Books are to be returned on or before

LIVERPOOL
JOHN MOORES UNIVERSITY
AVRIL ROBARTS LRC
TITHEBARN STREET
LIVERPOOL L2 2ER
TEL. 0151 231 4022

IVERPOOL JMU LIBRARY

3 1111 01184 0848

SECOND EDITION

TRANSFORM METHODS FOR SOLVING PARTIAL DIFFERENTIAL EQUATIONS

SECOND EDITION

TRANSFORM METHODS FOR SOLVING PARTIAL DIFFERENTIAL EQUATIONS

Dean G. Duffy

CHAPMAN & HALL/CRC

A CRC Press Company

Boca Raton   London   New York   Washington, D.C.

Library of Congress Cataloging-in-Publication Data

Duffy, Dean G.
  Transform methods for solving partial differential equations / Dean
G. Duffy.— 2nd ed.
    p. cm.
  Includes bibliographical references and index.
  ISBN 1-58488-451-7 (alk. paper)
  1. Fourier transformations. 2. Differential equations, Partial—
Numerical solutions.   I. Title.

  QA403.5.D84 2004
  515'.353--dc22   2004049448

This book contains information obtained from authentic and highly regarded sources. Reprinted material is quoted with permission, and sources are indicated. A wide variety of references are listed. Reasonable efforts have been made to publish reliable data and information, but the author and the publisher cannot assume responsibility for the validity of all materials or for the consequences of their use.

Neither this book nor any part may be reproduced or transmitted in any form or by any means, electronic or mechanical, including photocopying, microfilming, and recording, or by any information storage or retrieval system, without prior permission in writing from the publisher.

The consent of CRC Press LLC does not extend to copying for general distribution, for promotion, for creating new works, or for resale. Specific permission must be obtained in writing from CRC Press LLC for such copying.

Direct all inquiries to CRC Press LLC, 2000 N.W. Corporate Blvd., Boca Raton, Florida 33431.

Trademark Notice: Product or corporate names may be trademarks or registered trademarks, and are used only for identification and explanation, without intent to infringe.

Visit the CRC Press Web site at www.crcpress.com

© 2004 by Chapman & Hall/CRC

No claim to original U.S. Government works
International Standard Book Number 1-58488-451-7
Library of Congress Card Number 2004049448
Printed in the United States of America 1 2 3 4 5 6 7 8 9 0
Printed on acid-free paper

Dedicated to the Brigade of Midshipmen

Acknowledgments

I would like to thank the many midshipmen who have taken Engineering Mathematics II from me. They have been willing or unwilling guinea pigs in testing out many of the ideas and problems in this book. I would especially like to thank Dr. Gordon W. Inverarity for his many useful and insightful suggestions for improving the section on the numerical inversion of Fourier transforms. My appreciation goes to all those authors and publishers who allowed me the use of their material from the scientific and engineering literature. Finally, Many of the plots and calculations were done using MATLAB.

MATLAB is a registered trademark of
The MathWorks Inc.
24 Prime Park Way
Natick, MA 01760–1500
Phone: (508) 647–7000
Email: info@mathworks.com
www.mathworks.com

Introduction

Purpose. This book illustrates the use of Laplace, Fourier and Hankel transforms for solving linear partial differential equations that are encountered in engineering and sciences. To this end, this new edition features updated references as well as many new examples and exercises taken from a wide variety of sources. Of particular importance is the inclusion of numerical methods and asymptotic techniques for inverting particularly complicated transforms. Of course, typos from the first edition have been corrected.

Transform methods provide an alternative and bridge between the commonly employed methods of separation of variables and numerical methods in solving linear partial differential equations. The relationship between these techniques might be pictured as follows:

```
                    ┌────────────────┬────────────────┐
                    │                │                │
                    │                │   Numerical    │
                    │                │   Techniques   │
          ┌─────────┤                │                │
          │         │   Transform    ├────────────────┤
          │ Separation │  Methods     │                │
          │    of    │                │   Asymptotic   │
          │ Variables │               │   Analysis     │
          │         │                │                │
          └─────────┴────────────────┴────────────────┘
```

Transform methods are similar to separation of variables because they often yield closed form solutions via the powerful method of contour integration. Indeed, all of the results from separation of variables could also be derived using transform methods. Moreover, transform methods can handle a wider class of problems, such as those involving time-dependent boundary conditions, where separation of variables would fail.

Even in those cases when the inverse of the transform cannot be found analytically, transform methods can still be used profitably. A wide variety of numerical and asymptotic methods now exist for their inversion. Because some analytic aspects of the problem are retained, it is easier to obtain greater physical insight than from a purely numerical approach.

Prerequisites. The book assumes the usual undergraduate sequence of mathematics in engineering or the sciences: the traditional calculus and differential equations. A course in complex variables and Fourier and Laplace transforms is also essential. Finally some knowledge of Bessel functions is desirable to completely understand the book.

Audience. This book may be used as either a textbook or a reference book for applied physicists, geophysicists, civil, mechanical or electrical engineers and applied mathematicians.

Chapter Overview. The purpose of Chapter 1 is two-fold. The first four sections (and Section 1.7) serve as a refresher on the background material: linear ordinary differential equations, transform methods and complex variables. The amount of time spent with this material depends upon the background of the class. At least one class period should be spent on each section. Section 1.5 and Section 1.6 cover multivalued complex functions. These sections can be omitted if you only plan to teach Chapter 2 and Chapter 3. Otherwise, several class periods will be necessary to master this material because most students have never seen it. Due to the complexity of the problems, it is suggested that take-home problems are given to test the student's knowledge.

Chapter 2 through Chapter 5 are the meat-and-potatoes of the book. We subdivide the material according to whether we invert a single-valued or multivalued transform. Each chapter is then subdivided into two parts. The first part deals with simply the mechanics of how to invert the transform while the second part actually applies the transform methods to solving partial differential equations. Undergraduates with a strong mathematics background should be able to handle Chapter 2 and Chapter 3 while Chapter 4 and Chapter 5 are really graduate-level material. The constant theme is the repeated application of the residue theorem to invert Fourier and Laplace transforms.

In Chapter 6 we solve partial differential equations by repeated applications of transform methods. We are now in advanced topics and this material is really only suitable for graduate students. The first two sections are straightforward, brute-force applications of Laplace and Fourier transforms in solving partial differential equations where we hope that we can invert both transforms to find the solution. Section 6.3 and Section 6.4 are devoted to the very clever inversion techniques of Cagniard and De Hoop. For too long this interesting work has been restricted to the seismic, acoustic and electrodynamic communities.

In Chapter 7 we treat the classic Wiener–Hopf problem. This is very difficult material because of the very complicated analysis that is usually involved. I have tried to break the chapter into two parts: Section 7.1 deals with finite domains while Section 7.2 applies to infinite and semi-infinite domains.

Features. This is an unabashedly applied book because the intended audience is problem solvers in engineering and the applied sciences. However, references are given to other books that do cover any unproven point, should the reader be interested. Also I have tried to give some human touch to this field by including references to the original works and photographs of some of the leading figures.

It is always difficult to write a book that satisfies both the student and the researcher. The student usually wants all of the gory details while the researcher wants the answer now. I have tried to accommodate both by centering most of the text around examples of increasing difficulty. There are plenty of details for the student, but the researcher may quickly leaf through the examples to find the material that interests him.

As anyone who has taken a course knows, the only way that you know a subject is by working the problems. For that reason I have included several hundred well-crafted problems, most of which were taken from the scientific and engineering literature. When possible, these problems are grouped according to some common property – such as a cylindrical domain. Because many of these problems are difficult, I have included detailed solutions to most of them. The student is asked however to refrain from looking at the solution before he has really tried to solve it on his own. No pain; no gain. The remaining problems have intermediate results so that the student has confidence that he is on the right track. The researcher also might look at these problems because his problem might already have been solved.

A new feature of this book is the inclusion of sections on the numerical inversion of Laplace, Hankel and Fourier transforms. I have included MATLAB code for the reader's use. A quick glance at the scripts reveals their "Fortran"-like structure. This was done for a reason. For those who know MATLAB well, it is easy to optimize the scripts using MATLAB syntax. For the Fortran and C crowd, the scripts are easily convertible into those languages.

Finally, an important aspect of this book is the numerous references that can serve as further grist for the student or point the researcher toward a solution of his problem. Of course, we must strike a balance between having a book of references and leaving out some interesting papers. The criteria for inclusion were three-fold. First, the paper had to have used the technique and not merely chanted the magic words; quoting results was unacceptable. Second, the papers had to compute both the forward and inverse transforms. The use of asymptotic or numerical methods to invert the transform excluded the reference. Finally, we required some details of the process.

List of Definitions

Function	Definition
$\delta(t-a)$	$= \begin{cases} \infty, & t = 0, \\ 0, & t \neq 0, \end{cases} \qquad \int_{-\infty}^{\infty} \delta(t)\, dt = 1$
$\text{erfc}(x)$	$= 1 - \dfrac{2}{\sqrt{\pi}} \displaystyle\int_0^x e^{-\xi^2}\, d\xi$
$H(t-a)$	$= \begin{cases} 1, & t > a, \\ 0, & t < a. \end{cases}$
$H_n^{(1)}(x), H_n^{(2)}(x)$	Hankel functions of first and second kind of order n
$I_n(x)$	modified Bessel function of the first kind and order n
$J_n(x)$	Bessel function of the first kind and order n
$K_n(x)$	modified Bessel function of the second kind and order n
$\text{sgn}(t-a)$	$= \begin{cases} -1, & t < a, \\ 1, & t > a. \end{cases}$
$Y_n(x)$	Bessel function of the second kind and order n

Contents

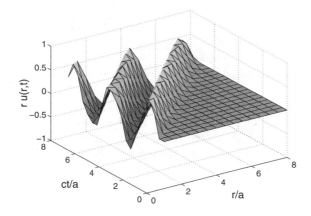

Chapter 1
The Fundamentals

Many physical processes in nature evolve with time in domains that may be treated as if they were of infinite or semi-infinite extent. For this reason, the use of Laplace and Fourier transforms has proven a powerful analytic technique for solving *linear* partial differential equations that occur in engineering and the sciences. The purpose of this book is to illustrate their use.

We have designed this first chapter to bring everyone to a common starting point. Section 1.1 to Section 1.4 provide a review for those who are a little rusty on their Laplace and Fourier transforms, ordinary differential equations and complex variables while Section 1.7 provides an overview of Bessel functions – special functions that will repeatedly appear in this book. We limit our discussion to those topics that we will use later. Section 1.5 and Section 1.6 give an in-depth examination of multivalued complex functions because most books on complex variables treat these functions in a rather perfunctory manner. Finally, Section 1.8 provides a glance into the power of transform methods.

1.1 FOURIER TRANSFORMS

The Fourier transform is the natural extension of Fourier series to a function $f(t)$ of infinite period. It is defined by the pair of integrals:

$$f(t) = \frac{1}{2\pi} \int_{-\infty}^{\infty} F(\omega)\, e^{i\omega t}\, d\omega, \qquad (1.1.1)$$

1

and

$$F(\omega) = \int_{-\infty}^{\infty} f(t)e^{-i\omega t}\, dt. \qquad (1.1.2)$$

Equation 1.1.2 is the *Fourier transform* of $f(t)$ while Equation 1.1.1 is the *inverse Fourier transform*, which converts a Fourier transform back to $f(t)$. If, following Hamming,[1] we imagine that $f(t)$ is a light beam, then the Fourier transform, like a glass prism, breaks up the function into its component frequencies ω, each of intensity $F(\omega)$. In optics, the various frequencies are called colors; by analogy the Fourier transform gives us the color spectrum of a function. On the other hand, the inverse Fourier transform blends a function's spectrum to give back the original function.

If $f(t)$ is an even function, we can replace Equation 1.1.1 with the *Fourier cosine transform*

$$F(\omega) = \int_{0}^{\infty} f(t)\cos(\omega t)\, dt \qquad (1.1.3)$$

with

$$f(t) = \frac{2}{\pi}\int_{0}^{\infty} F(\omega)\cos(\omega t)\, d\omega. \qquad (1.1.4)$$

On the other hand, if $f(t)$ is an odd function, then Equation 1.1.1 can be replaced with the *Fourier sine transform*

$$F(\omega) = \int_{0}^{\infty} f(t)\sin(\omega t)\, dt \qquad (1.1.5)$$

with

$$f(t) = \frac{2}{\pi}\int_{0}^{\infty} F(\omega)\sin(\omega t)\, d\omega. \qquad (1.1.6)$$

Fourier cosine and sine transforms are also useful when $f(t)$ is only defined on the semi-infinite interval $[0, \infty)$.

Clearly, for a Fourier transform to exist, then the integral in Equation 1.1.2 must also exist. A sufficient condition is that $f(t)$ is absolutely integrable on $(-\infty, \infty)$, or

$$\int_{-\infty}^{\infty} |f(t)|\, dt < \infty. \qquad (1.1.7)$$

Some of the simplest functions, such as $f(t) = \sin(at)$ and $f(t) = \cos(at)$, violate this condition and appear not to have a Fourier transform. Actually, they do exist but it requires the use of the Dirac delta function to express them.

To avoid the use of generalized functions, some investigators argue that all physical processes suffer a certain amount of dissipation. For that reason, the

---

[1] Hamming, R. W., 1977: *Digital Filters*. Prentice-Hall, p. 136.

Fourier transform should include a damping factor $e^{-\epsilon t}$ so that the definition of the Fourier transform becomes

$$F(\omega') = \int_{-\infty}^{\infty} f(t)e^{-\epsilon t}e^{-i\omega t}\, dt = \int_{-\infty}^{\infty} f(t)e^{-i\omega' t}\, dt, \qquad (1.1.8)$$

where $\omega' = \omega - i\epsilon$, $\epsilon > 0$. In this modified form, the inverse Fourier transform is

$$f(t) = \frac{1}{2\pi}\int_{-\infty-i\epsilon}^{\infty-i\epsilon} F(\omega')e^{i\omega' t}\, d\omega'. \qquad (1.1.9)$$

• **Example 1.1.1**

Let us find the Fourier transform of

$$f(t) = \begin{cases} 0, & \text{if} \quad -\infty < t < 0, \\ \cos(bt), & \text{if} \quad 0 < t < \infty, \end{cases} \qquad (1.1.10)$$

when we include a small amount of damping $\epsilon > 0$.

From the definition of the Fourier transform,

$$F(\omega') = \int_{0}^{\infty} \cos(bt)e^{-\epsilon t}e^{-i\omega t}\, dt. \qquad (1.1.11)$$

Direct integration yields

$$F(\omega') = \frac{i\omega'}{b^2 - \omega'^2}. \qquad (1.1.12)$$

For those familiar with Laplace transforms, this is the same answer as the Laplace transform of $\cos(bt)$ with s replaced by $i\omega'$. For this reason, Van der Pol and Bremmer[2] have called the transform pair Equation 1.1.8 and Equation 1.1.9, a two-sided Laplace or bilateral transform.

On the other hand, from the definition of the inverse Fourier transform

$$\frac{1}{2\pi}\int_{-\infty-i\epsilon}^{\infty-i\epsilon} \frac{i\omega'}{b^2 - \omega'^2}e^{i\omega' t}\, d\omega' = \begin{cases} 0, & \text{if} \quad -\infty < t < 0, \\ \cos(bt), & \text{if} \quad 0 < t < \infty. \end{cases} \qquad (1.1.13)$$

In principle, we can compute any Fourier transform from the definition. However, it is far more efficient to derive some simple relationships that relate

---

[2] Van der Pol, B., and H. Bremmer, 1955: *Operational Calculus Based on the Two-Sided Laplace Integral*. Cambridge University Press, 415 pp. See Equation 11 in Chapter 2.

Table 1.1.1: Some General Properties of Fourier Transforms

Property	Function, f(t)	Fourier Transform, F(ω)
1. Linearity	$c_1 f(t) + c_2 g(t)$	$c_1 F(\omega) + c_2 G(\omega)$
2. Complex conjugate	$f^*(t)$	$F^*(-\omega)$
3. Scaling	$f(\alpha t)$	$F(\omega/\alpha)/|\alpha|$
4. Delay	$f(t - \tau)$	$e^{-i\omega\tau} F(\omega)$
5. Frequency translation	$e^{i\omega_0 t} f(t)$	$F(\omega - \omega_0)$
6. Duality in time and frequency	$F(t)$	$2\pi f(-\omega)$
7. Time differentiation	$f'(t)$	$i\omega F(\omega)$

transforms to each other. Some of the most useful properties are stated in Table 1.1.1 without proof.

• **Example 1.1.2**

Repeatedly in this book, we must find the Fourier transform of the derivative of a function $f(t)$ that is differentiable for all t and vanishes as $t \to \pm\infty$. From the definition of the Fourier transform,

$$\mathcal{F}[f'(t)] = \int_{-\infty}^{\infty} f'(t) e^{-i\omega t}\, dt \tag{1.1.14}$$

$$= f(t) e^{-i\omega t}\Big|_{-\infty}^{\infty} + i\omega \int_{-\infty}^{\infty} f(t) e^{-i\omega t}\, dt \tag{1.1.15}$$

$$= i\omega F(\omega), \tag{1.1.16}$$

where $F(\omega)$ is the Fourier transform of $f(t)$. Similarly,

$$\mathcal{F}[f''(t)] = -\omega^2 F(\omega). \tag{1.1.17}$$

The most important property of Fourier transforms is convolution:

$$f(t) * g(t) = \int_{-\infty}^{\infty} f(x) g(t-x)\, dx = \int_{-\infty}^{\infty} f(t-x) g(x)\, dx. \tag{1.1.18}$$

Then,

$$\mathcal{F}[f(t) * g(t)] = \int_{-\infty}^{\infty} f(x)e^{-i\omega x} \left[\int_{-\infty}^{\infty} g(t-x)e^{-i\omega(t-x)} \, dt \right] dx \quad (1.1.19)$$

$$= \int_{-\infty}^{\infty} f(x)G(\omega)e^{-i\omega x} \, dx = F(\omega)G(\omega). \quad (1.1.20)$$

Thus, the Fourier transform of the convolution of two functions equals the product of the Fourier transforms of each of the functions.

● **Example 1.1.3**

Verify the convolution theorem using the functions

$$f(t) = \begin{cases} 1, & \text{if} \quad |t| < a, \\ 0, & \text{if} \quad |t| > a, \end{cases} \quad (1.1.21)$$

and

$$g(t) = \begin{cases} e^{-t}, & \text{if} \quad t > 0, \\ 0, & \text{if} \quad t < 0, \end{cases} \quad (1.1.22)$$

where $a > 0$.
The convolution of $f(t)$ with $g(t)$ is

$$f(t) * g(t) = \int_{-\infty}^{\infty} f(x)g(t-x) \, dx = \int_{-a}^{a} g(t-x) \, dx. \quad (1.1.23)$$

If $t < -a$, then $g(t-x)$ always equals zero for $|x| < a$, and $f(t) * g(t) = 0$. If $t > a$,

$$f(t) * g(t) = e^{-t} \int_{-a}^{a} e^{x} \, dx = e^{-(t-a)} - e^{-(t+a)}. \quad (1.1.24)$$

Finally, for $-a < t < a$,

$$f(t) * g(t) = e^{-t} \int_{-a}^{t} e^{x} \, dx = 1 - e^{-(t+a)}. \quad (1.1.25)$$

In summary,

$$f(t) * g(t) = \begin{cases} 0, & \text{if} \quad t < -a, \\ 1 - e^{-(t+a)}, & \text{if} \quad -a < t < a, \\ e^{-(t-a)} - e^{-(t+a)}, & \text{if} \quad a < t. \end{cases} \quad (1.1.26)$$

The Fourier transform of $f(t) * g(t)$ is

$$\mathcal{F}[f(t) * g(t)] = \int_{-a}^{a} \left[1 - e^{-(t+a)}\right] e^{-i\omega t} \, dt$$

$$+ \int_{a}^{\infty} \left[e^{-(t-a)} - e^{-(t+a)}\right] e^{-i\omega t} \, dt \qquad (1.1.27)$$

$$= \frac{2\sin(\omega a)}{\omega} - \frac{2i\sin(\omega a)}{1 + \omega i} \qquad (1.1.28)$$

$$= \frac{2\sin(\omega a)}{\omega} \left(\frac{1}{1 + \omega i}\right) = F(\omega)G(\omega), \qquad (1.1.29)$$

and the convolution theorem holds.

• Example 1.1.4

For our final example,[3] let us find the inverse $f(x)$ of the Fourier transform

$$F(k) = \int_{-\infty}^{\infty} f(x)e^{-ikx} \, dx = \frac{e^{ikct} - e^{-ikct}}{ki} \exp\left(-\frac{k^2 c^2 \tau t}{2}\right) \qquad (1.1.30)$$

using the convolution theorem where c, t and τ are positive and real.

We begin by noting that

$$\mathcal{F}\left[\frac{1}{c\sqrt{\tau t}} \exp\left(-\frac{x^2}{2c^2\tau t}\right)\right] = \exp\left(-\frac{k^2 c^2 \tau t}{2}\right) \qquad (1.1.31)$$

and

$$\mathcal{F}[H(ct + x) - H(ct - x)] = \frac{e^{ikct} - e^{-ikct}}{ki}. \qquad (1.1.32)$$

From the convolution theorem,

$$f(x) = \frac{1}{c\sqrt{\tau t}} \int_{-ct}^{ct} \exp\left[-\frac{(x-\eta)^2}{2c^2\tau t}\right] d\eta = \sqrt{2} \int_{(x-ct)/c\sqrt{2\tau t}}^{(x+ct)/c\sqrt{2\tau t}} e^{-r^2} \, dr, \quad (1.1.33)$$

if $r = (x - \eta)/\left(c\sqrt{2\tau t}\right)$. Then, if $x > ct$,

$$f(x) = \sqrt{\frac{\pi}{2}} \left[\text{erf}\left(\frac{x+ct}{c\sqrt{2\tau t}}\right) - \text{erf}\left(\frac{x-ct}{c\sqrt{2\tau t}}\right)\right]. \qquad (1.1.34)$$

For $|x| \le ct$,

$$f(x) = \sqrt{\frac{\pi}{2}} \left[\text{erf}\left(\frac{x+ct}{c\sqrt{2\tau t}}\right) + \text{erf}\left(\frac{ct-x}{c\sqrt{2\tau t}}\right)\right]. \qquad (1.1.35)$$

---

[3] Taken from Tanaka, K., and T. Kurokawa, 1973: Viscous property of steel and its effect on strain wave front. *Bull. JSME*, **16**, 188–193.

Finally, if $x < -ct$,

$$f(x) = \sqrt{\frac{\pi}{2}} \left[\text{erf}\left(\frac{ct - x}{c\sqrt{2\tau t}}\right) - \text{erf}\left(\frac{-ct - x}{c\sqrt{2\tau t}}\right) \right]. \tag{1.1.36}$$

In Section 3.1 and Section 5.1 we will discuss the inversion of Fourier transforms by complex variables.

In this section, we have given a quick overview of Fourier transforms. For greater detail, as well as drill exercises, the reader is referred to Chapter 5 of the author's *Advanced Engineering Mathematics with MATLAB*.[4]

1.2 LAPLACE TRANSFORMS

Consider a function $f(t)$ such that $f(t) = 0$ for $t < 0$. Then the *Laplace integral*

$$\mathcal{L}[f(t)] = F(s) = \int_0^\infty f(t)e^{-st}\,dt \tag{1.2.1}$$

defines the Laplace transform of $f(t)$, which we write $\mathcal{L}[f(t)]$ or $F(s)$. The Laplace transform of $f(t)$ exists, for sufficiently large s, provided $f(t)$ satisfies the following conditions:

- $f(t) = 0$ for $t < 0$,
- $f(t)$ is continuous or piece-wise continuous in every interval,
- $t^n|f(t)| < \infty$ as $t \to 0$ for some number n, where $n < 1$,
- $e^{-s_0 t}|f(t)| < \infty$ as $t \to \infty$, for some number s_0. The quantity s_0 is called the *abscissa of convergence*.

- **Example 1.2.1**

Let us find the Laplace transform for the *Heaviside step function*:

$$H(t - a) = \begin{cases} 1, & t > a, \\ 0, & t < a. \end{cases} \tag{1.2.2}$$

The Heaviside step function is essentially a bookkeeping device that gives us the ability to "switch on" and "switch off" a given function. For example, if we want a function $f(t)$ to become nonzero at time $t = a$, we represent this process by the product $f(t)H(t - a)$.

From the definition of the Laplace transform,

$$\mathcal{L}[H(t - a)] = \int_a^\infty e^{-st}\,dt = \frac{e^{-as}}{s}, \qquad s > 0. \tag{1.2.3}$$

---

[4] Duffy, D. G., 2003: *Advanced Engineering Mathematics with MATLAB*. Chapman & Hall/CRC, 818 pp.

• Example 1.2.2

The *Dirac delta function* or *impulse function*, often defined for computational purposes by

$$\delta(t) = \lim_{n \to \infty} \delta_n(t) = \lim_{n \to \infty} \begin{cases} n/2, & |t| < \frac{1}{2}, \\ \\ 0, & |t| > \frac{1}{2}, \end{cases} \tag{1.2.4}$$

plays an especially important role in transform methods because its Laplace transform is

$$\mathcal{L}[\delta(t-a)] = \int_0^\infty \delta(t-a)e^{-st}\,dt = \lim_{n \to \infty} \frac{n}{2} \int_{a-1/n}^{a+1/n} e^{-st}\,dt \tag{1.2.5}$$

$$= \lim_{n \to \infty} \frac{n}{2s}\left(e^{-as+s/n} - e^{-as-s/n}\right) \tag{1.2.6}$$

$$= \lim_{n \to \infty} \frac{n\,e^{-as}}{2s}\left(1 + \frac{s}{n} + \frac{s^2}{2n^2} + \cdots - 1 + \frac{s}{n} - \frac{s^2}{2n^2} + \cdots\right) \tag{1.2.7}$$

$$= e^{-as}. \tag{1.2.8}$$

A special case is $\mathcal{L}[\delta(t)] = 1$. The Fourier transform of $\delta(t-a)$ is similar, namely $e^{-ia\omega}$.

• Example 1.2.3

Although we could compute Equation 1.2.1 for every function that has a Laplace transform, these results have already been tabulated and are given in many excellent tables.[5] However, there are four basic transforms that the reader should memorize. They are

$$\mathcal{L}(e^{at}) = \int_0^\infty e^{at}e^{-st}\,dt = \int_0^\infty e^{-(s-a)t}\,dt \tag{1.2.9}$$

$$= -\frac{e^{-(s-a)t}}{s-a}\bigg|_0^\infty = \frac{1}{s-a}, \qquad s > a, \tag{1.2.10}$$

$$\mathcal{L}[\sin(at)] = \int_0^\infty \sin(at)e^{-st}\,dt = -\frac{e^{-st}}{s^2+a^2}[s\sin(at) + a\cos(at)]\bigg|_0^\infty \tag{1.1.11}$$

$$= \frac{a}{s^2+a^2}, \qquad s > 0, \tag{1.2.12}$$

---

[5] The most complete set is given by Erdélyi, A., W. Magnus, F. Oberhettinger, and F. G. Tricomi, 1954: *Tables of Integral Transforms, Vol I*. McGraw-Hill Co., 391 pp.

Table 1.2.1: Some General Properties of Laplace Transforms with $a > 0$

Property	Function, f(t)	Laplace Transform, F(s)
1. Linearity	$c_1 f(t) + c_2 g(t)$	$c_1 F(s) + c_2 G(s)$
2. Scaling	$f(t/a)/a$	$F(as)$
3. Multiplication by e^{bt}	$e^{bt} f(t)$	$F(s - b)$
4. Translation	$f(t - a)H(t - a)$	$e^{-as} F(s)$
5. Differentiation	$f^{(n)}(t)$	$s^n F(s) - s^{n-1} f(0)$ $-s^{n-2} f'(0) - \cdots$ $-f^{(n-1)}(0)$
6. Integration	$\int_0^t f(\tau)\,d\tau$	$F(s)/s$
7. Convolution	$\int_0^t f(t - \tau)g(\tau)\,d\tau$	$F(s)G(s)$

$$\mathcal{L}[\cos(at)] = \int_0^\infty \cos(at)e^{-st}dt = \left.\frac{e^{-st}}{s^2 + a^2}[-s\cos(at) + a\sin(at)]\right|_0^\infty \quad (1.1.13)$$

$$= \frac{s}{s^2 + a^2}, \qquad s > 0, \quad (1.2.14)$$

and

$$\mathcal{L}(t^n) = \int_0^\infty t^n e^{-st}dt = \left.-n!e^{-st}\sum_{m=0}^n \frac{t^{n-m}}{(n-m)!s^{m+1}}\right|_0^\infty = \frac{n!}{s^{n+1}}, \qquad s > 0, \quad (1.2.15)$$

where n is a positive integer.

The integral definition of the Laplace transform can be used to derive many important properties. For example, the transform of a sum equals the sum of the transforms:

$$\mathcal{L}[c_1 f(t) + c_2 g(t)] = c_1 \mathcal{L}[f(t)] + c_2 \mathcal{L}[g(t)]. \quad (1.2.16)$$

This linearity property holds with complex numbers and functions as well.

Another important property deals with derivatives. Suppose $f(t)$ is continuous and has a piece-wise continuous derivative $f'(t)$. Then,

$$\mathcal{L}[f'(t)] = \int_0^\infty f'(t)e^{-st}dt = \left.e^{-st}f(t)\right|_0^\infty + s\int_0^\infty f(t)e^{-st}dt \quad (1.2.17)$$

by integration by parts. If $f(t)$ is of exponential order,[6] $e^{-st}f(t)$ tends to zero as $t \to \infty$, for large enough s, so that

$$\mathcal{L}[f'(t)] = sF(s) - f(0). \tag{1.2.18}$$

Similarly, if $f(t)$ and $f'(t)$ are continuous, $f''(t)$ is piece-wise continuous, and all three functions are of exponential order, then

$$\mathcal{L}[f''(t)] = s\mathcal{L}[f'(t)] - f'(0) = s^2 F(s) - sf(0) - f'(0). \tag{1.2.19}$$

In general,

$$\mathcal{L}[f^{(n)}(t)] = s^n F(s) - s^{n-1}f(0) - \cdots - sf^{(n-2)}(0) - f^{(n-1)}(0) \tag{1.2.20}$$

on the assumption that $f(t)$ and its first $n - 1$ derivatives are continuous, $f^{(n)}(t)$ is piece-wise continuous, and all are of exponential order so that the Laplace transform exists.

Consider now the transform of the function $e^{-at}f(t)$, where a is any real number. Then, by definition,

$$\mathcal{L}\left[e^{-at}f(t)\right] = \int_0^\infty e^{-st}e^{-at}f(t)\,dt = \int_0^\infty e^{-(s+a)t}f(t)\,dt, \tag{1.2.21}$$

or

$$\mathcal{L}\left[e^{-at}f(t)\right] = F(s + a). \tag{1.2.22}$$

Equation 1.2.22 is known as the *first shifting theorem* and states that if $F(s)$ is the transform of $f(t)$ and a is a constant, then $F(s + a)$ is the transform of $e^{-at}f(t)$.

● **Example 1.2.4**

Let us find the Laplace transform of $f(t) = e^{-at}\sin(bt)$. Because the Laplace transform of $\sin(bt)$ is $b/(s^2 + b^2)$,

$$\mathcal{L}\left[e^{-at}\sin(bt)\right] = \frac{b}{(s + a)^2 + b^2}, \tag{1.2.23}$$

where we have simply replaced s by $s + a$ in the transform for $\sin(bt)$.

● **Example 1.2.5**

Let us find the inverse of the Laplace transform

$$F(s) = \frac{s + 2}{s^2 + 6s + 1}. \tag{1.2.24}$$

---

[6] By exponential order we mean that there exist some constants, M and k, for which $|f(t)| \le Me^{kt}$ for all $t > 0$.

Rearranging terms,

$$F(s) = \frac{s+2}{s^2 + 6s + 1} = \frac{s+2}{(s+3)^2 - 8} \tag{1.2.25}$$

$$= \frac{s+3}{(s+3)^2 - 8} - \frac{1}{2\sqrt{2}} \frac{2\sqrt{2}}{(s+3)^2 - 8}. \tag{1.2.26}$$

Immediately, from the first shifting theorem,

$$f(t) = e^{-3t} \cosh\left(2\sqrt{2}t\right) - \frac{e^{-3t}}{2\sqrt{2}} \sinh\left(2\sqrt{2}t\right). \tag{1.2.27}$$

The *second shifting theorem* states that if $F(s)$ is the transform of $f(t)$, then $e^{-bs}F(s)$ is the transform of $f(t-b)H(t-b)$, where b is real and positive. To show this, consider the Laplace transform of $f(t-b)H(t-b)$. Then, from the definition,

$$\mathcal{L}[f(t-b)H(t-b)] = \int_0^\infty f(t-b)H(t-b)e^{-st}\,dt \tag{1.2.28}$$

$$= \int_b^\infty f(t-b)e^{-st}\,dt = \int_0^\infty e^{-bs}e^{-sx}f(x)\,dx \tag{1.2.29}$$

$$= e^{-bs} \int_0^\infty e^{-sx}f(x)\,dx, \tag{1.2.30}$$

or

$$\mathcal{L}[f(t-b)H(t-b)] = e^{-bs}F(s), \tag{1.2.31}$$

where we have set $x = t - b$. This theorem is of fundamental importance because it allows us to write down the transforms for "delayed" time functions. These functions "turn on" b units after the initial time.

Let us find the inverse of the transform $e^{-\pi s}/[s^2(s^2 + 1)]$. Because

$$\frac{e^{-\pi s}}{s^2(s^2 + 1)} = \frac{e^{-\pi s}}{s^2} - \frac{e^{-\pi s}}{s^2 + 1}, \tag{1.2.32}$$

$$\mathcal{L}^{-1}\left[\frac{e^{-\pi s}}{s^2(s^2 + 1)}\right] = \mathcal{L}^{-1}\left(\frac{e^{-\pi s}}{s^2}\right) - \mathcal{L}^{-1}\left(\frac{e^{-\pi s}}{s^2 + 1}\right) \tag{1.2.33}$$

$$= (t - \pi)H(t - \pi) - \sin(t - \pi)H(t - \pi) \tag{1.2.34}$$

$$= (t - \pi)H(t - \pi) + \sin(t)H(t - \pi), \tag{1.2.35}$$

since $\mathcal{L}^{-1}(1/s^2) = t$ and $\mathcal{L}^{-1}[1/(s^2 + 1)] = \sin(t)$.

Finally, we consider a fundamental concept in Laplace transforms: convolution. We begin by formally introducing the mathematical operation of the *convolution product*:

$$f(t) * g(t) = \int_0^t f(t-x)g(x)\,dx = \int_0^t f(x)g(t-x)\,dx. \tag{1.2.36}$$

• **Example 1.2.6**

Let us find the convolution between $\cos(t)$ and $\sin(t)$.

$$\cos(t) * \sin(t) = \int_0^t \sin(t-x)\cos(x)\,dx \qquad (1.2.37)$$

$$= \tfrac{1}{2}\int_0^t [\sin(t) + \sin(t-2x)]\,dx \qquad (1.2.38)$$

$$= \tfrac{1}{2}\int_0^t \sin(t)\,dx + \tfrac{1}{2}\int_0^t \sin(t-2x)\,dx \qquad (1.2.39)$$

$$= \tfrac{1}{2}\sin(t)\,x\Big|_0^t + \tfrac{1}{4}\cos(t-2x)\Big|_0^t = \tfrac{1}{2}t\sin(t). \qquad (1.2.40)$$

The reason why we introduced convolution derives from the following fundamental theorem (often called *Borel's theorem*[7]). If

$$w(t) = u(t) * v(t), \qquad (1.2.41)$$

then

$$W(s) = U(s)V(s). \qquad (1.2.42)$$

In other words, we can invert a complicated transform by convoluting the inverses to two simpler functions.

• **Example 1.2.7**

Let us find the inverse of the transform

$$\frac{1}{(s^2+a^2)^2} = \frac{1}{a^2}\left(\frac{a}{s^2+a^2} \times \frac{a}{s^2+a^2}\right) \qquad (1.2.43)$$

$$= \frac{1}{a^2}\mathcal{L}[\sin(at)]\mathcal{L}[\sin(at)]. \qquad (1.2.44)$$

Therefore,

$$\mathcal{L}^{-1}\left[\frac{1}{(s^2+a^2)^2}\right] = \frac{1}{a^2}\int_0^t \sin[a(t-x)]\sin(ax)\,dx \qquad (1.2.45)$$

$$= \frac{1}{2a^2}\int_0^t \cos[a(t-2x)]\,dx - \frac{1}{2a^2}\int_0^t \cos(at)\,dx \qquad (1.2.46)$$

$$= -\frac{1}{4a^3}\sin[a(t-2x)]\Big|_0^t - \frac{1}{2a^2}\cos(at)\,x\Big|_0^t \qquad (1.2.47)$$

$$= \frac{1}{2a^3}[\sin(at) - at\cos(at)]. \qquad (1.2.48)$$

---

[7] Borel, É., 1901: *Leçons sur les séries divergentes.* Gauthier-Villars, p. 104.

In this section, we have given a quick overview of Laplace transforms. For greater detail, as well as drill exercises, the reader is referred to Chapter 6 of the author's *Advanced Engineering Mathematics with MATLAB.*[8]

1.3 LINEAR ORDINARY DIFFERENTIAL EQUATIONS

Most analytic techniques for solving a partial differential equation involve reducing it down to an ordinary differential equation or a set of ordinary differential equations that is hopefully easier to solve than the original partial differential equation. From the vast number of possible ordinary differential equations, we focus on second-order equations. All of the following techniques extend to higher-order equations.

Consider the ordinary differential equation

$$a\frac{d^2y}{dx^2} + b\frac{dy}{dx} + cy = f(x), \tag{1.3.1}$$

where a, b and c are real. For the moment let us take $f(x) = 0$. Assuming a solution of the form $y(x) = Ae^{mx}$ and substituting into Equation 1.3.1,

$$am^2 + bm + c = 0. \tag{1.3.2}$$

This purely algebraic equation is the *characteristic* or *auxiliary equation*. Because Equation 1.3.2 is quadratic, there are either two real roots, or else a repeated real root, or else conjugate complex roots. At this point, let us consider each case separately and state the solution. Any undergraduate book on ordinary differential equations will provide the details for obtaining these general solutions.

Case I: *Two distinct real roots m_1 and m_2,*

$$y(x) = c_1 e^{m_1 x} + c_2 e^{m_2 x}. \tag{1.3.3}$$

Case II: *A repeated real root m_1,*

$$y(x) = c_1 e^{m_1 x} + c_2 x e^{m_1 x}. \tag{1.3.4}$$

Case III: *Conjugate complex roots $m_1 = p + qi$ and $m_2 = p - qi$,*

$$y(x) = c_1 e^{px} \cos(qx) + c_2 e^{px} \sin(qx). \tag{1.3.5}$$

---

[8] Duffy, *op. cit.*

• Example 1.3.1

One of the most commonly encountered differential equations is

$$\frac{d^2y}{dx^2} - m^2 y = 0, \qquad (1.3.6)$$

where m is real and positive. Because there are two distinct roots, $m_{1,2} = \pm m$, the general solution is

$$y(x) = Ae^{mx} + Be^{-mx}. \qquad (1.3.7)$$

Although this solution is perfectly correct, it is most useful in semi-infinite domains. For finite domains, such as $0 < x < L$, we introduce the mathematical functions of hyperbolic sine and cosine. A little algebra shows that Equation 1.3.7 also equals

$$y(x) = C\cosh(mx) + D\sinh(mx), \qquad (1.3.8)$$

where

$$\cosh(mx) = \tfrac{1}{2}\left(e^{mx} + e^{-mx}\right) \qquad (1.3.9)$$

and

$$\sinh(mx) = \tfrac{1}{2}\left(e^{mx} - e^{-mx}\right). \qquad (1.3.10)$$

The advantage of using Equation 1.3.8 follows from the facts that $\sinh(0) = 0$ and $\cosh(0) = 1$.

So far we have only found the solution when $f(x) = 0$, the so-called *homogeneous* or *complementary solution* to Equation 1.3.1. When $f(x)$ is nonzero, we must add a *particular solution* to the complementary solution that yields $f(x)$ upon substitution into Equation 1.3.1. The most common technique for determining this particular solution is the *method of undetermined coefficients*. This method is as follows: (1) assume a particular solution $y_p(x)$ that consists of an arbitrary linear combination of all of the linearly independent functions which arise from repeated differentiations of $f(x)$, (2) substitute $y_p(x)$ into the differential equation and (3) determine the arbitrary constants of $y_p(x)$ so that the equation resulting from the substitution yields $f(x)$.

• Example 1.3.2

Let us find the particular solution for the equation

$$\frac{d^2y}{dx^2} + 2\frac{dy}{dx} + y = \cos^2(x). \qquad (1.3.11)$$

Our guess for the particular solution is then

$$y_p(x) = A + B\cos(2x) + C\sin(2x), \qquad (1.3.12)$$

because $\cos^2(x) = [1+\cos(2x)]/2$. Substituting Equation 1.3.12 into Equation 1.3.11 and equating coefficients for the constant, cosine and sine terms, we find that $A = 1/2$, $B = -3/50$ and $C = 2/25$.

The remaining task is to compute the arbitrary constants in the homogeneous solution. In this book we always have conditions at both ends of a given domain, even if one of these points is at infinity. We now illustrate the procedure used in solving these *boundary-value problems*.

● **Example 1.3.3**

Solve the boundary-value problem

$$\frac{d^2y}{dx^2} - sy = -\frac{1}{s}, \qquad y(0) = y(1) = 0, \tag{1.3.13}$$

where $s > 0$. The general solution to Equation 1.3.13 is

$$y(x) = A\sinh(x\sqrt{s}) + B\cosh(x\sqrt{s}) + \frac{1}{s^2}. \tag{1.3.14}$$

We have chosen to use hyperbolic functions because the domain lies between $x = 0$ and $x = 1$. Now,

$$y(0) = B + \frac{1}{s^2} = 0 \tag{1.3.15}$$

and

$$y(1) = A\sinh(\sqrt{s}) + B\cosh(\sqrt{s}) + \frac{1}{s^2} = 0. \tag{1.3.16}$$

Solving for A and B,

$$A = \frac{\cosh(\sqrt{s}) - 1}{s^2\sinh(\sqrt{s})} \qquad \text{and} \qquad B = -\frac{1}{s^2}. \tag{1.3.17}$$

Therefore,

$$y(x) = \frac{1 - \cosh(x\sqrt{s})}{s^2} + \frac{\cosh(\sqrt{s}) - 1}{s^2\sinh(\sqrt{s})}\sinh(x\sqrt{s}). \tag{1.3.18}$$

Problems

1. Solve the boundary-value problem

$$\frac{d^2y}{dx^2} - (a^2 + s)y = 0, \qquad y(-1) = \frac{1}{s}, \qquad y(1) = 0,$$

where a and s are real and positive.

1.4   COMPLEX VARIABLES

Complex variables provide analytic tools for the evaluation of integrals with an ease that rarely occurs with real functions. The power of integration on the complex plane has its roots in the basic three C's: the Cauchy-Riemann equations, the Cauchy-Goursat theorem and Cauchy's residue theorem.

The Cauchy-Riemann equations have their origin in the definition of the derivative in the complex plane. Just as we have the concept of the function in real variables, where for a given value of x we can compute a corresponding value of $y = f(x)$, we can define a complex function $w = f(z)$ where for a given value of $z = x + iy$ ($i = \sqrt{-1}$) we may compute $w = f(z) = u(x, y) + iv(x, y)$. In order for $f'(z)$ to exist in some region, $u(x, y)$ and $v(x, y)$ must satisfy the Cauchy-Riemann equations:

$$\frac{\partial u}{\partial x} = \frac{\partial v}{\partial y} \quad \text{and} \quad \frac{\partial u}{\partial y} = -\frac{\partial v}{\partial x}. \tag{1.4.1}$$

If u_x, u_y, v_x and v_y are continuous in some region surrounding a point z_0 and satisfy Equation 1.4.1 there, then $f(z)$ is *analytic* there. If a function is analytic everywhere in the complex plane, then it is an *entire* function. Alternatively, if the function is analytic everywhere except at some isolated singularities, then it is *meromorphic*. Note that $f(z)$ must satisfy the Cauchy-Riemann equations in a region and not just at a point. For example, $f(z) = |z|$ satisfies the Cauchy-Riemann equations at $z = 0$ and nowhere else. Consequently, this function is not analytic anywhere on the complex plane.

Integration on the complex plane is more involved than in real, single variables because $dz = dx + i\,dy$. We must specify a path or contour as we integrate from one point to another. Of all of the possible contour integrals, a closed contour is the best. To see why, we introduce the following results:

Cauchy-Goursat theorem:[9] *If $f(z)$ is an analytic function at each point within and on a closed contour C, then $\oint_C f(z)\,dz = 0$.*

This theorem leads immediately to

The principle of deformation of contours: *The value of a line integral of an analytic function around any simple closed contour remains unchanged if we deform the contour in such a manner that we do not pass over a point where $f(z)$ is not analytic.*

Consequently we can evaluate difficult integrals by deforming the contour so that the actual evaluation is along a simpler contour or the computations are made easier. See Example 1.4.1.

---

[9] See Goursat, E., 1900: Sur la définition générale des fonctions analytiques, d'après Cauchy. *Trans. Am. Math. Soc.*, **1**, 14–16.

Most integrations on the complex plane, however, deal with meromorphic functions. Our next theorem involves these functions; it is

Cauchy's residue theorem:[10] *If $f(z)$ is analytic inside a closed contour C (taken in the positive sense) except at points $z_1, z_2, \ldots, z_n$ where $f(z)$ has singularities, then*

$$\oint_C f(z)\,dz = 2\pi i \sum_{j=1}^{n} \text{Res}\,[f(z); z_j], \qquad (1.4.2)$$

where $\text{Res}[f(z); z_j]$ denotes the residue of $f(z)$ for the singularity located at z_j.

The question now turns to what is a residue and how do we compute it. The answer involves the nature of the singularity and an extension of the Taylor expansion, called a *Laurent expansion*:

$$f(z) = \sum_{n=0}^{\infty} a_n (z - z_j)^n + \sum_{n=1}^{\infty} a_{-n}(z - z_j)^{-n} \qquad (1.4.3)$$

for $0 < |z - z_j| < a$. The first summation is merely the familiar Taylor expansion; the second summation involves negative powers of $z - z_j$ and gives the behavior at singularity. The *residue* equals the coefficient a_{-1}.

The construction of a Laurent expansion for a given singularity has two practical purposes: (1) it gives the nature of the singularity and (2) we will occasionally use it to give the actual value of the residue. Turning to the nature of the singularity, there are three types:

• *Essential Singularity*: Consider the function $f(z) = \cos(1/z)$. Using the expansion for cosine,

$$\cos\left(\frac{1}{z}\right) = 1 - \frac{1}{2!z^2} + \frac{1}{4!z^4} - \frac{1}{6!z^6} + \cdots \qquad (1.4.4)$$

for $0 < |z| < \infty$. Note that this series never truncates in the inverse powers of z. Essential singularities have Laurent expansions which have an infinite number of inverse powers of $z - z_j$. The value of the residue for this essential singularity at $z = 0$ is zero.

• *Removable Singularity*: Consider the function $f(z) = \sin(z)/z$. This function appears, at first blush, to have a singularity at $z = 0$. Upon applying the expansion for sine, we see that

$$\frac{\sin(z)}{z} = 1 - \frac{z^2}{3!} + \frac{z^4}{5!} - \frac{z^6}{7!} + \frac{z^8}{9!} - \cdots \qquad (1.4.5)$$

---

[10] See Mitrinović, D. S., and J. K. Kečkić, 1984: *The Cauchy Method of Residues*. D. Reidel Publishing Co., 361 pp. Section 10.3 gives the historical development of the residue theorem.

for all z and we actually have no singularity at all. This is an example of a removable singularity; its residue is zero.

- *Pole of order n*: Consider the function

$$f(z) = \frac{1}{(z-1)^3(z+1)}. \tag{1.4.6}$$

This function has two singularities: one at $z = 1$ and the other at $z = -1$. We shall only consider the case $z = 1$. After a little algebra,

$$f(z) = \frac{1}{(z-1)^3} \frac{1}{2 + (z-1)} \tag{1.4.7}$$

$$= \frac{1}{2} \frac{1}{(z-1)^3} \frac{1}{1 + (z-1)/2} \tag{1.4.8}$$

$$= \frac{1}{2} \frac{1}{(z-1)^3} \left[1 - \frac{z-1}{2} + \frac{(z-1)^2}{4} - \frac{(z-1)^3}{8} + \cdots \right] \tag{1.4.9}$$

$$= \frac{1}{2(z-1)^3} - \frac{1}{4(z-1)^2} + \frac{1}{8(z-1)} - \frac{1}{16} + \cdots \tag{1.4.10}$$

for $0 < |z - 1| < 2$. Because the largest inverse (negative) power is three, the singularity at $z = 1$ is called a third-order pole; the value of the residue is $1/8$. Generally, we refer to a first-order pole as a *simple* pole.

The construction of a Laurent expansion is not the method of choice in computing a residue. (For an essential singularity, it is the only method; however, essential singularities are very rare in applications.) The common method for a pole of order n is

$$\text{Res}[f(z); z_j] = \frac{1}{(n-1)!} \lim_{z \to z_j} \frac{d^{n-1}}{dz^{n-1}} \left[(z - z_j)^n f(z) \right]. \tag{1.4.11}$$

For a simple pole Equation 1.4.11 simplifies to

$$\text{Res}[f(z); z_j] = \lim_{z \to z_j} (z - z_j) f(z). \tag{1.4.12}$$

Quite often, $f(z) = p(z)/q(z)$. From l'Hospital's rule, it follows that

$$\text{Res}[f(z); z_j] = \frac{p(z_j)}{q'(z_j)}. \tag{1.4.13}$$

The desirability of dealing with closed contour integrals should be clear by now. This is true to such an extent that mathematicians have devised several theorems that allow us to change a line integral into a closed one by adding an arc at infinity. The one of greatest relevance to us is by C. Jordan:[11]

Jordan's lemma: *Suppose that, on a circular arc C_R with radius R and center at the origin, $f(z) \to 0$ uniformly as $R \to \infty$. Then*

$$\lim_{R \to \infty} \int_{C_R} f(z) e^{imz} \, dz = 0, \qquad m > 0, \qquad (1.4.14)$$

if C_R is in the first and/or second quadrant;

$$\lim_{R \to \infty} \int_{C_R} f(z) e^{-imz} \, dz = 0, \qquad m > 0, \qquad (1.4.15)$$

if C_R is in the third and/or fourth quadrant;

$$\lim_{R \to \infty} \int_{C_R} f(z) e^{mz} \, dz = 0, \qquad m > 0, \qquad (1.4.16)$$

if C_R is in the second and/or third quadrant; and

$$\lim_{R \to \infty} \int_{C_R} f(z) e^{-mz} \, dz = 0, \qquad m > 0, \qquad (1.4.17)$$

if C_R is in the first and/or fourth quadrant. Technically, only Equation 1.4.14 is actually Jordan's lemma while the remaining points are variations.

Proof: We shall prove the first part; the remaining portions follow by analog. We begin by noting that

$$|I_R| = \left| \int_{C_R} f(z) e^{imz} dz \right| \le \int_{C_R} |f(z)| \left| e^{imz} \right| |dz|. \qquad (1.4.18)$$

Now,
$$|dz| = R \, d\theta, \quad |f(z)| \le M_R, \qquad (1.4.19)$$
$$\left| e^{imz} \right| = \left| \exp(imRe^{\theta i}) \right| = \left| \exp\{imR[\cos(\theta) + i\sin(\theta)]\} \right| = e^{-mR\sin(\theta)}. \qquad (1.4.20)$$

Therefore,
$$|I_R| \le RM_R \int_{\theta_0}^{\theta_1} \exp[-mR\sin(\theta)] \, d\theta, \qquad (1.4.21)$$

---

[11] Jordan, C., 1894: *Cours D'Analyse de l'École Polytechnique. Vol. 2.* Gauthier-Villars, pp. 285–286. See also Whittaker, E. T., and G. N. Watson, 1963: *A Course of Modern Analysis.* Cambridge University Press, p. 115.

where $0 \leq \theta_0 < \theta_1 \leq \pi$. Because the integrand is positive, the right side of Equation 1.4.21 is largest if we take $\theta_0 = 0$ and $\theta_1 = \pi$. Then

$$|I_R| \leq RM_R \int_0^\pi e^{-mR\sin(\theta)} d\theta = 2RM_R \int_0^{\pi/2} e^{-mR\sin(\theta)} d\theta. \qquad (1.4.22)$$

We cannot evaluate the integrals in Equation 1.4.22 as they stand. However, because $\sin(\theta) \geq 2\theta/\pi$, we can bound the value of the integral by

$$|I_R| \leq 2RM_R \int_0^{\pi/2} e^{-2mR\theta/\pi} d\theta = \frac{\pi}{m} M_R \left(1 - e^{-mR}\right). \qquad (1.4.23)$$

If $m > 0$, $|I_R|$ tends to zero with M_R as $R \to \infty$.   □

• Example 1.4.1

To illustrate how useful distorting the original contour may be in evaluating an integral, consider[12]

$$\frac{1}{2\pi i} \oint_C f(z)\, dz = \frac{1}{2\pi i} \oint_C \frac{z^{-n-1}}{1 - ae^z}\, dz, \qquad 0 < a < 1, \qquad (1.4.24)$$

where the contour C is the circle $|z| < -\ln(a)$ and $1 \leq n$. The integrand has an infinite number of simple poles at $z_m = -\ln(a) + 2m\pi i$ with $m = 0, \pm 1, \pm 2, \ldots$ (which lie outside the original contour) and a $(n+1)$th-order pole at $z = 0$. Because the straightforward evaluation of this integral by the residue theorem would require differentiating the denominator n times, we choose to evaluate Equation 1.4.24 by expanding the contour so that it is a circle of infinite radius with a cut that excludes the simple poles at z_m. See Figure 1.4.1. Then, by the residue theorem,

$$\frac{1}{2\pi i} \oint_C f(z)\, dz = I - \sum_{m=-\infty}^{\infty} \text{Res}\left[f(z); z_m\right], \qquad (1.4.25)$$

where I is the contribution from the circle at infinity. Because the residue of $f(z)$ at z_m is $-z_m^{-1-n}$,

$$\frac{1}{2\pi i} \oint_C f(z)\, dz = I + \sum_{m=-\infty}^{\infty} \left[2m\pi i - \ln(a)\right]^{-n-1}. \qquad (1.4.26)$$

---

[12] Based upon Götze, F., and H. Friedrich, 1980: Berechnungs- und Abschätzungsformeln für verallgemeinerte geometrische Reihen. *Z. Angew. Math. Mech.*, **60**, 737–739.

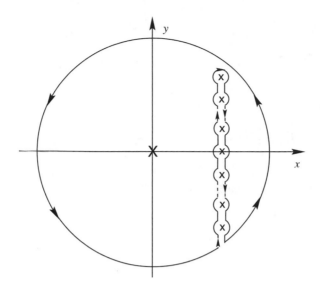

Figure 1.4.1: Modified contour used to evaluate Equation 1.4.24.

Turning to the evaluation of I,

$$|I| = \left| \frac{1}{2\pi} \int_0^{2\pi} \frac{\left(Re^{\theta i}\right)^{-n}}{1 - a\exp\{R[\cos(\theta) + i\sin(\theta)]\}} \, d\theta \right| \qquad (1.4.27)$$

$$\leq \frac{R^{-n}}{2\pi} \int_0^{2\pi} \frac{d\theta}{|1 - a\exp[R\cos(\theta)]|}. \qquad (1.4.28)$$

As $R \to \infty$, we note that

$$|1 - a\exp[R\cos(\theta)]|^{-1} \sim \begin{cases} O[e^{-R\cos(\theta)}], & 0 \leq \theta < \pi/2, \, 3\pi/2 < \theta \leq 2\pi, \\ O(1), & \pi/2 < \theta < 3\pi/2, \end{cases} \qquad (1.4.29)$$

so that the integral in Equation 1.4.28 is finite and equals M, say. Consequently,

$$|I| \leq \lim_{R \to \infty} \frac{MR^{-n}}{2\pi} \to 0. \qquad (1.4.30)$$

Therefore,

$$\frac{1}{2\pi i} \oint_C f(z) \, dz = -\frac{1}{[\ln(a)]^{n+1}} + \sum_{m=1}^{\infty} \frac{1}{[2m\pi i - \ln(a)]^{n+1}} + \frac{1}{[-2m\pi i - \ln(a)]^{n+1}}. \qquad (1.4.31)$$

• **Example 1.4.2**

Let us evaluate

$$\int_0^\infty \frac{\cos(kx)}{x^2 + a^2}\, dx,$$

where a and k are real and positive. First,

$$\int_0^\infty \frac{\cos(kx)}{x^2 + a^2}\, dx = \frac{1}{2} \int_{-\infty}^\infty \frac{\cos(kx)}{x^2 + a^2}\, dx = \frac{1}{2} \operatorname{Re}\!\left(\int_{-\infty}^\infty \frac{e^{ikx}}{x^2 + a^2}\, dx \right). \quad (1.4.32)$$

We close the line integral along the real axis by introducing an infinite semi-circle in the upper half-plane as dictated by Jordan's lemma. Therefore,

$$\int_0^\infty \frac{\cos(kx)}{x^2 + a^2}\, dx = \frac{1}{2} \operatorname{Re}\!\left(\oint_C \frac{e^{ikz}}{z^2 + a^2}\, dz \right) = \operatorname{Re}\!\left[\pi i \operatorname{Res}\!\left(\frac{e^{ikz}}{z^2 + a^2}; ai \right) \right] \quad (1.4.33)$$

$$= \operatorname{Re}\!\left[\pi i \lim_{z \to ia} \frac{(z - ia)e^{ikz}}{z^2 + a^2} \right] = \frac{\pi}{2a} e^{-ka}. \quad (1.4.34)$$

• **Example 1.4.3**

When the definite integral involves hyperbolic functions, a rectangular closed contour is generally the best one to use. For example, consider the contour integral[13]

$$\oint_C \frac{e^{2\lambda z}}{\{\cosh(z) + \cos[\pi/(2\lambda)]\}^2}\, dz, \qquad \tfrac{1}{2} < \lambda < 1, \quad (1.4.35)$$

where C is the closed rectangular contour $ABCD$ shown in Figure 1.4.2. Along AD as $R \to \infty$,

$$\frac{e^{2\lambda z}}{\{\cosh(z) + \cos[\pi/(2\lambda)]\}^2} \to e^{-2(1-\lambda)R} \to 0. \quad (1.4.36)$$

Similarly, along BC

$$\frac{e^{2\lambda z}}{\{\cosh(z) + \cos[\pi/(2\lambda)]\}^2} \to e^{-2(1+\lambda)R} \to 0. \quad (1.4.37)$$

Along AB

$$\int_\infty^{-\infty} \frac{e^{2\lambda z}}{\{\cosh(z) + \cos[\pi/(2\lambda)]\}^2}\, dz = -\int_0^\infty \frac{e^{2\lambda(-x+\pi i)}}{\{\cos[\pi/(2\lambda)] - \cosh(x)\}^2}\, dx$$

$$- \int_0^\infty \frac{e^{2\lambda(x+\pi i)}}{\{\cos[\pi/(2\lambda)] - \cosh(x)\}^2}\, dx, \quad (1.4.38)$$

---

[13] Taken from Hawthorne, W. R., 1954: The secondary flow about struts and airfoils. *J. Aeronaut. Sci.*, **21**, 588–608.

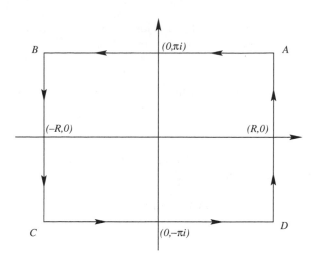

Figure 1.4.2: Closed contour used in evaluating Equation 1.4.35.

while along CD

$$\int_{-\infty}^{\infty} \frac{e^{2\lambda z}}{\{\cosh(z) + \cos[\pi/(2\lambda)]\}^2} \, dz = \int_{0}^{\infty} \frac{e^{2\lambda(x-\pi i)}}{\{\cos[\pi/(2\lambda)] - \cosh(x)\}^2} \, dx$$

$$+ \int_{0}^{\infty} \frac{e^{2\lambda(-x-\pi i)}}{\{\cos[\pi/(2\lambda)] - \cosh(x)\}^2} \, dx. \quad (\mathbf{1.4.39})$$

Therefore, summing the four segments,

$$-4i\sin(2\lambda\pi) \int_{0}^{\infty} \frac{\cosh(2\lambda x) \, dx}{\{\cosh(x) - \cos[\pi/(2\lambda)]\}^2}$$

$$= 2\pi i \sum_{j=1}^{n} \mathrm{Res}\left(\frac{e^{2\lambda z}}{\{\cosh(z) + \cos[\pi/(2\lambda)]\}^2} ; z_j \right), \quad (\mathbf{1.4.40})$$

or

$$\int_{0}^{\infty} \frac{\cosh^2(\lambda x)}{\{\cosh(x) - \cos[\pi/(2\lambda)]\}^2} \, dx$$

$$= -\frac{\pi}{4\sin(2\lambda\pi)} \sum_{j=1}^{n} \mathrm{Res}\left(\frac{e^{2\lambda z}}{\{\cosh(z) + \cos[\pi/(2\lambda)]\}^2} ; z_j \right)$$

$$+ \frac{1}{2} \int_{0}^{\infty} \frac{dx}{\{\cosh(x) - \cos[\pi/(2\lambda)]\}^2}. \quad (\mathbf{1.4.41})$$

From a table of integrals,

$$\int_{0}^{\infty} \frac{dx}{\{\cosh(x) - \cos[\pi/(2\lambda)]\}^2} = \int_{1}^{\infty} \frac{4y \, dy}{\{y^2 - 2y\cos[\pi/(2\lambda)] + 1\}^2} \quad (\mathbf{1.4.42})$$

$$= \frac{\pi[1 - 1/(2\lambda)]\cot[\pi/(2\lambda)] + 1}{\sin^2(2\pi/\lambda)}. \quad (\mathbf{1.4.43})$$

The only poles located inside the closed contour occur at $z_\pm = \pm\pi i[1 - 1/(2\lambda)]$. To compute their residues, we note that

$$\frac{e^{2\lambda z}}{[\cosh(z) - \cosh(z_\pm)]^2} = \frac{e^{2\lambda z_\pm} e^{2\lambda\zeta}}{\{\cosh(z_\pm)[\cosh(\zeta) - 1] + \sinh(z_\pm)\sinh(\zeta)\}^2} \quad (1.4.44)$$

$$= \frac{e^{2\lambda z_\pm}}{\cosh^2(z_\pm)} \left\{ \frac{1}{\zeta^2} + \frac{[2\lambda - \coth(z_\pm)]}{\zeta} + \cdots \right\}, \quad (1.4.45)$$

because $\cos[\pi/(2\lambda)] = -\cosh(z_\pm)$ and $\zeta = z - z_\pm$. Therefore, the poles are second order and the residues equal

$$-e^{\pm\pi i(2\lambda-1)}\{2\lambda \mp i\cot[\pi/(2\lambda)]\}\csc^2[\pi/(2\lambda)].$$

Hence the sum of residues is

$$\{4\lambda\cos(2\lambda\pi) + 2\sin(2\lambda\pi)\cot[\pi/(2\lambda)]\}\csc^2[\pi/(2\lambda)].$$

Substituting this sum and Equation 1.4.43 into Equation 1.4.41, we finally have

$$\int_0^\infty \frac{\cosh^2(\lambda x)}{\{\cosh(x) - \cos[\pi/(2\lambda)]\}^2}\,dx = \csc^2[\pi/(2\lambda)](1 - \pi\{2\lambda\cot(2\pi\lambda)$$

$$+ \cot[\pi/(2\lambda)]/(2\lambda)\})/2. \quad (1.4.46)$$

In this section, we have given a quick overview of complex variables as it applies to single-valued functions. For greater detail, as well as drill exercises, the reader is referred to Chapter 1 of the author's *Advanced Engineering Mathematics with MATLAB*.[14] In those instances where there are multivalued functions due to the presence of z raised to some rational power, inverse functions or logarithms, we must make them single-valued. This is the subject of the next two sections.

1.5 MULTIVALUED FUNCTIONS, BRANCH POINTS, BRANCH CUTS AND RIEMANN SURFACES

In this section, we introduce functions that yield several different values of w for a given z, i.e., multivalued functions. We must make these functions single-valued so that we can apply the techniques from the previous section. Furthermore, this condition is also necessary for a well-posed physical problem.

Consider the complex function $w = z^{1/2}$. For each value of z there are two possible values of w. For example, if $z = -i$, then w equals either $w_1 = -1/\sqrt{2} + i/\sqrt{2}$ or $w_2 = 1/\sqrt{2} - i/\sqrt{2}$. The points w_1 and w_2 are

---

[14] Duffy, *op. cit.*

distinct members from two *branches* of the same function that "branch off" from the same point, $z = 0$. The number of branches depends upon the nature of $f(z)$. For example, $\log(z)$ has an infinite number of branches, namely, $\ln(r) + \theta i + n\pi i, n = 0, 1, 2, \ldots$.

In the case of real variables, the branches easily separate. For example, the square root of a positive, real number has two distinct branches: a and $-a$, where a is a real, positive number. However, in complex functions the two branches are hardly distinguishable because they do not separate at all. Therefore, if we wish to keep them separate, we must do it artificially.

Consider again the complex function $w = z^{1/2} = r^{1/2}e^{\theta i/2}$. If we move around a closed path that does *not* encircle the origin, the values of r and θ vary continuously. At the end, the final value of θ equals our initial value, θ_0. Consequently, we can say that all of the values of w along this contour belong to the same branch, $w_1 = r^{1/2}e^{i\theta_0/2}$.

Let our closed contour now enclose the origin. Because the final value of $\theta = \theta_0 + 2\pi$, the final value of w now equals $w_2 = -r^{1/2}e^{i\theta_0/2} = -w_1$. We have reached the other branch of w in a continuous manner. Consequently, $z = 0$ appears to be a special point; one branch ties to the other there because they have the same value. We reach each branch from the other after a complete turn around the origin. A *branch point* is any point having this property. In this example, infinity is also a branch point. We can show this by the substitution $z = 1/z'$ and an examination of the transformed function about $z' = 0$.

In our example with $w = z^{1/2}$, we reached the other branch by completing a closed contour around the origin. However, we cannot say exactly when, in our journey, we crossed the boundary from w_1 and w_2. For example, it might have been when we crossed the positive real axis or when we crossed the negative real axis. The conclusions are the same either way. This ambiguity leads to the concept of the *branch cut*. The reason why we must define a branch cut between branch points lies in the fact that we cannot make a multivalued function single-valued by excluding the branch point and a small neighborhood around it. Its multivaluedness does not depend on the mere existence of the branch point itself, but on the possibility of encircling it.

In summary, a branch cut is a line that *we* choose that connects two branch points. Furthermore, it defines the separation between the branches. For this reason the branch cut is a barrier that we may not cross.[15]

A geometrical interpretation of this process is to limit each branch to a particular *Riemann surface*. We can view each Riemann surface as a floor in a large department store. On each floor (Riemann surface) you can only obtain one type of branch (for example, square roots with positive real parts). However, we can reach other floors (surfaces), if desired, through a set of escalators, located at the branch cut, that take you up to the next higher

---

[15] The Russian mathematical term for the edge along a branch cut is *bereg*, which commonly means "the bank of a river."

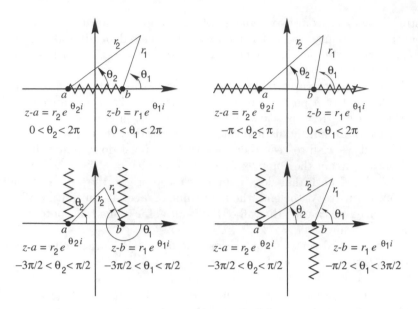

Figure 1.5.1: Some popular branch cut configurations for $\sqrt{(z-a)(z-b)}$. The branch cuts are denoted by wavy lines.

Riemann surface (if it exists) or down to the next lower Riemann surface (if it exists). These (very thin) escalators extend between the branch points. As the architect of your Riemann surface, there is great flexibility in choosing where your branch cuts lie. Figure 1.5.1 and Figure 1.5.2 show some of the more popular choices for $\sqrt{(z-a)(z-b)}$ and $\sqrt{(z-ai)(z-bi)}$, respectively.

The advantage of introducing a Riemann surface is that every continuous curve on the z-plane maps the multivalued function into a continuous curve on the w-plane. This relationship between the Riemann surfaces allows us to apply to multivalued analytic functions all of the techniques of integration, analytic continuation and so forth, which depend on a continuous path being drawn from one point to another.

In the next section, we illustrate the mechanics of integration involving multivalued functions.

1.6 SOME EXAMPLES OF INTEGRATION THAT INVOLVE MULTIVALUED FUNCTIONS

In this section we perform contour integration with multivalued functions. Essentially, the introduction of branch points and cuts requires careful bookkeeping of phases (or arguments). Once we set up our bookkeeping, the evaluation follows directly.

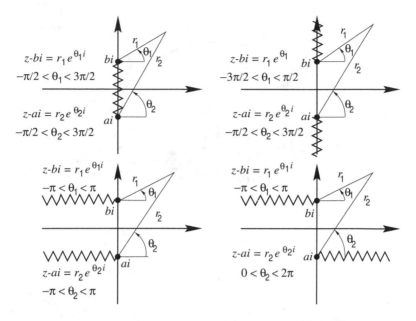

Figure 1.5.2: Some popular branch cut configurations for $\sqrt{(z-ai)(z-bi)}$. The branch cuts are denoted by wavy lines.

- **Example 1.6.1**

Let us evaluate the integral[16]

$$\oint_C f(z)\,dz = \oint_C \left(\frac{a+z}{c-z}\right)^{1/2} \frac{dz}{(d-z)(z-b)}, \qquad (1.6.1)$$

where $-a < b < c < d$. Figure 1.6.1 shows the contour $C = C_1 \cup C_2 \cup \ldots \cup C_8$. The integrand is a multivalued function with branch points at $z = -a$ and $z = c$ and simple poles at $z = b$ and $z = d$.

We begin by expanding our original contour C without bound so that it includes all of the singularities. Figure 1.6.1 shows this enlarged contour Γ. Next we employ the following theorem:

Theorem:[17] *If $zf(z)$ tends uniformly to a limit k as $|z|$ increases indefinitely, the value of $\oint f(z)\,dz$, taken around a very large circle, center the origin, tends toward $2\pi i k$.*

---

[16] This example is taken from Glagolev, N. I., 1945: Resistance of cylindrical bodies in rolling (in Russian). *Prikl. Mat. Mek.*, **9**, 318–333.

[17] Forsyth, A. R., 1965: *Theory of Functions of a Complex Variable.* Dover Publications, Inc., p. 41.

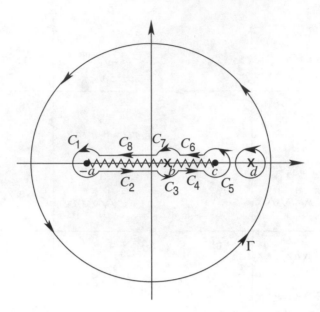

Figure 1.6.1: The contour used to evaluate Equation 1.6.1.

Because $zf(z)$ tends to zero,

$$\oint_\Gamma \left(\frac{a+z}{c-z}\right)^{1/2} \frac{dz}{(d-z)(z-b)} = 0. \tag{1.6.2}$$

We evaluate the integral around the contour C by noting that an integration around the contour Γ is equivalent to integrations along C and around the pole at $z = d$. If the phase of the functions $z + a$ and $z - c$ lies between 0 and 2π, which is consistent with our choice for the branch cut, then the residue theorem yields

$$\oint_C \left(\frac{a+z}{c-z}\right)^{1/2} \frac{dz}{(d-z)(z-b)} = \oint_\Gamma f(z)\, dz - 2\pi i \operatorname{Res}\left[\left(\frac{a+z}{c-z}\right)^{1/2} \frac{1}{(d-z)(z-b)}; d\right]$$

$$= \frac{2\pi}{d-b}\left(\frac{a+d}{d-c}\right)^{1/2}, \tag{1.6.3}$$

because $(a + z)/(c - z)$ always has the phase of $-\pi$ at the point $z = d$.

Integrals similar to those in Equation 1.6.3 are important because we can use them to evaluate definite integrals where the integrand contains a square root. In the present problem, for example, we can convert this contour integral into a real, definite integral by first noting that

$$\oint_C f(z)\, dz = \int_{C_1} f(z)\, dz + \int_{C_2} f(z)\, dz + \int_{C_3} f(z)\, dz + \int_{C_4} f(z)\, dz$$

$$+ \int_{C_5} f(z)\, dz + \int_{C_6} f(z)\, dz + \int_{C_7} f(z)\, dz + \int_{C_8} f(z)\, dz. \tag{1.6.4}$$

Next we express $z - c$ and $z + a$ in terms of $re^{\theta i}$ with the phase $0 < \theta < 2\pi$. To evaluate the integral on the contour C_1, for example, $z + a = \epsilon e^{\theta i}$ and $z - c = (a + c)e^{\pi i}$, so that

$$\int_{C_1} f(z)\, dz = \lim_{\epsilon \to 0} \int_0^{2\pi} \frac{\epsilon^{1/2} e^{\theta i/2}}{\sqrt{a + c}} \frac{i\epsilon e^{\theta i}}{(a + d)(-a - b)}\, d\theta = 0, \qquad (1.6.5)$$

whereas for C_5, $z - c = \epsilon e^{\theta i}$ and $z + a = (a + c)e^{0i}$ or $(a + c)e^{2\pi i}$, resulting in

$$\int_{C_5} f(z)\, dz = -\lim_{\epsilon \to 0} \int_{-\pi}^0 \frac{(a + c)^{1/2}}{i\epsilon^{1/2} e^{\theta i/2}} \frac{i\epsilon e^{\theta i}}{(d - c)(c - b)}\, d\theta$$

$$+ \lim_{\epsilon \to 0} \int_0^{\pi} \frac{(a + c)^{1/2}}{i\epsilon^{1/2} e^{\theta i/2}} \frac{i\epsilon e^{\theta i}}{(d - c)(c - b)}\, d\theta = 0. \qquad (1.6.6)$$

At $z = b$, $z - b = \epsilon e^{\theta i}$, $z - c = (c - b)e^{\pi i}$, $z + a = (a + b)e^{0i}$ on C_7 and $(a + b)e^{2\pi i}$ on C_3, so that

$$\int_{C_3} f(z)\, dz = \lim_{\epsilon \to 0} \left(\frac{a + b}{c - b}\right)^{1/2} \left(\frac{1}{d - b}\right) \int_{-\pi}^0 \frac{i\epsilon e^{\theta i}}{\epsilon e^{\theta i}}\, d\theta \qquad (1.6.7)$$

$$= \left(\frac{a + b}{c - b}\right)^{1/2} \frac{\pi i}{d - b} \qquad (1.6.8)$$

and

$$\int_{C_7} f(z)\, dz = -\lim_{\epsilon \to 0} \left(\frac{a + b}{c - b}\right)^{1/2} \left(\frac{1}{d - b}\right) \int_0^{\pi} \frac{i\epsilon e^{\theta i}}{\epsilon e^{\theta i}}\, d\theta \qquad (1.6.9)$$

$$= -\left(\frac{a + b}{c - b}\right)^{1/2} \frac{\pi i}{d - b}. \qquad (1.6.10)$$

We evaluate the remaining contour integrals by noting that $z - c = (c - x)e^{\pi i}$, $z + a = (a + x)e^{2\pi i}$ along C_2 and C_4 and $z + a = (a + x)e^{0i}$ along C_6 and C_8, where $-a < x < c$. Upon substituting these definitions into the integrals and simplifying,

$$\int_{C_2 \cup C_4} f(z)\, dz = \int_{-a}^c \left(\frac{a + x}{c - x}\right)^{1/2} \frac{dx}{(d - x)(x - b)} \qquad (1.6.11)$$

and

$$\int_{C_6 \cup C_8} f(z)\, dz = -\int_c^{-a} \left(\frac{a + x}{c - x}\right)^{1/2} \frac{dx}{(d - x)(x - b)}. \qquad (1.6.12)$$

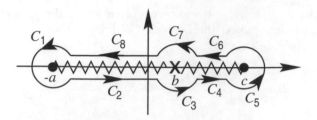

Figure 1.6.2: The contour used to evaluate Equation 1.6.15.

Substituting these results into Equation 1.6.4, our final answer is

$$\int_{-a}^{c} \left(\frac{a+x}{c-x}\right)^{1/2} \frac{dx}{(d-x)(x-b)} - \int_{c}^{-a} \left(\frac{a+x}{c-x}\right)^{1/2} \frac{dx}{(d-x)(x-b)}$$

$$= \frac{2\pi}{d-b}\left(\frac{a+d}{d-c}\right)^{1/2}, \tag{1.6.13}$$

or

$$\int_{-a}^{c} \left(\frac{a+x}{c-x}\right)^{1/2} \frac{dx}{(d-x)(x-b)} = \frac{\pi}{d-b}\left(\frac{a+d}{d-c}\right)^{1/2}. \tag{1.6.14}$$

A similar problem involves the integral

$$\oint_{C} f(z)\,dz = \oint_{C} \left(\frac{a+z}{c-z}\right)^{\delta} \frac{dz}{z-b}, \tag{1.6.15}$$

where $-a < b < c$ and $|\delta| < 1$. This multivalued function has branch points at $z = -a$ and $z = c$. Figure 1.6.2 shows the contour of integration. To evaluate the contour integral, we again make use of the theorem stated after Equation 1.6.1. Thus,

$$\oint_{C} f(z)\,dz = 2\pi i(e^{-\pi i})^{\delta} = 2\pi i e^{-\pi \delta i}. \tag{1.6.16}$$

Once again, we can reduce our contour integral to a real definite integral by breaking the closed contour into eight segments:

$$\oint_{C} f(z)\,dz = \int_{C_1} f(z)\,dz + \int_{C_2} f(z)\,dz + \int_{C_3} f(z)\,dz + \int_{C_4} f(z)\,dz$$

$$+ \int_{C_5} f(z)\,dz + \int_{C_6} f(z)\,dz + \int_{C_7} f(z)\,dz + \int_{C_8} f(z)\,dz. \tag{1.6.17}$$

Proceeding as before, the contribution from the circles at $z = -a$ and $z = c$ vanish. We evaluate the remaining integrals by noting that $z - c = (c - x)e^{\pi i}$,

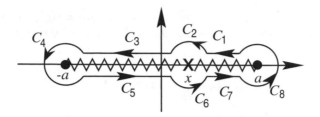

Figure 1.6.3: The contour used to evaluate Equation 1.6.20.

$z + a = (x+a)e^{2\pi i}$ along C_2, C_3 and C_4, and $z + a = (x+a)e^{0i}$ along C_6, C_7 and C_8 with $-a < x < c$. After substituting into Equation 1.6.17,

$$
\int_{-a}^{c} \left(\frac{a+x}{c-x}\right)^{\delta} \frac{dx}{x-b} + e^{-2\pi\delta i} \int_{c}^{-a} \left(\frac{a+x}{c-x}\right)^{\delta} \frac{dx}{x-b}
$$
$$
+ \pi i \left(\frac{a+b}{c-b}\right)^{\delta} + e^{-2\pi\delta i} \pi i \left(\frac{a+b}{c-b}\right)^{\delta} = 2\pi i e^{-\pi\delta i}, \quad (1.6.18)
$$

or

$$
\int_{-a}^{c} \left(\frac{a+x}{c-x}\right)^{\delta} \frac{dx}{x-b} = \frac{\pi}{\sin(\delta\pi)} - \pi\cot(\delta\pi) \left(\frac{a+b}{c-b}\right)^{\delta}. \quad (1.6.19)
$$

• **Example 1.6.2**

Let us evaluate

$$
\oint_C f(z)\,dz = \oint_C \frac{dz}{(z^2 - a^2)^{\mu/2}(z-x)^{1-\mu}}, \quad (1.6.20)
$$

where $-a < x < a$ and $0 < \mu < 1$. Figure 1.6.3 shows that the contour runs just below the real axis from $-a$ to a and then back to $-a$ just above the real axis.[18] The integrand is a multivalued function with three branch points at $z = a$, x and $-a$. The branch cut runs along the real axis from $-a$ to a.

The value of Equation 1.6.20 follows from the limit of $zf(z)$ as $|z| \to \infty$. (See the previous example.) If the argument of $z - a$, $z + a$ and $z - x$ lies between 0 and 2π, this limit equals one and

$$
\oint_C f(z)\,dz = 2\pi i. \quad (1.6.21)
$$

---

[18] Reprinted from *J. Appl. Math. Mech.*, **23**, V. Kh. Arutiunian, The plane contact problem of the theory of creep, 1283–1313, ©1959, with kind permission from Pergamon Press Ltd., Headington Hill Hall, Oxford OX3 0BX, UK.

Once again we can use our results to evaluate a real, definite integral by evaluating Equation 1.6.21 along the various legs of the contour shown in Figure 1.6.3,

$$
\int_{C_1} f(z)\,dz + \int_{C_2} f(z)\,dz + \int_{C_3} f(z)\,dz + \int_{C_4} f(z)\,dz
$$
$$
+ \int_{C_5} f(z)\,dz + \int_{C_6} f(z)\,dz + \int_{C_7} f(z)\,dz + \int_{C_8} f(z)\,dz = 2\pi i. \quad (1.6.22)
$$

At $z = -a$, $z + a = \epsilon e^{\theta i}$, $z - x = (x+a)e^{\pi i}$ and $z - a = 2ae^{\pi i}$ so that

$$
\int_{C_4} f(z)\,dz = \lim_{\epsilon \to 0} \int_0^{2\pi} \frac{i\epsilon e^{\theta i}\,d\theta}{(2a\epsilon e^{\theta i + \pi i})^{\mu/2}(x+a)^{1-\mu}e^{\pi i - \mu\pi i}} = 0; \quad (1.6.23)
$$

while at $z = a$, $z - a = \epsilon e^{\theta i}$, $z + a = 2ae^{0i}$ or $2ae^{2\pi i}$, and $z - x = (a-x)e^{0i}$ or $(a-x)e^{2\pi i}$,

$$
\int_{C_8} f(z)\,dz = \lim_{\epsilon \to 0} \int_{-\pi}^0 \frac{i\epsilon e^{\theta i}\,d\theta}{(2a\epsilon e^{\theta i + 2\pi i})^{\mu/2}(a-x)^{1-\mu}e^{2\pi i - 2\mu\pi i}}
$$
$$
+ \lim_{\epsilon \to 0} \int_0^\pi \frac{i\epsilon e^{\theta i}\,d\theta}{(2a\epsilon e^{\theta i})^{\mu/2}(a-x)^{1-\mu}} = 0. \quad (1.6.24)
$$

At $z = x$, $z - x = \epsilon e^{\theta i}$, $z - a = (a-x)e^{\pi i}$, and $z + a = (x+a)e^{0i}$ on C_2 and $(x+a)e^{2\pi i}$ on C_6 so that

$$
\int_{C_2} f(z)\,dz = \lim_{\epsilon \to 0} \int_0^\pi \frac{i\epsilon e^{\theta i}\,d\theta}{(a^2 - x^2)^{\mu/2}e^{\pi\mu i/2}(\epsilon e^{\theta i})^{1-\mu}} = 0 \quad (1.6.25)
$$

and

$$
\int_{C_6} f(z)\,dz = \lim_{\epsilon \to 0} \int_\pi^{2\pi} \frac{i\epsilon e^{\theta i}\,d\theta}{(a^2 - x^2)^{\mu/2}e^{3\pi\mu i/2}(\epsilon e^{\theta i})^{1-\mu}} = 0. \quad (1.6.26)
$$

Finally, for the straight line segments, $z - a = (a-s)e^{\pi i}$, $z + a = (a+s)e^{0i}$ along C_1 and C_3 and $(a+s)e^{2\pi i}$ on C_5 and C_7, and $z - x = |s-x|e^{0i}$ on C_1, $|x-s|e^{\pi i}$ on C_3 and C_5 and $|s-x|e^{2\pi i}$ on C_7. Therefore,

$$
\int_{C_1} f(z)\,dz = \int_a^x \frac{e^{-\mu\pi i/2}\,ds}{(a^2 - s^2)^{\mu/2}|s - x|^{1-\mu}} = -\int_x^a \frac{e^{-\mu\pi i/2}\,ds}{(a^2 - s^2)^{\mu/2}|s - x|^{1-\mu}}, \quad (1.6.27)
$$

$$
\int_{C_3} f(z)\,dz = \int_x^{-a} \frac{e^{-\pi i + \mu\pi i/2}\,ds}{(a^2 - s^2)^{\mu/2}|s - x|^{1-\mu}} = \int_{-a}^x \frac{e^{\mu\pi i/2}\,ds}{(a^2 - s^2)^{\mu/2}|s - x|^{1-\mu}}, \quad (1.6.28)
$$

$$
\int_{C_5} f(z)\,dz = \int_{-a}^x \frac{e^{-\pi i - \mu\pi i/2}\,ds}{(a^2 - s^2)^{\mu/2}|s - x|^{1-\mu}}, \quad (1.6.29)
$$

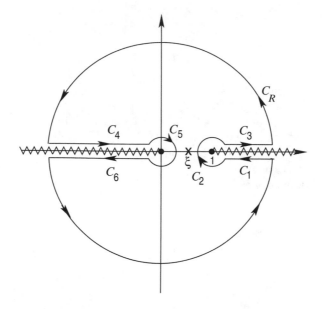

Figure 1.6.4: The contour used to evaluate Equation 1.6.34.

and

$$\int_{C_7} f(z)\, dz = \int_x^a \frac{e^{-2\pi i + \mu \pi i/2}\, ds}{(a^2 - s^2)^{\mu/2}|s - x|^{1-\mu}}. \tag{1.6.30}$$

Substituting these integrals into Equation 1.6.22 and simplifying,

$$\int_{-a}^a \sin\!\left(\frac{\mu\pi}{2}\right) \frac{(a^2 - s^2)^{-\mu/2}}{|s - x|^{1-\mu}}\, ds = \frac{\pi}{2}. \tag{1.6.31}$$

• Example 1.6.3

Let us simplify the integral with a singular kernel[19]

$$\int_1^\infty t^{\alpha-1}(t - 1)^{\delta-1}\frac{t^{1/2}}{t - \xi}\, dt, \qquad 0 < \xi < 1, \tag{1.6.32}$$

by evaluating the complex integral

$$\oint_C z^{\alpha-1}(z - 1)^{\delta-1}\frac{z^{1/2}}{z - \xi}\, dz, \tag{1.6.33}$$

where α and δ are nonintegers. The contour C is a circle of infinite radius with appropriate branch cuts as shown in Figure 1.6.4. From the residue theorem,

---

[19] Taken from Liu, P. L.-F., 1986: Hydrodynamic pressures on rigid dams during earthquakes. *J. Fluid Mech.*, **165**, 131–145. Reprinted with the permission of Cambridge University Press.

$$\oint_C z^{\alpha-1}(z-1)^{\delta-1}\frac{z^{1/2}}{z-\xi}\,dz = 2\pi i \xi^{\alpha-1}(1-\xi)^{\delta-1}\xi^{1/2}e^{(\delta-1)\pi i}, \qquad \textbf{(1.6.34)}$$

because there is a simple pole at $z = \xi$.

Because the integrals vanish along C_R as $R \to \infty$ and along the arcs C_2 and C_5 as the circles about $z = 0$ and $z = 1$ become infinitesimally small, we need only evaluate the line integrals along C_1, C_3, C_4 and C_6. The branch cuts associated with $z^{\alpha-1}$ and $z^{1/2}$ lie along the negative real axis. Consequently the argument for z runs from $-\pi$ and π. The argument for $z - 1$ used in $(z-1)^{\delta-1}$, however, lies between 0 and 2π. Taking these arguments into account,

$$\int_{C_3} z^{\alpha-1}(z-1)^{\delta-1}\frac{z^{1/2}}{z-\xi}\,dz = \int_1^\infty x^{\alpha-1}(x-1)^{\delta-1}\frac{x^{1/2}}{x-\xi}\,dx, \qquad \textbf{(1.6.35)}$$

$$\int_{C_4} z^{\alpha-1}(z-1)^{\delta-1}\frac{z^{1/2}}{z-\xi}\,dz$$

$$= \int_\infty^0 \frac{x^{\alpha-1}e^{(\alpha-1)\pi i}(x+1)^{\delta-1}e^{(\delta-1)\pi i}x^{1/2}e^{\pi i/2}}{x+\xi}\,dx \qquad \textbf{(1.6.36)}$$

$$= -ie^{(\alpha+\delta)\pi i}\int_0^\infty \frac{x^{\alpha-1}(x+1)^{\delta-1}x^{1/2}}{x+\xi}\,dx, \qquad \textbf{(1.6.37)}$$

$$\int_{C_6} z^{\alpha-1}(z-1)^{\delta-1}\frac{z^{1/2}}{z-\xi}\,dz$$

$$= \int_0^\infty \frac{x^{\alpha-1}e^{-(\alpha-1)\pi i}(x+1)^{\delta-1}e^{(\delta-1)\pi i}x^{1/2}e^{-\pi i/2}}{x+\xi}\,dx \qquad \textbf{(1.6.38)}$$

$$= -ie^{(\delta-\alpha)\pi i}\int_0^\infty \frac{x^{\alpha-1}(x+1)^{\delta-1}x^{1/2}}{x+\xi}\,dx, \qquad \textbf{(1.6.39)}$$

and

$$\int_{C_1} z^{\alpha-1}(z-1)^{\delta-1}\frac{z^{1/2}}{z-\xi}\,dz = \int_\infty^1 \frac{x^{\alpha-1}(x-1)^{\delta-1}e^{2\pi(\delta-1)i}x^{1/2}}{x-\xi}\,dx \qquad \textbf{(1.6.40)}$$

$$= -e^{2\delta\pi i}\int_1^\infty \frac{x^{\alpha-1}(x-1)^{\delta-1}x^{1/2}}{x-\xi}\,dx. \qquad \textbf{(1.6.41)}$$

Substituting these results into Equation 1.6.34,

$$\int_1^\infty \frac{x^{\alpha-\frac{1}{2}}(x-1)^{\delta-1}}{x-\xi}\,dx = \frac{\xi^{\alpha-\frac{1}{2}}(1-\xi)^{\delta-1}}{\sin(\pi\delta)}\pi$$

$$- \frac{\cos(\pi\alpha)}{\sin(\pi\delta)}\int_0^\infty \frac{x^{\alpha-1/2}(x+1)^{\delta-1}}{x+\xi}\,dx. \qquad \textbf{(1.6.42)}$$

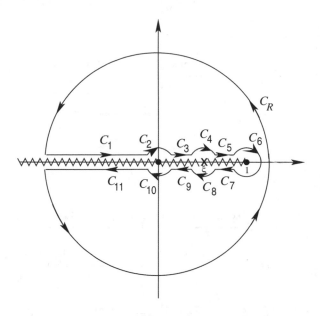

Figure 1.6.5: The contour used in redoing Equation 1.6.34.

Although we must still evaluate an integral numerically, the integrand is no longer singular.

Let us now redo this integral with the contour shown in Figure 1.6.5. The phase of both z and $z-1$ now run from $-\pi$ to π. The contributions from C_R, C_2 and C_{10} are zero. However, for the other contours,

$$\int_{C_1} \frac{z^{\alpha-1}(z-1)^{\delta-1}z^{1/2}}{z-\xi}\, dz$$

$$= \int_\infty^0 \frac{x^{\alpha-1}e^{(\alpha-1)\pi i}(1+x)^{\delta-1}e^{(\delta-1)\pi i}x^{1/2}e^{\pi i/2}}{x+\xi}\, dx, \quad (\mathbf{1.6.43})$$

$$= -ie^{(\alpha+\delta)\pi i}\int_0^\infty \frac{x^{\alpha-1}(1+x)^{\delta-1}x^{1/2}}{x+\xi}\, dx, \quad (\mathbf{1.6.44})$$

$$\int_{C_3\cup C_5} \frac{z^{\alpha-1}(z-1)^{\delta-1}z^{1/2}}{z-\xi}\, dz = \int_0^1 \frac{x^{\alpha-1}(1-x)^{\delta-1}e^{(\delta-1)\pi i}x^{1/2}}{x-\xi}\, dx \quad (\mathbf{1.6.45})$$

$$= -e^{\delta\pi i}\int_0^1 \frac{x^{\alpha-1}(1-x)^{\delta-1}x^{1/2}}{x-\xi}\, dx, \quad (\mathbf{1.6.46})$$

$$\int_{C_7\cup C_9} \frac{z^{\alpha-1}(z-1)^{\delta-1}z^{1/2}}{z-\xi}\, dz = \int_1^0 \frac{x^{\alpha-1}(1-x)^{\delta-1}e^{-(\delta-1)\pi i}x^{1/2}}{x-\xi}\, dx \quad (\mathbf{1.6.47})$$

$$= e^{-\delta\pi i}\int_0^1 \frac{x^{\alpha-1}(1-x)^{\delta-1}x^{1/2}}{x-\xi}\, dx, \quad (\mathbf{1.6.48})$$

$$\int_{C_{11}} \frac{z^{\alpha-1}(z-1)^{\delta-1}z^{1/2}}{z-\xi}\, dz$$

$$= \int_0^\infty \frac{x^{\alpha-1}e^{-(\alpha-1)\pi i}(1+x)^{\delta-1}e^{-(\delta-1)\pi i}x^{1/2}e^{-\pi i/2}}{x+\xi}\, dx \qquad (1.6.49)$$

$$= -ie^{-(\alpha+\delta)\pi i}\int_0^\infty \frac{x^{\alpha-1}(1+x)^{\delta-1}x^{1/2}}{x+\xi}\, dx, \qquad (1.6.50)$$

and

$$\int_{C_4\cup C_8} \frac{z^{\alpha-1}(z-1)^{\delta-1}z^{1/2}}{z-\xi}\, dz$$

$$= \lim_{\epsilon\to 0}\int_\pi^0 i\epsilon e^{\theta i}d\theta \frac{(\xi+\epsilon e^{\theta i})^{\alpha-1}(1-\xi-\epsilon e^{\theta i})^{\delta-1}e^{(\delta-1)\pi i}(\xi+\epsilon e^{\theta i})^{1/2}}{\epsilon e^{\theta i}}$$

$$+ \lim_{\epsilon\to 0}\int_0^{-\pi} i\epsilon e^{\theta i}d\theta \frac{(\xi+\epsilon e^{\theta i})^{\alpha-1}(1-\xi-\epsilon e^{\theta i})^{\delta-1}e^{-(\delta-1)\pi i}(\xi+\epsilon e^{\theta i})^{1/2}}{\epsilon e^{\theta i}}$$

$$\qquad (1.6.51)$$

$$= -i\pi\xi^{\alpha-1}(1-\xi)^{\delta-1}\xi^{1/2}e^{(\delta-1)\pi i} - i\pi\xi^{\alpha-1}(1-\xi)^{\delta-1}\xi^{1/2}e^{-(\delta-1)\pi i}$$

$$\qquad (1.6.52)$$

$$= 2i\pi\xi^{\alpha-1}(1-\xi)^{\delta-1}\xi^{1/2}\cos(\pi\delta). \qquad (1.6.53)$$

Bringing these results together,

$$\int_0^1 \frac{x^{\alpha-\frac{1}{2}}(1-x)^{\delta-1}}{x-\xi}\, dx = \pi\xi^{\alpha-1/2}(1-\xi)^{\delta-1}\frac{\cos(\pi\delta)}{\sin(\pi\delta)}$$

$$- \frac{\cos[\pi(\alpha+\delta)]}{\sin(\pi\delta)}\int_0^\infty \frac{x^{\alpha-1/2}(1+x)^{\delta-1}}{x+\xi}\, dx. \qquad (1.6.54)$$

• Example 1.6.4

We now turn to some examples that involve logarithms. For example, consider the contour integral of $2z\log(z)/(z^2+2iaz-1)$ around the contour shown in Figure 1.6.6 where $a > 1$. The branch cut lies along the negative real axis. The only singularity within the contour is a pole located at $z_1 = i(\sqrt{a^2-1}-a)$. For the contour C_1, $z = xe^{-\pi i}$ and $dz = dx\,e^{-\pi i}$. Therefore,

$$2\int_{C_1}\frac{z\log(z)}{z^2+2iaz-1}\, dz = 2\int_0^1 \frac{x[\ln(x)-i\pi]}{x^2-2iax-1}\, dx, \qquad (1.6.55)$$

whereas, along C_3, $z = xe^{\pi i}$ and $dz = dx\,e^{\pi i}$, so that

$$2\int_{C_3}\frac{z\log(z)}{z^2+2iaz-1}\, dz = 2\int_1^0 \frac{x[\ln(x)+i\pi]}{x^2-2iax-1}\, dx. \qquad (1.6.56)$$

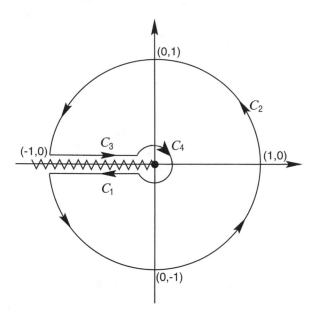

Figure 1.6.6: The contour used to derive Equation 1.6.65 and Equation 1.6.66.

Along the unit circle, $z = e^{\theta i}$, $dz = ie^{\theta i}\,d\theta$ and

$$2\int_{C_2} \frac{z\log(z)}{z^2 + 2iaz - 1}\,dz = 2\int_{-\pi}^{\pi} \frac{i\theta e^{\theta i}}{e^{2\theta i} + 2iae^{\theta i} - 1}ie^{\theta i}\,d\theta \qquad (1.6.57)$$

$$= 2\int_{-\pi}^{\pi} \frac{i\theta e^{\theta i}}{e^{\theta i} + 2ia - e^{-\theta i}}i\,d\theta \qquad (1.6.58)$$

$$= \int_{-\pi}^{\pi} \frac{\cos(\theta) + i\sin(\theta)}{\sin(\theta) + a}i\theta\,d\theta. \qquad (1.6.59)$$

Finally,

$$2\int_{C_4} \frac{z\log(z)}{z^2 + 2iaz - 1}\,dz = \lim_{\epsilon \to 0}\int_{\pi}^{-\pi} \frac{\epsilon e^{\theta i}[\ln(\epsilon) + i\theta]i\epsilon e^{\theta i}}{\epsilon^2 e^{2\theta i} + 2ia\epsilon e^{\theta i} - 1}\,d\theta = 0. \qquad (1.6.60)$$

The sum of the integrals given by Equation 1.6.55, Equation 1.6.56, Equation 1.6.59 and Equation 1.6.60 must equal $2\pi i$ times the residues at the pole z_1, or

$$-4\pi i\int_0^1 \frac{x}{x^2 - 2iax - 1}\,dx + \int_{-\pi}^{\pi} \frac{\cos(\theta) + i\sin(\theta)}{a + \sin(\theta)}i\theta\,d\theta$$

$$= 4\pi i\frac{z_1[\ln(|z_1|) + i\arg(z_1)]}{2z_1 + 2ia} \qquad (1.6.61)$$

$$= 2\pi i\left(\sqrt{a^2 - 1} - a\right)\frac{\ln\left(a - \sqrt{a^2 - 1}\right) - i\pi/2}{\sqrt{a^2 - 1}}. \qquad (1.6.62)$$

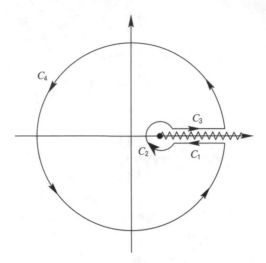

Figure 1.6.7: The contour used to evaluate Equation 1.6.67.

Now,

$$
\int_0^1 \frac{x}{x^2 - 2iax - 1}\,dx = \frac{1}{2}\int_0^1 \frac{(x^2-1)\,d(x^2-1)}{(x^2-1)^2 + 4a^2(x^2-1) + 4a^2}
$$

$$
+\, 2ia \int_0^1 \frac{x^2}{(x^2-1)^2 + 4a^2 x^2}\,dx \tag{1.6.63}
$$

$$
= \frac{a}{2\sqrt{a^2-1}} \ln\!\left(a - \sqrt{a^2-1}\right) + \frac{1}{2}\ln(2a)
$$

$$
+\, \frac{i}{2\sqrt{a^2-1}} \left[\frac{\pi}{2}\sqrt{a^2-1} - a\arctan\!\left(\sqrt{a^2-1}\right)\right]. \tag{1.6.64}
$$

Substituting Equation 1.6.64 into Equation 1.6.62 and separating the real and imaginary parts,

$$
\int_{-\pi}^{\pi} \frac{\theta\sin(\theta)}{a + \sin(\theta)}\,d\theta = \frac{2\pi a}{\sqrt{a^2-1}} \left[\frac{\pi}{2} - \arctan\!\left(\sqrt{a^2-1}\right)\right] \tag{1.6.65}
$$

and

$$
\int_{-\pi}^{\pi} \frac{\theta\cos(\theta)}{a + \sin(\theta)}\,d\theta = 2\pi \ln\!\left(2a^2 - 2a\sqrt{a^2-1}\right). \tag{1.6.66}
$$

• **Example 1.6.6**

Let us evaluate the expression

$$
\oint_C f(z)\log(c - z)\,dz, \tag{1.6.67}
$$

where Figure 1.6.7 shows the contour C, $f(z)$ is a single-valued function that vanishes along $|z| = R \to \infty$. Our branch cut has real values for z lying along the real axis from $-\infty$ and c. Evaluating the various line integrals,

$$\int_{C_1} f(z) \log(c - z)\, dz = \int_{\infty}^{c} f(x)[\ln(x - c) + \pi i]\, dx, \qquad (1.6.68)$$

$$\int_{C_2} f(z) \log(c - z)\, dz = 0, \qquad (1.6.69)$$

$$\int_{C_3} f(z) \log(c - z)\, dz = \int_{c}^{\infty} f(x)[\ln(x - c) - \pi i]\, dx, \qquad (1.6.70)$$

and

$$\int_{C_4} f(z) \log(c - z)\, dz = 0. \qquad (1.6.71)$$

Therefore,

$$\oint_{C} f(z) \log(c - z)\, dz = 2\pi i \sum_{j=1}^{n} \mathrm{Res}[f(z) \log(c - z); z_j], \qquad (1.6.72)$$

$$-2\pi i \int_{c}^{\infty} f(x)\, dx = 2\pi i \sum_{j=1}^{n} \mathrm{Res}[f(z) \log(c - z); z_j], \qquad (1.6.73)$$

or

$$\int_{c}^{\infty} f(x)\, dx = -\sum_{j=1}^{n} \mathrm{Res}[f(z) \log(c - z); z_j], \qquad (1.6.74)$$

where z_j denotes the jth singularity of $f(z) \log(c - z)$ located within the contour C; there are n singularites. Consequently, the value of $\int_{c}^{\infty} f(x)\, dx$ equals the negative of the sum of the residues of $f(z) \log(c - z)$ where we only use the poles associated with $f(z)$. This nice result was first published in an obscure paper by Neville.[20] Since then, Boas and Schoenfeld[21] extended Neville's results to include integrals of the form

$$PV \int_{0}^{\infty} f(t)\, dt,$$

$$PV \int_{a}^{b} f(t)\, dt, \qquad -\infty < a < b < \infty,$$

and

$$PV \int_{\alpha}^{\beta} f(e^{i\theta})\, d\theta, \qquad \alpha < \beta < \alpha + 2\pi.$$

---

[20] Neville, E. H., 1945: Indefinite integration by means of residues. *Math. Student*, **13**, 16–25.

[21] Boas, R. P., and L. Schoenfeld, 1966: Indefinite integration by residues. *SIAM Review*, **8**, 173–183.

Here PV denotes the Cauchy principal value, in case the integrand has a singularity in the interval of integration.

To illustrate this result, consider the case of

$$I = \int_c^\infty \frac{dx}{x^2 - a^2} \tag{1.6.75}$$

with $c > a$. The function $f(z) = 1/(z^2 - a^2)$ has simple poles at $z = \pm a$. Our analysis shows that

$$\int_c^\infty \frac{dx}{x^2 - a^2} = -\left\{ \text{Res}\left[\frac{\log(c-z)}{z^2 - a^2}; a \right] + \text{Res}\left[\frac{\log(c-z)}{z^2 - a^2}; -a \right] \right\} \tag{1.6.76}$$

$$= -\frac{\log(c-a)}{2a} + \frac{\log(c+a)}{2a} = \frac{1}{2a} \ln\left(\frac{c+a}{c-a} \right), \tag{1.6.77}$$

if $0 < a < c$.

Problems

1. Prove that

$$\oint_{|z|=1} a^z \, dz = 0,$$

for any single-valued branch of the function a^z. Hint: $a^z = \exp[z \log(a)]$.

2. Show that

$$\oint_{|z|=1} z^\alpha \, dz = \begin{cases} \left(e^{2\alpha\pi i} - 1 \right)/(1+\alpha), & \text{if} \quad \alpha \neq -1, \\ 0, & \text{if} \quad \alpha = -1, \end{cases}$$

where α is an arbitrary complex number, and the branch cut associated with z^α lies along the positive real axis.

3. Evaluate $\oint_C dz/[z \log(z)]$ around the illustrated contour and show that

$$\int_0^\infty \frac{dx}{x[\ln(x)^2 + \pi^2]} = 1.$$

Use the principal value of the logarithm, $\log(z) = \ln(r) + \theta i$.

4. Use the illustrated contour to evaluate $\oint_C dz/[(z+1)\sqrt{z}]$ and show that

$$\int_0^\infty \frac{dx}{(x+1)\sqrt{x}} = \pi.$$

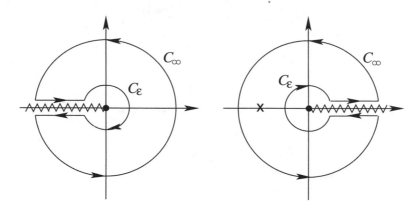

Problem 3   Problem 4

5. Evaluate $\oint_C dz/\left(z\sqrt{z^2-1}\right)$ around the illustrated contour and verify that

$$\int_1^\infty \frac{dx}{x\sqrt{x^2-1}} = \frac{\pi}{2}.$$

6. Use the illustrated contour and two different complex functions to show that

$$\int_{-1}^1 \frac{dx}{(x^2+1)\sqrt{1-x^2}} = \frac{\pi}{\sqrt{2}}$$

and

$$\int_{-1}^1 \frac{dx}{(x^2+1)^2\sqrt{1-x^2}} = \frac{3\pi}{4\sqrt{2}}.$$

Hint: The square root will have different signs at the two poles.

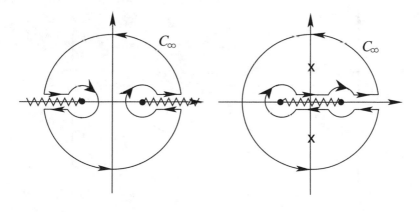

Problem 5   Problem 6

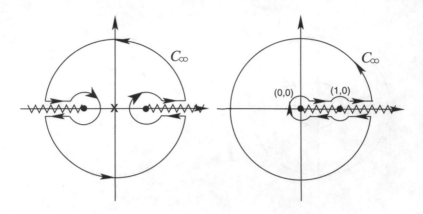

Problem 7   **Problem 8**

7. Use the illustrated contour and show that

$$\int_1^\infty \frac{dx}{x^2} \sqrt{\frac{x-1}{x+1}} = \pi - \int_1^\infty \frac{dx}{x^2} \sqrt{\frac{x+1}{x-1}}.$$

8. Use the illustrated contour and the complex function $f(z) = z^{\alpha-1}/(1-z) = \exp[(\alpha-1)\log(z)]/(1-z)$ to show that

$$\int_0^\infty \frac{x^{\alpha-1}}{1-x} \, dx = \pi \cot(\alpha\pi), \qquad 0 < \alpha < 1.$$

9. Use the illustrated contour and show that

$$\int_0^1 \frac{dx}{(x^2 - x^3)^{1/3}} = \frac{2\pi}{\sqrt{3}}.$$

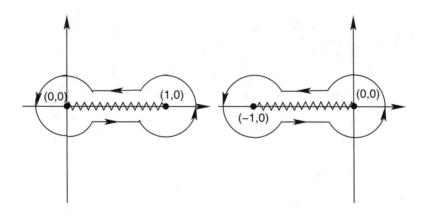

Problem 9   **Problem 10**

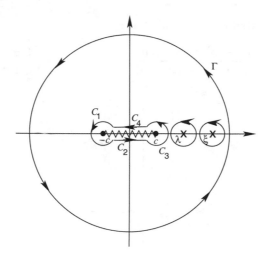

Problem 11

10. Show[22] that

$$\frac{1}{2\pi i} \oint_C \left(\frac{z+1}{z}\right)^\alpha z^{n-1}\, dz = \frac{\sin[(n-\alpha)\pi]}{\pi} \int_{0+}^1 x^{n-1-\alpha}(1-x)^\alpha\, dx,$$

where $0 \le n$, α is real and noninteger and C is a dumbbell-shaped contour lying along the negative real axis from $z = -1$ and $z = 0$.

11. Evaluate

$$\oint_\Gamma \left(\frac{z-c}{z+c}\right)^{1/2} \frac{dz}{(z-\lambda)(\xi-z)}, \qquad c \le \lambda, \xi,$$

around the illustrated contour and show[23]

$$\int_{-c}^c \left(\frac{c-x}{c+x}\right)^{1/2} \frac{dx}{(x-\lambda)(\xi-x)} = \frac{\pi}{\lambda-\xi}\left[\left(\frac{\xi-c}{\xi+c}\right)^{1/2} - \left(\frac{\lambda-c}{\lambda+c}\right)^{1/2}\right].$$

12. Evaluate

$$\oint_\Gamma \left(\frac{z-c}{z+c}\right)^{1/2} \frac{dz}{(z-\sigma)(\xi-z)}, \qquad |\sigma| \le c \le \xi,$$

[22] Taken from Jury, E. I., and C. A. Galtieri, 1961: A note on the inverse z transformation. *IRE Trans. Circuit Theory*, **CT-8**, 371–374. ©1961 IEEE.

[23] Taken from Ahmadi, A. R., and S. E. Widnall, 1994: Energetics of oscillating lifting surfaces by the use of integral conservation laws. *J. Fluid Mech.*, **266**, 347–370. Reprinted with the permission of Cambridge University Press.

around the illustrated contour and show[24]

$$PV \int_{-c}^{c} \left(\frac{c-x}{c+x}\right)^{1/2} \frac{dx}{(x-\sigma)(\xi-x)} = \frac{\pi}{\sigma-\xi} \left(\frac{\xi-c}{\xi+c}\right)^{1/2}.$$

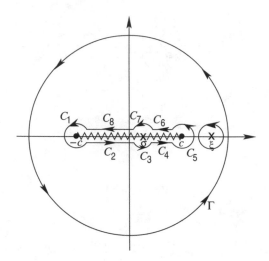

Problem 12

13. Evaluate

$$\oint_{C} \frac{z^{p-1}}{z^2 - 2\alpha z + 1}\, dz, \qquad 0 \le \alpha < 1,$$

around the contour used in Problem 4, where $p = \pi/[2\pi - 2\arccos(\alpha)]$ with the branch $0 \le \arccos(\alpha) < 2\pi$. Show[25] that

$$\int_{0}^{\infty} \frac{x^{p-1}}{x^2 - 2\alpha x + 1}\, dx = \frac{\pi}{\sqrt{1-\alpha^2}} \big\{ \cos[(p-1)\arccos(\alpha)]$$
$$- \cot[\pi(p-1)]\sin[(p-1)\arccos(\alpha)]\big\}.$$

Hint: $\alpha \pm i\sqrt{1-\alpha^2} = \exp[\pm i\arccos(\alpha)]$.

14. By first showing that

$$\int_{0}^{2\pi} \frac{e^{-i\omega\theta}}{\cos(\theta) - a}\, d\theta = \frac{2}{i} \oint_{C_2} \frac{z^{|\omega|}}{z^2 - 2az + 1}\, dz, \qquad 0 < a,$$

---

[24] *Ibid.*

[25] Taken from Greenwell, R. N., and C. Y. Wang, 1980: Fluid flow through a partially filled cylinder. *Appl. Sci. Res.*, **36**, 61–75. Reprinted by permission of Kluwer Academic Publishers.

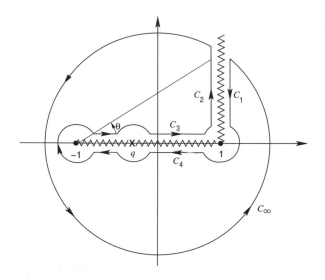

Figure 1.6.8: Contour used in solving Problem 15. Redrawn from Lewis, P. A., and G. R. Wickham, *Philos. Trans. R. Soc. London, Ser. A*, **340**, 503–529 (1992).

and C_2 is the contour shown in Figure 1.6.6, prove that

$$\int_0^{2\pi} \frac{e^{-i\omega\theta}}{\cos(\theta) - a}\, d\theta = 4\pi \begin{cases} \dfrac{\sin[|\omega|\arccos(a)]}{\sqrt{1 - a^2}}, & 0 < a < 1, \\[2ex] -\dfrac{\left(a - \sqrt{a^2 - 1}\right)^{|\omega|}}{2\sqrt{a^2 - 1}}, & 1 < a < \infty, \end{cases}$$
$$- 4\sin(|\omega|\pi) \int_0^1 \frac{x^{|\omega|}}{x^2 + 2ax + 1}\, dx.$$

15. Show[26] that

$$\int_{-1}^1 \frac{\ln|t - p|}{(t - q)\sqrt{1 - t^2}}\, dt = -2\pi \frac{\operatorname{sgn}(p - q)}{\sqrt{1 - q^2}} \arctan\!\left(\sqrt{\frac{1 \pm q}{1 \mp q}}\,\right),$$

where $-1 < q < 1$ and the plus sign applies when $p > q$ while the negative sign is for $p < q$.

Step 1: Consider first the case $p = 1$. From the contours shown in Figure 1.6.8, show that

$$\int_{C_3 \cup C_4} \frac{\log(z - 1)}{(z - q)\sqrt{1 - z^2}}\, dz = -\int_{C_1 \cup C_2} \frac{\log(z - 1)}{(z - q)\sqrt{1 - z^2}}\, dz.$$

---

[26] Taken from Lewis, P. A., and G. R. Wickham, 1992: The diffraction of SH waves by an arbitrary shaped crack in two dimensions. *Phil. Trans. R. Soc. London, Ser. A*, **340**, 503–529.

Step 2: By parameterizing each contour as follows:

$$C_1 : z - 1 = \rho e^{\pi i/2}, \quad z + 1 = \sqrt{4 + \rho^2} e^{\theta i}, \quad \log(z - 1) = \ln(\rho) + \pi i/2,$$

$$C_2 : z - 1 = \rho e^{\pi i/2}, \quad z + 1 = \sqrt{4 + \rho^2} e^{\theta i}, \quad \log(z - 1) = \ln(\rho) - 3\pi i/2,$$

where $0 < \rho < \infty$ and $\tan(\theta) = \rho/2$, and

$$C_3 : z - 1 = (1 - \rho) e^{\pi i}, \quad z + 1 = (1 + \rho) e^{0i}, \quad \log(z - 1) = \ln(1 - \rho) - \pi i,$$

$$C_4 : z - 1 = (1 - \rho) e^{\pi i}, \quad z + 1 = (1 + \rho) e^{2\pi i}, \quad \log(z - 1) = \ln(1 - \rho) - \pi i,$$

where $-1 < \rho < 1$, show that

$$\int_{C_1 \cup C_2} \frac{\log(z - 1)}{(z - q)\sqrt{1 - z^2}} \, dz = \int_0^{\pi/2} \frac{2\pi e^{-\pi i/4} e^{-\theta i/2}}{\sqrt{\sin(\theta)}[(1 - q)\cos(\theta) + 2i\sin(\theta)]} \, d\theta,$$

and

$$\int_{C_3 \cup C_4} \frac{\log(z - 1)}{(z - q)\sqrt{1 - z^2}} \, dz = -2i \int_{-1}^1 \frac{\ln(1 - t)}{(t - q)\sqrt{1 - t^2}} \, dt.$$

Note that the simple pole at $z = q$ makes no contribution.

Step 3: Evaluate the integral for $C_1 \cup C_2$ by noting that $(1 + q) + (1 - q) = 2$, replacing cosine and sine by their exponential equivalents, and letting $1 - e^{2\theta i} = \eta^2$. You should find that

$$\int_{C_1 \cup C_2} \frac{\log(z - 1)}{(z - q)\sqrt{1 - z^2}} \, dz = -\frac{4\pi i}{\sqrt{1 - q^2}} \arctan\left(\sqrt{\frac{1 + q}{1 - q}}\right).$$

Finally, by matching imaginary parts, show that

$$\int_{-1}^1 \frac{\ln(1 - t)}{(t - q)\sqrt{1 - t^2}} \, dt = -\frac{2\pi}{\sqrt{1 - q^2}} \arctan\left(\sqrt{\frac{1 + q}{1 - q}}\right).$$

Step 4: Now redo the first three steps with $p = -1$ and show that

$$\int_{-1}^1 \frac{\ln(t - 1)}{(t - q)\sqrt{1 - t^2}} \, dt = \frac{2\pi}{\sqrt{1 - q^2}} \arctan\left(\sqrt{\frac{1 - q}{1 + q}}\right).$$

Then, use analytic continuity to argue for the more general result.

16. Evaluate[27]

$$\int_{-1}^1 \frac{\ln(x + a)}{(x + b)\sqrt{1 - x^2}} \, dx, \quad 1 < a, \quad |b| \le 1.$$

---

[27] Taken from Chen, C. F., 1962: Linearized theory for supercavitating hydrofoils with spoiler-flaps. *J. Ship Res.*, **6, No. 3**, 1–9. Reprinted with the permission of the Society of Naval Architects and Marine Engineers (SNAME). Material originally appearing in SNAME publications cannot be reprinted without written permission from the Society, 601 Pavonia Ave., Jersey City, NJ 07306.

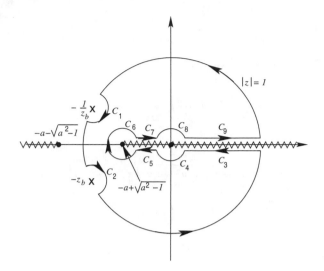

Problem 16

Step 1: By introducing $x = \cos(\theta)$, show that

$$\int_{-1}^{1} \frac{\ln(x+a)}{(x+b)\sqrt{1-x^2}}\, dx = \int_{0}^{\pi} \frac{\ln[a+\cos(\theta)]}{\cos(\theta)+\cos(\theta_b)}\, d\theta = \frac{1}{2}\int_{0}^{2\pi} \frac{\ln[a+\cos(\theta)]}{\cos(\theta)+\cos(\theta_b)}\, d\theta,$$

where $\theta_b = \arccos(b)$.

Step 2: Setting $z = e^{\theta i}$, show that

$$\int_{-1}^{1} \frac{\ln(x+a)}{(x+b)\sqrt{1-x^2}}\, dx$$

$$= \frac{1}{i}\oint_{|z|=1} \frac{\log(z^2+2az+1) - \log(z) - \ln(2)}{\left(z+z_b\right)\left(z+z_b^{-1}\right)}\, dz$$

$$= \frac{1}{i}\oint_{|z|=1} \frac{\log\!\left[\left(z+a-\sqrt{a^2-1}\right)\left(z+a+\sqrt{a^2-1}\right)\right] - \log(z) - \ln(2)}{\left(z+z_b\right)\left(z+z_b^{-1}\right)}\, dz.$$

Step 3: If

$$f(z) = \frac{\log\!\left[\left(z+a-\sqrt{a^2-1}\right)\left(z+a+\sqrt{a^2-1}\right)\right] - \log(z) - \ln(2)}{\left(z+z_b\right)\left(z+z_b^{-1}\right)}$$

and using the accompanying figure, show that

$$\oint_{|z|=1} f(z)\, dz = \pi i\,\mathrm{Res}[f(z); -z_b] + \pi i\,\mathrm{Res}[f(z); -z_b^{-1}] - \int_C f(z)\, dz,$$

where $C = C_3 \cup C_4 \cup \cdots \cup C_8 \cup C_9$.

Step 4: Show that

$$\text{Res}[f(z); -z_b] + \text{Res}[f(z); -z_b^{-1}] = 0.$$

Hint: $-z_b = e^{\theta_b i - \pi i}$ and $-z_b^{-1} = e^{-\theta_b i - \pi i}$.

Step 5: Show that

$$\int_{C_6} f(z)\, dz = \int_{C_4 \cup C_8} f(z)\, dz = 0.$$

Step 6: Since along C_3,

$$z = x e^{2\pi i}, \qquad z + a - \sqrt{a^2 - 1} = \left(x + a - \sqrt{a^2 - 1}\right) e^{2\pi i},$$

$$z + a + \sqrt{a^2 - 1} = \left(x + a + \sqrt{a^2 - 1}\right) e^{0i},$$

and

$$z = x e^{0i}, \qquad z + a - \sqrt{a^2 - 1} = \left(x + a - \sqrt{a^2 - 1}\right) e^{0i},$$

$$z + a + \sqrt{a^2 - 1} = \left(x + a + \sqrt{a^2 - 1}\right) e^{0i}$$

along C_9, show that

$$\int_{C_3 \cup C_9} f(z)\, dz = 0.$$

Step 7: Since along C_5,

$$z = x e^{-\pi i}, \qquad z + a - \sqrt{a^2 - 1} = \left(x + a - \sqrt{a^2 - 1}\right) e^{2\pi i},$$

$$z + a + \sqrt{a^2 - 1} = \left(x + a + \sqrt{a^2 - 1}\right) e^{0i},$$

and

$$z = x e^{-\pi i}, \qquad z + a - \sqrt{a^2 - 1} = \left(x + a - \sqrt{a^2 - 1}\right) e^{0i},$$

$$z + a + \sqrt{a^2 - 1} = \left(x + a + \sqrt{a^2 - 1}\right) e^{0i}$$

along C_7, show that

$$\int_{C_5 \cup C_7} f(z)\, dz = -2\pi i \int_0^{a - \sqrt{a^2 - 1}} \frac{dx}{\left(x - z_b\right)\left(x - z_b^{-1}\right)}$$

$$= -\frac{\pi}{\sqrt{1 - b^2}} \log\left[\frac{1 - \left(a - \sqrt{a^2 - 1}\right) z_b^{-1}}{1 - \left(a - \sqrt{a^2 - 1}\right) z_b}\right]$$

$$= -\frac{2\pi i}{\sqrt{1 - b^2}} \arctan\left[\frac{\sqrt{1 - b^2}\left(a - \sqrt{a^2 - 1}\right)}{1 - b\left(a - \sqrt{a^2 - 1}\right)}\right].$$

Hint:

$$\arctan(z) = \frac{1}{2i} \log\left(\frac{1 + iz}{1 - iz}\right).$$

Step 8: Conclude the problem by showing that

$$\int_{-1}^{1} \frac{\ln(x+a)}{(x+b)\sqrt{1-x^2}}\,dx = \frac{2\pi}{\sqrt{1-b^2}}\arctan\left[\frac{\sqrt{1-b^2}\left(a-\sqrt{a^2-1}\right)}{1-b\left(a-\sqrt{a^2-1}\right)}\right].$$

1.7 BESSEL FUNCTIONS

In Section 1.3 we dealt only with ordinary differential equations that have constant coefficients. In problems involving cylindrical coordinates, we will solve the equation

$$r^2\frac{d^2y}{dr^2} + r\frac{dy}{dr} + (\lambda^2 r^2 - n^2)y = 0, \tag{1.7.1}$$

commonly known as *Bessel's equation of order n with a parameter* λ. The general solution to Equation 1.7.1 is

$$y(r) = c_1 J_n(\lambda r) + c_2 Y_n(\lambda r), \tag{1.7.2}$$

where $J_n(\cdot)$ and $Y_n(\cdot)$ are nth order Bessel functions of the first and second kind, respectively. Bessel functions have been exhaustively studied and a vast literature now exists on them.[28] The Bessel function $J_n(z)$ is an entire function, has no complex zeros, and has an infinite number of real zeros symmetrically located with respect to the point $z = 0$, which is itself a zero if $n > 0$. All of the zeros are simple, except the point $z = 0$, which is a zero of order n if $n > 0$. On the other hand, $Y_n(z)$ is analytic in the complex plane with a branch cut along the segment $(-\infty, 0]$ and becomes infinite as $z \to 0$.

Considerable insight into the nature of Bessel functions is gained from their asymptotic expansions. These expansions are

$$J_n(z) \sim \left(\frac{2}{\pi z}\right)^{1/2}\cos\left(z - \tfrac{1}{2}n\pi - \tfrac{1}{4}\pi\right), \quad |\arg(z)| \le \pi - \epsilon, \quad |z| \to \infty, \tag{1.7.3}$$

and

$$Y_n(z) \sim \left(\frac{2}{\pi z}\right)^{1/2}\sin\left(z - \tfrac{1}{2}n\pi - \tfrac{1}{4}\pi\right), \quad |\arg(z)| \le \pi - \epsilon, \quad |z| \to \infty, \tag{1.7.4}$$

where ϵ denotes an arbitrarily small positive number. Therefore, Bessel functions are sinusoidal in nature and decay as $z^{-1/2}$.

---

[28] *The* standard reference is Watson, G. N., 1966: *A Treatise on the Theory of Bessel Functions*. Cambridge University Press, 804 pp.

LIVERPOOL JOHN MOORES UNIVERSITY
LEARNING SERVICES

A closely related differential equation is the *modified Bessel equation*

$$r^2 \frac{d^2 y}{dr^2} + r \frac{dy}{dr} - (\lambda^2 r^2 + \nu^2) y = 0. \tag{1.7.5}$$

Its general solution is

$$y(r) = c_1 I_\nu(\lambda r) + c_2 K_\nu(\lambda r), \tag{1.7.6}$$

where $I_\nu(\cdot)$ and $K_\nu(\cdot)$ are ν-th order, modified Bessel functions of the first and second kind, respectively. Both $I_\nu(\cdot)$ and $K_\nu(\cdot)$ are analytic functions in the complex z-plane provided that we introduce a branch cut along the segment $(-\infty, 0]$. As $z \to 0$, $K_\nu(z)$ becomes infinite.

Turning to the zeros, $I_\nu(z)$ has zeros that are purely imaginary for $\nu > -1$. On the other hand, $K_\nu(z)$ has no zeros in the region $|\arg(z)| \leq \pi/2$. In the remaining portion of the cut z-plane, it has a finite number of zeros.

The modified Bessel functions also have asymptotic representations:

$$I_\nu(z) \sim \frac{e^z}{\sqrt{2\pi z}} + \frac{e^{-z \pm \pi(\nu + \frac{1}{2})}}{\sqrt{2\pi z}}, \quad |\arg(z)| \leq \pi - \epsilon, \quad |z| \to \infty, \tag{1.7.7}$$

and

$$K_n(z) \sim \frac{\pi e^{-z}}{\sqrt{2\pi z}}, \quad |\arg(z)| \leq \pi - \epsilon, \quad |z| \to \infty, \tag{1.7.8}$$

where we chose the plus sign if $\text{Im}(z) > 0$, and the minus sign if $\text{Im}(z) < 0$. Note that $K_n(z)$ decrease exponentially as $x \to \infty$, while $I_n(z)$ increases exponentially as $x \to \infty$ *and* $x \to -\infty$.

These results just scratch the known relationships concerning Bessel functions; we will introduce additional results as we need them.

• Example 1.7.1

Solve the boundary-value problem

$$\frac{d^2 y}{dr^2} + \frac{1}{r} \frac{dy}{dr} - sy = 0, \quad y'(a) = -\frac{1}{s}, \quad \lim_{r \to \infty} y(r) \to 0, \tag{1.7.9}$$

where $s > 0$. The general solution to Equation 1.7.9 is

$$y(r) = A I_0(r\sqrt{s}) + B K_0(r\sqrt{s}) \tag{1.7.10}$$

from Equation 1.7.5 and Equation 1.7.6. Because $I_0(r\sqrt{s}) \to \infty$ as $r \to \infty$, $A = 0$. At $r = a$,

$$y'(a) = B\sqrt{s} K_0'(a\sqrt{s}) = -B\sqrt{s} K_1(a\sqrt{s}) = -\frac{1}{s} \tag{1.7.11}$$

LIVERPOOL JOHN MOORES UNIVERSITY
LEARNING SERVICES

Table 1.7.1: Some Useful Relationships Involving Bessel Functions of Integer Order

$$J_{n-1}(z) + J_{n+1}(z) = \frac{2n}{z} J_n(z), \qquad n = 1, 2, 3, \cdots$$

$$J_{n-1}(z) - J_{n+1}(z) = 2J_n'(z), \; n = 1, 2, 3, \cdots; \quad J_0'(z) = -J_1(z)$$

$$I_{n-1}(z) - I_{n+1}(z) = \frac{2n}{z} I_n(z), \qquad n = 1, 2, 3, \cdots$$

$$I_{n-1}(z) + I_{n+1}(z) = 2I_n'(z), \; n = 1, 2, 3, \cdots; \quad I_0'(z) = I_1(z)$$

$$K_{n-1}(z) - K_{n+1}(z) = -\frac{2n}{z} K_n(z), \qquad n = 1, 2, 3, \cdots$$

$$K_{n-1}(z) + K_{n+1}(z) = -2K_n'(z), \; n = 1, 2, 3, \cdots; \quad K_0'(z) = -K_1(z)$$

$$J_n(ze^{m\pi i}) = e^{nm\pi i} J_n(z)$$

$$I_n(ze^{m\pi i}) = e^{nm\pi i} I_n(z)$$

$$K_n(ze^{m\pi i}) = e^{-mn\pi i} K_n(z) - m\pi i \frac{\cos(mn\pi)}{\cos(n\pi)} I_n(z)$$

$$I_n(z) = e^{-n\pi i/2} J_n(ze^{\pi i/2}), \qquad -\pi < \arg(z) \le \pi/2$$

$$I_n(z) = e^{3n\pi i/2} J_n(ze^{-3\pi i/2}), \qquad \pi/2 < \arg(z) \le \pi$$

because $K_0'(z) = -K_1(z)$ or

$$B = \frac{1}{s^{3/2} K_1(a\sqrt{s})}. \tag{1.7.12}$$

Therefore, the solution to the boundary-value problem is

$$y(r) = \frac{K_0(r\sqrt{s})}{s^{3/2} K_1(a\sqrt{s})}. \tag{1.7.13}$$

We have repeatedly noted how Bessel functions are very similar in nature to sine and cosine. This suggests that for axisymmetric problems we could develop an alternative transform based on Bessel functions. To examine this possibility, let us write the two-dimensional Fourier transform pair as

$$f(x, y) = \frac{1}{2\pi} \int_{-\infty}^{\infty} \int_{-\infty}^{\infty} F(k, \ell) e^{i(kx+\ell y)} \, dk \, d\ell, \tag{1.7.14}$$

where

$$F(k,\ell) = \frac{1}{2\pi} \int_{-\infty}^{\infty} \int_{-\infty}^{\infty} f(x,y) \, e^{-i(kx+\ell y)} \, dx \, dy. \qquad (1.7.15)$$

Consider now the special case where $f(x,y)$ is only a function of $r = \sqrt{x^2 + y^2}$, so that $f(x,y) = g(r)$. Then, changing to polar coordinates through the substitution $x = r\cos(\theta)$, $y = r\sin(\theta)$, $k = \rho\cos(\varphi)$ and $\ell = \rho\sin(\varphi)$, we have that

$$kx + \ell y = r\rho[\cos(\theta)\cos(\varphi) + \sin(\theta)\sin(\varphi)] = r\rho\cos(\theta - \varphi) \qquad (1.7.16)$$

and

$$dA = dx \, dy = r \, dr \, d\theta. \qquad (1.7.17)$$

Therefore, the integral in Equation 1.7.15 becomes

$$F(k,\ell) = \frac{1}{2\pi} \int_{0}^{\infty} \int_{0}^{2\pi} g(r) \, e^{-ir\rho\cos(\theta-\varphi)} r \, dr \, d\theta \qquad (1.7.18)$$

$$= \frac{1}{2\pi} \int_{0}^{\infty} r \, g(r) \left[\int_{0}^{2\pi} e^{-ir\rho\cos(\theta-\varphi)} \, d\theta \right] dr. \qquad (1.7.19)$$

If we introduce $\lambda = \theta - \varphi$, the integral inside the square brackets can be evaluated as follows:

$$\int_{0}^{2\pi} e^{-ir\rho\cos(\theta-\varphi)} \, d\theta = \int_{-\varphi}^{2\pi-\varphi} e^{-ir\rho\cos(\lambda)} \, d\lambda \qquad (1.7.20)$$

$$= \int_{0}^{2\pi} e^{-ir\rho\cos(\lambda)} \, d\lambda \qquad (1.7.21)$$

$$= 2\pi J_0(\rho r). \qquad (1.7.22)$$

Equation 1.7.21 is equivalent to Equation 1.7.20 because the integral of a periodic function over one full period is the same regardless of where the integration begins. Equation 1.7.22 follows from the integral definition of the Bessel function.[29] Therefore,

$$F(k,\ell) = \int_{0}^{\infty} r \, g(r) \, J_0(\rho r) \, dr. \qquad (1.7.23)$$

Finally, because Equation 1.7.23 is clearly a function of $\rho = \sqrt{k^2 + \ell^2}$, $F(k,\ell) = G(\rho)$ and

$$G(\rho) = \int_{0}^{\infty} r \, g(r) \, J_0(\rho r) \, dr. \qquad (1.7.24)$$

---

[29] *Ibid.*, Section 2.2, Equation 5.

Conversely, if we begin with Equation 1.7.14, make the same substitution, and integrate over the $k\ell$-plane, we have

$$f(x,y) = g(r) = \frac{1}{2\pi} \int_0^\infty \int_0^{2\pi} F(k,\ell) \, e^{ir\rho\cos(\theta-\varphi)} \rho \, d\rho \, d\varphi \qquad (1.7.25)$$

$$= \frac{1}{2\pi} \int_0^\infty \rho \, G(\rho) \left[\int_0^{2\pi} e^{ir\rho\cos(\theta-\varphi)} \, d\varphi \right] d\rho \qquad (1.7.26)$$

$$= \int_0^\infty \rho \, G(\rho) \, J_0(\rho r) \, d\rho. \qquad (1.7.27)$$

Thus, we obtain the result that if $\int_0^\infty |F(r)| \, dr$ exists, then

$$g(r) = \int_0^\infty \rho \, G(\rho) \, J_0(\rho r) \, d\rho, \qquad (1.7.28)$$

where

$$G(\rho) = \int_0^\infty r \, g(r) \, J_0(\rho r) \, dr. \qquad (1.7.29)$$

Taken together, Equation 1.7.28 and Equation 1.7.29 constitute the *Hankel transform pair for Bessel function of order 0*, named after the German mathematician Hermann Hankel (1839–1873). The function $G(\rho)$ is called the Hankel transform of $g(r)$.

For asymmetric problems, we can generalize our results to Hankel transforms of order ν

$$F(k) = \int_0^\infty f(r) J_\nu(kr) \, r \, dr, \qquad -\tfrac{1}{2} < \nu, \qquad (1.7.30)$$

and its inverse[30]

$$f(r) = \int_0^\infty F(k) J_\nu(kr) \, k \, dk. \qquad (1.7.31)$$

Finally, it is well know that $\sin(\theta)$ and $\cos(\theta)$ can be expressed in terms of the complex exponential $e^{\theta i}$ and $e^{-\theta i}$. In the case of Bessel functions $J_\nu(z)$ and $Y_\nu(z)$, the corresponding representations are called Hankel functions (or Bessel functions of the third kind)

$$H_\nu^{(1)}(z) = J_\nu(z) + iY_\nu(z) \quad \text{and} \quad H_\nu^{(2)}(z) = J_\nu(z) - iY_\nu(z), \qquad (1.7.32)$$

where ν is arbitrary and z is any point in the z-plane cut along the segment $(-\infty, 0]$. The analogy is most clearly seen in the asymptotic expansions for these functions:

$$H_\nu^{(1)}(z) = \sqrt{\frac{2}{\pi z}} e^{i(z-\nu\pi/2-\pi/4)} \quad \text{and} \quad H_\nu^{(2)}(z) = \sqrt{\frac{2}{\pi z}} e^{-i(z-\nu\pi/2-\pi/4)}$$

$$(1.7.33)$$

---

[30] *Ibid.*, Section 14.4. See also MacRobert, T. M., 1931: Fourier integrals. *Proc. R. Soc. Edinburgh*, **51**, 116–126.

Table 1.7.2: Some Useful Recurrence Relations for Hankel Functions

---

$$\frac{d}{dx}\left[x^n H_n^{(p)}(x)\right] = x^n H_{n-1}^{(p)}(x), \;\; n = 1, 2, \ldots; \; \frac{d}{dx}\left[H_0^{(p)}(x)\right] = -H_1^{(p)}(x)$$

$$\frac{d}{dx}\left[x^{-n} H_n^{(p)}(x)\right] = -x^{-n} H_{n+1}^{(p)}(x), \qquad n = 0, 1, 2, 3, \ldots$$

$$H_{n-1}^{(p)}(x) + H_{n+1}^{(p)}(x) = \frac{2n}{x} H_n^{(p)}(x), \qquad n = 1, 2, 3, \ldots$$

$$H_{n-1}^{(p)}(x) - H_{n+1}^{(p)}(x) = 2\frac{dH_n^{(p)}(x)}{dx}, \qquad n = 1, 2, 3, \ldots$$

---

for $|z| \to \infty$ with $|\arg(z)| \le \pi - \epsilon$, where ϵ is an arbitrarily small positive number. These functions are linearly independent solutions of

$$\frac{d^2 u}{dz^2} + \frac{1}{z}\frac{du}{dz} + \left(1 - \frac{\nu^2}{z^2}\right)u = 0. \tag{1.7.34}$$

Table 1.7.2 gives additional relationships involving Hankel functions.

Problems

1. Solve the boundary-value problem

$$\frac{d^2 y}{dr^2} + \frac{1}{r}\frac{dy}{dr} - sy = 0, \qquad |y(0)| < \infty, \qquad y'(a) = -\frac{1}{s},$$

where a and s are real and positive.

2. Show that the Hankel transform of $\delta(r)$ is $1/(2\pi)$.

1.8 WHAT ARE TRANSFORM METHODS?

For many students of engineering and the sciences, Laplace and Fourier transforms are an important tool in solving linear ordinary differential equations. When it comes to *linear* partial differential equations, such as the wave, heat or Laplace's equations, similar considerations hold true. The purpose of this book is to show you how this method works.

For many years, scholars attributed the invention of operational mathematics (another name for transform methods) to Oliver Heaviside. Modern research,[31] however, has shown that during the early nineteenth century both English and French mathematicians developed both symbolic calculus and operational methods. For example, Cauchy used operational methods, applying the Fourier transform to solve the wave equation. Later in the nineteenth century, this knowledge was apparently forgotten until Heaviside rediscovered the Laplace transform.[32] However, because of Heaviside's lack of mathematical rigor, a controversy of legendary proportions[33] developed between him and the mathematical "establishment." It remained for the English mathematician T. J. I'a. Bromwich[34] and the German electrical engineer K. W. Wagner[35] to justify Heaviside's work. Although each was ignorant of the other's work, both used function theory. Wagner concentrated on the expansion formula, whereas Bromwich gave a broader explanation of the operational calculus.

Having given a brief history of transform methods, let us explore the relationship of transform methods to the traditional methods of separation of variables and numerical methods by studying the sound waves that arise when a sphere of radius a begins to pulsate at time $t = 0$. The symmetric wave equation in spherical coordinates is

$$\frac{1}{r}\frac{\partial^2(ru)}{\partial r^2} = \frac{1}{c^2}\frac{\partial^2 u}{\partial t^2}, \tag{1.8.1}$$

where c is the speed of sound, $u(r,t)$ is the velocity potential and $-\partial u/\partial r$ gives the velocity of the parcel of air. At the surface of the sphere $r = a$, the radial velocity must equal the velocity of the pulsating sphere

$$-\frac{\partial u}{\partial r} = \frac{d\xi}{dt}, \tag{1.8.2}$$

where $\xi(t)$, the displacement of the surface of the pulsating sphere, equals $B\sin(\omega t)H(t)$. The air is initially at rest.

---

[31] Cooper, J. L. B., 1952: Heaviside and the operational calculus. *Math. Gaz.*, **36**, 5–19; Lützen, J., 1979: Heaviside's operational calculus and the attempts to rigorise it. *Arch. Hist. Exact Sci.*, **21**, 161–200; Deakin, M. A. B., 1981: The development of the Laplace transform, 1737–1937: I. Euler to Spitzer, 1737–1880. *Arch. Hist. Exact Sci.*, **25**, 343–390; Deakin, M. A. B., 1982: The development of the Laplace transform, 1737–1937: II. Poincaré to Doetsch, 1880 1937. *Arch. Hist. Exact Sci.*, **26**, 351–381; Petrova, S. S., 1986: Heaviside and the development of the symbolic calculus. *Arch. Hist. Exact Sci.*, **37**, 1–23; Deakin, M. A. B., 1992: The ascendancy of the Laplace transform and how it came about. *Arch. Hist. Exact Sci.*, **44**, 265–286.

[32] Heaviside, O., 1893: On operators in physical mathematics. *Proc. R. Soc. London, Ser. A*, **52**, 504–529.

[33] Nahin, P., 1988: *Oliver Heaviside: Sage in Solitude*. IEEE Press, Chapter 10.

[34] Bromwich, T. J. I'a., 1916: Normal coordinates in dynamical systems. *Proc. London Math. Soc., Ser. 2*, **15**, 401–448.

[35] Wagner, K. W., 1915/16: Über eine Formel von Heaviside zur Berechnung von Einschaltvorgängen. *Arch. Electrotechnik*, **4**, 159–193.

Figure 1.8.1: Largely a self-educated man, Oliver Heaviside (1850–1925) lived the life of a recluse. It was during his studies of the implications of Maxwell's theory of electricity and magnetism that he reinvented Laplace transforms. Initially rejected, it would require the work of Bromwich to justify its use. (Portrait courtesy of the Institution of Electrical Engineers, London.)

Most students, when faced with this problem, would try separation of variables where you guess the product solution $ru(r,t) = R(r)T(t)$. The partial differential equation then reduces to the ordinary differential equations

$$\frac{R''(r)}{R(r)} = \frac{1}{c^2}\frac{T''(t)}{T(t)} = -k^2. \qquad (1.8.3)$$

Solving for $R(r)$ first, there are three possible solutions:

$$R(r) = A_1 e^{-mr}, \qquad k^2 = -m^2 < 0, \qquad (1.8.4)$$

$$R(r) = B_1, \qquad k^2 = 0, \qquad (1.8.5)$$

and

$$R(r) = C_1 \cos(kr) + D_1 \sin(kr), \qquad 0 < k^2. \qquad (1.8.6)$$

All of these solutions satisfy the condition that $|R(r)|$ remains finite as $r \to \infty$. From Equation 1.8.3 it follows that the temporal solutions are

$$T(t) = A_2 e^{-mct}, \qquad (1.8.7)$$

$$T(t) = B_2, \tag{1.8.8}$$

and

$$T(t) = C_2 \cos(kct) + D_2 \sin(kct), \tag{1.8.9}$$

respectively. Here we have only retained those solutions that remain finite at $t \to \infty$. The initial conditions require that $T(0) = T'(0) = 0$. Applying this condition to Equation 1.8.7 through Equation 1.8.9, we obtain $T(t) = 0$ in all of the cases. Clearly separation of variables has failed and we must try something else.

One alternative is numerical methods. If we introduce the new dependent variables $v(r,t) = ru(r,t)$ and then apply centered-in-space and centered-in-time finite differences, Equation 1.8.1 becomes

$$\frac{v_m^{n+1} - 2v_m^n + v_m^{n-1}}{(\Delta t)^2} = \frac{v_{m+1}^n - 2v_m^n + v_{m-1}^n}{(\Delta r)^2}, \tag{1.8.10}$$

where v_m^n is the value of $v(r,t)$ at time $t_n = na\Delta t/c$, $n = 0, 1, 2, \ldots$, and location $r_m = a + ma\Delta r$, $m = 0, 1, 2, \ldots$. Here, Δr and Δt denote the *nondimensional* intervals between the grid points in the radial and time directions, respectively. Equation 1.8.10, with the initial values $v_m^0 = v_m^1 = 0$ if $m > 0$, can then be used to predict the values of $v(r,t)$. At the surface $r = a$, we have from the boundary condition:

$$\frac{v_{-1}^n - v_1^n}{2\Delta r} + v_0^n = Bwa^2 \sin(\omega t_n). \tag{1.8.11}$$

Combining Equation 1.8.10 and Equation 1.8.11, we find that

$$\frac{v_0^{n+1} - 2v_0^n + v_0^{n-1}}{(\Delta t)^2} = 2\frac{v_1^n - v_0^n}{(\Delta r)^2} - \frac{2v_0^n}{\Delta r} + \frac{2B\omega a^2}{\Delta r} \sin(\omega t_n). \tag{1.8.12}$$

Figure 1.8.2 illustrates this numerical solution.

Although numerical methods provide a straightforward procedure for computing the solution, this technique does suffer from several drawbacks. First, there is no universal form for the solution. The numerical computations must be redone for each new value of $\omega a/c$. Second, the accuracy of the solution depends upon the magnitude of Δr and Δt; the smaller Δr and Δt are, the more accurate the results are. Greater accuracy therefore demands greater computational resources and storage.

These considerations beg the question whether or not there still might be a way to find an exact solution to this problem. In the case of linear problems, transform methods frequently provide the answer.

How would transform methods handle this problem? To answer this question, we must first address another question: What is the Laplace transform

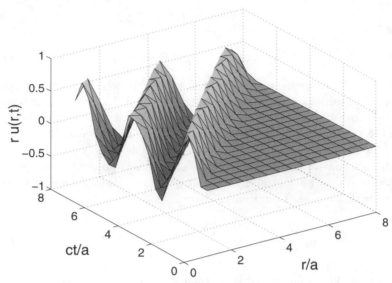

Figure 1.8.2: Numerical solution of Equation 1.8.1 and Equation 1.8.2 where we have plotted $ru(r,t)$ as functions of time ct/a and distance r/a. Here $wa/c = 2$, $wBa^2 = 1$, $\Delta r = 0.5$ and $\Delta t = 0.4$.

of a multivariable function? Perhaps surprisingly, the answer to this question is quite simple. From the fundamental definition of the Laplace transform

$$U(r,s) = \int_0^\infty u(r,t)e^{-st}\,dt, \qquad (1.8.13)$$

where $U(r,s)$ denotes the Laplace transform of the multivariable function $u(r,t)$, it then follows that

$$\mathcal{L}[u_t(r,t)] = sU(r,s) - u(r,0) \qquad (1.8.14)$$

and

$$\mathcal{L}[u_{tt}(r,t)] = s^2U(r,s) - su(r,0) - u_t(r,0). \qquad (1.8.15)$$

This is very similar to the derivative rules for single variable functions; the difference here is that the Laplace transform depends not only on the transform variable s but also the independent variable r.

Turning to the derivatives involving r, we have

$$\mathcal{L}\left\{\frac{\partial[ru(r,t)]}{\partial r}\right\} = \frac{d}{dr}\left\{r\mathcal{L}[u(r,t)]\right\} = \frac{d}{dr}\left[rU(r,s)\right] \qquad (1.8.16)$$

and

$$\mathcal{L}\left\{\frac{\partial^2[ru(r,t)]}{\partial r^2}\right\} = \frac{d^2}{dr^2}\left\{r\mathcal{L}[u(r,t)]\right\} = \frac{d^2}{dr^2}\left[rU(r,s)\right]. \qquad (1.8.17)$$

This is quite new and introduces ordinary derivatives of the Laplace transform $U(r, s)$.

Having derived the derivative rules, we are ready to take the Laplace transform of Equation 1.8.1. It is

$$\frac{d^2}{dr^2}\left[rU(r, s)\right] - \frac{s^2}{c^2}rU(r, s) = 0. \tag{1.8.18}$$

The solution to Equation 1.8.18 is

$$rU(r, s) = A\exp(-rs/c). \tag{1.8.19}$$

We have discarded the $\exp(rs/c)$ solution because it becomes infinite in the limit of $r \to \infty$. After substituting Equation 1.8.19 into the Laplace transformed Equation 1.8.2, we find

$$-\frac{d}{dr}\left(A\frac{e^{-sr/c}}{r}\right)\Bigg|_{r=a} = \frac{\omega Bs}{s^2 + \omega^2} = Ae^{-as/c}\left(\frac{1}{a^2} + \frac{s}{ac}\right). \tag{1.8.20}$$

Therefore,

$$rU(r, s) = \frac{\omega Ba^2 cs}{(s^2 + \omega^2)(as + c)}e^{-s(r-a)/c} \tag{1.8.21}$$

$$= \frac{\omega Ba^2 c}{a^2\omega^2 + c^2}e^{-s(r-a)/c}\left(\frac{cs + \omega^2 a}{s^2 + \omega^2} - \frac{c}{s + c/a}\right). \tag{1.8.22}$$

Applying the second shifting theorem, the inversion of Equation 1.8.22 follows directly, namely,

$$ru(r, t) = \frac{\omega Ba^2 c^2}{a^2\omega^2 + c^2}\left\{\cos\left[\omega\left(t - \frac{r-a}{c}\right)\right] + \frac{\omega a}{c}\sin\left[\omega\left(t - \frac{r-a}{c}\right)\right]\right.$$
$$\left. - \exp\left[-\frac{c}{a}\left(t - \frac{r-a}{c}\right)\right]\right\}H\left(t - \frac{r-a}{c}\right). \tag{1.8.23}$$

Equation 1.8.23 clarifies the numerical solution presented in Figure 1.8.2. First, there is a sharp demarcation between the quiescent region ahead of the wave front located at $t = (r - a)/c$ and the oscillations that occur behind it. Second, at a given fixed point r, the frequency of the wave motion rapidly approaches ω after the passage of the wave front. Both of these results are captured in the numerical solution although the exact solution has a universality that the numerical solution lacks.

This problem highlights all of the advantages of transform methods. Once we generalized our concept of Laplace transforms to include multivariable functions, we could immediately take the Laplace transform of the partial differential equation and use our knowledge of ordinary differential equations to find the Laplace transform $U(r, s)$. Finally, we can use the same inversion techniques as those employed in single variable problems since the independent variable r acts as a parameter.

With the straightforward nature of this procedure, a natural question is "Why isn't this technique used for all linear partial differential equations?" There are essentially three difficulties. First, taking the Laplace transform of the partial differential equation may be difficult. For example, for most partial differential equations with nonconstant coefficients, we do not know how to take their transform. Second, having taken the Laplace transform of the partial differential equation, we may be unable to solve the resulting ordinary differential equation. Finally, even if we can solve the differential equation, we may be unable to invert the transform analytically.

Of these three difficulties, inversion seems the greatest stumbling block, probably because transform methods would be immediately abandoned if we could not find analytically the transform of the dependent variable. For this reason, we will spend considerable time on various inversion techniques. First, we will apply integration techniques on the complex plane to evaluate the inversion integral. Indeed, we shall do this so often that this book will read like a book on applied complex variables.[36] When this technique fails, we still have the option of inverting the transform using asymptotic or numerical methods. Using numerical methods for only the inversion is preferable to using numerical methods to solve the entire problem since we still have an analytic solution in the other independent variable.

In addition to Laplace transforms, we may also use Fourier transforms to solve partial differential equations. Again, the most difficult aspect of the analysis is the inversion and, again, we may apply complex variables and numerical methods to find its inverse. We illustrate this in Chapter 3 for single-valued transforms while we treat multivalued transforms in Chapter 5.

---

[36] "Had Heaviside been able to make full use of Cauchy's method of complex integration, then (to quote a well-known saying) 'we should have learned something'." Quote taken from Bromwich, T. J. I'a., 1928: Note on Prof. Carslaw paper. *Math. Gaz.*, **14**, p. 227.

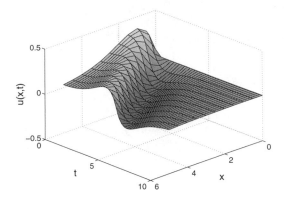

Chapter 2
Methods Involving
Single-Valued Laplace Transforms

Having given an overview of transform techniques, we begin with this chapter to examine in greater detail how Laplace transforms may be used to solve linear partial differential equations. We limit ourselves to the situation where the transform does not contain a branch point or cut. We address that question in Chapter 4.

2.1 INVERSION OF LAPLACE TRANSFORMS BY CONTOUR INTEGRATION

Laplace transforms are a popular tool for solving initial-value problems. In most undergraduate courses, the use of tables, special theorems, partial fractions and convolution are the methods taught for finding the inverse. In most problems involving partial differential equations, these techniques fail us.

In 1916, Bromwich[1] showed that we can express the inverse of a Laplace transform (*Bromwich's integral*) as the contour integral

$$f(t) = \frac{1}{2\pi i} \int_{c-\infty i}^{c+\infty i} F(z)e^{tz}\, dz, \qquad (2.1.1)$$

---

[1] Bromwich, T. J. I'a., 1916: Normal coordinates in dynamical systems. *Proc. London Math. Soc., Ser. 2*, **15**, 401–448.

where $F(z)$ is the Laplace transform

$$F(s) = \int_0^\infty f(t)e^{-st}\, dt, \qquad (2.1.2)$$

and c is the Laplace convergence abscissa. The value of c must be greater than the real part of any of the singularities of $F(s)$.

Proof: Consider the piecewise differentiable function $f(x)$ which vanishes for $x < 0$. We can express the function $e^{-cx}f(x)$ by the complex Fourier representation

$$f(x)e^{-cx} = \frac{1}{2\pi}\int_{-\infty}^\infty e^{i\omega x}\left[\int_0^\infty e^{-i\omega t}e^{-ct}f(t)\, dt\right] d\omega, \qquad (2.1.3)$$

where we choose c so that $\int_0^\infty e^{-ct}|f(t)|\, dt$ exists. We now multiply Equation 2.1.3 by e^{cx} and then bring this factor inside the first integral. Then

$$f(x) = \frac{1}{2\pi}\int_{-\infty}^\infty e^{(c+i\omega)x}\left[\int_0^\infty e^{-(c+i\omega)t}f(t)\, dt\right] d\omega. \qquad (2.1.4)$$

Let $s = c + i\omega$, where s is a new, complex variable of integration. Then,

$$f(x) = \frac{1}{2\pi i}\int_{c-\infty i}^{c+\infty i} e^{sx}\left[\int_0^\infty e^{-st}f(t)\, dt\right] ds. \qquad (2.1.5)$$

The path of integration from $c - \infty i$ to $c + \infty i$ is called *Bromwich's contour*. Because the quantity inside the square brackets is equal to the Laplace transform $F(s)$,

$$f(t) = \frac{1}{2\pi i}\int_{c-\infty i}^{c+\infty i} F(z)e^{zt}\, dz. \qquad (2.1.6)$$

$\square$

Of course, this brings us no closer to actually finding the inverse unless we can evaluate the integral. Fortunately, complex variables are particularly adept at doing this. In the following examples, we illustrate how these techniques apply when the transform is single-valued.

● **Example 2.1.1**

For our first example of the inversion of Laplace transforms by complex integration, let us find the inverse of

$$F(s) = \frac{1}{s\sinh(as)}, \qquad (2.1.7)$$

Figure 2.1.1: An outstanding mathematician at Cambridge University at the turn of the twentieth century, Thomas John I'Anson Bromwich (1875–1929) came to Heaviside's operational calculus through his interest in divergent series. Beginning a correspondence with Heaviside, Bromwich was able to justify operational calculus through the use of contour integrals by 1915. After his premature death, individuals such as J. R. Carson and Sir H. Jeffreys brought Laplace transforms to the increasing attention of scientists and engineers. (Portrait courtesy of ©The Royal Society.)

where a is real. From Bromwich's integral,

$$f(t) = \frac{1}{2\pi i} \int_{c-\infty i}^{c+\infty i} \frac{e^{tz}}{z \sinh(az)} \, dz. \qquad (2.1.8)$$

Here c is greater than the real part of any of the singularities in Equation 2.1.8.

Our first task in the inversion of $F(s)$ is the classification of the singularities of the integrand of Equation 2.1.8. Using the infinite product for the hyperbolic sine,[2]

$$\frac{e^{tz}}{z \sinh(az)} = \frac{e^{tz}}{az^2[1 + a^2z^2/\pi^2][1 + a^2z^2/(4\pi^2)][1 + a^2z^2/(9\pi^2)]\cdots}. \qquad (2.1.9)$$

Thus, we have a second-order pole at $z = 0$ and simple poles at $z_n = \pm n\pi i/a$, where $n = 1, 2, 3, \ldots$.

---

[2] Gradshteyn, I. S., and I. M. Ryzhik, 1965: *Table of Integrals, Series and Products.* Academic Press, Section 1.431, Formula 2.

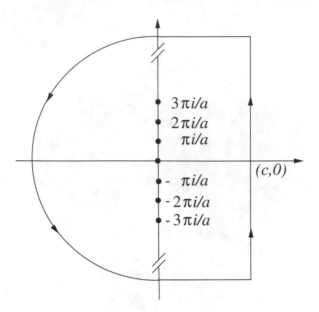

Figure 2.1.2: Contours used in the inversion of Laplace transform given by Equation 2.1.7.

We can convert the line integral in Equation 2.1.8, with the Bromwich contour lying parallel and slightly to the right of the imaginary axis, into a closed contour using Jordan's lemma through the addition of an infinite semicircle joining $i\infty$ to $-i\infty$ as shown in Figure 2.1.2. We now apply the residue theorem. For the second-order pole at $z = 0$,

$$\text{Res}\left[\frac{e^{tz}}{z\sinh(az)}; 0\right] = \frac{1}{1!}\lim_{z\to 0}\frac{d}{dz}\left[\frac{(z-0)^2 e^{tz}}{z\sinh(az)}\right] \tag{2.1.10}$$

$$= \lim_{z\to 0}\frac{d}{dz}\left[\frac{ze^{tz}}{\sinh(az)}\right] \tag{2.1.11}$$

$$= \lim_{z\to 0}\left[\frac{e^{tz}}{\sinh(az)} + \frac{zte^{tz}}{\sinh(az)} - \frac{az\cosh(az)e^{tz}}{\sinh^2(az)}\right] \tag{2.1.12}$$

$$= \frac{t}{a} \tag{2.1.13}$$

after using $\sinh(az) = az + O(z^3)$. For the simple poles $z_n = \pm n\pi i/a$,

$$\text{Res}\left[\frac{e^{tz}}{z\sinh(az)}; z_n\right] = \lim_{z\to z_n}\frac{(z-z_n)e^{tz}}{z\sinh(az)} \tag{2.1.14}$$

$$= \lim_{z\to z_n}\frac{e^{tz}}{\sinh(az) + az\cosh(az)} \tag{2.1.15}$$

$$= \frac{\exp(\pm n\pi it/a)}{(-1)^n(\pm n\pi i)}, \tag{2.1.16}$$

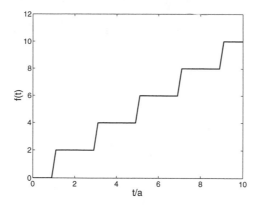

Figure 2.1.3: The inverse of the Laplace transform given by Equation 2.1.7.

because $\cosh(\pm n\pi i) = \cos(n\pi) = (-1)^n$. Thus, summing the residues gives

$$f(t) = \frac{t}{a} + \sum_{n=1}^{\infty} \frac{(-1)^n \exp(n\pi it/a)}{n\pi i} - \sum_{n=1}^{\infty} \frac{(-1)^n \exp(-n\pi it/a)}{n\pi i} \qquad (2.1.17)$$

$$= \frac{t}{a} + \frac{2}{\pi} \sum_{n=1}^{\infty} \frac{(-1)^n}{n} \sin\left(\frac{n\pi t}{a}\right). \qquad (2.1.18)$$

Figure 2.1.3 illustrates this inverse at various times t.

• **Example 2.1.2**

For our second example, we invert

$$F(s) = \frac{\cosh(qx)}{sq\sinh(qL)}, \qquad (2.1.19)$$

where $q = s^{1/2}/a$, and the constants a, L and x are real. One immediate concern is the presence of $s^{1/2}$ because this is a multivalued function. However, when we replace the hyperbolic cosine and sine functions with their Taylor expansions, $F(s)$ contains only powers of s and is, in fact, single-valued.

From Bromwich's integral,

$$f(t) = \frac{1}{2\pi i} \int_{c-\infty i}^{c+\infty i} \frac{\cosh(qx)\, e^{tz}}{zq\sinh(qL)}\, dz, \qquad (2.1.20)$$

where $q = z^{1/2}/a$. Using the Taylor expansions for the hyperbolic cosine and sine, we find that $z = 0$ is a second-order pole. The remaining poles are located where $\sinh(qL) = -i\sin(iqL) = 0$. Therefore, $z_n^{1/2} L/a = n\pi i$ or $z_n = -n^2\pi^2 a^2/L^2$, where $n = 1, 2, 3, \ldots$. We have chosen the positive sign

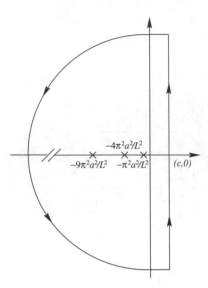

Figure 2.1.4: Contour used in the inversion of Laplace transform given by Equation 2.1.19.

because $z^{1/2}$ must be single-valued; a negative sign would lead to the same result. Further analysis reveals that these poles are simple.

Having classified the poles, we now close the line contour which lies slightly to the right of the imaginary axis with an infinite semicircle in the left half-plane and use the residue theorem. See Figure 2.1.4. The values of the residues are

$$\text{Res}\left[\frac{\cosh(qx)e^{tz}}{zq\sinh(qL)};0\right] = \frac{1}{1!}\lim_{z\to 0}\frac{d}{dz}\left[\frac{(z-0)^2\cosh(qx)e^{tz}}{zq\sinh(qL)}\right] \tag{2.1.21}$$

$$= \lim_{z\to 0}\frac{d}{dz}\left[\frac{z\cosh(qx)e^{tz}}{q\sinh(qL)}\right] \tag{2.1.22}$$

$$= \frac{a^2}{L}\lim_{z\to 0}\frac{d}{dz}\left[\frac{z\left(1+\frac{zx^2}{2!a^2}+\cdots\right)\left(1+tz+\frac{t^2z^2}{2!}+\cdots\right)}{z+\frac{L^2z^2}{3!a^2}+\cdots}\right] \tag{2.1.23}$$

$$= \frac{a^2}{L}\lim_{z\to 0}\frac{d}{dz}\left(1+tz+\frac{zx^2}{2a^2}-\frac{zL^2}{3!a^2}+\cdots\right) \tag{2.1.24}$$

$$= \frac{a^2}{L}\left(t+\frac{x^2}{2a^2}-\frac{L^2}{6a^2}\right), \tag{2.1.25}$$

and

$$\text{Res}\left[\frac{\cosh(qx)e^{tz}}{zq\sinh(qL)};z_n\right] = \left[\lim_{z\to z_n}\frac{\cosh(qx)}{zq}e^{tz}\right]\left[\lim_{z\to z_n}\frac{z-z_n}{\sinh(qL)}\right] \tag{2.1.26}$$

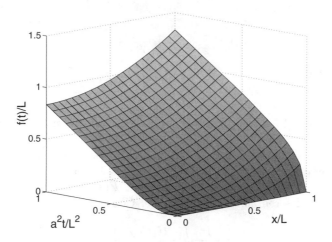

Figure 2.1.5: The inverse of the Laplace transform given by Equation 2.1.19.

$$\text{Res}\left[\frac{\cosh(qx)e^{tz}}{zq\sinh(qL)}; z_n\right] = \lim_{z\to z_n} \frac{\cosh(qx)e^{tz}}{zq\cosh(qL)L/(2a^2q)} \tag{2.1.27}$$

$$= \frac{\cosh(n\pi xi/L)\exp(-n^2\pi^2a^2t/L^2)}{(-n^2\pi^2a^2/L^2)\cosh(n\pi i)L/(2a^2)} \tag{2.1.28}$$

$$= -\frac{2L(-1)^n}{n^2\pi^2}\cos\left(\frac{n\pi x}{L}\right)\exp\left(-\frac{n^2\pi^2a^2t}{L^2}\right). \tag{2.1.29}$$

Summing the residues,

$$f(t) = \frac{a^2}{L}\left(t + \frac{x^2}{2a^2} - \frac{L^2}{6a^2}\right) - \frac{2L}{\pi^2}\sum_{n=1}^{\infty}\frac{(-1)^n}{n^2}\cos\left(\frac{n\pi x}{L}\right)\exp\left(-\frac{n^2\pi^2a^2t}{L^2}\right). \tag{2.1.30}$$

Figure 2.1.5 illustrates this inverse at various times t and values of x.

• Example 2.1.3

To illustrate inversion by Bromwich's integral when Bessel functions are present, consider the transform[3]

$$F(s) = \frac{2I_1(\lambda\sqrt{s})}{\lambda s^{3/2}I_0(\lambda\sqrt{s})}. \tag{2.1.31}$$

The power series representations for $I_0(z)$ and $I_1(z)$ are

$$I_0(z) = 1 + \frac{z^2}{4} + \frac{z^4}{64} + O(z^6) \tag{2.1.32}$$

---

[3] Taken from Raval, U., 1972: Quasi-static transient response of a covered permeable inhomogeneous cylinder to a line current source. *Pure Appl. Geophys.*, **96**, 140–156. Published by Birkhäuser Verlag, Basel, Switzerland.

and

$$I_1(z) = \frac{z}{2} + \frac{z^3}{16} + \frac{z^5}{384} + O(z^7). \qquad (2.1.33)$$

Consequently,

$$F(s) = \frac{1 + \lambda^2 s/8 + \lambda^4 s^2/192 + \cdots}{s(1 + \lambda^2 s/4 + \lambda^4 s^2/64 + \cdots)}. \qquad (2.1.34)$$

From Equation 2.1.34, we see that $F(s)$ is single-valued and has a simple pole at $s = 0$. The remaining poles occur where $I_0(\lambda\sqrt{s}) = 0$. If $\lambda\sqrt{s} = -i\alpha$, then these poles are located where $I_0(-i\alpha) = J_0(\alpha) = 0$. Denoting the nth root of $J_0(\alpha) = 0$ by α_n, then $s_n = -\alpha_n^2/\lambda^2$, where $n = 1, 2, 3, \ldots$.

As a result of this analysis, we see that $F(s)$ has simple poles that lie along the negative real axis of the z-plane. Consequently,

$$f(t) = \frac{1}{2\pi i} \oint_C \frac{2I_1(\lambda\sqrt{z})e^{tz}}{\lambda z^{3/2} I_0(\lambda\sqrt{z})} \, dz, \qquad (2.1.35)$$

where C consists of a path along, and just to the right of, the imaginary axis and a semicircle of infinite radius in the left side of the complex plane. To evaluate the contour integral, we apply the residue theorem where

$$\text{Res}\left[\frac{2I_1(\lambda\sqrt{z})e^{tz}}{\lambda z^{3/2} I_0(\lambda\sqrt{z})}; 0\right] = \lim_{z \to 0} \frac{(1 + \lambda^2 z/8 + \lambda^4 z^2/192 + \cdots)(1 + tz + \cdots)}{1 + \lambda^2 z/4 + \lambda^4 z^2/64 + \cdots} = 1$$
$$(2.1.36)$$

and

$$\text{Res}\left[\frac{2I_1(\lambda\sqrt{z})e^{tz}}{\lambda z^{3/2} I_0(\lambda\sqrt{z})}; -\frac{\alpha_n^2}{\lambda^2}\right] = \lim_{z \to -\alpha_n^2/\lambda^2} \frac{2(z + \alpha_n^2/\lambda^2)I_1(\lambda\sqrt{z})e^{tz}}{\lambda z^{3/2} I_0(\lambda\sqrt{z})} \qquad (2.1.37)$$

$$= \lim_{z \to -\alpha_n^2/\lambda^2} \frac{2I_1(\lambda\sqrt{z})e^{tz}}{\lambda z^{3/2} I_0'(\lambda\sqrt{z})[\lambda/(2\sqrt{z})]} \qquad (2.1.38)$$

$$= -\frac{4\exp(-\alpha_n^2 t/\lambda^2)}{\alpha_n^2}, \qquad (2.1.39)$$

because $I_0'(z) = I_1(z)$. Adding the residues together, we find that the inverse of $F(s)$ is

$$f(t) = 1 - 4 \sum_{n=1}^{\infty} \frac{\exp(-\alpha_n^2 t/\lambda^2)}{\alpha_n^2}, \qquad (2.1.40)$$

where $J_0(\alpha_n) = 0$ for $n = 1, 2, 3, \ldots$. Figure 2.1.6 illustrates this inverse at various times t.

• **Example 2.1.4**

Consider now the inversion of

$$F(s) = K_0\left(r\sqrt{s^2 - a^2}\right), \qquad (2.1.41)$$

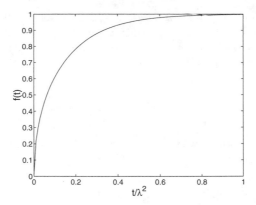

Figure 2.1.6: The inverse of the Laplace transform given by Equation 2.1.31.

where a and r are real, positive. One of the integral representations of $K_0(\cdot)$ is

$$K_0(rz) = \int_0^\infty \frac{\cos(r\eta)}{\sqrt{\eta^2 + z^2}}\, d\eta, \qquad 0 < r, \quad |\arg(z)| < \pi/2. \qquad (2.1.42)$$

Therefore,

$$f(t) = \frac{1}{2\pi i} \int_{c-\infty i}^{c+\infty i} \int_0^\infty \frac{\cos(r\eta)e^{tz}}{\sqrt{\eta^2 - a^2 + z^2}}\, d\eta\, dz \qquad (2.1.43)$$

$$= \int_0^\infty \cos(r\eta)\left[\frac{1}{2\pi i}\int_{c-\infty i}^{c+\infty i}\frac{e^{tz}}{\sqrt{\eta^2 - a^2 + z^2}}\, dz\right]d\eta \qquad (2.1.44)$$

$$= \int_0^\infty \cos(r\eta)\, J_0\left(t\sqrt{\eta^2 - a^2}\right)d\eta \qquad (2.1.45)$$

$$= \frac{\cosh(a\sqrt{t^2 - r^2})}{\sqrt{t^2 - r^2}}H(t - r), \qquad (2.1.46)$$

because $\mathcal{L}[J_0(bt)] = 1/\sqrt{b^2 + s^2}$.

● **Example 2.1.5**

In the previous example, we used an integral representation of the modified Bessel function $K_0(\cdot)$ to reexpress the transform so that we could take its inverse even though we had to write it as an integral. See Equation 2.1.45. Then we carried out the integration and obtained Equation 2.1.46. Yang, Latychev and Edwards[4] used a similar trick to invert the transform

$$F(s) = \frac{\exp\left[-c\sqrt{1 + b\left(s + \sqrt{s}/a\right)}\right]}{s}, \qquad (2.1.47)$$

[4] Yang, J.-W., K. Latychev, and R. N. Edwards, 1998: Numerical computation of hydrothermal fluid circulation in fractured Earth structures. *Geophys. J. Int.*, **135**, 627–649. Published by Blackwell Publishing.

where a, b and c are constants with $a > 0$. In Chapter 4, we will show how to deal with Laplace transforms that contain multivalued functions. Even then, Equation 2.1.47 is tricky because it contains a square root of a square root. Let's see how Yang et al. solved this problem.

They began by noting that

$$\int_0^\infty e^{-\eta^2 - r^2/\eta^2}\, d\eta = \frac{\sqrt{\pi}}{2} e^{-2r}. \tag{2.1.48}$$

We may view Equation 2.1.48 as an integral representation of e^{-2r}. Consequently, we can reexpress Equation 2.1.47 as

$$F(s) = \frac{2}{s\sqrt{\pi}} \int_0^\infty \exp\left\{-\eta^2 - \frac{c^2}{4\eta^2}\left[1 + b\left(s + \sqrt{s}/a\right)\right]\right\} d\eta \tag{2.1.49}$$

$$= \frac{2}{s\sqrt{\pi}} \int_0^\infty \exp\left(-\eta^2 - \frac{c^2}{4\eta^2}\right) \exp\left[-\frac{bc^2}{4a\eta^2}\left(as + \sqrt{s}\right)\right] d\eta. \tag{2.1.50}$$

Using the linearity property of Laplace transforms,

$$f(t) = \frac{2}{\sqrt{\pi}} \int_0^\infty \exp\left(-\eta^2 - \frac{c^2}{4\eta^2}\right) \mathcal{L}^{-1}\left\{\frac{1}{s}\exp\left[-\frac{bc^2}{4a\eta^2}\left(as + \sqrt{s}\right)\right]\right\} d\eta. \tag{2.1.51}$$

Because

$$\mathcal{L}^{-1}\left(\frac{e^{-2b\sqrt{s}}}{s}\right) = \operatorname{erfc}\left(\frac{b}{\sqrt{t}}\right), \qquad 0 < b, \tag{2.1.52}$$

and

$$\mathcal{L}^{-1}\left(\frac{e^{-cs - 2b\sqrt{s}}}{s}\right) = \operatorname{erfc}\left(\frac{b}{\sqrt{t-c}}\right) H(t-c), \qquad 0 < b, c, \tag{2.1.53}$$

we obtain the final result that

$$f(t) = \frac{2}{\sqrt{\pi}} \int_{c\sqrt{b/t}/2}^\infty \exp\left(-\eta^2 - \frac{c^2}{4\eta^2}\right) \operatorname{erfc}\left[\frac{bc^2}{8a\eta^2\sqrt{t - bc^2/(4\eta^2)}}\right] d\eta. \tag{2.1.54}$$

Figure 2.1.7 illustrates this inverse at various times t.

- **Example 2.1.6**

Let us find the inverse[5] of the Laplace transform

$$F(s) = \frac{e^{\gamma s^2 + s^3/3}}{s}, \qquad 0 < \gamma, \tag{2.1.55}$$

---

[5] Reprinted from *Int. J. Solids Struct.*, **16**, T. C. T. Ting, The effects of dispersion and dissipation on wave propagation in viscoelastic layered composites, pp. 903–911, ©1980, with kind permission from Pergamon Press Ltd., Headington Hill Hall, Oxford OX3 0BW, UK. See also Cole, J. D., and T. Y. Wu, 1952: Heat conduction in a compressible fluid. *J. Appl. Mech.*, **19**, 209–213.

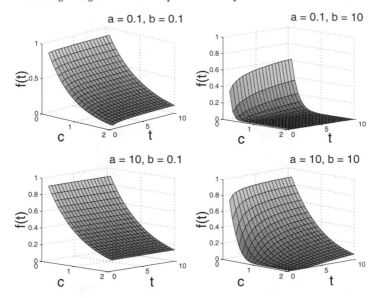

Figure 2.1.7: The inverse of the Laplace transform given by Equation 2.1.47.

or

$$f(t) = \frac{1}{2\pi i} \int_{c-\infty i}^{c+\infty i} e^{zt+\gamma z^2+z^3/3} \frac{dz}{z}.$$ (2.1.56)

We begin by noting that

$$\frac{e^{zt}}{z} = \frac{1}{z} + \int_0^t e^{z\eta}\, d\eta,$$ (2.1.57)

so that

$$f(t) = \frac{1}{2\pi i} \int_{c-\infty t}^{c+\infty i} e^{\gamma z^2+z^3/3} \frac{dz}{z} + \int_0^t \left(\frac{1}{2\pi i} \int_{c-\infty i}^{c+\infty i} e^{\eta z+\gamma z^2+z^3/3}\, dz \right) d\eta.$$ (2.1.58)

To evaluate the first integral in Equation 2.1.58, we now deform the original Bromwich integral to the contour shown in Figure 2.1.8. The contribution from the integrals along the arcs AB and EF vanish as $R \to \infty$. Along BC, $z = re^{-\pi i/3}$ and $dz = dr\, e^{-\pi i/3}$, while $z = re^{\pi i/3}$ and $dz = dr\, e^{\pi i/3}$ along DE. Then,

$$\frac{1}{2\pi i} \int_{BC} e^{\gamma z^2+z^3/3} \frac{dz}{z} = \frac{1}{2\pi i} \int_{\infty}^{0} \exp\left(\gamma r^2 e^{-2\pi i/3} - \frac{r^3}{3} \right) \frac{dr}{r},$$ (2.1.59)

$$\frac{1}{2\pi i} \int_{C_\epsilon} e^{\gamma z^2+z^3/3} \frac{dz}{z} = \frac{1}{3},$$ (2.1.60)

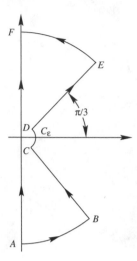

Figure 2.1.8: Contour used in the inversion of the Laplace transform given by Equation 2.1.55.

and

$$\frac{1}{2\pi i} \int_{DE} e^{\gamma z^2 + z^3/3} \frac{dz}{z} = \frac{1}{2\pi i} \int_0^\infty \exp\left(\gamma r^2 e^{2\pi i/3} - \frac{r^3}{3}\right) \frac{dr}{r}. \qquad (2.1.61)$$

Combining Equation 2.1.59 through Equation 2.1.61,

$$\frac{1}{2\pi i} \int_{BE} e^{\gamma z^2 + z^3/3} \frac{dz}{z} = \frac{1}{3} + \frac{1}{2\pi i} \int_0^\infty e^{-r^3/3} \frac{dr}{r} \left[\exp\left(-\frac{\gamma r^2}{2} + \frac{\sqrt{3}i\gamma r^2}{2}\right) \right.$$
$$\left. - \exp\left(-\frac{\gamma r^2}{2} - \frac{\sqrt{3}i\gamma r^2}{2}\right) \right] \qquad (2.1.62)$$

$$= \frac{1}{3} + \frac{1}{\pi} \int_0^\infty e^{-\gamma r^2/2 - r^3/3} \sin\left(\frac{\sqrt{3}}{2}\gamma r^2\right) \frac{dr}{r}. \qquad (2.1.63)$$

Consider now the integral

$$\int_{c-\infty i}^{c+\infty i} e^{\eta z + \gamma z^2 + z^3/3} \, dz = e^{-\eta\gamma + 2\gamma^3/3} \int_{c'-\infty i}^{c'+\infty i} e^{(\eta-\gamma^2)\chi + \chi^3/3} \, d\chi, \qquad (2.1.64)$$

where $z = \chi - \gamma$. If we set $b = \gamma^2 - \eta$, then

$$\frac{1}{2\pi i} \int_{c'-\infty i}^{c'+\infty i} e^{-b\chi + \chi^3/3} \, d\chi = \frac{1}{2\pi i} \int_{AF} e^{-b\chi + \chi^3/3} \, d\chi \qquad (2.1.65)$$

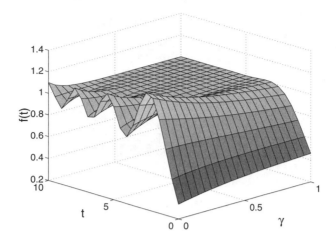

Figure 2.1.9: The inverse of the Laplace transform given by Equation 2.1.55.

$$\frac{1}{2\pi i} \int_{c'-\infty i}^{c'+\infty i} e^{-b\chi+\chi^3/3}\, d\chi = \frac{1}{2\pi i} \int_0^\infty e^{-ibr-r^3i/3}i\, dr - \frac{1}{2\pi i} \int_\infty^0 e^{ibr+r^3i/3}i\, dr$$

$$(2.1.66)$$

$$= \frac{1}{\pi} \int_0^\infty \cos(br+r^3/3)\, dr = \text{Ai}(\gamma^2 - \eta), \quad (2.1.67)$$

where Ai($\cdot$) is the Airy function of the first kind. Therefore, the inverse is

$$f(t) = \frac{1}{3} + \frac{1}{\pi} \int_0^\infty e^{-\gamma r^2/2-r^3/3} \sin\left(\frac{\sqrt{3}}{2}\gamma r^2\right) \frac{dr}{r} + \int_0^t e^{-\eta\gamma+2\gamma^3/3}\text{Ai}(\gamma^2 - \eta)\, d\eta.$$

$$(2.1.68)$$

In this form the inverse is now more amenable to physical interpretation or further asymptotic analysis. Figure 2.1.9 illustrates this inverse for $0 \le \gamma \le 1$.

Problems

For the following transforms, use the inversion integral to find the inverse Laplace transform for constant a, M, R and α.

1. $\dfrac{1}{s^3(s+1)^2}$

2. $\dfrac{s+1}{(s+2)^2(s+3)}$

3. $\dfrac{s+2}{s(s-a)(s^2+4)}$

4. $\dfrac{1}{s\,\sinh(as)}$

5. $\dfrac{\sinh(as/2)}{s\,\cosh(as/2)}$

6. $\dfrac{1}{s^2\,\cosh(\sqrt{as})}$

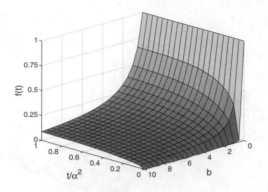

Problem 11

7. $\dfrac{\sinh(a\sqrt{s}\,)}{s\,\sinh(\sqrt{s}\,)}$

8. $\dfrac{\sinh(s)}{s^2\cosh(s)}$

9. $\dfrac{1}{Rs+M^2/\alpha}\left[1-\dfrac{\cosh\!\left(a\sqrt{Rs+M^2/\alpha}\,\right)}{\cosh\!\left(\sqrt{Rs+M^2/\alpha}\,\right)}\right]$

10. $\dfrac{1}{2sa-\sqrt{s}\,\tanh(\sqrt{s}\,)}$

11. Show that the inverse[6] of the Laplace transform

$$F(s)=\frac{\tanh(\alpha\sqrt{s}\,)}{s[m\sqrt{s}+\tanh(\alpha\sqrt{s}\,)]}$$

is

$$f(t)=\frac{1}{1+b}+\frac{1}{b}\sum_{n=1}^{\infty}\frac{\sin(2\lambda_n)\exp(-\lambda_n^2 t/\alpha^2)}{\lambda_n[\cos^2(\lambda_n)+1/b]},$$

where λ_n is the nth root of $\tan(\lambda)=-b\lambda$, $b=m/\alpha$, and α and m are real. This inverse is illustrated in the figure labeled Problem 11.

12. Show that the inverse of the Laplace transform

$$F(s)=\frac{I_0\!\left(r\sqrt{s/\kappa}\,\right)}{I_0\!\left(a\sqrt{s/\kappa}\,\right)},\qquad 0<a,r,\kappa,$$

is

$$f(t)=\frac{2\kappa}{a}\sum_{n=1}^{\infty}\frac{\alpha_n J_0(\alpha_n r)}{J_1(\alpha_n a)}e^{-\kappa\alpha_n^2 t},$$

[6] Taken from Arutunyan, N. H., 1949: On the research of statically indeterminate systems with vibrating support columns (in Russian). *Prikl. Mat. Mek.*, **13**, 399–500.

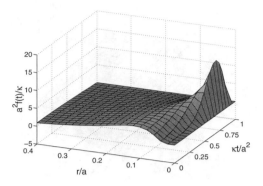

Problem 12

where α_n is the nth positive root of $J_0(\alpha a) = 0$. This inverse is illustrated in the figure labeled Problem 12.

13. Show that the inverse[7] of the Laplace transform

$$F(s) = \frac{I_0\left(r\sqrt{s/\kappa}\right)}{s\ I_0\left(a\sqrt{s/\kappa}\right)(k_1^2/\kappa - s)}, \qquad 0 \le r < a, \qquad 0 < \kappa,$$

is

$$f(t) = \frac{\kappa}{k_1^2} - \frac{\kappa}{k_1^2}\frac{I_0(k_1 r/\kappa)}{I_0(k_1 a/\kappa)}e^{k_1^2 t/\kappa} + \frac{2\kappa}{a}\sum_{n=1}^{\infty}\frac{J_0(\alpha_n r)\,e^{-\kappa\alpha_n^2 t}}{\alpha_n J_1(\alpha_n a)(k_1^2 + \kappa^2\alpha_n^2)},$$

where α_n is the nth positive root of $J_0(\alpha a) = 0$. This inverse is illustrated in the figure labeled Problem 13.

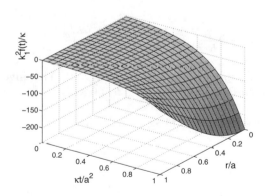

Problem 13

---

[7] Taken from Wadhawan, M. C., 1974: Dynamic thermoelastic response of a cylinder. *Pure Appl. Geophys.*, **112**, 73–82. Published by Birkhäuser Verlag, Basel, Switzerland.

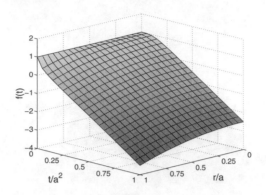

Problem 14

14. Show that the inverse of the Laplace transform

$$F(s) = \frac{1}{s} - \frac{I_0(r\sqrt{s})}{s^{3/2}\, I_1(a\sqrt{s})}, \qquad\qquad 0 \le r < a,$$

is

$$f(t) = 1 - a\left[\frac{2t}{a^2} + \frac{r^2}{2a^2} - \frac{1}{4} - 2\sum_{n=1}^{\infty} \frac{J_0(\alpha_n r/a)}{\alpha_n^2\, J_0(\alpha_n)} e^{-\alpha_n^2 t/a^2}\right],$$

where α_n is the nth positive root of $J_1(\alpha) = 0$. This inverse is illustrated in the figure labeled Problem 14 with $a = 2$.

15. Show that the inverse of the Laplace transform

$$F(s) = \frac{I_n(a\sqrt{s})}{\sqrt{s}\, I_{n+1}(b\sqrt{s})}, \qquad\qquad 0 \le n, \quad 0 < a < b,$$

is

$$f(t) = 2(n+1)\frac{a^n}{b^{n+1}} + \frac{4}{b}\sum_{m=1}^{\infty} \frac{J_n(a\alpha_m/b)}{J_n(\alpha_m) - J_{n+2}(\alpha_m)} e^{-\alpha_m^2 t/b^2},$$

where α_m is the mth positive root of $J_{n+1}(\alpha) = 0$. This inverse is illustrated in the figure labeled Problem 15 with $a/b = 0.8$.

16. Find the inverse[8] of the Laplace transform

$$F(s) = \frac{\exp[-r\sqrt{s}\,\tanh(\sqrt{s})]}{s}, \qquad\qquad 0 < r.$$

This transform has a simple pole at $s = 0$ and an infinite number of essential singularities at $s_n = -(2n-1)^2\pi^2/4$, where $n = 1, 2, 3, \ldots.$. The most

[8] Taken from Roshal', A. A., 1969: Mass transfer in a two-layer porous medium. *J. Appl. Mech. Tech. Phys.*, **10**, 551–558.

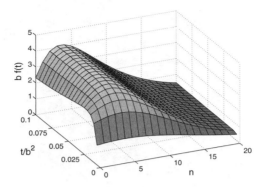

Problem 15

convenient method for finding the inverse is to deform Bromwich's contour along the imaginary axis of the s-plane, except for an infinitesimally small semicircle around the simple pole. If we use this contour, show that

$$f(t) = \frac{1}{2} + \frac{2}{\pi} \int_{0+}^{\infty} \exp\left[-\frac{r\eta}{2} \frac{\sinh(\eta) - \sin(\eta)}{\cosh(\eta) + \cos(\eta)}\right]$$
$$\times \sin\left[\frac{\eta^2 t}{2} - \frac{r\eta}{2} \frac{\sinh(\eta) + \sin(\eta)}{\cosh(\eta) + \cos(\eta)}\right] \frac{d\eta}{\eta}.$$

This inverse is illustrated in the figure labeled Problem 16.

17. Consider a function $f(t)$ which has the Laplace transform $F(z)$ which is analytic in the half-plane $\operatorname{Re}(z) > s_0$. Can we use this knowledge to find $g(t)$ whose Laplace transform $G(z)$ equals $F[\varphi(z)]$, where $\varphi(z)$ is also analytic for $\operatorname{Re}(z) > s_0$? The answer to this question leads to the Schouten[9]–Van der Pol[10] theorem.

Step 1: Show that the following relationships hold true:

$$G(z) = F[\varphi(z)] = \int_0^{\infty} f(\tau) e^{-\varphi(z)\tau} \, d\tau$$

and

$$g(t) = \frac{1}{2\pi i} \int_{c-\infty i}^{c+\infty i} F[\varphi(z)] e^{tz} \, dz.$$

Step 2: Using the results from Step 1, show that

$$g(t) = \int_0^{\infty} f(\tau) \left[\frac{1}{2\pi i} \int_{c-\infty i}^{c+\infty i} e^{-\varphi(z)\tau} e^{tz} \, dz\right] d\tau.$$

---

[9] Schouten, J. P., 1935: A new theorem in operational calculus together with an application of it. *Physica*, **2**, 75–80.

[10] Van der Pol, B., 1934: A theorem on electrical networks with applications to filters. *Physica*, **1**, 521–530.

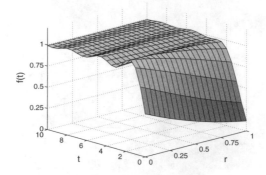

Problem 16

This is the Schouten–Van der Pol theorem.

Step 3: If $G(z) = F(\sqrt{z})$ show that

$$g(t) = \frac{1}{2\sqrt{\pi t^3}} \int_0^\infty \tau f(\tau) \exp\left(-\frac{\tau^2}{4t}\right) d\tau.$$

Hint: Do not evaluate the contour integral. Instead, ask yourself: What function of time has a Laplace transform that equals $e^{-\varphi(z)\tau}$, where τ is a parameter? Then use tables.

2.2 THE HEAT EQUATION

The solution of linear partial differential equations by Laplace transforms is the most commonly employed analytic technique after separation of variables. One of the key elements of this method involves finding the Laplace transform of $u(x,t)$ and its partial derivatives, where $u(x,t)$ denotes the solution to the partial differential equation. For example, because the Laplace transform of $u(x,t)$ is defined by the integral

$$U(x,s) = \int_0^\infty u(x,t)e^{-st}\,dt, \qquad (\mathbf{2.2.1})$$

the Laplace transform of u_t follows directly from integration by parts, or

$$\mathcal{L}[u_t(x,t)] = \int_0^\infty u_t(x,t)e^{-st}\,dt = sU(x,s) - u(x,0), \qquad (\mathbf{2.2.2})$$

which introduces the initial condition $u(x,0)$. Note that these Laplace transforms depend upon the variable x as well as the parameter s. On the other hand, derivatives involving x become

$$\mathcal{L}[u_x(x,t)] = \frac{d}{dx}\{\mathcal{L}[u(x,t)]\} = \frac{dU(x,s)}{dx}, \qquad (\mathbf{2.2.3})$$

and

$$\mathcal{L}[u_{xx}(x,t)] = \frac{d^2}{dx^2}\{\mathcal{L}[u(x,t)]\} = \frac{d^2U(x,s)}{dx^2}. \qquad (2.2.4)$$

Because the transformation eliminates the time variable, only $U(x,s)$ and its derivatives remain in the equation. Consequently, we transform the partial differential equation into a boundary-value problem involving an ordinary differential equation. Because this equation is often easier to solve than a partial differential equation, the use of Laplace transforms considerably simplifies the original problem. Of course, the Laplace transforms must exist for this technique to work.

The following schematic summarizes the Laplace transform method:

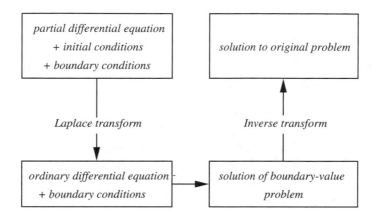

The following examples illustrate how Laplace transforms can be used to solve the heat equation in various types of domains and coordinate systems.

• **Example 2.2.1**

For our first example of how to solve the heat equation via Laplace transforms, consider the heat equation

$$\frac{\partial u}{\partial t} = \frac{\partial^2 u}{\partial x^2}, \qquad 0 < x < \infty, \quad 0 < t, \qquad (2.2.5)$$

with the initial condition

$$u(x,0) = 1, \qquad 0 < x < \infty, \qquad (2.2.6)$$

and boundary conditions

$$u_x(0,t) = 1, \qquad \lim_{x \to \infty}|u(x,t)| < \infty, \qquad 0 < t. \qquad (2.2.7)$$

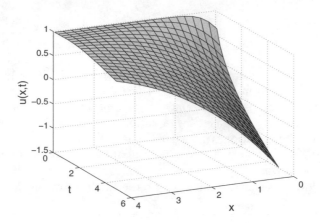

Figure 2.2.1: A plot of Equation 2.2.11 as a function of distance x and time t.

Taking the Laplace transform of Equation 2.2.5 and Equation 2.2.7 and substituting the initial condition, we obtain

$$\frac{d^2 U(x,s)}{dx^2} = sU(x,s) - 1 \qquad (\mathbf{2.2.8})$$

with the boundary conditions

$$U'(0,s) = \frac{1}{s} \qquad \text{and} \qquad \lim_{x \to \infty} |U(x,s)| < \infty. \qquad (\mathbf{2.2.9})$$

The solution that satisfies this boundary-value problem is

$$U(x,s) = \frac{1}{s} - \frac{e^{-x\sqrt{s}}}{s^{3/2}}. \qquad (\mathbf{2.2.10})$$

Using tables, the inverse of the Laplace transform given by Equation 2.2.10 is

$$u(x,t) = 1 - 2\sqrt{\frac{t}{\pi}} \exp\left(-\frac{x^2}{4t}\right) + x \operatorname{erfc}\left(\frac{x}{2\sqrt{t}}\right). \qquad (\mathbf{2.2.11})$$

Figure 2.2.1 illustrates Equation 2.2.11 as a function of distance x and time t.

• Example 2.2.2

Several slight modifications to Example 2.2.1 yield a more challenging problem, namely,

$$\frac{\partial u}{\partial t} = \frac{\partial^2 u}{\partial x^2} - u, \qquad 0 < x < \infty, \quad 0 < t, \qquad (\mathbf{2.2.12})$$

with the initial condition

$$u(x,0) = 0, \qquad 0 < x < \infty, \tag{2.2.13}$$

and boundary conditions

$$u(0,t) = 1, \qquad \lim_{x \to \infty} |u(x,t)| < \infty, \qquad 0 < t. \tag{2.2.14}$$

Taking the Laplace transform of Equation 2.2.12 and Equation 2.2.14 and substituting the initial condition, we obtain the boundary-value problem

$$\frac{d^2U(x,s)}{dx^2} = (s+1)U(x,s) \tag{2.2.15}$$

with

$$U(0,s) = \frac{1}{s} \qquad \text{and} \qquad \lim_{x \to \infty} |U(x,s)| < \infty. \tag{2.2.16}$$

The solution that satisfies this boundary-value problem is

$$U(x,s) = \frac{e^{-x\sqrt{s+1}}}{s}. \tag{2.2.17}$$

Using tables, the first shifting theorem and convolution, the inverse of the Laplace transform given by Equation 2.2.17 is

$$u(x,t) = \int_0^t e^{-\xi} \frac{x}{2\sqrt{\pi\xi^3}} \exp\left(-\frac{x^2}{4\xi}\right) d\xi \tag{2.2.18}$$

$$= \frac{2}{\sqrt{\pi}} \int_{x/(2\sqrt{t})}^{\infty} \exp\left(-\eta^2 - \frac{x^2}{4\eta^2}\right) d\eta, \tag{2.2.19}$$

where $\xi = x^2/(4\eta^2)$.

An alternative method for inverting Equation 2.2.17 is to apply the technique shown in Example 2.1.5. Using Equation 2.1.48 to replace the exponential in Equation 2.2.17, we obtain

$$U(x,s) = \frac{2}{\sqrt{\pi}} \int_0^\infty \exp\left(-\eta^2 - \frac{x^2}{4\eta^2}\right) \frac{e^{-x^2 s/(4\eta^2)}}{s} d\eta. \tag{2.2.20}$$

Applying the second shifting theorem,

$$u(x,t) = \frac{2}{\sqrt{\pi}} \int_0^\infty \exp\left(-\eta^2 - \frac{x^2}{4\eta^2}\right) H\left(t - \frac{x^2}{4\eta^2}\right) d\eta. \tag{2.2.21}$$

Eliminating the Heaviside function from Equation 2.2.21 leads directly to Equation 2.2.19.

• Example 2.2.3

To illustrate how the Laplace transform technique applies to problems over a finite domain, we solve a heat conduction problem[11] in a plane slab of thickness $2L$. Initially the slab has a constant temperature of unity. For $0 < t$, we allow both faces of the slab to radiatively cool in a medium which has a temperature of zero.

If $u(x, t)$ denotes the temperature, a^2 is the thermal diffusivity, h is the relative emissivity, t is the time, and x is the distance perpendicular to the face of the slab and measured from the middle of the slab, then the governing equation is

$$\frac{\partial u}{\partial t} = a^2 \frac{\partial^2 u}{\partial x^2}, \qquad -L < x < L, \quad 0 < t, \tag{2.2.22}$$

with the initial condition

$$u(x, 0) = 1, \qquad -L < x < L, \tag{2.2.23}$$

and boundary conditions

$$\frac{\partial u(L, t)}{\partial x} + hu(L, t) = 0, \quad \frac{\partial u(-L, t)}{\partial x} + hu(-L, t) = 0, \quad 0 < t. \tag{2.2.24}$$

Taking the Laplace transform of Equation 2.2.22 and substituting the initial condition,

$$a^2 \frac{d^2 U(x, s)}{dx^2} - sU(x, s) = -1. \tag{2.2.25}$$

If we set $s = a^2 q^2$, Equation 2.2.25 becomes

$$\frac{d^2 U(x, s)}{dx^2} - q^2 U(x, s) = -\frac{1}{a^2}. \tag{2.2.26}$$

From the boundary conditions, $U(x, s)$ is an even function in x and we may conveniently write the solution as

$$U(x, s) = \frac{1}{s} + A \cosh(qx). \tag{2.2.27}$$

From Equation 2.2.24,

$$qA \sinh(qL) + \frac{h}{s} + hA \cosh(qL) = 0, \tag{2.2.28}$$

and

$$U(x, s) = \frac{1}{s} - \frac{h \cosh(qx)}{s[q \sinh(qL) + h \cosh(qL)]}. \tag{2.2.29}$$

---

[11] Goldstein, S., 1932: The application of Heaviside's operational method to the solution of a problem in heat conduction. *Z. Angew. Math. Mech.*, **12**, 234–243.

The inverse of $U(x, s)$ consists of two terms. The first term is simply unity. We will invert the second term by contour integration.

We begin by examining the nature and location of the singularities in the second term. Using the product formulas for the hyperbolic cosine and sine functions, the second term equals

$$\frac{(h/s)\left[1 + 4q^2x^2/\pi^2\right]\left[1 + 4q^2x^2/(9\pi^2)\right]\cdots}{q^2L\left[1 + (qL/\pi)^2\right]\left[1 + (qL/2\pi)^2\right]\cdots + h\left[1 + (2qL/\pi)^2\right]\left[1 + (2qL/3\pi)^2\right]\cdots}. \tag{2.2.30}$$

Because $q^2 = s/a^2$, Equation 2.2.30 shows that we do not have any $\sqrt{s}$ in the transform and we need not concern ourselves with branch points and cuts. Furthermore, we have only simple poles: one located at $s = 0$ and the others where

$$q\sinh(qL) + h\cosh(qL) = 0. \tag{2.2.31}$$

If we set $q = i\lambda$, Equation 2.2.31 becomes

$$h\cos(\lambda L) - \lambda\sin(\lambda L) = 0, \tag{2.2.32}$$

or

$$\lambda L\tan(\lambda L) = hL. \tag{2.2.33}$$

From Bromwich's integral,

$$\mathcal{L}^{-1}\left\{\frac{h\cosh(qx)}{s[q\sinh(qL) + h\cosh(qL)]}\right\} = \frac{1}{2\pi i}\oint_C \frac{h\cosh(qx)e^{tz}}{z[q\sinh(qL) + h\cosh(qL)]}\,dz, \tag{2.2.34}$$

where $q = z^{1/2}/a$ and the closed contour C consists of Bromwich's contour plus a semicircle of infinite radius in the left half of the z-plane. The residue at $z = 0$ is 1 while at $z_n = -a^2\lambda_n^2$,

$$\mathrm{Res}\left\{\frac{h\cosh(qx)e^{tz}}{z[q\sinh(qL) + h\cosh(qL)]}; z_n\right\}$$

$$= \lim_{z \to z_n}\frac{h(z + a^2\lambda_n^2)\cosh(qx)e^{tz}}{z[q\sinh(qL) + h\cosh(qL)]} \tag{2.2.35}$$

$$= \lim_{z \to z_n}\frac{h\cosh(qx)e^{tz}}{z[(1 + hL)\sinh(qL) + qL\cosh(qL)]/(2a^2q)} \tag{2.2.36}$$

$$= \frac{2ha^2\lambda_n i\cosh(i\lambda_n x)\exp(-\lambda_n^2 a^2 t)}{(-a^2\lambda_n^2)[(1 + hL)i\sin(\lambda_n L) + i\lambda_n L\cos(\lambda_n L)]} \tag{2.2.37}$$

$$= -\frac{2h\cos(\lambda_n x)\exp(-a^2\lambda_n^2 t)}{\lambda_n[(1 + hL)\sin(\lambda_n L) + \lambda_n L\cos(\lambda_n L)]}. \tag{2.2.38}$$

Therefore, the inversion of $U(x, t)$ is

$$u(x, t) = 1 - \left\{1 - 2h\sum_{n=1}^{\infty}\frac{\cos(\lambda_n x)\exp(-a^2\lambda_n^2 t)}{\lambda_n[(1 + hL)\sin(\lambda_n L) + \lambda_n L\cos(\lambda_n L)]}\right\}, \tag{2.2.39}$$

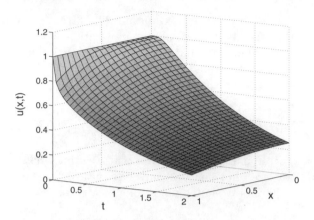

Figure 2.2.2: The temperature within a slab $0 < x/L < 1$ at various times a^2t/L^2 if the faces of the slab radiate to free space at temperature zero and the slab initially has the temperature 1. The parameter $hL = 1$.

or

$$u(x,t) = 2h \sum_{n=1}^{\infty} \frac{\cos(\lambda_n x)\exp(-a^2\lambda_n^2 t)}{\lambda_n[(1+hL)\sin(\lambda_n L) + \lambda_n L\cos(\lambda_n L)]}. \qquad (2.2.40)$$

We can further simplify Equation 2.2.40 by using $h/\lambda_n = \tan(\lambda_n L)$ and $hL = \lambda_n L\tan(\lambda_n L)$. Substituting these relationships into Equation 2.2.40 and simplifying,

$$u(x,t) = 2\sum_{n=1}^{\infty} \frac{\sin(\lambda_n L)\cos(\lambda_n x)\exp(-a^2\lambda_n^2 t)}{\lambda_n L + \sin(\lambda_n L)\cos(\lambda_n L)}. \qquad (2.2.41)$$

Figure 2.2.2 illustrates Equation 2.2.41.

● **Example 2.2.4**

For this example, we use Laplace transforms to solve a partial differential equation in cylindrical coordinates. It differs from our previous problems in its use of Bessel functions.

Let us solve

$$\frac{\partial u}{\partial t} = \frac{\partial^2 u}{\partial r^2} + \frac{1}{r}\frac{\partial u}{\partial r} - \frac{u}{r^2}, \qquad 0 < r < 1, \quad 0 < t, \qquad (2.2.42)$$

subject to the boundary conditions that $u(0,t) = 0$ and $u(1,t) = 1$, and the initial condition that $u(r,0) = 0$. Introducing the Laplace transform of $u(r,t)$,

$$U(r,s) = \int_0^{\infty} u(r,t)e^{-st}\,dt, \qquad (2.2.43)$$

Equation 2.2.42 becomes

$$r^2 \frac{d^2 U(r,s)}{dr^2} + r \frac{dU(r,s)}{dr} - (r^2 s + 1)U(r,s) = 0 \qquad (2.2.44)$$

with the associated boundary conditions $U(0,s) = 0$ and $U(1,s) = 1/s$. The general solution to Equation 2.2.44 is

$$U(r,s) = AI_1(r\sqrt{s}) + BK_1(r\sqrt{s}). \qquad (2.2.45)$$

Because $K_1(\cdot)$ becomes unbounded at $r = 0$, we discard it and the boundary condition $U(1,s) = 1/s$ leads to

$$U(r,s) = \frac{I_1(r\sqrt{s})}{s\, I_1(\sqrt{s})}. \qquad (2.2.46)$$

We now invert $U(r,s)$. Because the first few terms of the power series for $I_1(z)$ are

$$I_1(z) = \frac{z}{2} + \frac{z^3}{16} + \frac{z^5}{384} + \cdots, \qquad (2.2.47)$$

we find upon substituting this power series into Equation 2.2.46 that

$$U(r,s) = \frac{r(1/2 + r^2 s/16 + r^4 s^2/384 + \cdots)}{s(1/2 + s/16 + s^2/384 + \cdots)}. \qquad (2.2.48)$$

Consequently, $U(r,s)$ is a single-valued function and has a simple pole at $s = 0$. The remaining poles are at $I_1(\sqrt{s}) = 0$. If $\sqrt{s} = -\alpha i$, then $I_1(-\alpha i) = -iJ_1(\alpha) = 0$. Denoting the zeros of $J_1(\cdot)$ by α_n, then additional poles are located along the negative real axis at $s_n = -\alpha_n^2$ with $n = 1, 2, 3, \ldots$.

Having located the poles of $U(r,s)$, the inverse $u(r,t)$ is

$$u(r,t) = \frac{1}{2\pi i} \oint_C \frac{I_1(r\sqrt{z})e^{tz}}{z\, I_1(\sqrt{z})}\, dz, \qquad (2.2.49)$$

where C consists of a line parallel to the imaginary axis, but slightly to the right of it, and a semicircle of infinite radius in the left side of the complex plane. From Equation 2.2.48, the residue at $z = 0$ is r. The remaining residues are

$$\mathrm{Res}\left[\frac{I_1(r\sqrt{z})e^{tz}}{z\, I_1(\sqrt{z})}; -\alpha_n^2\right] = \lim_{z \to -\alpha_n^2} \frac{(z + \alpha_n^2)I_1(r\sqrt{z})e^{tz}}{z I_1(\sqrt{z})} \qquad (2.2.50)$$

$$= \frac{2I_1(-\alpha_n ri)}{(-\alpha_n i)I_1'(\alpha_n i)} e^{-\alpha_n^2 t} \qquad (2.2.51)$$

$$= \frac{2J_1(\alpha_n r)}{\alpha_n J_0(\alpha_n)} e^{-\alpha_n^2 t}, \qquad (2.2.52)$$

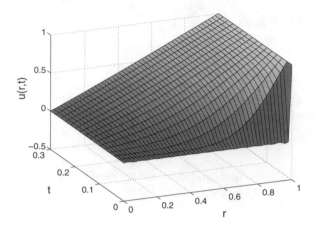

Figure 2.2.3: A plot of $u(r,t)$ given by Equation 2.2.53 as a function of distance r and time t.

because $2I_1'(z) = I_0(z) + I_2(z)$, $I_0(-i\alpha_n) = J_0(\alpha_n)$ and $I_2(-i\alpha_n) = -J_2(\alpha_n)$. Furthermore, we have $J_2(\alpha_n) = -J_0(\alpha_n)$ because $J_2(\alpha_n) + J_0(\alpha_n) = 2J_1(\alpha_n)/\alpha_n = 0$. Consequently, upon summing the residues, the final solution is

$$u(r,t) = r + 2\sum_{n=1}^{\infty} \frac{J_1(\alpha_n r)}{\alpha_n J_0(\alpha_n)} e^{-\alpha_n^2 t}, \tag{2.2.53}$$

where $J_1(\alpha_n) = 0$ with $n = 1, 2, 3, \ldots$. Figure 2.2.3 illustrates Equation 2.2.53 as a function of distance r and time t.

• **Example 2.2.5**

Let us solve the heat equation within an infinitely long cylindrical shell

$$\frac{\partial u}{\partial t} = \frac{1}{r}\frac{\partial}{\partial r}\left(r\frac{\partial u}{\partial r}\right), \qquad a < r < b, \quad 0 < t, \tag{2.2.54}$$

where the interior surface is maintained at a constant temperature, $u(a,t) = 1$, while the outside surface is kept at zero, $u(b,t) = 0$. Initially, the shell has the temperature of zero, $u(r,0) = 0$.

We begin by taking the Laplace transform of Equation 2.2.54 and find that

$$\frac{1}{r}\frac{d}{dr}\left[r\frac{dU(r,s)}{dr}\right] - sU(r,s) = -u(r,0) = 0 \tag{2.2.55}$$

with $U(a,s) = 1/s$ and $U(b,s) = 0$. The general solution to Equation 2.2.55 is

$$U(r,s) = AI_0(r\sqrt{s}) + BK_0(r\sqrt{s}). \tag{2.2.56}$$

The constants A and B may be found from the boundary conditions. Upon substituting these values, we have

$$U(r, s) = \frac{K_0(b\sqrt{s})I_0(r\sqrt{s}) - K_0(r\sqrt{s})I_0(b\sqrt{s})}{s\,[K_0(b\sqrt{s})I_0(a\sqrt{s}) - K_0(a\sqrt{s})I_0(b\sqrt{s})]}. \qquad (2.2.57)$$

To find $u(r, t)$, we use the inversion integral

$$u(r, t) = \frac{1}{2\pi i} \oint_C \frac{K_0(b\sqrt{z})I_0(r\sqrt{z}) - K_0(r\sqrt{z})I_0(b\sqrt{z})}{z\,[K_0(b\sqrt{z})I_0(a\sqrt{z}) - K_0(a\sqrt{z})I_0(b\sqrt{z})]} e^{tz}\, dz, \quad (2.2.58)$$

where the closed contour C includes all of the singularities of the integrand of Equation 2.2.58. These singularities are simple poles at $z = 0$ and $z_n = -\alpha_n^2$, where α_n is the nth root of $J_0(\alpha a)Y_0(\alpha b) - J_0(\alpha b)Y_0(\alpha a) = 0$.

What about the possibility that we have branch points? After all, there are square roots in Equation 2.2.58 and the power series for $K_0(\cdot)$ contains a logarithm. When the power series for $I_0(\cdot)$ and $K_0(\cdot)$ are substituted into Equation 2.2.58 and the expression simplified, the logarithms subtract out and square roots disappear. Consequently, we only have the simple poles $z = 0$ and $z = -\alpha_n^2$.

For the pole at $z = 0$,

$$\text{Res}\left\{\frac{K_0(b\sqrt{z})I_0(r\sqrt{z}) - K_0(r\sqrt{z})I_0(b\sqrt{z})}{z\,[K_0(b\sqrt{z})I_0(a\sqrt{z}) - K_0(a\sqrt{z})I_0(b\sqrt{z})]} e^{tz}; 0\right\}$$

$$= \lim_{z \to 0} \frac{K_0(b\sqrt{z})I_0(r\sqrt{z}) - K_0(r\sqrt{z})I_0(b\sqrt{z})}{K_0(b\sqrt{z})I_0(a\sqrt{z}) - K_0(a\sqrt{z})I_0(b\sqrt{z})} e^{tz} = \frac{\ln(b/r)}{\ln(b/a)}. \quad (2.2.59)$$

We have used the asymptotic formulas that $I_0(z) \approx 1$ and $K_0(z) \approx \log(2/z)$ as $|z| \to 0$.

For the pole at $z = z_n = -\alpha_n^2$, we have

$$\text{Res}\left\{\frac{K_0(b\sqrt{z})I_0(r\sqrt{z}) - K_0(r\sqrt{z})I_0(b\sqrt{z})}{z\,[K_0(b\sqrt{z})I_0(a\sqrt{z}) - K_0(a\sqrt{z})I_0(b\sqrt{z})]} e^{tz}; z_n\right\}$$

$$= \lim_{z \to z_n} \frac{K_0(b\sqrt{z})I_0(r\sqrt{z}) - K_0(r\sqrt{z})I_0(b\sqrt{z})}{z} e^{tz}$$

$$\times \lim_{z \to z_n} \frac{z - z_n}{K_0(b\sqrt{z})I_0(a\sqrt{z}) - K_0(a\sqrt{z})I_0(b\sqrt{z})} \qquad (2.2.60)$$

$$= -\frac{2\alpha_n i[K_0(\alpha_n bi)I_0(\alpha_n ri) - I_0(\alpha_n bi)K_0(\alpha_n ri)]}{\alpha_n^2 [aK_0(\alpha_n bi)I_1(\alpha_n ai) + aI_0(\alpha_n bi)K_1(\alpha_n ai)} e^{-\alpha_n^2 t} \qquad (2.2.61)$$
$$\hspace{2.5cm} {\scriptstyle -bK_1(\alpha_n bi)I_0(\alpha_n ai) - bI_1(\alpha_n bi)K_0(\alpha_n ai)]}$$

$$= 2\frac{I_0(\alpha_n ai)I_0(\alpha_n bi)[K_0(\alpha_n bi)I_0(\alpha_n ri) - I_0(\alpha_n bi)K_0(\alpha_n ri)]}{I_0(\alpha_n bi) - I_0^2(\alpha_n ai)} e^{-\alpha_n^2 t}$$

$$\hspace{11cm} (2.2.62)$$

$$- \pi \frac{J_0(\alpha_n a)J_0(\alpha_n bi)[Y_0(\alpha_n b)J_0(\alpha_n r) - J_0(\alpha_n b)Y_0(\alpha_n r)]}{J_0^2(\alpha_n a) - J_0^2(\alpha_n b)} e^{-\alpha_n^2 t}, \quad (2.2.63)$$

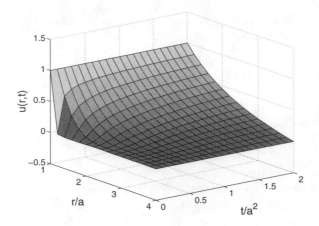

Figure 2.2.4: The temperature within a cylindrical shell $a < r < b$ with $b/a = 4$ at various times t if the inner surface is held at the temperature of 1 and outer surface has the temperature 0. Initially, the shell has the temperature of 0.

where we have used the properties that $I_0'(z) = I_1(z)$, $K_0'(z) = -K_1(z)$, $I_0(z)K_1(z) + I_1(z)K_0(z) = 1/z$, $I_0(xi) = J_0(x)$ and $K_0(xi) = -\pi i[J_0(x) - iY_0(x)]/2$.

Summing the residues leads to

$$u(r,t) = \frac{\ln(b/r)}{\ln(b/a)} + \pi \sum_{n=1}^{\infty} J_0(\alpha_n a) J_0(\alpha_n b) e^{-\alpha_n^2 t}$$
$$\times \frac{Y_0(\alpha_n b) J_0(\alpha_n r) - J_0(\alpha_n b) Y_0(\alpha_n r)}{J_0^2(\alpha_n a) - J_0^2(\alpha_n b)}. \quad (2.2.64)$$

This solution is illustrated in Figure 2.2.4 when $b/a = 4$.

• Example 2.2.6: Moving boundary

In the previous examples, we showed that Laplace transforms are particularly useful when the boundary conditions are time dependent.[12] Consider now the case when one of the boundaries is moving.

We wish to solve the heat equation

$$\frac{\partial u}{\partial t} = a^2 \frac{\partial^2 u}{\partial x^2}, \qquad \beta t < x < \infty, \quad 0 < t, \qquad (2.2.65)$$

subject to the boundary conditions

$$u(x,t)\big|_{x=\beta t} = f(t) \qquad \text{and} \qquad \lim_{x\to\infty} |u(x,t)| < \infty, \qquad 0 < t, \quad (2.2.66)$$

---

[12] Taken from Redozubov, D. B., 1960: The solution of linear thermal problems with a uniformly moving boundary in a semiinfinite region. *Sov. Phys. Tech. Phys.*, **5**, 570–574.

and the initial condition

$$u(x,0) = 0, \qquad 0 < x < \infty. \tag{2.2.67}$$

This type of problem arises in combustion problems where the boundary moves due to the burning of the fuel.

We begin by introducing the coordinate $\eta = x - \beta t$. Then the problem can be reformulated as

$$\frac{\partial u}{\partial t} - \beta \frac{\partial u}{\partial \eta} = a^2 \frac{\partial^2 u}{\partial \eta^2}, \qquad 0 < \eta < \infty, \quad 0 < t, \tag{2.2.68}$$

subject to the boundary conditions

$$u(0,t) = f(t) \qquad \text{and} \qquad \lim_{\eta \to \infty} |u(\eta,t)| < \infty, \qquad 0 < t, \tag{2.2.69}$$

and the initial condition

$$u(\eta,0) = 0, \qquad 0 < \eta < \infty. \tag{2.2.70}$$

Taking the Laplace transform of Equation 2.2.68 and Equation 2.2.69, we have

$$\frac{d^2 U(\eta,s)}{d\eta^2} + \frac{\beta}{a^2} \frac{dU(\eta,s)}{d\eta} - \frac{s}{a^2} U(\eta,s) = 0, \tag{2.2.71}$$

with

$$U(0,s) = F(s) \qquad \text{and} \qquad \lim_{\eta \to \infty} |U(\eta,s)| < \infty. \tag{2.2.72}$$

The solution to Equation 2.2.71 and Equation 2.2.72 is

$$U(\eta,s) = F(s) \exp\left(-\frac{\beta \eta}{2a^2} - \frac{\eta}{a} \sqrt{s + \frac{\beta^2}{4a^2}} \right). \tag{2.2.73}$$

Because

$$\mathcal{L}\left[\Phi(\eta,t) \right] = \exp\left(-\frac{\eta}{a} \sqrt{s + \frac{\beta^2}{4a^2}} \right), \tag{2.2.74}$$

where

$$\Phi(\eta,t) = \frac{1}{2} \left[e^{-\beta\eta/(2a^2)} \operatorname{erfc}\left(\frac{\eta}{2a\sqrt{t}} - \frac{\beta\sqrt{t}}{2a} \right) + e^{\beta\eta/(2a^2)} \operatorname{erfc}\left(\frac{\eta}{2a\sqrt{t}} + \frac{\beta\sqrt{t}}{2a} \right) \right], \tag{2.2.75}$$

we have by the convolution theorem that

$$u(\eta,t) = e^{-\beta\eta/2a^2} \int_0^t f(t-\tau) \Phi(\eta,\tau) \, d\tau, \tag{2.2.76}$$

or

$$u(x,t) = e^{-\beta(x-\beta t)/2a^2} \int_0^t f(t-\tau)\Phi(x-\beta\tau,\tau)\,d\tau. \qquad (2.2.77)$$

• **Example 2.2.7**

So far we have only found the particular solution for the differential equation governing $U(x,s)$ by the method of undetermined coefficients. In this problem we introduce an alternative method where we expand both the solution and inhomogeneous terms as eigenfunction expansions that satisfy the boundary conditions.

Let us solve the heat equation

$$\frac{\partial u}{\partial t} = \frac{a^2}{r}\frac{\partial}{\partial r}\left(r\frac{\partial u}{\partial r}\right), \qquad 0 \le r < b, \quad 0 < t, \qquad (2.2.78)$$

subject to the boundary conditions

$$\lim_{r\to 0}|u(r,t)| < \infty, \qquad u(b,t) = 0, \qquad 0 < t, \qquad (2.2.79)$$

and the initial condition

$$u(r,0) = 1 - cr^2, \qquad 0 \le r < b. \qquad (2.2.80)$$

Taking the Laplace transform of differential equation and boundary conditions, we find that

$$\frac{a^2}{r}\frac{d}{dr}\left[r\frac{dU(r,s)}{dr}\right] - sU(r,s) = -u(r,0) = cr^2 - 1, \qquad (2.2.81)$$

with

$$\lim_{r\to 0}|U(r,s)| < \infty \quad \text{and} \quad U(b,s) = 0. \qquad (2.2.82)$$

If we solved Equation 2.2.81 and Equation 2.2.82 using the method of undetermined coefficients, we would first find the particular solution by assuming

$$U_p(r,s) = Ar^2 + Br + C. \qquad (2.2.83)$$

Substituting $U_p(r,s)$ into Equation 2.2.81, we have

$$U_p(r,s) = \frac{1 - cr^2}{s} - \frac{4ca^2}{s^2}. \qquad (2.2.84)$$

This particular solution is then added to the homogeneous solution to construct the general solution, or

$$U(r,s) = C_1 I_0(qr) + C_2 K_0(qr) + \frac{1}{s} - \frac{cr^2}{s} - \frac{4ca^2}{s^2}, \qquad (2.2.85)$$

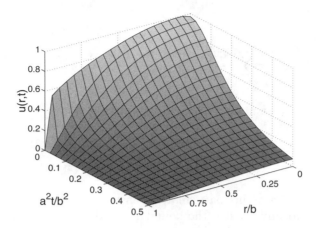

Figure 2.2.5: The temperature within a cylinder $0 \leq r < b$ at various times t if the surface is held at the temperature of 0 and the initial temperature distribution is $1 - cr^2$. Here we have chosen $cb^2 = 0.5$.

where $q^2 = s/a^2$. Finally, the coefficients C_1 and C_2 are chosen so that the general solution satisfies the boundary conditions given by Equation 2.2.82, yielding

$$U(r, s) = \frac{1}{s}\left[1 - \frac{I_0(qr)}{I_0(qb)}\right] - \frac{cr^2}{s} + \frac{cb^2}{s}\frac{I_0(qr)}{I_0(qb)} - \frac{4ca^2}{s^2} + \frac{4ca^2}{s^2}\frac{I_0(qr)}{I_0(qb)}. \quad (2.2.86)$$

Using Bromwich's integral, Equation 2.2.86 may be inverted to yield

$$u(r, t) = \sum_{n=1}^{\infty} \frac{8cb^2 + 2(1 - cb^2)k_n^2 b^2}{k_n^3 b^3 \, J_1(k_n b)} J_0(k_n r) e^{-a^2 k_n^2 t}, \quad (2.2.87)$$

where k_n denotes the nth root of $J_0(kb) = 0$. This solution is illustrated in Figure 2.2.5 when $cb^2 = 0.5$.

An alternative to the method of undetermined coefficients consists of expanding the nonhomogeneous term as a Fourier-Bessel expansion:[13]

$$1 - cr^2 = \sum_{n=1}^{\infty} A_n J_0(k_n r), \quad (2.2.88)$$

where k_n is again the nth root of $J_0(kb) = 0$, and

$$A_n = \frac{2}{b^2 J_1^2(k_n b)} \int_0^b (r - cr^3) J_0(k_n r) \, r \, dr = \frac{8cb^2 + 2(1 - cb^2)k_n^2 b^2}{k_n^3 b^3 \, J_1(k_n b)}. \quad (2.2.89)$$

---

[13] If you are not familiar with Fourier-Bessel series, see Duffy, D. G., 2003: *Advanced Engineering Mathematics with MATLAB*. Chapman & Hall/CRC, 818 pp. See Section 9.5.

Therefore,

$$1 - cr^2 = \sum_{n=1}^{\infty} \frac{8cb^2 + 2(1 - cb^2)k_n^2 b^2}{k_n^3 b^3 J_1(k_n b)} J_0(k_n r). \qquad (2.2.90)$$

Next, let us assume that

$$U(r, s) = \sum_{n=1}^{\infty} B_n J_0(k_n r). \qquad (2.2.91)$$

Why have we chosen this particular expansion for $U(r, s)$? One reason is that it automatically satisfies the boundary conditions because $J_0(k_n b) = 0$. The second reason becomes clear when we substitute Equation 2.2.91 into Equation 2.2.81. For example, the first term on the left side equals

$$\frac{a^2}{r} \frac{d}{dr}\left[r \frac{dU(r, s)}{dr}\right] = a^2 \sum_{n=1}^{\infty} B_n \frac{1}{r} \frac{d}{dr}\left[r \frac{dJ_0(k_n r)}{dr}\right] = - \sum_{n=1}^{\infty} a^2 k_n^2 B_n J_0(k_n r).$$

$$(2.2.92)$$

Substituting the power series for the remaining terms, Equation 2.2.81 becomes

$$\sum_{n=1}^{\infty}(s + a^2 k_n^2) B_n J_0(k_n r) = \sum_{n=1}^{\infty} \frac{8cb^2 + 2(1 - cb^2)k_n^2 b^2}{k_n^3 b^3 J_1(k_n b)} J_0(k_n r). \qquad (2.2.93)$$

Because Equation 2.2.93 must hold for any r, we immediately have

$$(s + a^2 k_n^2) B_n = \frac{8cb^2 + 2(1 - cb^2)k_n^2 b^2}{k_n^3 b^3 J_1(k_n b)}. \qquad (2.2.94)$$

Solving for B_n and substituting B_n into Equation 2.2.91, we find

$$U(r, s) = \sum_{n=1}^{\infty} \frac{8cb^2 + 2(1 - cb^2)k_n^2 b^2}{(s + a^2 k_n^2)k_n^3 b^3 J_1(k_n b)} J_0(k_n r). \qquad (2.2.95)$$

Inverting Equation 2.2.95 term by term, we obtain the final result, Equation 2.2.87, as we expect.

• **Example 2.2.8: Heat dissipation in disc brakes**

Disc brakes consist of two blocks of frictional material known as pads which press against each side of a rotating annulus, usually made of a ferrous material. In this problem, we determine the transient temperatures reached in a disc brake during a single brake application.[14] If we ignore the errors introduced by replacing the cylindrical portion of the drum by a rectangular plate, we can model our disc brakes as a one-dimensional solid which friction heats at both ends. Assuming symmetry about $x = 0$, the boundary condition there is $u_x(0, t) = 0$. To model the heat flux from the pads, we assume a uniform disc deceleration that generates heat from the frictional surfaces at the rate $N(1 - Mt)$, where M and N are experimentally determined constants.

If $u(x, t)$, κ and a^2 denote the temperature, thermal conductivity and diffusivity of the rotating annulus, respectively, then the heat equation is

$$\frac{\partial u}{\partial t} = a^2 \frac{\partial^2 u}{\partial x^2}, \qquad 0 < x < L, \quad 0 < t, \qquad (2.2.96)$$

with the boundary conditions

$$\frac{\partial u(0, t)}{\partial x} = 0, \quad \kappa \frac{\partial u(L, t)}{\partial x} = N(1 - Mt), \qquad 0 < t. \qquad (2.2.97)$$

The boundary condition at $x = L$ gives the frictional heating of the disc pads. Introducing the Laplace transform of $u(x, t)$, defined as

$$U(x, s) = \int_0^\infty u(x, t) e^{-st} \, dt, \qquad (2.2.98)$$

the equation to be solved becomes

$$\frac{d^2 U(x, s)}{dx^2} - \frac{s}{a^2} U(x, s) = 0, \qquad (2.2.99)$$

subject to the boundary conditions that

$$\frac{dU(0, s)}{dx} = 0 \quad \text{and} \quad \frac{dU(L, s)}{dx} = \frac{N}{\kappa} \left(\frac{1}{s} - \frac{M}{s^2} \right). \qquad (2.2.100)$$

The solution to Equation 2.2.99 is

$$U(x, s) = A \cosh(qx) + B \sinh(qx), \qquad (2.2.101)$$

where $q = s^{1/2}/a$. Using the boundary conditions, Equation 2.2.101 becomes

$$U(x, s) = \frac{N}{\kappa} \left(\frac{1}{s} - \frac{M}{s^2} \right) \frac{\cosh(qx)}{q \sinh(qL)}. \qquad (2.2.102)$$

---

[14] From Newcomb, T. P., 1958: The flow of heat in a parallel-faced infinite solid. *Brit. J. Appl. Phys.*, **9**, 370–372. See also Newcomb, T. P., 1958/59: Transient temperatures in brake drums and linings. *Proc. Inst. Mech. Eng.*, *Auto. Div.*, 227–237; Newcomb, T. P., 1959: Transient temperatures attained in disk brakes. *Brit. J. Appl. Phys.*, **10**, 339–340.

We must now find $u(x,t)$. We will invert the $\cosh(qx)/[sq\sinh(qL)]$ term in Equation 2.2.102; the inversion of the second term follows by analog.

Our first concern is the presence of $\sqrt{s}$ because this is a multivalued function. However, when we replace the hyperbolic cosine and sine functions with their Taylor expansions, $\cosh(qx)/[sq\sinh(qL)]$ contains only powers of s and is, in fact, single-valued.

From Bromwich's integral,

$$\mathcal{L}^{-1}\left[\frac{\cosh(qx)}{sq\sinh(qL)}\right] = \frac{1}{2\pi i}\int_{c-\infty i}^{c+\infty i}\frac{\cosh(qx)e^{tz}}{zq\sinh(qL)}\,dz, \qquad (2.2.103)$$

where $q = z^{1/2}/a$. Just as in the previous example, we replace the hyperbolic cosine and sine with their product expansion to determine the nature of the singularities. The point $z = 0$ is a second-order pole. The remaining poles are located where $z_n^{1/2}L/a = n\pi i$, or $z_n = -n^2\pi^2a^2/L^2$, where $n = 1, 2, 3, \ldots$. We have chosen the positive sign because $z^{1/2}$ must be single-valued; if we had chosen the negative sign, the answer would have been the same. Our expansion also shows that the poles are simple.

Having classified the poles, we now close Bromwich's contour, which lies slightly to the right of the imaginary axis, with an infinite semicircle in the left half-plane, and use the residue theorem. The values of the residues are

$$\text{Res}\left[\frac{\cosh(qx)e^{tz}}{zq\sinh(qL)};0\right] = \frac{1}{1!}\lim_{z\to 0}\frac{d}{dz}\left[\frac{(z-0)^2\cosh(qx)e^{tz}}{zq\sinh(qL)}\right] \qquad (2.2.104)$$

$$= \lim_{z\to 0}\frac{d}{dz}\left[\frac{z\cosh(qx)e^{tz}}{q\sinh(qL)}\right] \qquad (2.2.105)$$

$$= \frac{a^2}{L}\lim_{z\to 0}\frac{d}{dz}\left[\frac{z\left(1+\frac{zx^2}{2!a^2}+\cdots\right)\left(1+tz+\frac{t^2z^2}{2!}+\cdots\right)}{z+\frac{L^2z^2}{3!a^2}+\cdots}\right] \qquad (2.2.106)$$

$$= \frac{a^2}{L}\lim_{z\to 0}\frac{d}{dz}\left(1+tz+\frac{zx^2}{2a^2}-\frac{zL^2}{3!a^2}+\cdots\right) \qquad (2.2.107)$$

$$= \frac{a^2}{L}\left(t+\frac{x^2}{2a^2}-\frac{L^2}{6a^2}\right), \qquad (2.2.108)$$

and

$$\text{Res}\left[\frac{\cosh(qx)e^{tz}}{zq\sinh(qL)};z_n\right] = \left[\lim_{z\to z_n}\frac{\cosh(qx)}{zq}e^{tz}\right]\left[\lim_{z\to z_n}\frac{z-z_n}{\sinh(qL)}\right] \qquad (2.2.109)$$

$$= \lim_{z\to z_n}\frac{\cosh(qx)e^{tz}}{zq\cosh(qL)L/(2a^2q)} \qquad (2.2.110)$$

$$= \frac{\cosh(n\pi xi/L)\exp(-n^2\pi^2a^2t/L^2)}{(-n^2\pi^2a^2/L^2)\cosh(n\pi i)L/(2a^2)} \qquad (2.2.111)$$

$$= -\frac{2L(-1)^n}{n^2\pi^2}\cos(n\pi x/L)e^{-n^2\pi^2a^2t/L^2}. \qquad (2.2.112)$$

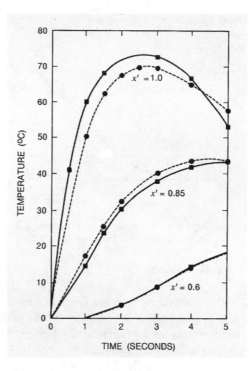

Figure 2.2.6: Transient temperature at different locations in a brake lining. Circles denote computed values while squares are experimental measurements. (From Newcomb, T. P., 1958: The flow of heat in a parallel-faced infinite solid. *Brit. J. Appl. Phys.*, **9**, 372 with permission.)

When we sum the residues from both inversions,

$$
u(x,t) = \frac{a^2 N}{\kappa L}\left(t + \frac{x^2}{2a^2} - \frac{L^2}{6a^2}\right) - \frac{2LN}{\kappa \pi^2}\sum_{n=1}^{\infty}\frac{(-1)^n}{n^2}\cos(n\pi x/L)e^{-n^2\pi^2 a^2 t/L^2}
$$
$$
- \frac{a^2 NM}{\kappa L}\left(\frac{t^2}{2} + \frac{tx^2}{2a^2} - \frac{tL^2}{6a^2} + \frac{x^4}{24a^4} - \frac{x^2 L^2}{12a^4} + \frac{7L^4}{360a^4}\right)
$$
$$
- \frac{2L^3 NM}{a^2 \kappa \pi^4}\sum_{n=1}^{\infty}\frac{(-1)^n}{n^4}\cos(n\pi x/L)e^{-n^2\pi^2 a^2 t/L^2}. \tag{2.2.113}
$$

Figure 2.2.6 shows the temperature in the brake lining at various places within the lining [$x' = x/L$] if $a^2 = 3.3 \times 10^{-3}$ cm^2/sec, $\kappa = 1.8 \times 10^{-3}$ cal/(cm sec °C), $L = 0.48$ cm and $N = 1.96$ cal/(cm^2 sec). Initially, the frictional heating results in an increase in the disc brake's temperature. As time increases, the heating rate decreases and radiative cooling becomes sufficiently large that the temperature begins to fall.

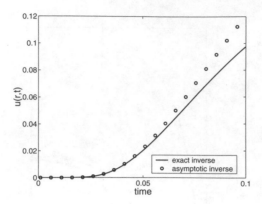

Figure 2.2.7: Comparison between the solution to Equation 2.2.42 given by the exact solution (solid line) and the asymptotic expansion, Equation 2.2.115, when $r = 0.2$.

• Example 2.2.9: Asymptotic Analysis

In Example 2.2.4 we solved a heat conduction problem and found the temperature, Equation 2.2.53, within a cylinder of unit radius whose surface is maintained at a temperature of 1 after initially having a temperature of 0. For small time, this series is not very satisfactory because it converges slowly. In this example, we introduce a method for finding an approximate solution which is valid as $t \to 0$.

Starting with the Laplace transform of $u(r,t)$, Equation 2.2.48, we apply the asymptotic expansion for $I_1(\cdot)$ for large s and find that

$$U(r,s) = \frac{e^{-(1-r)\sqrt{s}}}{s\sqrt{r}} - \frac{3e^{-(1-r)\sqrt{s}}}{8r^{3/2}s^{3/2}} + \frac{3e^{-(1-r)\sqrt{s}}}{8\sqrt{r}s^{3/2}} + \cdots . \qquad (2.2.114)$$

The reason behind our choice of expanding in large s will soon become clear. Taking the inverse Laplace transform of Equation 2.2.114 term by term,

$$u(r,t) \sim \frac{\operatorname{erfc}\left[r/\left(2\sqrt{t}\right)\right]}{\sqrt{r}} - \frac{3}{4}\left(\frac{1}{r} - 1\right)\sqrt{\frac{t}{r}}\operatorname{ierfc}\left(\frac{1-r}{2\sqrt{t}}\right) + \cdots , \qquad (2.2.115)$$

where

$$\operatorname{ierfc}(x) = \frac{e^{-x^2}}{\sqrt{\pi}} - x\operatorname{erfc}(x). \qquad (2.2.116)$$

Figure 2.2.7 compares the exact solution, Equation 2.2.53, with the asymptotic expansion, Equation 2.2.115, when $r = 0.2$. Clearly, as $t \to 0$, the asymptotic expansion gives a very accurate result.

We see now why we wrote $U(r,s)$ as an expansion in large s where each new term contains increasing inverse powers of s. Its inverse contains ever increasing powers of t. In the case of $t \to 0$, the lead term dominates.

Let us now devise a scheme for finding an asymptotic expansion for large t. To this end, we translate Bromwich's contour toward the left. In the process we will cross some of the poles in $F(z)$. Therefore, we can write the inverse as

$$f(t) = \sum_j \text{Res}\left[F(z)e^{tz}; z_j\right] + \frac{1}{2\pi i} \int_\Gamma F(z)e^{tz} \, dz, \qquad (\textbf{2.2.117})$$

where the j summation is over the singularities that we have crossed. As the contour Γ moves farther and farther to the left, the integral in Equation 2.2.117 vanishes. Clearly, the dominate term in the sum of the residues is the residue associated with the singularity that has the *largest* real part.

To illustrate this, consider again the transform $U(r, s)$ from Example 2.2.4. The right-most singularity occurs at $z = 0$ and its residue equals r. Therefore, as $t \to \infty$, $u(r, t) \sim r$, which is confirmed upon examining Equation 2.2.53.

Problems

Homogeneous Heat Equation on a Semi-Infinite, Cartesian Domain

1. Solve[15] the heat equation

$$\frac{\partial u}{\partial t} = a^2 \frac{\partial^2 u}{\partial x^2}, \qquad 0 < x < \infty, \quad 0 < t,$$

with the boundary conditions $u(0, t) = 1$ and $\lim_{x \to \infty} |u(x, t)| < \infty$, and the initial condition $u(x, 0) = 0$.

Step 1: Show that the Laplace transform of the partial differential equation and boundary conditions yields the boundary-value problem

$$\frac{d^2 U(x, s)}{dx^2} - \frac{s}{a^2} U(x, s) = 0, \qquad U(0, s) = 1/s, \quad \lim_{x \to \infty} |U(x, s)| < \infty.$$

Step 2: Show that the solution to the boundary-value problem is $U(x, s) = e^{-x\sqrt{s}/a}/s$.

Step 3: Use tables to invert $U(x, s)$ and show that $u(x, t) = \text{erfc}\left[x/(2a\sqrt{t})\right]$. This solution is illustrated in the figure labeled Problem 1.

2. Solve the heat equation

$$\frac{\partial u}{\partial t} = a^2 \left(\frac{\partial^2 u}{\partial r^2} + \frac{2}{r} \frac{\partial u}{\partial r} \right), \qquad b < r < \infty, \quad 0 < t,$$

---

[15] If $u(x, t)$ denotes the Eulerian velocity of a viscous fluid in the half space $x > 0$ and parallel to the wall located at $x = 0$, then this problem was first solved by Stokes, G. G., 1850: On the effect of the internal friction of fluids on the motions of pendulums. *Proc. Cambridge Philos. Soc.*, **9, Part II**, [8]–[106].

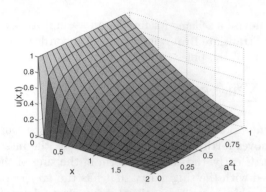

Problem 1

with the boundary conditions $u(b,t) = 1$, $\lim_{r\to\infty} |u(r,t)| < \infty$, $0 < t$, and the initial condition $u(r,0) = 0$, $b < r < \infty$.

Step 1: Show that the Laplace transform of the partial differential equation and boundary conditions yields the boundary-value problem

$$a^2 \left[\frac{d^2 U(r,s)}{dr^2} + \frac{2}{r} \frac{dU(r,s)}{dr} \right] - sU(r,s) = 0, \qquad U(b,s) = \frac{1}{s}, \ \lim_{r\to\infty} |U(r,s)| < \infty.$$

Step 2: Show that the solution to the boundary-value problem is

$$U(r,s) = \frac{b}{rs} e^{-(r-b)\sqrt{s}/a}.$$

Step 3: Use tables to invert $U(r,s)$ and show that

$$u(r,t) = \frac{b}{r} \operatorname{erfc}\left(\frac{r-b}{2a\sqrt{t}} \right).$$

This solution is illustrated in the figure labeled Problem 2.

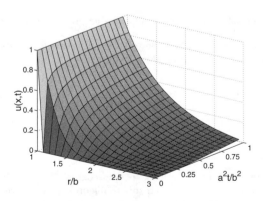

Problem 2

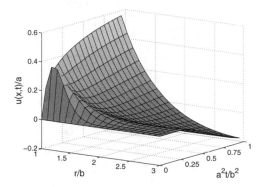

Problem 3

3. Solve the heat equation

$$\frac{\partial u}{\partial t} = a^2 \left(\frac{\partial^2 u}{\partial r^2} + \frac{2}{r} \frac{\partial u}{\partial r} \right), \qquad b < r < \infty, \quad 0 < t,$$

with the boundary conditions $u_r(b, t) = -1$, $\lim_{r \to \infty} |u(r, t)| < \infty$, $0 < t$, and the initial condition $u(r, 0) = 0$, $b < r < \infty$.

Step 1: Show that the Laplace transform of the partial differential equation and boundary conditions yields the boundary-value problem

$$a^2 \left[\frac{d^2 U(r, s)}{dr^2} + \frac{2}{r} \frac{dU(r, s)}{dr} \right] - sU(r, s) = 0,$$

with $U'(b, s) = -1/s$ and $\lim_{r \to \infty} |U(r, s)| < \infty$.

Step 2: Show that the solution to the boundary-value problem is

$$U(r, s) = \frac{ab^2}{rs} \frac{e^{-(r-b)\sqrt{s}/a}}{a + b\sqrt{s}}.$$

Step 3: Use tables to invert $U(r, s)$ and show that

$$u(r, t) = \frac{ab}{r} \left[\mathrm{erfc}\left(\frac{r-b}{2a\sqrt{t}} \right) - \exp\left(\frac{r-b}{b} + \frac{a^2 t}{b^2} \right) \mathrm{erfc}\left(\frac{a\sqrt{t}}{b} + \frac{r-b}{2a\sqrt{t}} \right) \right].$$

This solution is illustrated in the figure labeled Problem 3.

4. Solve the heat equation

$$\frac{\partial u}{\partial t} = \frac{\partial^2 u}{\partial x^2}, \qquad 0 < x < \infty, \quad 0 < t,$$

with the boundary conditions $u_x(0, t) = 1$ and $\lim_{x \to \infty} |u(x, t)| < \infty$, and the initial condition $u(x, 0) = 0$.

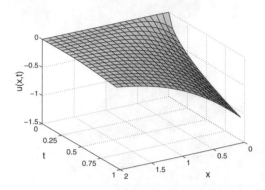

Problem 4

Step 1: Show that the Laplace transform of the partial differential equation and boundary conditions yields the boundary-value problem

$$\frac{d^2U(x,s)}{dx^2} - sU(x,s) = 0, \qquad U'(0,s) = 1/s, \qquad \lim_{x\to\infty} |U(x,s)| < \infty.$$

Step 2: Show that the solution to the boundary-value problem is $U(x,s) = -e^{-x\sqrt{s}}/s^{3/2}$.

Step 3: Use tables to invert $U(x,s)$ and show that

$$u(x,t) = x \operatorname{erfc}\left(\frac{x}{2\sqrt{t}}\right) - 2\sqrt{\frac{t}{\pi}} \exp\left(-\frac{x^2}{4t}\right).$$

This solution is illustrated in the figure labeled Problem 4.

5. Solve the heat equation

$$\frac{\partial u}{\partial t} = \frac{\partial^2 u}{\partial x^2}, \qquad 0 < x < \infty, \quad 0 < t,$$

with the boundary conditions $u(0,t) = 1$ and $\lim_{x\to\infty} |u(x,t)| < \infty$, and the initial condition $u(x,0) = e^{-x}$.

Step 1: Show that the Laplace transform of the partial differential equation and boundary conditions yields the boundary-value problem

$$\frac{d^2U(x,s)}{dx^2} - sU(x,s) = -e^{-x}, \qquad U(0,s) = 1/s, \qquad \lim_{x\to\infty} |U(x,s)| < \infty.$$

Step 2: Show that the solution to the boundary-value problem is

$$U(x,s) = \frac{e^{-x}}{s-1} + \left(\frac{1}{s} - \frac{1}{s-1}\right) e^{-x\sqrt{s}}.$$

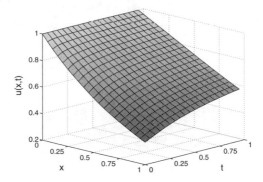

Problem 5

Step 3: Use tables to invert $U(x, s)$ and show that

$$u(x,t) = e^{t-x} + \text{erfc}\left(\frac{x}{2\sqrt{t}}\right) - \tfrac{1}{2}e^t\left[e^{-x}\text{erfc}\left(\frac{x}{2\sqrt{t}} - \sqrt{t}\right) + e^x\text{erfc}\left(\frac{x}{2\sqrt{t}} + \sqrt{t}\right)\right].$$

Hint:

$$\frac{e^{-x\sqrt{s}}}{s-1} = \frac{1}{2}\left(\frac{e^{-x\sqrt{s}}}{\sqrt{s}-1} - \frac{e^{-x\sqrt{s}}}{\sqrt{s}+1}\right).$$

This solution is illustrated in the figure labeled Problem 5.

6. Solve the heat equation

$$\frac{\partial u}{\partial t} = \frac{\partial^2 u}{\partial x^2}, \qquad 0 < x < \infty, \quad 0 < t,$$

with the boundary conditions $u(0, t) - u_x(0, t) = 1$ and $\lim_{x\to\infty}|u(x,t)| < \infty$, and the initial condition $u(x, 0) = 0$.

Step 1: Show that the Laplace transform of the partial differential equation and boundary conditions yields the boundary-value problem

$$\frac{d^2U(x,s)}{dx^2} - sU(x,s) = 0, \qquad U(0,s) - U'(0,s) = 1/s, \quad \lim_{x\to\infty}|U(x,s)| < \infty.$$

Step 2: Show that the solution to the boundary-value problem is

$$U(x,s) = \frac{e^{-x\sqrt{s}}}{s(1+\sqrt{s})}.$$

Step 3: Use tables to invert $U(x, s)$ and show that

$$u(x,t) = \text{erfc}\left(\frac{x}{2\sqrt{t}}\right) - e^{x+t}\text{erfc}\left(\frac{x}{2\sqrt{t}} + \sqrt{t}\right).$$

This solution is illustrated in the figure labeled Problem 6.

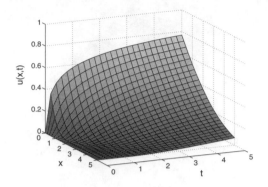

Problem 6

7. Solve the partial differential equation[16]

$$\frac{\partial u}{\partial t} = a^2 \frac{\partial^2 u}{\partial x^2} + c \frac{\partial u}{\partial x}, \qquad 0 < x < \infty, \quad 0 < t,$$

with the boundary conditions $a^2 u_x(0, t) + (c-d)u(0, t) = f(t)$, $\lim_{x\to\infty} |u(x, t)| < \infty$, $0 < t$, and the initial condition $u(x, 0) = 0$, $0 < x < \infty$. Here, a, c and d are constants.

Step 1: Show that the Laplace transform of the partial differential equation and boundary conditions yields the boundary-value problem

$$a^2 \frac{d^2 U(x, s)}{dx^2} + c \frac{dU(x, s)}{dx} - sU(x, s) = 0,$$

with $a^2 U'(0, s) + (c - d)U(0, s) = F(s)$ and $\lim_{x\to\infty} |U(x, s)| < \infty$.

Step 2: Show that the solution to the boundary-value problem is

$$U(x, s) = F(s) \frac{\exp\left(-\frac{cx}{2a^2} - \frac{x}{2}\sqrt{\frac{c^2}{a^4} + \frac{4s}{a^2}}\right)}{\frac{c}{2} - d - \frac{a^2}{2}\sqrt{\frac{c^2}{a^4} + \frac{4s}{a^2}}}.$$

Step 3: Use tables and the convolution theorem to invert $U(x, s)$ and show that

$$u(x, t) = e^{-cx/(2a^2)} \int_0^t f(t - \xi)h(x, \xi) \, d\xi,$$

where

$$h(x, \xi) = -\frac{\exp\left[-c^2\xi/(4a^2) - x^2/(4a^2\xi)\right]}{\sqrt{\pi a^2 \xi}}$$

$$+ \left(\frac{c}{2a} - \frac{d}{a}\right) \exp\left[-\frac{x}{a^2}\left(\frac{c}{2} - d\right) + \frac{\xi}{a^2}\left(\frac{c}{2} - d\right)^2\right]$$

$$\times \operatorname{erfc}\left[-\left(\frac{c}{2} - d\right)\sqrt{\frac{\xi}{a^2}} + \frac{x}{\sqrt{4a^2\xi}}\right].$$

[16] Taken from Duggan, G., and F. Berz, 1982: The theory of the Shockley-Haynes experiments: Contact effects. *J. Appl. Phys.*, **53**, 470–476.

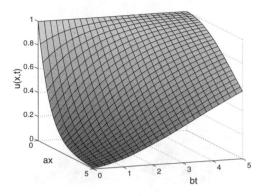

Problem 8

8. Solve the partial differential equation[17]

$$\frac{\partial^2 u}{\partial x \partial t} + a\frac{\partial u}{\partial t} + b\frac{\partial u}{\partial x} = 0, \qquad 0 < x < \infty, \quad 0 < a, b, t,$$

with the boundary conditions $u(0,t) = 1$, $\lim_{x \to \infty} |u(x,t)| < \infty$, $0 < t$, and the initial condition $u_x(x,0) + au(x,0) = 0$, $0 < x < \infty$.

Step 1: Show that the Laplace transform of the partial differential equation and boundary conditions yields the boundary-value problem

$$(s+b)\frac{dU(x,s)}{dx} + asU(x,s) = 0, \qquad 0 < x < \infty,$$

with $U(0,s) = 1/s$ and $\lim_{x \to \infty} |U(x,s)| < \infty$.

Step 2: Show that the solution to the boundary-value problem is

$$U(x,s) = \frac{1}{s}\exp\left(-\frac{axs}{s+b}\right).$$

Step 3: Using the fact that

$$e^{-c\xi} = 1 - c\int_0^\xi e^{-c\eta}\, d\eta,$$

show that

$$U(x,s) = \frac{1}{s} - \int_0^{ax} \frac{e^{-\eta}}{s+b}\exp\left(\frac{b\eta}{s+b}\right)\, d\eta.$$

Step 4: Using tables, show that

$$u(x,t) = 1 - e^{-bt}\int_0^{ax} e^{-\eta}I_0\left(2\sqrt{bt\eta}\right)\, d\eta.$$

This solution is illustrated in the figure labeled Problem 8.

[17] Taken from Liaw, C. H., J. S. P. Wang, R. A. Greenkorn, and K. P. Chao, 1979: Kinetics of fixed-bed adsorption: A new solution. *AICHE J.*, **25**, 376–381. This particular form of the solution is due to a letter by R. G. Rice published in *AICHE J.*, **26**, 334.

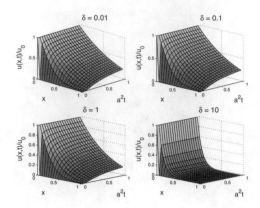

Problem 9

9. Solve the partial differential equation

$$\frac{\partial u}{\partial t} = a^2 \left[\frac{\partial^2 u}{\partial x^2} + (1 + \delta)\frac{\partial u}{\partial x} + \delta u \right], \qquad 0 < x < \infty, \quad 0 < t,$$

where δ is a constant, with the boundary conditions $u(0, t) = u_0$ and $\lim_{x \to \infty} |u(x, t)| < \infty$, and the initial condition $u(x, 0) = 0$.

Step 1: Show that the Laplace transform of the partial differential equation and boundary conditions yields the boundary-value problem

$$\frac{d^2 U(x, s)}{dx^2} + (1 + \delta)\frac{dU(x, s)}{dx} + \left(\delta - \frac{s}{a^2} \right) U(x, s) = 0,$$

with $U(0, s) = u_0/s$ and $\lim_{x \to \infty} |U(x, s)| < \infty$.

Step 2: Show that the solution to the boundary-value problem is

$$U(x, s) = \frac{u_0}{s} \exp\left[-\frac{(1 + \delta)x}{2} - \frac{x}{a}\sqrt{\frac{a^2(1 - \delta)^2}{4} + s} \right].$$

Step 3: Use tables to invert $U(x, s)$ and show that

$$u(x, t) = \tfrac{1}{2}u_0 e^{-\delta x}\,\mathrm{erfc}\left[\frac{x}{2a\sqrt{t}} + \frac{a(1 - \delta)\sqrt{t}}{2} \right] + \tfrac{1}{2}u_0 e^{-x}\,\mathrm{erfc}\left[\frac{x}{2a\sqrt{t}} - \frac{a(1 - \delta)\sqrt{t}}{2} \right].$$

This solution is illustrated in the figure labeled Problem 9.

10. Solve the partial differential equation[18]

$$\frac{\partial u}{\partial t} - \nu\frac{\partial^2 u}{\partial x^2} - \ell^2\frac{\partial^3 u}{\partial x^2 \partial t} + \nu\ell^2\frac{\partial^4 u}{\partial x^4} = 0, \qquad 0 < x < \infty, \quad 0 < t,$$

---

[18] Taken from Guram, G. S., 1983: Flow of a dipolar fluid due to suddenly accelerated flat plate. *Acta Mech.*, **49**, 133–138.

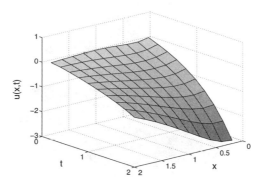

Problem 10

with the boundary conditions $u(0,t) = V_0$, $\mu\ell^2 u_{xx}(0,t) = M_0$, $\lim_{x\to\infty} u(x,t)$ $\to 0$, $\lim_{x\to\infty} \mu\ell^2 u_{xx}(x,t) \to 0$, and the initial condition $u(x,0) = \ell^2 u_{xx}(x,0)$.

Step 1: Show that the Laplace transform of the partial differential equation yields the boundary-value problem

$$\nu\ell^2 \frac{d^4 U(x,s)}{dx^4} - (\nu + \ell^2 s)\frac{d^2 U(x,s)}{dx^2} + sU(x,s) = 0,$$

with $\lim_{x\to\infty} U(x,s) \to 0$, $\lim_{x\to\infty} \mu\ell^2 U''(x,s) \to 0$, $U(0,s) = V_0/s$ and $\mu\ell^2 U''(0,s) = M_0/s$.

Step 2: Show that the solution to the boundary-value problem is

$$U(x,s) = \left(\frac{V_0}{s} + \frac{a}{s} - \frac{a}{s - \nu/\ell^2}\right)e^{-x/\ell} - \left(\frac{a}{s} - \frac{a}{s - \nu/\ell^2}\right)e^{-x\sqrt{s/\nu}},$$

where $a = M_0/\mu - V_0$.

Step 3: Using a table of Laplace transforms and the convolution theorem, show that

$$u(x,t) = \left(V_0 + a - ae^{\nu t/\ell^2}\right)\exp(-x/\ell) - a\,\mathrm{erfc}\!\left[x/\left(2\sqrt{\nu t}\right)\right]$$

$$+ \frac{ax}{2\sqrt{\nu\pi}}\exp(\nu t/\ell^2)\int_0^t \exp\!\left(-\frac{\nu\beta}{\ell^2} - \frac{x^2}{4\nu\beta}\right)\frac{d\beta}{\beta^{3/2}}.$$

This solution is illustrated in the figure labeled Problem 10 when $V_0/a = 1$.

Homogeneous Heat Equation on a Finite, Cartesian Domain

11. Solve the heat equation

$$\frac{\partial u}{\partial t} = \frac{\partial^2 u}{\partial x^2}, \qquad 0 < x < 1, \quad 0 < t,$$

with the boundary conditions $u(0,t) = 0$ and $u(1,t) = 1$, and the initial condition $u(x,0) = 0$.

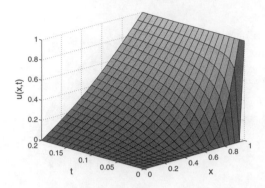

Problem 11

Step 1: Show that the Laplace transform of the partial differential equation and boundary conditions yields the boundary-value problem

$$\frac{d^2 U(x,s)}{dx^2} - sU(x,s) = 0, \qquad U(0,s) = 0, \quad U(1,s) = \frac{1}{s}.$$

Step 2: Show that the solution to the boundary-value problem is

$$U(x,s) = \frac{\sinh\left(x\sqrt{s}\right)}{s\,\sinh\left(\sqrt{s}\right)}.$$

Step 3: Show that the singularities in the Laplace transform are located at $s_n = -n^2\pi^2$, where $n = 0, 1, 2, 3, \ldots$, and are *all* simple poles.

Step 4: Use Bromwich's integral to invert $U(x,s)$ and show that

$$u(x,t) = x + \frac{2}{\pi} \sum_{n=1}^{\infty} \frac{(-1)^n}{n} \sin(n\pi x) e^{-n^2\pi^2 t}.$$

This solution is illustrated in the figure labeled Problem 11.

12. Solve the heat equation

$$\frac{\partial u}{\partial t} = \frac{\partial^2 u}{\partial x^2}, \qquad 0 < x < 1, \quad 0 < t,$$

with the boundary conditions $u(0,t) = 0$ and $u_x(1,t) = 1$, and the initial condition $u(x,0) = 0$.

Step 1: Show that the Laplace transform of the partial differential equation and boundary conditions yields the boundary-value problem

$$\frac{d^2 U(x,s)}{dx^2} - sU(x,s) = 0, \qquad U(0,s) = 0, \quad U'(1,s) = \frac{1}{s}.$$

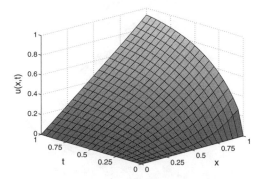

Problem 12

Step 2: Show that the solution to the boundary-value problem is

$$U(x, s) = \frac{\sinh\left(x\sqrt{s}\right)}{s\sqrt{s}\,\cosh\left(\sqrt{s}\right)}.$$

Step 3: Show that the singularities in the Laplace transform are located at $s = 0$ and $s_n = -(2n-1)^2\pi^2/4$, where $n = 0, 1, 2, 3, \ldots$, and are *all* simple poles.

Step 4: Use Bromwich's integral to invert $U(x, s)$ and show that

$$u(x, t) = x + \frac{8}{\pi^2}\sum_{n=1}^{\infty}\frac{(-1)^n}{(2n-1)^2}\sin[(2n-1)\pi x/2]e^{-(2n-1)^2\pi^2 t/4}.$$

This solution is illustrated in the figure labeled Problem 12.

13. Solve the heat equation

$$\frac{\partial u}{\partial t} = \frac{\partial^2 u}{\partial x^2}, \qquad 0 < x < 1, \quad 0 < t,$$

with the boundary conditions $u_x(0, t) = 0$ and $u(1, t) = t$, and the initial condition $u(x, 0) = 0$.

Step 1: Show that the Laplace transform of the partial differential equation and boundary conditions yields the boundary-value problem

$$\frac{d^2 U(x, s)}{dx^2} - sU(x, s) = 0, \qquad U'(0, s) = 0, \quad U(1, s) = \frac{1}{s^2}.$$

Step 2: Show that the solution to the boundary-value problem is

$$U(x, s) = \frac{\cosh\left(x\sqrt{s}\right)}{s^2\,\cosh\left(\sqrt{s}\right)}.$$

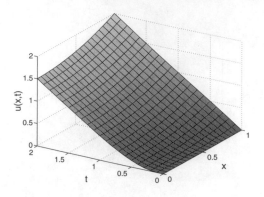

Problem 13

Step 3: Show that the singularities in the Laplace transform consist of a second-order pole located at $s = 0$ and simple poles located at $s_n = -(2n-1)^2\pi^2/4$, where $n = 1, 2, 3, \ldots$.

Step 4: Use Bromwich's integral to invert $U(x, s)$ and show that

$$u(x,t) = t + \frac{x^2 - 1}{2} - \frac{16}{\pi^3} \sum_{n=1}^{\infty} \frac{(-1)^n}{(2n-1)^3} \cos\left[\frac{(2n-1)\pi x}{2}\right] \exp\left[-\frac{(2n-1)^2\pi^2 t}{4}\right].$$

This solution is illustrated in the figure labeled Problem 13.

14. Solve the heat equation

$$\frac{\partial u}{\partial t} = \frac{\partial^2 u}{\partial x^2}, \qquad -\tfrac{1}{2} < x < \tfrac{1}{2}, \quad 0 < t,$$

with the boundary conditions $u_x\left(-\tfrac{1}{2}, t\right) = 0$ and $u_x\left(\tfrac{1}{2}, t\right) = \delta(t)$, and the initial condition $u(x, 0) = 0$.

Step 1: Show that the Laplace transform of the partial differential equation and boundary conditions yields the boundary-value problem

$$\frac{d^2 U(x, s)}{dx^2} - sU(x, s) = 0, \qquad U'\left(-\tfrac{1}{2}, s\right) = 0, \quad U'\left(\tfrac{1}{2}, s\right) = 1.$$

Step 2: Show that the solution to the boundary-value problem is

$$U(x, s) = \frac{\cosh\left[(x + \tfrac{1}{2})\sqrt{s}\right]}{\sqrt{s}\,\sinh\left(\sqrt{s}\right)}.$$

Step 3: Show that the singularities in the Laplace transform consist of simple poles at $s = 0$ and $s_n = -n^2\pi^2$ with $\sqrt{s_n} = n\pi i$, where $n = 1, 2, 3, \ldots$.

Step 4: Use Bromwich's integral to invert $U(x, s)$ and show that

$$u(x,t) = 1 + 2\sum_{n=1}^{\infty}(-1)^n \cos\left[n\pi\left(x + \tfrac{1}{2}\right)\right] e^{-n^2\pi^2 t}.$$

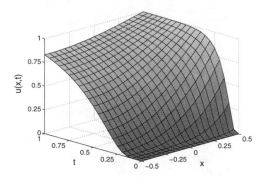

Problem 14

This solution is illustrated in the figure labeled Problem 14.

15. Solve the heat equation

$$\frac{\partial u}{\partial t} = a^2 \frac{\partial^2 u}{\partial x^2}, \qquad 0 < x < L, \quad 0 < t,$$

with the boundary conditions $u_x(0, t) = -A$ and $u_x(L, t) = 0$, and the initial condition $u(x, 0) = 0$.

Step 1: Show that the Laplace transform of the partial differential equation and boundary conditions yields the boundary-value problem

$$\frac{d^2U(x, s)}{dx^2} - q^2 U(x, s) = 0, \qquad U'(0, s) = -A/s, \quad U'(L, s) = 0,$$

where $q^2 = s/a^2$.

Step 2: Show that the solution to the boundary-value problem is

$$U(x, s) = \frac{A \cosh[q(x - L)]}{sq \sinh(qL)}.$$

Step 3: Show that the singularities in the Laplace transform consist of a second-order pole at $s = 0$ and simple poles located at $s_n = -n^2\pi^2a^2/L^2$ with $\sqrt{s_n} = n\pi ai/L$, where $n = 1, 2, 3, \ldots$.

Step 4: Use Bromwich's integral to invert $U(x, s)$ and show that

$$u(x, t) = \frac{a^2 At}{L} + \frac{A}{L}\left[\frac{(x - L)^2}{2} - \frac{L^2}{6}\right] - \frac{2AL}{\pi^2}\sum_{n=1}^{\infty}\frac{\cos(n\pi x/L)}{n^2}\exp\left(-\frac{a^2n^2\pi^2t}{L^2}\right).$$

This solution is illustrated in the figure labeled Problem 15.

16. Solve the heat equation

$$\frac{\partial u}{\partial t} = \frac{\partial^2 u}{\partial x^2}, \qquad 0 < x < 1, \quad 0 < t,$$

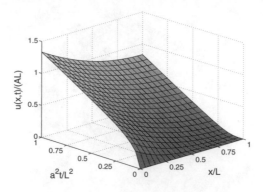

Problem 15

with the boundary conditions $u(0,t) = 0$ and $3a\,[u_x(1,t) - u(1,t)] + u_t(1,t)$ $= \delta(t)$, $0 < a$, and the initial condition $u(x,0) = 0$.

Step 1: Show that the Laplace transform of the partial differential equation and boundary conditions yields the boundary-value problem

$$\frac{d^2 U(x,s)}{dx^2} - sU(x,s) = 0,$$

with $U(0,s) = 0$ and $3a[U'(1,s) - U(1,s)] + sU(1,s) = 1$.

Step 2: Show that the solution to the boundary-value problem is

$$U(x,s) = \frac{\sinh\left(x\sqrt{s}\right)}{3a\left[\sqrt{s}\cosh\left(\sqrt{s}\right) - \sinh\left(\sqrt{s}\right)\right] + s\,\sinh\left(\sqrt{s}\right)}.$$

Step 3: Show that the singularities in the Laplace transform consist of simple poles at $s = 0$ and $s_n = -\lambda_n^2$, where $\lambda_n \cot(\lambda_n) = (3a + \lambda_n^2)/3a$ and $n = 1, 2, 3, \ldots$.

Step 4: Use Bromwich's integral to invert $U(x,s)$ and show that

$$u(x,t) = \frac{x}{a+1} + 2\sum_{n=1}^{\infty} \frac{\sin(\lambda_n x)\exp(-\lambda_n^2 t)}{[3a + 3 + \lambda_n^2/(3a)]\sin(\lambda_n)}.$$

This solution is illustrated in the figure labeled Problem 16.

17. Solve the heat equation[19]

$$\frac{\partial u}{\partial t} = \frac{\partial^2 u}{\partial x^2}, \qquad 0 < x < 1, \quad 0 < t,$$

---

[19] Reprinted from *Chem. Eng. Sci.*, **40**, J. C. Merchuk, Z. Tsur and E. Horn, Oxygen transfer resistance as a criterion of blood ageing, 1101–1107, ©1985, with permission from Elsevier.

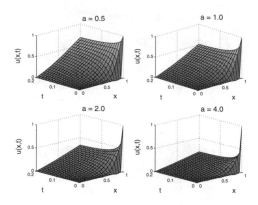

Problem 16

with the boundary conditions $u_x(0, t) = 0$ and

$$u_x(1, t) = ab \left[1 - \int_0^1 u(\xi, t) \, d\xi \right] - bu(1, t), \qquad 0 < a, b, t,$$

and the initial condition $u(x, 0) = 0$, $0 < x < 1$.

Step 1: Show that the Laplace transform of the partial differential equation and boundary conditions yields the boundary-value problem

$$\frac{d^2 U(x, s)}{dx^2} - sU(x, s) = 0,$$

with $U(0, s) = 0$ and

$$U'(1, s) = ab \left[\frac{1}{s} - \int_0^1 U(\xi, s) \, d\xi \right] - bU(1, s).$$

Step 2: Show that the solution to the boundary-value problem is

$$U(x, s) = \frac{a \cosh\left(x\sqrt{s} \right)}{s \left[\left(a/\sqrt{s} + \sqrt{s}/b \right) \sinh\left(\sqrt{s} \right) + \cosh\left(\sqrt{s} \right) \right]}.$$

Step 3: Show that the singularities in the Laplace transform consist of simple poles at $s = 0$ and $s_n = -\lambda_n^2$, where $\cot(\lambda_n) = (\lambda_n/b - a/\lambda_n)$ and $n = 1, 2, 3, \ldots$.

Step 4: Use Bromwich's integral to invert $U(x, s)$ and show that

$$u(x, t) = \frac{a}{1 + a} - 2a \sum_{n=1}^{\infty} \frac{\cos(\lambda_n x) e^{-\lambda_n^2 t}}{\lambda_n \left[(1 + 1/b + a/\lambda_n^2) \sin(\lambda_n) + \cot(\lambda_n) \cos(\lambda_n) \right]}.$$

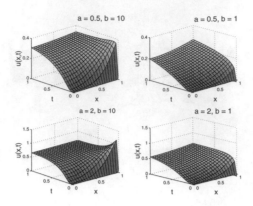

Problem 17

This solution is illustrated in the figure labeled Problem 17.

18. Solve the partial differential equation[20]

$$\frac{\partial u}{\partial t} = \frac{\partial^2 u}{\partial x^2} - a^2 u, \qquad -1 < x < 1, \quad 0 < t,$$

with the boundary conditions $u(-1,t) = 1$ and $u(1,t) = 0$, and the initial condition $u(x,0) = 0$.

Step 1: Show that the Laplace transform of the partial differential equation and boundary conditions yields the boundary-value problem

$$\frac{d^2 U(x,s)}{dx^2} - (a^2 + s)U(x,s) = 0, \qquad U(-1,s) = \frac{1}{s}, \quad U(1,s) = 0.$$

Step 2: Show that the solution to the boundary-value problem is

$$U(x,s) = \frac{\sinh\left[(1 - x)\sqrt{a^2 + s}\,\right]}{s\,\sinh\left(2\sqrt{a^2 + s}\right)}.$$

Step 3: Show that the singularities in the Laplace transform consist of simple poles located at $s = 0$ and $s_n = -a^2 - n^2\pi^2/4$, where $n = 1, 2, 3, \ldots$.

Step 4: Use Bromwich's integral to invert $U(x,s)$ and show that

$$u(x,t) = \frac{\sinh[a(1 - x)]}{\sinh(2a)} + \frac{\pi}{2}\sum_{n=1}^{\infty} \frac{(-1)^n\, n}{a^2 + n^2\pi^2/4} \sin\left[\frac{n\pi(1 - x)}{2}\right] e^{-(a^2 + n^2\pi^2/4)t}.$$

---

[20] Taken from Jacobs, C., 1971: Transient motions produced by disks oscillating torsionally about a state of rigid rotation. *Quart. J. Mech. Appl. Math.*, **24**, 221–236. By permission of Oxford University Press.

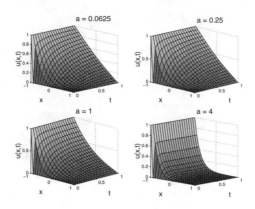

Problem 18

This solution is illustrated in the figure labeled Problem 18.

19. Solve[21] the partial differential equation

$$\frac{\partial u}{\partial t} + V\frac{\partial u}{\partial x} = \frac{\partial^2 u}{\partial x^2}, \qquad 0 < x < 1, \quad 0 < t,$$

where V is a constant, with the boundary conditions $u(0,t) = 1$ and $u_x(1,t) = 0$, and the initial condition $u(x,0) = 0$.

Step 1: Show that the Laplace transform of the partial differential equation and boundary conditions yields the boundary-value problem

$$\frac{d^2 U(x,s)}{dx^2} - V\frac{dU(x,s)}{dx} - sU(x,s) = 0, \qquad U(0,s) = 1/s, \quad U'(1,s) = 0.$$

Step 2: Show that the solution to the boundary-value problem is

$$U(x,s) = e^{Vx/2}\frac{\sqrt{s'}\cosh\left[(1-x)\sqrt{s'}\right] + (V/2)\sinh\left[(1-x)\sqrt{s'}\right]}{(s' - V^2/4)\left[\sqrt{s'}\cosh\left(\sqrt{s'}\right) + (V/2)\sinh\left(\sqrt{s'}\right)\right]},$$

where $s' = s + V^2/4$.

Step 3: Show that the singularities in the Laplace transform consist of simple poles $s' = V^2/4$ and $s'_n = -\lambda_n^2$ with $\lambda_n\cot(\lambda_n) = -V/2$, where $n = 1,2,3,\ldots$.

Step 4: Use Bromwich's integral to invert $U(x,s)$ and show that

$$u(x,t) = 1 - 2e^{Vx/2 - V^2t/4}\sum_{n=1}^{\infty}\frac{\lambda_n\{(V/2)\sin[\lambda_n(1-x)] + \lambda_n\cos[\lambda_n(1-x)]\}}{(\lambda_n^2 + V^2/4)[\lambda_n\sin(\lambda_n) - (1+V/2)\cos(\lambda_n)]}e^{-\lambda_n^2 t}.$$

---

[21] Reprinted from *Solar Energy*, **56**, H. Yoo and E.-T. Pak, Analytical solutions to a one-dimensional finite-domain model for stratified thermal storage tanks, 315–322, ©1996, with permission from Elsevier.

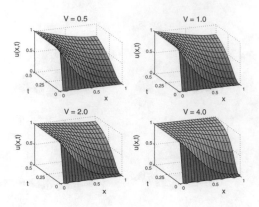

Problem 19

This solution is illustrated in the figure labeled Problem 19.

20. Solve the partial differential equation

$$\frac{\partial u}{\partial t} + V\frac{\partial u}{\partial x} = a^2\frac{\partial^2 u}{\partial x^2}, \qquad 0 < x < L, \quad 0 < t,$$

where V is a constant, with the boundary conditions $u(0,t) = 1$ and $u(L,t) = 0$, and the initial condition $u(x,0) = 0$.

Step 1: Show that the Laplace transform of the partial differential equation and boundary conditions yields the boundary-value problem

$$a^2\frac{d^2 U(x,s)}{dx^2} - V\frac{dU(x,s)}{dx} - sU(x,s) = 0, \qquad U(0,s) = 1/s, \quad U(L,s) = 0.$$

Step 2: Show that the solution to the boundary-value problem is

$$U(x,s) = e^{\mu x}\frac{\sinh[q(L-x)]}{s\,\sinh(qL)} = \frac{e^{\mu x}}{s}\left\{1 - \frac{x}{L} + \frac{2}{\pi}\sum_{n=1}^{\infty}\frac{(-1)^n}{n}\frac{q^2}{q^2 + \lambda_n^2}\sin[\lambda_n(L-x)]\right\},$$

where $q^2 = s/a^2 + \mu^2$, $\lambda_n = n\pi/L$ and $\mu = V/(2a^2)$.

Step 3: Inverting the Laplace transforms given by the series term by term, show that

$$u(x,t) = e^{\mu x}\left\{1 - \frac{x}{L} - \frac{2}{\pi}\sum_{n=1}^{\infty}\frac{\sin(\lambda_n x)}{n}\frac{\mu^2 + \lambda_n^2\exp[-a^2(\mu^2 + \lambda_n^2)t]}{\mu^2 + \lambda_n^2}\right\}.$$

This solution is illustrated in the figure labeled Problem 20.

21. Solve the partial differential equation[22]

$$\frac{\partial u}{\partial t} + V\frac{\partial u}{\partial x} = a^2\frac{\partial^2 u}{\partial x^2}, \qquad 0 < x < L, \quad 0 < t,$$

---

[22] Leij and Toride [Leij, F. J., and N. Toride, 1998: Analytical solutions for solute transport in finite soil columns with arbitrary initial distributions. *Soil Sci. Soc. Am. J.*, **62**, 855–864.] employed Laplace transforms in their study of this problem but solved the boundary-value problem in a different manner.

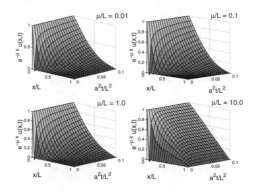

Problem 20

where V is a constant, with the boundary conditions $u(0, t) = 0$ and $u_x(L, t) = 0$, $0 < t$, and the initial condition $u(x, 0) = f(x)$, $0 < x < L$.

Step 1: If $\mu = V/(2a^2)$ and $u(x, t) = v(x, t)e^{\mu x - \mu^2 a^2 t}$, show that the problem becomes

$$\frac{\partial v}{\partial t} = a^2 \frac{\partial^2 v}{\partial x^2}, \qquad 0 < x < L, \quad 0 < t,$$

with the boundary conditions $v(0, t) = 0$ and $v_x(L, t) + \mu v(L, t) = 0$, $0 < t$, and the initial condition $u(x, 0) = f(x)e^{-\mu x} = g(x)$, $0 < x < L$.

Step 2: Show that the Laplace transform of Step 1 yields the boundary-value problem

$$a^2 \frac{d^2 V(x, s)}{dx^2} - sV(x, s) = -g(x), \qquad V(0, s) = 0, \quad V'(L, s) + \mu V(L, s) = 0.$$

Step 3: Show that $g(x)$ can be reexpressed by the orthogonal expansion

$$g(x) = \sum_{n=1}^{\infty} A_n \sin(k_n x), \qquad 0 < x < L,$$

where k_n is the nth root of $k \cos(kL) + \mu \sin(kL) = 0$, and

$$A_n = 2 \frac{\mu^2 + k_n^2}{\mu + L(\mu^2 + k_n^2)} \int_0^L g(\xi) \sin(k_n \xi) \, d\xi.$$

Hint: Show that $\mu^2 = (\mu^2 + k_n^2) \cos^2(k_n L)$.

Step 4: Show that the solution to the boundary-value problem is

$$V(x, s) = 2 \sum_{n=1}^{\infty} \frac{(\mu^2 + k_n^2) \sin(k_n x)}{[\mu + L(\mu^2 + k_n^2)] (s + a^2 k_n^2)} \int_0^L g(\xi) \sin(k_n \xi) \, d\xi.$$

Note that this series satisfies the boundary conditions.

Step 5: Inverting the Laplace transforms given by the series term by term, show that

$$v(x,t) = 2\sum_{n=1}^{\infty} \frac{\mu^2 + k_n^2}{\mu + L(\mu^2 + k_n^2)} \sin(k_n x) e^{-a^2 k_n^2 t} \int_0^L g(\xi) \sin(k_n \xi)\, d\xi,$$

or

$$u(x,t) = 2\sum_{n=1}^{\infty} \frac{(\mu^2 + k_n^2)e^{-a^2(\mu^2 + k_n^2)t}}{\mu + L(\mu^2 + k_n^2)} \sin(k_n x) \int_0^L f(\xi) e^{\mu(x-\xi)} \sin(k_n \xi)\, d\xi.$$

Homogeneous Heat Equation on a Cylindrical Domain

22. Solve the heat equation[23]

$$\frac{\partial u}{\partial t} = \frac{\partial^2 u}{\partial r^2} + \frac{1}{r}\frac{\partial u}{\partial r}, \qquad 0 \le r < a, \quad 0 < t,$$

with the boundary conditions $\lim_{r\to 0} |u(r,t)| < \infty$ and $u_r(a,t) = -1$, and the initial condition $u(r,0) = 1$.

Step 1: Show that the Laplace transform of the partial differential equation and boundary conditions yields the boundary-value problem

$$\frac{d^2 U(r,s)}{dr^2} + \frac{1}{r}\frac{dU(r,s)}{dr} - sU(r,s) = -1, \qquad 0 \le r < a,$$

with $\lim_{r\to 0} |U(r,s)| < \infty$ and $U'(a,s) = -1/s$.

Step 2: Show that the solution to the boundary-value problem is

$$U(r,s) = \frac{1}{s} - \frac{I_0\left(r\sqrt{s}\right)}{s^{3/2} I_1\left(a\sqrt{s}\right)}.$$

Step 3: Show that the singularities in the Laplace transform are located at $s = 0$ and $s_n = -\alpha_n^2/a^2$ with $a\sqrt{s_n} = -\alpha_n i$, where α_n is the nth positive zero of $J_1(\alpha) = 0$ and $n = 0, 1, 2, 3, \ldots$. The pole at $s = 0$ is second order while the remaining poles are simple.

Step 4. Use Bromwich's integral to invert $U(r,s)$ and show that

$$u(r,t) = 1 - a\left[\frac{2t}{a^2} + \frac{r^2}{2a^2} - \frac{1}{4} - 2\sum_{n=1}^{\infty} \frac{J_0(\alpha_n r/a)}{\alpha_n^2 J_0(\alpha_n)} e^{-\alpha_n^2 t/a^2}\right].$$

---

[23] Taken from Jaeger, J. C., 1944: Note on a problem in radial flow. *Proc. Phys. Soc.*, **56**, 197–203.

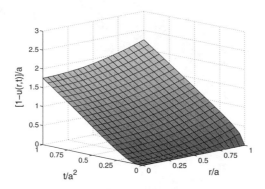

Problem 22

This solution is illustrated in the figure labeled Problem 22.

23. Solve the heat equation

$$\frac{\partial u}{\partial t} = \frac{1}{r}\frac{\partial}{\partial r}\left(r\frac{\partial u}{\partial r}\right), \qquad 0 \le r < a, \quad 0 < t,$$

with the boundary conditions $\lim_{r\to 0}|u(r,t)| < \infty$ and $u(a,t) = e^{-t/\tau_0}$, and the initial condition $u(r,0) = 1$.

Step 1: Show that the Laplace transform of the partial differential equation and boundary conditions yields the boundary-value problem

$$\frac{1}{r}\frac{d}{dr}\left[r\frac{dU(r,s)}{dr}\right] - sU(r,s) = -1,$$

with $\lim_{r\to 0}|U(r,s)| < \infty$ and $U(a,s) = 1/(s + 1/\tau_0)$.

Step 2: Show that the solution to the boundary-value problem is

$$U(r,s) = \frac{1}{s} + \left(\frac{1}{s + 1/\tau_0} - \frac{1}{s}\right)\frac{I_0\left(r\sqrt{s}\right)}{I_0\left(a\sqrt{s}\right)}.$$

Step 3: Show that the Laplace transform has simple poles at $s = 0$, $s = -1/\tau_0$ and $s_n = -k_n^2/a^2$, where $J_0(k_n) = 0$ and $n = 1, 2, 3, \ldots$.

Step 4: Use Bromwich's integral to invert $U(r,s)$ and show that

$$u(r,t) = e^{-t/\tau_0} + 2a^2 \sum_{n=1}^{\infty} \frac{J_0(k_n r/a)}{k_n(a^2 - k_n^2\tau_0)J_1(k_n)}\left(e^{-k_n^2 t/a^2} - e^{-t/\tau_0}\right).$$

This solution is illustrated in the figure labeled Problem 23 with $\tau_0/a^2 = 1$.

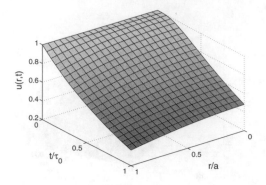

Problem 23

24. Solve the heat equation[24]

$$\frac{\partial u}{\partial t} = \frac{\partial^2 u}{\partial r^2} + \frac{1}{r}\frac{\partial u}{\partial r}, \qquad 0 \le r < a, \quad 0 < t,$$

with $\lim_{r\to 0} |u(r,t)| < \infty$ and $u(a,t) = \cos(\omega t)$, and the initial condition $u(r,0) = 0$.

Step 1: By taking the Laplace transform of the partial differential equation and boundary conditions, show that they reduce to the boundary-value problem

$$\frac{d^2 U(r,s)}{dr^2} + \frac{1}{r}\frac{dU(r,s)}{dr} - sU(r,s) = 0, \qquad 0 \le r < a,$$

with $\lim_{r\to 0} |U(r,s)| < \infty$ and $U(a,s) = s/(s^2 + \omega^2)$.

Step 2: Show that the solution to the boundary-value problem is

$$U(r,s) = \frac{s}{s^2 + \omega^2}\frac{I_0\!\left(r\sqrt{s}\right)}{I_0\!\left(a\sqrt{s}\right)}.$$

Step 3: Show that the singularities of $U(r,s)$ are the simple poles at $s = \pm\omega i$ and $s = -\alpha_n^2$ with $n = 1,2,3,\ldots$, where α_n is the nth zero of $J_0(\alpha a) = 0$.

Step 4: Use Bromwich's integral to invert $U(r,s)$ and show that

$$u(r,t) = \frac{I_0(r\sqrt{\omega i})}{2I_0(a\sqrt{\omega i})}e^{i\omega t} + \frac{I_0(r\sqrt{-\omega i})}{2I_0(a\sqrt{-\omega i})}e^{-i\omega t} - \frac{2}{a}\sum_{n=1}^{\infty}\frac{\alpha_n^3}{\alpha_n^4 + \omega^2}\frac{J_0(\alpha_n r)}{J_1(\alpha_n a)}e^{-\alpha_n^2 t}.$$

---

[24] Reprinted from *Int. J. Mech. Sci.*, **8**, P. G. Bhuta and L. R. Koval, A viscous ring damper for a freely precessing satellite, pp. 383–395, ©1966, with kind permission from Pergamon Press Ltd., Headington Hill Hall, Oxford OX3 0BW, UK.

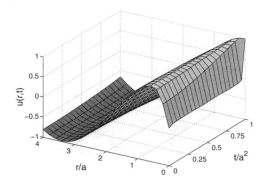

Problem 24

This solution is illustrated in the figure labeled Problem 24.

25. Solve the heat equation[25]

$$\frac{\partial u}{\partial t} = \frac{\partial^2 u}{\partial r^2} + \frac{1}{r}\frac{\partial u}{\partial r} - \frac{u}{r^2}, \qquad 0 \le r < 1, \quad 0 < t,$$

with the boundary conditions $\lim_{r \to 0} |u(r,t)| < \infty$ and $u(1,t) = \sin(\omega t)$, and the initial condition $u(r,0) = 0$.

Step 1: By taking the Laplace transform of the partial differential equation and boundary conditions, show that they reduce to the boundary-value problem

$$\frac{d^2 U(r,s)}{dr^2} + \frac{1}{r}\frac{dU(r,s)}{dr} - \left(s + \frac{1}{r^2}\right)U(r,s) = 0, \qquad 0 \le r < 1,$$

with $\lim_{r \to 0} |U(r,s)| < \infty$ and $U(1,s) = \omega/(s^2 + \omega^2)$.

Step 2: Show that the solution to the boundary-value problem is

$$U(r,s) = \frac{\omega}{s^2 + \omega^2} \frac{I_1(r\sqrt{s})}{I_1(\sqrt{s})}.$$

Step 3: Show that the singularities of $U(r,s)$ are simple poles at $s = \pm\omega i$ and $s = -\alpha_n^2$ with $n = 1, 2, 3, \ldots$, where α_n is the nth root of $J_1(\alpha) = 0$.

Step 4: Use Bromwich's integral to invert $U(r,s)$ and show that

$$u(r,t) = \frac{M_1(r\sqrt{\omega})}{M_1(\sqrt{\omega})} \sin[\omega t + \theta_1(r\sqrt{\omega}) - \theta_1(\sqrt{\omega})] - 2\omega \sum_{n=1}^{\infty} \frac{\alpha_n}{\alpha_n^4 + \omega^2} \frac{J_1(\alpha_n r)}{J_0(\alpha_n)} e^{-\alpha_n^2 t},$$

[25] Taken from Schwarz, W. H., 1963: The unsteady motion of an infinite oscillating cylinder in an incompressible Newtonian fluid at rest. *Appl. Sci. Res., Ser. A,* **11**, 115–124. Reprinted by permission of Kluwer Academic Publishers.

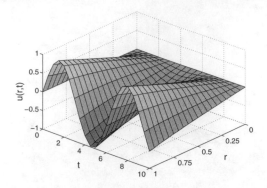

Problem 25

where

$$M_1(z) = \sqrt{\text{ber}_1^2(z) + \text{bei}_1^2(z)}, \qquad \theta_1(z) = \arctan[\text{bei}_1(z)/\text{ber}_1(z)],$$

and $\text{ber}_1(z)$ and $\text{bei}_1(z)$ are Kelvin's ber and bei functions, respectively. This solution is illustrated in the figure labeled Problem 25 when $\omega = 1$.

26. Solve the heat equation

$$\frac{\partial u}{\partial t} = a^2 \left(\frac{\partial^2 u}{\partial r^2} + \frac{1}{r} \frac{\partial u}{\partial r} \right), \qquad 0 \le r < b, \quad 0 < t,$$

with the boundary conditions

$$\lim_{r \to 0} |u(r,t)| < \infty, \qquad b^2 u(b,t) - 2 \int_0^b u(r,t)\, r\, dr = b^2 \delta(t), \qquad 0 < t,$$

and the initial condition $u(r,0) = 1$, $0 \le r < b$.

Step 1: By taking the Laplace transform of the partial differential equation and boundary conditions, show that they reduce to the boundary-value problem

$$\frac{d^2 U(r,s)}{dr^2} + \frac{1}{r} \frac{dU(r,s)}{dr} - q^2 U(r,s) = -\frac{1}{a^2}, \qquad 0 \le r < b,$$

with

$$\lim_{r \to 0} |U(r,s)| < \infty, \qquad b^2 U(b,s) - 2 \int_0^b U(r,s)\, r\, dr = b^2,$$

where $q^2 = s/a^2$.

Step 2: Show that the solution to the boundary-value problem is

$$U(r,s) = \frac{1}{s} + \frac{I_0(qr)}{I_2(qb)}.$$

Hint:

$$I_{\nu-1}(z) + I_{\nu+1}(z) = 2\nu I_\nu(z)/z \quad \text{and} \quad \frac{d}{dz}[z^\nu I_\nu(z)] = z^\nu I_{\nu-1}(z).$$

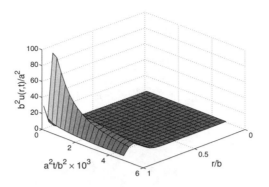

Problem 26

Step 3: Show that the singularities of $U(r,s)$ are simple poles at $s = 0$ and $s_n = -a^2\alpha_n^2/b^2$ with $bq_n = -i\alpha_n$ and $n = 1, 2, 3, \ldots$, where $J_0(\alpha_n) - 2J_1(\alpha_n)/\alpha_n = -J_2(\alpha_n) = 0$.

Step 4: Use Bromwich's integral to invert $U(r,s)$ and show that

$$u(r,t) = 1 + \frac{8a^2}{b^2} + \frac{2a^2}{b^2} \sum_{n=1}^{\infty} \frac{\alpha_n}{J_1(\alpha_n)} J_0\left(\frac{\alpha_n r}{b}\right) \exp\left(-\frac{a^2\alpha_n^2 t}{b^2}\right).$$

This solution is illustrated in the figure labeled Problem 26 when $a/b = 0.5$.

27. Solve the heat equation

$$\frac{\partial u}{\partial t} = \frac{\partial^2 u}{\partial r^2} + \frac{1}{r}\frac{\partial u}{\partial r} - \frac{u}{r^2}, \qquad a < r < b, \quad 0 < t,$$

with the boundary conditions $u(a,t) = 1$, $u(b,t) = 0$, $0 < t$, and the initial condition $u(r,0) = 0$, $a < r < b$.

Step 1: By taking the Laplace transform of the partial differential equation and boundary conditions, show that they reduce to the boundary-value problem

$$\frac{d^2U(r,s)}{dr^2} + \frac{1}{r}\frac{dU(r,s)}{dr} - \frac{U(r,s)}{r^2} - sU(r,s) = 0, \qquad a < r < b,$$

with $U(a,s) = 1/s$ and $U(b,s) = 0$.

Step 2: Show that the solution to the boundary-value problem is

$$U(r,s) = \frac{K_1(b\sqrt{s})I_1(r\sqrt{s}) - K_1(r\sqrt{s})I_1(b\sqrt{s})}{s\left[K_1(b\sqrt{s})I_1(a\sqrt{s}) - K_1(a\sqrt{s})I_1(b\sqrt{s})\right]}.$$

Step 3: Show that the singularities of $U(r,s)$ are simple poles at $s = 0$ and $s_n = -\alpha_n^2$, $n = 1, 2, 3, \ldots$, where $J_1(\alpha a)Y_1(\alpha b) - J_1(\alpha b)Y_1(\alpha a) = 0$.

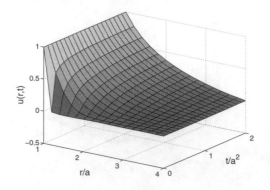

Problem 27

Step 4: Use Bromwich's integral to invert $U(r, s)$ and show that

$$u(r,t) = \frac{ab^2}{(b^2 - a^2)r} - \frac{ar}{b^2 - a^2}$$

$$+ \pi \sum_{n=1}^{\infty} J_1(\alpha_n a) J_1(\alpha_n b) e^{-\alpha_n^2 t} \frac{Y_1(\alpha_n b) J_1(\alpha_n r) - J_1(\alpha_n b) Y_1(\alpha_n r)}{J_1^2(\alpha_n a) - J_1^2(\alpha_n b)}.$$

This solution is illustrated in the figure labeled Problem 27 when $b/a = 4$.

Homogeneous Heat Equation on a Spherical Domain

28. Solve[26]

$$\frac{\partial u}{\partial t} = \frac{\partial^2 u}{\partial r^2} + \frac{2}{r} \frac{\partial u}{\partial r}, \qquad 0 \le r < 1, \quad 0 < t,$$

with the boundary conditions $\lim_{r \to 0} |u(r,t)| < \infty$ and $u_r(1,t) = 1$. The initial condition is $u(r, 0) = 0$.

Step 1: By introducing the new variable $v(r, t) = r\, u(r, t)$, show that the problem now becomes

$$\frac{\partial v}{\partial t} = \frac{\partial^2 v}{\partial r^2}, \qquad 0 \le r < 1, \quad 0 < t,$$

with the boundary conditions $\lim_{r \to 0} v(r, t) \to 0$ and $v_r(1,t) - v(1,t) = 1$. The initial condition is $v(r, 0) = 0$.

Step 2: Take the Laplace transform of the partial differential equation and boundary conditions and show that you obtain the boundary-value problem

$$\frac{d^2 V(r, s)}{dr^2} - s V(r, s) = 0,$$

---

[26] From Reismann, H., 1962: Temperature distribution in a spinning sphere during atmospheric entry. *J. Aerosp. Sci.*, **29**, 151–159 with permission.

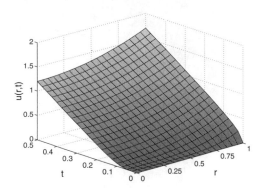

Problem 28

with $\lim_{r \to 0} V(r, s) \to 0$ and $V'(1, s) - V(1, s) = 1/s$.

Step 3: Show that the solution to the boundary-value problem is

$$V(r, s) = \frac{\sinh\left(r\sqrt{s}\right)}{s\left[\sqrt{s}\cosh\left(\sqrt{s}\right) - \sinh\left(\sqrt{s}\right)\right]}.$$

Step 4: Show that the Laplace transform has a second-order pole at $s = 0$, and simple poles at $\sqrt{s_n} = i\lambda_n$ or $s_n = -\lambda_n^2$, where $\tan(\lambda_n) = \lambda_n$ and $n = 1, 2, 3, \ldots$.

Step 5: Use Bromwich's integral to invert $V(x, s)$ and show that

$$u(r, t) = \frac{r^2}{2} + 3t - \frac{3}{10} - \frac{2}{r}\sum_{n=1}^{\infty} \frac{\sin(\lambda_n r)}{\lambda_n^2 \sin(\lambda_n)} e^{-\lambda_n^2 t}.$$

This solution is illustrated in the figure labeled Problem 28.

29. Solve[27]

$$\frac{\partial u}{\partial t} = \frac{\partial^2 u}{\partial r^2} + \frac{2}{r}\frac{\partial u}{\partial r}, \qquad 1 \le r < b, \quad 0 < t,$$

with the boundary conditions $u(1, t) = 1$, $u_r(b, t) = 0$, $0 < t$, and the initial condition $u(r, 0) = 0$, $1 < r < b$.

Step 1: By introducing the new variable $v(r, t) = r\,u(r, t)$, show that the problem now becomes

$$\frac{\partial v}{\partial t} = \frac{\partial^2 v}{\partial r^2}, \qquad 1 \le r < b, \quad 0 < t,$$

[27] Taken from Jordan, P. M., and P. Puri, 2001: Thermal stresses in a spherical shell under three thermoelastic models. *J. Therm. Stresses*, **24**, 47–70. ©2001 from "Thermal stresses in a spherical shell under three thermoelastic models" by P. M. Jordan and P. Puri. Reproduced by permission of Taylor & Francis, Inc., http://www.routledge-ny.com.

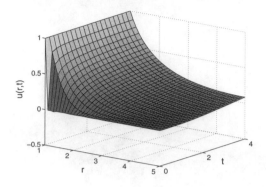

Problem 29

with the boundary conditions $v(1,t) = 1$ and $v_r(b,t) = v(b,t)/b$. The initial condition is $v(r,0) = 0$.

Step 2: Take the Laplace transform of the partial differential equation and boundary conditions and show that you obtain the boundary-value problem

$$\frac{d^2V(r,s)}{dr^2} - sV(r,s) = 0,$$

with $V(1,s) = 1/s$ and $V'(1,s) = V(1,s)/b$.

Step 3: Show that the solution to the boundary-value problem is

$$V(r,s) = \frac{\sinh\left[(b-r)\sqrt{s}\right] - b\sqrt{s}\,\cosh\left[(b-r)\sqrt{s}\right]}{s\left\{\sinh\left[(b-1)\sqrt{s}\right] - b\sqrt{s}\,\cosh\left[(b-1)\sqrt{s}\right]\right\}}.$$

Step 4: Show that the Laplace transform has a simple pole at $s = 0$, and simple poles at $\sqrt{s_n} = i\lambda_n$ or $s_n = -\lambda_n^2$, where $\tan[\lambda_n(b-1)] = b\,\lambda_n$ and $n = 1,2,3,\ldots$.

Step 5: Use Bromwich's integral to invert $V(r,s)$ and show that

$$u(r,t) = 1 - \sum_{n=1}^{\infty} \frac{\sin[\lambda_n(r-1)]}{\lambda_n\{b\sin^2[\lambda_n(b-1)] - 1\}}e^{-\lambda_n^2 t}.$$

This solution is illustrated in the figure labeled Problem 29.

Nonhomogeneous Heat Equation on a Semi-Infinite Cartesian Domain

30. Solve the nonhomogeneous heat equation

$$\frac{\partial u}{\partial t} = a^2\frac{\partial^2 u}{\partial x^2} + Ae^{-kx}, \qquad\qquad 0 < x < \infty, \quad 0 < t,$$

with the boundary conditions $u_x(0,t) = 0$ and $\lim_{x\to\infty} u(x,t) = u_0$, and the initial condition $u(x,0) = u_0$.

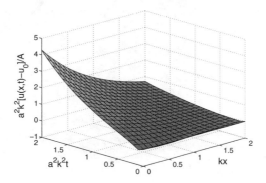

Problem 30

Step 1: Show that the Laplace transform of the partial differential equation and boundary conditions yields the boundary-value problem

$$\frac{d^2U(x,s)}{dx^2} - \frac{s}{a^2}U(x,s) = -\frac{u_0}{a^2} - \frac{A}{a^2}\frac{1}{s}e^{-kx},$$

with $U'(0,s) = 0$ and $\lim_{x\to\infty} U(x,s) = u_0/s$.

Step 2: Show that the solution to the boundary-value problem is

$$U(x,s) = \frac{u_0}{s} + \frac{Ae^{-kx}}{a^2k^2}\left(\frac{1}{s - a^2k^2} - \frac{1}{s}\right) + \frac{Ae^{-qx}}{aks\sqrt{s}} - \frac{Ae^{-qx}}{ak\sqrt{s}(s - a^2k^2)}.$$

Step 3: Use tables and convolution to invert $U(x,s)$ and show that

$$u(x,t) = u_0 + \frac{Ae^{-kx}}{a^2k^2}\left(e^{a^2k^2t} - 1\right) + \frac{A}{ak}\left[2\sqrt{\frac{t}{\pi}}\exp\left(-\frac{x^2}{4a^2t}\right) - \frac{x}{a}\operatorname{erfc}\left(\frac{x}{2a\sqrt{t}}\right)\right]$$

$$- \frac{2Ae^{a^2k^2t}}{ak\sqrt{\pi}}\int_0^t e^{-a^2k^2\eta^2}\exp\left(-\frac{x^2}{4a^2\eta^2}\right)d\eta.$$

This solution is illustrated in the figure labeled Problem 30.

Nonhomogeneous Heat Equation on a Finite Cartesian Domain

31. Solve the nonhomogeneous heat equation[28]

$$\frac{\partial u}{\partial t} - \frac{\partial^2 u}{\partial x^2} = 1, \qquad 0 < x < 1, \quad 0 < t,$$

with the boundary conditions $u(0,t) = u(1,t) = 0$, and the initial condition $u(x,0) = 0$.

---

[28] For the three-dimensional case, see Webster, A. G., 1966: *Partial Differential Equations of Mathematical Physics*. Dover, Section 45.

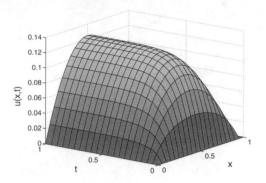

Problem 31

Step 1: Show that the Laplace transform of the partial differential equation and boundary conditions yields the boundary-value problem

$$\frac{d^2 U(x,s)}{dx^2} - sU(x,s) = -\frac{1}{s}, \qquad U(0,s) = U(1,s) = 0.$$

Step 2: Show that the solution to the boundary-value problem is

$$U(x,s) = \frac{1 - \cosh(x\sqrt{s})}{s^2} + \frac{[\cosh(\sqrt{s}) - 1]\sinh(x\sqrt{s})}{s^2 \sinh(\sqrt{s})}.$$

Step 3: Show that the singularities in the Laplace transform are located at $s_n = -n^2\pi^2$, where $n = 0, 1, 2, 3, \ldots$, and are *all* simple poles.

Step 4: Use Bromwich's integral to invert $U(x,s)$ and show that

$$u(x,t) = \frac{x(1-x)}{2} - \frac{4}{\pi^3} \sum_{n=1}^{\infty} \frac{\sin[(2n-1)\pi x]}{(2n-1)^3} e^{-(2n-1)^2\pi^2 t}.$$

This solution is illustrated in the figure labeled Problem 31.

32. Solve the nonhomogeneous partial differential equation[29]

$$\frac{\partial u}{\partial t} = b + \frac{\partial^2 u}{\partial x^2} - m^2 u, \qquad -1 < x < 1, \quad 0 < t,$$

with the boundary conditions $u(-1,t) = u(1,t) = 0$, and the initial condition $u(x,0) = 0$.

Step 1: Show that the Laplace transform of the partial differential equation and boundary conditions yields the boundary-value problem

$$\frac{d^2 U(x,s)}{dx^2} - (m^2 + s)U(x,s) = -\frac{b}{s}, \qquad U(-1,s) = U(1,s) = 0.$$

---

[29] Taken from Ram, G., and R. S. Mishra, 1977: Unsteady flow through magnetohydrodynamic porous media. *Indian J. Pure Appl. Math.*, **8**, 637–647.

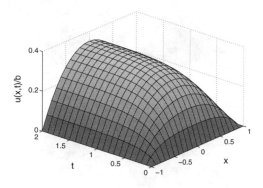

Problem 32

Step 2: Show that the solution to the boundary-value problem is

$$U(x,s) = \frac{b}{s(s+m^2)}\left[1 - \frac{\cosh\left(x\sqrt{s+m^2}\right)}{\cosh\left(\sqrt{s+m^2}\right)}\right].$$

Step 3: Show that the singularities in the Laplace transform consist of simple poles located at $s = 0$, $s = -m^2$ and $s_n = -m^2 - (2n-1)^2\pi^2/4$, where $n = 1, 2, 3, \ldots$.

Step 4: Use Bromwich's integral to invert $U(x,s)$ and show that

$$u(x,t) = \frac{b}{m^2}\left[1 - \frac{\cosh(mx)}{\cosh(m)}\right]$$

$$+ \frac{16b}{\pi}\sum_{n=1}^{\infty}\frac{(-1)^n \cos[(2n-1)\pi x/2]}{(2n-1)[(2n-1)^2\pi^2 + 4m^2]}e^{-[m^2+(2n-1)^2\pi^2/4]t}.$$

This solution is illustrated in the figure labeled Problem 32 with $m = 1$.

33. Solve the nonhomogeneous heat equation

$$\frac{\partial u}{\partial t} = a^2\frac{\partial^2 u}{\partial x^2} - P, \qquad 0 < x < L, \quad 0 < t,$$

with the boundary conditions $u(0,t) = t$ and $u(L,t) = 0$, and the initial condition $u(x,0) = 0$.

Step 1: Show that the Laplace transform of the partial differential equation and boundary conditions yields the boundary-value problem

$$\frac{d^2 U(x,s)}{dx^2} - \frac{s}{a^2}U(x,s) = \frac{P}{sa^2},$$

with $U(0,s) = 1/s^2$ and $U(L,s) = 0$.

Step 2: Show that the solution to the boundary-value problem is

$$U(x,s) = \frac{P}{s^2}\left[\frac{\sinh(qx)}{\sinh(qL)} - 1\right] + (P+1)\frac{\sinh[q(L-x)]}{s^2\sinh(qL)}, \qquad q^2 = s/a^2.$$

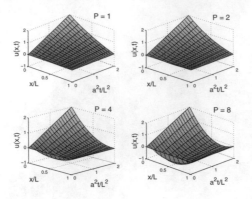

Problem 33

Step 3: Show that the singularities in the Laplace transform consist of simple poles located at $s = 0$, $s = -m^2$ and $s_n = -n^2\pi^2a^2/L^2$, where $n = 1, 2, 3, \ldots$.

Step 4: Use Bromwich's integral to invert $U(x, s)$ and show that

$$u(x, t) = \frac{t(L - x)}{L} + \frac{Px(x - L)}{2a^2} - \frac{x(x - L)(x - 2L)}{6a^2L}$$

$$- \frac{2PL^2}{a^2\pi^3} \sum_{n=1}^{\infty} \frac{(-1)^n}{n^3} \sin\left(\frac{n\pi x}{L}\right) \exp\left(-\frac{a^2n^2\pi^2t}{L^2}\right)$$

$$+ \frac{2(P + 1)L^2}{a^2\pi^3} \sum_{n=1}^{\infty} \frac{1}{n^3} \sin\left(\frac{n\pi x}{L}\right) \exp\left(-\frac{a^2n^2\pi^2t}{L^2}\right).$$

This solution is illustrated in the figure labeled Problem 33.

34. Solve the nonhomogeneous heat equation

$$\frac{\partial u}{\partial t} = a^2 \frac{\partial^2 u}{\partial x^2} - g, \qquad\qquad 0 < x < L, \quad 0 < t,$$

with the boundary conditions $u(0, t) = f(t)$ and $u_x(L, t) = 0$, and the initial condition $u(x, 0) = 0$.

Step 1: Show that the Laplace transform of the partial differential equation and boundary conditions yields the boundary-value problem

$$\frac{d^2U(x, s)}{dx^2} - \frac{s}{a^2}U(x, s) = \frac{g}{sa^2},$$

with $U(0, s) = F(s)$ and $U'(L, s) = 0$.

Step 2: Show that the solution to the boundary-value problem is

$$U(x, s) = \left[F(s) + \frac{g}{s^2}\right] \frac{\cosh[q(x - L)]}{\cosh(qL)} - \frac{g}{s^2}, \qquad q^2 = s/a^2.$$

Step 3: Show that the singularities in the Laplace transform consist of a second-order pole located at $s = 0$, and simple poles at $s_n = -(2n + 1)^2\pi^2 a^2/(4L^2)$, where $n = 0, 1, 2, 3, \ldots$.

Step 4: Use Bromwich's integral to invert $U(x, s)$ and show that

$$u(x, t) = \frac{g\left(x^2 - 2xL\right)}{2a^2} + \frac{16gL^2}{a^2\pi^3}\sum_{n=0}^{\infty}\frac{1}{(2n + 1)^3}\sin\left[\frac{(2n + 1)\pi x}{2L}\right]\exp\left[-\frac{a^2(2n + 1)^2\pi^2 t}{4L^2}\right]$$

$$+ \frac{\pi a^2}{L^2}\sum_{n=0}^{\infty}(2n + 1)\sin\left[\frac{(2n + 1)\pi x}{2L}\right]\exp\left[-\frac{a^2(2n + 1)^2\pi^2 t}{4L^2}\right]$$

$$\times \int_0^t f(\tau)\exp\left[\frac{a^2(2n + 1)^2\pi^2\tau}{4L^2}\right]d\tau.$$

35. Solve the nonhomogeneous partial differential equation

$$\frac{\partial u}{\partial t} = \frac{\partial^2 u}{\partial x^2} - b^2 u + \cos^2\left(\frac{\pi x}{a}\right), \qquad -a/2 < x < a/2, \quad 0 < t,$$

with the boundary conditions $u(-a/2, t) = u(a/2, t) = 0$, and the initial condition $u(x, 0) = 0$.

Step 1: Show that the Laplace transform of the partial differential equation and boundary conditions yields the boundary-value problem

$$\frac{d^2 U(x, s)}{dx^2} - (b^2 + s)U(x, s) = -\frac{1}{2s}\left[1 + \cos\left(\frac{2\pi x}{a}\right)\right],$$

with $U(-a/2, s) = U(a/2, s) = 0$.

Step 2: Show that the solution to the boundary-value problem is

$$U(x, s) = \frac{1}{2s(s + b^2)}\left[1 - \frac{\cosh\left(x\sqrt{s + b^2}\right)}{\cosh\left(a\sqrt{s + b^2}/2\right)}\right]$$

$$+ \frac{1}{2s(s + b^2 + 4\pi^2/a^2)}\left[\cos\left(\frac{2\pi x}{a}\right) + \frac{\cosh\left(x\sqrt{s + b^2}\right)}{\cosh\left(a\sqrt{s + b^2}/2\right)}\right].$$

Step 3: Show that the singularities in the Laplace transform consist of removable poles at $s = -b^2$ and $s = -b^2 - 4\pi^2/a^2$ and simple poles at $s = 0$ and $s_n = -b^2 - (2n - 1)^2\pi^2/a^2$, where $n = 1, 2, 3, \ldots$.

Step 4: Use Bromwich's integral to invert $U(x, s)$ and show that

$$u(x, t) = \frac{a^2(a^2 b^2 + 2\pi^2)}{b^2(a^2 b^2 + 4\pi^2)}\cos^2\left(\frac{\pi x}{a}\right) + \frac{2\pi^2 a^2}{a^2 b^2(a^2 b^2 + 4\pi^2)}\left[\sin^2\left(\frac{\pi x}{a}\right) - \frac{\cosh(bx)}{\cosh(ab/2)}\right]$$

$$- \frac{8a^2}{\pi}\sum_{n=1}^{\infty}(-1)^n\frac{\cos[(2n - 1)\pi x/a]\exp\{-[b^2 + (2n - 1)^2\pi^2/a^2]t\}}{(2n - 1)[a^2 b^2 + (2n - 1)^2\pi^2][(2n - 1)^2 - 4]}.$$

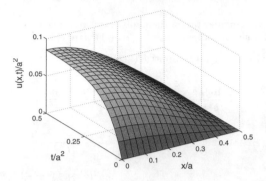

Problem 35

This solution is illustrated in the figure labeled Problem 35 with $ab = 0.5$.

36. Solve the nonhomogeneous heat equation[30]

$$\frac{\partial u}{\partial t} = \frac{1}{R}\frac{\partial^2 u}{\partial x^2} + P(1 - e^{-\alpha t}), \qquad -1 < x < 1, \quad 0 < t,$$

with the boundary conditions $u(-1,t) = 0$ and $u(1,t) = 0$, and the initial condition $u(x,0) = 0$.

Step 1: Show that the Laplace transform of the partial differential equation yields the boundary-value problem

$$\frac{d^2U(x,s)}{dx^2} - RsU(x,s) = -\frac{\alpha RP}{s(s+\alpha)}, \qquad U(-1,s) = U(1,s) = 0.$$

Step 2: Show that the solution the boundary-value problem is

$$U(x,s) = \frac{\alpha P}{s^2(s+\alpha)}\left[1 - \frac{\cosh\left(x\sqrt{Rs}\right)}{\cosh\left(\sqrt{Rs}\right)}\right].$$

Step 3: Show that the singularities in the Laplace transform are located at $s = 0$, $s = -\alpha$ and $s_n = -(2n-1)^2\pi^2/4R$, where $n = 1, 2, 3, \ldots$. All of the poles are simple.

Step 4 Use Bromwich's integral to invert $U(x,s)$ and show that

$$u(x,t) = \frac{PR}{2}(1 - x^2) + \frac{P}{\alpha}\left[1 - \frac{\cos\left(x\sqrt{R\alpha}\right)}{\cos\left(\sqrt{R\alpha}\right)}\right]e^{-\alpha t}$$

$$+ \frac{64\alpha RP}{\pi^3}\sum_{n=1}^{\infty}\frac{(-1)^n\cos[(2n-1)\pi x/2]}{(2n-1)^3[4\alpha - (2n-1)^2\pi^2/R]}e^{-(2n-1)^2\pi^2 t/(4R)}.$$

---

[30] Taken from Prakash, S., 1967: Nonsteady parallel viscous flow through a straight channel. *Bull. Calcutta Math. Soc.*, **59**, 55–59.

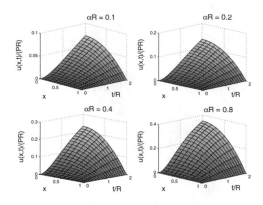

Problem 36

This solution is illustrated in the figure labeled Problem 36.

Nonhomogeneous Heat Equation on a Cylindrical Domain

37. Solve the nonhomogeneous heat equation[31] in cylindrical coordinates

$$\frac{\partial u}{\partial t} - \frac{1}{r}\frac{\partial}{\partial r}\left(r\frac{\partial u}{\partial r}\right) = 1, \qquad 0 < r < 1, \quad 0 < t,$$

with the boundary conditions $\lim_{r \to 0} |u(r,t)| < \infty$ and $u(1,t) = 0$, and the initial condition $u(r,0) = 0$.

Step 1: Show that the Laplace transform of the partial differential equation and boundary conditions yields the boundary-value problem

$$\frac{1}{r}\frac{d}{dr}\left[r\frac{dU(r,s)}{dr}\right] - sU(r,s) = -\frac{1}{s}$$

with $\lim_{r \to 0} |U(r,s)| < \infty$ and $U(1,s) = 0$.

Step 2: Show that the solution the boundary-value problem is

$$U(r,s) = \frac{1}{s^2}\left[1 - \frac{I_0(r\sqrt{s})}{I_0(\sqrt{s})}\right].$$

Step 3: Show that the singularities in the Laplace transform are located at $s = 0$ and $s_n = -\alpha_n^2$ with $\sqrt{s_n} = -\alpha_n i$, where α_n is the nth positive zero of the Bessel function $J_0(\cdot)$ and $n = 0, 1, 2, 3, \ldots$. All of the singularities are simple poles.

---

[31] Taken from Achard, J. L., and G. M. Lespinard, 1981: Structure of the transient wall-friction law in one-dimensional models of laminar pipe flows. *J. Fluid Mech.*, **113**, 283–298. Reprinted with the permission of Cambridge University Press.

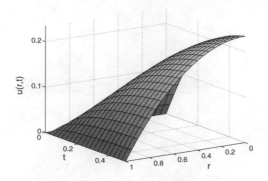

Problem 37

Step 4: Use Bromwich's integral to invert $U(r, s)$ and show that

$$u(r,t) = \tfrac{1}{4}(1 - r^2) - 2\sum_{n=1}^{\infty} \frac{J_0(\alpha_n r)}{\alpha_n^3 J_1(\alpha_n)}e^{-\alpha_n^2 t}.$$

This solution is illustrated in the figure labeled Problem 37.

38. Solve the nonhomogeneous heat equation

$$\frac{1}{r}\frac{\partial}{\partial r}\left(r\frac{\partial u}{\partial r}\right) - \frac{\partial u}{\partial t} = \delta(t), \qquad 0 \leq r < a, \quad 0 < t,$$

with the boundary conditions $\lim_{r \to 0}|u(r,t)| < \infty$ and $u(a,t) = 0$, and the initial condition $u(r,0) = 1$.

Step 1: Show that the Laplace transform of the partial differential equation and boundary conditions yields the boundary-value problem

$$\frac{1}{r}\frac{d}{dr}\left[r\frac{dU(r,s)}{dr}\right] - sU(r,s) = 1$$

with $\lim_{r \to 0}|U(r,s)| < \infty$ and $U(a,s) = 0$.

Step 2: Show that the solution to the boundary-value problem is

$$U(r,s) = \frac{I_0\!\left(r\sqrt{s}\right) - I_0\!\left(a\sqrt{s}\right)}{s\, I_0\!\left(a\sqrt{s}\right)}.$$

Step 3: Show that the Laplace transform has a removable singularity at $s = 0$ and simple poles at $s_n = -k_n^2/a^2$, where k_n is the nth positive zero of $J_0(\cdot)$ and $n = 1, 2, 3, \ldots$.

Step 4: Use Bromwich's integral to invert $U(r, s)$ and show that

$$u(r,t) = -2\sum_{n=1}^{\infty} \frac{J_0(k_n r/a)}{k_n J_1(k_n)}e^{-k_n^2 t/a^2}.$$

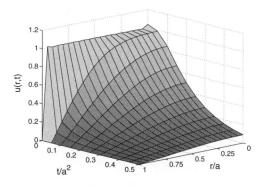

Problem 38

This solution is illustrated in the figure labeled Problem 38.

39. Solve the nonhomogeneous partial differential equation[32]

$$\frac{\partial u}{\partial t} = 2a^2\left(\frac{\partial^2 u}{\partial r^2} + \frac{1}{r}\frac{\partial u}{\partial r} - \frac{u}{r^2}\right) - bu - r, \qquad 0 \le r < 1, \quad 0 < t,$$

with the boundary conditions $\lim_{r\to 0}|u(r,t)| < \infty$ and $u(1,t) = 0$, $0 < t$, and the initial condition $u(r,0) = 0$, $0 \le r < 1$.

Step 1: Show that the Laplace transform of the partial differential equation and boundary conditions yields the boundary-value problem

$$\frac{d^2 U(r,s)}{dr^2} + \frac{1}{r}\frac{dU(r,s)}{dr} - \frac{U(r,s)}{r^2} - \frac{s+b}{2a^2}U(r,s) = \frac{r}{2a^2 s},$$

with $\lim_{r\to 0}|U(r,s)| < \infty$ and $U(1,s) = 0$.

Step 2: Show that the solution to the boundary-value problem is

$$U(r,s) = \frac{I_1\left(r\sqrt{s+b}/\sqrt{2a^2}\right)}{s(s+b)\,I_1\left(\sqrt{s+b}/\sqrt{2a^2}\right)} - \frac{r}{s(s+b)}.$$

Step 3: Show that the Laplace transform has simple poles at $s = 0$, $s = -b$ and $s_n = -b - 2a^2\alpha_n^2$, where α_n is the nth root of $J_1(\alpha) = 0$.

Step 4: Use Bromwich's integral to invert $U(r,s)$ and show that

$$u(r,t) = \frac{I_1\left[r\sqrt{b/(2a^2)}\right]}{b\,I_1\left[\sqrt{b/(2a^2)}\right]} - \frac{r}{b} + 2e^{-bt}\sum_{n=1}^{\infty}\frac{J_1(\alpha_n r)\,e^{-2a^2\alpha_n^2 t}}{\alpha_n(b + 2a^2\alpha_n^2)J_2(\alpha_n)}.$$

---

[32] Taken from Ungarish, M., 1997: Some spin-up effects on the geostrophic and quasi-geostrophic drag on a slowly rising particle or drop in a rotating fluid. *Phys. Fluids, Ser. A*, **9**, 325–336.

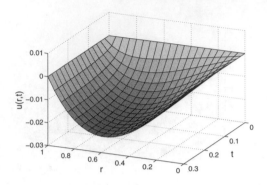

Problem 39

This solution is illustrated in the figure labeled Problem 39 when $a = b = 1$.

Systems of Heat Equations

40. Solve the set of partial differential equations[33]

$$\frac{\partial u}{\partial t} = a^2 \frac{\partial^2 u}{\partial x^2}, \qquad 0 < x < \infty, \quad 0 < t,$$

$$\frac{\partial v}{\partial t} = a^2 \frac{\partial^2 v}{\partial x^2} + \beta \frac{\partial u}{\partial t}, \qquad 0 < x < \infty, \quad 0 < t,$$

with the boundary conditions $u(0,t) = 1$, $\lim_{x\to\infty} |u(x,t)| < \infty$, $v_x(0,t) = -1$, $\lim_{x\to\infty} |v(x,t)| < \infty$, $0 < t$. The initial conditions are $u(x,0) = v(x,0) = 0$, $-\infty < x < \infty$.

Step 1: Take the Laplace transform of this system of partial differential equations and show that

$$a^2 \frac{d^2 U(x,s)}{dx^2} - sU(x,s) = 0, \qquad U(0,s) = \frac{1}{s}, \qquad \lim_{x\to\infty} |U(x,s)| < \infty,$$

and

$$a^2 \frac{d^2 V(x,s)}{dx^2} - sV(x,s) = -\beta s U(x,s), \qquad V'(0,s) = -\frac{1}{s}, \qquad \lim_{x\to\infty} |V(x,s)| < \infty.$$

Step 2: Show that the solutions to Step 1 are

$$U(x,s) = \frac{1}{s} \exp\left(-\frac{x\sqrt{s}}{a}\right)$$

---

[33] Simplified version of a problem solved by Wang, Y., and E. Papamichos, 1999: Thermal effects on fluid flow and hydraulic fracturing from wellbores and cavities in low-permeability formations. *Int. J. Numer. Anal. Meth. Geomech.*, **23**, 1819–1834.

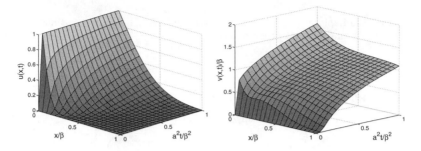

Problem 40

and

$$V(x, s) = \left(\frac{a}{s\sqrt{s}} + \frac{\beta}{2s} + \frac{\beta x}{2a\sqrt{s}} \right) \exp\left(-\frac{x\sqrt{s}}{a} \right).$$

Step 3: Using tables to invert the transforms in Step 2, show that

$$u(x, t) = \text{erfc}\left(\frac{x}{2a\sqrt{t}} \right)$$

and

$$v(x, t) = \left(\frac{\beta}{2} - x \right) \text{erfc}\left(\frac{x}{2a\sqrt{t}} \right) + \left(2a\sqrt{\frac{t}{\pi}} + \frac{\beta x}{2a\sqrt{\pi t}} \right) \exp\left(-\frac{x^2}{4a^2 t} \right).$$

This solution is illustrated in the figure labeled Problem 40.

41. Solve the partial differential equation[34]

$$\frac{\partial u}{\partial t} - fv = \nu \frac{\partial^2 u}{\partial z^2}, \qquad 0 < z < \infty, \quad 0 < t,$$

$$\frac{\partial v}{\partial t} + fu = \nu \frac{\partial^2 v}{\partial z^2}, \qquad 0 < z < \infty, \quad 0 < t,$$

with the boundary conditions $\lim_{z \to \infty} u(z, t) \to 0$, $\lim_{z \to \infty} v(z, t) \to 0$, $u_z(0, t)$ $= -g(t)$ and $v_z(0, t) = 0$. The initial conditions are $u(z, 0) = v(z, 0) = 0$, where f and ν are real and positive constants.

Step 1: If you introduce $w = u + iv$, show that we can rewrite the system of partial differential equations as

$$\frac{\partial w}{\partial t} + ifw = \nu \frac{\partial^2 w}{\partial z^2}, \qquad 0 < z < \infty, \quad 0 < t,$$

with the boundary conditions $\lim_{z \to \infty} w(z, t) \to 0$ and $w_z(0, t) = -g(t)$, and the initial condition $w(z, 0) = 0$.

---

[34] First posed and solved by Ekman, V. W., 1905: On the influence of the earth's rotation on ocean-currents. *Ark. Math. Astr. Fys.*, **2, No. 11**, 52 pp.

Step 2: Take the Laplace transform of the partial differential equation and boundary conditions in Step 1 and show that they reduce to

$$\nu \frac{d^2 W(z,s)}{dz^2} - (s+if)W(z,s) = 0, \qquad 0 < z < \infty,$$

with $\lim_{z \to \infty} W(z,s) \to 0$ and $W'(0,s) = -G(s)$.

Step 3: Show that the solution to Step 2 is

$$W(z,s) = \frac{G(s)}{\sqrt{(s+if)/\nu}} \exp\left(-z\sqrt{\frac{s+if}{\nu}}\right).$$

Step 4: Use the convolution theorem and the first shifting theorem to invert the Laplace transform and show that

$$w(z,t) = \sqrt{\frac{\nu}{\pi}} e^{-ift} \int_0^t \frac{g(\eta)}{\sqrt{t-\eta}} \exp\left[-\frac{z^2}{4\nu(t-\eta)}\right] d\eta,$$

or

$$u(z,t) = \sqrt{\frac{\nu}{\pi}} \cos(ft) \int_0^t \frac{g(\eta)}{\sqrt{t-\eta}} \exp\left[-\frac{z^2}{4\nu(t-\eta)}\right] d\eta$$

and

$$v(z,t) = \sqrt{\frac{\nu}{\pi}} \sin(ft) \int_0^t \frac{g(\eta)}{\sqrt{t-\eta}} \exp\left[-\frac{z^2}{4\nu(t-\eta)}\right] d\eta.$$

42. Solve the partial differential equations

$$\frac{\partial u}{\partial t} = \frac{\partial^2 u}{\partial x^2} - v, \qquad 0 < x < \infty, \quad 0 < t,$$

and

$$\frac{\partial v}{\partial t} = \frac{\partial^2 v}{\partial x^2} + u, \qquad 0 < x < \infty, \quad 0 < t,$$

with the boundary conditions $u(0,t) = 1$, $v(0,t) = 0$, $\lim_{x \to \infty} u(x,t) \to 0$ and $\lim_{x \to \infty} v(x,t) \to 0$. The initial conditions are $u(x,0) = v(x,0) = 0$.

Step 1: By defining the new variable $w(x,t) = u(x,t) + iv(x,t)$, show that you can write the system of partial differential equations as

$$\frac{\partial w}{\partial t} = \frac{\partial^2 w}{\partial x^2} + iw, \qquad 0 < x < \infty, \quad 0 < t,$$

with the boundary conditions $w(0,t) = 1$ and $\lim_{x \to \infty} w(x,t) \to 0$. The initial condition becomes $w(x,0) = 0$.

Step 2: By taking the Laplace transform of the partial differential equation and boundary conditions in Step 1, show that they become

$$\frac{d^2 W(x,s)}{dx^2} - (s-i)W(x,s) = 0, \qquad W(0,s) = \frac{1}{s}, \qquad \lim_{x \to \infty} W(x,s) \to 0.$$

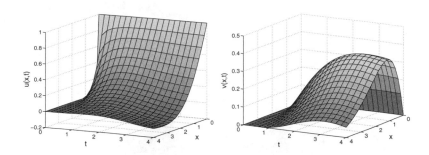

Problem 42

Step 3: Show that the solution to Step 2 is $W(x,s) = e^{-x\sqrt{s-i}}/s$.

Step 4: From a table of Laplace transforms, show that

$$u(x,t) = \frac{1}{2}\,\text{Re}\left\{ e^{(1-i)x/\sqrt{2}}\,\text{erfc}\left[\frac{x}{2\sqrt{t}} + (1-i)\sqrt{\frac{t}{2}}\right] \right.$$
$$\left. + e^{-(1-i)x/\sqrt{2}}\,\text{erfc}\left[\frac{x}{2\sqrt{t}} - (1-i)\sqrt{\frac{t}{2}}\right] \right\}$$

and

$$v(x,t) = \frac{1}{2}\,\text{Im}\left\{ e^{(1-i)x/\sqrt{2}}\,\text{erfc}\left[\frac{x}{2\sqrt{t}} + (1-i)\sqrt{\frac{t}{2}}\right] \right.$$
$$\left. + e^{-(1-i)x/\sqrt{2}}\,\text{erfc}\left[\frac{x}{2\sqrt{t}} - (1-i)\sqrt{\frac{t}{2}}\right] \right\}.$$

This solution is illustrated in the figure labeled Problem 42.

Applications

43. An electric fuse protects electrical devices by using resistance heating to melt an enclosed wire when excessive current passes through it. A knowledge of the distribution of temperature along the wire is important in the design of the fuse. If the temperature rises to the melting point only over a small interval of the element, the melt will produce a small gap, resulting in an unnecessary prolongation of the fault and a considerable release of energy. Therefore, the desirable temperature distribution should melt most of the wire. For this reason, Guile and Carne[35] solved the heat conduction equation

$$\frac{\partial u}{\partial t} = a^2 \frac{\partial^2 u}{\partial x^2} + q(1 + \alpha u), \qquad -L < x < L, \quad 0 < t,$$

---

[35] From Guile, A. E., and Carne, E. B., 1954: An analysis of an analogue solution applied to the heat conduction problem in a cartridge fuse. *AIEE Trans., Part 1*, **72**, 861–868. ©AIEE (now IEEE).

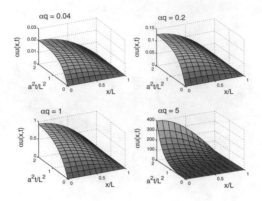

Problem 43

to model the temperature structure within a fuse just before meltdown. The second term on the right side of the heat conduction equation gives the resistance heating which is assumed to vary linearly with temperature. If the terminals at $x = \pm L$ remain at a constant temperature, which we can take to be zero, the boundary conditions are $u(-L, t) = u(L, t) = 0$. The initial condition is $u(x, 0) = 0$. Find the temperature field as a function of the parameters a, q and α.

Step 1: Take the Laplace transform of the partial differential equation and boundary conditions and show that you obtain the boundary-value problem

$$\frac{d^2 U(x, s)}{dx^2} + \frac{\alpha q - s}{a^2} U(x, s) = -\frac{q}{a^2 s}, \qquad U(-L, s) = U(L, s) = 0.$$

Step 2: Show that the solution to the boundary-value problem is

$$sU(x, s) = \frac{q}{s - \alpha q} - \frac{q \cosh(x\sqrt{s - \alpha q}/a)}{(s - \alpha q) \cosh(L\sqrt{s - \alpha q}/a)}.$$

Step 3: Show that the Laplace transform has a removable singularity at $s = \alpha q$ and simple poles at $s_n = \alpha q - (2n - 1)^2 \pi^2 a^2 / 4L^2$ with $\sqrt{s_n - \alpha q} = (2n - 1)\pi a i / 2L$, where $n = 1, 2, 3, \ldots$.

Step 4: Use Bromwich's integral to invert $sU(x, s)$ and show that

$$u_t(x, t) = \frac{4q}{\pi} \sum_{n=1}^{\infty} \frac{(-1)^{n+1}}{2n - 1} \cos\left[\frac{(2n - 1)\pi x}{2L}\right] \exp\left[\alpha q t - \frac{(2n - 1)^2 \pi^2 a^2 t}{4L^2}\right].$$

Step 5: Integrate $u_t(x, t)$ and show that

$$u(x, t) = \frac{4q}{\pi} \sum_{n=1}^{\infty} \frac{(-1)^n}{2n - 1} \frac{\cos[(2n - 1)\pi x/(2L)]}{\alpha q - (2n - 1)^2 \pi^2 a^2 /(4L^2)} \left[1 - e^{\alpha q t - (2n - 1)^2 \pi^2 a^2 t /(4L^2)}\right],$$

because $u(x, 0) = 0$. This solution is illustrated in the figure labeled Problem 43.

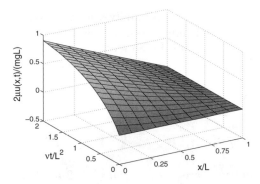

Problem 44

44. Consider[36] a viscous fluid located between two fixed walls $x = \pm L$. At $x = 0$, we introduce a thin, infinitely long rigid barrier of mass m per unit area and let it fall under the force of gravity which points in the direction of positive x. We wish to find the velocity of the fluid $u(x,t)$. The fluid is governed by the partial differential equation

$$\frac{\partial u}{\partial t} = \nu \frac{\partial^2 u}{\partial x^2}, \qquad 0 < x < L, \quad 0 < t,$$

subject to the boundary conditions $u(L,t) = 0$ and $u_t(0,t) - 2\mu u_x(0,t)/m = g$, $0 < t$, and the initial condition $u(x,0) = 0$, $0 < x < L$.

Step 1: Take the Laplace transform of the partial differential equation and boundary conditions and show that you obtain the boundary-value problem

$$\frac{d^2 U(x,s)}{dx^2} - \frac{s}{\nu} U(x,s) = 0$$

with $U(L,s) = 0$ and $sU(0,s) - 2\mu U'(0,s)/m = g/s$.

Step 2: Show that the solution to the boundary-value problem is

$$U(x,s) = \frac{g \sinh\left[(L-x)\sqrt{s/\nu}\right]}{s\left[s \sinh\left(L\sqrt{s/\nu}\right) + 2\mu\sqrt{s} \cosh\left(L\sqrt{s/\nu}\right)/(m\sqrt{\nu})\right]}.$$

Step 3: Show that the Laplace transform has simple poles at $s = 0$ and $s_n = -\nu\lambda_n^2/L^2$, where $\lambda_n \tan(\lambda_n) = 2\mu L/(m\nu) \equiv k$ and $n = 1, 2, 3, \ldots$.

Step 4: Use Bromwich's integral to invert $U(x,s)$ and show that

$$u(x,t) = \frac{mg(L-x)}{2\mu} - \frac{4g\mu L^3}{m\nu^2} \sum_{n=1}^{\infty} \frac{\sin[\lambda_n(L-x)/L]\exp(-\nu\lambda_n^2 t/L^2)}{\lambda_n^2[\lambda_n^2 + k(1+k)]\sin(\lambda_n)}.$$

This solution is illustrated in the figure labeled Problem 44 when $k = 2$.

[36] Havelock, T. H., 1921: The solution of an integral equation occurring in certain problems of viscous fluid motion. *Philos. Mag., Ser. 6*, **42**, 620–628.

45. Consider[37] a viscous fluid located between two fixed walls at $x = \pm L$. At $x = 0$, we introduce a thin, infinitely long rigid barrier of mass m per unit area. The barrier is acted upon by an elastic force in such a manner that it would vibrate with a frequency ω if the liquid were absent. We wish to find the barrier's deviation from equilibrium, $y(t)$. The fluid is governed by the partial differential equation

$$\frac{\partial u}{\partial t} = \nu \frac{\partial^2 u}{\partial x^2}, \qquad 0 < x < L, \quad 0 < t.$$

The boundary conditions are $u(L, t) = my'' - 2\mu u_x(0, t) + m\omega^2 y = 0$ and $y' = u(0, t)$, $0 < t$, and the initial conditions are $u(x, 0) = 0$, $0 < x < L$, and $y(0) = A$, $y'(0) = 0$.

Step 1: Take the Laplace transform of the partial differential equation and boundary conditions and show that

$$\frac{d^2 U(x, s)}{dx^2} - \frac{s}{\nu} U(x, s) = 0, \qquad U(L, s) = 0,$$

$$ms^2 Y(s) - 2\mu U'(0, s) + m\omega^2 Y(s) = -msA, \quad \text{and} \quad sY(s) - A = U(0, s).$$

Step 2: Show that the solution to Step 1 is

$$Y(s) = A \frac{ms + 2\mu\sqrt{s/\nu} \coth\left(L\sqrt{s/\nu}\right)}{ms^2 + 2\mu s \sqrt{s/\nu} \coth\left(L\sqrt{s/\nu}\right) + m\omega^2}.$$

Step 3: Show that the Laplace transform has simple poles at $s_n = \lambda_n$ which are the roots of

$$\lambda_n^2 + 2\mu \lambda_n^{3/2} \coth\left(L\sqrt{\lambda_n/\nu}\right) / (m\sqrt{\nu}) + \omega^2 = 0.$$

This equation yields two classes of solutions. The first class consists of a conjugate pair; in the limit of $\mu \to 0$, $\lambda_n = \omega i$. The second class consists of λ_n's along the negative real axis and represent disturbances created at the plate and radiated into the fluid.

Step 4: Use Bromwich's integral to invert $Y(s)$ and show that

$$y(t) = 2\mu' A\omega^2 \sum_{n=1}^{\infty} \frac{\lambda_n e^{\lambda_n t}}{\lambda_n^4 - \mu'(1 + \mu' L^2/\nu)\lambda_n^3 + 2\omega^2 \lambda_n^2 + 3\mu' \omega^2 \lambda_n + \omega^4},$$

where $\mu' = 2\mu/(mL)$. This solution is illustrated in the figure labeled Problem 45 when $\omega = 2$ and $L^2/\nu = 10$.

---

[37] Havelock, T. H., 1921: On the decay of oscillation of a solid body in a viscous fluid. *Philos. Mag., Ser. 6*, **42**, 628–634.

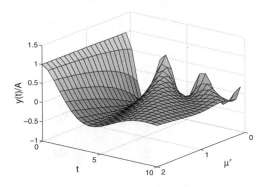

Problem 45

2.3 THE WAVE EQUATION

In the previous section we showed how the heat equation can be solved by the method of Laplace transforms. This is also true for the wave equation. Once again, we take the Laplace transform with respect to time. From the definition of Laplace transforms,

$$\mathcal{L}[u(x,t)] = U(x,s), \qquad (2.3.1)$$

$$\mathcal{L}[u_t(x,t)] = sU(x,s) - u(x,0), \qquad (2.3.2)$$

$$\mathcal{L}[u_{tt}(x,t)] = s^2 U(x,s) - su(x,0) - u_t(x,0), \qquad (2.3.3)$$

and

$$\mathcal{L}[u_{xx}(x,t)] = \frac{d^2 U(x,s)}{dx^2}. \qquad (2.3.4)$$

We next solve the resulting ordinary differential equation, known as the *auxiliary equation*, with the corresponding Laplace transformed boundary conditions. The initial condition gives us the required value of $u(x,0)$ and $u_t(x,0)$. The final step is the inversion of the Laplace transform $U(x,s)$. We typically use the inversion integral.

● **Example 2.3.1**

For our first example on how to use Laplace transforms to solve the wave equation, we solve

$$\frac{\partial^2 u}{\partial t^2} = \frac{\partial^2 u}{\partial x^2}, \qquad 0 < x < \infty, \quad 0 < t, \qquad (2.3.5)$$

subject to the initial conditions

$$u(x,0) = xe^{-x}, \qquad u_t(x,0) = 0, \qquad 0 < x < \infty, \qquad (2.3.6)$$

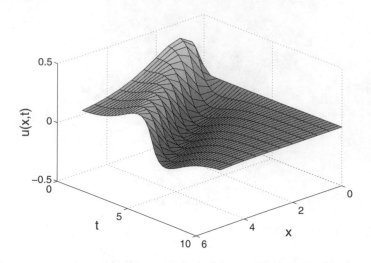

Figure 2.3.1: A plot of Equation 2.3.12 as a function of distance x and time t.

and the boundary conditions

$$u(0,t) = 0, \qquad \lim_{x \to \infty} |u(x,t)| < \infty, \qquad 0 < t. \tag{2.3.7}$$

Taking the Laplace transform of Equation 2.3.5 and Equation 2.3.7 and substituting the initial condition, we obtain

$$\frac{d^2 U(x,s)}{dx^2} = s^2 U(x,s) - sxe^{-x}, \qquad 0 < x < \infty, \tag{2.3.8}$$

with the boundary conditions

$$U(0,s) = 0, \qquad \lim_{x \to \infty} |U(x,s)| < \infty. \tag{2.3.9}$$

Solving this ordinary differential equation,

$$U(x,s) = \frac{2s}{(s^2-1)^2} e^{-sx} + \frac{sx}{s^2-1} e^{-x} - \frac{2s}{(s^2-1)^2} e^{-x} \tag{2.3.10}$$

$$= \frac{e^{-sx}}{2(s-1)^2} - \frac{e^{-sx}}{2(s+1)^2} - \frac{e^{-x}}{2(s-1)^2}$$

$$+ \frac{e^{-x}}{2(s+1)^2} + \frac{xe^{-x}}{2(s-1)} + \frac{xe^{-x}}{2(s+1)}. \tag{2.3.11}$$

A straightforward application of tables and the second shifting theorem yields

$$u(x,t) = \tfrac{1}{2}(t-x)e^{t-x} H(t-x) - \tfrac{1}{2}(t-x)e^{x-t} H(t-x)$$
$$- \tfrac{1}{2}te^{t-x} + \tfrac{1}{2}te^{-t-x} + \tfrac{1}{2}xe^{t-x} + \tfrac{1}{2}xe^{-t-x}. \tag{2.3.12}$$

Figure 2.3.1 illustrates Equation 2.3.12 as a function of distance x and time t.

• Example 2.3.2

To illustrate how these concepts apply to the wave equation on a finite domain, let us solve the wave equation

$$\frac{\partial^2 u}{\partial t^2} = \frac{\partial^2 u}{\partial x^2}, \qquad 0 < x < 1, \quad 0 < t, \qquad (2.3.13)$$

subject to the initial conditions

$$u(x,0) = 0, \qquad u_t(x,0) = x, \qquad 0 < x < 1, \qquad (2.3.14)$$

and boundary conditions

$$u(0,t) = u_x(1,t) = 0, \qquad 0 < t. \qquad (2.3.15)$$

Taking the Laplace transform of Equation 2.3.13 and Equation 2.3.15 and substituting the initial condition, we obtain

$$\frac{d^2 U(x,s)}{dx^2} = s^2 U(x,s) - x, \qquad 0 < x < 1, \qquad (2.3.16)$$

with the boundary conditions $U(0,s) = U'(1,s) = 0$. Solving this ordinary differential equation,

$$U(x,s) = \frac{xs \, \cosh(s) - \sinh(sx)}{s^3 \, \cosh(s)}. \qquad (2.3.17)$$

To invert $U(x,s)$, we will apply contour integration. From Bromwich's integral,

$$u(x,t) = \frac{1}{2\pi i} \int_{c-\infty i}^{c+\infty i} \frac{xz \, \cosh(z) - \sinh(xz)}{z^3 \, \cosh(z)} e^{tz} \, dz, \qquad (2.3.18)$$

where the line integral lies to the right of any singularities. Presently it appears that we have poles at $z = 0$ and $z_n = \pm(2n - 1)\pi i/2$, $n = 1, 2, 3, \ldots$.

Let us first examine the integrand around $z = 0$. Using the Taylor expansions for the hyperbolic functions, we have

$$\frac{xz \, \cosh(z) - \sinh(xz)}{z^3 \, \cosh(z)} e^{tz} = \frac{xz \left(1 + z^2/2 + \cdots\right) - \left(xz + x^3 z^3/6 + \cdots\right)}{z^3 \left(1 + z^2/2 + z^4/24 + \cdots\right)}$$
$$\times \left(1 + tz + t^2 z^2/2 + \cdots\right) \qquad (2.3.19)$$

$$= \frac{x}{2} - \frac{x^3}{6} + O(z). \qquad (2.3.20)$$

Equation 2.3.20 clearly shows that $z = 0$ is a removable singularity. Because its residue equals zero, we need not consider it further.

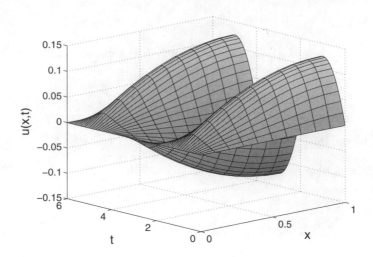

Figure 2.3.2: A plot of Equation 2.3.24 as a function of distance x and time t.

Turning to the simple poles at $z = z_n$, we find that

$$\text{Res}\left[\frac{xz\,\cosh(z) - \sinh(xz)}{z^3\,\cosh(z)}e^{tz}; z_n\right]$$

$$= \lim_{z \to z_n} \frac{xz\,\cosh(z) - \sinh(xz)}{z^3}e^{tz}\lim_{z \to z_n}\frac{z - z_n}{\cosh(z)} \quad \textbf{(2.3.21)}$$

$$= \frac{xz_n\,\cosh(z_n) - \sinh(xz_n)}{z_n^3\,\sinh(z_n)}e^{tz_n} \quad \textbf{(2.3.22)}$$

$$= \frac{8\sinh[\pm(2n-1)\pi xi/2]\exp[\pm(2n-1)\pi ti/2]}{\pm i(2n-1)^3\pi^3\sinh[\pm(2n-1)\pi i/2]}. \quad \textbf{(2.3.23)}$$

Noting that $\sinh(iz) = i\sin(z)$, $\sin[(2n-1)\pi/2] = (-1)^{n+1}$ and summing the residues, we obtain the final answer, namely,

$$u(x,t) = \frac{4}{\pi^3}\sum_{n=1}^{\infty}\frac{(-1)^{n+1}}{(2n-1)^3}\sin\left[\frac{(2n-1)\pi x}{2}\right]\sin\left[\frac{(2n-1)\pi t}{2}\right]. \quad \textbf{(2.3.24)}$$

We could also obtain Equation 2.3.24 by employing separation of variables. Figure 2.3.2 illustrates Equation 2.3.24 as a function of distance x and time t.

• **Example 2.3.3**

For our next example, we use Laplace transforms to solve the wave equation[38] in a semi-infinite domain $0 < z < \infty$ consisting of two layers

---

[38] Taken from Section 2 of Veselov, G. I., A. I. Kirpa, and N. I. Platonov, 1986: Transient-field representation in terms of steady-improper solutions. *IEE Proc., Part H*, **133**, 21–25.

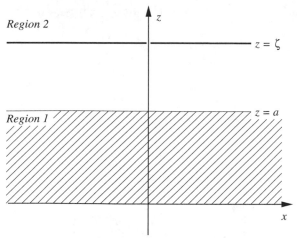

Figure 2.3.3: Schematic of a dielectric of thickness a lying above a perfect conductor at $z = 0$ and below some external medium $a < z < \infty$.

in which the physical properties differ, as shown in Figure 2.3.3. Within the slab $0 < z < a$, waves travel at the speed of c/n, where c is the speed of light in free-space $a < z < \infty$. There is an impulse forcing in the free space at $z = \zeta$.

The governing equations for each layer are

$$\frac{\partial^2 g_1}{\partial z^2} - \frac{n^2}{c^2} \frac{\partial^2 g_1}{\partial t^2} = 0, \qquad 0 < z < a, \quad 0 < t, \qquad (\mathbf{2.3.25})$$

and

$$\frac{\partial^2 g_2}{\partial z^2} - \frac{1}{c^2} \frac{\partial^2 g_2}{\partial t^2} = -\delta(z - \zeta)\delta(t - \tau), \qquad a < z < \infty, \quad 0 < t, \quad (\mathbf{2.3.26})$$

where $\zeta > a$, z is the vertical distance and t is time. At $z = 0$, we assume a perfect conductor and

$$g_1(0, t|\zeta, \tau) = 0, \qquad 0 < t. \qquad (\mathbf{2.3.27})$$

At the interface $z = a$, we require continuity of tangential components, or

$$g_1(a, t|\zeta, \tau) = g_2(a, t|\zeta, \tau), \qquad 0 < t, \qquad (\mathbf{2.3.28})$$

and

$$\frac{\partial g_1(a, t|\zeta, \tau)}{\partial z} = \frac{\partial g_2(a, t|\zeta, \tau)}{\partial z}, \qquad 0 < t. \qquad (\mathbf{2.3.29})$$

Initially, the system is undisturbed so that

$$g_1(z, 0|\zeta, \tau) = \frac{\partial g_1(z, 0|\zeta, \tau)}{\partial t} = 0, \qquad 0 < z < a, \qquad (\mathbf{2.3.30})$$

and

$$g_2(z, 0|\zeta, \tau) = \frac{\partial g_2(z, 0|\zeta, \tau)}{\partial t} = 0, \qquad a < z < \infty. \qquad (\mathbf{2.3.31})$$

At infinity, $\lim_{z \to \infty} |g_2(z, t|\zeta, \tau)| < \infty$.

Taking the Laplace transform of Equation 2.3.25 through Equation 2.3.29,

$$\frac{d^2 G_1(z,s)}{dz^2} - \frac{n^2 s^2}{c^2} G_1(z,s) = 0, \tag{2.3.32}$$

$$\frac{d^2 G_2(z,s)}{dz^2} - \frac{s^2}{c^2} G_2(z,s) = -\delta(z - \zeta)e^{-s\tau}, \tag{2.3.33}$$

with

$$G_1(0, s|\zeta, \tau) = 0, \tag{2.3.34}$$

$$G_1(a, s|\zeta, \tau) = G_2(a, s|\zeta, \tau), \tag{2.3.35}$$

and

$$G_1'(a, s|\zeta, \tau) = G_2'(a, s|\zeta, \tau). \tag{2.3.36}$$

The difficulty in solving this set of differential equations is the presence of the delta function in Equation 2.3.33. We solve this problem by further subdividing the region $a < z < \infty$ into two separate regions $a < z < \zeta$ and $\zeta < z < \infty$. At $z = \zeta$, we integrate Equation 2.3.33 across a very narrow strip from ζ^- to ζ^+. This gives the additional conditions that

$$G_2(\zeta^-, s|\zeta, \tau) = G_2(\zeta^+, s|\zeta, \tau) \tag{2.3.37}$$

and

$$G_2'(\zeta^+, s|\zeta, \tau) - G_2'(\zeta^-, s|\zeta, \tau) = -e^{-s\tau}. \tag{2.3.38}$$

Solutions that satisfy the differential equations stated in Equations 2.3.32 and Equation 2.3.33 plus the boundary conditions given by Equation 2.3.34 to Equation 2.3.38 are

$$G_1(z, s|\zeta, \tau) = \frac{\sinh(nsz/c)}{\Delta(s)} e^{-s(\zeta-a)/c - s\tau}, \qquad 0 < z < a, \tag{2.3.39}$$

and

$$\begin{aligned} G_2(z, s|\zeta, \tau) = {} & \frac{c}{2s} \exp\left(-\frac{s}{c}|z - \zeta| - s\tau\right) \\ & - \frac{c}{2s} \exp\left[-\frac{s(\zeta - a)}{c} - \frac{s(z - a)}{c} - s\tau\right] \\ & + \frac{\sinh(nsa/c)}{\Delta(s)} \exp\left[-\frac{s(\zeta - a)}{c} - \frac{s(z - a)}{c} - s\tau\right], \end{aligned} \tag{2.3.40}$$

where $a < z < \infty$, and

$$\Delta(s) = \frac{s}{c} \sinh\left(\frac{nsa}{c}\right) + \frac{ns}{c} \cosh\left(\frac{nsa}{c}\right). \tag{2.3.41}$$

We invert the first term of Equation 2.3.40 by inspection and discover that

$$g_2^{(d)}(z,t|\zeta,\tau) = \frac{c}{2}H\left(t - \tau - \frac{|z-\zeta|}{c}\right). \tag{2.3.42}$$

This portion of the solution represents the direct wave emitted from $z = \zeta$. The second term in $G_2(z,s|\zeta,\tau)$ contains a simple pole at $s = 0$. Performing the inversion by inspection,

$$g_2^{(s)}(z,t|\zeta,\tau) = -\frac{c}{2}H\left(t - \tau - \frac{z-a}{c} - \frac{\zeta-a}{c}\right). \tag{2.3.43}$$

This is a steady-state field set up in Region 2 due to the reflection of the direct wave from the interface at $z = a$.

The inversion of the last term in Equation 2.3.40 depends upon the value of n. For $n > 1$, the most interesting case, this term contains simple poles that are the zeros of $s^{-1}\Delta(s)$. A little algebra gives

$$\frac{as_m}{c} = \frac{1}{2n}\ln\left(\frac{n-1}{n+1}\right) + \frac{m\pi}{2n}i, \tag{2.3.44}$$

where $m = \pm 1, \pm 3, \ldots$. These poles at $s = s_m$ give the transient solutions to the problem. Applying the residue theorem in conjunction with Bromwich's integral,

$$g_1^{(t)}(z,t|\zeta,\tau) = \frac{c}{n^2-1}H\left(t - \tau + \frac{n(z-a)}{c} - \frac{\zeta-a}{c}\right)$$
$$\times \left\{\sum_m \frac{\sinh(ns_mz/c)\exp[s_m(t-\tau) - s_m(\zeta-a)/c]}{(as_m/c)\sinh(ns_ma/c)}\right\} \tag{2.3.45}$$

and

$$g_2^{(t)}(z,t|\zeta,\tau) = \frac{c}{n^2-1}H\left(t - \tau - \frac{z-a}{c} - \frac{\zeta-a}{c}\right)$$
$$\times \left\{\sum_m \frac{\exp[s_m(t-\tau) - s_m(\zeta-a)/c - s_m(z-a)/c]}{as_m/c}\right\}. \tag{2.3.46}$$

The solution $g_1^{(t)}(z,t|\zeta,\tau)$ gives the total field in Region 1 while the total field within Region 2 equals the sum of $g_2^{(d)}(z,t|\zeta,\tau)$, $g_2^{(s)}(z,t|\zeta,\tau)$ and $g_2^{(t)}(z,t|\zeta,\tau)$.

Let us examine the transient solution $g_2^{(t)}(z,t|\zeta,\tau)$ more closely. Its origins are in Equation 2.3.40, which contain terms that behave as $e^{-s_m z/c}$. Because $\text{Re}(s_m) < 0$ for $n > 1$, this solution grows exponentially as $z \to \infty$. Have we made a mistake in deriving Equation 2.3.40? No, because there is an exponential *decay* $e^{s_m t}$ in time for each fixed space point at which the wave front passed. This exponential time decay is sufficiently large to counteract the exponential growth in z so that the net effect yields solutions that are always finite. We see this in a plot of the solution given in Figure 2.3.4.

In the present case, each term in Equation 2.3.45 and Equation 2.3.46 decays with time at the same rate if $n > 1$. In many scattering problems, the singularities have a larger negative real part as m increases. For this reason, only a few terms of the residue expansion are needed to compute the temporal response at large time. This fact has given birth to a numerical tool for finding the temporal response in antenna scattering problems, called *singularity expansion method* or *SEM*.[39]

Individual episodes are illustrated in Figure 2.3.5. Initially, a square pulse emanates from $z = 2a$, propagating both inward and outward from the source. When the inwardly propagating wave strikes the interface, a portion enters the dielectric while the remaining energy is reflected, as shown in Frame $c(t - \tau)/a = 2$. After reflecting from the conductor at $z = 0$, a portion of the energy escapes the dielectric while the remaining portion is reflected back into the dielectric, as Frame $c(t - \tau)/a = 5$ shows. This repeated process of reflection and transmission continues until all of the energy escapes the dielectric.

• Example 2.3.4: Ray Expansion

So far we have concentrated on solving problems where we could analytically invert the Laplace transforms. In many cases this is not possible and we must be satisfied with an approximate solution or perform numerical inversion. Here, we show a technique, which was originally developed by Heaviside,[40] for expressing the inversion as a *ray expansion*. This expansion derives its name from the geometric theory of light which follows the propagation of a light pulse along its ray path. Even in those cases where we can find inverses exactly, ray expansions may be insightful in understanding the physical processes that are occurring.

Consider the wave problem:

$$\frac{\partial^2 u}{\partial t^2} = c^2 \frac{\partial^2 u}{\partial x^2}, \qquad 0 < x < L, \quad 0 < t, \qquad (2.3.47)$$

subject to the initial conditions $u(x,0) = u_t(x,0) = 0$, $0 < x < L$, and boundary conditions $u(0,t) = 0$, $u(L,t) = E$, $0 < t$.

The transformed partial differential equation is

$$\frac{d^2 U(x,s)}{dx^2} - \frac{s^2}{c^2} U(x,s) = 0, \qquad 0 < x < L. \qquad (2.3.48)$$

---

[39] A nice example is presented in Chen, K.-M., and D. Westmoreland, 1981: Impulse response of a conducting sphere based on singularity expansion method. *Proc. IEEE*, **69**, 747–750.

[40] Heaviside, O., 1950: *Electromagnetic Theory*. Dover, Section 256.

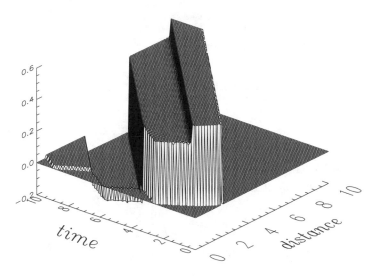

Figure 2.3.4: Plot of the Green's function (divided by c) at various distances z/a and times $c(t - \tau)/a$ for $\zeta = 2a$, and $n = 1.5811$.

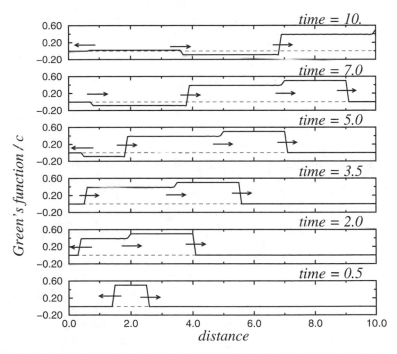

Figure 2.3.5: Snapshots of the Green's function (divided by c) at various distances z/a and times $c(t - \tau)/a$ for $\zeta = 2a$ and $n = 1.5811$.

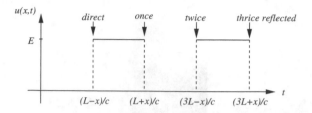

Figure 2.3.6: The solution to the wave equation given by Equation 2.3.47 with boundary conditions $u(0, t) = 0$ and $u(L, t) = E$.

The Laplace transform of the boundary conditions is $U(0, s) = 0$ and $U(L, s) = E/s$. The solution of Equation 2.3.48 which satisfies these boundary conditions is

$$U(x, s) = \frac{E \sinh(sx/c)}{s \sinh(sL/c)}. \tag{2.3.49}$$

Using Bromwich's integral, we could invert $U(x, s)$ and would find that

$$u(x, t) = \frac{Ex}{L} + \frac{2E}{\pi} \sum_{n=1}^{\infty} \frac{(-1)^n}{n} \sin\left(\frac{n\pi x}{L}\right) \cos\left(\frac{n\pi ct}{L}\right). \tag{2.3.50}$$

Although Equation 2.3.50 is absolutely correct, it is not particularly insightful. As an alternative, let us rewrite Equation 2.3.49 in a form involving negative exponentials and expand the denominator by the binomial theorem,

$$U(x, s) = \frac{E}{s} e^{-s(L-x)/c} \frac{1 - e^{-2sx/c}}{1 - e^{-2sL/c}} \tag{2.3.51}$$

$$= \frac{E}{s} e^{-s(L-x)/c} \left(1 - e^{-2sx/c}\right) \left(1 + e^{-2sL/c} + e^{-4sL/c} + \cdots\right) \tag{2.3.52}$$

$$= \frac{E}{s} \left[e^{-s(L-x)/c} - e^{-s(L+x)/c} + e^{-s(3L-x)/c} - e^{-s(3L+x)/c} + \cdots\right]. \tag{2.3.53}$$

Taking the inverse of Equation 2.3.53 term by term, we obtain

$$u(x, t) = E\left[H\left(t - \frac{L-x}{c}\right) - H\left(t - \frac{L+x}{c}\right)\right.$$

$$\left. + H\left(t - \frac{3L-x}{c}\right) - H\left(t - \frac{3L+x}{c}\right) + \cdots\right]. \tag{2.3.54}$$

We illustrate Equation 2.3.54 in Figure 2.3.6. The solution at x is zero up to the time $(L-x)/c$, at which time a wave traveling directly from the end $x = L$ would reach the point x. The solution then has the constant value E up to the time $(L+x)/c$, at which time a wave traveling from the end $x = L$ and reflected back from the end $x = 0$ would arrive. From this time up to the time of arrival of a twice-reflected wave, it has the value zero, and so on. However, as t becomes large, the bookkeeping effort necessary to account for all of the wave frontal passages becomes too cumbersome to use. For that reason, ray expansions are effective only when the solution is required for small t.

As a second example, consider the wave problem:

$$\frac{\partial^2 u}{\partial t^2} + a\frac{\partial u}{\partial t} = c^2\frac{\partial^2 u}{\partial x^2}, \qquad 0 < x < L, \quad 0 < a, t, \qquad (2.3.55)$$

subject to the initial conditions $u(x,0) = u_t(x,0) = 0$, $0 < x < L$, and boundary conditions $u_x(0,t) = \delta(t)$, $u(L,t) = 0$, $0 < t$.

The transformed partial differential equation is

$$\frac{d^2 U(x,s)}{dx^2} - \left(\frac{s^2 + as}{c^2}\right) U(x,s) = 0, \qquad 0 < x < L. \qquad (2.3.56)$$

The Laplace transform of the boundary conditions is $U'(0,s) = 1$ and $U(L,s) = 0$. The solution to Equation 2.3.56 which satisfies the boundary conditions is

$$U(x,s) = -\frac{\sinh[q(L-x)]}{q\cosh(qL)} = -\frac{e^{(L-x)q} - e^{-(L-x)q}}{q\left(e^{qL} + e^{-qL}\right)} \qquad (2.3.57)$$

$$= -\frac{1}{q}\sum_{n=0}^{\infty}(-1)^n \left\{ e^{-(2nL+x)q} - e^{-[(2n+1)L-x]q}\right\}, \qquad (2.3.58)$$

where $q = \sqrt{s^2 + as}/c$. Inverting Equation 2.3.58 term by term, we obtain the solution

$$u(x,t) = -ce^{-at/2}\sum_{n=0}^{\infty}(-1)^n \left\{ f(2nL + x, t) - f[(2n+1)L - x, t]\right\}, \qquad (2.3.59)$$

where

$$f(x,t) = I_0\left(a\sqrt{t^2 - x^2/c^2}/2\right) H(t - x/c). \qquad (2.3.60)$$

Problems

Homogeneous Wave Equation on a Semi-Infinite or Infinite Cartesian Domain

1. Use transform methods to solve the wave equation

$$\frac{\partial^2 u}{\partial t^2} - \frac{\partial^2 u}{\partial x^2} = te^{-x}, \qquad 0 < x < \infty, \quad 0 < t,$$

with the boundary conditions $u(0,t) = 1 - e^{-t}$ and $\lim_{x\to\infty}|u(x,t)| \sim x^n$, n finite, and the initial conditions $u(x,0) = 0$ and $u_t(x,0) = x$.

Step 1: Show that the Laplace transform of the partial differential equation and boundary conditions yields the boundary-value problem

$$\frac{d^2 U(x,s)}{dx^2} - s^2 U(x,s) = -x - \frac{e^{-x}}{s^2}, \qquad U(0,s) = \frac{1}{s} - \frac{1}{s+1}, \qquad \lim_{x\to\infty}|U(x,s)| \sim x^n.$$

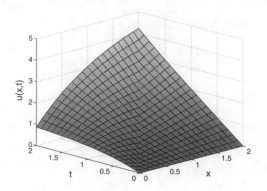

Problem 1

Step 2: Show that the solution to the boundary-value problem is

$$U(x,s) = \left(\frac{1}{s} - \frac{1}{s+1} + \frac{1}{s^2} - \frac{1}{s^2-1}\right)e^{-sx} + \frac{x}{s^2} - \frac{e^{-x}}{s^2} + \frac{e^{-x}}{s^2-1}.$$

Step 3: Invert $U(x,s)$ and show that

$$u(x,t) = xt - te^{-x} + \sinh(t)e^{-x} + \left[1 - e^{-(t-x)} + t - x - \sinh(t-x)\right]H(t-x).$$

This solution is illustrated in the figure labeled Problem 1.

2. Use transform methods to solve the wave equation

$$\frac{\partial^2 u}{\partial t^2} - \frac{\partial^2 u}{\partial x^2} = xe^{-t}, \qquad 0 < x < \infty, \quad 0 < t,$$

with the boundary conditions $u(0,t) = \cos(t)$ and $\lim_{x\to\infty} |u(x,t)| \sim x^n$, n finite, and the initial conditions $u(x,0) = 1$ and $u_t(x,0) = 0$.

Step 1: Show that the Laplace transform of the partial differential equation and boundary conditions yields the boundary-value problem

$$\frac{d^2 U(x,s)}{dx^2} - s^2 U(x,s) = -s - \frac{x}{s+1},$$

with

$$U(0,s) = \frac{s}{s^2+1} \qquad \text{and} \qquad \lim_{x\to\infty} |U(x,s)| \sim x^n.$$

Step 2: Show that the solution to the boundary-value problem is

$$U(x,s) = \left(\frac{s}{s^2+1} - \frac{1}{s}\right)e^{-sx} + \frac{1}{s} + \frac{x}{s^2} - \frac{x}{s} + \frac{x}{s+1}.$$

Step 3: Invert $U(x,s)$ and show that

$$u(x,t) = 1 + xt - x + xe^{-t} + [\cos(t-x) - 1]H(t-x).$$

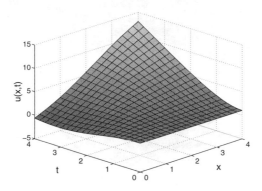

Problem 2

This solution is illustrated in the figure labeled Problem 2.

3. Solve the partial differential equation[41]

$$\epsilon\frac{\partial^2 u}{\partial t^2} + \frac{\partial u}{\partial t} = \frac{\partial^2 u}{\partial x^2}, \qquad 0 < \epsilon, \quad 0 < x < \infty, \quad 0 < t,$$

with the boundary conditions $\lim_{x\to\infty} u(x,t) \to 0$ and $u_x(0,t) = -1 - \epsilon\delta(t)$, and the initial conditions $u(x,0) = u_t(x,0) = 0$.

Step 1: By taking the Laplace transform of the partial differential equation and boundary conditions, show that they reduce to the boundary-value problem

$$\frac{d^2 U(x,s)}{dx^2} - s(1+\epsilon s)U(x,s) = 0, \qquad \lim_{x\to\infty} U(x,s) \to 0, \quad U'(0,s) = -\epsilon - \frac{1}{s}.$$

Step 2: Show that the solution to the boundary-value problem is

$$U(x,s) = \sqrt{\epsilon}\left(1 + \frac{1}{\epsilon s}\right)\frac{\exp\left[-\sqrt{\epsilon}x\sqrt{s(s+1/\epsilon)}\right]}{\sqrt{s(s+1/\epsilon)}}.$$

Step 3: Use tables to invert the Laplace transform in Step 2 and show that

$$u(x,t) = \sqrt{\epsilon}H(t - \sqrt{\epsilon}x)\left\{e^{-t/2\epsilon}I_0\left[\sqrt{\left(\frac{t}{2\epsilon}\right)^2 - \frac{x^2}{4\epsilon}}\right]\right.$$
$$\left. + \frac{1}{\epsilon}\int_{\sqrt{\epsilon}x}^{t}e^{-\eta/2\epsilon}I_0\left[\sqrt{\left(\frac{\eta}{2\epsilon}\right)^2 - \frac{x^2}{4\epsilon}}\right]d\eta\right\}.$$

[41] Taken from Maurer, M. J., and H. A. Thompson, 1973: Non-Fourier effects at high heat flux. *J. Heat Transfer*, **95**, 284–286.

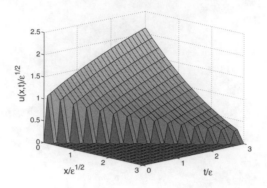

Problem 3

This solution is illustrated in the figure labeled Problem 3.

4. Solve the partial differential equation[42]

$$\frac{\partial^2 u}{\partial t^2} + 2\frac{\partial u}{\partial t} = \frac{\partial^2 u}{\partial x^2}, \qquad 0 < x < \infty, \quad 0 < t,$$

with the boundary conditions $u(0, t) = 1$ and $\lim_{x \to \infty} u(x, t) \to 0$, and the initial conditions $u(x, 0) = u_t(x, 0) = 0$.

Step 1: Taking the Laplace transform of the partial differential equation and boundary conditions, show that they reduce to the boundary-value problem

$$\frac{d^2 U(x, s)}{dx^2} - s(s + 2)U(x, s) = 0, \qquad U(0, s) = \frac{1}{s}, \quad \lim_{x \to \infty} U(x, s) \to 0.$$

Step 2: Show that the solution to boundary-value problem is

$$U(x, s) = \frac{\exp\left[-x\sqrt{s(s + 2)}\right]}{s}.$$

Step 3: Show that

$$e^{-x\sqrt{s(s+2)}} = -\frac{d}{dx}\left\{\mathcal{L}\left[e^{-t}I_0\left(\sqrt{t^2 - x^2}\right)H(t - x)\right]\right\}$$

$$= \mathcal{L}\left[xe^{-t}\frac{I_1\left(\sqrt{t^2 - x^2}\right)}{\sqrt{t^2 - x^2}}H(t - x) + e^{-t}I_0\left(\sqrt{t^2 - x^2}\right)\delta(t - x)\right].$$

Step 4: Finish the problem by noting that

$$u(x, t) = \int_0^t \mathcal{L}^{-1}\left[e^{-x\sqrt{s(s+2)}}\right]d\tau = \left[e^{-x} + x\int_x^t e^{-\eta}\frac{I_1\left(\sqrt{\eta^2 - x^2}\right)}{\sqrt{\eta^2 - x^2}}d\eta\right]H(t - x).$$

---

[42] Taken from Baumeister, K. J., and T. D. Hamill, 1969: Hyperbolic heat-conduction equation – a solution for the semi-infinite body problem. *J. Heat Transfer*, **91**, 543–548.

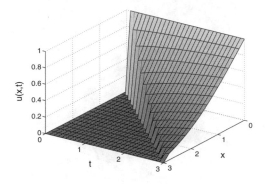

Problem 4

This solution is illustrated in the figure labeled Problem 4.

5. Solve the telegraph-like equation[43]

$$\frac{\partial^2 u}{\partial t^2} + k\frac{\partial u}{\partial t} = c^2\left(\frac{\partial^2 u}{\partial x^2} + \alpha\frac{\partial u}{\partial x}\right), \qquad 0 < x < \infty, \quad 0 < t,$$

subject to the boundary conditions $u_x(0,t) = -u_0\delta(t)$ and $\lim_{x\to\infty}|u(x,t)| < \infty$, and the initial conditions $u(x,0) = u_0$ and $u_t(x,0) = 0$ with $\alpha c > k$.

Step 1: Taking the Laplace transform of the partial differential equation and boundary conditions, show that they become the boundary-value problem

$$\frac{d^2 U(x,s)}{dx^2} + \alpha\frac{dU(x,s)}{dx} - \left(\frac{s^2 + ks}{c^2}\right)U(x,s) = -\left(\frac{s+k}{c^2}\right)u_0$$

with $U'(0,s) = -u_0$ and $\lim_{x\to\infty}|U(x,s)| < \infty$.

Step 2: Show that the solution to boundary-value problem is

$$U(x,s) = \frac{u_0}{s} + u_0 e^{-\alpha x/2}\,\frac{\exp\left[-x\sqrt{\left(s+\frac{k}{2}\right)^2 + a^2}/c\right]}{\frac{\alpha}{2} + \sqrt{(s+\frac{k}{2})^2 + a^2}/c},$$

where $4a^2 = \alpha^2 c^2 - k^2 > 0$.

Step 3: Using the first and second shifting theorems and the property that

$$F\left(\sqrt{s^2 + a^2}\right) = \mathcal{L}\left[f(t) - a\int_0^t \frac{J_1\left(a\sqrt{t^2 - \tau^2}\right)}{\sqrt{t^2 - \tau^2}}\tau f(\tau)\,d\tau\right],$$

---

[43] Taken from Abbott, M. R., 1959: The downstream effect of closing a barrier across an estuary with particular reference to the Thames. *Proc. R. Soc. London, Ser. A*, **251**, 426–439 with permission.

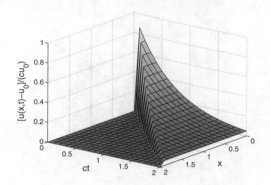

Problem 5

show that

$$u(x,t) = u_0 + u_0 c e^{-kt/2} H(t - x/c) \left[e^{-\alpha ct/2} - a \int_{x/c}^{t} \frac{J_1\left(a\sqrt{t^2 - \tau^2}\right)}{\sqrt{t^2 - \tau^2}} \tau e^{-\alpha c\tau/2} \, d\tau \right].$$

This solution is illustrated in the figure labeled Problem 5 when $\alpha = 2$ and $k/c = 1$.

6. Solve the partial differential equation[44]

$$\frac{\partial^2 u}{\partial x^2} - \frac{1}{c^2} \frac{\partial^2 u}{\partial t^2} - \frac{\omega_0^2}{c^2} u = 0, \qquad -\infty < x < \infty, \quad 0 < t,$$

subject to the boundary conditions $\lim_{|x| \to \infty} u(x,t) \to 0$, and the initial conditions $u(x,0) = u_0(x)$ and $u_t(x,0) = u_1(x)$.

Step 1: By taking the Laplace transform of the partial differential equation and boundary conditions, show that they reduce to the boundary-value problem

$$\frac{d^2 U(x,s)}{dx^2} - \left(\frac{s^2 + \omega_0^2}{c^2} \right) U(x,s) = -\frac{s}{c^2} u_0(x) - \frac{1}{c^2} u_1(x), \qquad \lim_{|x| \to \infty} U(x,s) \to 0.$$

Step 2: Show that the solution to the boundary-value problem is

$$U(x,s) = \frac{1}{2c^2 q} \int_{-\infty}^{x} e^{-q(x-\eta)} [s u_0(\eta) + u_1(\eta)] \, d\eta + \frac{1}{2c^2 q} \int_{x}^{\infty} e^{-q(\eta-x)} [s u_0(\eta) + u_1(\eta)] \, d\eta,$$

where $q^2 = (s^2 + \omega_0^2)/c^2$.

Step 3: Using the transforms that

$$\mathcal{L}\left[c J_0\left(\omega_0 \sqrt{t^2 - x^2/c^2} \right) H(t - x/c) \right] = e^{-qx}/q, \qquad x > 0,$$

---

[44] Taken from Jaeger, J. C., and K. C. Westfold, 1949: Transients in an ionized medium with applications to bursts of solar noise. *Aust. J. Sci. Res.*, Ser. A, **2**, 322–334.

and

$$\mathcal{L}\left[c\delta(t - x/c) - \frac{c\omega_0 t J_1\left(\omega_0\sqrt{t^2 - x^2/c^2}\right)}{\sqrt{t^2 - x^2/c^2}} H(t - x/c) \right] = se^{-qx}/q, \qquad x > 0,$$

show that

$$u(x, t) = \tfrac{1}{2}\left[u_0(x - ct) + u_0(x + ct) \right]$$

$$+ \frac{1}{2c} \int_{x-ct}^{x+ct} \left\{ u_1(\eta) J_0\left[\omega_0 \sqrt{t^2 - (x - \eta)^2/c^2} \right] \right.$$

$$\left. - u_0(\eta) \frac{\omega_0 t J_1\left[\omega_0 \sqrt{t^2 - (x - \eta)^2/c^2} \right]}{\sqrt{t^2 - (x - \eta)^2/c^2}} \right\} d\eta.$$

This corresponds to disturbances, each of half the original waveform, moving in opposite directions, and each followed by a tail.

7. Solve the partial differential equation[45]

$$\frac{\partial^3 u}{\partial t^3} + 4\frac{\partial^2 u}{\partial t^2} + 4(1 - \alpha)\frac{\partial u}{\partial t} = \frac{\partial^3 u}{\partial t \partial x^2} + 2\frac{\partial^2 u}{\partial x^2} + 8\alpha u, \qquad 0 < x < \infty, \quad 0 < t,$$

subject to the boundary conditions $u(0, t) = f(t)$, $\lim_{x \to \infty} |u(x, t)| < \infty$, $0 < t$, and the initial conditions $u(x, 0) = u_t(x, 0) = 0$ and $u_{tt}(x, 0) = u_{xx}(x, 0)$, $0 < x < \infty$.

Step 1: By taking the Laplace transform of the partial differential equation and boundary conditions, show that they reduce to the boundary-value problem

$$\frac{d^2 U(x, s)}{dx^2} - (s^2 + 2s - 4\alpha)U(x, s) = 0, \qquad 0 < x < \infty,$$

with $U(x, 0) = F(s)$ and $\lim_{x \to \infty} |U(x, s)| < \infty$.

Step 2: Show that the solution to the boundary-value problem is

$$U(x, s) = F(s) \exp\left[-x\sqrt{(s + 1)^2 - (1 + 4\alpha)} \right].$$

Step 3: Using tables and the convolution theorem, show that

$$u(x, t) = e^{-x} f(t - x) H(t - x)$$

$$+ x\sqrt{1 + 4\alpha}\, H(t - x) \int_x^t f(t - \xi) e^{-\xi} I_1\left[\frac{\sqrt{(1 + 4\alpha)(\xi^2 - x^2)}}{\sqrt{\xi^2 - x^2}} \right] d\xi.$$

---

[45] Reprinted from *Int. J. Heat Mass Transfer*, **39**, L. Malinowski, Relaxation heat conduction and generation: An analysis of the semi-infinite body case by method of Laplace transforms, 1543–1549, ©1966, with permission of Elsevier.

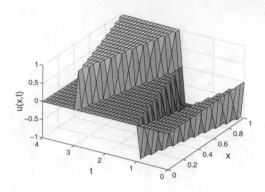

Problem 8

Homogeneous Wave Equation on a Finite Cartesian Domain

8. Solve the wave equation

$$\frac{\partial^2 u}{\partial t^2} = \frac{\partial^2 u}{\partial x^2}, \qquad 0 < x < 1, \quad 0 < t,$$

with the boundary conditions $u(0,t) = 0$ and $u_x(1,t) = \delta(t)$, and the initial conditions $u(x,0) = u_t(x,0) = 0$.

Step 1: Show that the Laplace transform of the partial differential equation and boundary conditions yields the boundary-value problem

$$\frac{d^2 U(x,s)}{dx^2} - s^2 U(x,s) = 0, \qquad U(0,s) = 0, \quad U'(1,s) = 1.$$

Step 2: Show that the solution to the boundary-value problem is

$$U(x,s) = \frac{\sinh(sx)}{s \, \cosh(s)}.$$

Step 3: Show that the singularities in the Laplace transform are located at $s_n = \pm(2n - 1)\pi i/2$, where $n = 1, 2, 3, \ldots$, and are *all* simple poles.

Step 4: Use Bromwich's integral to invert $U(x,s)$ and show that

$$u(x,t) = \frac{4}{\pi} \sum_{n=1}^{\infty} \frac{(-1)^{n+1}}{2n - 1} \sin\left[\frac{(2n-1)\pi x}{2}\right] \sin\left[\frac{(2n-1)\pi t}{2}\right].$$

This solution is illustrated in the figure labeled Problem 8.

9. Solve the wave equation

$$\frac{\partial^2 u}{\partial t^2} = \frac{\partial^2 u}{\partial x^2}, \qquad 0 < x < 1, \quad 0 < t,$$

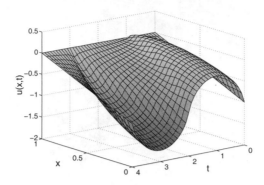

Problem 9

with the boundary conditions $u_x(0, t) = 1$ and $u(1, t) = 0$, and the initial conditions $u(x, 0) = u_t(x, 0) = 0$.

Step 1: Show that the Laplace transform of the partial differential equation and boundary conditions yields the boundary-value problem

$$\frac{d^2 U(x, s)}{dx^2} - s^2 U(x, s) = 0, \qquad U'(0, s) = \frac{1}{s}, \quad U(1, s) = 0.$$

Step 2: Show that the solution to the boundary-value problem is

$$U(x, s) = \frac{\sinh[s(x - 1)]}{s^2 \cosh(s)}.$$

Step 3: Show that the singularities in the Laplace transform are located at $s = 0$ and $s_n = \pm(2n - 1)\pi i/2$, where $n = 1, 2, 3, \dots$ and are *all* simple poles.

Step 4: Use Bromwich's integral to invert $U(x, s)$ and show that

$$u(x, t) = x - 1 + \frac{8}{\pi^2} \sum_{n=1}^{\infty} \frac{(-1)^n}{(2n - 1)^2} \sin\left[\frac{(2n - 1)\pi(x - 1)}{2}\right] \cos\left[\frac{(2n - 1)\pi t}{2}\right].$$

This solution is illustrated in the figure labeled Problem 9.

10. Solve the wave equation

$$\frac{\partial^2 u}{\partial t^2} = \frac{\partial^2 u}{\partial x^2}, \qquad 0 < x < 1, \quad 0 < t,$$

with the boundary conditions $u_x(0, t) = 0$ and $u(1, t) = 1$, and the initial conditions $u(x, 0) = x$ and $u_t(x, 0) = 0$.

Step 1: Show that the Laplace transform of the partial differential equation and boundary conditions yields the boundary-value problem

$$\frac{d^2 U(x, s)}{dx^2} - s^2 U(x, s) = -sx, \qquad U'(0, s) = 0, \quad U(1, s) = \frac{1}{s}.$$

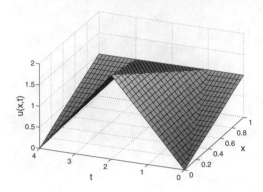

Problem 10

Step 2: Show that the solution to the boundary-value problem is

$$U(x,s) = \frac{x}{s} + \frac{\sinh[s(1-x)]}{s^2 \cosh(s)}.$$

Step 3: Show that the singularities in the Laplace transform are located at $s = 0$ and $s_n = \pm(2n-1)\pi i/2$, where $n = 1, 2, 3, \ldots$, and are *all* simple poles.

Step 4. Use Bromwich's integral to invert $U(x,s)$ and show that

$$u(x,t) = 1 + \frac{8}{\pi^2} \sum_{n=1}^{\infty} \frac{(-1)^n}{(2n-1)^2} \sin\left[\frac{(2n-1)\pi(1-x)}{2}\right] \cos\left[\frac{(2n-1)\pi t}{2}\right].$$

This solution is illustrated in the figure labeled Problem 10.

11. Use transform methods to solve the wave equation

$$\frac{\partial^2 u}{\partial t^2} = \frac{\partial^2 u}{\partial x^2}, \qquad 0 < x < 1, \quad 0 < t,$$

for the boundary conditions $u(0,t) = u(1,t) = 0$, and the initial conditions $u(x,0) = 0$ and $u_t(0,t) = 1$.

Step 1: Show that the Laplace transform of the partial differential equation and boundary conditions yields the boundary-value problem

$$\frac{d^2 U(x,s)}{dx^2} - s^2 U(x,s) = -1, \qquad U(0,s) = U(1,s) = 0.$$

Step 2: Show that the solution to the boundary-value problem is

$$U(x,s) = \frac{1 - \cosh(sx)}{s^2} + \frac{[\cosh(s) - 1]\sinh(sx)}{s^2 \sinh(s)}.$$

Step 3: Show that the singularities in the Laplace transform are the simple poles at $s_n = \pm n\pi i$, with $n = 1, 2, 3, \ldots$, and a removable pole at $s = 0$.

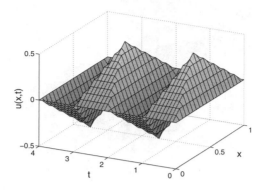

Problem 11

Step 4. Use Bromwich's integral to invert $U(x, s)$ and show that

$$u(x,t) = \frac{4}{\pi^2} \sum_{m=1}^{\infty} \frac{\sin[(2m-1)\pi x]\sin[(2m-1)\pi t]}{(2m-1)^2}.$$

This solution is illustrated in the figure labeled Problem 11.

12. Use transform methods to solve the wave equation

$$\frac{\partial^2 u}{\partial t^2} = \frac{\partial^2 u}{\partial x^2}, \qquad 0 < x < 1, \quad 0 < t,$$

for the boundary conditions $u(0,t) = u_x(1,t) = 0$, and the initial conditions $u(x,0) = 0$ and $u_t(0,t) = x$.

Step 1: Show that the Laplace transform of the partial differential equation and boundary conditions yields the boundary-value problem

$$\frac{d^2 U(x,s)}{dx^2} - s^2 U(x,s) = -x, \qquad U(0,s) = U'(1,s) = 0.$$

Step 2 Show that the solution to the boundary-value problem is

$$U(x,s) = \frac{xs\,\cosh(s) - \sinh(sx)}{s^3\,\cosh(s)}.$$

Step 3: Show that the Laplace transform has simple poles at $s_n = \pm(2n-1)\pi i/2$, where $n = 1, 2, 3, \ldots$, and a removable pole at $s = 0$.

Step 4 Use Bromwich's integral to invert $U(x,s)$ and show that

$$u(x,t) = \frac{16}{\pi^3} \sum_{n=1}^{\infty} \frac{(-1)^{n+1}}{(2n-1)^3} \sin\left[\frac{(2n-1)\pi x}{2}\right] \sin\left[\frac{(2n-1)\pi t}{2}\right].$$

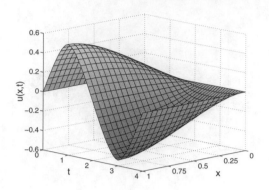

Problem 12

This solution is illustrated in the figure labeled Problem 12.

13. Use transform methods to solve the wave equation

$$\frac{\partial^2 u}{\partial t^2} = \frac{\partial^2 u}{\partial x^2}, \qquad 0 < x < 1, \quad 0 < t,$$

with the boundary conditions $u(0, t) = u(1, t) = 0$, and the initial conditions $u(x, 0) = \sin(\pi x)$ and $u_t(x, 0) = -\sin(\pi x)$.

Step 1: Show that the Laplace transform of the partial differential equation and boundary conditions yields the boundary-value problem

$$\frac{d^2 U(x, s)}{dx^2} - s^2 U(x, s) = -s \sin(\pi x) + \sin(\pi x), \qquad U(0, s) = U(1, s) = 0.$$

Step 2: Show that the solution to the boundary-value problem is

$$U(x, s) = \frac{s - 1}{s^2 + \pi^2} \sin(\pi x).$$

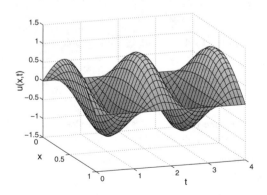

Problem 13

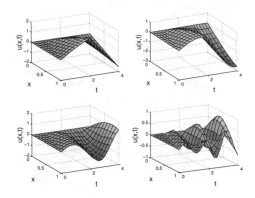

Problem 14

Step 3: Invert by inspection and show that

$$u(x,t) = \sin(\pi x)\cos(\pi t) - \sin(\pi x)\sin(\pi t)/\pi.$$

This solution is illustrated in the figure labeled Problem 13.

14. Solve the wave equation

$$\frac{\partial^2 u}{\partial t^2} = \frac{\partial^2 u}{\partial x^2}, \qquad 0 < x < 1, \quad 0 < t,$$

with the boundary conditions $u(0,t) = 0$ and $u_x(1,t) = \cos(\omega t)$, and the initial conditions $u(x,0) = u_t(x,0) = 0$. Assume that $\omega \neq (2n-1)\pi/2$, where $n = 1, 2, 3, \ldots$.

Step 1: Show that the Laplace transform of the partial differential equation and boundary conditions yields the boundary-value problem

$$\frac{d^2 U(x,s)}{dx^2} - s^2 U(x,s) = 0, \qquad U(0,s) = 0, \quad U'(1,s) = \frac{s}{s^2 + \omega^2}.$$

Step 2: Show that the solution to the boundary-value problem is

$$U(x,s) = \frac{\sinh(sx)}{(s^2 + \omega^2)\cosh(s)}.$$

Step 3: Show that the singularities in the Laplace transform are located at $s = \pm\omega i$ and $s_n = \pm(2n-1)\pi i/2$, where $n = 1, 2, 3, \ldots$. All of the singularities are simple poles.

Step 4: Use Bromwich's integral to invert $U(x,s)$ and show that

$$u(x,t) = \frac{\sin(\omega x)\cos(\omega t)}{\omega \cos(\omega)} - 2\sum_{n=1}^{\infty}(-1)^n \frac{\sin[(2n-1)\pi x/2]}{\omega^2 - (2n-1)^2\pi^2/4}\cos\left[\frac{(2n-1)\pi t}{2}\right].$$

This solution is illustrated in the figure labeled Problem 14.

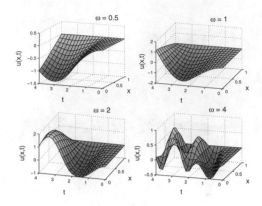

Problem 15

15. Solve the wave equation

$$\frac{\partial^2 u}{\partial t^2} = \frac{\partial^2 u}{\partial x^2}, \qquad 0 < x < 1, \quad 0 < t,$$

with the boundary conditions $u_x(0, t) = \sin(\omega t)$ and $u(1, t) = 0$, and the initial conditions $u(x, 0) = u_t(x, 0) = 0$. Assume that $\omega \neq (2n - 1)\pi/2$, where $n = 1, 2, 3, \ldots$.

Step 1: Show that the Laplace transform of the partial differential equation and boundary conditions yields the boundary-value problem

$$\frac{d^2 U(x, s)}{dx^2} - s^2 U(x, s) = 0, \qquad U'(0, s) = \frac{\omega}{s^2 + \omega^2}, \quad U(1, s) = 0.$$

Step 2: Show that the solution to the boundary-value problem is

$$U(x, s) = \frac{\omega \sinh[s(x - 1)]}{s(s^2 + \omega^2) \cosh(s)}.$$

Step 3: Show that the singularities in the Laplace transform are located at $s = \pm \omega i$ and $s_n = \pm(2n - 1)\pi i/2$, where $n = 1, 2, 3, \ldots$. All of the singularities are simple poles.

Step 4: Use Bromwich's integral to invert $U(x, s)$ and show that

$$u(x, t) = \frac{\sin[\omega(x - 1)] \sin(\omega t)}{\omega \cos(\omega)}$$

$$- \frac{4\omega}{\pi} \sum_{n=1}^{\infty} (-1)^n \frac{\sin[(2n - 1)\pi(x - 1)/2]}{(2n - 1)[\omega^2 - (2n - 1)^2 \pi^2/4]} \sin\left[\frac{(2n - 1)\pi t}{2}\right]$$

$$= \frac{4}{\pi} \sum_{n=1}^{\infty} (-1)^{n+1} \frac{\sin[(2n - 1)\pi(x - 1)/2]}{(2n - 1)[\omega^2 - (2n - 1)^2 \pi^2/4]}$$

$$\times \left\{ \omega \sin\left[\frac{(2n - 1)\pi t}{2}\right] - \frac{(2n - 1)\pi}{2} \sin(\omega t) \right\}.$$

This solution is illustrated in the figure labeled Problem 15.

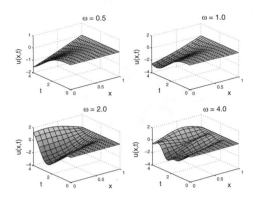

Problem 16

16. Solve the wave equation

$$\frac{\partial^2 u}{\partial t^2} = \frac{\partial^2 u}{\partial x^2}, \qquad 0 < x < 1, \quad 0 < t,$$

with the boundary conditions $u_x(0,t) = 1 - \cos(\omega t)$ and $u(1,t) = 0$, and the initial conditions $u(x,0) = u_t(x,0) = 0$. Assume that $\omega \neq (2n-1)\pi/2$, where $n = 1,2,3,\ldots$.

Step 1: Show that the Laplace transform of the partial differential equation and boundary conditions yields the boundary-value problem

$$\frac{d^2 U(x,s)}{dx^2} - s^2 U(x,s) = 0, \qquad U'(0,s) = \frac{1}{s} - \frac{s}{s^2 + \omega^2}, \quad U(1,s) = 0.$$

Step 2: Show that the solution to the boundary-value problem is

$$U(x,s) = \left(\frac{1}{s} - \frac{s}{s^2 + \omega^2}\right) \frac{\sinh[s(x-1)]}{s\cosh(s)}.$$

Step 3: Show that the singularities in the Laplace transform are located at $s = 0$, $s = \pm\omega i$ and $s_n = \pm(2n-1)\pi i/2$, where $n = 1,2,3,\ldots$. All of the singularities are simple poles.

Step 4: Use Bromwich's integral to invert $U(x,s)$ and show that

$$u(x,t) = x - 1 - \frac{\sin[\omega(x-1)]\cos(\omega t)}{\omega\cos(\omega)}$$

$$+ \frac{8}{\pi^2}\sum_{n=1}^{\infty}\frac{(-1)^n}{(2n-1)^2}\sin\left[\frac{(2n-1)\pi(x-1)}{2}\right]\cos\left[\frac{(2n-1)\pi t}{2}\right]$$

$$+ 2\sum_{n=1}^{\infty}\frac{(-1)^n}{\omega^2 - (2n-1)^2\pi^2/4}\sin\left[\frac{(2n-1)\pi(x-1)}{2}\right]\cos\left[\frac{(2n-1)\pi t}{2}\right],$$

or

$$u(x,t) = x - 1 - \frac{\sin[\omega(x-1)]\cos(\omega t)}{\omega\cos(\omega)}$$

$$+ \frac{8\omega^2}{\pi^2}\sum_{n=1}^{\infty}(-1)^n\frac{\sin[(2n-1)\pi(x-1)/2]\cos[(2n-1)\pi t/2]}{(2n-1)^2\left[\omega^2 - (2n-1)^2\pi^2/4\right]}.$$

This solution is illustrated in the figure labeled Problem 16.

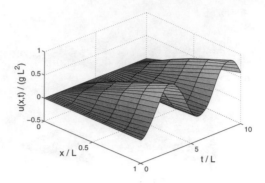

Problem 17

17. Use transform methods to solve the wave equation

$$\frac{\partial^2 u}{\partial t^2} = \frac{\partial^2 u}{\partial x^2}, \qquad 0 < x < L, \quad 0 < t,$$

for the boundary conditions $u(0,t) = 0$ and $u_{tt}(L,t) + ku_x(L,t)/m = g$, and the initial conditions $u(x,0) = u_t(x,0) = 0$, where k, m and g are constants.

Step 1: Show that the Laplace transform of the partial differential equation and boundary conditions yields the boundary-value problem

$$\frac{d^2 U(x,s)}{dx^2} - s^2 U(x,s) = 0, \qquad U(0,s) = 0, \quad s^2 U(L,s) + \omega^2 U'(L,s) = \frac{g}{s}.$$

Step 2: Show that the solution to the boundary-value problem is

$$U(x,s) = \frac{g \, \sinh(sx)}{s^2[s \, \sinh(sL) + \omega^2 \cosh(sL)]}.$$

Step 3: Show that the Laplace transform has simple poles at $s = 0$ and $s_n = \pm \lambda_n i$, where $\tan(\lambda_n L)\lambda_n L = \omega^2 L$ with $n = 1, 2, 3, \ldots$.

Step 4: Invert $U(x,s)$ and show that

$$u(x,t) = \frac{gx}{\omega^2} - \frac{2g\omega^2}{L} \sum_{n=1}^{\infty} \frac{\sin(\lambda_n x) \cos(\lambda_n t)}{\lambda_n^2 (\omega^4 + \omega^2/L + \lambda_n^2) \sin(\lambda_n L)}.$$

This solution is illustrated in the figure labeled Problem 17 when $\omega^2 L = 2$.

18. Solve the wave equation[46]

$$\frac{\partial^2 u}{\partial t^2} = c^2 \frac{\partial^2 u}{\partial x^2}, \qquad 0 < x < L, \quad 0 < t,$$

---

[46] Taken from Durant, N. J., 1960: Stress in a dynamically loaded helical spring. *Quart. J. Mech. Appl. Math.*, **13**, 251–256. By permission of Oxford University Press.

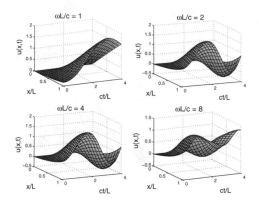

Problem 18

with the boundary conditions $u(0,t) = 0$ and

$$M\frac{\partial^2 u(L,t)}{\partial t^2} - kL\frac{\partial u(L,t)}{\partial x} = -k[1 - \cos(\omega t)],$$

and the initial conditions $u(x,0) = u_t(x,0) = 0$. Assume that $\alpha_n \neq \omega L/c$.

Step 1: Show that the Laplace transform of the partial differential equation yields the boundary-value problem

$$\frac{d^2 U(x,s)}{dx^2} - \frac{s^2}{c^2}U(x,s) = 0,$$

with $U(0,s) = 0$ and

$$s^2 MU(L,s) - kL\frac{dU(L,s)}{dx} = -\frac{k\omega^2}{s(s^2 + \omega^2)}.$$

Step 2: Show that the solution to the boundary-value problem is

$$U(x,s) = \frac{\beta c\omega^2}{L}\frac{\sinh(sx/c)}{s^2(s^2 + \omega^2)[\beta \cosh(sL/c) - (sL/c)\sinh(sL/c)]},$$

where $\beta = kL^2/Mc^2$.

Step 3: Show that the singularities in the Laplace transform are located at $s = 0$, $s = \pm\omega i$ and $s_n = \pm ic\alpha_n/L$ with $\alpha_n \tan(\alpha_n) = -\beta$, where $n = 0, 1, 2, 3, \ldots$. All of these singularities are simple poles.

Step 4: Use Bromwich's integral to invert $U(x,s)$ and show that

$$u(x,t) = \frac{x}{L} - \frac{\beta c}{L\omega}\frac{\sin(\omega x/c)\cos(\omega t)}{\beta\cos(\omega L/c) + (\omega L/c)\sin(\omega L/c)}$$

$$- \frac{2\beta\omega^2 L^2}{c^2}\sum_{n=1}^{\infty}\frac{\sin(\alpha_n x/L)\cos(\alpha_n ct/L)}{\alpha_n[\alpha_n^2 - (\omega L/c)^2](\beta^2 - \beta + \alpha_n^2)\cos(\alpha_n)}.$$

This solution is illustrated in the figure labeled Problem 18 when $\beta = 2$.

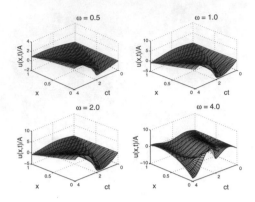

Problem 19

19. Use transform methods[47] to solve the wave equation

$$\frac{\partial^2 u}{\partial t^2} = c^2 \frac{\partial}{\partial x}\left(x \frac{\partial u}{\partial x}\right), \qquad 0 < x < 1, \quad 0 < t,$$

for the boundary conditions $\lim_{x \to 0} |u(x,t)| < \infty$ and $u(1,t) = A\sin(\omega t)$, and the initial conditions $u(x,0) = u_t(x,0) = 0$. Assume that $2\omega \neq c\beta_n$, where $J_0(\beta_n) = 0$.

Step 1: Show that the Laplace transform of the partial differential equation yields the boundary-value problem

$$\frac{d}{dx}\left[x \frac{dU(x,s)}{dx}\right] - \frac{s^2}{c^2}U(x,s) = 0, \qquad \lim_{x \to 0}|U(x,s)| < \infty, \quad U(1,s) = \frac{A\omega}{s^2 + \omega^2}.$$

Step 2: Show that the solution to the boundary-value problem is

$$U(x,s) = \frac{A\omega}{s^2 + \omega^2} \frac{I_0\left(2s\sqrt{x}/c\right)}{I_0(2s/c)}.$$

Step 3: Show that the Laplace transform has simple poles at $s = \pm\omega i$ and $s_n = \pm c\beta_n i/2$, where $J_0(\beta_n) = 0$ with $n = 1, 2, 3, \ldots$.

Step 4: Use Bromwich's integral to invert $U(x,s)$ and show that

$$u(x,t) = A\frac{J_0(2\omega\sqrt{x}/c)}{J_0(2\omega/c)} \sin(\omega t) + Ac\omega \sum_{n=1}^{\infty} \frac{J_0(\beta_n\sqrt{x})\sin(\beta_n ct/2)}{(\omega^2 - c^2\beta_n^2/4)J_1(\beta_n)}.$$

This solution is illustrated in the figure labeled Problem 19.

---

[47] Suggested by a problem solved by Brown, J., 1975: Stresses in towed cables during re-entry. *J. Spacecr. Rockets*, **12**, 524–527.

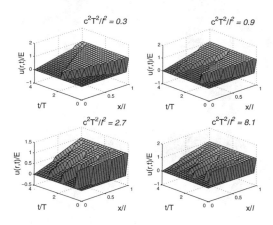

Problem 20

20. Solve the equation of telegraphy without leakage

$$\frac{\partial^2 u}{\partial x^2} = CR\frac{\partial u}{\partial t} + CL\frac{\partial^2 u}{\partial t^2}, \qquad 0 < x < \ell, \quad 0 < t,$$

with the boundary conditions $u(0,t) = 0$ and $u(\ell,t) = E$, and the initial conditions $u(x,0) = u_t(x,0) = 0$.

Step 1: By taking the Laplace transform of the partial differential equation and boundary conditions, show that they reduce to the boundary-value problem

$$\frac{d^2 U(x,s)}{dx^2} - s(CR + CLs)U(x,s) = 0, \qquad U(0,s) = 0, \quad U(\ell,s) = \frac{E}{s}.$$

Step 2: Show that the solution to the boundary-value problem is

$$U(x,s) = \frac{E \sinh\left[x\sqrt{s(1/T+s)}/c\right]}{s \sinh\left[\ell\sqrt{s(1/T+s)}/c\right]},$$

where $T = L/R$ and $c = 1/\sqrt{LC}$.

Step 3: Show that the Laplace transform has simple poles at $s = 0$ and the points where $\sinh\left[\ell\sqrt{s(1/T+s)}/c\right] = 0$, or $s_n = \left(-1 \pm \sqrt{1 - 4n^2\pi^2 c^2 T^2/\ell^2}\right)/(2T)$.

Step 4: Use Bromwich's integral to invert $U(x,s)$ and show that

$$\frac{u(x,t)}{E} = \frac{x}{\ell} - \frac{2}{\pi}e^{-t/(2T)}\sum_{n=1}^{N}\frac{(-1)^n}{n}\sin\left(\frac{n\pi x}{\ell}\right)\left\{\cosh[\alpha_n t/(2T)] + \frac{\sinh[\alpha_n t/(2T)]}{\alpha_n}\right\}$$

$$- \frac{2}{\pi}e^{-t/(2T)}\sum_{n=N+1}^{\infty}\frac{(-1)^n}{n}\sin\left(\frac{n\pi x}{\ell}\right)\left\{\cos[\beta_n t/(2T)] + \frac{\sin[\beta_n t/(2T)]}{\beta_n}\right\},$$

where

$$\alpha_n = \sqrt{1 - 4n^2\pi^2 c^2 T^2/\ell^2} \qquad \text{and} \qquad \beta_n = \sqrt{4n^2\pi^2 c^2 T^2/\ell^2 - 1}.$$

Here, N is the largest value of n such that α_n is real. This solution is illustrated in the figure labeled Problem 20.

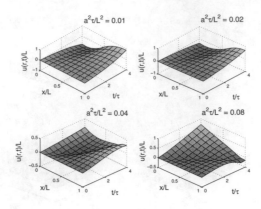

Problem 21

21. Solve the wave equation

$$\tau \frac{\partial^2 u}{\partial t^2} + \frac{\partial u}{\partial t} = a^2 \frac{\partial^2 u}{\partial x^2}, \qquad 0 < x < L, \quad 0 < t, \tau,$$

with the boundary conditions $u_x(0,t) = -1$, $u_x(L,t) = 0$, $0 < t$, and the initial conditions $u(x,0) = u_t(x,0) = 0$, $0 < x < L$.

Step 1: By taking the Laplace transform of the partial differential equation and boundary conditions, show that they reduce to the boundary-value problem

$$a^2 \frac{d^2 U(x,s)}{dx^2} - (\tau s^2 + s)U(x,s) = 0, \qquad U'(0,s) = -1/s, \quad U'(L,s) = 0.$$

Step 2: Show that the solution to the boundary-value problem is

$$U(x,s) = \frac{\cosh[q(x-L)]}{sq \, \sinh(qL)}, \qquad q^2 = (\tau s^2 + s)/a^2.$$

Step 3: Show that the Laplace transform has simple poles at $s = 0$ and the points where $\sinh(qL) = 0$ or $s_n = -\left(1 \pm \sqrt{1 - 4n^2\pi^2 a^2\tau/L^2}\right)/(2\tau)$.

Step 4: Use Bromwich's integral to invert $U(x,s)$ and show that

$$u(x,t) = \frac{a^2(t-\tau)}{L} + \frac{a^2\tau}{L}e^{-t/\tau} + \frac{(x-L)^2}{2L} - \frac{L}{6}$$

$$- 2Le^{-t/(2\tau)} \sum_{n=1}^{N} \frac{\cos(n\pi x/L)}{n^2\pi^2} \left\{\cosh[\alpha_n t/(2\tau)] + \sin[\alpha_n t/(2\tau)]/\alpha_n\right\}$$

$$- 2Le^{-t/(2\tau)} \sum_{n=N+1}^{\infty} \frac{\cos(n\pi x/L)}{n^2\pi^2} \left\{\cos[\beta_n t/(2\tau)] + \sin[\beta_n t/(2\tau)]/\beta_n\right\},$$

where

$$\alpha_n = \sqrt{1 - 4n^2\pi^2 a^2\tau/L^2} \qquad \text{and} \qquad \beta_n = \sqrt{4n^2\pi^2 a^2\tau/L^2 - 1}.$$

Here, N is the largest value of n such that α_n is real. This solution is illustrated in the figure labeled Problem 21.

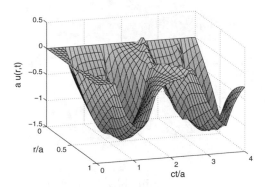

Problem 22

Homogeneous Wave Equation on a Cylindrical Domain

22. Solve the wave equation

$$\frac{1}{c^2}\frac{\partial^2 u}{\partial t^2} - \frac{\partial^2 u}{\partial r^2} - \frac{1}{r}\frac{\partial u}{\partial r} + \frac{u}{r^2} = \frac{\delta(r-\alpha)}{\alpha^2}, \qquad 0 \leq r < a, \quad 0 < t, \quad 0 < \alpha < a,$$

subject to the boundary conditions

$$\lim_{r \to 0} |u(r,t)| < \infty \qquad \text{and} \qquad \frac{\partial u(a,t)}{\partial r} + \frac{h}{a}u(a,t) = 0,$$

and the initial conditions $u(r,0) = u_t(r,0) = 0$.

Step 1: Take the Laplace transform of the partial differential equation and boundary conditions and show that you obtain the boundary-value problem

$$\frac{d^2 U(r,s)}{dr^2} + \frac{1}{r}\frac{dU(r,s)}{dr} - \left(\frac{s^2}{c^2} + \frac{1}{r^2}\right)U(r,s) = -\frac{\delta(r-\alpha)}{s\alpha^2}, \qquad 0 \leq r < a,$$

with

$$\lim_{r \to 0} |U(r,s)| < \infty \qquad \text{and} \qquad \frac{dU(a,s)}{dr} + \frac{h}{a}U(a,s) = 0.$$

Step 2: Show that the Dirac delta function can be reexpressed as the Fourier-Bessel series:

$$\delta(r-\alpha) = \frac{2\alpha}{a^2}\sum_{n=1}^{\infty}\frac{\beta_n^2 J_1(\beta_n\alpha/a)}{(\beta_n^2 + h^2 - 1)J_1^2(\beta_n)}J_1(\beta_n r/a), \qquad 0 \leq r < a,$$

where β_n is the nth root of $\beta J_1'(\beta) + h J_1(\beta) = \beta J_0(\beta) + (h-1)J_1(\beta) = 0$.

Step 3: Show that solution to the boundary-value problem is

$$U(r,s) = \frac{2}{\alpha}\sum_{n=1}^{\infty}\frac{J_1(\beta_n\alpha/a)J_1(\beta_n r/a)}{(\beta_n^2 + h^2 - 1)\,J_1^2(\beta_n)}\left(\frac{1}{s} - \frac{s}{s^2 + c^2\beta_n^2/a^2}\right).$$

Note that this solution satisfies the boundary conditions.

Step 4: Taking the inverse of the Laplace transform in Step 3, show that the solution to the partial differential equation is

$$u(r,t) = \frac{2}{\alpha}\sum_{n=1}^{\infty}\frac{J_1(\beta_n\alpha/a)J_1(\beta_n r/a)}{(\beta_n^2 + h^2 - 1)\,J_1^2(\beta_n)}\left[1 - \cos\left(\frac{c\beta_n t}{a}\right)\right].$$

This solution is illustrated in the figure labeled Problem 22 with $h = 2$ and $\alpha/a = 0.5$.

23. Solve the wave equation[48]

$$\frac{\partial^2 u}{\partial t^2} = \frac{\partial^2 u}{\partial r^2} + \frac{1}{r}\frac{\partial u}{\partial r} - \frac{u}{r^2} + \frac{\partial^2 u}{\partial z^2}, \qquad 0 \leq r < a, \quad 0 < z < \infty, \quad 0 < t,$$

subject to the boundary conditions

$$\lim_{r \to 0} |u(r, z, t)| < \infty, \qquad \frac{\partial u(a, z, t)}{\partial r} - \frac{u(a, z, t)}{a} = 0, \quad 0 < z < \infty, \quad 0 < t,$$

$$u_z(r, 0, t) = \begin{cases} 0, & 0 \leq r < b, \\ -r, & b < r < a, \end{cases} \qquad \lim_{z \to \infty} |u(r, z, t)| < \infty, \quad 0 \leq r < a, \quad 0 < t,$$

and the initial conditions

$$u(r, z, 0) = u_t(r, z, 0) = 0, \qquad 0 \leq r < a, \quad 0 < z < \infty.$$

Step 1: Take the Laplace transform of the partial differential equation and boundary conditions. Show that

$$\frac{\partial^2 U(r, z, s)}{\partial z^2} + \frac{\partial^2 U(r, z, s)}{\partial r^2} + \frac{1}{r}\frac{\partial U(r, z, s)}{\partial r} - \left(s^2 + \frac{1}{r^2}\right) U(r, z, s) = 0,$$

for $0 \leq r < a$ and $0 < z < \infty$, with

$$\lim_{r \to 0} |U(r, z, s)| < \infty, \qquad \frac{\partial U(a, z, s)}{\partial r} - \frac{U(a, z, s)}{a} = 0, \qquad 0 < z < \infty,$$

and

$$\frac{\partial U(r, 0, s)}{\partial z} = \begin{cases} 0, & 0 \leq r < b, \\ -r/s, & b < r < a, \end{cases} \qquad \text{and} \qquad \lim_{z \to \infty} |U(r, z, s)| < \infty.$$

Step 2: Show that the solution to the boundary-value problem can be expressed as Fourier-Bessel series:

$$U(r, z, s) = Z_0(z, s) r/a + \sum_{n=1}^{\infty} Z_n(z, s) J_1(k_n r/a),$$

where k_n is the nth nonzero root of $J_2(k) = 0$.

Step 3: Show that

$$Z_0(z, s) = A_0(s) e^{-sz} \qquad \text{and} \qquad Z_n(z, s) = A_n(s) e^{-z\sqrt{(k_n/a)^2 + s^2}},$$

where $\mathrm{Re}\left[\sqrt{(k_n/a)^2 + s^2}\right] \geq 0$.

---

[48] Taken from Ohyoshi, T., 1980: Transient responses of a finite elastic cylinder to a torsional end shear (for estimation of applicability of simple elementary theory) (in Japanese). *Nihon Kikai Gakkai Rombunshu (Trans. Japan Soc. Mech. Engrs.), Ser. A*, **46**, 907–913.

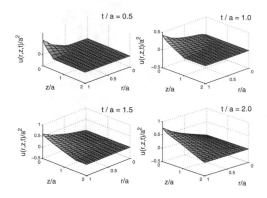

Problem 23

Step 4: Using the boundary condition at $U_z(r, 0, s)$, show that

$$U(r, z, s) = \frac{a^4 - b^4}{s^2\,a^4}re^{-sz} - \frac{2b^2}{a}\sum_{n=1}^{\infty}\frac{J_2(bk_n/a)\,J_1(k_nr/a)}{k_n\,J_1^2(k_n)}\;\frac{e^{-z\sqrt{(k_n/a)^2+s^2}}}{s\,\sqrt{(k_n/a)^2 + s^2}}.$$

Hint: To find A_0, multiply $U_z(r, 0, s)$ by $r^2\,dr$ and integrate from 0 to a. Similarly, to find A_n, multiply $U_z(r, 0, s)$ by $r\,J_1(k_nr/a)\,dr$ and integrate from 0 to a, using orthogonality to simplify the expressions.

Step 5: Taking the inverse of the Laplace transform in Step 4 term by term, show that the solution is

$$u(r, z, t) = \frac{a^4 - b^4}{a^4}r(t - z)H(t - z)$$

$$- \frac{2b^2}{a}H(t - z)\sum_{n=1}^{\infty}\frac{J_2(bk_n/a)\,J_1(k_nr/a)}{k_n\,J_1^2(k_n)}\int_z^t J_0\!\left(k_n\sqrt{\tau^2 - z^2}/a\right)d\tau.$$

This solution is illustrated in the figure labeled Problem 23 when $b/a = 0.9$.

Homogeneous Wave Equation on a Spherical Domain

24. Solve the wave equation[49]

$$\frac{\partial^2(ru)}{\partial t^2} = c^2\frac{\partial^2(ru)}{\partial r^2}, \qquad a < r < \infty, \quad 0 < t,$$

with the boundary condition that $\lim_{r\to\infty} u(r, t) \to 0$ and

$$-\rho c^2\left(\frac{\partial^2 u}{\partial r^2} + \frac{2}{3r}\frac{\partial u}{\partial r}\right)\bigg|_{r=a} = p_0 e^{-\alpha t}, \qquad 0 < \alpha.$$

---

[49] See Sharpe, J. A., 1942: The production of elastic waves by explosion pressures. I. Theory and empirical field observations. *Geophysics*, **7**, 144–154.

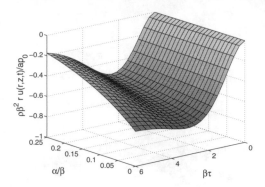

Problem 24

The initial conditions are $u(r,0) = u_t(r,0) = 0$.

Step 1: Show that the Laplace transform of the partial differential equation yields the boundary-value problem

$$\frac{d^2[rU(r,s)]}{dr^2} - \frac{s^2}{c^2}[rU(r,s)] = 0, \qquad a < r < \infty,$$

with

$$-\rho c^2\left[\frac{d^2U(a,s)}{dr^2} + \frac{2}{3a}\frac{dU(a,s)}{dr}\right] = \frac{p_0}{s+\alpha} \qquad \text{and} \qquad \lim_{r\to\infty} U(r,s) \to 0.$$

Step 2: Show that the solution to the boundary-value problem is

$$U(r,s) = -\frac{ap_0\exp[-s(r-a)/c]}{\rho r(s+\alpha)[s^2 + 4sc/(3a) + 4c^2/(3a^2)]}.$$

Step 3: Show that the singularities in the Laplace transform are simple poles located at $s = -\alpha$ and $s = -2c/(3a) \pm 2\sqrt{2}ci/(3a)$.

Step 4: Use Bromwich's integral to invert $U(r,s)$ and show that

$$u(r,t) = \frac{ap_0}{\rho r[(\beta/\sqrt{2} - \alpha)^2 + \beta^2]}\left\{e^{-\beta\tau/\sqrt{2}}\left[\left(\frac{1}{\sqrt{2}} - \frac{\alpha}{\beta}\right)\sin(\beta\tau) + \cos(\beta\tau)\right] - e^{-\alpha\tau}\right\} H(\tau),$$

where $\tau = t - (r-a)/c$ and $\beta = 2\sqrt{2}c/(3a)$. This solution is illustrated in the figure labeled Problem 24.

25. Solve the wave equation[50]

$$\frac{\partial^2(ru)}{\partial t^2} = c^2\frac{\partial^2(ru)}{\partial r^2}, \qquad a < r < \infty, \quad 0 < t,$$

---

[50] See Ghosh, S. S., 1969: On the disturbances in a viscoelastic medium due to blast inside a spherical cavity. *Pure Appl. Geophys*, **75**, 93–97.

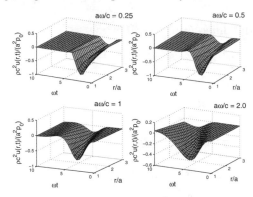

Problem 25

with the boundary condition that

$$-\rho c^2 \left(\frac{\partial^2 u}{\partial r^2} + \frac{2}{3r} \frac{\partial u}{\partial r} \right) \bigg|_{r=a} = p_0 \sin(\omega t)[1 - H(t - \pi/\omega)]$$

and $\lim_{r \to \infty} u(r,t) \to 0$. The initial conditions are $u(r,0) = u_t(r,0) = 0$.

Step 1: Show that the Laplace transform of the partial differential equation yields the boundary-value problem

$$\frac{d^2[rU(r,s)]}{dr^2} - \frac{s^2}{c^2}[rU(r,s)] = 0, \qquad a < r < \infty,$$

with

$$\rho c^2 \left[\frac{d^2 U(a,s)}{dr^2} + \frac{2}{3a} \frac{dU(a,s)}{dr} \right] = -p_0 \frac{\omega}{s^2 + \omega^2} \left(1 + e^{-s\pi/\omega}\right)$$

and $\lim_{r \to \infty} U(r,s) \to 0$.

Step 2: Show that the solution to the boundary-value problem is

$$U(r,s) = -\frac{a\omega p_0 \exp[-s(r-a)/c]}{\rho r (s^2 + \omega^2)[s^2 + 4sc/(3a) + 4c^2/(3a^2)]} \left(1 + e^{-s\pi/\omega}\right).$$

Step 3: Show that the singularities in the Laplace transform are simple poles located at $s = \pm \omega i$ and $s = -\delta \pm \beta i$, where $\delta = 2c/(3a)$ and $\beta = 2\sqrt{2}c/(3a)$.

Step 4: Show that the inverse of

$$F(s) = -\frac{a\omega p_0}{\rho r (s^2 + \omega^2)[s^2 + 4sc/(3a) + 4c^2/(3a^2)]}$$

is

$$f(t) = \frac{ap_0 \omega}{\rho r[\omega^2 - 4c^2/(3a^2)]^2 + 16c^2\omega^2/(9a^2)}$$

$$\times \left[\left(\omega^2 - \frac{4c^2}{3a^2} \right) \frac{\sin(\omega t)}{\omega} + \frac{4c}{3a} \cos(\omega t) \right.$$

$$\left. - \left(\omega^2 - \frac{4c^2}{9a^2} \right) e^{-\delta t} \frac{\sin(\beta t)}{\beta} - \frac{4c}{3a} e^{-\delta t} \cos(\beta t) \right].$$

Step 5: Use the second shifting theorem to show that

$$u(r,t) = f(\tau)H(\tau) + f(\tau - \pi/\omega)H(\tau - \pi/\omega),$$

where $\tau = t - (r-a)/c$. This solution is illustrated in the figure labeled Problem 25.

26. Solve the partial differential equation[51]

$$\left(1 + \frac{1}{\omega_0}\frac{\partial}{\partial t}\right)\frac{\partial^2(ru)}{\partial r^2} = \frac{1}{c^2}\frac{\partial^2(ru)}{\partial t^2}, \qquad a < r < \infty, \quad 0 < t,$$

with the boundary conditions

$$\mu\left(1 + \frac{1}{\omega_0}\frac{\partial}{\partial t}\right)\left[3\frac{\partial^2 u(a,t)}{\partial r^2} + \frac{2}{a}\frac{\partial u(a,t)}{\partial r}\right] = -\delta(t)$$

and $\lim_{r\to\infty}[r\,u(r,t)] \to 0$. The initial conditions are $u(r,0) = u_t(r,0) = 0$.

Step 1: Show that the partial differential equation and boundary conditions reduce to the boundary-value problem

$$\frac{d^2[rU(r,s)]}{dr^2} - \frac{s^2}{c^2(1+s/\omega_0)}rU(r,s) = 0$$

with

$$\mu\left(1 + \frac{s}{\omega_0}\right)\left[3\frac{d^2U(a,s)}{dr^2} + \frac{2}{a}\frac{dU(a,s)}{dr}\right] = -1$$

and $\lim_{r\to\infty}[r\,U(r,s)] \to 0$.

Step 2: Show that the solution to the boundary-value problem is

$$rU(r,s) = -\frac{a\exp\left[-s\,(r-a)\,/\,\left(c\sqrt{1+s/\omega_0}\right)\right]}{\mu(1+s/\omega_0)\left\{4/a^2 + 4s/\left(ac\sqrt{1+s/\omega_0}\right) + 3s^2/\left[c^2\left(1+s/\omega_0\right)\right]\right\}}.$$

Step 3: Show that

$$\mathcal{L}\left[\frac{\mu r}{a\omega_0}e^{-t}u(r,t/\omega_0)\right] = -\frac{\exp[-\alpha(s-1)/\sqrt{s}]}{s[A(s-1)^2/s + B(s-1)/\sqrt{s} + C]},$$

where $A = 3\omega_0^2/c^2$, $B = 4\omega_0/(ac)$, $C = 4/a^2$ and $\alpha = \omega_0(r-a)/c$.

Step 4: Use[52]

$$F\left(\sqrt{s}\right) = \mathcal{L}\left[\int_0^\infty \frac{\tau}{2\sqrt{\pi}\,t^{3/2}}e^{-\tau^2/(4t)}f(\tau)\,d\tau\right]$$

to show that

$$\frac{\mu r}{a\omega_0}e^{-t}u(r,t/\omega_0) = -\frac{1}{2\sqrt{\pi}\,t^{3/2}}\int_0^\infty \tau e^{-\tau^2/(4t)}\mathcal{L}^{-1}[G(s)]\,d\tau,$$

---

[51] Taken from Clark, G. B., and G. B. Rupert, 1966: Plane and spherical waves in a Voigt medium. *J. Geophys. Res.*, **71**, 2047–2053. ©1966 American Geophysical Union. Reproduced/modified by permission of American Geophysical Union.

[52] Doetsch, G., 1961: *Guide to the Application of Laplace Transforms*. D. van Nostrand Co., Ltd., p. 229.

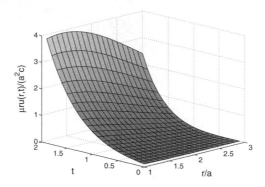

Problem 26

where

$$G(s) = \frac{\exp[-\alpha(s - 1/s)]}{s^2[A(s - 1/s)^2 + B(s - 1/s) + C]}.$$

Step 5: Use[53]

$$\mathcal{L}\left[\int_0^t \left(\frac{t - \tau}{a\tau}\right)^\nu J_{2\nu}\left(2\sqrt{a\tau t - a\tau^2}\right) f(\tau)\,d\tau\right] = s^{-2\nu - 1} F\left(s + \frac{a}{s}\right)$$

to show that

$$g(t) = \frac{2}{\sqrt{4AC - B^2}} \int_0^t \sqrt{\frac{\eta - t}{\eta}}\, J_1\left(2\sqrt{\eta^2 - \eta t}\right) \exp\left[-\frac{B(\eta - \alpha)}{2A}\right]$$

$$\times \sin\left[\frac{\sqrt{4AC - B^2}\,(\eta - \alpha)}{2A}\right] H(\eta - \alpha)\,d\eta.$$

Therefore, the solution is

$$\frac{\mu r}{a^2 c} u(r, t/\omega_0) = \frac{e^t}{4\sqrt{2\pi t^3}} \int_\alpha^\infty \tau e^{-\tau^2/(4t)} \left\{\int_\alpha^\tau \sqrt{\frac{\tau - \eta}{\eta}}\, I_1\left(2\sqrt{\eta\tau - \eta^2}\right) \exp\left[-\frac{2c(\eta - \alpha)}{3a\omega_0}\right]\right.$$

$$\left.\times \sin\left[\frac{2\sqrt{2}c(\eta - \alpha)}{3a\omega_0}\right] d\eta\right\} d\tau.$$

This solution is illustrated in the figure labeled Problem 26 when $a\omega_0/c = 1$.

Systems of Wave Equations

27. Solve the partial differential equations

$$\frac{\partial^2 u_1}{\partial t^2} = a_1^2 \frac{\partial^2 u_1}{\partial x^2}, \qquad 0 < x < \infty, \quad 0 < t,$$

---

[53] Erdélyi, A., W. Magnus, F. Oberhettinger, and F. G. Tricomi, 1954: *Table of Integral Transforms. Volume 1*. McGraw-Hill Book Co., Inc., p. 133.

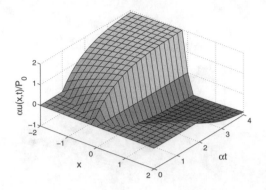

Problem 27

and

$$\frac{\partial^2 u_2}{\partial t^2} = a_2^2 \frac{\partial^2 u_2}{\partial x^2}, \qquad -\infty < x < 0, \quad 0 < t,$$

with the boundary conditions $\lim_{x \to \infty} u_1(x,t) \to 0$, $\lim_{x \to -\infty} u_2(x,\,t) \to 0$, $\partial u_1(0,t)/\partial x = \partial u_2(0,t)/\partial x$ and $\partial u_1(0,t)/\partial x = P_0 e^{-\alpha t}$ with $\alpha > 0$. The initial conditions are $u_1(x,0) = u_2(x,0) = 0$.

Step 1: Show that the Laplace transform of the partial differential equations and boundary conditions yields the boundary-value problems

$$\frac{d^2 U_1(x,s)}{dx^2} - \frac{s^2}{a_1^2} U_1(x,s) = 0, \qquad 0 < x < \infty,$$

and

$$\frac{d^2 U_2(x,s)}{dx^2} - \frac{s^2}{a_2^2} U_2(x,s) = 0, \qquad -\infty < x < 0,$$

with $\lim_{x \to \infty} U_1(x,s) \to 0$, $\lim_{x \to -\infty} U_2(x,s) \to 0$, $U_1'(0,s) = U_2'(0,s)$ and $U_1'(0,s) = P_0/(s+\alpha)$.

Step 2: Show that the solutions to the boundary-value problems are

$$U_1(x,s) = -\frac{a_1 P_0}{s(s+\alpha)} \exp(-sx/a_1) \quad \text{and} \quad U_2(x,s) = \frac{a_2 P_0}{s(s+\alpha)} \exp(sx/a_2).$$

Step 3: By taking the inverse Laplace transform of Step 2, show that

$$u_1(x,t) = -\frac{a_1 P_0}{\alpha} \left[1 - e^{-\alpha(t-x/a_1)} \right] H(t - x/a_1)$$

and

$$u_2(x,t) = \frac{a_2 P_0}{\alpha} \left[1 - e^{-\alpha(t+x/a_2)} \right] H(t + x/a_2).$$

A portion of this solution is illustrated in the figure labeled Problem 27 when $a_1 = 1$ and $a_2 = 2$. Note that when $x > 0$, we have $u_1(x,t)$ while $u_2(x,t)$ is given in the domain $x < 0$.

Applications

28. The pressure and velocity oscillations from water hammer in a pipe without friction[54] are given by the equations

$$\frac{\partial p}{\partial t} = -\rho c^2 \frac{\partial u}{\partial x} \quad \text{and} \quad \frac{\partial u}{\partial t} = -\frac{1}{\rho}\frac{\partial p}{\partial x},$$

where $p(x,t)$ denotes the pressure perturbation, $u(x,t)$ is the velocity perturbation, c is the speed of sound in water and ρ is the density of water. These two first-order partial differential equations may be combined to yield

$$\frac{\partial^2 p}{\partial t^2} = c^2 \frac{\partial^2 p}{\partial x^2}.$$

Find the solution to this partial differential equation if $p(0,t) = p_0$ and $u(L,t) = 0$. The initial conditions are $p(x,0) = p_0$, $p_t(x,0) = 0$ and $u(x,0) = u_0$.

Step 1: Show that the partial differential equation and boundary conditions reduce to the boundary-value problem

$$\frac{d^2 P(x,s)}{dx^2} - \frac{s^2}{c^2} P(x,s) = -\frac{s}{c^2} p_0,$$

with $P(0,s) = p_0/s$ and $P'(L,s) = -\rho U(L,s) + \rho u(L,0) = \rho u_0$.

Step 2: Show that the solution to the boundary-value problem is

$$P(x,s) = \frac{p_0}{s} + \frac{\rho u_0 c \, \sinh(sx/c)}{s \, \cosh(sL/c)}.$$

Step 3: Show that the Laplace transform has simple poles at $s_n = \pm(2n-1)c\pi i/(2L)$ with $n = 1, 2, 3, \ldots$.

Step 4: Use Bromwich's integral to invert $P(x,s)$ and show that

$$p(x,t) = p_0 - \frac{4\rho u_0 c}{\pi} \sum_{n=1}^{\infty} \frac{(-1)^n}{2n-1} \sin\left[\frac{(2n-1)\pi x}{2L}\right] \sin\left[\frac{(2n-1)c\pi t}{2L}\right].$$

This solution is illustrated in the figure labeled Problem 28.

29. Consider a vertical rod or column of length L that is supported at both ends. The elastic waves that arise when the support at the bottom is suddenly removed are governed by the wave equation[55]

$$\frac{\partial^2 u}{\partial t^2} = c^2 \frac{\partial^2 u}{\partial x^2} + g, \qquad 0 < x < L, \quad 0 < t,$$

---

[54] See Rich, G. R., 1945: Water-hammer analysis by the Laplace-Mellin transformation. *Trans. ASME*, **67**, 361–376.

[55] Abstracted with permission from Hall, L. H., 1953: Longitudinal vibrations of a vertical column by the method of Laplace transform. *Am. J. Phys.*, **21**, 287–292. ©1953 American Association of Physics Teachers.

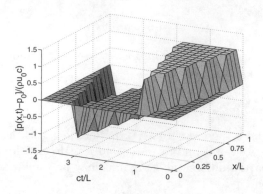

Problem 28

where g denotes the gravitational acceleration, $c^2 = E/\rho$, E is Young's modulus and ρ is the mass density. Find the wave solution if the boundary conditions are $u_x(0,t) = u_x(L,t) = 0$ and the initial conditions are

$$u(x,0) = -\frac{gx^2}{2c^2} \qquad \text{and} \qquad \frac{\partial u(x,0)}{\partial t} = 0.$$

Step 1: Show that the partial differential equation and boundary conditions reduce to the boundary-value problem

$$\frac{d^2 U(x,s)}{dx^2} - \frac{s^2}{c^2} U(x,s) = \frac{sgx^2}{2c^4} - \frac{g}{sc^2}, \qquad U'(0,s) = U'(L,s) = 0.$$

Step 2: Show that the solution to the boundary-value problem is

$$U(x,s) = \frac{gL \cosh(sx/c)}{cs^2 \sinh(sL/c)} - \frac{gx^2}{2sc^2}.$$

Step 3: Show that the Laplace transform has simple poles at $s = 0$ and $s_n = \pm n\pi ci/L$, where $n = 1, 2, 3, \ldots$.

Step 4: Use Bromwich's integral to invert $U(x,s)$ and show that

$$u(x,t) = \frac{gt^2}{2} - \frac{gL^2}{6c^2} - \frac{2gL^2}{c^2\pi^2} \sum_{n=1}^{\infty} \frac{(-1)^n}{n^2} \cos\left(\frac{n\pi x}{L}\right) \cos\left(\frac{n\pi ct}{L}\right).$$

Step 5: Show that ray expansion for $u(x,t)$ is

$$u(x,t) = -\frac{gx^2}{2c^2} + \frac{gL}{c} \sum_{m=0}^{\infty} \left[t - \frac{(2m+1)L - x}{c} \right] H\left[t - \frac{(2m+1)L - x}{c} \right]$$

$$+ \frac{gL}{c} \sum_{m=0}^{\infty} \left[t - \frac{(2m+1)L + x}{c} \right] H\left[t - \frac{(2m+1)L + x}{c} \right].$$

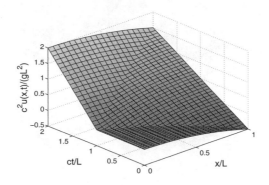

Problem 29

This solution is illustrated in the figure labeled Problem 29.

30. As an electric locomotive travels down a track at the speed V, the pantograph (the metallic framework that connects the overhead power lines to the locomotive) pushes up the line with a force P. Let us find the behavior[56] of the overhead wire as a pantograph passes between two supports of the electrical cable that are located a distance L apart. We model this system as a vibrating string with a point load:

$$\frac{\partial^2 u}{\partial t^2} = c^2 \frac{\partial^2 u}{\partial x^2} + \frac{P}{\rho V}\delta\left(t - \frac{x}{V}\right), \qquad 0 < x < L, \quad 0 < t.$$

Let us assume that the wire is initially at rest $[u(x,0) = u_t(x,0) = 0$ for $0 < x < L]$ and fixed at both ends $[u(0,t) = u(L,t) = 0$ for $0 < t]$.

Step 1: Take the Laplace transform of the partial differential equation and show that

$$s^2 U(x,s) = c^2 \frac{d^2 U(x,s)}{dx^2} + \frac{P}{\rho V}e^{-xs/V}.$$

Step 2: Solve the ordinary differential equation in Step 1 as a Fourier half-range sine series:

$$U(x,s) = \sum_{n=1}^{\infty} B_n(s)\sin\left(\frac{n\pi x}{L}\right),$$

where

$$B_n(s) = \frac{2P\beta_n}{\rho L(\beta_n^2 - \alpha_n^2)}\left(\frac{1}{s^2 + \alpha_n^2} - \frac{1}{s^2 + \beta_n^2}\right)\left[1 - (-1)^n e^{-Ls/V}\right],$$

$\alpha_n = n\pi c/L$ and $\beta_n = n\pi V/L$. This solution satisfies the boundary conditions.

---

[56] From Oda, O., and Ooura, Y., 1976: Vibrations of catenary overhead wire. *Q. Rep.*, *(Tokyo) Railway Tech. Res. Inst.*, **17**, 134–135 with permission. Material originally appearing in RTRI publications cannot be reprinted without written permission from the Institute.

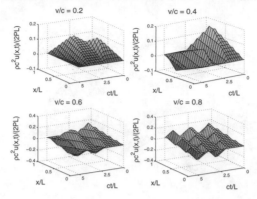

Problem 30

Step 3: By inverting the solution in Step 2, show that

$$u(x,t) = \frac{2P}{\rho L} \sum_{n=1}^{\infty} \left[\frac{\sin(\beta_n t)}{\alpha_n^2 - \beta_n^2} - \frac{V}{c} \frac{\sin(\alpha_n t)}{\alpha_n^2 - \beta_n^2} \right] \sin\left(\frac{n\pi x}{L}\right)$$

$$- \frac{2P}{\rho L} H\left(t - \frac{L}{V}\right) \sum_{n=1}^{\infty} (-1)^n \sin\left(\frac{n\pi x}{L}\right) \left\{ \frac{\sin[\beta_n(t - L/V)]}{\alpha_n^2 - \beta_n^2} - \frac{V}{c} \frac{\sin[\alpha_n(t - L/V)]}{\alpha_n^2 - \beta_n^2} \right\}$$

$$= \frac{2P}{\rho L} \sum_{n=1}^{\infty} \left[\frac{\sin(\beta_n t)}{\alpha_n^2 - \beta_n^2} - \frac{V}{c} \frac{\sin(\alpha_n t)}{\alpha_n^2 - \beta_n^2} \right] \sin\left(\frac{n\pi x}{L}\right)$$

$$- \frac{2P}{\rho L} H\left(t - \frac{L}{V}\right) \sum_{n=1}^{\infty} \sin\left(\frac{n\pi x}{L}\right) \left\{ \frac{\sin(\beta_n t)}{\alpha_n^2 - \beta_n^2} - \frac{V}{c}(-1)^n \frac{\sin[\alpha_n(t - L/V)]}{\alpha_n^2 - \beta_n^2} \right\}.$$

The first term in both summations represents the static uplift on the line; this term disappears after the pantograph has passed. The second term in both summations represents the vibrations excited by the traveling force. Even after the pantograph passes, they will continue to exist. This solution is illustrated in the figure labeled Problem 30.

2.4 LAPLACE'S AND POISSON'S EQUATIONS

Using Laplace transforms to solve Laplace's or Poisson's equations would appear to be a strange choice since there are no initial conditions. However, for half and quarter plane problems, one (or both) of the independent variables can act as the "time" variable. The tricky part is satisfying the boundary conditions. The following example shows how this is done.

• **Example 2.4.1**

Let us solve Poisson's equation:

$$\frac{\partial^2 u}{\partial x^2} + \frac{\partial^2 u}{\partial y^2} = xe^{-x}, \qquad 0 < x < \infty, \quad 0 < y < a, \tag{2.4.1}$$

subject to the boundary conditions

$$u(0, y) = 0, \qquad \lim_{x \to \infty} |u(x, y)| < \infty, \qquad 0 < y < a, \qquad (2.4.2)$$

and

$$u(x, 0) = 0, \qquad u_x(x, a) = 0, \qquad 0 < x < \infty. \qquad (2.4.3)$$

This problem gives the electrostatic potential within a semi-infinite slab of thickness a with a charge density xe^{-x}.

Because the domain is semi-infinite in the x-direction, we introduce the Laplace transform

$$U(s, y) = \int_0^\infty u(x, y) \, e^{-sx} \, dx. \qquad (2.4.4)$$

Thus, taking the Laplace transform of Equation 2.4.1, we have

$$\frac{d^2 U(s, y)}{dy^2} + s^2 U(s, y) - su(0, y) - u_x(0, y) = \frac{1}{(s+1)^2}. \qquad (2.4.5)$$

Although $u(0, y) = 0$, $u_x(0, y)$ is unknown and we denote its value by $f(y)$. Therefore, Equation 2.4.5 becomes

$$\frac{d^2 U(s, y)}{dy^2} + s^2 U(s, y) = f(y) + \frac{1}{(s+1)^2}, \qquad 0 < y < a, \qquad (2.4.6)$$

with $U(s, 0) = U'(s, a) = 0$.

To solve Equation 2.4.6, we first assume that we can rewrite $f(y)$ as the Fourier series

$$f(y) = \sum_{n=1}^{\infty} A_n \sin(m_n y), \qquad (2.4.7)$$

where

$$A_n = \frac{2}{a} \int_0^a f(y) \sin(m_n y) \, dy, \qquad (2.4.8)$$

and $m_n = (2n - 1)\pi/(2a)$. Furthermore,

$$\frac{1}{(s+1)^2} = \frac{2}{a(s+1)^2} \sum_{n=1}^{\infty} \frac{\sin(m_n y)}{m_n}. \qquad (2.4.9)$$

The reason behind this particular choice for our expansion will become clear shortly. Thus, Equation 2.4.6 may be rewritten as

$$\frac{d^2 U(s, y)}{dy^2} + s^2 U(s, y) = \sum_{n=1}^{\infty} \left[A_n + \frac{2}{a m_n (s+1)^2} \right] \sin(m_n y). \qquad (2.4.10)$$

The form of the right side of Equation 2.4.10 suggests that we seek solutions of the form

$$U(s,y) = \sum_{n=1}^{\infty} B_n \sin(m_n y), \qquad 0 < y < a. \tag{2.4.11}$$

We now understand why we rewrote the right side of Equation 2.4.6 as a Fourier series; the solution $U(s,y)$ automatically satisfies the boundary condition $U(s,0) = U'(s,a) = 0$. Substituting Equation 2.4.11 into Equation 2.4.10, we find that

$$\sum_{n=1}^{\infty} \left(s^2 - m_n^2\right) B_n \sin(m_n y) = \sum_{n=1}^{\infty} \left[A_n + \frac{2}{am_n(s+1)^2}\right] \frac{\sin(m_n y)}{s^2 - m_n^2}. \tag{2.4.12}$$

Because Equation 2.4.12 must be true for any x,

$$B_n = \frac{\{A_n + 2/[am_n(s+1)^2]\}}{s^2 - m_n^2} \tag{2.4.13}$$

and

$$U(s,y) = \sum_{n=1}^{\infty} \left[A_n + \frac{2}{am_n(s+1)^2}\right] \frac{\sin(m_n y)}{s^2 - m_n^2}. \tag{2.4.14}$$

We have not yet determined A_n. Note, however, that in order for the inverse of Equation 2.4.14 *not* to grow as $e^{m_n x}$, the numerator must vanish when $s = m_n$ and $s = m_n$ is a removable pole. Thus, $A_n = -2/[am_n(1+m_n)^2]$ and

$$U(s,y) = -\frac{2}{a} \sum_{n=1}^{\infty} \frac{(s + m_n + 2)\sin(m_n y)}{m_n(1+m_n)^2(s+1)^2(s+m_n)}, \qquad 0 < y < a. \tag{2.4.15}$$

The inverse of $U(s,y)$ then follows directly from partial fraction and equals

$$u(x,y) = \frac{4}{a} \sum_{n=1}^{\infty} \frac{\sin(m_n y)}{m_n(1 - m_n^2)^2} \left(e^{-x} - e^{-m_n x}\right) + \frac{2x}{a} e^{-x} \sum_{n=1}^{\infty} \frac{\sin(m_n y)}{m_n(1 - m_n^2)}. \tag{2.4.16}$$

Figure 2.4.1 illustrates Equation 2.4.16 when $a = 4$.

• Example 2.4.2

Let us now solve a similar problem to the previous one but in cylindrical coordinates. Here,

$$\frac{1}{r}\frac{\partial}{\partial r}\left(r\frac{\partial u}{\partial r}\right) + \frac{\partial^2 u}{\partial z^2} = \frac{2}{b}n(z)\delta(r - b), \qquad 0 \leq r < a, \quad 0 < z < \infty, \tag{2.4.17}$$

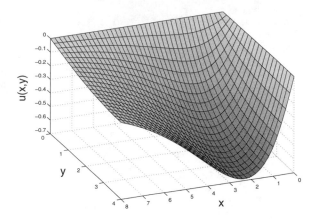

Figure 2.4.1: A plot of Equation 2.4.16 as a function of x and y when $a = 4$.

subject to the boundary conditions

$$u(r, 0) = 0, \qquad \lim_{z \to \infty} |u(r, z)| < \infty, \qquad 0 \le r < a, \qquad (2.4.18)$$

and

$$u(a, z) = 0, \qquad 0 < z < \infty, \qquad (2.4.19)$$

where $0 < b < a$. This problem gives the electrostatic potential within a semi infinite cylinder of radius a that is grounded and has the charge density of $n(z)$ within an infinitesimally thin shell located at $r = b$.

Because the domain is semi-infinite in the z direction, we introduce the Laplace transform

$$U(r, s) = \int_0^\infty u(r, z) \, e^{-sz} \, dz. \qquad (2.4.20)$$

Thus, taking the Laplace transform of Equation 2.4.17, we have

$$\frac{1}{r} \frac{d}{dr} \left[r \frac{dU(r, s)}{dr} \right] + s^2 U(r, s) - su(r, 0) - u_z(r, 0) = \frac{2}{b} N(s) \delta(r - b). \quad (2.4.21)$$

Although $u(r, 0) = 0$, $u_z(r, 0)$ is unknown and we denote its value by $f(r)$. Therefore, Equation 2.4.21 becomes

$$\frac{1}{r} \frac{d}{dr} \left[r \frac{dU(r, s)}{dr} \right] + s^2 U(r, s) = f(r) + \frac{2}{b} N(s) \delta(r - b), \quad 0 \le r < a, \quad (2.4.22)$$

with $\lim_{r \to 0} |U(r, s)| < \infty$ and $U(a, s) = 0$.

To solve Equation 2.4.22, we first assume that we can rewrite $f(r)$ as the Fourier-Bessel series

$$f(r) = \sum_{n=1}^\infty A_n J_0(k_n r/a), \qquad (2.4.23)$$

where k_n is the nth root of the $J_0(k) = 0$, and

$$A_n = \frac{2}{a^2 J_1^2(k_n)} \int_0^a f(r) J_0(k_n r/a) r \, dr. \qquad (2.4.24)$$

Similarly, the expansion for the delta function is

$$\delta(r - b) = \frac{2b}{a^2} \sum_{n=1}^{\infty} \frac{J_0(k_n b/a) J_0(k_n r/a)}{J_1^2(k_n)}, \qquad (2.4.25)$$

because

$$\int_0^a \delta(r - b) J_0(k_n r/a) r \, dr = b J_0(k_n b/a). \qquad (2.4.26)$$

Thus, Equation 2.4.22 may be rewritten as

$$\frac{1}{r} \frac{d}{dr}\left[r \frac{dU(r, s)}{dr} \right] + s^2 U(r, s) = \frac{2}{a^2} \sum_{n=1}^{\infty} \frac{2N(s) J_0(k_n b/a) + a_k}{J_1^2(k_n)} J_0(k_n r/a),$$
$$(2.4.27)$$

where $a_k = \int_0^a f(r) J_0(k_n r/a) r \, dr$.

The form of the right side of Equation 2.4.22 suggests that we seek solutions of the form

$$U(r, s) = \sum_{n=1}^{\infty} B_n J_0(k_n r/a), \qquad 0 \le r < a. \qquad (2.4.28)$$

This Fourier-Bessel representation of $U(r, s)$ automatically satisfies the boundary condition $U(a, s) = 0$. Substituting Equation 2.4.28 into Equation 2.4.27, we find that

$$U(r, s) = \frac{2}{a^2} \sum_{n=1}^{\infty} \frac{2N(s) J_0(k_n b/a) + a_k}{(s^2 - k_n^2/a^2) J_1^2(k_n)} J_0(k_n r/a), \qquad 0 \le r < a. \quad (2.4.29)$$

We have not yet determined a_k. Note, however, that in order for the inverse of Equation 2.4.29 *not* to grow as $e^{k_n z/a}$, the numerator must vanish when $s = k_n/a$, and $s = k_n/a$ is a removable pole. Thus, $a_k = -2N(k_n/a) \times J_0(k_n b/a)$, and

$$U(r, s) = \frac{4}{a^2} \sum_{n=1}^{\infty} \frac{[N(s) - N(k_n/a)] J_0(k_n b/a)}{(s^2 - k_n^2/a^2) J_1^2(k_n)} J_0(k_n r/a), \qquad 0 \le r < a.$$
$$(2.4.30)$$

The inverse of $U(r, s)$ then follows directly from simple inversions, the convolution theorem and the definition of the Laplace transform. The complete solution is

$$u(r,z) = \frac{2}{a} \sum_{n=1}^{\infty} \frac{J_0(k_n b/a) J_0(k_n r/a)}{k_n J_1^2(k_n)}$$

$$\times \left[\int_0^z n(\tau) e^{k_n(z-\tau)/a} \, d\tau - \int_0^z n(\tau) e^{-k_n(z-\tau)/a} \, d\tau \right.$$

$$\left. - \int_0^{\infty} n(\tau) e^{-k_n \tau/a} e^{k_n z/a} \, d\tau + \int_0^{\infty} n(\tau) e^{-k_n \tau/a} e^{-k_n z/a} \, d\tau \right] \quad (\mathbf{2.4.31})$$

$$= \frac{2}{a} \sum_{n=1}^{\infty} \frac{J_0(k_n b/a) J_0(k_n r/a)}{k_n J_1^2(k_n)}$$

$$\times \left[\int_0^{\infty} n(\tau) e^{-k_n(z+\tau)/a} \, d\tau - \int_0^z n(\tau) e^{-k_n(z-\tau)/a} \, d\tau \right.$$

$$\left. - \int_z^{\infty} n(\tau) e^{-k_n(\tau-z)/a} \, d\tau \right]. \quad (\mathbf{2.4.32})$$

Problems

1. Use Laplace transforms to solve Laplace's equation

$$\frac{\partial^2 u}{\partial x^2} + \frac{\partial^2 u}{\partial y^2} = 0, \qquad 0 < x < \infty, \quad 0 < y < a,$$

subject to the boundary conditions $u(0,y) = 1$, $\lim_{x\to\infty} |u(x,y)| < \infty$ and $u(x,0) = u(x,a) = 0$.

Step 1: By taking the Laplace transform of the partial differential equation and boundary conditions, show that they reduce to the boundary-value problem

$$\frac{d^2 U(s,y)}{dy^2} + s^2 U(s,y) = s + f(y),$$

or

$$\frac{d^2 U(s,y)}{dy^2} + s^2 U(s,y) = \sum_{n=1}^{\infty} \left[\frac{4s}{(2n-1)\pi} + A_n \right] \sin\left[\frac{(2n-1)\pi y}{a} \right],$$

where $U(s,0) = U(s,a) = 0$, $u_x(0,y) = f(y)$ and

$$A_n = \frac{2}{a} \int_0^a f(y) \sin\left[\frac{(2n-1)\pi y}{a} \right] dy.$$

Step 2: Show that the solution to the boundary-value problem is

$$U(s,y) = \sum_{n=1}^{\infty} \frac{4s + (2n-1)\pi A_n}{(2n-1)\pi [s^2 - (2n-1)^2 \pi^2/a^2]} \sin\left[\frac{(2n-1)\pi y}{a} \right].$$

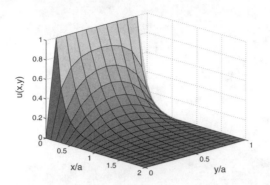

Problem 1

Step 3: Show that $s = (2n - 1)\pi/a$ cannot be a pole of $U(s, y)$ so that

$$U(s, y) = \sum_{n=1}^{\infty} \frac{4s}{(2n - 1)\pi[s + (2n - 1)\pi/a]} \sin\left[\frac{(2n - 1)\pi y}{a}\right].$$

Step 4: Invert $U(s, y)$ and show that

$$u(x, y) = \frac{4}{\pi} \sum_{n=1}^{\infty} \frac{1}{2n - 1} \exp\left[-\frac{(2n - 1)\pi x}{a}\right] \sin\left[\frac{(2n - 1)\pi y}{a}\right].$$

This solution is illustrated in the figure labeled Problem 1.

2. Use Laplace transforms to solve Laplace's equation

$$\frac{\partial^2 u}{\partial x^2} + \frac{\partial^2 u}{\partial y^2} = 0, \qquad 0 < x < \infty, \quad 0 < y < a,$$

subject to the boundary conditions $u_x(0, y) = ay$, $\lim_{x \to \infty} |u(x, y)| < \infty$, $0 < y < a$ and $u(x, 0) = u_y(x, a) = 0$, $0 < x < \infty$.

Step 1: By taking the Laplace transform of the partial differential equation and boundary conditions, show that they reduce to the boundary-value problem

$$\frac{d^2 U(s, y)}{dy^2} + s^2 U(s, y) = ay + sf(y),$$

or

$$\frac{d^2 U(s, y)}{dy^2} + s^2 U(s, y) = \sum_{n=1}^{\infty} \left[sA_n + \frac{2(-1)^n}{m_n^2} \right] \sin(m_n y),$$

where $U(s, 0) = U(s, a) = 0$, $m_n = (2n - 1)\pi/(2a)$, $u(0, y) = f(y)$ and

$$A_n = \frac{2}{a} \int_0^a f(y) \sin(m_n y) \, dy.$$

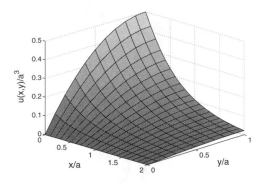

Problem 2

Step 2: Show that the solution to the boundary-value problem is

$$U(s,y) = \sum_{n=1}^{\infty} \frac{sm_n^2 A_n + 2(-1)^n}{m_n^2(s^2 - m_n^2)} \sin(m_n y).$$

Step 3: Show that $s = m_n$ cannot be a pole of $U(s,y)$ so that

$$U(s,y) = -2\sum_{n=1}^{\infty} \frac{(-1)^n}{m_n^3(s+m_n)} \sin(m_n y).$$

Step 4: Invert $U(s,y)$ and show that

$$u(x,y) = -2\sum_{n=1}^{\infty}(-1)^n \frac{\sin(m_n y)}{m_n^3} e^{-m_n x}.$$

This solution is illustrated in the figure labeled Problem 2.

3. Use Laplace transforms to solve Laplace's equation

$$\frac{\partial^2 u}{\partial x^2} + \frac{\partial^2 u}{\partial y^2} = H(x - \pi), \qquad 0 < x < \infty, \quad 0 < y < a,$$

subject to the boundary conditions $u(0,y) = 0$, $\lim_{x\to\infty} |u(x,y)| < \infty$, $0 < y < a$ and $u(x,0) = u(x,a) = 0$, $0 < x < \infty$.

Step 1: By taking the Laplace transform of the partial differential equation and boundary conditions, show that they reduce to the boundary-value problem

$$\frac{d^2 U(s,y)}{dy^2} + s^2 U(s,y) = \frac{e^{-s\pi}}{s} + f(y),$$

or

$$\frac{d^2 U(s,y)}{dy^2} + s^2 U(s,y) = \sum_{n=1}^{\infty}\left[A_n + \frac{4\,e^{-s\pi}}{asm_n}\right]\sin(m_n y),$$

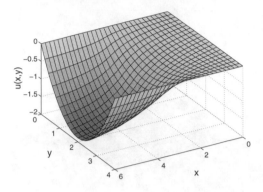

Problem 3

where $U(s,0) = U(s,a) = 0$, $m_n = (2n-1)\pi/a$, $u_x(0,y) = f(y)$ and

$$A_n = \frac{2}{a} \int_0^a f(y) \sin(m_n y)\, dy.$$

Step 2: Show that the solution to the boundary-value problem is

$$U(s,y) = \sum_{n=1}^{\infty} \left[A_n + \frac{4\,e^{-s\pi}}{asm_n} \right] \frac{\sin(m_n y)}{s^2 - m_n^2}.$$

Step 3: Show that $s = m_n$ cannot be a pole of $U(s,y)$ so that

$$U(s,y) = \frac{2}{a} \sum_{n=1}^{\infty} \left[\frac{e^{-s\pi}}{m_n^3(s - m_m)} + \frac{e^{-s\pi}}{m_n^3(s + m_m)} - 2\frac{e^{-s\pi}}{m_n^3 s} \right] \sin(m_n y)$$

$$- \frac{2}{a} \sum_{n=1}^{\infty} \left[\frac{e^{-m_n \pi}}{m_n^3(s - m_m)} - \frac{e^{-m_n \pi}}{m_n^3(s + m_m)} \right] \sin(m_n y).$$

Step 4: Invert $U(s,y)$ and show that

$$u(x,y) = \frac{2\,H(x-\pi)}{a} \sum_{n=1}^{\infty} \left[e^{m_n(x-\pi)} + e^{-m_n(x-\pi)} - 2 \right] \frac{\sin(m_n y)}{m_n^3}$$

$$- \frac{2}{a} \sum_{n=1}^{\infty} \left[e^{m_n(x-\pi)} - e^{-m_n(x+\pi)} \right] \frac{\sin(m_n y)}{m_n^3}.$$

This solution is illustrated in the figure labeled Problem 3 when $a = 4$.

Papers Using Laplace Transforms
to Solve Partial Differential Equations

Aquifers, Reservoirs and Porous Media

Abdel-Salam, A., and C. V. Chrysikopoulos, 1994: Analytical solutions for one-dimensional colloid transport in saturated fractures. *Adv. Water Resour.*, **17**, 283–296.

Ansari, M. A. A., 1978: Longitudinal dispersion in saturated porous media. *Indian J. Pure Appl. Math.*, **9**, 436–446.

Ansari, M. A. A., 1978: Longitudinal dispersion in an isotropic porous media flow. *Indian J. Pure Appl. Math.*, **9**, 588–599.

Batu, V., 1982: Time-dependent, linearized two-dimensional infiltration and evaporation from nonuniform and nonperiodic strip sources. *Water Resour. Res.*, **18**, 1725–1733.

Batu, V., 1983: Time-dependent linearized two-dimensional analytical infiltration and evaporation from nonuniform and periodic strip sources. *Water Resour. Res.*, **19**, 1523–1529.

Batu, V., 1989: A generalized two-dimensional analytical solution for hydrodynamic dispersion in bounded media with the first-type boundary condition at the source. *Water Resour. Res.*, **25**, 1125–1132.

Batu, V., and M. T. van Genuchten, 1990: First- and third-type boundary conditions in two-dimensional solute transport modeling. *Water Resour. Res.*, **26**, 339–350.

Bödvarsson, G. S., and C. F. Tsang, 1982: Injection and thermal breakthrough in fractured geothermal reservoirs. *J. Geophys. Res.*, **87**, 1031–1048.

Bödvarsson, G. S., S. M. Benson, and P. A. Witherspoon, 1982: Theory of the development of geothermal systems charged by vertical faults. *J. Geophys. Res.*, **87**, 9317–9328.

Bremer, R. E., H. Winston, and S. Vela, 1985: Analytical model for vertical interference tests across low-permeability zones. *Soc. Pet. Eng. J.*, **25**, 407–418.

Chen, C.-S., 1985: Analytical and approximate solutions to radial dispersion from an injection well to a geological unit with simultaneous diffusion into adjacent strata. *Water Resour. Res.*, **21**, 1069–1076.

Chen, C.-S., 1986: Solutions for radionuclide transport from an injection well into a single fracture in a porous formation. *Water Resour. Res.*, **22**, 508–518.

Clegg, M. W., and M. Mills, 1969: A study of the behavior of partially penetrating wells. *Soc. Pet. Eng. J.*, **9**, 189–203.

Coats, K. H., and B. D. Smith, 1964: Dead-end pore volume and dispersion in porous medium. *Soc. Pet. Eng. J.*, **4**, 73–84.

Elrick, D. E., 1961: Transient two-phase capillary flow in porous media. *Phys. Fluids*, **4**, 572–575.

Everdingen, A. F. van, and W. Hurst, 1949: The application of the Laplace transform to flow problems in reservoirs. *Trans. Am. Inst. Min. Metall. Pet. Eng.*, **186**, 305–324.

Fellah, Z. E. A., M. Fellah, W. Lauriks, C. Depollier, J.-Y. Chapelon, and Y. C. Angel, 2003: Solution in time domain of ultrasonic propagation equation in a porous material. *Wave Motion*, **38**, 151–163.

Grader, A. S., and R. N. Horne, 1988: Interference testing: Detecting a circular impermeable or compressible subregion. *SPE Form. Eval.*, **3**, 420–428.

Hantush, M. S., 1960: Modification of the theory of leaky aquifers. *J. Geophys. Res.*, **65**, 3713–3725.

Kazemi, H., M. S. Seth, and G. W. Thomas, 1969: The interpretation of interference tests in naturally fractured reservoirs with uniform fracture distribution. *Soc. Pet. Eng. J.*, **9**, 463–472.

Kumar, R., and A. Miglani, 1997: Radial displacements of an infinite liquid saturated porous medium with spherical cavity. *Proc. Indian Acad. Sci., Math. Sci.*, **107**, 57–70.

Lee, S.-T., and J. R. Brockenbrough, 1986: A new approximate analytic solution for finite-conductivity vertical fractures. *SPE Form. Eval.*, **1**, 75–88.

Lefkovits, H. C., P. Hazebroek, E. E. Allen, and C. S. Mathews, 1961: A study of the behavior of bounded reservoirs composed of stratified layers. *Soc. Pet. Eng. J.*, **1**, 43–58.

Lindstrom, F. T., and L. Boersma, 1989: Analytical solutions for convective-dispersive transport in confined aquifers with different initial and boundary conditions. *Water Resour. Res.*, **25**, 241–256.

Liu, C.-X., and W. P. Ball, 1998: Analytical modeling of diffusion-limited contamination and decontamination in a two-layer porous medium. *Adv. Water Resour.*, **21**, 297–313.

Luthin, J. N., and J. W. Holmer, 1960: An analysis of the flow of water in a shallow, linear aquifer, and of the approach to a new equilibrium after intake. *J. Geophys. Res.*, **65**, 1573–1576.

Marino, M. A., 1967: Hele-Shaw model study of the growth and decay of groundwater ridges. *J. Geophys. Res.*, **72**, 1195–1205.

Mason, D. P., A. Solomon, and L. O. Nicolaysen, 1991: Evolution of stress and strain during the consolidation of a fluid-saturated porous elastic sphere. *J. Appl. Phys.*, **70**, 4724–4740.

Mohammad, H. K., T. Oroveanu, and A. Stan, 1981: Temperature distribution in an oil layer due to hot water injection. *Rev. Roum. Sci. Tech., Ser. Mec. Appl.*, **26**, 673–685.

Mustafa, S., 1984: Unsteady flow to a ditch from a semi-confined and leaky aquifer. *Adv. Water Resour.*, **7**, 81–84.

Oguztoreli, M., and S. M. Faroug Ali, 1986: A mathematical model for the solvent leaching of tar sand. *SPE Reservior Eng.*, **1**, 545–555.

Pascal, H., 1973: Non-steady multiphase flow through a porous medium. *Rev. Roum. Sci. Tech., Ser. Mec. Appl.*, **18**, 329–343.

Polubarinova-Kochina, P. Ia., 1959: Ground water movements at water level fluctuations in a reservoir with a vertical boundary. *J. Appl. Math. Mech.*, **23**, 762–769.

Prats, M., P. Hazebroek, and W. R. Strickler, 1962: Effect of vertical fractures on reservoir behavior – compressible-fluid cases. *Soc. Pet. Eng. J.*, **2**, 87–94.

Roshal', A. A., 1969: Mass transfer in a two-layer porous medium. *J. Appl. Mech. Tech. Phys.*, **10**, 551–558.

Sacheti, N. C., and B. S. Bhatt, 1984: Stokes and Rayleigh layers in presence of naturally permeable boundaries. *J. Eng. Mech. Div. (Am. Soc. Civ. Eng.)*, **110**, 713–722.

Srivastava, R., and T.-C. J. Yeh, 1991: Analytical solutions for one-dimensional, transient infiltration toward the water table in homogeneous and layered soils. *Water Resour. Res.*, **27**, 753–762.

Temeng, K. O., and R. N. Horne, 1984: Pressure distributions in eccentric circular systems. *Soc. Pet. Eng. J.*, **24**, 677–684.

Wang, Y.-L., and E. Papamichos, 1999: Thermal effects on fluid flow and hydraulic fracturing from wellbores and cavities in low-permeability formations. *Int. J. Numer. Anal. Meth. Geomech.*, **23**, 1819–1834.

Diffusion

Agmon, N., 1984: Diffusion with back reaction. *J. Chem. Phys.*, **81**, 2811–2817.

Aksel'rud, G. A., 1959: Theory of the diffusion extraction of substances from porous bodies. I. Equations for the extraction kinetics. *Russ. J. Phys. Chem.*, **33**, 406–410.

Bochever, F. M., A. E. Oradovskaya, and V. I. Pagurova, 1966: Convective diffusion of salts in a radial flow of groundwater. *J. Appl. Mech. Tech. Phys.*, **7**, **No. 2**, 87–88.

Brenner, H., 1962: The diffusion model of longitudinal mixing in beds of finite length. Numerical values. *Chem. Eng. Sci.*, **17**, 229–243.

Chun, D. H., 1970: Distribution of concentration in flow through a circular pipe. *Int. J. Heat Mass Transfer*, **13**, 717–723.

Chun, K. R., 1972: Evaporation from a semi-infinite region with a nonvolatile solute. *J. Heat Transfer*, **94**, 238–240.

Cooper, R. S., 1965: Slow particle diffusion in ion exchange columns. *Indust. Eng. Chem. Fund.*, **4**, 308–313.

Crank, J., 1952: Simultaneous diffusion and reversible chemical reaction. *Philos. Mag., Ser.* 7, **43**, 811–826.

Deverall, L. I., 1958: Solution of the time-dependent thermal neutron diffusion equation. *Nucl. Sci. Eng.*, **4**, 495–498.

Gill, M. A., 1983: Diffusion model for aggrading channels. *J. Hydraul. Res.*, **21**, 355–367.

Gill, M. A., 1983: Diffusion model for degrading channels. *J. Hydraul. Res.*, **21**, 369–378.

Goldner, R. B., K. K. Wong, and T. E. Haas, 1992: One-dimensional diffusion into a multilayer structure: An exact solution for a bilayer. *J. Appl. Phys.*, **72**, 4674–4676.

Goldstein, S., 1932: Some two-dimensional diffusion problems with circular symmetry. *Proc. London Math. Soc., Ser. 2*, **34**, 51–88.

Hsieh, D.-Y., and M. S. Plesset, 1961: Theory of rectified diffusion of mass into gas bubbles. *J. Acoust. Soc. Am.*, **33**, 206–215.

Hsu, J. T., and J. S. Dranoff, 1986: On initial condition problems for reactor dispersion model. *Chem. Eng. Sci.*, **41**, 1930–1934.

Hurwitz, H., and L. Gierst, 1961: Théorie et applications de la méthode chronopotentiométrique avec courant imposé croissant proportionnellement à la racine carrée du temps. I. Introduction générale et cas des réactions de décharge directe. *J. Electranal. Chem.*, **2**, 128–141.

Hurwitz, H., 1961: Théorie et applications de la méthode chronopotentiométrique avec courant imposé croissant proportionnellement à la racine carrée du temps. II. Cas des réactions de décharge directe procédant en régime de diffusion sphérique et cylindrique. *J. Electranal. Chem.*, **2**, 142–151.

Hurwitz, H., 1961: Théorie et applications de la méthode chronopotentiométrique avec courant imposé croissant proportionnellement à la racine carrée du temps. III. Cas de plusieurs réactions de décharge directe successives et cas d'une réaction de décharge en plusieurs étapes consécutives. *J. Elcctranal. Chem.*, **2**, 328–339.

Kim, H., and K. J. Shin, 1999: Exact solution of the reversible diffusion-influenced reaction for an isolated pair in three dimensions. *Phys. Rev. Lett.*, **82**, 1578–1581.

Kim, H., and K. J. Shin, 2000: On the diffusion-influenced reversible trapping problem in one dimension. *J. Chem. Phys.*, **112**, 8312–8317.

Lozgachev, V. I., 1960: Isotope-exchange method for measuring saturated-vapor pressures and diffusion coefficients. II. The solution of equations for diffusion with exchange. *Russ. J. Phys. Chem.*, **34**, 144–149.

Malkovich, R. Sh., 2003: Diffusion with the formation and decay of immobile complexes in solids. *Tech. Phys. Lett.*, **29**, 422–425.

Mancy, K. H., D. A. Okun, and C. N. Reilley, 1962: A galvanic cell oxygen analyzer. *J. Electroanal. Chem.*, **4**, 65–92.

McKay, A. T., 1930: Diffusion into an infinite plane sheet subject to a surface condition, with a method of application to experimental data. *Proc. Phys. Soc.*, **42**, 547–555.

McKay, A. T., 1932: Diffusion for the infinite plane sheet. *Proc. Phys. Soc.*, **44**, 17–24.

Merchuk, J. C., Z. Tsur, and E. Horn, 1985: Oxygen transfer resistance as a criterion of blood ageing. *Chem. Eng. Sci.*, **40**, 1101–1107.

Neogi, P., 1983: Anomalous diffusion of vapors through solid polymers. Part II: Anomalous sorption. *AICHE J.*, **29**, 833–839.

Ogawa, S., and N. Shiono, 1995: Generalized diffusion-reaction model for the low-field charge-buildup instability at the Si-SiO$_2$ interface. *Phys. Rev. B*, **51**, 4218–4230.

O'Neill, K., 1982: One-dimensional transport from a highly concentrated, transfer type source. *Int. J. Heat Mass Transfer*, **25**, 27–36.

Rangarajan, S. K., and K. S. G. Doss, 1963: Faradaic admittance, a diffusion model. II. *J. Electroanal. Chem*, **5**, 114–123.

Rangarajan, S. K., 1963: Migration in the diffuse layer studies on the steady state. *J. Electroanal. Chem*, **5**, 350–361.

Scott, E. J., L. H. Tung, and H. G. Drickamer, 1951: Diffusion through an interface. *J. Chem. Phys.*, **19**, 1075–1078.

Shair, F. H., and D. S. Cohen, 1969: Transient ordinary and forced diffusion in a tube connecting stirred-tank end bulbs of finite size. *Chem. Eng. Sci.*, **24**, 39–48.

Shibata, T., and M. Kugo, 1983: Generalization and application of Laplace transformation formulas for diffusion. *Int. J. Heat Mass Transfer*, **26**, 1017–1027.

Singh, K. D., and S. N. Dube, 1986: Exact solution of the mass transfer equation for time variation of inlet concentration in a circular pipe. *J. Math. Phys. Sci.*, **20**, 335–342.

Stevenson, J. F., 1974: Unsteady mass transfer in a long composite cylinder with interfacial resistances. *AICHE J.*, **20**, 461–466.

Sun, L. M., and F. Meunier, 1987: A detailed model for nonisothermal sorption in porous adsorbents. *Chem. Eng. Sci.*, **42**, 1585–1593.

Varelas, C. G., D. G. Dixon, and C. A. Steiner, 1995: Mathematical model of mass transport through dispersed-phase polymer networks. *AIChE*, **41**, 805–811.

Vesala, T., 1993: On droplet evaporation in the presence of a condensing substance: The effect of internal diffusion. *Int. J. Heat Mass Transfer*, **36**, 695–703.

Wilson, J. E., 1954: Diffusion effects in the photochemistry of solid films. *J. Chem. Phys.*, **22**, 334–343.

Yu, K., and M. M. Klein, 1964: Diffusion of small particles in a nonuniform atmosphere. *Phys. Fluids*, **7**, 651–657.

Zhdanov, S. K., A. S. Chikhachev, and Yu. N. Yavlinskii, 1976: Diffusion boundary-value problem for regions with moving boundaries and conservation of particles. *Sov. Phys. Tech. Phys.*, **21**, 883–884.

Electromagnetism

Aleksin, V. F., and S. S. Romanov, 1973: Penetration of quasistationary magnetic fields into a conducting chamber. I. Uniform magnetic field. *Sov. Phys. Tech. Phys.*, **18**, 734–738.

Berger, H., and J. W. E. Griemsmann, 1968: Transient electromagnetic guided wave propagation in moving media. *IEEE Trans. Microwave Theory Tech.*, **MTT-16**, 842–849.

Bhattacharyya, B. K., 1957: Propagation of an electric pulse through a homogeneous and isotropic medium. *Geophysics*, **22**, 905–921.

Kapetanakos, C. A., P. Sprangle, D. P. Chernin, S. J. Marsh, and I. Haber, 1983: Equilibrium of a high-current electron ring in a modified-betatron accelerator. *Phys. Fluids*, **26**, 1634–1648.

Klevans, E. H., and S. P. Gary, 1980: Particle and energy end loss from a collisionless plasma. *Phys. Fluids*, **23**, 2219–2224.

Kuznetszov, P. E., 1947: The propagation of electromagnetic waves along a line (in Russian). *Prikl. Mat. Mek.*, **11**, 615–620.

Kuznetszov, P. E., 1948: The propagation of electromagnetic waves along two parallel, single conductor lines (in Russian). *Prikl. Mat. Mek.*, **12**, 141–148.

Maksiejewski, J. L., 1993: Thermal response of bimetallic conductors under surge currents taking the skin effect into account. *IEE Proc. A*, **140**, 338–344.

Maksiejewski, J. L., 1993: Losses in tubular cylindrical conductors due to current surges, taking skin effect into account. *IEE Proc. A*, **140**, 496–500.

Payne, W. T., and L. N. Zadoff, 1965: Electromagnetic diffusion into a moving conductor. *AIAA J.*, **3**, 1294–1297.

Raval, U., and K. N. N. Rao, 1973: Quasi-stationary electromagnetic response of covered permeable conductors to some pulses of spatially uniform magnetic field. *Pure Appl. Geophys.*, **104**, 553–565.

Schatz, E. R., and E. M. Williams, 1950: Pulse transients in exponential transmission lines. *Proc. IRE*, **38**, 1208–1212.

Verma, S. K., and R. N. Singh, 1970: Transient electromagnetic response of an inhomogeneous conducting sphere. *Geophysics*, **35**, 331–336.

Verma, S. K., 1972: Quasi-static time-domain electromagnetic response of a homogeneous conducting infinite cylinder. *Geophysics*, **37**, 92–97.

Verma, S. K., 1973: Time-dependent electromagnetic fields of an infinite, conducting cylinder excited by a long current-carrying cable. *Geophysics*, **38**, 369–379.

Wait, J. R., 1953: A transient magnetic dipole source in a dissipative medium. *J. Appl. Phys.*, **24**, 341–343.

Fluid Dynamics

Abu-Abdou, K., 1982: Decay of laminar flow inside a retarding rotating cylindrical vessel. *Acta Mech.*, **45**, 197–204.

Achard, J. L., and G. M. Lespinard, 1981: Structure of the transient wall-friction law in one-dimensional models of laminar pipe flows. *J. Fluid Mech.*, **113**, 283–298.

Agarwal, J. P., and S. K. Roy, 1977: On the effect of magnetic field on viscous lifting and drainage of conducting fluid. *Appl. Sci. Res.*, **33**, 141–149.

Annapurna, N., and G. Ramanaiah, 1976: A unified treatment of drainage, withdrawal, and postwithdrawal drainage with inertial effects. *AICHE J.*, **22**, 940–942.

Ansari, J. S., and R. Oldenburger, 1967: Propagation of disturbance in fluid lines. *J. Basic Eng.*, **89**, 415–422.

Atabek, H. B., 1964: Start-up flow of a Bingham plastic in a circular tube. *Z. Angew. Math. Mech.*, **44**, 332–333.

Baines, P. G., 1967: Forced oscillations of an enclosed rotating fluid. *J. Fluid Mech.*, **30**, 533–546.

Bandelli, R., K. R. Rajagopal, and G. P. Galdi, 1995: On some unsteady motions of fluids of second grade. *Arch. Mech.*, **47**, 661–676.

Bandelli, R., and K. R. Rajagopal, 1995: Start-up flows of second grade fluids in domains with one finite dimension. *Int. J. Nonlinear Mech.*, **30**, 817–839.

Baral, M. C., 1969: On the unsteady flow of conducting liquid between two parallel plates. *Proc. Natl. Inst. Sci. India*, **A35**, 422–427.

Bathaiah, D., 1978: Forced oscillations of an enclosed rotating fluid under a uniform magnetic field. *Indian J. Pure Appl. Math.*, **9**, 996–1003.

Bhattacharyya, P., 1962: Note on the flow of a viscous fluid standing on a rigid base due to a single periodic pulse of tangential force on the surface. *Indian J. Theoret. Phys.*, **10**, 73–76.

Bhattacharyya, P., 1966: Note on the flow of two incompressible immiscible viscous fluids due to the pulses of tangential force on the upper surface. *Indian J. Theoret. Phys.*, **14**, 45–58.

Binnie, A. M., 1951: The effect of friction on surges in long pipe-lines. *Quart. J. Mech. Appl. Math.*, **4**, 330–343.

Cerda, E. A., and E. L. Tirapegui, 1998: Faraday's instability in viscous fluid. *J. Fluid Mech.*, **368**, 195–228.

Chang, C. C., and H. B. Atabek, 1962: Flow between two co-axial tubes near the entry. *Z. Angew. Math. Mech.*, **42**, 425–430.

Chang, S. J., K. J. DeWitt, and J. G. Fikioris, 1973: Annular Couette flow of a suddenly pressurized viscoelastic fluid. *Ind. Eng. Chem. Fund.*, **12**, 31–33.

Chu, C., 1972: A note on "cocurrent" oblique crossflow transfer. *Chem. Eng. Sci.*, **27**, 1173–1175.

Cooper, F., and P. Blewett, 1978: Pressure generation due to a temperature discontinuity at a liquid-liquid plane interface. *Phys. Fluids*, **21**, 334 346.

Crossley, A. F., 1928: Operational solution of some problems in viscous fluid motion. *Proc. Cambridge Philos. Soc.*, **24**, 231–235.

Das, D., and J. H. Arakeri, 2000: Unsteady laminar duct flow with a given volume flow rate variation. *J. Appl. Mech.*, **67**, 274–281.

Datta, N., and S. K. Mishra, 1988: Couette flow of a dusty fluid in a rotating frame of reference. *J. Math. Phys. Sci.*, **22**, 421–430.

Datta, N., and D. C. Dalal, 1992: Unsteady flow of a dusty fluid through a circular pipe with impulsive pressure gradient. *Acta Mech.*, **95**, 51–57.

Datta, S., 1958: A note on the motion of viscous fluid subjected to uniform or periodic body force acting for a finite time. *J. Technol.*, **3**, 73–79.

Datta, S. K., 1963: Flow formation in Couette motion of an elastico-viscous Maxwell fluid in the presence of a transverse magnetic field. *J. Phys. Soc. Japan*, **18**, 1667–1671.

Debnath, L., 1973: On Ekman and Hartmann boundary layers in a rotating fluid. *Acta Mech.*, **18**, 333–341.

Devanathan, R., and A. R. Rao, 1973: Forced oscillations of a contained rotating stratified fluid. *Z. Angew. Math. Mech.*, **53**, 617–623.

Dube, S. K., 1970: Temperature distribution in Poiseuille flow between two parallel flat plates. *Indian J. Pure Appl. Math.*, **1**, 277–283.

Dube, S. N., 1967: On the temperature distribution of a viscous incompressible fluid flowing between two parallel flat plates, one in uniform motion and the other at rest with unsteady rate of heat generation. *J. Technol.*, **12**, 91–100.

Dube, S. N., 1972: On the flow of Rivlin-Ericksen fluids in a channel bounded by two parallel flat plates. *Indian J. Pure Appl. Math.*, **3**, 396–401.

Dube, S. N., and J. Singh, 1972: Unsteady flow of a dusty fluid between two parallel flat plates. *Indian J. Pure Appl. Math.*, **3**, 1175–1182.

Dube, S. N., and C. L. Sharma, 1976: Exact solution of the transient forced convection energy equation for timewise variation of inlet temperature in a circular pipe. *Indian J. Pure Appl. Math.*, **7**, 610–615.

Dutta, S., 1972: The effects of Hall current on unsteady slip flow over a flat plate under transverse magnetic field. *Arch. Mech.*, **24**, 269–276.

Dźygadło, Z., 1968: Pressure on a cylindrical shell performing unsteady oscillation in external or internal linearized supersonic flow. *Proc. Vib. Prob.*, **9**, 129–146.

Ehlers, F. E., 1955: The lift and moment on a ring concentric to a cylindrical body in supersonic flow. *J. Aeronaut. Sci.*, **22**, 239–248.

Erdoğan, M. E., 1997: Unsteady flow of a viscous fluid over a plane wall. *Z. Angew. Math. Mech.*, **77**, 733–740.

Francisco, S. F., J. W. Taunton, and E. N. Lightfoot, 1970: Transient creeping flow around spheres. *AIChE J.*, **16**, 386–391.

Fukumoto, Y., 1990: General unsteady circulatory flow outside a porous circular cylinder with suction or injection. *J. Phys. Soc. Japan*, **59**, 918–926.

Garg, P. C., 1978: Flow of a conducting dusty gas past an accelerated plate. *Bull. Calcutta Math. Soc.*, **70**, 265–270.

Gerbes, W., 1951: Zur instationären, laminaren Strömung einer inkompressiblen, zähen Flüssigkeit in kreiszylindrischen Rohren. *Z. Angew. Phys.*, **3**, 267–271.

Gershunov, E. M., 1972: Hydrodynamic pressure of a liquid on a shell with hydraulic shock. *Sov. Appl. Mech.*, **8**, 715–721.

Gershunov, E. M., 1975: Hydrodynamic pressure of a liquid against a shell in the case of hydraulic shock. *Sov. Appl. Mech.*, **11**, 405–409.

Ghosh, A. K., 1967: Flow of a viscous liquid between two coaxial circular porous cylinders due to longitudinal motion of the inner cylinder. *Indian J. Pure Appl. Phys.*, **5**, 179–181.

Ghosh, A. K., 1968: Note on the temperature distribution in a viscous fluid flowing through a channel bounded by two coaxial pipes. *Rev. Roum. Sci. Tech., Ser. Mec. Appl.*, **13**, 1073–1084.

Gopinath, M., and L. Debnath, 1973: On the growth of unsteady boundary layers on porous flat plates. *Pure Appl. Geophys.*, **109**, 1810–1818.

Gopinath, M., and L. Debnath, 1973: On unsteady motion of a rotating fluid bounded by flat porous plates. *Pure Appl. Geophys.*, **110**, 1996–2004.

Guha, D. K., 1982: Flow of a dusty viscous gas through a circular tube with pressure gradient of any function of time. *Bull. Calcutta Math. Soc.*, **74**, 333–338.

Gupta, M., R. Kant, and H. S. Sharma, 1979: Unsteady Hele-Shaw flow of non-Newtonian fluid. *J. Math. Phys. Sci.*, **13**, 189–197.

Gupta, M., 1980: Unsteady flow of dusty viscous fluid through equilateral triangular ducts with pressure gradient as any function of time. *Indian J. Theoret. Phys.*, **28**, 151–159.

Gupta, M., and S. L. Khandpur, 1980: Unsteady flow of non-Newtonian fluid through elliptical long ducts with pressure gradient as any function of time. *Indian J. Theoret. Phys.*, **28**, 319–327.

Gupta, M., and N. K. Varshneya, 1981: Unsteady flow of non-Newtonian fluid through equilateral triangular ducts with pressure gradient as any function of time. *Indian J. Theoret. Phys.*, **29**, 57–64.

Gupta, M., and H. S. Sharma, 1981: Unsteady flow of a dusty viscous fluid through confocal elliptical ducts. *Indian J. Theoret. Phys.*, **29**, 227–237.

Guram, G. S., 1983: Flow of a dipolar fluid due to suddenly accelerated flat plate. *Acta Mech.*, **49**, 133–138.

Gurchenkov, A. A., 2001: Steady motion of a viscous liquid between rotating parallel walls in the presense of a crossflow. *J. Appl. Mech. Tech. Phys.*, **42**, 603–606.

Hall, M., and L. Debnath, 1973: Some exact solutions of unsteady boundary layer equations – I. *Pure Appl. Geophys.*, **102**, 167–174.

Hanin, M., 1957: Propagation of an aperiodic wave in a compressible viscous medium. *J. Math. Phys. (Cambridge, MA)*, **36**, 234–249.

Hasimoto, H., 1956: Note on Rayleigh's problem for a circular cylinder with uniform suction and related flow problem. *J. Phys. Soc. Japan*, **11**, 611–612, 721.

Hershey, D., and G. Song, 1967: Friction factors and pressure drop for sinusoidal laminar flow of water and blood in rigid tubes. *AICHE J.*, **13**, 491–496.

Howarth, L, 1949: Rayleigh's problem for a semi-infinite plate. *Proc. Cambridge Philos. Soc.*, **45**, 127–140.

Jacobs, C., 1972: Transient and steady state vorticity generated by horizontal temperature gradients. *Quart. J. Mech. Appl. Math.*, **25**, 303–318.

Jaeger, J. C., 1945: On thermal stresses in circular cylinders. *Philos. Mag.*, Ser. 7, **36**, 418–428.

Jahagirdar, M. D., and R. M. Lahurikar, 1989: Transient forced and free convection flow past an infinite vertical plate. *Indian J. Pure Appl. Math.*, **20**, 711–715.

Jayasinghe, D. A. P., and H. J. Leutheusser, 1977: Pulsatile waterhammer subject to laminar friction. *J. Basic Eng.*, **94**, 467–472.

Jha, P. K., 1980: Unsteady flow of a dusty viscous fluid through a long duct whose cross-section is a cardioid. *Indian J. Theoret. Phys.*, **28**, 89–94.

Johri, A. K., 1978: Unsteady channel flow of an elastico-viscous liquid. *Indian J. Pure Appl. Math.*, **9**, 481–489.

Kantola, R., 1971: Transient response of fluid lines including frequency modulated inputs. *J. Basic Eng.*, **93**, 274–281.

Khare, Km. S., 1980: Flow of perfect gas over porous flat plate. *Indian J. Theoret. Phys.*, **28**, 191–195.

Kurosaka, M., 1974: On the unsteady supersonic cascade with a subsonic leading edge – An exact first order theory – Part 1. *J. Eng. Power*, **96**, 13–22.

Lahiri, S., and S. K. Dhar, 1984: On the motion of conducting liquid down an inclined plane. *Indian J. Theoret. Phys.*, **32**, 255–260.

Lal, G., P. C. Gupta, and R. G. Sharma, 1982: Laminar flow of a viscous fluid between two parallel plates, the upper plate being laid with a charge density and the lower plate moving with a constant velocity. *Indian J. Theoret. Phys.*, **30**, 271–278.

Majumder, S. R., 1962: Impulsive rotatory motion of a sphere in a viscous fluid. *Rev. Roum. Sci. Tech., Ser. Mec. Appl.*, **7**, 893–897.

Marcus, D. L., and S. A. Berger, 1989: The interaction between a counter-rotating vortex pair in vertical ascent and a free surface. *Phys. Fluids A*, **1**, 1988–2000.

Mithal, K. G., 1960: Unsteady flow of a viscous homogeneous incompressible fluid in a circular pipe of uniform cross-section. *Bull. Calcutta Math. Soc.*, **52**, 147–154.

Mitra, P., and P. Bhattacharyya, 1982: On the unsteady flow of a dusty gas between two parallel plates, one being at rest and the other oscillating. *Rev. Roum. Sci. Tech., Ser. Mec. Appl.*, **27**, 57–68.

Mitra, P., 1984: Note on the problem of unsteady dusty viscous fluid flow past a flat plate. *Bull. Calcutta Math. Soc.*, **76**, 162–166.

Morgenthaler, G. W., and H. Reismann, 1963: Temperature in a rotating cylinder in a high-speed gas flow. *Astronaut. Acta*, **9**, 351–365.

Mukherjee, S., and S. Mukherjee, 1983: Unsteady axisymmetric rotational flow of elastico-viscous liquid. *Indian J. Pure Appl. Math.*, **14**, 1534–1541.

Nag, S. K., R. N. Jana, and N. Datta, 1979: Couette flow of a dusty gas. *Acta Mech.*, **33**, 179–187.

Naidu, K. B., 1974: Stratified viscous flow between two oscillating cylinders. *Indian J. Pure Appl. Math.*, **5**, 1127–1136.

Nandy, S. C., 1972: On the unsteady flow of fluid between two parallel plates acted upon by an electric field. *Indian J. Pure Appl. Math.*, **3**, 890–895.

Ogibalov, P. M., 1941: On the spread of plastico-viscous flow about a rotating cylinder (in Russian). *Prikl. Mat. Mek.*, **5**, 13–29.

Pack, D. C., 1956: The oscillations of a supersonic gas jet embedded in a supersonic stream. *J. Aeronaut. Sci.*, **23**, 747–753.

Pal, S. K., and P. R. Sengupta, 1986: On the motion of a visco-elastic Maxwell fluid subjected to uniform or periodic body force acting for a finite times. *Indian J. Theoret. Phys.*, **34**, 349–358.

Park, J. H., P. Bahukudumbi, and A. Beskok, 2004: Rarefaction effects on shear driven oscillatory gas flows: A direct simulation Monte Carlo study in the entire Knudsen regime. *Phys. Fluids*, **16**, 317–330.

Pascal, H., 1973: Nonsteady gas flow in a pipeline network. *Rev. Roum. Sci. Tech., Ser. Mec. Appl.*, **18**, 491–510.

Pascal, H., 1973: Nonsteady gas flow in interconnected pipelines. *Rev. Roum. Sci. Tech., Ser. Mec. Appl.*, **18**, 851–871.

Pascal, H., 1982: Nonsteady gas flow through pipeline systems. *Acta Mech.*, **42**, 49–69.

Prakash, S., 1966: Unsteady viscous flow past a flat plate with suction. *Proc. Natl. Inst. Sci. India*, **A32**, 481–485.

Prakash, S., 1967: Non-steady parallel viscous flow through a straight channel. *Bull. Calcutta Math. Soc.*, **59**, 55–59.

Prakash, S., 1969: An exact solution of the problem of unsteady viscous flow through a porous straight channel. *Proc. Natl. Inst. Sci. India, Ser. A*, **35, Suppl. 2**, 123–129.

Prakash, S., 1971: Note on the problem of unsteady viscous flow past a flat plate. *Indian J. Pure Appl. Math.*, **2**, 283–289.

Puri, P., and P. M. Jordan, 1999: Stokes's first problem for a dipolar fluid with nonclassical heat conduction. *J. Eng. Math.*, **36**, 219–240.

Purohit, G. N., 1967: Temperature distribution in Covette flow between two parallel flat plates. *Proc. Natl. Sci. India*, **A33**, 142–149.

Ramamurthy, V., 1990: Free convection effects on the Stokes problem for an infinite vertical plate in a dusty fluid. *J. Math. Phys. Sci.*, **24**, 297–312.

Rao, P. B., 1967: Motion of a viscous fluid through a tube subjected to a series of longitudinal pulses. *Indian J. Pure Appl. Phys.*, **5**, 1–5.

Rich, G. R., 1945: Water-hammer analysis by the Laplace-Mellin transformation. *Trans. ASME*, **67**, 361–376.

Sacheti, N. C., and B. S. Bhatt, 1975: Unsteady motion of a second order fluid between parallel plates. *Indian J. Pure Appl. Math.*, **6**, 996–1006.

Sacheti, N. C., and B. S. Bhatt, 1984: Stokes and Rayleigh layers in presence of naturally permeable boundaries. *J. Eng. Mech. Div. (Am. Soc. Civ. Eng.)*, **110**, 713–722.

Schetz, J. A., and R. Eichhorn, 1962: Unsteady natural convection in the vicinity of a doubly infinite vertical plate. *J. Heat Transfer*, **84**, 334–338.

Schwarz, W. H., 1963: The unsteady motion of an infinite oscillating cylinder in an incompressible Newtonian fluid at rest. *Appl. Sci. Res., Ser. A*, **11**, 115–124.

Sen, S. K., 1983: On the unsteady flow of conducting liquid between two co-axial circular cylinders. *Indian J. Theoret. Phys.*, **31**, 225–232.

Sharma, H. S., 1972: Unsteady flow of viscoelastic fluids through circular and coaxial circular ducts with pressure gradients as any function of time. *Indian J. Pure Appl. Math.*, **3**, 535–542.

Sharma, R. S., 1975: Flow over an oscillating porous plate. *Arch. Mech.*, **27**, 115–123.

Sharma, R. S., 1979: Unsteady flow of an elastic-viscous fluid past an infinite porous plate. *Arch. Mech.*, **31**, 199–208.

Singh, J., and S. N. Dube, 1975: Unsteady flow of a dusty fluid through a circular pipe. *Indian J. Pure Appl. Math.*, **6**, 69–79.

Sinha, K. S., and V. Gupta, 1979: Slow motion of a spheroid on a rotating fluid. *Indian J. Pure Appl. Math.*, **10**, 1183–1195.

Srinivasan, V., and D. Bathoiath, 1978: The flow of a conducting viscous incompressible fluid between two parallel plates under a uniform transverse magnetic field. *Indian J. Pure Appl. Math.*, **9**, 511–517.

Stewartson, K., 1953: A weak spherical source in a rotating fluid. *Quart. J. Mech. Appl. Math.*, **6**, 45–49.

Subba, R., and R. Gorla, 1982: Unsteady heat transfer in laminar non-Newtonian boundary layer over a wedge. *AICHE J.*, **28**, 56–60.

Sucec, J., 1986: Transient heat transfer in the laminar thermal entry region of a pipe: An analytical solution. *Appl. Sci. Res.*, **43**, 115–125.

Tanahashi, T., T. Sawada, K. Shizawa, and T. Ando, 1985: Distorted pressure history due to the step responses in a linear tapered pipe (in Japanesse). *Nihon Kikai Gakkai Rombunshu (Trans. Japan Soc. Mech. Eng.), Ser. B*, **51**, 2313–2319.

Travelho, J. S., and W. F. N. Santos, 1991: Solution for transient conjugated forced convection in the thermal entrance region of a duct with periodically varying inlet temperature. *J. Heat Transfer*, **113**, 558–562.

Ungarish, M., 1989: Side wall effects in centrifugal separation of mixtures. *Phys. Fluids, Ser. A*, **1**, 810–818.

Wang, Y., and E. Papamichos, 1999: Thermal effects on fluid flow and hydraulic fracturing from wellbores and cavities in low-permeability formations. *Int. J. Numer. Anal. Meth. Geomech.*, **23**, 1819–1834.

Washio, S., and T. Konishi, 1983: Theoretical study of the end correction problem (in Japanese). *Nihon Kikai Gakkai Rombunshu (Trans. Japan Soc. Mech. Eng.), Ser. B*, **49**, 1162–1168.

Washio, S., and T. Konishi, 1984: Theoretical investigation on end correction problems (Part IV, Transient flow analysis by multi-mode wave equation). *Bull. JSME*, **27**, 196–203.

Wood, F. M., 1937: The application of Heaviside's operational calculus to the solution of problems in water hammer. *Trans. ASME*, **59**, 707–713.

LIVERPOOL JOHN MOORES UNIVERSITY
LEARNING SERVICES

General

Bouthillon, L., 1947: Oscillations et phénomènes transitoires. Leur étude par les transformations de Laplace et de Cauchy. *Ann. Radioélec.*, **2**, 283–328.

Bromwich, T. J. I'a., 1916: Normal coordinates in dynamical systems. *Proc. London Math. Soc.*, *Ser. 2*, **15**, 401–448.

Bromwich, T. J. I'a., 1919: Examples of operational methods in mathematical physics. *Philos. Mag.*, *Ser. 6*, **37**, 407–419.

Carslaw, H. S., 1928: Operational methods in mathematical physics. *Math. Gaz.*, **14**, 216–228.

Carslaw, H. S., 1938: Operational methods in mathematical physics. *Math. Gaz.*, **22**, 264–280.

Carslaw, H. S., 1940: A simple application of the Laplace transform. *Philos. Mag.*, *Ser. 7*, **30**, 414–417.

Carslaw, H. S., and J. C. Jaeger, 1963: *Operational Methods in Applied Mathematics.* Dover Publications, Inc., Chapters 5–10.

Carson, J. R., 1922: The Heaviside operational calculus. *Bell Syst. Tech. J.*, **1**, **No. 2**, 43–55.

Carson, J. R., 1925: Electric circuit theory and the operational calculus. *Bell Syst. Tech. J.*, **4**, 685–761.

Carson, J. R., 1926: The Heaviside operational calculus. *Bull. Am. Math. Soc.*, *Ser. 2*, **32**, 43–68.

Carson, J. R., 1926: Electric circuit theory and the operational calculus. *Bell Syst. Tech. J.*, **5**, 50–95, 336–384.

Churchill, R. V., 1936/37: The inversion of the Laplace transformation by a direct expansion in series and its application to boundary-value problems. *Math. Z.*, **42**, 567–579.

Cohen, L., 1923: Alternating current cable telegraphy. *J. Franklin Inst.*, **195**, 165–182.

Davies, B., 1978: *Integral Transforms and Their Applications.* Springer-Verlag, Chapter 4.

Goldman, S., 1949: *Laplace Transform Theory and Electrical Transients.* Dover Publications, Inc., Chapter 10.

Howell, W. H., 1939: A note on the solution of some partial differential equations in the finite domain. *Philos. Mag.*, *Ser. 7*, **28**, 396–402.

Jaeger, J. C., 1940: The solution of boundary value problems by a double Laplace transformation. *Bull. Am. Math. Soc.*, *Ser. 2*, **46**, 687–693.

Jaeger, J. C., 1949: *An Introduction to the Laplace Transform with Engineering Applications.* John Wiley & Sons, Chapter 4.

Jeffreys, H., 1964: *Operational Methods in Mathematical Physics.* Stechert-Hafner Service Agency, Chapters 4–6.

Jeffreys, H., and B. S. Jeffreys, 1972: *Methods of Mathematical Physics.* Cambridge University Press, Chapters 18 and 19.

Lowan, A. N., 1934: On the operational treatment of certain mechanical and electrical problems. *Philos. Mag.*, *Ser. 7*, **17**, 1134–1144.

McLachlan, N. W., 1939: *Complex Variables & Operational Calculus with Technical Applications.* Cambridge University Press, Chapters 13–15.

McLachlan, N. W., 1948: *Modern Operational Calculus with Applications in Technical Mathematics.* Macmillan and Co., Ltd., Chapter 4.

Miles, J. W., 1971: *Integral Transforms in Applied Mathematics.* Cambridge University Press, Chapter 2.

Pipes, L. A., 1939: The operational calculus. III. *J. Appl. Phys.*, **10**, 301–312.

Prager, W., 1933: Über die Verwendung symbolischer Methoden in der Mechanik. *Ing. Arch.*, **4**, 16–34.

Riess, K., 1947: Some applications of the Laplace transform. *Am. J. Phys.*, **15**, 45–48.

Smith, J. J., 1923: The solution of differential equations by a method similar to Heaviside's. *J. Franklin Inst.*, **195**, 815–850.

Thomson, W. T., 1950: *Laplace Transforms*. Prentice-Hall, Inc., Chapter 8.

Wagner, K. W., 1940: *Operatorenrechnung nebst Anwendungen in Physik und Technik*. Johann Ambrosius Berth Verlag, Chapter 7.

Walters, A. G., 1949: The solution of some transient differential equations by means of Green's function. *Proc. Cambridge Philos. Soc.*, **45**, 69–80.

Geophysical Sciences

Acheson, D. J., 1975: On hydromagnetic oscillations within the earth and core-mantle coupling. *Geophys. J. R. Astr. Soc.*, **43**, 253–268.

Abbott, M. R., 1959: The downstream effect of closing a barrier across an estuary with particular reference to the Thames. *Proc. R. Soc. London, Ser. A*, **251**, 426–439.

Belluigi, A., 1958: La non stazionarietà dei fenomeni elettrogeosmotici. *Geofis. Pura Appl.*, **40**, 97–119.

Birchfield, G. E., 1969: Response of a circular model Great Lake to a suddenly imposed wind stress. *J. Geophys. Res.*, **74**, 5547–5554.

Csanady, G. T., 1968: Motions in a model Great Lake due to a suddenly imposed wind. *J. Geophys. Res.*, **73**, 6435–6447.

Das Gupta, S. P., 1968: Effect of low velocity layer in earthquakes. *Z. Geophys.*, **34**, 1–8.

De, T. K., 1982: On the phase boundary motion in the earth due to pressure and temperature excitations at the earth's surface. *Z. Angew. Math. Mech.*, **62**, 249–255.

Heaps, N. S., 1965: Storm surges on a continental shelf. *Phil. Trans. R. Soc. London, Ser. A*, **257**, 351–383.

Hearn, C. J., 1993: Response of a uniform unbounded ocean to a moving tropical cyclone. *Appl. Math. Modell.*, **17**, 205–212.

Jeffreys, H., 1931: On the cause of oscillatory movement in seismograms. *Mon. Not. R. Astron. Soc., Geophys. Suppl.*, **2**, 407–416.

Jelesnianski, C. P., 1970: "Bottom stress time-history" in linearized equations of motion for storm surges. *Mon. Wea. Rev.*, **98**, 462–478.

Kalugin, V. M., G. I. Kolomiytseva, and N. M. Rotanova, 1979: Use of the operational method for solving the problem of electromagnetic induction in the lower mantle of the earth. *Geomagnet. Aeronom.*, **19**, 85–89.

Kamenkovich, V. M., 1989: Development of Rossby waves generated by localized effects. *Oceanology*, **29**, 1–11.

Mareschal, J.-C., 1983: Uplift and heat flow following the injection of magnas into the lithosphere. *Geophys. J. R. Astr. Soc.*, **73**, 109–127.

Mudford, B. S., 1988: Modeling the occurrence of overpressures on the Scotian Shelf, offshore eastern Canada. *J. Geophys. Res.*, **93**, 7845–7855.

Nakamura, K., 1961: Motion of water due to long waves in a rectangular bay of uniform depth. *Sci. Rep. Tohoku Univ., Ser. 5 Geophys.*, **12**, 191–213.

Nalesso, G. F., and A. R. Jacobson, 1988: Shaping of an ion cloud's velocity field by differential braking due to Alfvén wave dissipation in the ionosphere. 1. Coupling with an infinite ionosphere. *J. Geophys. Res.*, **93**, 5794–5802.

Nalesso, G. F., and A. R. Jacobson, 1988: Shaping of an ion cloud's velocity field by differential braking due to Alfvén wave dissipation in the ionosphere. 2. Reflections from the E layer. *J. Geophys. Res.*, **93**, 5803–5809.

Nikiforov, Ye. G., 1963: Some hydrodynamic effects in nonstationary, purely wind currents. *Dokl. Acad. Sci. USSR, Earth Sci. Section*, **140**, 1038–1040.

Omer, G. C., 1950: Volcanic tremor. Part two: The theory of volcanic tremor. *Bull. Seism. Soc. Am.*, **40**, 175–194.

Poppendiek, H. F., 1968: Two-dimensional transport models for the lower layers of the atmosphere. *Int. J. Heat Mass Transfer*, **11**, 67–79.

Salm, B., 1964: Anlage zur Untersuchung dynamischer Wirkungen von bewegtem Schnee. *Z. Angew. Math. Phys.*, **15**, 357–374.

Veronis, G., 1958: On the transient response of a β-plane ocean. *J. Oceanogr. Soc. Japan*, **14**, 1–5.

Yang, J., K. Latychev, and R. N. Edwards, 1998: Numerical computation of hydrothermal fluid circulation in fractured Earth structures. *Geophys. J. Int.*, **135**, 627–649.

Zdunkowski, W., and T. Kandlbinder, 1997: An analytic solution to nocturnal cooling. *Beitr. Phys. Atmos.*, **70**, 337–348.

Heat Conduction

Abarband, S. S., 1960: Time dependent temperature distribution in radiating solids. *J. Math. Phys. (Cambridge, MA)*, **39**, 246–257.

Baumeister, K. J., and T. D. Hamill, 1969: Hyperbolic heat-conduction equation – A solution for the semi-infinite body problem. *J. Heat Transfer*, **91**, 543–548.

Bell, R. P., 1945: A problem of heat conduction with spherical symmetry. *Proc. Phys. Soc.*, **57**, 45–48.

Benfield, A. E., 1951: The temperature in an accumulating snow field. *Mon. Not. R. Astron. Soc., Geophys. Suppl.*, **6**, 139–147.

Bhat, A. M., R. Prakash, and J. S. Saini, 1983: Heat transfer in nucleate pool boiling at high heat flux. *Int. J. Heat Mass Transfer*, **26**, 833–840.

Blackwell, B. F., 1990: Temperature profile in semi-infinite body with exponential source and convective boundary condition. *J. Heat Transfer*, **112**, 567–571.

Blackwell, J. H., 1953: Radial-axial heat flow in regions bounded internally by circular cylinders. *Can. J. Phys.*, **31**, 472–479.

Boehringer, J. C., and J. Spindler, 1963: Radiant heating of semitransparent materials. *AIAA J.*, **1**, 84–88.

Bonilla, C. F., J. S. Busch, H. G. Landau, and L. L. Lynn, 1961: Formal heat transfer solutions. *Nucl. Sci. Eng.*, **9**, 323–331.

Bromwich, T. J. I'a., 1921: Symbolical methods in the theory of conduction of heat. *Proc. Cambrigde Philos. Soc.*, **20**, 411–427.

Brown, A., 1965: Diffusion of heat from a sphere to a surrounding medium. *Aust. J. Phys.*, **18**, 483–489.

Burka, A. L., 1966: Asymmetric radiative-convective heating of an infinite plate. *J. Appl. Mech. Tech. Phys.*, **7**, **No. 2**, 85–86.

Carslaw, H. S., 1920: Bromwich's method of solving problems in the conduction of heat. *Philos. Mag., Ser. 6*, **39**, 603–611.

Carslaw, H. S., and J. C. Jaeger, 1938: Some problems in the mathematical theory of the conduction of heat. *Philos. Mag., Ser. 7*, **26**, 473–495.

Carslaw, H. S., and J. C. Jaeger, 1940: Some two-dimensional problems in conduction of heat with circular symmetry. *Proc. London Math. Soc., Ser. 2*, **46**, 361–388.

Carslaw, H. S., and J. C. Jaeger, 1940: The determination of Green's function for the equation of conduction of heat in cylindrical coordinates by the Laplace transformation. *J. London Math. Soc., Ser. 1*, **15**, 273–281.

Carslaw, H. S., and J. C. Jaeger, 1941: The determination of Green's function for line sources for the equation of conduction of heat in cylindrical coordinates by the Laplace transformation. *Philos. Mag., Ser. 7*, **31**, 204–208.

Carslaw, H. S., and J. C. Jaeger, 1957: *Conduction of Heat in Solids*. Second Ed. Oxford University Press, Chapters 12–15.

Chase, C. A., D. Gidaspow, and R. E. Peck, 1969: A regenerator–Prediction of Nusselt numbers. *Int. J. Heat Mass Transfer*, **12**, 727–736.

Chen, S.-Y., 1961: One-dimensional heat conduction with arbitrary heating rate. *J. Aerospace Sci.*, **28**, 336–337.

Chin, J. H., 1962: Effect of uncertainties in thermocouple location on computing surface heat fluxes. *Am. Rocket Soc. J.*, **32**, 273–274.

Choudhury, N. K. D., and Z. U. A. Warsi, 1964: Weighting function and transient thermal response of buildings. Part I – Homogeneous structure. *Int. J. Heat Mass Transfer*, **7**, 1309–1321.

Chu, H. S., C. K. Chen, and C. I. Weng, 1983: Transient response of circular pins. *J. Heat Transfer*, **105**, 205–208.

Chu, S. C., and S. G. Bankoff, 1964/65: Heat transfer to slug flows with finite wall thickness. *Appl. Sci. Res., Ser. A*, **14**, 379–395.

Churchill, R. V., 1941: A heat conduction problem introduced by C. J. Tranter. *Philos. Mag., Ser. 7*, **31**, 81–87.

Cole, J. D., and T. Y. Wu, 1952: Heat conduction in a compressible fluid. *J. Appl. Mech.*, **19**, 209–213.

Cooper, F., 1977: Heat transfer from a sphere to an infinite medium. *Int. J. Heat Mass Transfer*, **20**, 991–993.

Copley, J. A., and W. C. Thomas, 1974: Two-dimensional transient temperature distribution in cylindrical bodies with pulsating time and space-dependent boundary conditions. *J. Heat Transfer*, **96**, 300–306.

Craggs, J. W., 1945: Heat conduction in semi-infinite cylinders. *Philos. Mag., Ser. 7*, **36**, 220–222.

Davies, W., 1959: Thermal transients in graphite-copper contacts. *Brit. J. Appl. Phys.*, **10**, 516–522.

Domingos, H., and D. Voelker, 1976: Transient temperature rise in layered media. *J. Heat Transfer*, **98**, 329–330.

Doorly, J. E., and M. L. G. Oldfield, 1987: The theory of advanced multi-layer thin film heat transfer gauges. *Int. J. Heat Mass Transfer*, **30**, 1159–1168.

Duffy, D., 1985: The temperature distribution within a sphere placed in a directed uniform heat flux and allowed to radiatively cool. *J. Heat Transfer*, **107**, 28–32.

Edeskuty, F. J., and N. R. Amundson, 1952: Mathematics of adsorption. IV. Effects of intraparticle diffusion in agitated static systems. *J. Phys. Chem.*, **56**, 148–152.

Edwards, J. V., R. Evans, and S. D. Probert, 1971: Computation of transient temperatures in regenerators. *Int. J. Heat Mass Transfer*, **14**, 1175–1202.

El-Adawi, M. K., M. A. Abdel-Naby, and S. A. Shalaby, 1995: Laser heating of a two-layer system with constant surface absorption: An exact solution. *Int. J. Heat Mass Transfer*, **38**, 947–952.

Frank, I., 1958: Transient temperature distribution in aircraft structures. *J. Aeronaut. Sci.*, **25**, 265–267.

Gal-Or, B., and W. Resnick, 1964: Mass transfer from gas bubbles in an agitated vessel with and without simultaneous chemical reaction. *Chem. Eng. Sci.*, **19**, 653–663.

Gembarovič, J., and V. Majerník, 1987: Determination of thermal parameters of relaxation materials. *Int. J. Heat Mass Transfer*, **30**, 199–201.

Giedt, W. H., and D. R. Hornbaker, 1962: Transient temperature variation in a thermally orthotropic plate. *Am. Rocket Soc. J.*, **32**, 1902–1909.

Giere, A. C., 1964/65: Transient heat flow in a composite slab – constant flux, zero flux boundary conditions. *Appl. Sci. Res., Ser. A*, **14**, 191–198.

Goldenberg, H., 1951: A problem in radial heat flow. *Brit. J. Appl. Phys.*, **2**, 233–237.

Grannemann, W. W., and J. D. Reese, 1960: Transient junction temperatures in power transitors. *Electr. Eng. (Am. Inst. Electr. Eng.)*, **79**, 53–57.

Griffith, M. V., and G. K. Horton, 1946: The transient flow of heat through a two-layer wall. *Proc. Phys. Soc.*, **58**, 481–487.

Gupta, R. K., 1963: Two problems of heat flow in solids. *Proc. Natl. Inst. Sci. India*, **A29**, 84–89.

Hayasi, N., and K. Inouye, 1965: Transient heat transfer through a thin circular pipe due to unsteady flow in the pipe. *J. Heat Transfer*, **87**, 513–520.

Heasley, J. H., 1965: Transient heat flow between contacting solids. *Int. J. Heat Mass Transfer*, **8**, 146–154.

Hector, L. G., W.-S. Kim, and M. N. Özisik, 1992: Propagation and reflection of thermal waves in a finite medium due to axisymmetric surface sources. *Int. J. Heat Mass Transfer*, **35**, 897–912.

Hemmings, J. W., 1979: The non-adiabatic calorimeter problem and its application to transfer processes in suspensions of solid. *Int. J. Heat Mass Transfer*, **22**, 99–109.

Hume, J. R., and V. J. Skoglund, 1962: Theoretical solution of a transient, fluid-cooled heat generator. *J. Aerospace Sci.*, **29**, 1156–1163.

Imison, B. W., and R. G. Rice, 1975: The transient response of a CSTR to a spherical catalyst particle with mass transfer resistance. *Chem. Eng. Sci.*, **30**, 1421–1423.

Jaeger, J. C., 1941: Heat conduction in composite circular cylinders. *Philos. Mag., Ser. 7*, **32**, 324–335.

Jaeger, J. C., 1941: Conduction of heat in regions bounded by planes and cylinders. *Bull. Am. Math. Soc., Ser. 2*, **47**, 734–741.

Jaeger, J. C., 1942: Heat conduction in a wedge, or an infinite cylinder whose cross-section is a circle or a sector of a circle. *Philos. Mag., Ser. 7*, **33**, 527–536.

Jaeger, J. C., 1944: Some problems involving line sources in conduction of heat. *Philos. Mag., Ser. 7*, **35**, 169–179.

Jaeger, J. C., 1944: Note on a problem in radial flow. *Proc. Phys. Soc.*, **56**, 197–203.

Jaeger, J. C., 1945: Conduction of heat in a slab in contact with well-stirred fluid. *Proc. Cambridge Philos. Soc.*, **41**, 43–49.

Jaeger, J. C., 1953: Pulsed surface heating of a semi-infinite solid. *Quart. Appl. Math.*, **11**, 132–137.

Kahn, M. A., 1973: Cooling of an initially hot semi-infinite piezoelectric rod by radiation. *Rev. Roum. Sci. Tech., Ser. Mec. Appl.*, **18**, 683–697.

Kao, T., 1977: Non-Fourier heat conduction in thin surface layers. *J. Heat Transfer*, **99**, 343–345 and 501.

Kardas, A., 1966: On a problem in the theory of the unidirectional regenerator. *Int. J. Heat Mass Transfer*, **9**, 567–579.

Kartashov, É. M., and G. M. Bartenev, 1971: Method for solving heat-conduction boundary-value problems in a region with a boundary in uniform motion. *Sov. Phys. J.*, **14**, 45–52.

Kaushik, N. D., and A. Srivastava, 1980: Temperature distribution in ground: Response function technique. *Int. J. Heat Mass Transfer*, **23**, 903–906.

Kaye, J., and V. C. M. Yeh, 1955: Design charts for transient temperature distribution resulting from aerodynamic heating at supersonic speeds. *J. Aeronaut. Sci.*, **22**, 755–762.

Kumar, I. J., 1960: A problem in heat transfer. *Proc. Natl. Inst. Sci. India*, **A26**, 414–421.

Kumar, I. J., 1962: Heat flow induced in a hollow circular cylinder by a periodic surface source. *Proc. Natl. Sci. India*, **A28**, 325–335.

Kumar, I. J., and H. N. Narang, 1966: Drying of a moist capillary-porous body in moving air. *Int. J. Heat Mass Transfer*, **9**, 95–102.

Lebedev, N. N., and I. P. Skal'skaya, 1964: Some problems in the theory of heat conduction for wedge-shaped bodies. *Sov. Phys. Tech. Phys.*, **9**, 614–618, 1207–1213.

Lorenzini, E., and M. Spiga, 1982: Temperature in heat generating solids with memory. *Wärme Stoffübertrag.*, **16**, 113–118.

Lowan, A. N., 1934: On the problem of the heat recuperator. *Philos. Mag., Ser. 7*, **17**, 914–933.

Lowan, A. N., 1937: On the operational determination of Green's functions in the theory of heat conduction. *Philos. Mag., Ser. 7*, **24**, 62–70.

Lowan, A. N., 1937: On some two-dimensional problems in heat conduction. *Philos. Mag., Ser. 7*, **24**, 410–424.

Lowan, A. N., 1940: On some problems in the diffraction of heat. *Philos. Mag., Ser. 7*, **29**, 93–99.

Lowan, A. N., 1945: On the problem of heat conduction in thin plates. *J. Math. Phys. (Cambridge, MA)*, **24**, 22–29.

Luikov, A., 1936: The application of the Heaviside-Bromwich operational method to the solution of a problem in heat conduction. *Philos. Mag.*, *Ser. 7*, **22**, 239–248.

Madejski, J., 1967: Simultaneous mass and heat transfer on an absorbing porous sphere. *Arch. Mech. Stosow.*, **19**, 183–196.

Malkovich, R. Sh., 1977: Heating of a spherical shell by a radial current. *Sov. Phys. Tech. Phys.*, **22**, 636.

Mersman, W. A., 1942: Heat conduction in a semi-infinite slab. *Philos. Mag.*, *Ser. 7*, **33**, 303–309.

Minyatov, A. V., 1960: The heating of an infinite cylinder enclosed in a sheath. *Sov. Phys. Tech. Phys.*, **5**, 575–578.

Narin, F., and D. Langford, 1959: Analytic solutions for transient temperature distributions in two-region nuclear reactor fuel elements. *Nucl. Sci. Eng.*, **6**, 386–390.

Newcomb, T. P., 1958: The flow of heat in a parallel-faced infinite solid. *Brit. J. Appl. Phys.*, **9**, 370–372.

Newcomb, T. P., 1958: The radial flow of heat in an infinite cylinder. *Brit. J. Appl. Phys.*, **9**, 456–458.

Newcomb, T. P., 1959: Flow of heat in a composite solid. *Brit. J. Appl. Phys.*, **10**, 204–206.

Newcomb, T. P., 1959: Transient temperatures attained in disk brakes. *Brit. J. Appl. Phys.*, **10**, 339–340.

Newcomb, T. P., 1960: Temperatures reached in disc brakes. *J. Mech. Eng. Sci.*, **2**, 167–177.

Parkus, H., 1962: Wärmespannungen bei zufallsabhängiger Oberflächentemperatur. *Z. Angew. Math. Mech.*, **42**, 499–507.

Paterson, S., 1947: The heating or cooling of a solid sphere in a well-stirred fluid. *Proc. Phys. Soc.*, **59**, 50–58.

Petrenko, V. G., 1967: On heat conduction of an inhomogeneous plate. *Sov. Appl. Mech.*, **3, No. 11**, 38–40.

Phythian, J. E., 1963: Cylindrical heat flow with arbitrary heat rates. *AIAA J.*, **1**, 925–927.

Poppendick, H. F., 1953: Transient and steady-state heat transfer in irradiated citrus fruit. *ASME Trans.*, **75**, 421–425.

Pratt, A. W., and E. F. Ball, 1963: Transient cooling of a heated enclosure. *Int. J. Heat Mass Transfer*, **6**, 703–718.

Pratt, A. W., 1965: Fundamentals of heat transmission through the external walls of buildings. *J. Mech. Eng. Sci.*, **7**, 357–366.

Pratt, A. W., 1965: Variable heat flow through walls of cavity construction, naturally exposed. *Int. J. Heat Mass Transfer*, **8**, 861–872.

Pratt, A. W., and R. E. Lacy, 1966: Measurement of the thermal diffusivities of some single-layer walls in buildings. *Int. J. Heat Mass Transfer*, **9**, 345–353.

Pugh, H. Ll. D., and A. J. Harris, 1942: The temperature distribution around a spherical hole in an infinite conducting medium. *Philos. Mag.*, *Ser. 7*, **33**, 661–666.

Purohit, G. N., 1967: Temperature distribution in Covette flow between two parallel flat plates. *Proc. Natl. Sci. India*, **A33**, 142–149.

Redozubov, D. B., 1960: The solution of linear thermal problems with a uniformly moving boundary in a semiinfinite region. *Sov. Phys. Tech. Phys.*, **5**, 570–574.

Reinheimer, J., 1962: Ablation with volume distribution of heat sources. *Am. Rocket Soc. J.*, **32**, 1106–1107.

Reismann, H., and W. H. Jurney, 1961: Temperature distribution in a spinning spherical shell during atmospheric entry. *Astronaut. Acta*, **7**, 290–321.

Reismann, H., 1962: Temperature distribution in a spinning sphere during atmospheric entry. *J. Aerospace Sci.*, **29**, 151–159.

Rikenglaz, L. É., 1991: Solution of heat conduction boundary-value problems for a region with moving boundaries using the spatial Laplace transform. *Sov. Phys. Tech. Phys.*, **36**, 1415–1416.

Rotem, Z., J. Gildor, and A. Solan, 1963: Transient heat dissipation from storage reservoirs. *Int. J. Heat Mass Transfer*, **6**, 129–141.

Ruisseau, N. R. des, and R. D. Zerkle, 1970: Temperature in semi-infinite and cylindrical bodies subjected to moving heat sources and surface cooling. *J. Heat Transfer*, **92**, 456–464.

Sabherwal, K. C., 1965: An inverse problem of transient heat conduction. *Indian J. Pure Appl. Phys.*, **3**, 397–398.

Serroukh, M., 1992: Le transfert de chaleur pendant l'injection d'un fluide chaud dans un gisement de pétrole. *Rev. Roum. Sci. Tech., Ser. Mec. Appl.*, **37**, 401–420.

Shah, Y. T., 1972: Gas-liquid interface temperature rise for rapid absorption-reaction in a thin liquid film. *Chem. Eng. Sci.*, **27**, 1893–1895.

Służalec, A., and A. Służalec, 1993: Solution of thermal problems in friction welding – comparative study. *Int. J. Heat Mass Transfer*, **36**, 1583–1587.

Smith, E. G., 1941: A simple and rigorous method for the determination of the heat requirements of simple intermittently heated exterior walls. *J. Appl. Phys.*, **12**, 638–642.

Smith, E. G., 1941: The heat requirements of simple intermittently heated interior walls and furniture. *J. Appl. Phys.*, **12**, 642–644.

Soliman, M., and P. L. Chambré, 1967: On the time-dependent Lévêque problem. *Int. J. Heat Mass Transfer*, **10**, 169–180.

Spiga, G., and M. Spiga, 1986: Two-dimensional transient solutions for crossflow heat exchangers with neither gas mixed. *J. Heat Transfer*, **109**, 281–286.

Starr, A. T., 1930: Lag in a thermometer when the temperature of the external medium is varying. *Philos. Mag., Ser. 7*, **9**, 901–912.

Suryanarayana, N. V., 1975: Transient response of straight fins. *J. Heat Transfer*, **97**, 417–423.

Suryanarayana, N. V., 1976: Transient response of straight fins. Part II. *J. Heat Transfer*, **98**, 324–327.

Szekely, J., 1963: Notes on the transfer at the interface of two independently stirred liquids. *Int. J. Heat Mass Transfer*, **6**, 833–840.

Tang, D. W., and N. Araki, 1996: Non-Fourier heat conduction in a finite medium under periodic surface thermal disturbance. *Int. J. Heat Mass Transfer*, **39**, 1585–1590.

Touryan, K. J., 1964: Transient temperature variation in a thermally orthotropic cylindrical shell. *AIAA J.*, **2**, 124–126.

Towler, B. F., and R. G. Rice, 1974: A note on the response of a CSTR to a spherical catalyst pellet. *Chem. Eng. Sci.*, **29**, 1828–1832.

Tranter, C. J., 1944: On a problem in heat conduction. *Philos. Mag., Ser. 7*, **35**, 102–105.

Tranter, C. J., 1947: Heat flow in an infinite medium heated by a cylinder. *Philos. Mag., Ser. 7*, **38**, 131–134.

Tranter, C. J., 1947: Note on a problem in heat conduction. *Philos. Mag., Ser. 7*, **38**, 530–531.

Tsybin, A. M., 1975: Solution of the Stefan problem. *Sov. Phys. Tech. Phys.*, **19**, 1514–1515.

Van Zee, A. F., and C. L. Babcock, 1951: A method for the measurement of thermal diffusivity of molton glass. *J. Am. Ceram. Soc.*, **34**, 244–250.

Wah, T., 1971: Analysis of heat conduction in bridge slab. *Acta Mech.*, **11**, 9–26.

Warsi, Z. U. A., and N. K. D. Choudhury, 1964: Weighting function and transient response of buildings. Part II – Composite structure. *Int. J. Heat Mass Transfer*, **7**, 1323–1334.

Whitehead, S., 1944: An approximate method for calculating heat flow in an infinite medium heated by a cylinder. *Proc. Phys. Soc.*, **56**, 357–366.

Yang, W. J., J. A. Clark, and V. S. Arpaci, 1961: Dynamic response of heat exchangers having internal heat sources. Part IV. *J. Heat Transfer*, **83**, 321–338.

Yang, W.-J., and J. A. Clark, 1964: On the application of the source theory to the solution of problems involving phase change. Part 2. Transient interface heat and mass transfer in multi-component liquid-vapor system. *J. Heat Transfer*, **86**, 443–448.

Yoo, H., and E.-T. Pak, 1996: Analytical solutions to a one-dimensional finite-domain model for stratified thermal storage tanks. *Solar Energy*, **56**, 315–322.

Yuen, W. W., and S. C. Lee, 1989: Non-Fourier heat conduction in a semi-infinite solid subjected to oscillatory surface thermal disturbances. *J. Heat Transfer*, **111**, 178–181.

Zarubin, V. S., 1959: One problem in nonstationary heat conduction. *Am. Rocket Soc. J.*, **29**, 773–776.

Zatzkis, H., 1953: A certain problem in heat conduction. *J. Appl. Phys.*, **24**, 895–896.

Zhou, Z. W., 1995: Analytical solution for transient heat conduction in hollow cylinders containing well-stirred fluid with uniform heat sink. *Int. J. Heat Mass Transfer*, **38**, 2915–2919.

Zinsmeister, G. E., and J. R. Dixon, 1965: An extension of linear moving heat source solutions to a transient case in a composite system. *Int. J. Heat Mass Transfer*, **8**, 1–6.

Magnetohydrodynamics

Bathaiah, D., 1980: MHD flow through a porous straight channel. *Acta Mech.*, **35**, 223–229.

Bujurke, N. M., 1982: Effect of frequency of residual stresses of oscillating plate problem in MHD. *Indian J. Pure Appl. Math.*, **13**, 1492–1496.

Chandran, P., N. C. Sacheti, and A. K. Singh, 1993: Effect of rotation on unsteady hydromagnetic Couette flow. *Astrophys. Space Sci.*, **202**, 1–10.

Chandran, P., N. C. Sacheti, and A. K. Singh, 1998: Unsteady hydromagnetic free convection flow with heat flux and accelerated boundary motion. *J. Phys. Soc. Japan*, **67**, 124–129.

Chang, C. C., and J. T. Yen, 1959: Rayleigh's problem in magnetohydrodynamics. *Phys. Fluids*, **2**, 393–403.

Charles, M., and Ph. R. Smith, 1970: A general solution to unsteady coupled magnetohydrodynamic Couette flow. *Appl. Sci. Res.*, **22**, 44–59.

Chawla, S. S., 1967: Magnetohydrodynamic unsteady free convection. *Z. Angew. Math. Mech.*, **47**, 499–508.

Debnath, L., 1970: Unsteady hydromagnetic flow induced by an oscillating disk. *Bull. Calcutta Math. Soc.*, **62**, 173–182.

Debnath, L., 1972: On unsteady magnetohydrodynamic boundary layers in a rotating flow. *Z. Angew. Math. Mech.*, **52**, 623–626.

Debnath, L., 1972: On an unsteady hydromagnetic channel flow. *Bull. Calcutta Math. Soc.*, **64**, 143–149.

Debnath, L., 1973: On an unsteady hydromagnetic boundary layer flow. *Bull. Calcutta Math. Soc.*, **65**, 109–114.

Debnath, L., 1974: Resonant oscillations of a porous plate in an electrically conducting rotating viscous fluid. *Phys. Fluids*, **17**, 1704–1706.

Drake, D. G., 1960: Rayleigh's problem in magnetohydrodynamics for a non-perfect conductor. *Appl. Sci. Res., Ser. B*, **8**, 467–477.

Dube, S. K., and M. A. A. Khan, 1968: On unsteady hydromagnetic flow of an electrically conducting incompressible fluid in contact with a harmonically oscillating plate. *Bull. Cl. Sci., Acad. R. Belg., Ser. 5*, **54**, 732–744.

Dube, S. K., 1970: Hydromagnetic flow near an accelerated flat plate. *Indian J. Pure Appl. Math.*, **1**, 170–177.

Farn, C. L. S., and V. S. Arpaci, 1966: Suddenly pressurized elastomagnetohydrodynamic channel flow. *Phys. Fluids*, **9**, 1970–1973.

Filippov, G. V., and V. G. Shakov, 1967: The unsteady spatial laminar boundary layer on magnetohydrodynamics. *J. Appl. Mech. Tech. Phys.*, **8, No. 6**, 42–45.

Gambirasio, G., 1960: On the electrical behavior of an ideal plasma. *Phys. Fluids*, **3**, 299–302.

Groves, R. N., 1966: On the unsteady motion of a viscous hydromagnetic fluid contained in a cylindrical vessel. *J. Appl. Mech.*, **33**, 748–752.

Gupta, S. C., and B. Singh, 1970: Unsteady magnetohydrodynamic flow in a circular pipe under a transverse magnetic field. *Phys. Fluids*, **13**, 346–352.

Hata, T., 1996: Analysis of stress-focusing effect of magnetothermoelastic waves in a perfectly conducting solid cylinder (in Japanesse). *Nihon Kikai Gakkai Rombunshu (Trans. Japan Soc. Mech. Eng.), Ser. A*, **62**, 2104–2108.

Kabadi, S. A., and B. Siddappa, 1984: On the flow of an electrically conducting Maxwell fluid past an infinite flat plate in the presence of a transverse magnetic field. *Rev. Roum. Sci. Tech., Ser. Mec. Appl.*, **29**, 593–605.

Kant, R., 1980: Hydromagnetic flow of an electrically conducting viscous fluid near a time-varying accelerated porous plate with Hall effects. *Indian J. Theoret. Phys.*, **28**, 31–39.

Kasiviswanathan, S. R., and A. Ramachandra Rao, 1982: On exact solutions of unsteady MHD flow between eccentrically rotating disks. *Arch. Mech.*, **39**, 411–418.

Kishore, N., S. Tejpal, and H. K. Katiyar, 1981: On unsteady MHD flow through two parallel porous flat plates. *Indian J. Pure Appl. Math.*, **12**, 1372–1379.

Kumar, P., and N. P. Singh, 1989: MHD Hele-Shaw flow of an elasticoviscous fluid through porous media. *Bull. Calcutta Math. Soc.*, **81**, 32–41.

Ludford, G. S. S., 1959: Rayleigh's problem in hydromagnetics: The impulsive motion of a pole-piece. *Arch. Ration. Mech. Anal.*, **3**, 14–27.

Mahapatra, J. R., 1973: A note on the unsteady motion of a viscous conducting liquid between two porous concentric circular cylinders acted on by a radial magnetic field. *Appl. Sci. Res.*, **27**, 274–282.

Mazzucato, E., 1975: Exact equilibria of axisymmetric magnetic configurations. *Phys. Fluids*, **18**, 536–540.

Mishra, S. P., and P. Mohapatra, 1973: Flow near an accelerated plate in the presence of a magnetic field. *J. Appl. Phys.*, **44**, 1194–1199.

Mishra, S. P., and D. G. Sahoo, 1978: Magnetohydrodynamic unsteady free convection past a hot vertical plate. *Appl. Sci. Res.*, **34**, 1–16.

Mitra, P., 1981: Unsteady flow of a conducting dusty gas through a circular tube under time dependent pressure gradient in presence of a transverse magnetic field. *Rev. Roum. Sci. Tech., Ser. Mec. Appl.*, **26**, 795–803.

Mitra, P., and P. Bhattacharyya, 1981: Unsteady hydromagnetic laminar flow of a conducting dusty fluid between two parallel plates started impulsively from rest. *Acta Mech.*, **39**, 171–182.

Mohapatra, P., 1971: Magnetohydrodynamic flow near a time-varying accelerated plate. *Indian J. Phys.*, **45**, 421–431.

Nandy, S. C., 1972: On the unsteady flow of fluid between two parallel plates acted upon by an electric field. *Indian J. Pure Appl. Math.*, **3**, 890–895.

Ong, R. S., and J. A. Nicholls, 1959: On the flow of a hydromagnetic fluid near an oscillating flat plate. *J. Aerospace Sci.*, **26**, 313–314.

Prasada Rao, D. R. V., D. V. Krishna, and L. Debnath, 1981: A theory of convective heat transfer in a rotating hydromagnetic viscous flow. *Acta Mech.*, **39**, 225–240.

Puri, P., and P. K. Kulshrestha, 1976: Unsteady hydromagnetic boundary layer in a rotating medium. *J. Appl. Mech.*, **43**, 205–208.

Puri, P., and P. K. Kulshrestha, 1983: Structure of waves in a time dependent hydromagnetic plane Couette flow. *Z. Angew. Math. Mech.*, **63**, 489–495.

Ram, G., and R. S. Mishra, 1977: Unsteady flow through magnetohydrodynamic porous media. *Indian J. Pure Appl. Math.*, **8**, 637–647.

Rosciszewski, J., and A. Kritz, 1964: Magnetic field diffusion associated with an ionizing shock wave interacting with an electromagnetic field. *Phys. Fluids*, **7**, 1393–1394.

Sastry, D. V. S., and R. Seetharamaswamy, 1982: MHD dusty viscous flow through a circular pipe. *Indian J. Pure Appl. Math.*, **13**, 811–817.

Saxena, S., and G. C. Sharma, 1987: Unsteady flow of an electrically conducting dusty viscous liquid between two parallel plates. *Indian J. Pure Appl. Math.*, **18**, 1131–1138.

Seth, G. S., R. N. Jana, and M. K. Maiti, 1981: Unsteady hydromagnetic flow past a porous plate in a rotating medium with time-dependent free stream. *Rev. Roum. Sci. Tech., Ser. Mec. Appl.*, **26**, 383–400.

Singh, A. K., N. C. Sacheti, and P. Chandran, 1994: Transient effects on magnetohydrodynamic Couette flow with rotation: Accelerated motion. *Int. J. Eng. Sci.*, **32**, 133–139.

Singh, C. B., and P. C. Ram, 1977: Unsteady flow of an electrically conducting dusty viscous liquid through a channel. *Indian J. Pure Appl. Math.*, **8**, 1022–1028.

Singh, D., 1963/64: Unsteady motion of a viscous conducting liquid contained between two infinite coaxial cylinders in the presence of an axial magnetic field. *Appl. Sci. Res., Ser. B*, **10**, 412–416.

Sloan, D. M., 1971: An unsteady MHD duct flow. *Appl. Sci. Res.*, **25**, 126–136.

Soundalgekar, V. M., 1965: Hydrodynamic flow near an accelerated plate in the presence of a magnetic field. *Appl. Sci. Res., Ser. B*, **12**, 151–156.

Soundalgekar, V. M., S. Ravi, and S. B. Miremath, 1980: Hall effects on MHD flow past an accelerated plate. *J. Plasma Phys.*, **23**, 495–500.

Soundalgekar, V. M., and D. D. Haldavnekar, 1987: MHD boundary layer growth in a rotating liquid with suspended particles. *Proc. Indian Natl. Sci. Acad., Part A*, **53**, 597–601.

Sozou, C., 1972: The development of magnetohydrodynamic flow due to the passage of an electric current past a sphere immersed in a fluid. *J. Fluid Mech.*, **56**, 497–503.

Steketee, J. A., 1964: An application of the operational calculus to the equations of the Rayleigh problem in MHD. *Appl. Sci. Res., Ser. B*, **11**, 255–272.

Switick, D. M., and L. A. Kennedy, 1968: A solution for unsteady magnetohydrodynamic flow including ion-slip effects. *Z. Angew. Math. Phys.*, **19**, 145–148.

Tandon, P. N., 1970: Flow of a conducting viscoelastic fluid between two parallel plates under a transverse magnetic field. *Indian J. Theoret. Phys.*, **18**, 45–53.

Venkatasiva Murthy, K. N., 1979: MHD viscous flow between torsionally oscillating disks. *Appl. Sci. Res.*, **35**, 111–125.

Other

Adirovich, E. I., and V. G. Kolotilova, 1956: Propagation of a short pulse in a semiconductor bounded by two hole-electron transistors. *Sov. Phys. JETP*, **2**, 670–676.

Bachmann, K. J., and K. J. Vetter, 1966: Zur Theorie Potentiostatischer Ein- und Umschaltvorgänge bei der Elektrokristallisation. *Electrochim. Acta*, **11**, 1279–1299.

Berzins, T., and P. Delahay, 1955: Kinetics of fast electrode reactions. *J. Am. Chem. Soc.*, **77**, 6448–6453.

Bălă, C. V., 1965: Die Erwärmung des massiven Läufers beim asynchronen Anlauf der Wechselstrommachinen. *Rev. Roum. Sci. Tech., Ser. Electrotech. Energ.*, **10**, 479–490.

Baskoff, S. G., 1962: Bubble dynamics at the surface of an exponentially heat plate. *Indust. Eng. Chem. Fund.*, **1**, 257–259.

Bellman, R., R. E. Marshak, and G. W. Wing, 1949: Laplace transform solution of two-medium neutron ageing problem. *Philos. Mag., Ser. 7*, **40**, 297–308.

Borisov, V. V., 1970: Transition in the limit to the velocity of light in the problem of the incidence of a plane wave on a moving ionization front. *Radiophys. Quantum Electron.*, **13**, 1059–1061.

Brandani, S., 1996: Analytical solution for ZLC desorption curves with bi-porous adsorbent particles. *Chem. Eng. Sci.*, **51**, 3283–3288.

Braude, S. Ya., and I. L. Verbitskii, 1977: Derivation of the function describing the energy distribution of particles in a turbulent synchrotron pile. *Radiophys. Quantum Electr.*, **20**, 663–668.

Buck, R. P., 1963: Half-wave potentials for reversible processes with prior kinetic complexity. *J. Electroanal. Chem.*, **5**, 295–314.

Bucur, R. V., I. Covaci, and C. Miron, 1967: Voltametrie a tension lineairement variable en couche mince rigide: I. L'oxydation de l'hydrogene dissous dans une couche mince de palladium en solution agitee. *J. Electroanal. Chem.*, **13**, 263–274.

Carney, J. F., 1968: Dynamic response of a spherical structural system in fluid media. *Int. J. Mech. Sci.*, **10**, 583–591.

Carta, G., and R. L. Pigford, 1986: Analytical solution for cycling zone absorption. *Chem. Eng. Sci.*, **41**, 511–517.

Carta, G., 1988: Exact analytic solution of a mathematical model for chromatographic operations. *Chem. Eng. Sci.*, **43**, 2877–2883.

Carta, G., 1993: The linear driving force approximation for cyclic mass transfer in spherical particles. *Chem. Eng. Sci.*, **48**, 622–625.

Chakrabarty, A., 1971: Mechanical response in a piezoelectric transducer subjected to a current flowing in a semi-conducting boundary layer characterized by a polarization gradient. *Indian J. Theoret. Phys.*, **19**, 65–70.

Chekmareva, O. M., 1971: Novel integral equations in phase-transition problems. *Sov. Phys. Tech. Phys.*, **16**, 882–886.

Claesson, J., and J. Lindberg, 1972: Diffusional titration errors in coulometry. Part I. Analytical solutions for non-stationary diffusion through non-interacting membranes. *J. Electroanal. Chem.*, **40**, 255–263.

Compagnone, N. F., 1991: A new equation for the limiting capacity of the lead/acid cell. *J. Power Sources*, **35**, 97–111.

Daitch, P. B., and D. B. Ebeoglu, 1963: Asymptotic and transient analysis of pulsed moderators. *Nucl. Sci. Eng.*, **17**, 212–219.

Denison, M. R., and E. Baum, 1961: A simplified model of unstable burning in solid propellants. *Am. Rocket Soc. J.*, **31**, 1112–1122.

DeLevie, R., 1965: Uneven current distribution on the surface of a dropping mercury electrode and its possible relation to maxima of the first kind. *J. Electroanal. Chem.*, **9**, 311–320.

Duggan, G., and F. Berz, 1982: The theory of the Schockley-Haynes experiment: Contact effects. *J. Appl. Phys.*, **53**, 470–476.

Dvorkin, L. B., 1961: An accurate equation for dialysis. *Russ. J. Phys. Chem.*, **35**, 1378–1380.

Fleck, J. A., 1957: Transient pressures in nuclear reactors. *Nucl. Sci. Eng.*, **2**, 694–708.

Fleischmann, M., and H. R. Thirsk, 1960: Anodic electrocrystallization. *Electrochim. Acta*, **2**, 22–49.

Gaman, V. I., 1965: Transients in thin-base semiconductor diodes. *Sov. Phys. J.*, **8, No. 1**, 38–42.

Gaunaurd, G., 1977: One-dimensional model for acoustic absorption in a viscoelastic medium containing short cylindrical cavities. *J. Acoust. Soc. Am.*, **62**, 298–307.

Ghosh, S. K., S. K. Banerjee, and T. K. Banerjee, 1984: Transient response in a piezo-electric transducer. *Indian J. Theoret. Phys.*, **31**, 27–31, 199–211; **32**, 21–30, 113–128.

Ginsborg, B. L., and J. W. Searl, 1980: On the determination of the concentration over a sphere with a barrier layer. *J. Theoret. Biol.*, **82**, 521–524.

Gournay, L. S., 1966: Conversion of electromagnetic to acoustic energy by surface heating. *J. Acoust. Soc. Am.*, **40**, 1322–1330.

Guidelli, R., and D. Cozzi, 1967: Theory of first-order depolarizer regeneration in polarography and its application to a solid microelectrode with periodical renewal of the diffusion layer. *J. Electroanal. Chem.*, **14**, 245–259.

Guidelli, R., 1968: Theoretical current-time curves at constant potential with linear adsorption of the depolarizer on spherical and plane electrodes. *J. Electroanal. Chem.*, **18**, 5–19.

Hsieh, T. C., and R. Greif, 1972: Theoretical determination of the absorption coefficient and the total band absorptance including a specific application to carbon monoxide. *Int. J. Heat Mass Transfer*, **15**, 1477–1487.

Hsu, J. T., and J. S. Dranoff, 1986: On initial condition problems for reactor dispersion model. *Chem. Eng. Sci.*, **41**, 1930–1934.

Huang, C.-J., and C.-H. Kuo, 1963: General mathematical model for mass transfer accompanied by chemical reaction. *AICHE J.*, **9**, 161–167.

Jaeger, J. C., and K. C. Westfold, 1949: Transients in an ionized medium with applications to bursts of solar noise. *Aust. J. Sci. Res., Ser. A*, **2**, 322–334.

Kapoor, A., R. M. Mehra, V. K. Sharma, K. N. Tripathi, and P. C. Mathur, 1992: Theory of constant current phase of reverse recovery transient for measurement of minority carrier lifetime in a homojunction P$^+$N solar cell. *J. Instn. Electron. Telecommun. Eng.*, **38**, 289–293.

Kern, J., and J. W. Hemmings, 1978: On the analogy between the calorimeter problem and some granulate-fluid exchange processes. *J. Heat Transfer*, **100**, 319–323.

Kletskii, S. V., 1991: Fourier ring with a moving periodic delta-function source. *Sov. Phys. Tech. Phys.*, **36**, 1050–1051.

Knudsen, H. L., 1947: Pressure and oil flow in oil-filled cables at load variations. *J. Appl. Phys.*, **18**, 545–562.

Kubenko, V. D., and A. A. Babaev, 1997: Influence of a cable line on the transient operation of a piezoelectric transmitter. *Int. Appl. Mech.*, **33**, 888–894.

Kval'vasser, V. I., and Ya. F. Rutner, 1965: The problem of expansion of a neutral plasma in an external magnetic field. *Sov. Phys. Tech. Phys.*, **9**, 908–911.

Leibowitz, M. A., and R. C. Ackerberg, 1963: The vibration of a conducting wire in a magnetic field. *Quart. J. Mech. Appl. Math.*, **16**, 507–519.

Levinson, N., 1935: The Fourier transform solution of ordinary and partial differential equations. *J. Math. Phys. (Cambridge, MA)*, **14**, 195–227.

Li, C.-H., 1986: Exact transient solutions of parallel-current transfer processes. *J. Heat Transfer*, **108**, 365–369.

Macey, H. H., 1940: Clay-water relationships. *Proc. Phys. Soc.*, **52**, 625–656.

Masamune, S., and J. M. Smith, 1965: Adsorption rate studies – Interaction of diffusion and surface processes. *AICHE J.*, **11**, 34–40.

Mei, C. C., 1985: Gravity effects in consolidation of layer of soft soil. *J. Eng. Mech. Div. (Am. Soc. Civ. Eng.)*, **111**, 1038–1047.

Mintzer, D., and B. S. Tanenbaum, 1960: Spatial and temporal absorption in a viscous medium. *J. Acoust. Soc. Am.*, **32**, 67–71.

Papadopoulos, K. D., and R. V. Bailey, 1986: Reaction-diffusion in suspended particles with limited supply of reactant. *Ind. Eng. Chem. Fund.*, **25**, 303–305.

Peskin, E., and E. Weber, 1948: The d. c. thermal characteristics of microwave bolometers. *Rev. Sci. Instrum.*, **19**, 188–195.

Purves, R. D., 1977: The time course of cellular responses to iontophoretically applied drugs. *J. Theoret. Biol.*, **65**, 327–344.

Rasmuson, A., 1986: Exact solutions of some models for the dynamics of fixed beds using Danckwert's inlet condition. *Chem. Eng. Sci.*, **41**, 599–600.

Reddy, A. K. N., M. A. V. Devanathan, and J. O'M. Bockris, 1963: Chronoellipsometry: A new technique for the study of anodic processes of the dissolution – precipitation type. *J. Electroanal. Chem.*, **6**, 61–67.

Richardson, J. F., 1959: The evaporation of two-component liquid mixtures. *Chem. Eng. Sci.*, **10**, 234–242.

Roy, P., 1969: Note on responses in a piezoelectric crystal with divided electrodes. *Proc. Natl. Inst. Sci. India, Ser. A*, **35**, 612–618.

Ruckenstein, E., A. S. Vaidyanathan, and G. R. Youngquist, 1971: Sorption by solids with bidisperse pore structure. *Chem. Eng. Sci.*, **26**, 1305–1318.

Saidel, G. M., E. D. Morris, and G. M. Chisolm, 1987: Transport of macromolecules in arterial wall *in vivo*: A mathematical model and analytic solutions. *Bull. Math. Biol.*, **49**, 153–169.

Savastano, C. A., and E. S. Ortiz, 1991: Effect of changing hydrodynamic conditions on the rate of processes at liquid-liquid interfaces: Metal ion extraction and water solubilization by reversed micelles. *Chem. Eng. Sci.*, **46**, 741–749.

Schaaf, S. A., 1947: On the superposition of a heat source and contact resistance. *Quart. Appl. Math.*, **5**, 107–111.

Schatzman, E., 1969: Gravitational separation of the elements and turbulent transport. *Astron. Astrophys.*, **3**, 331–346.

Schuder, C. B., and R. C. Binder, 1959: The response of pneumatic transmission lines to step inputs. *J. Basic Eng.*, **81**, 578–583.

Shishov, V. I., 1966: Propagation of highly energetic solar protons in the interplanetary magnetic field. *Geomagnet. Aeronom.*, **6**, 174–180.

Siegel, R., and M. Perlmutter, 1962: Heat transfer for pulsating laminar duct flow. *J. Heat Transfer*, **84**, 111–122.

Sigrist, M. W., and F. K. Kneubühl, 1978: Laser-generated stress waves in liquids. *J. Acoust. Soc. Am.*, **64**, 1652–1663.

Siver, Yu. G., 1959: Non-stationary electrode processes in stirred media. I. Potential-current measurements at constant potential. *Russ. J. Phys. Chem.*, **33**, 533–539.

Snider, A. D., D. L. Akins, and R. L. Birke, 1977: Transient analysis of electrolytically initiated polymerization. *J. Electroanal. Chem.*, **79**, 31–47.

Teichmann, T., 1960: Slowing down of neutrons. *Nucl. Sci. Eng.*, **7**, 292–294.

Teng, H., C. M. Kinoshita, and S. M. Masutani, 1995: Hydrate formation on the surface of a CO_2 droplet in high-pressure, low-temperature water. *Chem. Eng. Sci.*, **50**, 559–564.

Überall, H., 1959: Neutron absorption by control rods of varying transparency. *Nucl. Sci. Eng.*, **7**, 228–234.

Venerus, E. R., and Ozisik, M. N., 1966: Theoretical investigations of fission product deposition from flowing gas streams. *Nucl. Sci. Eng.*, **26**, 122–130.

Walters, A. G., 1949: The solution of some transient differential equations by means of Green's function. *Proc. Cambridge Philos. Soc.*, **45**, 69–80.

Wang, J.-F., and H. Langemann, 1994: Unsteady two-film model for mass transfer accompanied by chemical reaction. *Chem. Eng. Sci.*, **49**, 3457–3463.

Yamamoto, H., T. Shiraishi, M. Ataka, and Y. Iwasaki, 1993: Experimental and numerical analyses of effects of thermal stability of lubricants on heatstreaks in cold rolling of stainless steel. *J. Tribol.*, **115**, 532–537.

Yang, W.-J., 1964: Transient heat transfer in a vapor-heated heat exchanger with arbitrary timewise-variant flow perturbation. *J. Heat Transfer*, **86**, 133–142.

Zaidel, R. M., 1968: Electrical phenomena on exposing a cable to neutrons and gamma rays. *J. Appl. Mech. Tech. Phys.*, **9**, 278–284.

Zil'bergleit, A. S., I. N. Zlatina, and I. B. Suslova, 1992: Two-dimensional time-dependent scalar diffraction problems in systems with nonintersecting semi-infinite screens. *Sov. Phys. Tech. Phys.*, **37**, 14–19.

Solid Mechanics

Alterman, Z., and P. Kornfeld, 1966: Effect of a fluid core on propagation of an SH-torque pulse from a point-source in a sphere. *Geophysics*, **31**, 741–763.

Alterman, Z., and F. Abramovici, 1967: The motion of a sphere caused by an impulsive force and by an explosive point-source. *Geophys. J. R. Astr. Soc.*, **13**, 117–148.

Alterman, Z. S., and J. Aboudi, 1969: Seismic pulse in a layered sphere: Normal modes and surface waves. *J. Geophys. Res.*, **74**, 2618–2636.

Amada, S., 1983: Dynamic stress analysis of a solid rotating disc (in Japanese). *Nihon Kikai Gakkai Rombunshu (Trans. Japan Soc. Mech. Engrs.), Ser. A*, **49**, 1540–1546.

Amada, S., 1984: Dynamic shear stress analysis of discs subjected to variable rotations (in Japanese). *Nihon Kikai Gakkai Rombunshu (Trans. Japan Soc. Mech. Engrs.), Ser. A*, **50**, 1719–1726.

Amada, S., 1985: Dynamic stress analysis of rotating hollow discs (in Japanese). *Nihon Kikai Gakkai Rombunshu (Trans. Japan Soc. Mech. Engrs.), Ser. A*, **51**, 2103–2110.

Amada, S., 1985: Dynamic shear stress analysis of discs subjected to variable rotations. *Bull. JSME*, **28**, 1029–1035.

Amada, S., 1986: Dynamic stress analysis of hollow rotating discs. *Bull. JSME*, **29**, 1383–1389.

Arenz, R. J., 1965: Two-dimensional wave propagation in realistic viscoelastic materials. *J. Appl. Mech.*, **32**, 303–314.

Banerjee, A., 1980: Torsional oscillation of a semi-infinite non-homogeneous transversely isotropic circular cylinder one of whose ends is acted on by an impulsive twist. *Bull. Calcutta Math. Soc.*, **72**, 309–314.

Barzukov, O. P., L. A. Vaisberg, B. V. Semkin, and V. A. Tsukerman, 1972: Energy relations in the case of impact action on a rod. *Sov. Phys. J.*, **15**, 1007–1011.

Bhaduri, S., and M. Kanoria, 1981: Forced vibration of an isotropic circular cylinder having a rigid cylindrical inclusion. *Rev. Roum. Sci. Tech., Ser. Mec. Appl.*, **26**, 475–480.

Bhaduri, S., and M. Kanoria, 1982: Forced vibration of an isotropic sphere having a rigid spherical inclusion. *Indian J. Theoret. Phys.*, **30**, 93–98.

Bhanja, N., 1971: Torsional vibration of a solid finite cylinder having transverse isotropy. *Pure Appl. Geophys.*, **85**, 182–188.

Bhattacharya, J., and S. C. Das Gupta, 1963: On leaking modes coupled with shear waves. *Z. Geophys.*, **29**, 101–114.

Bhattacharya, R., 1975: On forced vibration of anisotropic spherical shell of variable density. *Bull. Calcutta Math. Soc.*, **67**, 87–95.

Bhuta, P. G., and L. R. Koval, 1966: A viscous ring damper for a freely precessing satellite. *Int. J. Mech. Sci.*, **8**, 383–395.

Chakrabarti, R., 1974: Forced vibrations of a non-homogeneous isotropic elastic spherical shell. *Pure Appl. Geophys.*, **112**, 52–57.

Chakraborty, S. K., 1958: On disturbances generated by a pulse of pressure on the surface of a spherical cavity in an elastic medium. *Indian J. Theoret. Phys.*, **6**, 85–89.

Chakraborty, S. K., and A. B. Roy, 1965: Note on the propagation of waves from a spherical cavity in an infinite solid. *Pure Appl. Geophys.*, **61**, 23–28.

Chakraborty, S. K., and A. B. Roy, 1967: Propagation of waves in anisotropic visco-elastic solid from a spherical cavity. *Bull. Calcutta Math. Soc.*, **59**, 81–84.

Chakravorty, J. G., and P. K. Chaudhuri, 1983: On the propagation of waves due to a sudden impulse on the boundary of the spherical cavity. *Indian J. Pure Appl. Math.*, **14**, 965–973.

Chakravorty, M., and S. Bhaduri, 1981: Dynamic response of isotropic non-homogeneous spherical and cylindrical thick-walled shells to applied pressure on the inner boundary. *Bull. Calcutta Math. Soc.*, **73**, 232–236.

Chester, W., 1952: The reflection of a transient pulse by a parabolic cylinder and a paraboloid of revolution. *Quart. J. Mech. Appl. Math.*, **5**, 196–205.

Chonan, S., 1977: Response of an elastically supported finite beam to a moving load with consideration of the mass of the foundation. *Bull. JSME*, **20**, 1566–1571.

Chou, P. C., and P. F. Gordon, 1967: Radial propagation of axial shear waves in nonhomogeneous elastic media. *J. Acoust. Soc. Am.*, **42**, 36–41.

Clark, G. B., and G. B. Rupert, 1966: Plane and spherical waves in a Voigt medium. *J. Geophys. Res.*, **71**, 2047–2053.

Daimaruya, M., M. Naitoh, and K. Hamada, 1983: Propagation of elastic waves in a finite length bar with a variable cross section (in Japanese). *Nihon Kikai Gakkai Rombunshu (Trans. Japan Soc. Mech. Engrs.), Ser. A*, **49**, 1119–1125.

Daimaruya, M., M. Naitoh, and K. Hamada, 1984: Propagation of elastic waves in a finite length bar with a variable cross section. *Bull. JSME*, **27**, 872–878.

Daimaruya, M., M. Naitoh, and S. Onozaki, 1987: Dynamic behaviors of a finite length bar with a variable cross section colliding with a rigid wall (in Japanese). *Nihon Kikai Gakkai Rombunshu (Trans. Japan Soc. Mech. Engrs.), Ser. A*, **53**, 2324–2329.

Daimaruya, M., M. Naitoh, and S. Tanimura, 1988: Impact end stress and elastic response of a finite length bar with a variable cross-section colliding with a rigid wall. *J. Sound Vib.*, **121**, 105–115.

Das, A., and A. Ray, 1981: Stress response in a piezoelectric circular plate. *J. Math. Phys. Sci.*, **15**, 517–524.

Das, N. C., 1969: On mechanical response in a composite piezoelectric plate transducer characterized by a time-decaying polorization gradient. *Proc. Natl. Sci. India*, **A35**, **Suppl. II**, 173–181.

Das Gupta, S. C., 1954: Waves and stresses produced in an elastic medium due to impulse radial forces and twist on the surface of spherical cavity. *Geofis. Pura Appl.*, **27**, 1–6.

Datta, B. K., and P. R. Sengupta, 1984: Note on wave propagation in an infinite elastic space due to explosion at the cavity centre. *Rev. Roum. Sci. Tech., Ser. Mec. Appl.*, **29**, 487–492.

Datta, S., 1959: A note on stress waves in visco-elastic rod. *J. Technol.*, **4**, 41–46.

Davies, R. M., 1948: A critical study of the Hopkinson pressure bar. *Philos. Trans. R. Soc. London, Ser. A*, **240**, 376–457.

Davis, A. M. J., 1969: A semi-infinite elastic strip supporting a thin heavy overhanging beam. *Proc. Natl. Inst. Sci. India*, **A35**, 382–395.

De, S., 1970: Forced vibration of a thin non-homogeneous circular plate having a central hole. *Pure Appl. Geophys.*, **80**, 84–91.

De, S., 1971: Disturbances produced in an infinite cylindrically-aeolotropic medium. *Pure Appl. Geophys.*, **90**, 23–29.

Dörr, J., 1943: Der unendliche, federnd gebettete Balken unter dem Einfluss einer gleichförmig bewegten Last. *Ing.-Arch.*, **14**, 167–192.

Dubinkin, M. V., 1959: The propagation of waves in infinite plates. *J. Appl. Math. Mech.*, **23**, 1409–1413.

Duffey, T. A., and J. N. Johnson, 1981: Transient response of a pulsed spherical shell surrounded by an infinite elastic medium. *Int. J. Mech. Sci.*, **23**, 589–593.

Durant, N. J., 1960: Stress in a dynamically loaded helical spring. *Quart. J. Mech. Appl. Math.*, **13**, 251–256.

Dutta, K. L., and S. K. Chakraborty, 1990: Dynamic response of an elastic medium to random cylindrical sources. *Indian J. Pure Appl. Math.*, **21**, 867–877.

Eason, G., 1963: Propagation of waves from spherical and cylindrical cavities. *Z. Angew. Math. Phys.*, **14**, 12–23.

Esmeijer, W. L., 1949: On the dynamic behaviour of an elastically supported beam of infinite length, loaded by a concentrated force. *Appl. Sci. Res., Ser. A*, **1**, 151–168.

Ganguly, A., 1978: A note on the commencement of yielding due to suddenly applied pressure in a transversely isotropic medium. *Bull. Calcutta Math. Soc.*, **70**, 215–219.

Ghosh, S. L., 1966: Note on the longitudinal propagation of elastic disturbance in a thin inhomogeneous elastic rod. *Indian J. Theoret. Phys.*, **14**, 21–26.

Ghosh, S. L., 1966: Note on the longitudinal vibration of a paraboloidal bar. *Indian J. Theoret. Phys.*, **14**, 81–85.

Ghosh, S. L., 1966: Stress waves in a rod of viscoelastic linear Maxwell material. *Indian J. Pure Appl. Phys.*, **4**, 441–442.

Ghosh, S. S., 1969: On the disturbances in a viscoelastic medium due to blast inside a spherical cavity. *Pure Appl. Geophys.*, **75**, 93–97.

Gibson, R. E., R. L. Schiffman, and S. L. Pu, 1970: Plane strain and axially symmetric consolidation of a clay layer on a smooth impervious bar. *Quart. J. Mech. Appl. Math.*, **23**, 505–520.

Hall, L. H., 1953: Longitudinal vibrations of a vertical column by the method of Laplace transform. *Am. J. Phys.*, **21**, 287–292.

Hata, T., 1992: Stress-focusing effect in a solid cylinder caused by ramp unloading. *Nihon Kikai Gakkai Rombunshu (Trans. Japan Soc. Mech. Eng.), Ser. A*, **58**, 1684–1688.

Hill, J. L., 1966: Torsional-wave propagation from a rigid sphere semiembedded in an elastic half-space. *J. Acoust. Soc. Am.*, **40**, 376–379.

Huang, C.-L., 1969: On forced vibration of isotropic cylinders. *Appl. Sci. Res.*, **20**, 1–15.

Jacobs, C., 1971: Transient motions produced by disks oscillating torsionally about a state of rigid rotation. *Quart. J. Mech. Appl. Math.*, **24**, 221–236.

Jaeger, J. C., 1940: Magnetic screening by hollow circular cylinders. *Philos. Mag., Ser. 7*, **29**, 18–31.

Jeffreys, H., and E. R. Lapwood, 1957: The reflexion of a pulse within a sphere. *Proc. R. Soc. London, Ser. A*, **241**, 455–479.

Jiang, W., T. L. Wang, and W. K. Jones, 1991: Forced vibration of coupled extensional-torsional systems. *J. Eng. Mech. Div. (Am. Soc. Civ. Eng.)*, **117**, 1171–1190.

Jingu, T., K. Hisada, I. Nakahara, and S. Machida, 1985: Transient stress in a circular disk under diametrical impact loads. *Bull. JSME*, **28**, 13–19.

Jingu, T., and K. Nezu, 1985: Transient stress in an elastic sphere under diametrical concentrated impact loads. *Bull. JSME*, **28**, 2553–2561.

Jingu, T., and K. Nezu, 1985: Stress waves in an infinite medium under the diametrical concentrated impact loads on the spherical cavity. *Bull. JSME*, **28**, 2592–2598.

Jingu, T., H. Matsumoto, K. Nezu, and K. Sakamoto, 1987: The effect of local contact behavior on an impact load due to collision between a plate and a sphere (in Japanese). *Nihon Kikai Gakkai Rombunshu (Trans. Japan Soc. Mech. Engrs.), Ser. A*, **53**, 2331–2335.

Jones, R. P. N., 1954: The wave method for solving flexural vibration problems. *J. Appl. Mech.*, **21**, 75–80.

Kaliski, S., and E. Włodarczyk, 1967: Resonance of a longitudinal elastic-visco-plastic wave in a finite bar. *Proc. Vib. Prob.*, **8**, 113–128.

Kaliski, S., and E. Włodarczyk, 1967: On certain closed-form solutions of the propagation and reflection problem of an elastic-visco-plastic wave in a bar. *Arch. Mech. Stosow.*, **19**, 434–452.

Kecs, W. W., 1986: On the longitudinal vibration of the homogeneous, elastic rods. *Rev. Roum. Sci. Tech., Ser. Mec. Appl.*, **31**, 63–70.

Kubenko, V. D., 1997: Impact interaction of bodies with a medium. *Int. Appl. Mech.*, **33**, 933–957.

Lee, E. H., and I. Kanter, 1953: Wave propagation in finite rods of viscoelastic material. *J. Appl. Phys.*, **24**, 1115–1122.

Liu, T., and X.-Z. Zhang, 1994: Response of a disk mass on a half-space to dilatational waves. *J. Appl. Mech.*, **61**, 722–724.

Lowan, A. N., 1935: On transverse oscillations of beams under the action of moving variable loads. *Philos. Mag., Ser. 7*, **19**, 708–715.

Mandal, N. C., 1986: Torsional oscillation of a nonhomogeneous semiinfinite circular cylinder by an impulsive twist applied at one end. *Proc. Indian Natl. Sci. Acad.*, **A52**, 1218–1236.

Matsumoto, H., and S. Ujihashi, 1972: Deformations and stresses in a hollow sphere with spherical transversal isotropy under impulsive pressure. *Bull. JSME*, **15**, 1324–1332.

Matsumoto, H., E. Tsuchida, S. Miyao, and N. Tsunadu, 1976: Torsional stress wave propagation in a semi-infinite conical bar. *Bull. JSME*, **19**, 8–14.

Matsumoto, H., I. Nakahara, and H. Sekino, 1980: Identification of the dynamic visco-elastic properties under longitudinal impact. *Bull. JSME*, **23**, 1086–1091.

Mitra, A. K., 1961: Disturbance produced in a semi-infinite elastic medium by an impulsive twist applied on the plane surface. *Proc. Inst. Sci. India*, **A27**, 470–475.

Mitra, M., 1959: Solution of the buried source problem for an extended two-dimensional source in an elastic medium. *Proc. Inst. Sci. India*, **A25**, 236–242.

Morrison, J. A., 1956: Wave propagation in rods of Voigt material and visco-elastic materials with three-parameter models. *Quart. Appl. Math.*, **14**, 153–169.

Mukherjee, J., 1970: Waves produced in an infinite elastic medium due to normal forces and twists on the surface of a buried spherical source. *Pure Appl. Geophys.*, **82**, 11–18.

Mukherjee, S., 1968: Note on the propagation of waves in infinite spherically-aeolotropic medium produced by a blast in a spherical cavity. *Pure Appl. Geophys.*, **70**, 39–46.

Muthuswamy, V. P., and T. Raghavan, 1984: Forced vibrations of an isotropic, perfectly plastic circular cylinder having a rigid cylindrical inclusion. *Rev. Roum. Sci. Tech., Ser. Mec. Appl.*, **29**, 651–655.

Naitoh, M., and M. Daimaruya, 1985: The influence of a rise time of longitudinal impact on the propagation of elastic waves in a bar. *Bull. JSME*, **28**, 20–25.

Nakagawa, N., R. Kawai, and M. Akao, 1977: Behavior of impact waves at the interface of an elastic-viscoelastic bar. *Bull. JSME*, **20**, 1402–1408.

Negi, J. G., and T. Lal, 1969: Deformation of the shape of seismic pulses by a layer of non-uniform velocity distributions. *Z. Geophys.*, **35**, 589–610.

Newman, M. K., 1955: Effect of rotating inertia and shear on maximum strain in cantilever impact excitation. *J. Aeronaut. Sci.*, **22**, 313–320.

Odaka, T., and I. Nakahara, 1967: Stresses in an infinite beam impacted by an elastic bar. *Bull. JSME*, **10**, 863–872.

Olunloya, V. O. S., and K. Hutter, 1979: Forced vibration of a prestressed rectangular membrane: Near resonance response. *Acta Mech.*, **32**, 63–77.

Pal, B. R., 1987: Rotatory vibration of a thin non-homogeneous circular plate under shearing forces. *Rev. Roum. Sci. Tech., Ser. Mec. Appl.*, **32**, 389–396.

Pan, M., 1974: Forced vibrations of an non-homogeneous anisotropic elastic spherical shell. *Bull. Calcutta Math. Soc.*, **66**, 223–232.

Pan, M., 1975: Forced vibrations of an inhomogeneous anisotropic cylindrical shell. *Bull. Calcutta Math. Soc.*, **67**, 165–174.

Raghavan, T., and V. P. Muthuswamy, 1985: Radial vibrations in cylindrical soil massive in plastic yield state. *Rev. Roum. Sci. Tech., Ser. Mec. Appl.*, **30**, 31–36.

Reismann, H., and J. Gideon, 1971: Forced wave motion in an unbounded space surrounding a lined, spherical cavity. *Pure Appl. Geophys.*, **85**, 189–213.

Rizk, A. E.-F., and S. F. Radwan, 1993: Fracture of a plate under transient thermal stresses. *J. Therm. Stresses*, **16**, 79–102.

Sarker, A. K., 1965: Disturbance due to a shearing-stress applied to the boundary of a spherical cavity in an infinite viscoelastic medium. *Bull. Calcutta Math. Soc.*, **57**, 5–15.

Scott, W. E., 1967: Free-flight spin decay of a spin-stabilized shell containing a viscous liquid. *AIAA J.*, **5**, 2276–2277.

Sen, B., 1962: Note on the transient response of a linear visco-elastic plate in the form of an equilateral triangle. *Indian J. Theoret. Phys.*, **10**, 77–81.

Sheehan, J. P., and L. Debnath, 1972: Transient vibrations of an isotropic elastic sphere. *Pure Appl. Geophys.*, **99**, 37–48.

Sheehan, J. P., and L. Debnath, 1972: Forced vibrations of an anisotropic elastic sphere. *Arch. Mech.*, **24**, 117–125.

Shibuya, T., 1975: On the torsional impact of a thick elastic plate. *Int. J. Solids Struct.*, **11**, 803–811.

Shipitsina, E. M., 1972: Investigation of the strain of a hollow ball by a pulse loading according to three dimensional elasticity theory and the theory of shells. *Sov. Appl. Mech.*, **8**, 1194–1197.

Sinha, D. K., 1962: Note on responses in a piezoelectric plate transducer with a periodic step input. *Indian J. Theoret. Phys.*, **10**, 21–28.

Sinha, D. K., 1963: Note on electrical and mechanical responses with ramp-type input signal in piezoelectric plate transducer. *Indian J. Theoret. Phys.*, **11**, 93–99.

Sinha, D. K., 1965: A note on mechanical response in a piezoelectric transducer owing to an impulse voltage input. *Proc. Natl. Inst. Sci. India*, **A31**, 395–402.

Solarz, L., 1966: Aero-magneto-flutter of a plane duct of finite length. *Proc. Vib. Prob.*, **7**, 347–362.

Spillers, W. R., 1961: The quasi-static elastic and viscoelastic consolidating spherical cavity. *Int. J. Mech. Sci.*, **4**, 381–386.

Sur, S. P., 1961: A note on the longitudinal propagation of elastic disturbance in a thin inhomogeneous elastic rod. *Indian J. Theoret. Phys.*, **9**, 61–67.

Ting, T. C. T., and P. S. Symonds, 1964: Longitudinal impact on viscoplastic rods – Linear stress-strain rate laws. *J. Appl. Mech.*, **31**, 199–207.

Tomski, L., 1981: Longitudinal mass impact of hydraulic servo. *Z. Angew. Math. Mech.*, **61**, T191–T193.

Ujihashi, S., K. Tanaka, H. Matsumoto, and T. Adachi, 1997: Estimating method of the impact strength of FRP plates by drop weight tests (in Japanese). *Nihon Kikai Gakkai Rombunshu (Trans. Japan Soc. Mech. Eng.), Ser. A*, **63**, 2560–2567.

Zverev, I. N., 1950: Propagation of perturbations in a visco-elastic and visco-plastic rod (in Russian). *Prik. Mat. Mek.*, **14**, 295–302.

Thermoelasticity and Magnetoelasticity

Agarwal, M., 1978: Stresses and displacements in an infinite elastic medium due to some types of temperature distribution inside a spherical cavity. *Bull. Calcutta Math. Soc.*, **70**, 1–10.

Alzheimer, W. E., M. J. Forrestal, and W. B. Murfin, 1968: Transient response of cylindrical, shell-core systems. *AIAA J.*, **6**, 1861–1866.

Akkas, N., and U. Zakout, 1997: Transient response of an infinite elastic medium containing a spherical cavity with and without a shell embedment. *Int. J. Eng. Sci.*, **35**, 89–112.

Bakshi, S. K., 1969: Magnetoelastic disturbances in a perfectly conducting cylindrical shell in a constant magnetic field. *Pure Appl. Geophys.*, **76**, 56–64.

Barber, A. D., J. H. Weiner, and B. A. Boley, 1957: An analysis of the effect of thermal contact resistance in a sheet-stringer structure. *J. Aeronaut. Sci.*, **24**, 232–234.

Biswas, P., 1976: Large deflection in heated equilateral triangular plate. *Indian J. Pure Appl. Math.*, **7**, 257–264.

Chattopadhyay, N. C., 1967: Note on thermo-elastic stresses in a thin semi-infinite rod due to some time-dependent temperature applied to its free end. *Indian J. Theoret. Phys.*, **15**, 49–55.

Chaudhuri, B., 1969: On magnetoelastic disturbances in an aeolotropic elastic cylinder placed in a magnetic field. *Pure Appl. Geophys.*, **73**, 60–68.

Chaudhuri, B., 1969: On mechanical response in a piezoelectric annular disk transducer owing to a step input with a prescribed temperature field. *Proc. Natl. Inst. Sci. India*, **A35**, 709–718.

Chaudhuri, B. R., 1971: Dynamical problems of thermo-magnetoelasticity for cylindrical regions. *Indian J. Pure Appl. Math.*, **2**, 631–656.

Chen, S.-Y., 1958: Transient temperature and thermal stresses in skin of hypersonic vehicle with variable boundary conditions. *Trans. ASME*, **80**, 1389–1394.

Choudhuri, S. K. R., 1970: Note on the thermoelastic stress in a thin rod of finite length due to some constant temperature applied to its free end, the other end being fixed and insulated. *Indian J. Theoret. Phys.*, **18**, 99–106.

Choudhuri, S. K. R., 1972: Thermoelastic stress in a rod due to distributed time-dependent heat sources. *AIAA J.*, **10**, 531–533.

Chouhury, A. R., 1971: On disturbances in a piezoelectric layer permeated by a magnetic field. *Pure Appl. Geophys.*, **87**, 111–116.

Chouhury, A. R., 1971: On the electrical response of a non-homogeneous piezo-electric transducer, due to shock-loaded stress. *Pure Appl. Geophys.*, **87**, 117–123.

Chu, H. S., C. K. Chen, and C. I. Weng, 1983: Transient response of circular pins. *J. Heat Transfer*, **105**, 205–208.

Chu, J.-Y., 1963: Elasto-dynamic stresses produced by a source of heat located around a spherical cavity. *Arch. Mech. Stosow.*, **15**, 565–582.

Ciałkowski, M. J., and K. W. Grysa, 1980: On a certain inverse problem of temperature and thermal stress fields. *Acta Mech.*, **36**, 169–185.

Danilovskaya, V. I., and V. N. Zubchaninova, 1968: Temperature fields and stresses created in a plate by radiant energy. *Sov. Appl. Mech.*, **4, No. 1**, 63–66.

Das, B. R., 1961: Note on thermoelastic stresses in a thin semi-infinite rod due to some time dependent temperature applied to its free end. *Indian J. Theoret. Phys.*, **9**, 49–55.

Das, N. C., 1967: A note on disturbances in an inhomogeneous elastic bar acted upon by a magnetic field. *Arch. Mech. Stosow.*, **19**, 765–769.

Derski, W., 1958: The state of stress in a thin circular ring, due to a non-steady temperature field. *Arch. Mech. Stosow.*, **10**, 255–269.

Derski, W., 1958: On transient thermal stresses in a thin circular plate. *Arch. Mech. Stosow.*, **10**, 551–558.

Derski, W., 1959: Non-steady-state of thermal stresses in a layered elastic space with a spherical cavity. *Arch. Mech. Stosow.*, **11**, 303–316.

Derski, W., 1961: A dynamic problem of thermoelasticity concerning a thin circular plate. *Arch. Mech. Stosow.*, **13**, 177–184.

Dhaliwal, R. S., and K. L. Chowdhury, 1968: Dynamic problems of thermoelasticity for cylindrical regions. *Arch. Mech. Stosow.*, **20**, 46–66.

Dhar, A. K., 1982: Heat-source problem of thermoelasticity in a non-simple elastic medium. *Indian J. Pure Appl. Math.*, **13**, 1384–1392.

Dhar, A. K., 1983: Non-simple quasi-static thermoelastic deflection of a thin clamped circular plate. *Rev. Roum. Sci. Tech., Ser. Mec. Appl.*, **28**, 431–437.

Eason, G., 1962: Thermal stress in anisotropic cylinders. *Proc. Edinburgh Math. Soc., Ser. 2*, **13**, 159–164.

Ghosn, A. H., and M. Sabbaghian, 1982: Quasi-static coupled problems of thermoelasticity for cylindrical regions. *J. Therm. Stresses*, **5**, 299–313.

Giri, R. R., 1967: Note on the mechanical response of a composite piezoelectric transducer subjected to a constant heat flow and to a time-dependent electrical voltage. *Proc. Natl. Sci. India*, **A33**, 325–332.

Glauz, R. D., and E. H. Lee, 1954: Transient wave analysis in a linear time-dependent material. *J. Appl. Phys.*, **25**, 947–953.

Hata, T., 1988: A dynamic thermal stress wave in a hollow sphere caused by instantaneous uniform heating (in Japanese). *Nihon Kikai Gakkai Rombunshu (Trans. Japan Soc. Mech. Engrs.), Ser. A*, **54**, 2000–2005.

Hata, T., 1989: Stress-focusing effect in a solid sphere caused by instantaneous uniform heating (in Japanese). *Nihon Kikai Gakkai Rombunshu (Trans. Japan Soc. Mech. Engrs.), Ser. A*, **55**, 1112–1115.

Hata, T., 1990: Stress-focusing effect in a solid sphere caused by instanteous uniform heating. *JSME Int. J., Ser. 1*, **33**, 33–36.

Hata, T., 1991: Stress-focusing effect in a uniformly heated solid sphere. *J. Appl. Mech.*, **58**, 58–63.

Hata, T., 1991: A dynamic thermoelasto/viscoplastic solution in a transversely isotropic hollow sphere subjected to instantaneous uniform heating (in Japanese). *Nihon Kikai Gakkai Rombunshu (Trans. Japan Soc. Mech. Engrs.), Ser. A*, **57**, 838–844.

Hata, T., 1993: Analysis of thermal stress-focusing effect in a solid cylinder using ray series (in Japanese). *Nihon Kikai Gakkai Rombunshu (Trans. Japan Soc. Mech. Engrs.), Ser. A*, **59**, 2960–2964.

Hata, T., 1994: Reconsideration of the stress-focusing effect in a uniformly heated solid cylinder. *J. Appl. Mech.*, **61**, 676–680.

Hata, T., 1995: Stress-focusing effect in a transversely isotropic cylinder caused by instantaneous heating (in Japanese). *Nihon Kikai Gakkai Rombunshu (Trans. Japan Soc. Mech. Engrs.), Ser. A*, **61**, 1404–1408.

Hata, T., 1998: Analysis of the stress-focusing effect in a spherical inclusion embedded in an infinite elastic medium (in Japanese). *Nihon Kikai Gakkai Rombunshu (Trans. Japan Soc. Mech. Engrs.), Ser. A*, **64**, 942–948.

Horvay, G., 1954: Transient thermal stresses in circular disks and cylinders. *Trans. ASME*, **76**, 127–135.

Hsu, T. R., 1969: Thermal shock on a finite disk due to an instantaneous point heat source. *J. Appl. Mech.*, **36**, 113–120.

Hu, C.-L., 1969: Spherical model for an acoustical wave generated by rapid laser heating in a liquid. *J. Acoust. Soc. Am.*, **46**, 728–736.

Jaeger, J. C., 1945: On thermal stresses in circular cylinders. *Philos. Mag., Ser. 7*, **36**, 418–428.

Jingu, T., and K. Nezu, 1985: Stress waves in an infinite medium under the diametrical concentrated impact loads on the spherical cavity (in Japanese). *Nihon Kikai Gakkai Rombunshu (Trans. Japan Soc. Mech. Engrs.), Ser. A*, **51**, 942–949.

Jordan, P. M., and P. Puri, 2001: Thermal stresses in a spherical shell under three thermoelastic models. *J. Therm. Stresses*, **24**, 47–70.

Khan, M. A., 1972: Heating of a piezoelectric rod of semi-infinite length due to prescribed heat flux into it through radiation condition. *Pure Appl. Geophys.*, **94**, 66–73.

Mahalanabis, M., 1971: On disturbances in an inhomogeneous viscoelastic bar of Reiss type in a magnetic field. *Pure Appl. Geophys.*, **92**, 52–61.

Mahalanabis, R. K., 1968: On temperature changes due to application of impulsive pressure on the surface of a spherical cavity. *Rev. Roum. Sci. Tech., Ser. Mec. Appl.*, **13**, 121–125.

Maiti, P. C., 1978: Quasi-static thermal deflection of a thin clamped circular plate subjected to random temperature distribution on its upper face. *Indian J. Pure Appl. Math.*, **9**, 541–547.

Murthy, M. G., 1966: A dynamical problem of thermoelasticity concerning a circular disc. *Indian J. Pure Appl. Phys.*, **4**, 367–370.

Nowacki, W., 1959: Some three-dimensional problems of thermoelasticity. *J. Appl. Math. Mech.*, **23**, 651–665.

Nowacki, W., 1959: Two one-dimensional problems of thermoelasticity. *Arch. Mech. Stosow.*, **11**, 333–345.

Ohyoshi, T., 1980: Transient responses of a finite elastic cylinder to a torsional end shear (for estimation of applicability of simple elementary theory) (in Japanese). *Nihon Kikai Gakkai Rombunshu (Trans. Japan Soc. Mech. Engrs.), Ser. A*, **46**, 907–913.

Pal, R., 1972: Transient stresses in inhomogeneous elastic hollow circular cylinder due a thermal shock. *Bull. Calcutta Math. Soc.*, **64**, 107–119.

Paria, G., 1963: Study of the transient effect on a superconducting sphere under the influence of a varying magnetic field. *Indian J. Pure Appl. Phys.*, **1**, 87–90.

Paton, A., and W. Millar, 1964: Compression of magnetic field between two semi-infinite slabs of constant conductivity. *J. Appl. Phys.*, **35**, 1141–1146.

Pohle, F. V., and H. Oliver, 1954: Temperature distribution and thermal stresses in a model of a supersonic wing. *J. Aeronaut. Sci.*, **21**, 8–16.

Roy, P., 1970: On longitudinal disturbances in a viscoelastic solid of Reiss type placed in a magnetic field. *Pure Appl. Geophys.*, **82**, 29–33.

Sharma, J. N., 1997: Temperature distribution in a generalized thermoelastic solid due to heat sources. *Indian J. Pure Appl. Math.*, **28**, 1303–1316.

Sherief, H. H., and M. A. Ezzat, 1994: Solution of the generalized problem of thermoelasticity in the form of series of functions. *J. Therm. Stresses*, **17**, 75–95.

Sinha, D. K., 1965: A note on torsional disturbances in an elastic cylinder in a magnetic field. *Proc. Vib. Prob.*, **6**, 91–97.

Steketee, J. A., 1973: EM induction in a semi-infinite solid, impulsively moving in a uniform magnetic field. *Appl. Sci. Res.*, **27**, 307–320.

Sugano, Y., and J. Kimoto, 1991: A stochastic analysis of unaxisymmetric thermal stresses in a nonhomogeneous hollow circular plate (in Japanesse). *Nihon Kikai Gakkai Rombunshu (Trans. Japan Soc. Mech. Eng.), Ser. A*, **57**, 845–851.

Tsui, T., and H. Kraus, 1965: Thermal stress-wave propagation in hollow elastic spheres. *J. Acoust. Soc. Am.*, **37**, 730–737.

Wadhawan, M. C., 1974: Dynamic thermoelastic response of a cylinder. *Pure Appl. Geophys.*, **112**, 73–82.

Wheeler, L., 1973: Focusing of stress waves in an elastic sphere. *J. Acoust. Soc. Am.*, **53**, 521–524.

Wave Equation

Afanas'ev, E. F., 1962: Diffraction of a nonstationary pressure wave on a moving plate. *J. Appl. Math. Mech.*, **26**, 268–276.

Barakat, R. G., 1960: Transient diffraction of scalar waves by a fixed sphere. *J. Acoust. Soc. Am.*, **32**, 61–66.

Barakat, R. G., 1961: Propagation of acoustic pulses from a circular cylinder. *J. Acoust. Soc. Am.*, **33**, 1759–1764.

Bromwich, T. J. I'a., 1928: Some solutions of the electromagnetic equations, and of the elastic equations, with applications to the problem of secondary waves. *Proc. London Math. Soc., Ser. 2*, **28**, 438–475.

Cohen, L., 1923: Electrical oscillations on lines. *J. Franklin Inst.*, **195**, 45–58.

Dunn, H. S., and W. G. Dove, 1967: Focusing of aperiodic waves in a linear viscous medium. *J. Acoust. Soc. Am.*, **42**, 613–615.

Lowan, A. N., 1938: On wave-motion for infinite domains. *Philos. Mag., Ser. 7*, **26**, 340–360.

Lowan, A. N., 1939: On the problem of wave-motion for sub-infinite domains. *Philos. Mag., Ser. 7*, **27**, 182–194.

Lowan, A. N., 1941: On the problem of wave-motion for the wedge of an angle. *Philos. Mag., Ser. 7*, **31**, 373–381.

Masket, A. V., 1946: Forced vibrations of a whirling wire. *Philos. Mag., Ser. 7*, **37**, 426–432.

Masket, A. V., and A. C. Vastano, 1962: Interior value problems of mathematical physics. Part I. Wave propagation. *Am. J. Phys.*, **30**, 687–696.

Pipes, L. A., 1938: Operational solution of the wave equation. *Philos. Mag., Ser. 7*, **26**, 333–340.

Sattarov, R. M., 1977: Some cases of unsteady motion of viscoplastic media in an infinitely long viscoelastic tube. *J. Appl. Mech. Tech. Phys.*, **18**, 355–359.

Tranter, C. J., 1942: The application of the Laplace transformation to a problem on elastic vibrations. *Philos. Mag., Ser. 7*, **33**, 614–622.

Yen, D. H. Y., 2003: Green's function for the damped wave equation on a finite interval subject to two Robin boundary conditions. *Chinese J. Mech., Ser. A*, **19**, 257–262.

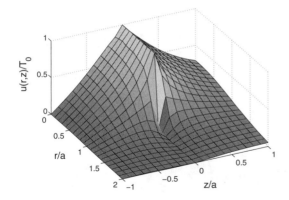

Chapter 3
Methods Involving Single-Valued Fourier and Hankel Transforms

In the previous chapter we used Laplace transforms to reduce the wave or heat equation to a boundary-value problem. When the spatial domain is of infinite or semi-infinite extent, an alternative possibility occurs if we use Fourier or Hankel transforms to reduce the partial differential equation to an initial-value problem. The purpose of this chapter is to work out the details of this alternative method.

3.1 INVERSION OF FOURIER TRANSFORMS BY CONTOUR INTEGRATION

In those instances when $f(t)$ exists for all t, is continuous and has a piecewise continuous derivative in any interval, the appropriate transform is the Fourier transform

$$F(\omega) = \int_{-\infty}^{\infty} f(t)e^{-i\omega t}\, dt. \qquad (3.1.1)$$

Given a Fourier transform, we compute its inverse from the inversion integral

$$f(t) = \frac{1}{2\pi} \int_{-\infty}^{\infty} F(\omega)e^{i\omega t}\, d\omega. \qquad (3.1.2)$$

Proof:[1] Consider the integral

$$I = \int_{-\infty}^{\infty} F(k)e^{ikx}\, dk \tag{3.1.3}$$

for all real values of k. Then,

$$I = \int_{-\infty}^{\infty} e^{ikx} \left[\int_{-\infty}^{\infty} f(x)e^{-ikx}\, dx \right] dk \tag{3.1.4}$$

$$= \int_{-\infty}^{0} e^{ikx} \left[\int_{C_1} f(z)e^{-ikz}\, dz \right] dk + \int_{0}^{\infty} e^{ikx} \left[\int_{C_2} f(z)e^{-ikz}\, dz \right] dk, \tag{3.1.5}$$

where C_1 is a contour that runs along and just above the real axis and C_2 is a contour that runs along and just below the real axis. In both integrals $\mathrm{Re}(ikz) > 0$. We now reverse the order of integration and obtain

$$I = \int_{C_1} f(z) \left[\int_{-\infty}^{0} e^{-ik(z-x)}\, dx \right] dz + \int_{C_2} f(z) \left[\int_{0}^{\infty} e^{-ik(z-x)}\, dx \right] dz \tag{3.1.6}$$

$$= -\int_{C_1} \frac{f(z)}{i(z-x)}\, dz + \int_{C_2} \frac{f(z)}{i(z-x)}\, dz = \oint_{C} \frac{f(z)}{i(z-x)}\, dz, \tag{3.1.7}$$

where C is a closed, counterclockwise contour between $-\infty + 0i$ and $\infty + 0i$. From Cauchy's integral formula, we find that

$$\int_{-\infty}^{\infty} F(k)e^{ikx}\, dk = 2\pi f(x). \tag{3.1.8}$$

$\square$

In Section 3.2 to Section 3.4 we will use Fourier transforms to solve partial differential equations where one or more of the independent variables extend from $-\infty$ to ∞. Although Equation 3.1.2 could be evaluated by direct integration, the most widely encountered technique is through contour integration. In this method we convert the integration along the real axis into a closed contour integration by the addition of an infinite semicircle in the upper or lower half-plane, as dictated by Jordan's lemma. The following examples illustrate the technique.

---

[1] Taken from MacRobert, T. M., 1931: Fourier integrals. *Proc. R. Soc. Edinburgh*, **51**, 116–126.

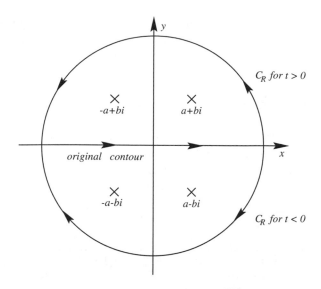

Figure 3.1.1: Contours used to find the inverse of the Fourier transform given by Equation 3.1.9.

• **Example 3.1.1**

For our first example, we find the inverse for

$$F(\omega) = \frac{1}{\omega^4 + 2(b^2 - a^2)\omega^2 + a^4 + 2a^2b^2 + b^4}, \qquad 0 < a, b, \qquad (3.1.9)$$

$$= \frac{1}{(\omega + a + bi)(\omega + a - bi)(\omega - a + bi)(\omega - a - bi)}. \qquad (3.1.10)$$

From the inversion integral,

$$f(t) = \frac{1}{2\pi} \int_{-\infty}^{\infty} \frac{e^{it\omega}}{\omega^4 + 2(b^2 - a^2)\omega^2 + a^4 + 2a^2b^2 + b^4} \, d\omega, \qquad (3.1.11)$$

or

$$f(t) = \frac{1}{2\pi} \oint_C \frac{e^{itz}}{z^4 + 2(b^2 - a^2)z^2 + a^4 + 2a^2b^2 + b^4} \, dz$$

$$- \frac{1}{2\pi} \int_{C_R} \frac{e^{itz}}{z^4 + 2(b^2 - a^2)z^2 + a^4 + 2a^2b^2 + b^4} \, dz, \qquad (3.1.12)$$

where C denotes a closed contour consisting of the entire real axis plus C_R. Because $f(z) = 1/[z^4 + 2(b^2 - a^2)z^2 + a^4 + 2a^2b^2 + b^4]$ tends to zero uniformly as $|z| \to \infty$ and $m = t$, the second integral in Equation 3.1.12 vanishes by Jordan's lemma if C_R is a semicircle of infinite radius in the upper half of the z-plane when $t > 0$ and a semicircle in the lower half of the z-plane when $t < 0$. See Figure 3.1.1.

Next, we must find the location and nature of the singularities. The denominator of the integrand in Equation 3.1.12 has four zeros, located at $z = \pm a \pm bi$. Therefore, we can rewrite Equation 3.1.12 as

$$f(t) = \frac{1}{2\pi} \oint_C \frac{e^{itz}}{(z+a+bi)(z+a-bi)(z-a+bi)(z-a-bi)} \, dz. \quad \textbf{(3.1.13)}$$

Consider now $t > 0$. As stated earlier, we close the line integral with an infinite semicircle in the upper half-plane. Inside this closed contour, there are two singularities: $z = \pm a + bi$. For these poles,

$$\operatorname{Res}\left[\frac{e^{itz}}{z^4 + 2(b^2 - a^2)z^2 + a^4 + 2a^2b^2 + b^4}; a + bi\right]$$

$$= \lim_{z \to a+bi} \frac{(z-a-bi)e^{itz}}{(z+a+bi)(z+a-bi)(z-a+bi)(z-a-bi)} \quad \textbf{(3.1.14)}$$

$$= \frac{e^{it(a+bi)}}{4(a+bi)^3 + 4(b^2 - a^2)(a+bi)} \quad \textbf{(3.1.15)}$$

$$= e^{-bt}\frac{[a\sin(at) - b\cos(at) - ia\cos(at) - ib\sin(at)]}{8ab(a^2 + b^2)}. \quad \textbf{(3.1.16)}$$

Similarly,

$$\operatorname{Res}\left[\frac{e^{itz}}{z^4 + 2(b^2 - a^2)z^2 + a^4 + 2a^2b^2 + b^4}; -a + bi\right]$$

$$= -e^{-bt}\frac{[a\sin(at) - b\cos(at) + ia\cos(at) + ib\sin(at)]}{8ab(a^2 + b^2)}. \quad \textbf{(3.1.17)}$$

Consequently, the inverse Fourier transform is

$$f(t) = e^{-bt}\frac{a\cos(at) + b\sin(at)}{4ab(a^2 + b^2)} \quad \textbf{(3.1.18)}$$

for $t > 0$. For $t < 0$, we take the semicircle in the lower half-plane because the contribution from the semicircle vanishes as $R \to \infty$. After evaluating the residues from the two poles at $z = \pm a - bi$ in the lower half-plane,

$$f(t) = e^{bt}\frac{a\cos(at) - b\sin(at)}{4ab(a^2 + b^2)} \quad \textbf{(3.1.19)}$$

for $t < 0$. Therefore, the inverse is

$$f(t) = e^{-b|t|}\frac{a\cos(a|t|) + b\sin(a|t|)}{4ab(a^2 + b^2)}. \quad \textbf{(3.1.20)}$$

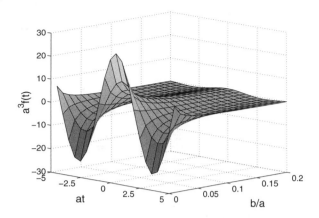

Figure 3.1.2: Plot of the inverse given by Equation 3.1.9.

Figure 3.1.2 illustrates this inverse as a function of at and b/a.

• **Example 3.1.2**

Let us find the inverse of the transform[2]

$$F(\omega) = -\frac{k + \omega}{\omega} \frac{I_n(\omega/\alpha)}{I_n'(\omega/\alpha)}, \tag{3.1.21}$$

where k and α are constants. We will assume that k and n are nonzero.

Once again we convert the inversion integral,

$$f(t) = -\frac{1}{2\pi} \int_{-\infty}^{\infty} \frac{k + \omega}{\omega} \frac{I_n(\omega/\alpha)}{I_n'(\omega/\alpha)} e^{i\omega t} \, d\omega, \tag{3.1.22}$$

into a closed contour by the addition of an infinite semicircle in either the upper or lower half-plane. To see which one is correct, we employ the asymptotic expansion[3] for $I_n(z)$ as $|z| \to \infty$. Because

$$\left| e^{izt} \frac{k + z}{z} \frac{I_n(z/\alpha)}{I_n'(z/\alpha)} \right| \to e^{-Rt \sin(\theta)} \to 0, \tag{3.1.23}$$

---

[2] Taken from Matsuzaki, Y., and Y. C. Fung, 1977: Unsteady fluid dynamic forces on a simply-supported circular cylinder of finite length conveying a flow, with applications to stability analysis. *J. Sound Vib.*, **54**, 317–330. Published by Academic Press Ltd., London, UK.

[3] See Watson, G. N., 1966: *Theory of Bessel Functions*. Cambridge University Press, p. 203.

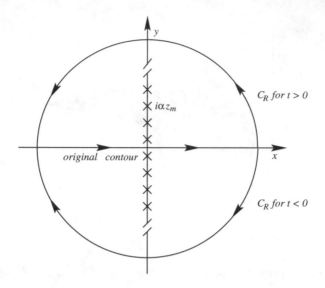

Figure 3.1.3: Contours used to find the inverse of the Fourier transform given by Equation 3.1.21.

as $R \to \infty$, $0 < t$ and $0 < \theta < \pi$, we close the contour in the upper half-plane. A similar argument shows that the contour should be closed in the lower half-plane for $t < 0$. See Figure 3.1.3.

One possible source of difficulty is the presence of z in the denominator of the integral because it might introduce a singularity along the contour of integration. However, from the power series expansion[4] of $I_n(z)$,

$$e^{izt}\frac{k+z}{z}\frac{I_n(z/\alpha)}{I_n'(z/\alpha)} \sim e^{izt}\frac{k}{n} \tag{3.1.24}$$

as $z \to 0$. Furthermore, because $I_n'(\pm ix) = (\pm i)^{n-1} J_n'(x)$, where x is real and positive, we conclude that the contour integral

$$\oint e^{izt}\frac{k+z}{z}\frac{I_n(z/\alpha)}{I_n'(z/\alpha)}\,dz \tag{3.1.25}$$

has only simple poles. They lie along the imaginary axis $\omega = \pm i\alpha z_m$, where z_m is the mth root of $J_n'(z_m) = 0$, excluding $z_0 = 0$, and $m = 1, 2, 3, \ldots$. Consequently, the value of the closed contour equals $2\pi i$ times the sum of the residues. This leads to

$$f(t) = \sum_{m=1}^{\infty}\frac{i\alpha z_m + k}{z_m}\frac{J_n(z_m)}{J_n''(z_m)}e^{-\alpha z_m t} \tag{3.1.26}$$

---

[4] *Ibid.*, p. 77

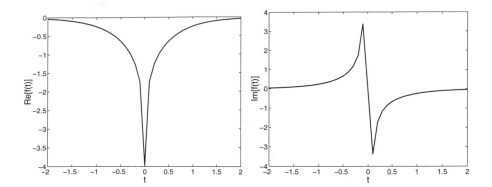

Figure 3.1.4: Plot of the inverse of the Fourier transform given by Equation 3.1.21 when $\alpha = 1$, $k = 2$ and $n = 1$.

for $t > 0$. A similar analysis yields the inverse

$$f(t) = \sum_{m=1}^{\infty} \frac{k - i\alpha z_m}{z_m} \frac{J_n(z_m)}{J_n''(z_m)} e^{\alpha z_m t} \qquad (3.1.27)$$

for $t < 0$ when we close the contour in the lower half-plane. Figure 3.1.4 illustrates this inverse as a function of t when $\alpha = 1$, $k = 2$ and $n = 1$.

• **Example 3.1.3**

Next, we find the inverse of

$$F(\omega) = \frac{\cos(\omega x)}{\cosh(\omega b)}, \qquad 0 < b, t, x. \qquad (3.1.28)$$

From the inversion integral,

$$f(t) = \frac{1}{2\pi} \int_{-\infty}^{\infty} \frac{\cos(\omega x)}{\cosh(\omega b)} e^{i\omega t} \, d\omega = \frac{1}{4\pi} \int_{-\infty}^{\infty} \left[\frac{e^{i\omega(t+x)}}{\cosh(\omega b)} + \frac{e^{i\omega(t-x)}}{\cosh(\omega b)} \right] d\omega \tag{3.1.29}$$

after rewriting $\cos(\omega x)$ in terms of complex exponentials. To convert Equation 3.1.29 into a contour integral, we close the line integral with a semicircle in either the upper or lower half-plane. See Figure 3.1.5. For x and $t > 0$, Jordan's lemma dictates that we choose a contour in the upper half-plane for all x and t for the first term in the second integral of Equation 3.1.29. On the other hand, Jordan's lemma requires the upper half-plane for $x < t$ and the lower half-plane for $t < x$ for the second term.

Regardless of how we close the contour, the poles of the integrand lie along the imaginary axis. Because $\cosh(\omega b) = \cos(i\omega b)$, the poles are located

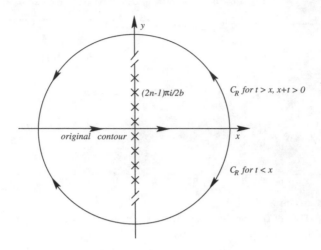

Figure 3.1.5: Contours used to find the inverse of the Fourier transform given by Equation 3.1.28.

at $z_n = \pm(2n + 1)\pi i/(2b)$, where $n = 0, 1, 2, 3, \ldots$. Consequently, for all positive x and t,

$$\frac{1}{2\pi i} \oint_C \frac{e^{iz(t+x)}}{\cosh(zb)}\, dz = \sum_{n=0}^{\infty} \frac{\exp[-(2n+1)\pi(x+t)/(2b)]}{b\sinh[i(2n+1)\pi/2]} \tag{3.1.30}$$

$$= \frac{1}{bi} \sum_{n=0}^{\infty} (-1)^n \exp\left[-\frac{(2n+1)\pi(t+x)}{2b}\right]. \tag{3.1.31}$$

Similarly for the last term in Equation 3.1.29,

$$\frac{1}{2\pi i} \oint_C \frac{e^{iz(t-x)}}{\cosh(zb)}\, dz = \frac{1}{bi} \sum_{n=0}^{\infty} (-1)^n \exp\left[-\frac{(2n+1)\pi|t-x|}{2b}\right]. \tag{3.1.32}$$

Therefore,

$$f(t) = \frac{1}{2b} \sum_{n=0}^{\infty} (-1)^n \left\{ \exp\left[-\frac{(2n+1)\pi(t+x)}{2b}\right] + \exp\left[-\frac{(2n+1)\pi|t-x|}{2b}\right] \right\}. \tag{3.1.33}$$

Figure 3.1.6 illustrates this inverse as a function of x and t.

• **Example 3.1.4**

For our next example, we find the inverse[5] of the bilateral Laplace transform

$$F(\omega) = \frac{ia}{\omega(\omega^2 - ia\omega - a)}, \qquad 0 < a < 4. \tag{3.1.34}$$

---

[5] The appearance of singularities along the real axis of the ω-plane indicates that we

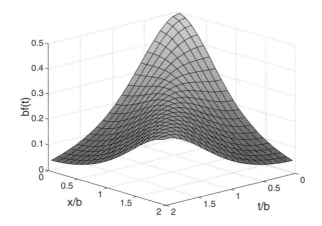

Figure 3.1.6: Plot of the inverse of the Fourier transform given by Equation 3.1.28.

We can convert the inversion integral

$$f(t) = \frac{1}{2\pi} \int_{-\infty-\epsilon i}^{\infty-\epsilon i} F(\omega)e^{i\omega t}\, d\omega, \qquad 0 < \epsilon \ll 1, \qquad (3.1.35)$$

into the closed contour integral

$$f(t) = \frac{1}{2\pi} \oint_C \frac{ia\, e^{izt}}{z(z^2 - iaz + a)}\, dz \qquad (3.1.36)$$

by adding an infinite semicircle in the upper half-plane as required by Jordan's lemma for $t > 0$ and then applying the residue theorem. See Figure 3.1.7. Three simple poles lie within the contour: $z = 0$ and $z = \pm\beta + ia/2$, where $\beta = \sqrt{a - a^2/4}$. Employing the residue theorem,

$$f(t) = 1 - \frac{ae^{-at/2}}{2\beta(\beta^2 + a^2/4)}\left[\left(\beta - \frac{ia}{2}\right)e^{i\beta t} + \left(\beta + \frac{ia}{2}\right)e^{-i\beta t}\right] \qquad (3.1.37)$$

$$= 1 - e^{-at/2}\left[\cos(\beta t) + \frac{2\beta}{4-a}\sin(\beta t)\right], \qquad (3.1.38)$$

where we have used the definition of β to simplify the resulting expression. Figure 3.1.8 illustrates this inverse as a function of a and t.

---

have a bilateral or two-sided Laplace transform. Usually the contour passes just below the real axis of the ω-plane [see Equation 1.1.4]. However, the only strictly valid method for determining whether we should pass above or below a singularity on the real axis consists of introducing a small amount of friction in the original formulation of the problem. This results in the singularity moving off the real axis. An alternative method tests various combinations of passing above and below the singularities and chooses the one which ensures solutions that radiate energy off to infinity.

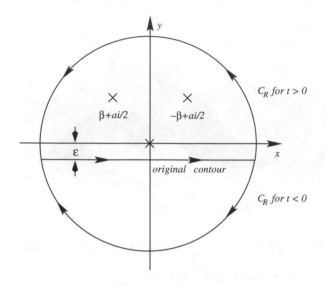

Figure 3.1.7: Contours used to find the inverse of the Fourier transform given by Equation 3.1.34.

• Example 3.1.5

Let us now invert the transform

$$F(\omega) = \frac{\sinh(\zeta)}{\zeta \cosh(\zeta)}, \tag{3.1.39}$$

where $\zeta^2 = (\omega - i\epsilon)^2 - k^2$ and $0 < \epsilon \ll 1$. The poles of the transform are located at $\zeta = 0$ and $\zeta_n = \pm(2n + 1)\pi i/2$, where $n = 0, 1, 2, 3, \ldots$. Closer examination reveals that $\zeta = 0$ is a removable singularity.

Solving for the ω_n associated with each ζ_n,

$$\omega_n^{\pm} = \pm\sqrt{k^2 + \zeta_n^2} + i\epsilon = \pm\sqrt{k^2 - \frac{(2n+1)^2\pi^2}{4}} + i\epsilon. \tag{3.1.40}$$

From Equation 3.1.40 we see that for sufficiently large n, say $N + 1$, ω_n becomes purely imaginary. Consequently, there are $N + 1$ poles that lie along and above the real axis; the vast majority lie along the imaginary axis.

To find the inverse for $t > 0$, we convert the line integral into a closed contour by adding an infinite semicircle in the upper half-plane. Therefore,

$$f(t) = \frac{1}{2\pi} \int_{-\infty}^{\infty} \frac{\sinh(\zeta)}{\zeta \cosh(\zeta)} e^{i\omega t} \, d\omega = \frac{1}{2\pi} \oint_C \frac{\sinh(\zeta)}{\zeta \cosh(\zeta)} e^{izt} \, dz. \tag{3.1.41}$$

Upon applying the residue theorem,

$$f(t) = i \sum_{n=0}^{N} \frac{\sinh(\zeta)e^{-\epsilon t}}{\frac{d}{d\zeta}[\zeta \cosh(\zeta)]\frac{d\zeta}{dz}\big|_{z=\omega_n^+}} \exp\left[it\sqrt{k^2 - \frac{(2n+1)^2\pi^2}{4}}\right]$$

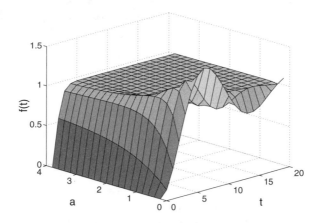

Figure 3.1.8: Plot of the inverse of Fourier transform given by Equation 3.1.34.

$$+ i \sum_{n=0}^{N} \frac{\sinh(\zeta)e^{-\epsilon t}}{\frac{d}{d\zeta}[\zeta \cosh(\zeta)]\frac{d\zeta}{dz}|_{z=\omega_n^-}} \exp\left[-it\sqrt{k^2 - \frac{(2n+1)^2\pi^2}{4}}\right]$$

$$+ i \sum_{n=N+1}^{\infty} \frac{\sinh(\zeta)e^{-\epsilon t}}{\frac{d}{d\zeta}[\zeta \cosh(\zeta)]\frac{d\zeta}{dz}|_{z=\omega_n^+}} \exp\left[-t\sqrt{\frac{(2n+1)^2\pi^2}{4} - k^2}\right], \quad (3.1.42)$$

or

$$f(t) = -2e^{-\epsilon t} \sum_{n=0}^{N} \frac{\sin\left[t\sqrt{k^2 - (2n+1)^2\pi^2/4}\right]}{\sqrt{k^2 - (2n+1)^2\pi^2/4}}$$

$$+ e^{-\epsilon t} \sum_{n=N+1}^{\infty} \frac{\exp\left[-t\sqrt{(2n+1)^2\pi^2/4 - k^2}\right]}{\sqrt{(2n+1)^2\pi^2/4 - k^2}}. \quad (3.1.43)$$

Problems

By taking the appropriate closed contour, find the inverse of the following Fourier transform by contour integration. The parameters a, h and x are real and positive. In Problem 8, replace the sine by its complex form and consider the three cases $t < -h/2$, $-h/2 < t < h/2$ and $h/2 < t$.

1. $\dfrac{1}{\omega^2 - ia^2\omega}$ 2. $\dfrac{1}{\omega^2 + a^2}$

3. $\dfrac{\omega}{\omega^2 + a^2}$ 3. $\dfrac{\omega}{(\omega^2 + a^2)^2}$

5. $\dfrac{\omega^2}{(\omega^2 + a^2)^2}$ 6. $\dfrac{1}{\omega^2 - 3i\omega - 3}$

7. $\dfrac{1}{(\omega - ia)^{2n+2}}$

8. $\dfrac{2i \sin(\omega h/2)}{\omega^2 + a^2}$

9. $\dfrac{\omega^2}{(\omega^2 - 1)^2 + 4a^2\omega^2}$

10. $\dfrac{1}{I_0(\omega)}$

11. $\dfrac{\cosh(\omega x)}{i \sinh(\omega h)}$

12. $\dfrac{\sin(\omega/a)}{\omega \cos(\omega h/a)}$

13. $\dfrac{m\pi \sinh(\omega a)}{2\omega(m^2\pi^2 - \omega^2)\cosh(\omega a)}[1 + (-1)^m \cos(\omega) - i(-1)^m \sin(\omega)]$

14. By replacing $\cos(\omega x)$ with its complex equivalent and taking the appropriate closed contours depending upon whether $0 < t \le x$ or $0 \le x < t$, find the inverse of Fourier transform

$$F(\omega) = \frac{2\cos(\omega x)}{\omega i \, \cosh(\omega a)} - \frac{1}{\omega i}.$$

15. Use the residue theorem to invert the Fourier transform[6]

$$F(\omega) = -\frac{Pa \exp[-\omega^2 b^2(r - a)/(2c^3)]}{r\rho c(\omega + B - Ai)(\omega - B - Ai)}, \qquad a < r,$$

where a, A, b, B, c, P, r and ρ are constant and positive.

16. If $\zeta = \sqrt{-\lambda^2 - i\omega}$ and λ is a real constant, show that the inverses[7] of the Fourier transforms

$$F(\omega) = \frac{\zeta \cot(\zeta) + \lambda}{2\lambda\zeta \cot(\zeta) - \zeta^2 + \lambda^2} \qquad \text{and} \qquad G(\omega) = \frac{1}{\zeta \cot(\zeta) + \lambda}$$

are

$$f(t) = 2H(t) \sum_{n=1}^{\infty} \frac{\overline{\zeta}_n - \lambda^2}{\overline{\zeta}_n + 2\lambda} e^{-\overline{\zeta}_n t},$$

where $\overline{\zeta}_n = \zeta_n^2 + \lambda^2$ and $2\lambda \cot(\zeta_n) - \zeta_n^2 + \lambda^2 = 0$, and

$$g(t) = 2H(t) \sum_{n=1}^{\infty} \frac{\overline{\zeta}_n - \lambda^2}{\overline{\zeta}_n + \lambda} e^{-\overline{\zeta}_n t},$$

[6] Taken from Chakraborty, S. K., 1961: Disturbances in a viscoelastic medium due to impulsive forces on a spherical cavity. *Geofis. Pura Appl.*, **48**, 23–26. Published by Birkhäuser Verlag, Basel, Switzerland.

[7] Taken from Goldman, M. M., and D. V. Fitterman, 1987: Direct time-domain calculation of the transient response for a rectangular loop over a two-layer medium. *Geophysics*, **52**, 997–1006.

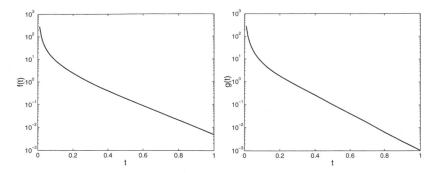

Problem 16

where $\overline{\zeta}_n = \zeta_n^2 + \lambda^2$ and $\zeta_n \cot(\zeta_n) + \lambda = 0$. These inverses $f(t)$ and $g(t)$ are illustrated in the figure labeled Problem 16 with $\lambda = 2$.

17. Find the inverse[8] of Fourier transform

$$F(\omega) = \frac{\cosh(\zeta)}{\zeta \sinh(\zeta)}$$

when $t > 0$, where $\zeta^2 = (\omega - i\epsilon)^2 - k^2$ and $0 < \epsilon \ll 1$.

18. Show that the inverse[9] of the Fourier transform

$$F(\omega) = \frac{\sqrt{\omega i}}{\sinh(d\sqrt{\omega i})}$$

is

$$f(t) = -\frac{H(t)}{2d^3} \sum_{n=1}^{\infty} (-1)^n n^2 \exp(-n^2 \pi^2 t/d^2).$$

This inverse is illustrated in the figure labeled Problem 18.

19. During the solution of the heat equation, Taitel et al.[10] inverted the Fourier transform

$$F(\omega) = \frac{\cosh(y\sqrt{\omega^2 + 1})}{\sqrt{\omega^2 + 1} \sinh(L\sqrt{\omega^2 + 1}/2)},$$

---

[8] A similar problem appeared in Abrahams, I. D., 1982: Scattering of sound by finite elastic surfaces bounding ducts or cavities near resonance. *Quart. J. Mech. Appl. Math.*, **35**, 91–101. Taken by permission of Oxford University Press.

[9] Taken from Stern, R. B., 1988: Time domain calculation of electric field penetration through metallic shields. *IEEE Trans. Electromag. Compat.*, **EC-30**, 307–311. ©1988 IEEE.

[10] Reprinted from *Int. J. Heat Mass Transfer*, **16**, Y. Taitel, M. Bentwich and A. Tamir, Effects of upstream and downstream boundary conditions on heat (mass) transfer with axial diffusion, 359–369, ©1973, with permission from Elsevier.

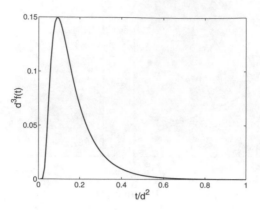

Problem 18

where y and L are real. Show that they should have found

$$f(t) = \frac{e^{-|t|}}{L} + \frac{2}{L} \sum_{n=1}^{\infty} \frac{(-1)^n}{\sqrt{1 + 4n^2\pi^2/L^2}} \cos\left(\frac{2n\pi y}{L}\right) e^{-\sqrt{1+4n^2\pi^2/L^2}\,|t|}.$$

In this case, our time variable t was their spatial variable $x - \xi$. This inverse is illustrated in the figure labeled Problem 19 with $L = 1$.

20. Find the inverse of the Fourier transform

$$F(\omega) = \left[\cos\left\{\frac{\omega L}{\beta[1 + i\gamma\,\operatorname{sgn}(\omega)]}\right\}\right]^{-1},$$

where L, β and γ are real and positive, and $\operatorname{sgn}(z) = 1$ if $\operatorname{Re}(z) > 0$ and -1 if $\operatorname{Re}(z) < 0$.

21. The concept of forced convection is normally associated with heat streaming through a duct or past an obstacle. Bentwich[11] wanted to show that a similar transport can exist when convection results from a wave traveling through an essentially stagnant fluid. In the process of computing the amount of heating, he had to prove the following identity:

$$\int_{-\infty}^{\infty} \frac{\cosh(hx) - 1}{x\,\sinh(hx)}\cos(ax)\,dx = \ln[\coth(|a|\pi/h)], \qquad 0 < h.$$

Confirm his result.

---

[11] Reprinted from *Int. J. Heat Mass Transfer*, **9**, M. Bentwich, Convection enforced by surface and tidal waves, 663–670, ©1966, with permission from Elsevier.

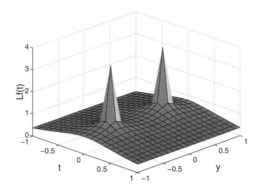

Problem 19

22. Find the inverse Fourier transform,[12] correct to $O(\epsilon)$, for

$$F(\omega) = \frac{\cosh(\omega h)}{(U\omega - i\epsilon)^2 \cosh(\omega h) - g\omega \sinh(\omega h)}$$

with $U^2 < gh$, $0 < \epsilon \ll 1$, and U and g are constants.

Step 1: Using a perturbation expansion, show that the poles are located at

$$\omega_1 = \frac{i\epsilon}{U - \sqrt{gh}} + O(\epsilon^2), \qquad \omega_2 = \frac{i\epsilon}{U + \sqrt{gh}} + O(\epsilon^2),$$

$$\omega_3 = \omega_0 - \frac{2i\epsilon U\omega_0}{g[\omega_0 h \, \mathrm{sech}^2(\omega_0 h) - \tanh(\omega_0 h)]} + O(\epsilon^2),$$

and

$$\omega_4 = -\omega_0 - \frac{2i\epsilon U\omega_0}{g[\omega_0 h \, \mathrm{sech}^2(\omega_0 h) - \tanh(\omega_0 h)]} + O(\epsilon^2),$$

where $\tanh(\omega_0 h) = U^2 \omega_0 / g$.

Step 2: Because $\omega_0 h \, \mathrm{sech}^2(\omega_0 h) - \tanh(\omega_0 h)$ is negative, the last three poles lie above the real axis whereas the first pole lies below it. By taking the appropriate contour, find the inverse correct to $O(\epsilon)$.

23. Find the inverse Fourier transform[13] of

$$F(\omega) = -\frac{iU\omega + \epsilon}{(U\omega - i\epsilon)^2[\omega^2 + (nN/U\kappa)^2] - N^2\omega^2}$$

---

[12] Taken from Newman, J. N., 1976: Blockage with a free surface. *J. Ship Res.*, **20**, 199–203.

[13] Taken from Janowitz, G. S., 1981: Stratified flow over a bounded obstacle in a channel. *J. Fluid Mech.*, **110**, 161–170. Reprinted with the permission of Cambridge University Press.

as $\epsilon \to 0$. The constants U, N, κ and n are real and positive. Consider the two cases of $\kappa < n$ and $n < \kappa$.

Step 1: Using a perturbation expansion, show that the poles are located at

$$\omega_1 = \frac{n\epsilon i/(U\kappa)}{n/\kappa - 1} + O(\epsilon^2), \qquad \omega_2 = \frac{n\epsilon i/(U\kappa)}{n/\kappa + 1} + O(\epsilon^2),$$

$$\omega_3 = \omega_0 - \frac{\epsilon i}{U(n^2/\kappa^2 - 1)} + O(\epsilon^2) \quad \text{and} \quad \omega_4 = -\omega_0 - \frac{\epsilon i}{U(n^2/\kappa^2 - 1)} + O(\epsilon^2),$$

where $\omega_0 = N\sqrt{1 - (n/\kappa)^2}/U$.

Step 2: Show that if $n/\kappa < 1$, the inverse is

$$f(t) = \frac{U}{2N^2}\left\{ \frac{1}{(n/\kappa)(1 + n/\kappa)} + \frac{2\cos\left[Nt\sqrt{1 - (n/\kappa)^2}/U\right]}{1 - (n/\kappa)^2} \right\}$$

for $t > 0$, and

$$f(t) = \frac{U}{2N^2(n/\kappa)(1 - n/\kappa)}$$

for $t < 0$.

Step 3: Show that if $n/\kappa > 1$, the inverse is

$$f(t) = \frac{U}{2N^2}\left\{ \frac{2 - \exp\left[-Nt\sqrt{(n/\kappa)^2 - 1}/U\right]}{(n/\kappa)^2 - 1} \right\}$$

for $t > 0$, and

$$f(t) = \frac{U}{2N^2}\frac{\exp\left[Nt\sqrt{(n/\kappa)^2 - 1}/U\right]}{(n/\kappa)^2 - 1}$$

for $t < 0$.

3.2 THE WAVE EQUATION

Having improved our ability to invert Fourier transforms, we are ready to show how we can use them to solve the wave equation in infinite and semi-infinite domains.

• Example 3.2.1

The simplest example of applying Fourier transforms to solve the wave equation is provided by the problem:

$$\frac{\partial^2 u}{\partial t^2} = c^2 \frac{\partial^2 u}{\partial x^2}, \qquad -\infty < x < \infty, \quad 0 < t, \qquad (\textbf{3.2.1})$$

subject to the boundary conditions

$$\lim_{|x| \to \infty} u(x,t) \to 0, \tag{3.2.2}$$

and the initial conditions

$$u(x,0) = f(x) \quad \text{and} \quad u_t(x,0) = g(x), \qquad -\infty < x < \infty. \tag{3.2.3}$$

We begin by multiplying Equation 3.2.1 with e^{-ikx} and integrating from $-\infty$ to ∞. This yields

$$\int_{-\infty}^{\infty} \frac{\partial^2 u}{\partial t^2} e^{-ikx} \, dx = \frac{d^2}{dt^2} \left[\int_{-\infty}^{\infty} u(x,t) e^{-ikx} \, dx \right] = c^2 \int_{-\infty}^{\infty} \frac{\partial^2 u}{\partial x^2} e^{-ikx} \, dx. \tag{3.2.4}$$

Integrating the right side of Equation 3.2.4 twice by parts,

$$\int_{-\infty}^{\infty} \frac{\partial^2 u}{\partial x^2} e^{-ikx} \, dx = \left(\frac{\partial u}{\partial x} + iku \right) e^{-ikx} \Big|_{x=-\infty}^{x=\infty} - k^2 \int_{-\infty}^{\infty} u(x,t) e^{-ikx} \, dx. \tag{3.2.5}$$

Because $u(x,t)$ and $u_x(x,t)$ vanish as $|x| \to \infty$, a combination of Equation 3.2.4 and Equation 3.2.5 yields the initial-value problem

$$\frac{d^2 U}{dt^2} + k^2 c^2 U(k,t) = 0, \tag{3.2.6}$$

with

$$U(k,0) = F(k) = \int_{-\infty}^{\infty} f(x) e^{-ikx} \, dx \tag{3.2.7}$$

and

$$U'(k,0) = G(k) = \int_{-\infty}^{\infty} g(x) e^{-ikx} \, dx, \tag{3.2.8}$$

where

$$U(k,t) = \int_{\infty}^{\infty} u(x,t) e^{-ikx} \, dx. \tag{3.2.9}$$

Here, the prime in Equation 3.2.8 denotes an ordinary derivative with respect to time.

The solution to Equation 3.2.6 satisfying the initial conditions given by Equation 3.2.7 and Equation 3.2.8 is

$$U(k,t) = \tfrac{1}{2} F(k) \left(e^{ikct} + e^{-ikct} \right) + \frac{G(k)}{2ikc} \left(e^{ikct} - e^{-ikct} \right). \tag{3.2.10}$$

Therefore,

$$u(x,t) = \frac{1}{2\pi} \int_{-\infty}^{\infty} U(k,t) e^{ikx} \, dk \tag{3.2.11}$$

$$= \frac{1}{4\pi} \int_{-\infty}^{\infty} F(k) \left[e^{ik(x+ct)} + e^{ik(x-ct)} \right] dk$$

$$+ \frac{1}{4\pi c} \int_{-\infty}^{\infty} \frac{G(k)}{ik} \left[e^{ik(x+ct)} - e^{ik(x-ct)} \right] dk. \tag{3.2.12}$$

Finally, noting that

$$f(x \pm ct) = \frac{1}{2\pi} \int_{-\infty}^{\infty} F(k) e^{ik(x \pm ct)} \, dk, \tag{3.2.13}$$

and

$$\int_{x-ct}^{x+ct} g(\xi) \, d\xi = \frac{1}{2\pi} \int_{-\infty}^{\infty} G(k) \left(\int_{x-ct}^{x+ct} e^{ik\xi} \, d\xi \right) dk \tag{3.2.14}$$

$$= \frac{1}{2\pi} \int_{-\infty}^{\infty} \frac{G(k)}{ik} \left[e^{ik(x+ct)} - e^{ik(x-ct)} \right] dk, \tag{3.2.15}$$

we can invert Equation 3.2.12 and find that

$$u(x,t) = \tfrac{1}{2} \left[f(x + ct) + f(x - ct) \right] + \frac{1}{2c} \int_{x-ct}^{x+ct} g(\xi) \, d\xi. \tag{3.2.16}$$

This classic result is known as D'Alembert's formula[14] and gives the wave motions within a homogeneous domain of infinite extent.

• Example 3.2.2

In the previous example we used Fourier transforms to eliminate the x dependence. Fourier transforms could also be used to eliminate the time dependence just as we used Laplace transforms in the previous chapter. Because some favor this method, we illustrate it here.

Let us calculate the sound waves[15] radiated by a sphere of radius a whose surface expands radially with an impulsive acceleration $v_0 \delta(t)$. This problem has a number of applications; one of them would be the sound waves generated by an exploding depth charge.[16]

If we assume radial symmetry, the corresponding wave equation is

$$\frac{1}{r^2} \frac{\partial}{\partial r} \left(r^2 \frac{\partial u}{\partial r} \right) = \frac{1}{c^2} \frac{\partial^2 u}{\partial t^2} \tag{3.2.17}$$

subject to the boundary condition

$$\frac{\partial u}{\partial r} = -\rho v_0 \delta(t) \tag{3.2.18}$$

---

[14] D'Alembert, J., 1747: Recherches sur la courbe que forme une corde tenduë mise en vibration. *Hist. Acad. R. Sci. Belles Lett., Berlin*, 214–219.

[15] Taken from Hodgson, D. C., and J. E. Bowcock, 1975: Billet expansion as a mechanism for noise production in impact forming machines. *J. Sound Vib.*, **42**, 325–335. Published by Academic Press Ltd., London, UK.

[16] Probst (Probst, W., 1972: Die Schallerzeugung durch eine expandierende Kugel. *Acustica*, **27**, 299–306.) used a complicated form of this problem to explain the high sound levels generated by an inflating air-bag system.

at $r = a$, where $u(r, t)$ is the pressure field, c is the speed of sound and ρ is the average density of the fluid. Assuming that the pressure field possesses a Fourier transform, we may reexpress it by the Fourier integral

$$u(r, t) = \frac{1}{2\pi} \int_{-\infty}^{\infty} U(r, \omega) e^{i\omega t} \, d\omega. \tag{3.2.19}$$

Substituting into Equation 3.2.17, it becomes

$$\frac{1}{r^2} \frac{d}{dr} \left(r^2 \frac{dU}{dr} \right) + k^2 U = 0, \tag{3.2.20}$$

and

$$\frac{dU(a, \omega)}{dr} = -\rho v_0 \tag{3.2.21}$$

with $k = \omega/c$. The most general solution of Equation 3.2.20 is

$$U(r, \omega) = A(\omega) \frac{e^{ikr}}{r} + B(\omega) \frac{e^{-ikr}}{r} \tag{3.2.22}$$

and

$$u(r, t) = \frac{1}{2\pi} \int_{-\infty}^{\infty} \left[A(\omega) \frac{e^{i\omega t + ikr}}{r} + B(\omega) \frac{e^{i\omega t - ikr}}{r} \right] d\omega. \tag{3.2.23}$$

At this point, we note that the first term on the right side of Equation 3.2.23 represents an inward propagating wave while the second term is an outward propagating wave. Because there is no source of energy at infinity, the inward propagating wave is aphysical and we discard it. This boundary condition is often referred to as the *Sommerfeld radiation condition* because Sommerfeld[17] used it first in his study of electromagnetic waves on the surface of a conducting earth.

Upon substituting Equation 3.2.22 into Equation 3.2.21 with $A(\omega) = 0$,

$$U(r, \omega) = \frac{\rho a^2 v_0}{2\pi r} \frac{e^{-i\omega(r-a)/c}}{1 + i\omega a/c}, \tag{3.2.24}$$

or

$$u(r, t) = \frac{1}{2\pi} \int_{-\infty}^{\infty} U(r, \omega) e^{i\omega t} \, d\omega = \frac{\rho a^2 v_0}{2\pi r} \int_{-\infty}^{\infty} \frac{e^{i\omega[t - (r-a)/c]}}{1 + i\omega a/c} \, d\omega. \tag{3.2.25}$$

To evaluate Equation 3.2.25, we employ the residue theorem. For $t < (r-a)/c$, Jordan's lemma dictates that we close the line integral by an infinite semicircle

---

[17] Sommerfeld, A., 1909: Über die Ausbreitung der Wellen in der draftlosen Telegraphie. *Ann. Phys., Vierte Folge*, **28**, 665–736.

Figure 3.2.1: Originally drawn to mathematics, Arnold Johannes Wilhelm Sommerfeld (1868–1951) migrated into physics due to Klein's interest in applying the theory of complex variables and other pure mathematics to a range of physical topics from astronomy to dynamics. Later on, Sommerfeld contributed to quantum mechanics and statistical mechanics. (Portrait, AIP Emilio Segrè Visual Archives, Margrethe Bohr Collection.)

in the lower half-plane. For $t > (r-a)/c$, we close the contour with a semicircle in the upper half-plane. The final result is

$$u(r,t) = \frac{\rho a c v_0}{r} \exp\left[-\frac{c}{a}\left(t - \frac{r-a}{c}\right)\right] H\left(t - \frac{r-a}{c}\right). \qquad (3.2.26)$$

This solution is illustrated in Figure 3.2.2.

● **Example 3.2.3**

For our final example,[18] we find the solution to the scalar, reduced (Helmholtz), inhomogeneous wave equation in free space:

$$\nabla^2 u(x,y,z) + \kappa_0^2 u(x,y,z) = -4\pi\delta(x)\delta(y)\delta(z), \qquad (3.2.27)$$

where $\kappa_0 = \omega/c$, ω is the frequency of the wave and c is the phase speed. By direct substitution, we find[19] that

$$u(x,y,z) = \frac{e^{i\omega R/c}}{R} \qquad (3.2.28)$$

---

[18] Patterned after Biggs, A. W., 1977: Fourier transforms in propagation and scattering problems. *IEEE Trans. Antennas Propagat.*, **AP-25**, 585–586. ©1977 IEEE.

[19] For a derivation, see Aki, K., and P. G. Richards, 1980: *Quantitative Seismology: Theory and Methods. Vol. 1.* W. H. Freeman and Co., Section 4.1.

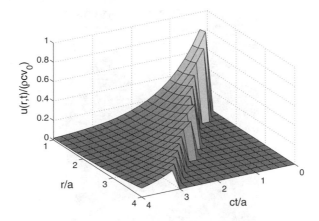

Figure 3.2.2: Plot of the wave solution given by Equation 3.2.26.

is a solution of Equation 3.2.27, where $R = \sqrt{x^2 + y^2 + z^2}$. However, in this section let us derive an alternative solution to Equation 3.2.27 using Fourier transforms.

We begin by assuming that a Fourier transform exists for $u(x, y, z)$. Therefore, we can write $u(x, y, z)$ as

$$u(x, y, z) = \frac{1}{(2\pi)^3} \int_{-\infty}^{\infty} \int_{-\infty}^{\infty} \int_{-\infty}^{\infty} U(k, \ell, m) e^{i(kx + \ell y + mz)} \, dk \, d\ell \, dm. \quad (3.2.29)$$

Direct substitution of Equation 3.2.29 into Equation 3.2.27 yields

$$U(k, \ell, m) = \frac{4\pi}{\kappa^2 - \kappa_0^2}, \quad (3.2.30)$$

where $\kappa^2 = k^2 + \ell^2 + m^2$, so that

$$u(x, y, z) = \frac{1}{2\pi^2} \int_{-\infty}^{\infty} \int_{-\infty}^{\infty} \int_{-\infty}^{\infty} \frac{e^{i(kx + \ell y + mz)}}{\kappa^2 - \kappa_0^2} \, dk \, d\ell \, dm. \quad (3.2.31)$$

To evaluate Equation 3.2.31 we first perform the integration in m by closing the line integral along the real axis as dictated by Jordan's lemma. For $z > 0$, this requires an infinite semicircle in the upper half-plane; for $z < 0$, an infinite semicircle in the lower half-plane. One source of difficulty is the presence of singularities along the real axis at $m = \pm\sqrt{\kappa_0^2 - k^2 - \ell^2}$. We avoid this difficulty by introducing some friction (making it a slightly lossy medium) so that $\kappa_0 = \kappa_0' + i\kappa_0''$, where $|\kappa_0'| \gg |\kappa_0''|$. This lifts the singularities off the real axis and a straightforward application of the residue theorem yields

$$u(x, y, z) = \frac{1}{2\pi} \int_{-\infty}^{\infty} \int_{-\infty}^{\infty} \frac{e^{ikx + i\ell y - |z|\sqrt{k^2 + \ell^2 - \omega^2/c^2}}}{\sqrt{k^2 + \ell^2 - \omega^2/c^2}} \, dk \, d\ell \quad (3.2.32)$$

Figure 2.3.3: Educated at Munich and Göttingen, Hermann Klaus Hugo Weyl (1885–1955) was Hilbert's greatest student. His primary contributions to mathematical physics are his attempt to incorporate electromagnetism into the geometric formalism of general relativity, the concept of continuous groups using matrix representations and the application of group theory to quantum mechanics. (Portrait, AIP Emilio Segrè Visual Archives, Nina Courant Collection.)

with the condition that the square root has a real part ≥ 0. Because the $u(x, y, z)$ representations given by Equation 3.2.28 and Equation 3.2.32 must be equivalent, we immediately obtain the *Weyl integral*:[20]

$$\frac{e^{i\omega R/c}}{R} = \frac{1}{2\pi} \int_{-\infty}^{\infty} \int_{-\infty}^{\infty} \frac{e^{ikx+i\ell y-|z|\sqrt{k^2+\ell^2-\omega^2/c^2}}}{\sqrt{k^2 + \ell^2 - \omega^2/c^2}} \, dk \, d\ell. \qquad (\mathbf{3.2.33})$$

We can further simplify Equation 3.2.33 by introducing $k = \rho\cos(\varphi)$, $\ell = \rho\sin(\varphi)$, $x = r\cos(\theta)$ and $y = r\sin(\theta)$. Then,

$$\frac{e^{i\omega R/c}}{R} = \frac{1}{2\pi} \int_{0}^{\infty} \int_{0}^{2\pi} \frac{e^{i\rho r \cos(\theta-\varphi)+i|z|\sqrt{\omega^2/c^2-\rho^2}}}{-i\sqrt{\omega^2/c^2 - \rho^2}} \, d\varphi \, \rho \, d\rho. \qquad (\mathbf{3.2.34})$$

If we now carry out the integration in φ, we obtain the *Sommerfeld integral*:[21]

$$\frac{e^{i\omega R/c}}{R} = i \int_{0}^{\infty} \frac{J_0(\rho r)e^{i|z|\sqrt{\omega^2/c^2-\rho^2}}}{\sqrt{\omega^2/c^2 - \rho^2}} \, \rho \, d\rho, \qquad (\mathbf{3.2.35})$$

---

[20] Weyl, H., 1919: Ausbreitung elektromagnetischer Wellen über einer ebenen Leiter. *Ann. Phys., Vierte Folge*, **60**, 481–500.

[21] Sommerfeld, *op. cit.*

where the imaginary part of the square root must be positive. Both the Weyl and Sommerfeld integrals are used extensively in electromagnetism and elasticity as an integral representation of spherical waves propagating from a point source.

Problems

1. The nondimensional shallow water equations in one spatial dimension

$$\frac{\partial u}{\partial t} - v = -\frac{\partial h}{\partial x}, \qquad \frac{\partial v}{\partial t} + u = 0, \quad \text{and} \quad \frac{\partial h}{\partial t} + c^2 \frac{\partial u}{\partial x} = 0$$

describe wave motions within a homogeneous ocean whose depth is small compared to the radius of the rotating earth. The zonal and meridional velocities are $u(x,t)$ and $v(x,t)$; the height of the free surface is $h(x,t)$. The phase speed c equals the square root of the depth of the ocean times gravity divided by the Coriolis parameter.

In this problem let us find the wave motion in a quiescent ocean after we raise its interface by h_0 in the region $-a < x < a$. Mathematically, this is equivalent to solving the shallow water equations subject to the initial condition that

$$u(x,0) = v(x,0) = 0 \qquad \text{and} \qquad h(x,0) = \begin{cases} h_0, & \text{if} \quad |x| < a, \\ 0, & \text{if} \quad |x| > a, \end{cases}$$

and assuming that $u(x,t)$, $v(x,t)$ and $h(x,t)$ vanish as $|x| \to \infty$.

Step 1: By defining the Fourier transform of $u(x,t)$, $v(x,t)$ and $h(x,t)$ as

$$U(k,t) = \int_{-\infty}^{\infty} u(x,t)e^{-ikx}\, dx, \qquad V(k,t) = \int_{-\infty}^{\infty} v(x,t)e^{-ikx}\, dx,$$

and

$$H(k,t) = \int_{-\infty}^{\infty} h(x,t)e^{-ikx}\, dx,$$

show that we can transform the shallow water equations into the initial-value problem

$$\frac{dU(k,t)}{dt} - kV(k,t) = -ikH(k,t), \qquad \frac{dV(k,t)}{dt} + U(k,t) = 0,$$

and

$$\frac{dH(k,t)}{dt} + ikc^2 U(k,t) = 0$$

with $U(k,0) = V(k,0) = 0$ and $H(k,0) = 2h_0 \sin(ka)/k$.

Step 2: Show that we can combine the three differential equations in Step 1 into the single equation

$$\frac{d^3 H(k,t)}{dt^3} + (1 + k^2 c^2)\frac{dH(k,t)}{dt} = 0.$$

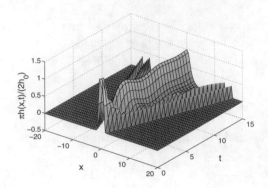

Problem 1

Step 3: Show that the solution to the initial-value problem is

$$H(k,t) = \frac{2h_0 \sin(ka)}{k}\left[\frac{1 + k^2c^2 \cos\left(t\sqrt{1 + k^2c^2}\right)}{1 + k^2c^2}\right].$$

Step 4: Show that $h(x,t)$ is

$$h(x,t) = \frac{h_0}{\pi}\int_{-\infty}^{\infty}\frac{\sin(ka)}{k}\left[\frac{1 + k^2c^2 \cos\left(t\sqrt{1 + k^2c^2}\right)}{1 + k^2c^2}\right]e^{ikx}\,dk,$$

or

$$h(x,t) = \frac{2h_0}{\pi}\int_{0}^{\infty}\frac{\sin(ka)}{k}\left[\frac{1 + k^2c^2 \cos\left(t\sqrt{1 + k^2c^2}\right)}{1 + k^2c^2}\right]\cos(kx)\,dk.$$

This solution $\pi h(x,t)/(2h_0)$ is illustrated in the figure labeled Problem 1 with $a = c = 1$.

2. Solve the nonhomogeneous equation of telegraphy[22]

$$\frac{\partial^2 u}{\partial x^2} - \frac{1}{c^2}\frac{\partial^2 u}{\partial t^2} - \frac{4\pi\sigma}{c^2}\frac{\partial u}{\partial t} = \frac{4\pi}{c}f(x), \qquad -\infty < x < \infty, \quad 0 < t,$$

where the initial conditions are $u(x,0) = u_t(x,0) = 0$.

Step 1: Assuming that $f(x)$ has a Fourier transform and

$$U(k,t) = \int_{-\infty}^{\infty} u(x,t)e^{-ikx}\,dx,$$

show that we can reduce the partial differential equation to the initial-value problem

$$\frac{d^2U(k,t)}{dt^2} + 4\pi\sigma\frac{dU(k,t)}{dt} + k^2c^2U(k,t) = -4\pi c F(k), \qquad U(k,0) = U'(k,0) = 0,$$

[22] Karzas, W. J., and R. Latter, 1962: The electromagnetic signal due to the interaction of nuclear explosions with the earth's magnetic field. *J. Geophys. Res.*, **67**, 4635–4640. ©1962 American Geophysical Union. Reproduced/modified by permission of American Geophysical Union.

where $F(k)$ is the Fourier transform of $f(x)$.

Step 2: Show that the solution to the initial-value problem is

$$U(k,t) = -4\pi c F(k) \int_0^t e^{-2\pi\sigma(t-\tau)} \frac{\sin\left[(t-\tau)\sqrt{k^2c^2 - 4\pi^2\sigma^2}\right]}{\sqrt{k^2c^2 - 4\pi^2\sigma^2}} \, d\tau.$$

[Hint: Use Laplace transforms.]

Step 3: Show that we can write the solution $u(x,t)$ as

$$u(x,t) = -2 \int_0^t e^{-2\pi\sigma(t-\tau)} \left\{ \int_{-\infty}^{\infty} F(k) \frac{c\sin\left[(t-\tau)\sqrt{k^2c^2 - 4\pi^2\sigma^2}\right]}{\sqrt{k^2c^2 - 4\pi^2\sigma^2}} e^{ikx} \, dk \right\} d\tau$$

$$= -4 \int_0^t \left[\int_{-\infty}^{\infty} f(\chi) I(|x - \chi|, t - \tau) \, d\chi \right] e^{-2\pi\sigma(t-\tau)} \, d\tau,$$

where

$$I(x,t) = \int_0^{\infty} \frac{\sin\left[ct\sqrt{k^2 - (2\pi\sigma/c)^2}\right]}{\sqrt{k^2 - (2\pi\sigma/c)^2}} \cos(kx) \, dk = \frac{\pi}{2} I_0\left[2\pi\sigma\sqrt{t^2 - (x/c)^2}\right] H(t - x/c).$$

3.3 THE HEAT EQUATION

As the following examples show, Fourier transforms are particularly useful in solving the heat equation in domains of infinite extent.

• Example 3.3.1

A simple application of solving the heat equation by Fourier transforms is provided by the following problem:

$$\frac{\partial u}{\partial t} = a^2 \frac{\partial^2 u}{\partial x^2}, \qquad -\infty < x < \infty, \quad 0 < t, \qquad (3.3.1)$$

subject to the boundary conditions

$$\lim_{|x|\to\infty} u(x,t) \to 0, \qquad (3.3.2)$$

and the initial condition

$$u(x,0) = e^{-\beta^2 x^2}, \qquad -\infty < x < \infty. \qquad (3.3.3)$$

We begin by multiplying Equation 3.3.1 with e^{-ikx} and integrating from $-\infty$ to ∞. This yields

$$\int_{-\infty}^{\infty} \frac{\partial u}{\partial t} e^{-ikx} \, dx = \frac{d}{dt}\left[\int_{-\infty}^{\infty} u(x,t)e^{-ikx} \, dx \right] = a^2 \int_{-\infty}^{\infty} \frac{\partial^2 u}{\partial x^2} e^{-ikx} \, dx. \quad (3.3.4)$$

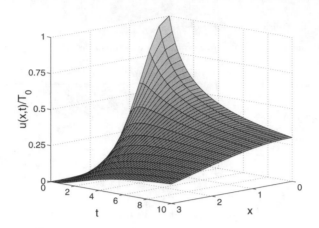

Figure 3.3.1: A plot $u(x,t)/T_0$ given by Equation 3.3.11 as a function of distance x and time t when $\beta = 1$.

Integrating the right side of Equation 3.3.4 twice by parts,

$$\int_{-\infty}^{\infty} \frac{\partial^2 u}{\partial x^2} e^{-ikx}\, dx = \left(\frac{\partial u}{\partial x} + iku \right) e^{-ikx} \Big|_{x=-\infty}^{x=\infty} - k^2 \int_{-\infty}^{\infty} u(x,t) e^{-ikx}\, dx. \tag{3.3.5}$$

Because $u(x,t)$ and $u_x(x,t)$ vanish as $|x| \to \infty$, a combination of Equation 3.3.4 and Equation 3.3.5 gives

$$\frac{dU}{dt} + a^2 k^2 U(k,t) = 0, \qquad U(k,0) = \frac{\sqrt{\pi}\, T_0}{\beta} e^{-k^2/(4\beta^2)}, \tag{3.3.6}$$

where

$$U(k,t) = \int_{-\infty}^{\infty} u(x,t) e^{-ikx}\, dx. \tag{3.3.7}$$

The solution to Equation 3.3.6 is

$$U(k,t) = \frac{\sqrt{\pi}\, T_0}{\beta} e^{-k^2/(4\beta^2) - a^2 k^2 t}. \tag{3.3.8}$$

Therefore,

$$u(x,t) = \frac{T_0}{2\sqrt{\pi}\beta} \int_{-\infty}^{\infty} e^{-(1+4a^2\beta^2 t)k^2/(4\beta^2)} e^{ikx}\, dk \tag{3.3.9}$$

$$= \frac{T_0}{\sqrt{\pi}\beta} \int_0^{\infty} e^{-(1+4a^2\beta^2 t)k^2/(4\beta^2)} \cos(kx)\, dk \tag{3.3.10}$$

$$= \frac{T_0}{\sqrt{1+a^2\beta^2 t}} \exp\left[-\frac{\beta^2 x^2}{1 + 4a^2\beta^2 t} \right]. \tag{3.3.11}$$

Figure 3.3.1 illustrates Equation 3.3.11 as a function of distance x and time t.

● **Example 3.3.2**

Fourier transforms can also be applied to problems on a semi-infinite domain. For example, let us solve[23]

$$\frac{\partial u}{\partial t} = \alpha \frac{\partial^3 u}{\partial t \partial x^2} + a^2 \frac{\partial^2 u}{\partial x^2}, \qquad 0 < x < \infty, \quad 0 < t, \qquad (3.3.12)$$

subject to the boundary conditions

$$u(0, t) = f(t), \qquad \lim_{x \to 0} |u(x, t)| < \infty, \qquad 0 < t, \qquad (3.3.13)$$

$$\lim_{x \to \infty} u(x, t) \to 0, \qquad \lim_{x \to \infty} u_x(x, t) \to 0, \qquad 0 < t, \qquad (3.3.14)$$

and the initial condition

$$u(x, 0) = 0, \qquad 0 < x < \infty. \qquad (3.3.15)$$

We begin by multiplying Equation 3.3.12 by $\sin(kx)$ and integrating over x from 0 to ∞:

$$\alpha \int_0^\infty u_{txx} \sin(kx)\, dx + a^2 \int_0^\infty u_{xx} \sin(kx)\, dx = \int_0^\infty u_t \sin(kx)\, dx. \quad (3.3.16)$$

Next, we integrate by parts. For example,

$$\int_0^\infty u_{xx} \sin(kx)\, dx = u_x \sin(kx)\Big|_{x=0}^{x=\infty} - k \int_0^\infty u_x \cos(kx)\, dx \qquad (3.3.17)$$

$$= -k \int_0^\infty u_x \cos(kx)\, dx \qquad (3.3.18)$$

$$= -ku(x, t) \cos(kx)\Big|_{x-0}^{x=\infty} - k^2 \int_0^\infty u(x, t) \sin(kx)\, dx \qquad (3.3.19)$$

$$= kf(t) - k^2 U(k, t), \qquad (3.3.20)$$

where

$$U(k, t) = \int_0^\infty u(x, t) \sin(kx)\, dx. \qquad (3.3.21)$$

We now see why we multiplied Equation 3.3.12 by $\sin(kx)$ rather than $\cos(kx)$. Although we do not know the value of $u_x(0, t)$, it does not matter because $u_x(0, t)$ is multiplied by $\sin(0)$ and the product equals zero. On the

---

[23] Taken from Fetecău, C., and J. Zierep, 2001: On a class of exact solutions of the equations of motion of a second grade fluid. *Acta Mech.*, **150**, 135–138.

LIVERPOOL
JOHN MOORES UNIVERSITY
AVRIL ROBARTS LRC
TITHEBARN STREET
LIVERPOOL L2 2ER
TEL 0151 231 4022

other hand, if we had multiplied by $\cos(kx)$, integration of parts would have given $u_x(0,t)\cos(0) = u_x(0,t)$, an unknown quantity.

Equation 3.3.21 is the definition of the *Fourier sine transform*. It and its mathematical cousin, the *Fourier cosine transform* $\int_0^\infty u(x,t)\cos(kx)\,dx$, are analogous to the half-range sine and cosine expansions that appear in solving the heat equation over the finite interval $(0, L)$. The difference here is that our range runs from 0 to ∞.

Applying the same technique to the other terms, we obtain

$$\alpha[kf'(t) - k^2 U'(k,t)] + a^2[kf(t) - k^2 U(k,t)] = U'(k,t) \qquad (3.3.22)$$

with $U(k,0) = 0$, where the primes denote differentiation with respect to time. Solving Equation 3.3.22,

$$e^{a^2 k^2 t/(1+\alpha k^2)} U(k,t) = \frac{\alpha k}{1+\alpha k^2} \int_0^t f'(\tau) e^{a^2 k^2 \tau/(1+\alpha k^2)}\, d\tau$$

$$+ \frac{a^2 k}{1+\alpha k^2} \int_0^t f(\tau) e^{a^2 k^2 \tau/(1+\alpha k^2)}\, d\tau. \qquad (3.3.23)$$

Using integration by parts on the second integral in Equation 3.3.23, we find that

$$U(k,t) = \frac{1}{k}\left[f(t) - f(0)e^{-a^2 k^2 t/(1+\alpha k^2)} \right.$$

$$\left. - \frac{1}{1+\alpha k^2} \int_0^t f'(\tau) e^{-a^2 k^2 (t-\tau)/(1+\alpha k^2)}\, d\tau \right]. \qquad (3.3.24)$$

Because

$$u(x,t) = \frac{2}{\pi} \int_0^\infty U(k,t)\sin(kx)\, dk, \qquad (3.3.25)$$

$$u(x,t) = \frac{2}{\pi} f(t) \int_0^\infty \frac{\sin(kx)}{k}\, dk$$

$$- \frac{2}{\pi} \int_0^\infty \frac{\sin(kx)}{k} e^{-a^2 k^2 t/(1+\alpha k^2)}\, dk$$

$$\times \left[f(0) + \frac{1}{1+\alpha k^2} \int_0^t f'(\tau) e^{a^2 k^2 \tau/(1+\alpha k^2)}\, d\tau \right] \qquad (3.3.26)$$

$$= f(t) - \frac{2}{\pi} \int_0^\infty \frac{\sin(kx)}{k} e^{-a^2 k^2 t/(1+\alpha k^2)}\, dk$$

$$\times \left[f(0) + \frac{1}{1+\alpha k^2} \int_0^t f'(\tau) e^{a^2 k^2 \tau/(1+\alpha k^2)}\, d\tau \right]. \qquad (3.3.27)$$

Problems

1. Solve the heat equation

$$\frac{\partial u}{\partial t} = \frac{\partial^2 u}{\partial x^2}, \qquad -\infty < x < \infty, \quad 0 < t,$$

subject to the boundary conditions $\lim_{|x| \to \infty} u(x, t) \to 0$, and the initial condition $u(x, 0) = f(x)$, $-\infty < x < \infty$.

Step 1: Assuming that the Fourier transform for $f(x)$ exists, show that partial differential equation reduces to the initial-value problem

$$\frac{dU(k, t)}{dt} + k^2 U(k, t) = 0, \qquad U(k, 0) = F(k),$$

where $F(k)$ is the Fourier transform of $f(x)$.

Step 2: Show that $U(k, t) = F(k)e^{-k^2 t}$.

Step 3: Using tables and the convolution theorem, show that

$$u(x, t) = \frac{1}{\sqrt{\pi}} \int_{-\infty}^{\infty} f\left(x - 2\eta\sqrt{t}\right) e^{-\eta^2} \, d\eta.$$

2. Find the forced solution to the partial differential equation[24]

$$\frac{\partial u}{\partial t} = \frac{\partial^2 u}{\partial r^2} + \frac{1}{r}\frac{\partial u}{\partial r} + \frac{\partial^2 u}{\partial z^2}, \qquad 0 < r < 1, \quad -\infty < z < \infty, \quad 0 < t,$$

subject to the boundary conditions $\lim_{|z| \to \infty} u(r, z, t) \to 0$, $|u(0, z, t)| < \infty$ and

$$u(1, z, t) = \begin{cases} 0, & \text{if} \quad z < 0, \\ 1, & \text{if} \quad 0 < z < t, \\ 0, & \text{if} \quad t < z. \end{cases}$$

Step 1: If

$$u(r, z, t) = \frac{1}{2\pi} \int_{-\infty}^{\infty} U(r, k, t)e^{ikz} \, dk,$$

show that the original partial differential equation reduces to the two-dimensional partial differential equation

$$\frac{\partial U(r, k, t)}{\partial t} = \frac{\partial^2 U(r, k, t)}{\partial r^2} + \frac{1}{r}\frac{\partial U(r, k, t)}{\partial r} - k^2 U(r, k, t)$$

with the boundary condition $U(1, k, t) = \left(1 - e^{-ikt}\right)/(ik)$.

---

[24] Taken from Singh, H., 1981: Thermal stresses in an infinite cylinder. *Indian J. Pure Appl. Math.*, **12**, 405–418.

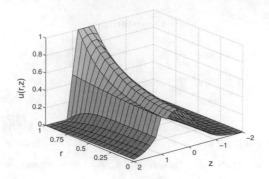

Problem 2

Step 2: Show that the forced solution to Step 1 is

$$U(r,k,t) = \frac{I_0(kr)}{ikI_0(k)} - \frac{I_0\left(r\sqrt{k^2-ik}\right)}{ikI_0\left(\sqrt{k^2-ik}\right)}e^{-ikt}.$$

Step 3: Using the inversion formula

$$f(t) = \frac{1}{4\pi}\left[\int_{-\infty-\epsilon i}^{\infty-\epsilon i} F(\omega)e^{i\omega t}\,d\omega + \int_{-\infty+\epsilon i}^{\infty+\epsilon i} F(\omega)e^{i\omega t}\,d\omega\right], \qquad 0 < \epsilon \ll 1,$$

and the residue theorem, show that

$$\mathcal{F}^{-1}\left[\frac{I_0(kr)}{ikI_0(k)}\right] = \tfrac{1}{2}\mathrm{sgn}(z) - \mathrm{sgn}(z)\sum_{n=1}^{\infty}\frac{J_0(\alpha_n r)}{\alpha_n J_1(\alpha_n)}e^{-\alpha_n|z|},$$

where α_n is the nth root of $J_0(\alpha) = 0$. Finally, show that

$$\mathcal{F}^{-1}\left[\frac{I_0(r\sqrt{k^2-ik})e^{-ikt}}{ikI_0(\sqrt{k^2-ik})}\right] = -\tfrac{1}{2}\mathrm{sgn}(t-z) + \mathrm{sgn}(t-z)$$

$$\times \sum_{n=1}^{\infty}\frac{2\alpha_n J_0(\alpha_n r)}{\lambda_n(2\lambda_n-1)J_1(\alpha_n)}e^{\lambda_n(t-z)},$$

where

$$\lambda_n = \begin{cases} \left(1 - \sqrt{1+4\alpha_n^2}\right)/2, & \text{if} \quad t - z > 0, \\[2mm] \left(1 + \sqrt{1+4\alpha_n^2}\right)/2, & \text{if} \quad t - z < 0. \end{cases}$$

This solution, $u(r,z,t)$, is illustrated in the figure labeled Problem 2 when $t = 1$.

3.4 LAPLACE'S EQUATION

Fourier transforms are a very popular method for solving Laplace's equation when the domain is of infinite extent in one direction. In the case of semi-infinite domains, an important question is whether we should use Fourier cosine or sine transform. The answer lies in the nature of the boundary conditions. The choice must be such that repeated integration by parts can be evaluated with known quantities. The following examples show the correct usage.

• **Example 3.4.1**

Let us solve Laplace's equation on the semi-infinite strip

$$\frac{\partial^2 u}{\partial x^2} + \frac{\partial^2 u}{\partial y^2} = 0, \qquad 0 < x < \infty, \quad 0 < y < b, \qquad (3.4.1)$$

subject to the boundary conditions

$$u_x(0, y) = 0, \qquad \lim_{x \to \infty} u(x, y) \to 0, \qquad 0 < y < b, \qquad (3.4.2)$$

and

$$u_y(x, 0) = 0, \qquad u(x, b) = f(x), \qquad 0 < x < \infty. \qquad (3.4.3)$$

We begin by multiplying Equation 3.4.1 with $\cos(kx)$ and integrating from 0 to ∞. This yields

$$\int_0^\infty \frac{\partial^2 u}{\partial y^2} \cos(kx) \, dx + \int_0^\infty \frac{\partial^2 u}{\partial x^2} \cos(kx) \, dx = 0, \qquad (3.4.4)$$

or

$$\frac{d^2}{dy^2} \left[\int_0^\infty u(x, y) \cos(kx) \, dx \right] + \int_0^\infty \frac{\partial^2 u}{\partial x^2} \cos(kx) \, dx = 0. \qquad (3.4.5)$$

Integrating the u_{xx} term of Equation 3.4.5 twice by parts,

$$\int_0^\infty \frac{\partial^2 u}{\partial x^2} \cos(kx) \, dx = \left. \frac{\partial u}{\partial x} \cos(kx) \right|_{x=0}^{x=\infty} - \left. ku(x, y) \sin(kx) \right|_{x=0}^{x=\infty}$$
$$- k^2 \int_0^\infty u(x, y) \cos(kx) \, dx. \qquad (3.4.6)$$

We now see why we multiplied Equation 3.4.1 by $\cos(kx)$ rather than $\sin(kx)$. As long as $u(0, y)$ is finite, the product $u(0, y) \sin(0)$ equals zero. On the other hand, if we had used $\sin(kx)$, integration by parts would yield $u(0, y) \cos(0)$, an unknown quantity.

Because $u_x(0, y) = 0$ and $u(x, y)$ vanishes as $x \to \infty$, a combination of Equation 3.4.5 and Equation 3.4.6 gives

$$\frac{d^2 U(k, y)}{dy^2} - k^2 U(k, y) = 0, \qquad 0 < y < b, \qquad (3.4.7)$$

with the boundary conditions $U'(k, 0) = 0$ and

$$U(k, b) = F(k) = \int_0^\infty f(x) \cos(kx)\, dx, \qquad (3.4.8)$$

where

$$U(k, y) = \int_0^\infty u(x, y) \cos(kx)\, dx. \qquad (3.4.9)$$

Here, the prime denotes an ordinary differentiation with respect to y.

The solution to Equation 3.4.6 satisfying the boundary conditions is

$$U(k, y) = F(k)\frac{\cosh(ky)}{\cosh(kb)}. \qquad (3.4.10)$$

Therefore,

$$u(x, y) = \frac{2}{\pi} \int_0^\infty U(k, y) \cos(kx)\, dk = \frac{2}{\pi} \int_0^\infty F(k)\frac{\cosh(ky)}{\cosh(kb)} \cos(kx)\, dk. \qquad (3.4.11)$$

• Example 3.4.2

Let us solve Laplace's equation inside a semi-infinite cylinder of radius a:

$$\frac{1}{r}\frac{\partial}{\partial r}\left(r\frac{\partial u}{\partial r}\right) + \frac{\partial^2 u}{\partial z^2} = 0, \qquad 0 \le r < a, \quad 0 < z < \infty, \qquad (3.4.12)$$

subject to the boundary conditions

$$\lim_{r \to 0} |u(r, z)| < \infty, \qquad u(a, z) = 0, \qquad 0 < z < \infty, \qquad (3.4.13)$$

and

$$u(r, 0) = T_0, \qquad \lim_{z \to \infty} u(r, z) \to 0, \qquad 0 \le r < a. \qquad (3.4.14)$$

We begin by introducing a new dependent variable $v(r, z)$, where $u(r, z) = T_0 + v(r, z)$. Thus, rewriting Equation 3.4.12 through Equation 3.4.14 in terms of $v(r, z)$,

$$\frac{1}{r}\frac{\partial}{\partial r}\left(r\frac{\partial v}{\partial r}\right) + \frac{\partial^2 v}{\partial z^2} = 0, \qquad 0 \le r < a, \quad 0 < z < \infty, \qquad (3.4.15)$$

subject to the boundary conditions

$$\lim_{r \to 0} |v(r, z)| < \infty, \qquad v(a, z) = -T_0, \qquad 0 < z < \infty, \qquad (3.4.16)$$

and

$$v(r, 0) = 0, \qquad \lim_{z \to \infty} v(r, z) \to 0, \qquad 0 \le r < a. \qquad (3.4.17)$$

Equation 3.4.15 through Equation 3.4.17 can now be solved using Fourier sine transforms. Multiplying Equation 3.4.15 by $\sin(mz)$ and integrating from 0 to ∞, we find

$$\int_0^\infty \frac{1}{r}\frac{\partial}{\partial r}\left(r\frac{\partial v}{\partial r}\right)\sin(mz)\,dz + \int_0^\infty \frac{\partial^2 v}{\partial z^2}\sin(mz)\,dz = 0, \qquad (3.4.18)$$

or

$$\frac{1}{r}\frac{d}{dr}\left\{r\frac{d}{dr}\left[\int_0^\infty v(r,z)\sin(mz)\,dz\right]\right\} + \int_0^\infty \frac{\partial^2 v}{\partial z^2}\sin(mz)\,dz = 0. \qquad (3.4.19)$$

Integrating the v_{zz} term twice by parts,

$$\int_0^\infty \frac{\partial^2 v}{\partial z^2}\sin(mz)\,dz = \frac{\partial v}{\partial z}\sin(mz)\Big|_{z=0}^{z=\infty} - mv(r,z)\cos(mz)\Big|_{z=0}^{z=\infty}$$
$$- m^2\int_0^\infty v(r,z)\sin(mz)\,dz. \qquad (3.4.20)$$

Because $v(r,0) = 0$ and $v(r,z)$ vanishes as $z \to \infty$, a combination of Equation 3.4.19 and Equation 3.4.20 yields the boundary-value problem

$$\frac{1}{r}\frac{d}{dr}\left[r\frac{dV(r,m)}{dr}\right] - m^2 V(r,m) = 0, \qquad 0 \le r < a, \qquad (3.4.21)$$

with

$$\lim_{r\to 0}|V(r,m)| < \infty \quad \text{and} \quad V(a,m) = -\frac{T_0}{m}, \qquad (3.4.22)$$

since

$$\frac{2}{\pi}\int_0^\infty \frac{\sin(kx)}{k}\,dk = 1, \qquad 0 < x. \qquad (3.4.23)$$

The solution to this boundary-value problem is

$$V(r,m) = \frac{T_0}{m}\frac{I_0(mr)}{I_0(ma)}. \qquad (3.4.24)$$

Therefore,

$$u(r,z) = T_0 - \frac{2T_0}{\pi}\int_0^\infty \frac{I_0(mr)}{I_0(ma)}\frac{\sin(mz)}{m}\,dm. \qquad (3.4.25)$$

Figure 3.4.1 illustrates Equation 3.4.25 as a function of r and z.

● **Example 3.4.3**

Let us solve Laplace's equation on the quarter plane:

$$\frac{\partial^2 u}{\partial x^2} + \frac{\partial^2 u}{\partial y^2} = 0, \qquad 0 < x, y < \infty, \qquad (3.4.26)$$

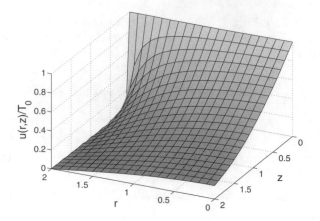

Figure 3.4.1: A plot of $u(r, z)/T_0$ given by Equation 3.4.25 as a function of r and z when $a = 2$.

subject to the boundary conditions

$$\lim_{x \to \infty} u(x, y) \to 0, \qquad \lim_{y \to \infty} u(x, y) \to 0, \qquad (3.4.27)$$

$$u(x, 0) = T_0, \qquad 0 < x < \infty, \qquad (3.4.28)$$

and

$$u_x(0, y) - hu(0, y) = 0, \qquad 0 < y < \infty. \qquad (3.4.29)$$

We begin by multiplying Equation 3.4.26 with $\sin(\ell y)$ and integrating from 0 to ∞. This yields

$$\int_0^\infty \frac{\partial^2 u}{\partial x^2} \sin(\ell y) \, dy + \int_0^\infty \frac{\partial^2 u}{\partial y^2} \sin(\ell y) \, dy = 0, \qquad (3.4.30)$$

or

$$\frac{d^2}{dx^2} \left[\int_0^\infty u(x, y) \sin(\ell y) \, dy \right] + \int_0^\infty \frac{\partial^2 u}{\partial y^2} \sin(\ell y) \, dy = 0. \qquad (3.4.31)$$

Integrating the second integral in Equation 3.4.31 twice by parts,

$$\int_0^\infty \frac{\partial^2 u}{\partial y^2} \sin(\ell y) \, dy = \frac{\partial u}{\partial y} \sin(\ell y) \Big|_{y=0}^{y=\infty} - \ell u(x, y) \cos(\ell y) \Big|_{y=0}^{y=\infty}$$
$$- \ell^2 \int_0^\infty u(x, y) \sin(\ell y) \, dy. \qquad (3.4.32)$$

We now see why we multiplied Equation 3.4.26 by $\sin(\ell y)$ rather than $\cos(\ell y)$. As long as $u_y(x, 0)$ is finite, the product $u_y(x, 0) \sin(0)$ equals zero. On the other hand, if we had used $\cos(\ell y)$, integration by parts would yield $u_y(x, 0) \cos(0)$, an unknown quantity.

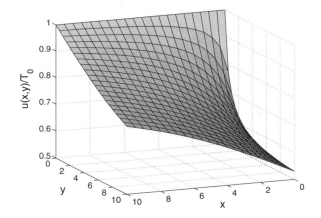

Figure 3.4.2: A plot of $u(x,y)/T_0$ given by Equation 3.4.38 as a function of x and y when $h = 1$.

Because $u(x,0) = T_0$ and $u(x,y)$ vanishes as $x \to \infty$, a combination of Equation 3.4.31 and Equation 3.4.32 gives

$$\frac{d^2 U(x,\ell)}{dx^2} - \ell^2 U(x,\ell) = -\ell T_0, \qquad 0 < x < \infty, \qquad (\textbf{3.4.33})$$

with the boundary conditions

$$U'(0,\ell) - hU(0,\ell) = 0, \qquad \lim_{x \to \infty} U(x,\ell) \to 0, \qquad (\textbf{3.4.34})$$

where

$$U(x,\ell) = \int_0^\infty u(x,y) \sin(\ell y)\, dy. \qquad (\textbf{3.4.35})$$

Here, the prime denotes an ordinary derivative with respect to x.

The solution to Equation 3.4.33 satisfying the boundary conditions is

$$U(x,\ell) = \frac{T_0}{\ell} - \frac{hT_0}{\ell(h+\ell)} e^{-\ell x}. \qquad (\textbf{3.4.36})$$

Therefore,

$$u(x,y) = \frac{2T_0}{\pi} \int_0^\infty \frac{\sin(\ell y)}{\ell}\, dy - \frac{2hT_0}{\pi} \int_0^\infty \frac{e^{-\ell x}}{\ell(h+\ell)} \sin(\ell y)\, d\ell \qquad (\textbf{3.4.37})$$

$$= T_0 \left[1 - \frac{2h}{\pi} \int_0^\infty \frac{e^{-\ell x}}{\ell(h+\ell)} \sin(\ell y)\, d\ell \right]. \qquad (\textbf{3.4.38})$$

Figure 3.4.2 illustrates Equation 3.4.38 as a function of x and y.

● **Example 3.4.4**

In the previous examples, we reduced the partial differential equation to an ordinary differential equation by formally taking the Fourier transform of the partial differential equation. Let's solve a problem similar to the previous example but do it by using the classic method of separation of variables.

The problem that we shall now solve is

$$\frac{\partial^2 u}{\partial x^2} + \frac{\partial^2 u}{\partial y^2} = 0, \qquad 0 < x, y < \infty, \qquad (3.4.39)$$

subject to the boundary conditions

$$\lim_{x \to \infty} u(x, y) \to 0, \qquad \lim_{y \to \infty} u(x, y) \to 0, \qquad (3.4.40)$$

$$u_y(x, 0) = hu(x, 0), \qquad 0 < x < \infty, \qquad (3.4.41)$$

and

$$u(0, y) = f(y), \qquad 0 < y < \infty. \qquad (3.4.42)$$

To solve Equation 3.4.39 through Equation 3.4.42, we assume that the solution consists of a product of two single-variable functions: $u(x, y) = X(x)Y(y)$. Substituting this solution into the partial differential equation and boundary conditions and separating the variables, we find that

$$\frac{X''}{X} = -\frac{Y''}{Y} = k^2, \qquad (3.4.43)$$

with

$$\lim_{x \to \infty} X(x) \to 0, \qquad \lim_{y \to \infty} Y(y) \to 0, \qquad Y'(0) = hY(0), \qquad (3.4.44)$$

where k is an arbitrary separation constant. A particular solution that satisfies Equation 3.4.39 to Equation 3.4.41 is

$$u_p(x, y) = A(k)e^{-kx} \left[\sin(ky) + k\cos(ky)/h\right]. \qquad (3.4.45)$$

Because Equation 3.4.45 is true for any k, the most general solution is a linear superposition of all of the particular solutions, or

$$u(x, y) = \int_0^\infty A(k)e^{-kx} \left[\sin(ky) + k\cos(ky)/h\right] dk. \qquad (3.4.46)$$

Our final task is to evaluate $A(k)$. Applying Equation 3.4.42, we find that

$$f(y) = \int_0^\infty A(k) \left[\sin(ky) + k\cos(ky)/h\right] dk. \qquad (3.4.47)$$

To find $A(k)$, we must solve the integral Equation 3.4.47. Following Karush,[25] we introduce the function

$$W(y) = \int_0^\infty A(k) \sin(ky) \, dk. \tag{3.4.48}$$

Therefore, we may rewrite Equation 3.4.47 as the ordinary differential equation

$$\frac{dW(y)}{dy} + hW(y) = hf(y), \qquad W(0) = 0. \tag{3.4.49}$$

Solving for $W(y)$,

$$W(y) = he^{-hy} \int_0^y f(\eta)e^{h\eta} \, d\eta. \tag{3.4.50}$$

Using Fourier integral theorem,

$$A(k) = \frac{2}{\pi} \int_0^\infty W(y) \sin(ky) \, dy \tag{3.4.50}$$

$$= \frac{2h}{\pi} \int_0^\infty e^{-hy} \left[\int_0^y f(\eta)e^{h\eta} \, d\eta \right] \sin(ky) \, dy \tag{3.4.51}$$

$$= \frac{2h}{\pi} \int_0^\infty f(\eta)e^{h\eta} \left[\int_y^\infty e^{-hy} \sin(ky) \, dy \right] d\eta \tag{3.4.52}$$

$$= \frac{2}{\pi \left[1 + (k/h)^2\right]} \int_0^\infty f(\eta) \left[\sin(k\eta) + k\cos(k\eta)/h\right] d\eta. \tag{3.4.53}$$

A combination of Equation 3.4.46 and Equation 3.4.53 is the solution to our problem.

• Example 3.4.5: Mixed Boundary-Value Problem

For our final example, let us solve Laplace's equation

$$\frac{\partial^2 u}{\partial x^2} + \frac{\partial^2 u}{\partial y^2} = 0, \qquad -\infty < x < \infty, \quad 0 < y < \infty, \tag{3.4.54}$$

subject to the boundary conditions

$$\lim_{|x| \to \infty} u(x, y) \to 0, \qquad 0 < y, \tag{3.4.55}$$

$$\lim_{y \to \infty} u(x, y) \to 0, \qquad -\infty < x < \infty, \tag{3.4.56}$$

$$u(x, 0) = \begin{cases} -1, & -b < x < -a, \\ 1, & a < x < b, \end{cases} \tag{3.4.57}$$

---

[25] Karush, W., 1952: A steady-state heat flow problem for a quarter infinite solid. *J. Appl. Phys.*, **23**, 492–494.

and

$$u_y(x,0) = 0, \qquad 0 < |x| < a, \quad b < |x| < \infty. \qquad (3.4.58)$$

The interesting aspect of this problem is the boundary condition along $y = 0$. For a portion of the boundary $(-b < x < -a$ and $a < x < b)$, it consists of a Dirichlet condition; otherwise, it is a Neumann condition. This is a simple example of a wide class of problems that appears in mathematical physics, commonly known as *mixed boundary-value problems*.

If we employ separation of variables as we did in the previous example, the most general solution is

$$u(x,y) = \int_0^\infty \frac{A(k)}{k} e^{-ky} \sin(kx)\, dk. \qquad (3.4.59)$$

Substituting Equation 3.4.59 into the boundary conditions given by Equation 3.4.57 and Equation 3.4.58, we obtain the following set of integral equations:

$$\int_0^\infty A(k) \sin(kx)\, dk = 0, \qquad 0 \le x < a, \qquad (3.4.60)$$

$$\int_0^\infty \frac{A(k)}{k} \sin(kx)\, dk = 1, \qquad a < x < b, \qquad (3.4.61)$$

and

$$\int_0^\infty A(k) \sin(kx)\, dk = 0, \qquad b < x < \infty. \qquad (3.4.62)$$

We must now solve for $A(k)$ which appears in a set of integral equations. In general, this is a rather formidable task. However, Tranter[26] has shown that triple integral equations of the form given by Equation 3.4.60 through Equation 3.4.62 have the solution

$$A(k) = 2 \sum_{n=1}^\infty (-1)^{n-1} A_n J_{2n-1}(bk), \qquad (3.4.63)$$

where the constants A_n are the solution of the dual series relationship

$$\sum_{n=1}^\infty (-1)^{n-1} A_n \sin\left[\left(n - \tfrac{1}{2}\right)\varphi\right] = 0, \qquad 0 \le \varphi < \gamma, \qquad (3.4.64)$$

$$\sum_{n=1}^\infty (-1)^{n-1} \frac{A_n}{n - \tfrac{1}{2}} \sin\left[\left(n - \tfrac{1}{2}\right)\varphi\right] = 1, \qquad \gamma < \varphi \le \pi, \qquad (3.4.65)$$

[26] Tranter, C. J., 1960: Some triple integral equations. *Proc. Glasgow Math. Assoc.*, **4**, 200–203. A more accessible analysis is given in Section 6.4 of Sneddon, I. N., 1966: *Mixed Boundary Value Problems in Potential Theory*. Wiley, 283 pp.

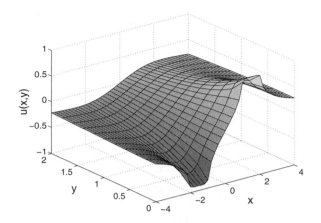

Figure 3.4.3: Plot of Equation 3.4.59 when $a = 1$ and $b = 2$.

and γ is defined by $a = b\sin(\gamma/2)$, $0 < \gamma \leq \pi$. If we now introduce the change of variables $\theta = \pi - \varphi$ and $c = \pi - \gamma$, we find that A_n is the solution of the following pair of dual series equations:

$$\sum_{n=1}^{\infty} \frac{A_n}{n - \frac{1}{2}} \cos\left[\left(n - \tfrac{1}{2}\right)\theta\right] = 1, \qquad 0 < \theta < c, \tag{3.4.66}$$

and

$$\sum_{n=1}^{\infty} A_n \cos\left[\left(n - \tfrac{1}{2}\right)\theta\right] = 0, \qquad c < \theta \leq \pi. \tag{3.4.67}$$

Consequently, we have reduced three integral equations to two dual trigonometric series. Tranter[27] has also analyzed dual trigonometric series of the form given by Equation 3.4.66 and Equation 3.4.67 and shown that in our particular case

$$A_n = \frac{P_{n-1}[\cos(c)]}{K(a/b)}, \tag{3.4.68}$$

where $K(\cdot)$ denotes the complete elliptic integral and $P_n(\cdot)$ is the Legendre polynomial of order n. Substituting Equation 3.4.68 into Equation 3.4.63 with $a = b\cos(c/2)$, we obtain

$$A(k) = \frac{2}{K(a/b)} \sum_{n=1}^{\infty} (-1)^{n-1} P_{n-1}[\cos(c)] J_{2n-1}(bk). \tag{3.4.69}$$

Figure 3.4.3 illustrates $u(x, y)$ when $a = 1$ and $b = 2$.

---

[27] Tranter, C. J., 1959: Dual trigonometric series. *Proc. Glasgow Math. Assoc.*, **4**, 49–57; Sneddon, *op. cit.*, Section 5.4.5

Problems

1. Find the solution to Laplace's equation

$$\frac{\partial^2 u}{\partial x^2} + \frac{\partial^2 u}{\partial y^2} = 0$$

on the upper half-plane $y > 0$. The solution should remain bounded over the entire domain and $u(x,0) = f(x)$ along the x-axis.

Step 1: Take the Fourier transform of Laplace's equation and the boundary condition and show that you obtain the boundary-value problem

$$\frac{d^2 U(k,y)}{dy^2} - k^2 U(k,y) = 0, \qquad 0 < y < \infty,$$

with

$$F(k) = U(k,0) = \int_{-\infty}^{\infty} f(x) e^{-ikx}\, dx,$$

where

$$U(k,y) = \int_{-\infty}^{\infty} u(x,y) e^{-ikx}\, dx.$$

Step 2: Solve the boundary-value problem and show that

$$U(k,y) = F(k) e^{-|k|y}.$$

Step 3: Find the inverse of the Fourier transform $e^{-|k|y}$ and use the convolution theorem to show that

$$u(x,y) = \frac{1}{\pi} \int_{-\infty}^{\infty} \frac{y f(\xi)}{(x-\xi)^2 + y^2}\, d\xi.$$

This classic result is called *Poisson's integral formula*[28] for the half-plane $y > 0$ or *Schwarz's integral formula*.[29]

Step 4: If $u(x,0) = 1$ if $|x| < 1$ and $u(x,0) = 0$ otherwise, then use Poisson's integral formula and show that

$$u(x,y) = \frac{1}{\pi} \left[\arctan\left(\frac{1-x}{y}\right) + \arctan\left(\frac{1+x}{y}\right) \right].$$

---

[28] Poisson, S. D., 1823: Suite du mémoire sur les intégrales définies et sur la sommation des séries. *J. École Polytech.*, **19**, 404–509. See pg. 462.

[29] Schwarz, H. A., 1870: Über die Integration der partiellen Differentialgleichung $\partial^2 u/\partial x^2 + \partial^2 u/\partial y^2 = 0$ für die Fläche eines Kreises. *Vierteljahrsschr. Naturforsch. Ges. Zürich*, **15**, 113–128.

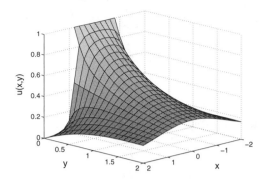

Problem 1

This solution $u(x, y)$ is illustrated in the figure labeled Problem 1.

2. Find the solution to Laplace's equation

$$\frac{\partial^2 u}{\partial x^2} + \frac{\partial^2 u}{\partial y^2} = 0$$

along the semi-infinite strip $0 < x < \infty$, $0 < y < b$, subject to the boundary conditions

$$u(0, y) = 0, \qquad \lim_{x \to \infty} u(x, y) \to 0, \qquad 0 < y < b,$$

and

$$u(x, 0) = f(x), \qquad u(x, b) = 0, \qquad 0 < x < \infty.$$

Step 1: Take the Fourier sine transform of Laplace's equation and the boundary conditions and show that you obtain the boundary-value problem

$$\frac{d^2 U(k, y)}{dy^2} - k^2 U(k, y) = 0, \qquad 0 < y < b,$$

with

$$F(k) = U(k, 0) = \int_0^\infty f(x) \sin(kx)\, dx \qquad \text{and} \qquad U(k, b) = 0,$$

where

$$U(k, y) = \int_0^\infty u(x, y) \sin(kx)\, dx.$$

Step 2: Solve the boundary-value problem and show that

$$U(k, y) = F(k) \frac{\sinh[k(b - y)]}{\sinh(kb)}.$$

Step 3: Invert the Fourier sine transform and show that

$$u(x,y) = \frac{2}{\pi} \int_0^\infty F(k) \frac{\sinh[k(b-y)]}{\sinh(kb)} \sin(kx) \, dk.$$

3. Find the solution to Laplace's equation

$$\frac{\partial^2 u}{\partial x^2} + \frac{\partial^2 u}{\partial y^2} = 0$$

in the strip $0 < x < \infty$, $-L < y < L$, subject to the boundary conditions

$$u_x(0,y) = 0, \qquad \lim_{x \to \infty} u(x,y) \to 0, \qquad -L < y < L,$$

and

$$u_y(x,L) = f(x), \qquad u_y(x,-L) = -f(x), \qquad 0 < x < \infty.$$

Step 1: Take the Fourier cosine transform of Laplace's equation and the boundary conditions and show that you obtain the boundary-value problem

$$\frac{d^2 U(k,y)}{dy^2} - k^2 U(k,y) = 0, \qquad -L < y < L,$$

with

$$U'(k,L) = -U'(k,-L) = F(k) = \int_0^\infty f(x) \cos(kx) \, dx,$$

where

$$U(k,y) = \int_0^\infty u(x,y) \cos(kx) \, dx.$$

Step 2: Solve the boundary-value problem and show that

$$U(k,y) = F(k) \frac{\cosh(ky)}{k \sinh(kL)}.$$

Step 3: Invert the Fourier cosine transform and show that

$$u(x,y) = \frac{2}{\pi} \int_0^\infty F(k) \frac{\cosh(ky)}{k \sinh(kL)} \cos(kx) \, dk.$$

4. Find the solution to Laplace's equation

$$\frac{\partial^2 u}{\partial x^2} + \frac{\partial^2 u}{\partial y^2} = 0$$

in the quarter plane $0 < x, y < \infty$, subject to the boundary conditions

$$u(0, y) = 0, \qquad \lim_{x \to \infty} u(x, y) \to 0, \qquad 0 < y < \infty,$$

and

$$hu_y(x, 0) = f(x), \qquad \lim_{y \to \infty} u(x, y) \to 0, \qquad 0 < x < \infty.$$

Step 1: Use separation of variables to show that

$$u(x, y) = \int_0^\infty A(k) \sin(kx) e^{-ky} \, dk.$$

Step 2: Use Fourier's integral theorem and show that

$$A(k) = -\frac{2}{\pi hk} \int_0^\infty f(\eta) \sin(k\eta) \, d\eta.$$

5. If $a > 0$, find the solution to Laplace's equation

$$\frac{\partial^2 u_1}{\partial x^2} + \frac{\partial^2 u_1}{\partial y^2} = 0, \qquad 0 < x < \infty, \quad -\infty < y < \infty,$$

$$\frac{\partial^2 u_2}{\partial x^2} + \frac{\partial^2 u_2}{\partial y^2} = 0, \qquad -a < x < 0, \quad -\infty < y < \infty,$$

subject to the boundary conditions

$$\lim_{x \to \infty} u_1(x, y) \to 0, \quad -\infty < y < \infty, \qquad \lim_{|y| \to \infty} u_i(x, y) \to 0, \quad -a < x < \infty,$$

$$u_1(0, y) = u_2(0, y) = \begin{cases} T_0, & |y| < b, \\ 0, & |y| > b, \end{cases} \quad \text{and} \quad u_2(-a, y) = 0, \quad -\infty < y < \infty.$$

Step 1: By taking the Fourier transform

$$U_i(x, \ell) = \int_{-\infty}^\infty u_i(x, y) e^{-i\ell y} \, dy$$

of the partial differential equations and boundary conditions, show that they reduce to the boundary-value problems

$$\frac{d^2 U_1(x, \ell)}{dx^2} - \ell^2 U_1(x, \ell) = 0, \qquad 0 < x < \infty,$$

and

$$\frac{d^2 U_2(x, \ell)}{dx^2} - \ell^2 U_2(x, \ell) = 0, \qquad -a < x < 0,$$

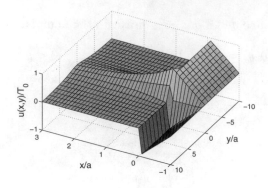

Problem 5

with $\lim_{x\to\infty} U_1(x,\ell) \to 0$, $U_2(-a,\ell) = 0$ and $U_1(0,\ell) = U_2(0,\ell) = T_0\sin(b\ell)/\ell$.

Step 2: Using the relationship

$$\int_0^\infty e^{-x\xi}\sin(y\xi)\,\frac{d\xi}{\xi} = \arctan\left(\frac{y}{x}\right), \qquad 0 < x, y,$$

invert the Fourier transform and show that

$$u_1(x,y) = \frac{2T_0}{\pi}\int_0^\infty e^{-x\ell}\sin(b\ell)\cos(y\ell)\,\frac{d\ell}{\ell} = \frac{2T_0}{\pi}\left[\arctan\left(\frac{b+y}{x}\right) + \arctan\left(\frac{b-y}{x}\right)\right].$$

Step 3: Using the relationship

$$\int_0^\infty \frac{\sinh(x\xi)}{\sinh(a\xi)}\sin(y\xi)\,\frac{d\xi}{\xi} = \arctan\left[\tan\left(\frac{\pi x}{2a}\right)\tanh\left(\frac{\pi y}{2a}\right)\right],$$

if $|\mathrm{Re}(x)| < \mathrm{Re}(a)$ and $0 < y$, invert the Fourier transform and show that

$$u_2(x,y) = \frac{2T_0}{\pi}\int_0^\infty \frac{\sinh[(x+a)\ell]}{\sinh(a\ell)}\sin(b\ell)\cos(y\ell)\,\frac{d\ell}{\ell}$$

$$= \frac{2T_0}{\pi}\left(\arctan\left\{\tan\left[\frac{\pi(x+a)}{2a}\right]\tanh\left[\frac{\pi(b+y)}{2a}\right]\right\}\right.$$

$$\left. + \arctan\left\{\tan\left[\frac{\pi(x+a)}{2a}\right]\tanh\left[\frac{\pi(b-y)}{2a}\right]\right\}\right).$$

This solution $u(x,y)$ is illustrated in the figure labeled Problem 5 when $b/a = 1$.

6. Solve Laplace's equation

$$\frac{\partial^2 u}{\partial x^2} + \frac{\partial^2 u}{\partial y^2} = 0, \qquad -\infty < x < \infty, \quad 0 < y < \infty,$$

with the boundary conditions

$$\lim_{|x|\to\infty} u(x,y) \to 0, \qquad 0 < y < \infty,$$

$$\lim_{y\to\infty} u(x,y) \to 0, \qquad -\infty < x < \infty,$$

$$u(x,0) = x, \quad 0 \le x < 1, \qquad \text{and} \qquad u_y(x,0) = 0, \quad 1 < x < \infty. \tag{1}$$

Step 1: Using separation of variables, show that the general solution to the problem is

$$u(x,y) = \int_0^\infty A(k)e^{-ky}\frac{\sin(kx)}{k}\,dk.$$

Step 2: Using boundary condition (1), show that $A(k)$ satisfies the dual integral equations

$$\int_0^\infty \frac{A(k)}{k}\sin(kx)\,dk = x, \qquad 0 \le x < 1,$$

and

$$\int_0^\infty A(k)\sin(kx)\,dk = 0, \qquad 1 < x < \infty.$$

Step 3: Fredricks[30] has shown that dual integral equations of the form

$$\int_0^\infty \frac{A(k)}{k}\sin(kx)\,dk = f(x), \qquad 0 \le x < a,$$

and

$$\int_0^\infty A(k)\sin(kx)\,dk = 0, \qquad a < x < \infty,$$

have the solution

$$A(k) = 2\sum_{n=1}^\infty \sum_{m=0}^\infty (2m+1)B_n J_{2m+1}(n\pi)J_{2m+1}(ka),$$

where B_n is given by the Fourier sine series

$$f(x) = \sum_{n=1}^\infty B_n \sin(n\pi x/a), \qquad 0 \le x < a.$$

Clearly $f(0) = 0$. Use this result to show that

$$A(k) = \frac{4}{\pi}\sum_{n=1}^\infty \sum_{m=0}^\infty \frac{(-1)^{n+1}}{n}(2m+1)B_n J_{2m+1}(n\pi)J_{2m+1}(k).$$

---

[30] Fredricks, R. W., 1958: Solution of a pair of integral equations from elastostatics. *Proc. Natl. Acad. Sci.*, **44**, 309–312.

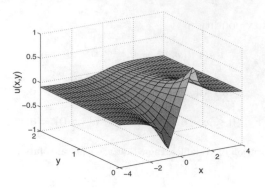

Problem 6

This solution $u(x, y)$ is illustrated in the figure labeled Problem 6.

7. Solve the partial differential equation[31]

$$\frac{\partial^2 u}{\partial x^2} + \frac{\partial^2 u}{\partial z^2} - u = 0, \qquad -\infty < x < \infty, \quad 0 < z < 1,$$

with the boundary conditions

$$\frac{\partial u(x, 0)}{\partial z} = 0, \qquad \text{and} \qquad u(x, 1) = e^{-x} H(x), \qquad -\infty < x < \infty.$$

Step 1: By taking the Fourier transform

$$U(k, z) = \int_{-\infty}^{\infty} u(x, z) e^{-ikx} \, dx$$

of the partial differential equation and boundary conditions, show that they reduce to the boundary-value problem

$$\frac{d^2 U(k, z)}{dz^2} - m^2 U(k, z) = 0, \qquad 0 < z < 1,$$

with $U'(k, 0) = 0$ and $U(k, 1) = 1/(1 + ki)$, where $m^2 = k^2 + 1$.

Step 2: Show that the solution to boundary-value problem is

$$U(k, z) = \frac{\cosh(mz)}{i(k - i) \cosh(m)}.$$

Step 3: Show that the singularities of $U(k, z)$ are simple poles at $k = i$ and $k_n = \pm i \sqrt{4 + (2n + 1)^2 \pi^2}/2$, where $n = 0, 1, 2, \ldots$.

---

[31] Taken from Horvay, G., 1961: Temperature distribution in a slab moving from a chamber at one temperature to a chamber at another temperature. *J. Heat Transfer*, **83**, 391–402.

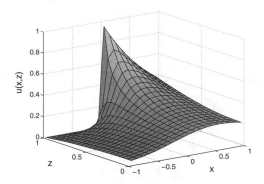

Problem 7

Step 4: Use the residue theorem to invert $U(k,z)$ and show that

$$u(x,z) = e^{-x}H(x) + 2\pi \sum_{n=0}^{\infty} \frac{(-1)^n(2n+1)}{\sigma_n[2 - \mathrm{sgn}(x)\sigma_n]} e^{-\sigma_n|x|/2} \cos\left[\frac{(2n+1)\pi z}{2}\right],$$

where $\sigma_n = \sqrt{4 + (2n+1)^2\pi^2}$. This solution $u(x,z)$ is illustrated in the figure labeled Problem 7.

8. Solve the Helmholtz's equation

$$\frac{\partial^2 u}{\partial x^2} + \frac{\partial^2 u}{\partial y^2} + k_0^2 u = 0, \qquad -\infty < x < \infty, \quad 0 < y < \infty,$$

with the boundary conditions $\lim_{|x|\to\infty} u(x,y) \to 0, 0 < y < \infty$, and

$$u_y(x,0) = \begin{cases} 1, & |x| \le a, \\ 0, & |x| > a, \end{cases} \qquad \lim_{y\to\infty} |u(x,y)| < \infty, \qquad -\infty < x < \infty.$$

Step 1: By taking the Fourier transform

$$U(k,y) = \int_{-\infty}^{\infty} u(x,y)e^{-ikx}\,dx$$

of the partial differential equation and boundary conditions, show that they reduce to the boundary-value problem

$$\frac{d^2 U(k,y)}{dy^2} - (k^2 - k_0^2)U(k,y) = 0, \qquad 0 < y < \infty,$$

with $U'(k,0) = 2\sin(ka)/k$ and $\lim_{y\to\infty} |U(k,y)| < \infty$.

Step 2: Show that the solution to boundary-value problem is

$$U(k,y) = -2\frac{\sin(ka)}{k}\frac{e^{-y\sqrt{k^2-k_0^2}}}{\sqrt{k^2-k_0^2}}, \qquad \mathrm{Re}\left(\sqrt{k^2 - k_0^2}\right) \ge 0.$$

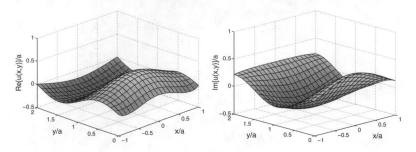

Problem 8

Step 3: Invert $U(k, y)$ and show that

$$u(x, y) = -\frac{2}{\pi} \int_0^\infty \frac{\sin(ka)}{k} \frac{e^{-y\sqrt{k^2 - k_0^2}}}{\sqrt{k^2 - k_0^2}} \cos(kx) \, dk.$$

This solution $u(x, y)$ is illustrated in the figure labeled Problem 8.

9. Find the particular solution to the partial differential equation[32]

$$\frac{\partial^2 u_p}{\partial x^2} + \frac{\partial^2 u_p}{\partial z^2} - i\omega\mu\sigma u_p = \sigma B_0[\delta(z + h/2) - \delta(z - h/2)] \sin(kx),$$

where $-\infty < x, z < \infty$ and $0 < h$.

Step 1: By taking the Fourier transform in the z-direction

$$U_p(x, m) = \int_{-\infty}^\infty u_p(x, z) e^{-imz} \, dz,$$

show that the partial differential equation reduces to the ordinary differential equation

$$\frac{d^2 U_p(x, m)}{dx^2} - (m^2 + i\omega\mu\sigma) U_p(x, m) = 2i\sigma B_0 \sin(mh/2) \sin(kx).$$

Step 2: Show that the particular solution $U_p(x, m)$ is

$$U_p(x, m) = -\frac{2i\sigma B_0 \sin(mh/2)}{m^2 + \alpha^2} \sin(kx),$$

where $\alpha^2 = k^2 + i\omega\mu\sigma$ with $\text{Re}(\alpha) > 0$.

Step 3: Invert $U_p(x, m)$ and show that

$$u_p(x, z) = \frac{\sigma B_0 \sin(kx)}{\alpha} \left\{ \begin{array}{ll} \sinh(\alpha z) e^{-\alpha h/2}, & |z| < h/2, \\ \text{sgn}(z) \sinh(\alpha h/2) e^{-\alpha|z|}, & |z| > h/2. \end{array} \right.$$

---

[32] Taken from Robey, D. H., 1953: Magnetic dispersion of sound in electrically conducting plates. *J. Acoust. Soc. Am.*, **25**, 603–609.

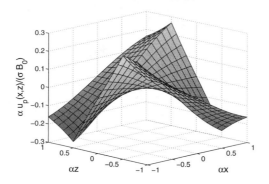

Problem 9

This solution $u_p(x, z)$ is illustrated in the figure labeled Problem 9 when $\alpha h = 1$.

10. Solve the partial differential equation

$$\left(\frac{\partial^2}{\partial r^2} - \frac{1}{r}\frac{\partial}{\partial r} + \frac{\partial^2}{\partial z^2}\right)^2 u = 0, \qquad 0 \le r < 1, \quad -\infty < z < \infty,$$

subject to the boundary conditions $\lim_{r \to 0} |u(r, z)| < \infty$, $u(1, z) = 0$, $u_r(1, z) = \delta(z)$, $-\infty < z < \infty$, and $\lim_{|z| \to \infty} u(r, z) \to 0$, $0 \le r < 1$.

Step 1: By taking the Fourier transform in the z-direction

$$U(r, k) = \int_{-\infty}^{\infty} u(r, z)e^{-ikz}\, dz,$$

show that the partial differential equation reduces to the boundary-value problem

$$\left(\frac{d^2}{dr^2} - \frac{1}{r}\frac{d}{dr} - k^2\right)^2 U(r, k) = 0,$$

with

$$\lim_{r \to 0} |U(r, k)| < \infty, \qquad U(1, k) = 0, \qquad U'(1, k) = 1.$$

Step 2: By direct substitution, show that

$$U(r, k) = \frac{r^2 I_1(k)I_0(kr) - r I_0(k)I_1(kr)}{k\left[I_1^2(k) - I_0(k)I_2(k)\right]}.$$

Step 3: Invert $U(r, k)$ and show that

$$u(r, z) = \frac{2}{\pi}\int_0^{\infty} \frac{r^2 I_1(k)I_0(kr) - r I_0(k)I_1(kr)}{k\left[I_1^2(k) - I_0(k)I_2(k)\right]}\cos(kz)\, dk.$$

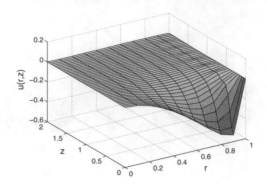

Problem 10

This solution $u(r, z)$ is illustrated in the figure labeled Problem 10.

3.5 THE SOLUTION OF PARTIAL DIFFERENTIAL EQUATIONS BY HANKEL TRANSFORMS

For many problems involving cylindrical geometries, the use of Hankel transforms is preferable to applying Fourier transforms, as the following examples illustrate.

• Example 3.5.1: Elastic Wave Equation

For our first example,[33] we find the free, small transverse oscillations of a thin elastic plate of infinite radius and of uniform density. Assuming symmetry about the z-axis, the differential equation in polar coordinates is

$$\frac{\partial^2 u}{\partial t^2} + b^2 \left(\frac{\partial^2}{\partial r^2} + \frac{1}{r} \frac{\partial}{\partial r} \right)^2 u = 0. \tag{3.5.1}$$

If $U(k, t)$ denotes the Hankel transform of $u(r, t)$ so that

$$U(k, t) = \int_0^\infty r\, u(r, t) J_0(kr)\, dr, \tag{3.5.2}$$

then, as the result of a series of integrations by parts,

$$\int_0^\infty r \left(\frac{\partial^2 u}{\partial r^2} + \frac{1}{r} \frac{\partial u}{\partial r} \right) J_0(kr)\, dr = r \frac{\partial u}{\partial r} J_0(kr) \bigg|_{r=0}^{r=\infty} - k \int_0^\infty r \frac{\partial u}{\partial r} J_0'(kr)\, dr \tag{3.5.3}$$

$$= k \int_0^\infty u(r, t) \left[J_0'(kr) + kr J_0''(kr) \right] dr. \tag{3.5.4}$$

---

[33] Sneddon, I. N., 1945: The Fourier transform solution of an elastic wave equation. *Proc. Cambridge Philos. Soc.*, **41**, 239–243. Reprinted with the permission of Cambridge University Press.

However, because $J_0(kr)$ satisfies the differential equation

$$kr J_0''(kr) + J_0'(kr) + kr J_0(kr) = 0, \tag{3.5.5}$$

$$\int_0^\infty r \left(\frac{\partial^2 u}{\partial r^2} + \frac{1}{r} \frac{\partial u}{\partial r} \right) J_0(kr) \, dr = -k^2 U(k,t). \tag{3.5.6}$$

Applying this result twice,

$$\int_0^\infty r \left(\frac{\partial^2}{\partial r^2} + \frac{1}{r} \frac{\partial}{\partial r} \right)^2 u(r,t) J_0(kr) \, dr = k^4 U(k,t). \tag{3.5.7}$$

Consequently, if we multiply Equation 3.5.1 by $r J_0(kr)$ and integrate with respect to r over the range $(0, \infty)$, it becomes the ordinary differential equation

$$\frac{d^2 U(k,t)}{dt^2} + b^2 k^4 U(k,t) = 0. \tag{3.5.8}$$

We then find the displacement $u(r,t)$ by the Hankel inversion formula

$$u(r,t) = \int_0^\infty k\, U(k,t) J_0(kr) \, dk. \tag{3.5.9}$$

If the initial conditions are $u(r,0) = f(r)$ and $u_t(r,0) = 0$, the solution of Equation 3.5.8 is

$$U(k,t) = A(k) \cos(bk^2 t) + B(k) \sin(bk^2 t). \tag{3.5.10}$$

From the definition of $U(k,t)$,

$$A(k) = U(k,0) = \int_0^\infty \eta f(\eta) J_0(k\eta) \, d\eta \tag{3.5.11}$$

and

$$B(k) = \frac{dU(k,0)}{dt} = 0. \tag{3.5.12}$$

Therefore,

$$u(r,t) = \int_0^\infty k \cos(bk^2 t) \left[\int_0^\infty \eta f(\eta) J_0(k\eta) \, d\eta \right] J_0(kr) \, dk. \tag{3.5.13}$$

By changing the order of integration,

$$u(r,t) = \int_0^\infty \eta f(\eta) \left[\int_0^\infty k J_0(k\eta) J_0(kr) \cos(bk^2 t) \, dk \right] d\eta \tag{3.5.14}$$

$$= \frac{1}{2bt} \int_0^\infty \eta f(\eta) J_0 \left(\frac{r\eta}{2bt} \right) \sin \left(\frac{r^2 + \eta^2}{4bt} \right) d\eta, \tag{3.5.15}$$

because

$$\int_0^\infty k J_0(k\eta) J_0(kr) \cos(btk^2) \, dk = \frac{1}{2bt} J_0\left(\frac{r\eta}{2bt}\right) \sin\left(\frac{r^2 + \eta^2}{4bt}\right). \quad (3.5.16)$$

• Example 3.5.2: Heat Equation

For our next example, let us solve the heat equation

$$\frac{\partial u}{\partial t} = \frac{a^2}{r} \frac{\partial}{\partial r}\left(r \frac{\partial u}{\partial r}\right), \qquad 0 \le r < \infty, \quad 0 < t, \quad (3.5.17)$$

subject to the boundary conditions

$$\lim_{r \to 0} |u(r,t)| < \infty, \qquad \lim_{r \to \infty} u(r,t) \to 0, \qquad 0 < t, \quad (3.5.18)$$

and the initial condition

$$u(r,0) = f(r), \qquad 0 < r < \infty. \quad (3.5.19)$$

We begin by multiplying Equation 3.5.17 by $r \, J_0(kr) \, dr$ and integrating from 0 to ∞:

$$\int_0^\infty \frac{\partial u}{\partial t} J_0(kr) \, r \, dr = \frac{d}{dt}\left[\int_0^\infty u(r,t) J_0(kr) \, r \, dr\right] \quad (3.5.20)$$

$$= a^2 \int_0^\infty \frac{\partial}{\partial r}\left(r \frac{\partial u}{\partial r}\right) J_0(kr) \, dr. \quad (3.5.21)$$

Integrating the right side of Equation 3.5.21 twice by parts,

$$\int_0^\infty \frac{\partial}{\partial r}\left(r \frac{\partial u}{\partial r}\right) J_0(kr) \, dr = r \frac{\partial u}{\partial r} J_0(kr) \Big|_{r=0}^{r=\infty} - k \int_0^\infty r \frac{\partial u}{\partial r} J_0'(kr) \, dr \quad (3.5.22)$$

$$= -kru(r,t) J_0'(kr) \Big|_{r=0}^{r=\infty}$$

$$+ k^2 \int_0^\infty u(r,t) \frac{d}{dr}\left[r \frac{dJ_0(kr)}{dr}\right] dr \quad (3.5.23)$$

$$= -k^2 \int_0^\infty u(r,t) J_0(kr) \, r \, dr, \quad (3.5.24)$$

because

$$\frac{1}{r} \frac{d}{dr}\left[r \frac{dJ_0(kr)}{dr}\right] = -k^2 J_0(kr). \quad (3.5.25)$$

Therefore, Equation 3.5.20 and Equation 3.5.21 become

$$\frac{dU(k,t)}{dt} + a^2 k^2 U(k,t) = 0, \quad (3.5.26)$$

with the initial condition

$$U(k,0) = F(k) = \int_0^\infty f(r)J_0(kr)\, r\, dr. \qquad (3.5.27)$$

The solution to this initial-value problem is

$$U(k,t) = F(k)e^{-a^2k^2t}. \qquad (3.5.28)$$

Therefore,

$$u(r,t) = \int_0^\infty F(k)J_0(kr)e^{-a^2k^2t}\, dk. \qquad (3.5.29)$$

Upon substituting Equation 3.5.27 into Equation 3.5.29 and reversing the order of integration,

$$u(r,t) = \int_0^\infty F(\xi)\left[\int_0^\infty J_0(kr)J_0(k\xi)e^{-a^2k^2t}\, dk\right]\xi\, d\xi \qquad (3.5.30)$$

$$= \frac{1}{2a^2t}\int_0^\infty F(\xi)\exp\left(-\frac{r^2+\xi^2}{4a^2t}\right)I_0\left(\frac{r\xi}{2a^2t}\right)\xi\, d\xi, \qquad (3.5.31)$$

because

$$\int_0^\infty e^{-\lambda^2\tau}J_0(\lambda s)J_0(\lambda r)\,\lambda\, d\lambda = \frac{1}{2\tau}\exp\left(-\frac{r^2+s^2}{4\tau}\right)I_0\left(\frac{rs}{2\tau}\right). \qquad (3.5.32)$$

• Example 3.5.3: Laplace's Equation

Let us find the steady-state temperature in the upper half-space

$$\frac{1}{r}\frac{\partial}{\partial r}\left(r\frac{\partial u}{\partial r}\right) + \frac{\partial^2 u}{\partial z^2} = 0, \qquad 0 \le r < \infty, \quad 0 < z < \infty, \qquad (3.5.33)$$

subject to the boundary conditions

$$\lim_{r \to 0}|u(r,z)| < \infty, \qquad \lim_{r \to \infty} u(r,z) \to 0, \qquad 0 < z < \infty, \qquad (3.5.34)$$

$$\lim_{z \to \infty} u(r,z) \to 0, \qquad 0 \le r < \infty, \qquad (3.5.35)$$

and

$$-\frac{\partial u(r,0)}{\partial z} + hu(r,0) = \begin{cases} Q/\kappa, & 0 \le r < a, \\ 0, & a < r < \infty. \end{cases} \qquad (3.5.36)$$

Physically, Equation 3.5.36 states that the surface $z = 0$ is heated at the rate Q, while the temperature is maintained at zero from $a < r < \infty$. The half space is then heated via radiative cooling of the surface.

Following the previous example, we take the Hankel transform of the differential equation

$$\int_0^\infty \frac{\partial^2 u}{\partial z^2} J_0(kr)\, r\, dr + \int_0^\infty \frac{\partial}{\partial r}\left(r\frac{\partial u}{\partial r}\right) J_0(kr)\, dr = 0. \qquad (3.5.37)$$

As before,

$$\int_0^\infty \frac{\partial}{\partial r}\left(r\frac{\partial u}{\partial r}\right) J_0(kr)\, dr = -k^2 \int_0^\infty u(r,z) J_0(kr)\, r\, dr, \qquad (3.5.38)$$

while

$$\int_0^\infty \frac{\partial^2 u}{\partial z^2} J_0(kr)\, r\, dr = \frac{d^2}{dz^2}\left[\int_0^\infty u(r,z) J_0(kr)\, r\, dr\right]. \qquad (3.5.39)$$

Therefore, Equation 3.5.37 becomes the boundary-value problem

$$\frac{d^2 U(k,z)}{dz^2} - k^2 U(k,z) = 0, \qquad 0 < z < \infty, \qquad (3.5.40)$$

with

$$\lim_{z\to\infty} U(k,z) \to 0, \qquad (3.5.41)$$

and

$$-U'(k,0) + hU(k,0) = \frac{Q}{\kappa}\int_0^a J_0(kr)\, r\, dr = \frac{Qa}{\kappa k} J_1(ka), \qquad (3.5.42)$$

where

$$U(k,z) = \int_0^\infty u(r,z)\, J_0(kr)\, r\, dr. \qquad (3.5.43)$$

The solution to Equation 3.5.40 with the boundary conditions given by Equation 3.5.41 and Equation 3.5.42 is

$$U(k,z) = \frac{Qa}{\kappa k(k+h)} J_1(ka)e^{-kz}. \qquad (3.5.44)$$

Finally, taking the inverse Hankel transform,

$$u(r,z) = \frac{Qa}{\kappa}\int_0^\infty \frac{e^{-kz}}{k+h} J_1(ka) J_0(kr)\, dk. \qquad (3.5.45)$$

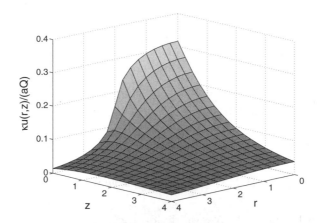

Figure 3.5.1: A plot of $\kappa u(r,z)/(Qa)$ given by Equation 3.5.45 as a function of r and z when $h = 1$.

Figure 3.5.1 illustrates Equation 3.5.45 as a function of r and z.

• Example 3.5.4: Poisson's Equation

Let us solve the Poisson equation

$$\frac{1}{r}\frac{\partial}{\partial r}\left(r\frac{\partial u}{\partial r}\right) + \frac{\partial^2 u}{\partial z^2} = -4\pi\delta(r)\delta(z), \qquad 0 \le r < \infty, \quad -a < z < a,$$

$$(3.5.46)$$

subject to the boundary conditions

$$\lim_{r\to 0}|u(r,z)| < \infty, \qquad \lim_{r\to\infty} u(r,z) \to 0, \qquad -a < z < a, \qquad (3.5.47)$$

and

$$u(r,-a) = u(r,a) = 0, \qquad 0 \le r < \infty. \qquad (3.5.48)$$

Mathematically we are finding the Green's function for Poisson's equation for the given domain. Physically we are computing the electrostatic potential in the free space between two grounded planes at $z = \pm a$ when a unit point charge is located at the origin.

Again, we begin by taking the Hankel transform of Equation 3.5.46 and obtain

$$\int_0^\infty \frac{\partial^2 u}{\partial z^2} J_0(kr)\,r\,dr + \int_0^\infty \frac{\partial}{\partial r}\left(r\frac{\partial u}{\partial r}\right) J_0(kr)\,dr = -4\pi\delta(z)\int_0^\infty \delta(r)J_0(kr)r\,dr.$$

$$(3.5.49)$$

Because

$$\int_0^\infty \frac{\partial}{\partial r}\left(r\frac{\partial u}{\partial r}\right) J_0(kr)\,dr = -k^2 \int_0^\infty u(r,z)J_0(kr)\,r\,dr, \qquad (3.5.50)$$

and

$$\int_0^\infty \frac{\partial^2 u}{\partial z^2} J_0(kr)\, r\, dr = \frac{d^2}{dz^2}\left[\int_0^\infty u(r,z) J_0(kr)\, r\, dr\right], \qquad (3.5.51)$$

Equation 3.5.49 becomes

$$\frac{d^2 U(k,z)}{dz^2} - k^2 U(k,z) = -2\delta(z), \qquad -a < z < a, \qquad (3.5.52)$$

with the boundary conditions $U(k,-a) = U(k,a) = 0$, where

$$U(k,z) = \int_0^\infty u(r,z)\, J_0(kr)\, r\, dr. \qquad (3.5.53)$$

To solve Equation 3.5.52, we divide the region $-a < z < a$ into two subregions, $-a < z < 0$ and $0 < z < a$. Within each region,

$$\frac{d^2 U_\pm(k,z)}{dz^2} - k^2 U_\pm(k,z) = 0. \qquad (3.5.54)$$

From the boundary conditions, we find that $U_+(k,a) = U_-(k,-a) = 0$. The corresponding solutions are

$$U_+(k,z) = A\sinh[k(z-a)], \qquad 0 < z < a, \qquad (3.5.55)$$

and

$$U_-(k,z) = B\sinh[k(z+a)], \qquad -a < z < 0. \qquad (3.5.56)$$

We now must evaluate A and B. Of course, $U(k,z)$ must be continuous and $U_+(k,0) = U_-(k,0)$. For the second condition, we integrate Equation 3.5.52 over the infinitesimal interval $[0^-, 0^+]$ and find that

$$\int_{0^-}^{0^+} U''(k,z)\, dz - k^2 \int_{0^-}^{0^+} U(k,z)\, dz = -2\int_{0^-}^{0^+} \delta(z)\, dz, \qquad (3.5.57)$$

or

$$U'(k,0^+) - U(k,0^-) = -2. \qquad (3.5.58)$$

Upon substituting Equation 3.5.55 and Equation 3.5.56 into Equation 3.5.58, we obtain

$$U_+(k,z) = \frac{e^{-kz}}{k} - \frac{\cosh(kz)}{\cosh(ka)}\frac{e^{-ka}}{k}, \qquad (3.5.59)$$

and

$$U_-(k,z) = \frac{e^{kz}}{k} - \frac{\cosh(kz)}{\cosh(ka)}\frac{e^{-ka}}{k}. \qquad (3.5.60)$$

Therefore,

$$U(k,z) = \frac{e^{-k|z|}}{k} - \frac{\cosh(kz)}{\cosh(ka)}\frac{e^{-ka}}{k}. \qquad (3.5.61)$$

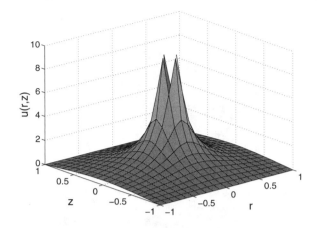

Figure 3.5.2: A plot of $u(r, z)$ given by Equation 3.5.63 as a function of r and z when $a = 1$.

Finally, taking the inverse Hankel transform,

$$u(r, z) = \int_0^\infty e^{-k|z|} J_0(kr)\, dk - \int_0^\infty \frac{\cosh(kz)}{\cosh(ka)} e^{-ka} J_0(kr)\, dk \quad \textbf{(3.5.62)}$$

$$= \frac{1}{\sqrt{r^2 + z^2}} - \int_0^\infty \frac{\cosh(kz)}{\cosh(ka)} e^{-ka} J_0(kr)\, dk, \quad \textbf{(3.5.63)}$$

since

$$\frac{1}{\sqrt{r^2 + z^2}} = \int_0^\infty e^{-\lambda z} J_0(\lambda r)\, d\lambda, \qquad 0 < z. \quad \textbf{(3.5.64)}$$

Figure 3.5.2 illustrates Equation 3.5.63 as a function of r and z.

• Example 3.5.5: Water Waves

Consider the surface waves[34] generated within an incompressible ocean of infinite depth due to an explosion above it that produces the pressure field $f(r, t)$. Assuming potential flow, the continuity equation is

$$\frac{1}{r} \frac{\partial}{\partial r}\left(r \frac{\partial u}{\partial r}\right) + \frac{\partial^2 u}{\partial z^2} = 0, \qquad 0 < r < \infty, \quad z < 0. \quad \textbf{(3.5.65)}$$

This velocity potential $u(r, z, t)$ must satisfy the boundary condition that

$$\frac{\partial^2 u}{\partial t^2} + g \frac{\partial u}{\partial z} = \frac{1}{\rho} \frac{\partial f}{\partial t}\{H(r) - H[r - r_0(t)]\} \quad \textbf{(3.5.66)}$$

---

[34] Sen, A. R., 1963: Surface waves due to blasts on and above liquids. *J. Fluid Mech.*, **16**, 65–81. Reprinted with the permission of Cambridge University Press.

at $z = 0$, where ρ is the density of the liquid and $r_0(t)$ is the extent of the blast. If the fluid is initially at rest, a general solution to Equation 3.5.65 is clearly

$$u(r, z, t) = \int_0^\infty A(k, t) e^{kz} J_0(kr) \, k \, dk. \tag{3.5.67}$$

Because the Hankel transform representation of $f(r, t)$ is

$$f(r, t) = \int_0^\infty \left[\int_0^{r_0(t)} F(\alpha, t) J_0(k\alpha) \, \alpha \, d\alpha \right] J_0(kr) \, k \, dk, \tag{3.5.68}$$

Equation 3.5.66 becomes

$$\frac{\partial^2 A(k, t)}{\partial t^2} + gk A(k, t) = \frac{1}{\rho} \frac{\partial}{\partial t} \int_0^{r_0(t)} F(\alpha, t) J_0(k\alpha) \, \alpha \, d\alpha. \tag{3.5.69}$$

The general solution to Equation 3.5.69 is

$$A(k, t) = A_0(k) \cos(\sigma t) + B_0(k) \sin(\sigma t) \tag{3.5.70}$$

$$+ \frac{1}{\rho \sigma} \int_0^t \frac{\partial}{\partial s} \left[\int_0^{r_0(s)} F(\alpha, s) J_0(k\alpha) \alpha \, d\alpha \right] \sin[\sigma(t - s)] \, ds,$$

where $\sigma^2 = gk$. From the initial conditions, $A_0(k) = B_0(k) = 0$. After an integration by parts, Equation 3.5.70 becomes

$$u(r, z, t) = \frac{1}{\rho} \int_0^\infty e^{kz} J_0(kr) \left\{ \int_0^t \left[\int_0^{r_0(s)} F(\alpha, s) J_0(k\alpha) \, \alpha \, d\alpha \right] \right. \tag{3.5.71}$$

$$\left. \times \cos[\sigma(t - s)] \, ds \right\} k \, dk.$$

• Example 3.5.6: Mixed Boundary-Value Problems

In all of our previous examples, the boundary condition along any specific boundary remained the same. In this example, we relax this condition and consider a *mixed boundary-value problem*.

Consider[35] the axisymmetric Laplace equation

$$\frac{1}{r} \frac{\partial}{\partial r} \left(r \frac{\partial u}{\partial r} \right) + \frac{\partial^2 u}{\partial z^2} = 0, \qquad 0 \leq r < \infty, \quad 0 < z < 1, \tag{3.5.72}$$

---

[35] Reprinted from *J. Theor. Biol.*, **81**, A. Nir and R. Pfeffer, Transport of macromolecules across arterial wall in the presence of local endothial injury, 685–711, ©1979, with permission from Elsevier.

subject to the boundary conditions

$$\lim_{r \to 0} |u(r,z)| < \infty, \qquad \lim_{r \to \infty} |u(r,z)| < \infty, \qquad u(r,0) = 0, \qquad \textbf{(3.5.73)}$$

and

$$\begin{cases} u(r,1) = 1, & 0 < r \le a, \\[2mm] u(r,1) + \dfrac{u_z(r,1)}{\sigma} = 1, & a < r < \infty. \end{cases} \qquad \textbf{(3.5.74)}$$

The interesting aspect of this example is the mixture of boundary conditions along the boundary $z = 1$. For $0 < r < a$, we have a Dirichlet boundary condition that becomes a Robin boundary condition when $a < r < \infty$.

Following the previous examples, we apply Hankel transforms to Equation 3.5.72 and the boundary conditions given by Equation 3.5.73. When the inverse is computed, the solution may be written

$$u(r,z) = \frac{\sigma z}{1+\sigma} + \frac{a}{1+\sigma} \int_0^\infty A(k,a) \sinh(kz) J_0(kr) \, dk. \qquad \textbf{(3.5.75)}$$

Substitution of Equation 3.5.75 into Equation 3.5.74 leads to the *dual integral equations*:

$$a \int_0^\infty A(k,a) \sinh(k) J_0(kr) \, dk = 1, \qquad \textbf{(3.5.76)}$$

if $0 < r \le a$, and

$$\int_0^\infty A(k,a) \left[\sinh(k) + \frac{k \cosh(k)}{\sigma} \right] J_0(kr) \, dk = 0, \qquad \textbf{(3.5.77)}$$

if $a < r < \infty$.

What sets this problem apart from others is the solution of dual integral equations;[36] in general, they are very difficult to solve. The process usually begins with finding a solution that satisfies Equation 3.5.77 via the orthogonality condition involving Bessel functions. This is the technique employed by Tranter[37] who proved that the dual integral equations

$$\int_0^\infty G(\lambda) f(\lambda) J_0(\lambda a) \, d\lambda = g(a) \qquad \textbf{(3.5.78)}$$

and

$$\int_0^\infty f(\lambda) J_0(\lambda a) \, d\lambda = 0 \qquad \textbf{(3.5.79)}$$

---

[36] The standard reference is Sneddon, *op. cit.*

[37] Tranter, C. J., 1950: On some dual integral equations occurring in potential problems with axial symmetry. *Quart. J. Mech. Appl. Math.*, **3**, 411–419.

have the solution

$$f(\lambda) = \lambda^{1-\kappa} \sum_{n=0}^{\infty} A_n J_{2m+\kappa}(\lambda), \qquad (3.5.80)$$

if $G(\lambda)$ and $g(a)$ are known. The value of κ is chosen so that the difference $G(\lambda) - \lambda^{2\kappa-2}$ is fairly small. In the present case, $f(\lambda) = \sinh(\lambda)A(\lambda, a)$, $g(a) = 1$ and $G(\lambda) = 1 + \lambda \coth(\lambda)/\sigma$.

What is the value of κ here? Clearly, we would like our solution to be valid for a wide range of σ. Because $G(\lambda) \to 1$ as $\sigma \to \infty$, a reasonable choice is $\kappa = 1$. Therefore, we take

$$\sinh(k)A(k, a) = \sum_{n=1}^{\infty} \frac{A_n}{1 + k \coth(k)/\sigma} J_{2n-1}(ka). \qquad (3.5.81)$$

Our final task remains to find A_n.

We begin by writing

$$\frac{A_n}{1 + k \coth(k)/\sigma} J_{2n-1}(ka) = \sum_{m=1}^{\infty} B_{mn} J_{2m-1}(ka), \qquad (3.5.82)$$

where B_{mn} depends only on a and σ. Multiplying Equation 3.5.82 by $dk/k \times J_{2p-1}(ka)$ and integrating,

$$\int_0^{\infty} \frac{A_n}{1 + k \coth(k)/\sigma} J_{2n-1}(ka) J_{2p-1}(ka) \frac{dk}{k}$$
$$= \int_0^{\infty} \sum_{m=1}^{\infty} B_{nm} J_{2m-1}(ka) J_{2p-1}(ka) \frac{dk}{k}. \qquad (3.5.83)$$

Because[38]

$$\int_0^{\infty} J_{2n-1}(ka) J_{2p-1}(ka) \frac{dk}{k} = \frac{\delta_{mp}}{2(2m-1)}, \qquad (3.5.84)$$

where δ_{mp} is the Kronecker delta:

$$\delta_{mp} = \begin{cases} 1, & m = p, \\ 0, & m \neq p, \end{cases} \qquad (3.5.85)$$

Equation 3.5.83 reduces to

$$A_n \int_0^{\infty} \frac{J_{2n-1}(ka) J_{2p-1}(ka)}{1 + k \coth(k)/\sigma} \frac{dk}{k} = \frac{B_{mn}}{2(2m-1)}. \qquad (3.5.86)$$

---

[38] Gradshteyn, I. S., and I. M. Ryzhik, 1965: *Table of Integrals, Series, and Products.* Academic Press, Section 6.538, Formula 2.

Table 3.5.1: The Convergence of the Coefficients A_n Given by Equation 3.5.88 Where S_{mn} Has Nonzero Values for $1 \leq m, n \leq N$

N	A_1	A_2	A_3	A_4	A_5	A_6	A_7	A_8
1	2.9980							
2	3.1573	−1.7181						
3	3.2084	−2.0329	1.5978					
4	3.2300	−2.1562	1.9813	−1.4517				
5	3.2411	−2.2174	2.1548	−1.8631	1.3347			
6	3.2475	−2.2521	2.2495	−2.0670	1.7549	−1.2399		
7	3.2515	−2.2738	2.3073	−2.1862	1.9770	−1.6597	1.1620	
8	3.2542	−2.2882	2.3452	−2.2626	2.1133	−1.8925	1.5772	−1.0972

If we define

$$S_{mn} = \int_0^\infty \frac{J_{2n-1}(ka) \, J_{2m-1}(ka)}{1 + k \coth(k)/\sigma} \, \frac{dk}{k}, \qquad (3.5.87)$$

then we can rewrite Equation 3.5.86 as

$$A_n S_{mn} = \frac{B_{mn}}{2(2m-1)}. \qquad (3.5.88)$$

Because[39]

$$a \int_0^\infty J_0(kr) \, J_{2m-1}(ka) \, dk = P_{m-1}\left(1 - \frac{2r^2}{a^2}\right), \qquad r < a, \qquad (3.5.89)$$

where $P_m(\cdot)$ is the Legendre polynomial of order m, Equation 3.5.76 can be rewritten

$$\sum_{n=1}^\infty \sum_{m=1}^\infty B_{mn} P_{m-1}\left(1 - \frac{2r^2}{a^2}\right) = 1. \qquad (3.5.90)$$

Equation 3.5.90 follows from the substitution of Equation 3.5.81 into Equation 3.5.76 and then using Equation 3.5.89. Multiplying Equation 3.5.90 by $P_{m-1}(\xi) \, d\xi$, integrating between −1 and 1, and using the orthogonality properties of the Legendre polynomial, we have

$$\sum_{n=1}^\infty B_{mn} \int_{-1}^1 [P_{m-1}(\xi)]^2 \, d\xi = \int_{-1}^1 P_{m-1}(\xi) \, d\xi \qquad (3.5.91)$$

$$= \int_{-1}^1 P_0(\xi) P_{m-1}(\xi) \, d\xi, \qquad (3.5.92)$$

which shows that only $m = 1$ yields a nontrivial sum. Thus,

$$\sum_{n=1}^\infty B_{mn} = 2(2m-1) \sum_{n=1}^\infty A_n S_{mn} = 0, \qquad 2 \leq m, \qquad (3.5.93)$$

---

[39] *Ibid.*, Section 6.512, Formula 4.

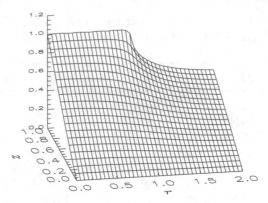

Figure 3.5.3: The solution of the axisymmetric Laplace's equation, Equation 3.5.72, with $u(r,0) = 0$ and the mixed boundary condition given by Equation 3.5.74. Here we have chosen $\sigma = 1$ and $a = 2$.

and

$$\sum_{n=1}^{\infty} B_{1n} = 2 \sum_{n=1}^{\infty} A_n S_{1n} = 1, \qquad (3.5.94)$$

or

$$\sum_{n=1}^{\infty} S_{mn} A_n = \tfrac{1}{2}\delta_{m1}. \qquad (3.5.95)$$

Thus, we have reduced the problem to the solution of an infinite number of linear equations which yield A_n — a common occurrence in the solution of dual integral equations. Selecting some maximum value for n and m, say N, each term in the matrix S_{mn}, $1 \leq m, n \leq N$, is evaluated numerically for a given value of a and σ. By inverting Equation 3.5.95, we obtain the coefficients A_n for $n = 1, \ldots, N$. Because we solved a truncated version of Equation 3.5.95, they will only be approximate. To find more accurate values, we can increase N by 1 and again invert Equation 3.5.95. In addition to the new A_{N+1}, the previous coefficients will become more accurate. We can repeat this process of increasing N until the coefficients converge to their correct value. This is illustrated in Table 3.5.1 when $\sigma = a = 1$.

Once we have computed the coefficients A_n necessary for the desired accuracy, we use Equation 3.5.81 to find $A(k, a)$ and then obtain $u(r, z)$ from Equation 3.5.75 via numerical integration. Figure 3.5.3 illustrates the solution when $\sigma = 1$ and $a = 2$.

• **Example 3.5.7: Heat Conduction Outside of a Cylinder**

So far we have solved problems where it is rather straightforward to apply Hankel transforms. Let's now show how to obtain a transform solution via the classic method of separation of variables.

Let us solve Laplace's equation in a domain *outside* of a cylinder of radius a and above the xy-plane:

$$\frac{1}{r}\frac{\partial}{\partial r}\left(r\frac{\partial u}{\partial r}\right) + \frac{\partial^2 u}{\partial z^2} = 0, \qquad a < r < \infty, \quad 0 < z < \infty, \qquad (3.5.96)$$

subject to the boundary conditions

$$u_r(a, z) = hu(a, z), \qquad \lim_{r\to\infty} u(r, z) \to 0, \qquad 0 < z < \infty, \qquad (3.5.97)$$

and

$$u(r, 0) = 1, \qquad \lim_{z\to\infty} |u(r, z)| < \infty, \qquad a < r < \infty. \qquad (3.5.98)$$

Physically, we are finding the steady-state temperature field within the half-space $z > 0$ outside of a cylinder of radius a. Heat radiates into the cylinder from the half-space while along the surface $z = 0$ the temperature is maintained at one.

We approach this solution[40] via the method of separation of variables. Setting $u(r, z) = R(r)Z(z)$, Equation 3.5.96 becomes

$$\frac{1}{rR}\frac{d}{dr}\left(r\frac{dR}{dr}\right) = -\frac{Z''}{Z} = -k^2, \qquad (3.5.99)$$

with

$$\lim_{r\to\infty} |R(r)| < \infty, \qquad \lim_{z\to\infty} |Z(z)| < \infty, \quad \text{and} \quad R'(a) = hR(a). \qquad (3.5.100)$$

Here, k is the separation constant.

A product solution that satisfies Equation 3.5.99 and Equation 3.5.100 is

$$u_p(r, z) = A(k)\left[g_Y(k)J_0(kr) - g_J(k)Y_0(kr)\right]e^{-kz}, \qquad (3.5.101)$$

where

$$g_J(k) = kJ_1(ka) + hJ_0(ka) \qquad \text{and} \qquad g_Y(k) = kY_1(ka) + hY_0(ka). \qquad (3.5.102)$$

Because there is no restriction on k, the most general solution is a linear superposition of these particular solutions, or

$$u(r, z) = \int_0^\infty A(k)\left[g_Y(k)J_0(kr) - g_J(k)Y_0(kr)\right]e^{-kz}\, dk. \qquad (3.5.103)$$

---

[40] Reprinted from *Chem. Eng. Sci.*, **24**, M. Manner, Steady-state temperature distributions in a quarter-infinite solid and in a semi-infinite solid internally bounded by a cylinder, 267–272, ©1969, with permission from Elsevier.

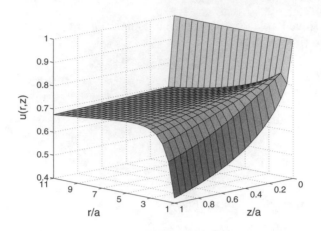

Figure 3.5.4: A plot of $u(r, z)$ given by Equation 3.5.106 as a function of r and z when $ah = 1$.

To determine $A(k)$, we substitute Equation 3.5.103 into the boundary condition at $z = 0$ and obtain the integral equation

$$\int_0^\infty A(k) \left[g_Y(k) J_0(kr) - g_J(k) Y_0(kr) \right] dk = 1. \tag{3.5.104}$$

Comparing Equation 3.5.104 with Equation 4.1.28 derived in the next chapter, we see that

$$A(k) = -\frac{2h}{\pi k [g_J^2(k) + g_Y^2(k)]}. \tag{3.5.105}$$

Therefore, the solution is

$$u(r, z) = -\frac{2h}{\pi} \int_0^\infty e^{-kz} \frac{g_Y(k) J_0(kr) - g_J(k) Y_0(kr)}{g_J^2(k) + g_Y^2(k)} \frac{dk}{k}. \tag{3.5.106}$$

Figure 3.5.4 illustrates Equation 3.5.106 as a function of r and z when $ah = 1$.

Problems

1. Solve the heat equation[41]

$$\frac{\partial u}{\partial t} = \frac{\partial^2 u}{\partial r^2} + \frac{1}{r} \frac{\partial u}{\partial r} - \lambda u + S(r), \qquad 0 \le r < \infty, \quad 0 < t,$$

subject to the boundary conditions $\lim_{r \to 0} |u(r, t)| < \infty$, $\lim_{r \to \infty} |u(r, t)| < \infty$, $0 < t$, and the initial condition $u(r, 0) = u_0(r)$, $0 \le r < \infty$.

---

[41] Taken from Hassan, M. H. A., 1988: Ion distribution functions during ion cyclotron resonance heating at the fundamental frequency. *Phys. Fluids*, **31**, 596–601.

Step 1: Defining the Hankel transform

$$U(k,t) = \int_0^\infty u(r,t) J_0(kr)\, r\, dr,$$

show that the partial differential equation and boundary conditions reduce to the initial-value problem

$$\frac{dU(k,t)}{dt} + (k^2 + \lambda) U(k,s) = \overline{S}(k), \qquad U(k,0) = \overline{u}_0(k),$$

where

$$\overline{u}_0(k) = \int_0^\infty u_0(r) J_0(kr)\, r\, dr \quad \text{and} \quad \overline{S}(k) = \int_0^\infty S(r) J_0(kr)\, r\, dr.$$

Step 2: Solve the initial-value problem and show that

$$U(k,t) = \overline{u}_0(k) e^{-(k^2+\lambda)t} + \frac{\overline{S}(k)}{k^2 + \lambda} \left[1 - e^{-(k^2+\lambda)t} \right].$$

Step 3: Complete the problem by inverting the Hankel transform and show that

$$
\begin{aligned}
u(r,t) &= \int_0^\infty \eta\, u_0(\eta) \int_0^\infty e^{-(k^2+\lambda)t} J_0(k\eta) J_0(kr)\, k\, dk\, d\eta \\
&\quad + \int_0^\infty \eta\, S(\eta) \int_0^\infty \left[1 - e^{-(k^2+\lambda)t} \right] \frac{J_0(k\eta) J_0(kr)}{k^2 + \lambda}\, k\, dk\, d\eta \\
&= \frac{e^{-r^2/(4t)-\lambda t}}{2t} \int_0^\infty \eta\, u_0(\eta)\, e^{-\eta^2/(4t)} I_0\!\left(\frac{r\eta}{2t} \right) d\eta \\
&\quad + \int_0^\infty \eta\, S(\eta) \int_0^\infty \left[1 - e^{-(k^2+\lambda)t} \right] \frac{J_0(k\eta) J_0(kr)}{k^2 + \lambda}\, k\, dk\, d\eta.
\end{aligned}
$$

2. Solve the partial differential equation

$$\frac{1}{r} \frac{\partial}{\partial r}\left(r \frac{\partial u}{\partial r} \right) + \frac{\partial^2 u}{\partial z^2} = 0, \qquad 0 < r, z < \infty,$$

by Hankel transforms with the boundary conditions $|u(0,z)| < \infty$, $\lim_{z\to\infty} u(r,z) \to 0$, $u(r,0) = u_0$ for $0 \le r < a$ and $u(r,0) = 0$ for $a < r < \infty$.

Step 1: By defining the Hankel transform

$$U(k,z) = \int_0^\infty u(r,z) J_0(kr)\, r\, dr,$$

show that we may convert the partial differential equation into the boundary-value problem

$$\frac{d^2 U(k,z)}{dz^2} - k^2 U(k,z) = 0$$

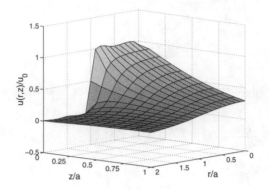

Problem 2

with $\lim_{z \to \infty} U(k, z) \to 0$.

Step 2: Show that the solution to Step 1 is $U(k, z) = A(k)e^{-kz}$.

Step 3: Using the remaining boundary condition, show that

$$U(k, z) = au_0 J_1(ka)e^{-kz}/k.$$

Step 4: Show that

$$u(r, z) = au_0 \int_0^\infty e^{-kz} J_1(ka) J_0(kr) \, dk.$$

This solution $u(r, z)/u_0$ is illustrated in the figure labeled Problem 2.

3. If $a > b > 0$, solve the partial differential equations

$$\frac{\partial^2 u_1}{\partial r^2} + \frac{1}{r} \frac{\partial u_1}{\partial r} + \frac{\partial^2 u_1}{\partial z^2} = 0, \qquad 0 \le r < \infty, \quad 0 < z < \infty,$$

and

$$\frac{\partial^2 u_2}{\partial r^2} + \frac{1}{r} \frac{\partial u_2}{\partial r} + \frac{\partial^2 u_2}{\partial z^2} = -4\pi\delta(r)\delta(z + b), \qquad 0 \le r < \infty, \quad -a < z < 0,$$

by Hankel transforms with the boundary conditions

$$|u_i(0, z)| < \infty, \qquad \lim_{r \to \infty} u_i(r, z) \to 0, \qquad -a < z < \infty,$$

$$\lim_{z \to \infty} u_1(r, z) \to 0, \qquad u_2(r, -a) = 0, \qquad 0 \le r < \infty,$$

$$u_1(r, 0) = u_2(r, 0), \qquad \text{and} \qquad \kappa_1 \frac{\partial u_1(r, 0)}{\partial z} = \kappa_2 \frac{\partial u_2(r, 0)}{\partial z}, \qquad 0 \le r < \infty.$$

Step 1: By defining the Hankel transform

$$U_i(k, z) = \int_0^\infty u_i(r, z) J_0(kr) \, r \, dr,$$

show that we may convert the partial differential equations into the boundary-value problems

$$\frac{d^2 U_1(k,z)}{dz^2} - k^2 U_1(k,z) = 0, \qquad 0 < z < \infty,$$

and

$$\frac{d^2 U_2(k,z)}{dz^2} - k^2 U_2(k,z) = -2\delta(z+b), \qquad -a < z < 0,$$

with

$$\lim_{z\to\infty} U_1(k,z) \to 0, \qquad U_2(k,-a) = 0,$$

$$U_1(k,0) = U_2(k,0), \qquad \text{and} \qquad \kappa_1 U_1'(k,0) = \kappa_2 U_2'(k,0).$$

Step 2: Show that the solutions to the boundary-value problems are

$$kU_1(k,z) = (1-\lambda)\frac{e^{-k(z+b)} - e^{-k(z+2a-b)}}{1 - \lambda e^{-2ka}}$$

$$= (1-\lambda)\sum_{m=0}^{\infty} \lambda^m \left[e^{-k(z+b+2ma)} - e^{-k(z+2a-b+2ma)} \right],$$

and

$$kU_2(k,z) = e^{-k|z+b|} - e^{-k(z+2a-b)}$$

$$- \lambda \frac{e^{-k(b-z)} + e^{-k(z+4a-b)} - e^{-k(2a-b-z)} - e^{-k(z+b+2a)}}{1 - \lambda e^{-2ka}}$$

$$= e^{-k|z+b|} - e^{-k(z+2a-b)} - \lambda \sum_{m=0}^{\infty} \lambda^m \left[e^{-k(b-z+2ma)} - e^{-k(2a-b-z+2ma)} \right]$$

$$+ \lambda \sum_{m=0}^{\infty} \lambda^m \left\{ e^{-k[z+b+2(m+1)a]} - e^{-k[z+2a-b+2(m+1)a]} \right\},$$

where $\lambda = (\kappa_1 - \kappa_2)/(\kappa_1 + \kappa_2)$.

Step 3: Using the relationship[42]

$$\int_0^\infty e^{-kz} J_0(kr)\, dk = \frac{1}{r^2 + z^2}, \qquad 0 < z,$$

show that

$$u_1(r,z) = (1-\lambda)\sum_{m=0}^{\infty} \lambda^m \left(\frac{1}{R_1} - \frac{1}{R_2} \right),$$

and

$$u_2(r,z) = \sum_{m=0}^{\infty} \lambda^m \left(\frac{1}{R_1} - \frac{1}{R_2} \right) - \sum_{m=0}^{\infty} \lambda^{m+1} \left(\frac{1}{R_1} - \frac{1}{R_2} \right),$$

---

[42] Gradshteyn and I. M. Ryzhik, *op. cit.*, Formula 6.611.1.

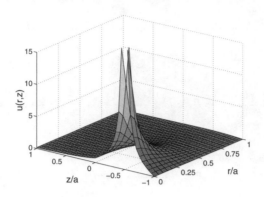

Problem 3

where

$$R_i^2 = (d_i + z)^2 + r^2, \qquad \overline{R}_i^2 = (d_i - z)^2 + r^2,$$

$$d_1 = 2ma + b, \quad \text{and} \quad d_2 = 2ma + 2a - b.$$

This solution $u_i(r,z)$ is illustrated in the figure labeled Problem 3 when $b/a = 0.5$ and $\lambda = 0.5$.

4. Let us solve Helmholtz's equation on the upper half-plane:

$$\frac{\partial^2 u}{\partial r^2} + \frac{1}{r}\frac{\partial u}{\partial r} + \frac{\partial^2 u}{\partial z^2} + k_0^2 u = 0, \qquad 0 \le r < \infty, \quad 0 < z < \infty,$$

subject to the boundary conditions

$$\lim_{r \to 0} |u(r,z)| < \infty, \qquad \lim_{r \to \infty} |u(r,z)| < \infty, \qquad 0 < z < \infty,$$

$$\lim_{z \to \infty} |u(r,z)| < \infty, \qquad 0 \le r < \infty,$$

and

$$u_z(r,0) = \begin{cases} -V, & 0 \le r < a, \\ 0, & a \le r < \infty. \end{cases}$$

Step 1: Take the Hankel transform of the governing equation and boundary conditions and show that the problem reduces to the boundary-value problem

$$\frac{d^2 U(k,z)}{dz} + (k_0^2 - k^2)U(k,z) = 0, \qquad 0 < z < \infty,$$

with

$$U'(k,0) = -aV\frac{J_1(ka)}{k}, \qquad \lim_{z \to \infty} |U(r,z)| < \infty.$$

Step 2: Show that the solution to the boundary-value problems is

$$U(k,z) = aV\frac{J_1(ka)}{k}\frac{\exp\left(-z\sqrt{k_0^2 - k^2}\right)}{\sqrt{k_0^2 - k^2}},$$

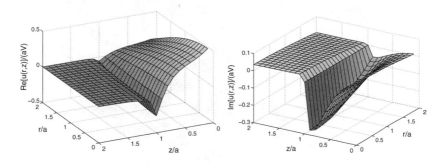

Problem 4

subject to $\mathrm{Re}\left(\sqrt{k_0^2 - k^2}\right) > 0$.

Step 3: Invert the Hankel transform and show that

$$u(r, z) = aV \int_0^\infty \frac{\exp\left(-z\sqrt{k_0^2 - k^2}\right)}{\sqrt{k_0^2 - k^2}} J_1(ka) J_0(kr) \, dk.$$

In acoustics, this result is known as the King integral.[43] This inverse $u(r, z)$ is illustrated in the figure labeled Problem 4 when $k_0 a = \pi$.

5. Solve[44]

$$\frac{\partial^2 u}{\partial r^2} + \frac{1}{r}\frac{\partial u}{\partial r} + \frac{\partial^2 u}{\partial z^2} = 0, \qquad 0 \le r < \infty, \quad 0 < z < \infty,$$

subject to the boundary conditions

$$\lim_{r \to 0} |u(r, z)| < \infty, \qquad \lim_{r \to \infty} |u(r, z)| < \infty, \quad 0 < z < \infty,$$

$$\lim_{z \to \infty} u(r, z) \to u_\infty, \qquad 0 \le r < \infty,$$

and

$$\begin{cases} u(r, 0) = u_\infty - \Delta u, & 0 \le r < a, \\ u_z(r, 0) = 0, & a \le r < \infty, \end{cases}$$

where u_∞ and Δu are constants.

Step 1: Show that

$$u(r, z) = u_\infty - \int_0^\infty A(k) e^{-kz} J_0(kr) \, dk$$

---

[43] King, L. V., 1934: On the acoustic radiation field of the piezo-electric oscillator and the effect of viscosity on transmission. *Can. J. Res.*, **11**, 135–155.

[44] Reprinted from *J. Electroanal. Chem.*, **222**, M. Fleischmann and S. Pons, The behavior of microdisk and microring electrodes, 107–115, ©1987, with permission from Elsevier.

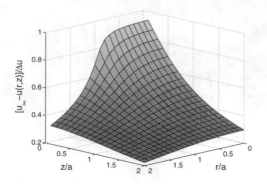

Problem 5

satisfies the partial differential equation and the boundary conditions as $r \to 0$, $r \to \infty$ and $z \to \infty$.

Step 2: Show that

$$\int_0^\infty k A(k) J_0(kr) \, dk = 0, \qquad a < r < \infty.$$

Step 3: Using the relationship[45]

$$\int_0^\infty \sin(ka) J_0(kr) \, dk = \begin{cases} (a^2 - r^2)^{-1/2}, & r < a, \\ 0, & r > a, \end{cases}$$

show that $k A(k) = C \sin(ka)$.

Step 4: Using the relationship[46]

$$\int_0^\infty \sin(ka) J_0(kr) \, \frac{dk}{k} = \begin{cases} \pi/2, & r \le a, \\ \arcsin(a/r), & r \ge a, \end{cases}$$

show that

$$u(r, z) = u_\infty - \frac{2\Delta u}{\pi} \int_0^\infty e^{-kz} \sin(ka) J_0(kr) \, \frac{dk}{k}.$$

The solution $u(r, z)$ is illustrated in the figure labeled Problem 5.

6. If $a > 0$, find the solution to Laplace's equation

$$\frac{\partial^2 u_1}{\partial r^2} + \frac{1}{r} \frac{\partial u_1}{\partial r} + \frac{\partial^2 u_1}{\partial z^2} = 0, \qquad 0 \le r < \infty, \quad 0 < z < \infty,$$

$$\frac{\partial^2 u_2}{\partial r^2} + \frac{1}{r} \frac{\partial u_2}{\partial r} + \frac{\partial^2 u_2}{\partial z^2} = 0, \qquad 0 \le r < \infty, \quad -a < z < 0,$$

---

[45] Gradshteyn and Ryzhik, *op. cit.*, Section 6.671, Formula 7.

[46] *Ibid.*, Section 6.693, Formula 1 with $\nu = 0$.

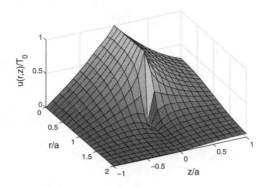

Problem 6

subject to the boundary conditions

$$\lim_{z \to \infty} u_1(r, z) \to 0, \quad 0 \le r < \infty, \qquad \lim_{r \to \infty} u_i(r, z) \to 0, \quad -a < z < \infty,$$

$$u_1(r, 0) = u_2(r, 0) = \begin{cases} T_0, & |r| < b, \\ 0, & |r| > b, \end{cases} \quad \text{and} \quad u_2(r, -a) = 0, \quad 0 \le r < \infty.$$

Step 1: By taking the Hankel transform

$$U_i(k, z) = \int_0^\infty u_i(r, z) J_0(kr) \, r \, dr$$

of the partial differential equations and boundary conditions, show that they reduce to the boundary-value problems

$$\frac{d^2 U_1(k, z)}{dz^2} - k^2 U_1(k, z) = 0, \qquad 0 < z < \infty,$$

and

$$\frac{d^2 U_2(k, z)}{dz^2} - k^2 U_2(k, z) = 0, \qquad -a < z < 0,$$

with

$$\lim_{z \to \infty} U_1(k, z) \to 0, \quad U_2(k, -a) = 0, \quad \text{and} \quad U_1(k, 0) = U_2(k, 0) = \frac{T_0 b}{k} J_1(bk).$$

Step 2: Solve the boundary-value problems and show that

$$U_1(k, z) = \frac{bT_0}{k} J_1(bk) e^{-kz} \quad \text{and} \quad U_2(k, z) = \frac{bT_0}{k} J_1(bk) \frac{\sinh[k(z + a)]}{\sinh(ka)}.$$

Step 3: Invert the Hankel transform and show that

$$u_1(r, z) = bT_0 \int_0^\infty J_1(bk) J_0(kr) e^{-kz} \, dk$$

and

$$u_2(r, z) = bT_0 \int_0^\infty \frac{\sinh[k(z + a)]}{\sinh(ka)} J_1(bk) J_0(kr) \, dk.$$

This solution $u(r, z)$ is illustrated in the figure labeled Problem 6 when $b/a = 1$.

7. Solve the partial differential equation[47]

$$
\left(\frac{\partial}{\partial t}+1\right)\left[\frac{1}{r}\frac{\partial}{\partial r}\left(r\frac{\partial u}{\partial r}\right)\right]-\left(\frac{1}{S}\frac{\partial u}{\partial t}+\gamma\right)u=\frac{1}{r}\frac{\partial m_0}{\partial r},
$$

where $0 < r < \infty$ and $0 < t$, by Hankel transforms with the boundary conditions $\lim_{r\to 0}|u(r,t)| < \infty$ and $\lim_{r\to\infty}u(r,t)\to 0$, and the initial condition $u(r,0)=0$.

Step 1: By defining the Hankel transforms

$$
U(k,t)=\int_0^\infty u(r,t)J_0(kr)\,r\,dr \quad \text{and} \quad M_0(k,t)=\int_0^\infty m_0(r,t)J_0(kr)\,r\,dr,
$$

show that we may convert the partial differential equation into the initial-value problem

$$
\left(k^2+\frac{1}{S}\right)\frac{dU(k,t)}{dt}+\left(k^2+\gamma\right)U(k,t)=-kM_0(k), \qquad U(k,0)=0.
$$

Step 2: Show that the solution to initial-value problem is

$$
U(k,t)=-\frac{kM_0(k)}{k^2+\gamma}\left[1-\exp\left(-\frac{k+\gamma}{k^2+1/S}t\right)\right].
$$

Step 3: Invert the Hankel transform and show that

$$
u(r,t)=-\int_0^\infty \frac{k^2 M_0(k)}{k^2+\gamma}\left[1-\exp\left(-\frac{k+\gamma}{k^2+1/S}t\right)\right]J_0(kr)\,dk.
$$

8. Solve the partial differential equation[48]

$$
\frac{\partial^2 u}{\partial z^2}+\frac{\partial^2 u}{\partial r^2}+\frac{1}{r}\frac{\partial u}{\partial r}-\frac{u}{r^2}=-\frac{f(z)\delta(r)}{r^2}, \qquad 0<r<\infty, \quad 0<z,
$$

by Hankel transforms with the boundary conditions $\lim_{r\to\infty}u(r,z)\to 0$, $\lim_{z\to\infty}u(r,z)\to 0$, $\lim_{r\to 0}|u(r,z)| < \infty$ and $u_z(r,0)=0$.

Step 1: By defining the Hankel transform

$$
U(k,z)=\int_0^\infty u(r,z)J_1(kr)\,r\,dr,
$$

[47] Taken from Ou, H. W., and A. L. Gordon, 1986: Spin-down of baroclinic eddies under sea ice. *J. Geophys. Res.*, **91**, 7623–7630. ©1986 American Geophysical Union. Reproduced/modified by permission of the American Geophysical Union.

[48] Taken from Agrawal, G. L., and W. G. Gottenberg, 1971: Response of a semi-infinite elastic solid to an arbitrary torque along the axis. *Pure Appl. Geophys.*, **91**, 34–39. Published by Birkhäuser Verlag, Basel, Switzerland.

show that we may convert the partial differential equation into the boundary-value problem

$$\frac{d^2 U(k,z)}{dz^2} - k^2 U(k,z) = -\frac{kf(z)}{4\pi}$$

with the boundary conditions $\lim_{z \to \infty} U(k,z) \to \infty$ and $U'(k,0) = 0$.

Step 2: Show that the solution to the boundary-value problem is

$$U(k,z) = \frac{1}{8\pi} \left[\int_0^\infty f(\tau) e^{-k(z+\tau)} \, d\tau + \int_z^\infty f(\tau) e^{-k(\tau-z)} \, d\tau + \int_0^z f(\tau) e^{-k(z-\tau)} \, d\tau \right].$$

Step 3: Invert $U(k,z)$ by noting that

$$\int_0^\infty \left[\int_0^L e^{-k(z+\tau)} f(\tau) \, d\tau \right] J_1(kr) k \, dk = \int_0^L f(\tau) \left[\int_0^\infty e^{-k(z+\tau)} J_1(kr) k \, dk \right] d\tau$$

$$= \int_0^L f(\tau) \frac{r}{[(z+\tau)^2 + r^2]^{3/2}} \, d\tau,$$

and show that

$$u(r,z) = \frac{1}{8\pi} \left\{ \int_0^\infty \frac{rf(\tau)}{[(z+\tau)^2 + r^2]^{3/2}} \, d\tau + \int_0^\infty \frac{rf(\tau)}{[(z-\tau)^2 + r^2]^{3/2}} \, d\tau \right\}.$$

3.6 NUMERICAL INVERSION OF HANKEL TRANSFORMS

Unlike Fourier and Laplace transforms, the analytic techniques available for the inversion of Hankel transforms are rather limited. For this reason, efficient numerical methods for inverting these transforms are important in applications. In this section, we examine several of these methods.

Regardless of the exact method used to invert the Hankel transform, the first step replaces the original definition of the inversion integral with

$$f(r) = \int_0^L F(k) J_0(kr) \, k \, dk. \tag{3.6.1}$$

Here L is chosen sufficiently large so that the approximation introduces a negligibly small error.

The simplest and most reliable method for inverting Hankel transforms is the trapezoidal rule. However, because this quadrature method requires many evaluations of the integrand which contains at least one Bessel function, this technique is computationally too expensive. For this reason, it is desirable to employ schemes that yield accurate results with a minimum number of evaluations of the integrand. One popular scheme is Gaussian quadrature which evaluates $\int_{-1}^1 f(x) \, dx$ using the sum $\sum_{j=1}^n A_j f(x_j)$, where x_j are the designated evaluation points (Gaussian abscissas) and A_j are the weights of each of the abscissa points. Because of the freedom in choosing the evaluation points as well as their weights, n-point Gaussian quadrature achieves the same accuracy as a scheme such as Simpson's rule that uses fixed equidistant points for about half the number of function evaluations. Because the limits of Equation 3.6.1 run from 0 to L rather than -1 to 1, we break the interval $[0, L]$ into M panels that run from $a = 2m\Delta k$ to $b = 2(m + 1)\Delta k$, where $m = 0, 1, 2, \ldots, M - 1$. Then Equation 3.6.1 may be rewritten as the sum of M integrals over M subdivisions:

$$f(r) = \Delta k \sum_{m=0}^{M-1} \int_{-1}^{1} kF(k)J_0(kr)\, d\eta, \qquad (3.6.2)$$

where $k = (2m + 1)\Delta k + \eta\Delta k$, and $2M\Delta k = L$.

Although Gaussian quadrature is quite good, Kronrod[49] showed in 1964 how the n-point Gaussian quadrature formulas may be augmented by a set of $n + 1$ abscissas to yield quadrature formulas of degree $3n + 1$ if n is even and $3n + 2$ if n is odd. The importance of this method is that the accuracy of a numerical integration can be considerably improved without wasting the integrand evaluation at the Gaussian abscissas.

Let us examine whether Gaussian and Gauss–Kronrod quadrature is useful in inverting Hankel transforms. To this end, we employ the transform pair:

$$F(k) = \frac{\sin(kc)}{k^2}, \qquad \text{and} \qquad f(r) = \begin{cases} \pi/2, & r \leq c, \\ \arcsin(c/r), & r \geq c. \end{cases} \qquad (3.6.3)$$

Using MATLAB, we first introduce the Gauss–Kronrod abscissas:

```
xgk( 1) = 0.9963696138895427; xgk( 2) = 0.9782286581460570;
xgk( 3) = 0.9416771085780681; xgk( 4) = 0.8870625997680953;
xgk( 5) = 0.8160574566562211; xgk( 6) = 0.7301520055740492;
xgk( 7) = 0.6305995201619651; xgk( 8) = 0.5190961292068118;
xgk( 9) = 0.3979441409523776; xgk(10) = 0.2695431559523450;
xgk(11) = 0.1361130007993617;
```

We have only listed the positive abscissas but there are abscissa points at -xgk(j) as well. Finally, there is an abscissa point at $\eta = 0$. If we were only using Gaussian quadrature, then we would only include points xgk(2), xgk(4), ..., xgk(10); the remaining abscissa points are introduced by Kronrod's modification. The weights for the Gaussian and Kronrod schemes are:

```
% weights for Gaussian quadrature
```

```
wg(1) = 0.05566856711617449; wg(2) = 0.1255803694649048;
wg(3) = 0.1862902109277352;  wg(4) = 0.2331937645919914;
wg(5) = 0.2628045445102478;
```

and

```
% weights for Kronrod quadrature
wgk( 1) = 0.009765441045961290; wgk( 2) = 0.02715655468210443;
wgk( 3) = 0.04582937856442671;  wgk( 4) = 0.06309742475037484;
wgk( 5) = 0.07866457193222764;  wgk( 6) = 0.09295309859690074;
wgk( 7) = 0.1058720744813894;   wgk( 8) = 0.1167395024610472;
wgk( 9) = 0.1251587991003195;   wgk(10) = 0.1312806842298057;
wgk(11) = 0.1351935727998845;
```

Having introduced the abscissa points and weights, the inversion for a specific r is as follows:

```
% initialize a and b, the beginning and end values
%           of the subdivision
a = 0; b = 2 * dk;
% initialize the integral values to zero
inverse_gauss = 0; inverse_kronrod = 0;
for m = 0:M-1 % compute the contribution from each subdivision
center = 0.5 * (a + b); % this corresponds to the point η = 0
% compute integrand at η = 0
fcenter = sin(c*center) * besselj(0,r*center) / center;
% for Gaussian quadrature, add in the contribution from η = 0
sum_gauss = 0.2729250867779007 * fcenter;
% for Gauss-Kronrod quadrature, add in the contribution
%           from η = 0
sum_kronrod = 0.1365777947111183 * fcenter;
% do the remaining quadrature points, dk = Δk
for j = 1:11
% for specific dk, find the location of k with respect to center
delta_k = dk * xgk(j);
% for a Kronrod abscissa point to the left of η = 0,
%           compute the integrand
k1 = center - delta_k;
if (k1 > 0)
f1 = sin(c*k1) * besselj(0,r*k1) / k1;
else
f1 = c * besselj(0,r*k1);
end
% for a Kronrod abscissa point to the right of η = 0,
%           compute the integrand
```

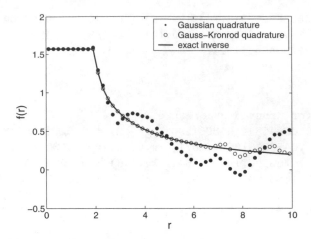

Figure 3.6.1: The numerical inversion of the Hankel transform $\sin(ka)/k$ when $c = 2$ using Gaussian and Gauss–Kronrod quadrature.

```
k2 = center + delta_k;
f2 = sin(c*k2) * besselj(0,r*k2) / k2;
% add together these two contributions
funsum = f1 + f2;
% add this contribution to the value of the integral
%      over [2m∆k, 2(m + 1)∆k]
sum_kronrod = sum_kronrod + wgk(j) * funsum;
% for Gaussian quadrature, do the same
if (mod(j,2) == 0) sum_gauss = sum_gauss + wg(j/2) * funsum; end
end
% add the contribution from all of the quadrature points within
% [a,b] to the value of the integral over the interval [0, L]
inverse_gauss = inverse_gauss + dk * sum_gauss;
inverse_kronrod = inverse_kronrod + dk * sum_kronrod;
% update a and b for the next subdivision's computation
a = b; b = b + 2 * dk;
end
```

Figure 3.6.1 illustrates these two methods when $L = 1000$ and $\Delta k = 5$. Both methods give excellent results for moderate values of r, but we note poorer results as r becomes large.

Our sample problem highlights an important problem in the inversion of Hankel transforms: If the integrand oscillates rapidly, poor accuracy results because the positive and negative areas under the curve are large and nearly equal. This is even true when $F(k)$ is nonoscillatory because $J_0(kr)$ will oscillate rapidly when r is large. In any case, the sampling interval of k, Δk, must be reduced if we are to achieve acceptable results. A similar problem occurs in the numerical inversion of Fourier transforms. There the oscillating integrand is due to $F(\omega)e^{it\omega}$. In 1928, Filon[50] derived a quadrature method that is similar to Simpson's rule but differs in fitting only $F(\omega)$ by a quadratic over $[a, b]$ rather than $F(\omega)e^{it\omega}$. Barakat and Parshall[51] have applied this same philosophy to the inversion of Hankel transforms of order zero; subsequently, Barakat and Sandler[52] developed a similar scheme for Hankel transform of the first kind.

Although implementing the Barakat–Parshall algorithm is straightforward, Markham and Conchello[53] have shown that one of the quadrature coefficients, $\$_1(x_1, x_2)$, is computationally very expensive as well as suffers from poor convergence for large x_1 and x_2. Even if these problems did not exist, we still could not use their algorithm because it assumes that $F(k)$ is finite for all k. Rather than attempting to improve the Barakat–Parshall scheme, we will implement a scheme devised by Chung, Evans and Webster[54] for oscillatory integrals.

One of the most widely used methods for dealing with integrals with singular or oscillatory functions is the quadrature rule:

$$\int_a^b g(x)w(x)\,dx \approx \sum_{n=0}^N w_n g(x_n). \qquad (3.6.4)$$

The weight function $w(x)$ is chosen so that it "takes out" the singular or oscillatory term. In our particular case, $w(x) = J_0(rx)$. The question then becomes how do you compute w_n? To answer this question, Chung et al. invented the function

$$g(x) = \frac{d}{dx}\left[x\frac{dv(x)}{dx}\right] + r^2 x v(x). \qquad (3.6.5)$$

Then, by integration by parts,

$$\int_a^b g(x)J_0(rx)\,dx = \int_a^b \left\{\frac{d}{dx}\left[x\frac{dv(x)}{dx}\right] + r^2 x v(x)\right\} J_0(rx)\,dx \qquad (3.6.6)$$

[50] Filon, L. N. G., 1928-29: On a quadrature formula for trigonometric integrals. *Proc. R. Soc. Edinburgh*, **49**, 38–47.

[51] Barakat, R., and E. Parshall, 1996: Numerical evaluation of the zero-order Hankel transform using Filon quadrature philosophy. *Appl. Math. Lett.*, **9, No. 5**, 21–26.

[52] Barakat, R., and B. H. Sandler, 1998: Evaluation of first-order Hankel transforms using Filon quadrature philosophy. *Appl. Math. Lett.*, **11, No. 1**, 127–131.

[53] Markham, J., and J. A. Conchello, 2003: Numerical evaluation of Hankel transforms for oscillating functions. *J. Opt. Soc. Am., Ser. A*, **20**, 621–630.

[54] Chung, K. C., G. A. Evans, and J. R. Webster, 2000: A method to generate generalized quadrature rules for oscillatory integrals. *Appl. Numer. Math.*, **34**, 85–93.

$$\int_a^b g(x) J_0(rx)\, dx = \left. x \frac{dv(x)}{dx} J_0(rx) \right|_a^b$$

$$- \int_a^b \left\{ x \frac{dJ_0(rx)}{dx} \frac{dv(x)}{dx} - r^2 x J_0(rx) v(x) \right\} dx \quad (3.6.7)$$

$$= \left. x \frac{dv(x)}{dx} J_0(rx) - x v(x) \frac{dJ_0(rx)}{dx} \right|_a^b$$

$$+ \int_a^b \left\{ \frac{d}{dx} \left[x \frac{dJ_0(rx)}{dx} \right] + r^2 x J_0(rx) \right\} v(x)\, dx \quad (3.6.8)$$

$$= \left. x \left[\frac{dv(x)}{dx} J_0(rx) + r\, v(x) J_1(rx) \right] \right|_a^b . \quad (3.6.9)$$

Equation 3.6.9 follows from Equation 3.6.8, because the term inside the wavy brackets vanishes since it is Bessel's equation.

Why did Chung et al. introduce $v(x)$? Recall that our goal is to compute w_n. A common technique is to require that Equation 3.6.4 becomes an identity for a set of functions $g_i(x)$ where $i = 0, 1, 2, \ldots, N$. If we choose $v(x) = p_i(x)$, where $p_i(x)$ is a polynomial of order i,

$$\sum_{n=0}^N w_n g_i(x_n) = \int_a^b g_i(x) J_0(rx)\, dx = \left. x \left[p_i'(x) J_0(rx) + r\, p_i(x) J_1(rx) \right] \right|_a^b . \quad (3.6.10)$$

Inverting Equation 3.6.10 then yields the quadrature coefficients w_n.

A few details need attention here. First, the original integral in Equation 3.6.1 runs from 0 to L rather than from a to b. We can rewrite Equation 3.6.1 in terms of Equation 3.6.4 as follows:

$$f(r) = \sum_{m=0}^{M-1} \int_{a_m}^{b_m} h(k) J_0(kr)\, dk, \quad (3.6.11)$$

where $h(k) = kF(k)$, $a_m = 2m\Delta k$, $b_m = 2(m+1)\Delta k$ and $L = 2M\Delta k$. Next, we need to specify $p_i(k)$. From previous experience, Chung et al. found that the best results occurred with Chebyshev polynomials T_i on a Clenshaw-Curtis grid $x_j = \cos(j\pi/N)$. Shifting the Chebyshev polynomials so that they lie in the range $[a_m, b_m]$, we have

$$g_i(\eta) = \frac{k}{\Delta k^2} \frac{d^2 T_i(\eta)}{d\eta^2} + \frac{1}{\Delta k} \frac{dT_i(\eta)}{d\eta} + r^2 k T_i(\eta), \quad (3.6.12)$$

where $\eta \in [-1, 1]$ and $k = (a_m + b_m)/2 + \Delta k \eta$.

To illustrate this technique, let us develop a MATLAB script that implements the Chung–Evans–Webster algorithm. We shall use the same test case employed by Barakat and Prashall, namely,

$$F(k) = \begin{cases} \arccos(k) - k\sqrt{1 - k^2}, & 0 \le k \le 1, \\ 0, & 1 \le k < \infty. \end{cases} \quad (3.6.13)$$

We start by computing the Chebyshev polynomials:

```
for n = 0:N
x(n+1) = cos(pi*n/N); % compute Clenshaw-Curtis points
% *** compute Tm(xn), T'm(xn), and T''m(xn) for m = 0,1
t(1,n+1) = 1; tp(1,n+1) = 0; tpp(1,n+1) = 0;
t(2,n+1) = x(n+1); tp(2,n+1) = 1; tpp(2,n+1) = 0;
% *** compute Tm(1), T'm(1), and T''m(1)
if (n == 0)
for m = 2:N
t(m+1,1) = 2 * x(1) * t(m,1) - t(m-1,1);
tp(m+1,1) = m * m;
tpp(m+1,1) = 2 * x(1) * tpp(m,1) + 4 * tp(m,1) - tpp(m-1,1);
end; end
% *** compute Tm(-1), T'm(-1), and T''m(-1)
if (n == N)
for m = 2:N
t(m+1,N+1) = 2 * x(N+1) * t(m,N+1) - t(m-1,N+1);
tp(m+1,N+1) = m * m;
tp(m+1,N+1) = (-1)^(m-1) * tp(m+1,N+1);
tpp(m+1,N+1) = 2 * x(N+1) * tpp(m,N+1) + 4 * tp(m,N+1) ...
                - tpp(m-1,N+1);
end; end
% *** compute Tm(xn), T'm(xn), and T''m(xn) otherwise
if ((n > 0) & (n < N))
for m = 2:N
t(m+1,n+1) = 2 * x(n+1) * t(m,n+1) - t(m-1,n+1);
tp(m+1,n+1) = m * t(m,n+1) - m * x(n+1) * t(m+1,n+1);
tp(m+1,n+1) = tp(m+1,n+1) / (1 - x(n+1) * x(n+1));
tpp(m+1,n+1) = 2 * x(n+1) * tpp(m,n+1) + 4 * tp(m,n+1) ...
                - tpp(m-1,n+1);
end; end
end
```

Having computed the Clenshaw-Curtis abscissa points and the Chebyshev polynomials, we are ready to perform the numerical integration. For a given r, this consists of two steps. First we set up the equations given by Equation 3.6.10 and then solve them to find w_n. Then the integration via Equation 3.6.11 is performed.

```
% compute some intermediate values, dk = Δk
for n = 0:N
for m = 0:N
term1(m+1,n+1) = tpp(m+1,N+1-n)/(dk*dk) + r*r*t(m+1,N+1-n);
term2(m+1,n+1) = tp(m+1,N+1-n)/dk;
```

```
end; end
inverse = 0;
%%%%%%%%%%%%%%%%%%%%%%%%%%%%%%%%%%%%%%%%%%%%%%%%%%%%%%%%%%%%%%%%%%
% for each subdivision, compute the weights w_n
%%%%%%%%%%%%%%%%%%%%%%%%%%%%%%%%%%%%%%%%%%%%%%%%%%%%%%%%%%%%%%%%%%
for m = 0:M-1
a_m = 2 * m * dk; b_m = 2 * (m+1) * dk; % a_m = a_m, b_m = b_m
clear A b LL UU w_n
% compute the left side of Equation 3.6.10.
% use Equation 3.6.12 to compute g_i(x_n)
for n = 0:N
k = 0.5 * (a_m+b_m) + dk * x(N+1-n);
for i = 0:N
A(i+1,n+1) = k * term1(i+1,n+1) + term2(i+1,n+1);
end; end
% compute the right side of Equation 3.6.10
% this equation is shifted from [-1,1] to the [a_m,b_m]
for i = 0:N
arg = r * a_m;
Z_a = r*besselj(1,arg)*t(i+1,N+1) ...
    + besselj(0,arg)*tp(i+1,N+1)/dk;
Z_a = a_m * Z_a;
arg = r * b_m;
Z_b = r*besselj(1,arg)*t(i+1,1)+besselj(0,arg)*tp(i+1,1)/dk;
Z_b = b_m * Z_b;
b(i+1,1) = Z_b - Z_a;
end
% Invert the set of equations to find weight w_n
[LL,UU] = lu(A);
w_n = UU\(LL\b);
%%%%%%%%%%%%%%%%%%%%%%%%%%%%%%%%%%%%%%%%%%%%%%%%%%%%%%%%%%%%%%%%%%
% now perform the numerical integration Equation 3.6.11
%%%%%%%%%%%%%%%%%%%%%%%%%%%%%%%%%%%%%%%%%%%%%%%%%%%%%%%%%%%%%%%%%%
for n = 0:N
k = 0.5 * (a_m+b_m) + dk * x(N+1-n);
% compute the Hankel transform F(k)
func = acos(k) - k * sqrt(1-k*k); % compute F(k), Eq.  3.6.13
func = k * F(k) % recall that h(k) = kF(k)
end
inverse = inverse + w_n(n+1) * func; % compute Equation 3.6.11
end
```

Figure 3.6.2 illustrates the Chung–Evans–Webster algorithm for two different N's. The accuracy is quite good, especially when $N = 20$.

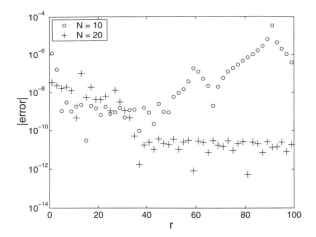

Figure 3.6.2: The difference between numerical inversion of the Hankel transform given by Equation 3.6.13 using the Chung–Evans–Webster algorithm with $\Delta k = 0.1$ and the exact inverse.

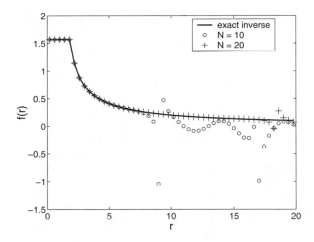

Figure 3.6.3: The numerical inversion of the Hankel transform given by Equation 3.6.3 using the Chung–Evans–Webster algorithm with $\Delta k = 1$.

With our success in inverting Equation 3.6.13, let us now apply this scheme to the Hankel transform given by Equation 3.6.3. The results are shown in Figure 3.6.3 when $L = 100$. The results are not as good as we might have hoped. What happened here? Recall that the transform contains two oscillatory functions: $J_0(kr)$ and $\sin(k)$. Although the Chung–Evans–Webster algorithm is designed to deal with the effect of a highly oscillatory Bessel function, it does nothing with regard to the $\sin(k)$ term and less than ideal results occur. In their paper, the authors show how to handle a case where the integrand includes both Bessel and trigonometric functions.

Finally, the present MATLAB code is rather inefficient because of the repeated inversion of a $(N + 1) \times (N + 1)$ set of equations. Other methods for inverting the matrix would make the code less expensive to run.

Papers Using Fourier Transforms
to Solve Partial Differential Equations

Diffusion

O'Neill, K., 1982: One-dimensional transport from a highly concentrated, transfer type source. *Int. J. Heat Mass Transfer*, **25**, 27–36.

Electromagnetism

Ignetik, R., Y.-C. Thio, and K. C. Westfold, 1985: Transient electromagnetic field above a permeable and conducting half-space. *Geophys. J. R. Astr. Soc.*, **81**, 623–639.

Seshadri, S. R., 1963: TEM mode in a parallel-plate waveguide filled with a gyrotropic dielectric. *IEEE Trans. Microwave Theory Tech.*, **MTT-11**, 436–437.

Sezginer, A., and W. C. Chew, 1984: Closed form expression of the Green's function for the time-domain wave equation for a lossy two-dimensional medium. *IEEE Trans. Antennas Propagat.*, **AP-32**, 527–528.

Fluid Dynamics

Chaudhuri, K. S., 1976: Unsteady wave motions on a sloping beach with application to under-water explosions. *Quart. J. Mech. Appl. Math.*, **29**, 89–100.

Lamb, H., 1905: On deep-water waves. *Proc. London Math. Soc.*, Ser. 2, **2**, 371–400.

Matsuzaki, Y., and Y. C. Fung, 1977: Unsteady fluid dynamic forces on a simply-supported circular cylinder of finite length conveying a flow, with applications to stability analysis. *J. Sound Vib.*, **54**, 317–330.

Miles, J. W., 1956: The compressible flow past an oscillating airfoil in a wind tunnel. *J. Aeronaut. Sci.*, **23**, 671–678.

Seeger, G., and H.-G. Stäblein, 1964: Über den Strom-Anlaufvorgang im zylindrischen Leiter. *Z. Angew. Phys.*, **16**, 419–424.

Sen, A. R., 1963: Surface waves due to blasts on and above liquids. *J. Fluid Mech.*, **16**, 65–81.

Takenouchi, K., and Z. Tang, 1990: Two-dimensional supersonic channel flow with the magnetic field due to a current perpendicular to the flow. *J. Phys. Soc. Japan*, **59**, 3167–3181.

Tsien, H. S., and M. Finston, 1949: Interaction between parallel streams of subsonic and supersonic velocities. *J. Aeronaut. Sci.*, **16**, 515–528.

General

Davies, B., 1978: *Integral Transforms and Their Applications*. Springer-Verlag, Chapter 8.

Miles, J. W., 1971: *Integral Transforms in Applied Mathematics*. Cambridge University Press, Chapter 3.

Geophysical Sciences

Chimonas, G., and C. O. Hines, 1970: Atmospheric gravity waves launched by auroral currents. *Planet. Space Sci.*, **18**, 565–582.

Janowitz, G. S., 1975: The effect of bottom topography on a stratified flow in the beta plane. *J. Geophys. Res.*, **80**, 4163–4168.

Kahalas, S. L., 1965: Excitation of extremely low frequency electromagnetic waves in the earth-ionosphere cavity by high-altitude nuclear detonations. *J. Geophys. Res.*, **70**, 3587–3594.

Kajiura, K., 1953: On the influence of bottom topography on ocean currents. *J. Oceanogr. Soc. Japan*, **8**, 1–14.

Karzas, W. J., and R. Latter, 1962: The electromagnetic signal due to the interaction of nuclear explosions with the earth's magnetic field. *J. Geophys. Res.*, **67**, 4635–4640.

Lamb, H., 1908: On the theory of waves propagated vertically in the atmosphere. *Proc. London Math. Soc.*, *Ser. 2*, **7**, 122–141.

Lyra, G., 1943: Theorie der stationären Leewellen Strömung in freier Atmosphäre. *Z. Angew. Math. Mech.*, **23**, 1–28.

Officer, C. B., W. S. Newman, J. M. Sullivan, and D. R. Lynch, 1988: Glacial isostatic adjustment and mantle viscosity. *J. Geophys. Res.*, **93**, 6397–6409.

Sezawa, K., 1929: Formation of shallow-water waves due to subaqueous shocks. *Bull. Earthq. Res. Inst.*, **7**, 15–40.

Sezawa, K., and K. Kanai, 1932: Possibility of free oscillations of strata excited by seismic waves. Part IV. *Bull. Earthq. Res. Inst.*, **10**, 273–299.

Heat Conduction

Delsante, A. E., A. N. Stokes, and P. J. Walsh, 1983: Application of Fourier transforms to periodic heat flow into the ground under a building. *Int. J. Heat Mass Transfer*, **26**, 121–132.

Horvay, G., 1961: Temperature distribution in a slab moving from a chamber at one temperature to a chamber at another temperature. *J. Heat Transfer*, **83**, 391–402.

Karush, W., 1952: A steady-state heat flow problem for a quarter infinite solid. *J. Appl. Phys.*, **23**, 492–494.

Manner, M., 1969: Steady-state temperature distribution in a quarter-infinite solid and in a semi-infinite solid internally bounded by a cylinder. *Chem. Eng. Sci.*, **24**, 267–272.

Magnetohydrodynamics

Ehlers, F. E., 1961: Linearized magnetogasdynamic channel flow with axial symmetry. *Am. Rocket Soc. J.*, **31**, 334–342.

Prasada Rao, D. R. V., and D. V. Krishna, 1977: Point source in a MHD rotating fluid. *Appl. Sci. Res.*, **33**, 177–185.

Solid Mechanics

Beaudet, P. R., 1970: Elastic wave propagation in heterogeneous media. *Bull. Seism. Soc. Am.*, **60**, 769–784.

Blake, F. G., 1952: Spherical wave propagation in solid media. *J. Acoust. Soc. Am.*, **24**, 211–215.

Bose, S. K., 1975: Transmission of SH waves across a rectangular step. *Bull. Seism. Soc. Am.*, **65**, 1779–1786.

Chabravorty, S. K., 1961: Disturbances in a viscoelastic medium due to impulsive forces on a spherical cavity. *Geofis. Pura Appl.*, **48**, 23–26.

Cheng, D. H., and J. E. Benveniste, 1966: Transient response of structural elements to traveling pressure waves of arbitrary shape. *Int. J. Mech. Sci.*, **8**, 607–618.

Herman, H., and J. M. Klosner, 1965: Transient response of a periodically supported cylindrical shell immersed in a fluid medium. *J. Appl. Mech.*, **32**, 562–568.

Jung, H., 1950: Über eine Anwendung der Fouriertransformation in der Elastizitätstheorie. *Ing.-Arch.*, **18**, 263–271.

Matthews, P. M., 1958: Vibrations of a beam on elastic foundation. *Z. Angew. Math. Mech.*, **38**, 105–115.

Matthews, P. M., 1959: Vibrations of a beam on elastic foundation. II. *Z. Angew. Math. Mech.*, **39**, 13–19.

Mattice, H. C., and P. Lieber, 1954: On attenuation of waves produced in viscoelastic materials. *Trans. Am. Geophys. Union*, **35**, 613–624.

Momoi, T., 1987: Scattering of Rayleigh waves by a semi-circular rough surface in layered media. *Bull. Earthq. Res. Inst.*, **62**, 163–200.

Mow, C. C., 1965: Transient response of a rigid spherical inclusion in an elastic medium. *J. Appl. Mech.*, **32**, 637–642.

Patil, S. P., 1987: Natural frequencies of a railroad track. *J. Appl. Mech.*, **54**, 299–304.

Sharma, R. K., and N. R. Garg, 1994: Stresses in an elastic plate lying over a base due to strip-loading. *Proc. Indian Acad. Sci., Math. Sci.*, **104**, 425–433.

Sharpe, J. A., 1942: The production of elastic waves by explosion pressures. I. Theory and empirical field observations. *Geophysics*, **7**, 144–154.

Sneddon, I. N., 1945: The symmetrical vibrations of a thin elastic plate. *Proc. Cambridge Philos. Soc.*, **41**, 27–43.

Tranter, C. J., and J. W. Craggs, 1945: The stress distribution in a long circular cylinder when a discontinuous pressure is applied to the curved surface. *Philos. Mag., Ser. 7*, **36**, 241–250.

Thermoelasticity

Jahanshahi, A., 1964: Thin plates and shallow cylindrical shells subjected to hot spots. *J. Appl. Mech.*, **31**, 79–82.

Manaker, A. M., and G. Horvay, 1975: Thermal response in laminated composites. *Z. Angew. Math. Mech.*, **55**, 503–513.

Singh, H., 1981: Thermal stresses in an infinite cylinder. *Indian J. Pure Appl. Math.*, **12**, 405–418.

Wave Equation

Bremmer, H., 1964: Long waves associated with disturbances produced in plasmas. *J. Res. NBS*, **68D**, 47–58.

Grigor'yan, F. E., 1974/75: Analysis of a plane waveguide and derivation of inseparable solutions of the Helmholtz equation. *Sov. Phys. Acoust.*, **20**, 132–136.

Leehey, P., and H. G. Davies, 1975: The direct and reverberant response of strings and membranes to convecting, random pressure fields. *J. Sound Vib.*, **38**, 163–184.

Robey, D. H., 1953: Magnetic dispersion of sound in electrically conducting plates. *J. Acoust. Soc. Am.*, **25**, 603–609.

Sawyers, K. N., 1968: Underwater sound pressure from sonic booms. *J. Acoust. Soc. Am.*, **44**, 523–524.

Schoenstadt, A. L., 1977: The effect of spatial discretization on the steady-state and transient solutions of a dispersive wave equation. *J. Comput. Phys.*, **23**, 364–379.

Tanaka, K., and T. Kurokawa, 1973: Viscous property of steel and its effect on strain wave front. *Bull. JSME*, **16**, 188–193.

Papers Using Hankel Transforms
to Solve Partial Differential Equations

Aquifers, Reservoirs and Porous Media

Coats, K. H., 1962: A mathematical model water movement about bottom-water-drive reservoir. *Soc. Pet. Eng. J.*, **2**, 44–52.

Electromagnetism

Chetaev, D. N., 1963: On the field of a low-frequency electric dipole situated on the surface of a uniform anisotropic conducting halfspace. *Sov. Phys. Tech. Phys.*, **7**, 991–995.

Paul, M. K., and B. Banerjee, 1970: Electrical potentials due to a point source upon models of continuously varying conductivity. *Pure Appl. Geophys.*, **80**, 218–237.

Fluid Dynamics

Lamb, H., 1905: On deep-water waves. *Proc. London Math. Soc., Ser. 2*, **2**, 371–400.

Lamb, H., 1922: On water waves due to disturbance beneath the surface. *Proc. London Math. Soc., Ser. 2*, **21**, 359–372.

Mackie, A. G., 1953: An application of Hankel transforms in axially symmetric potential flow. *Proc. Edinburgh Math. Soc., Ser. 2*, **9**, 128–131.

Sen, A. R., 1960: Problems of deep-water waves. Part I. The exact and asymptotic solutions. *Bull. Calcutta Math. Soc.*, **52**, 127–146.

Sen, A. R., 1963: Surface waves due to blasts on and above liquids. *J. Fluid Mech.*, **16**, 65–81.

Terazawa, K., 1915: On deep-sea water waves caused by a local disturbance on or beneath the surface. *Proc. R. Soc. London, Ser. A*, **92**, 57–81.

General

Miles, J. W., 1971: *Integral Transforms in Applied Mathematics*. Cambridge University Press, Chapter 3.

Geophysical Sciences

Ben-Menahem, A., 1961: Radiation of seismic surface-waves from finite moving sources. *Bull. Seism. Soc. Am.*, **51**, 401–435.

Kajiura, K., 1963: The leading wave of a tsunami. *Bull. Earthq. Res. Inst.*, **41**, 535–571.

Krimigis, S. M., 1965: Interplanetary diffusion model for the time behavior of intensity in a solar cosmic ray event. *J. Geophys. Res.*, **70**, 2943–2960.

Lamb, H., 1908: On the theory of waves propagated vertically in the atmosphere. *Proc. Lond. Math. Soc., Ser. 2*, **7**, 122–141.

Officer, C. B., W. S. Newman, J. M. Sullivan, and D. R. Lynch, 1988: Glacial isostatic adjustment and mantle viscosity. *J. Geophys. Res.*, **93**, 6397–6409.

Ou, H. W., and A. L. Gordon, 1986: Spin-down of baroclinic eddies under sea ice. *J. Geophys. Res.*, **91**, 7623–7630.

Sinha, A. K., and P. K. Bhattacharya, 1967: Electric dipole over an isotropic and inhomogeneous earth. *Geophysics*, **32**, 652–667.

Heat Conduction

Manner, M., 1969: Steady-state temperature distribution in a quarter-infinite solid and in a semi-infinite solid internally bounded by a cylinder. *Chem. Eng. Sci.*, **24**, 267–272.

Solid Mechanics

Agrawal, G. L., and W. G. Gottenberg, 1971: Response of a semi-infinite elastic solid to an arbitrary torque along the axis. *Pure Appl. Geophys.*, **91**, 34–39.

Das Gupta, S., 1966: Generation of transverse waves in an elastic medium due to distribution of surface forces. *Proc. Vib. Prob.*, **7**, 221–235.

Datta, S. K., 1961: Shear waves in a semi-infinite visco-elastic medium due to transient torsional couple applied on the circumference of a circle on the plane boundary. *Proc. Natl. Inst. Sci. India*, **A27**, 482–488.

Gürgözc, I. T., and M. S. Dokuz, 1998: An equilibrium solution for the Boussinesq problem for a mixture of an elastic solid and a fluid in an infinite half-space. *Int. J. Eng. Sci.*, **36**, 645–653.

Hron, M., 1970: Radiation of seismic surface waves from extending circular sources. *Bull. Seism. Soc. Am.*, **60**, 517–537.

Newlands, M., 1954: Lamb's problem with internal dissipation. I. *J. Acoust. Soc. Am.*, **26**, 434–448.

Nishida, Y., Y. Shindo, and A. Atsumi, 1984: Diffraction of horizontal shear waves by a moving interface crack. *Acta Mech.*, **54**, 23–34.

Olesiak, Z., and I. N. Sneddon, 1960: The distribution of thermal stress in an infinite elastic solid containing a penny-shaped crack. *Arch. Ration. Mech. Anal.*, **4**, 237–254.

Press, F., and M. Ewing, 1950: Propagation of explosive sound in a liquid layer overlying a semi-infinite elastic solid. *Geophysics*, **15**, 426–446.

Reißner, E., 1937: Freie und erzwungene Torsionsschwingungen des elastischen Halbraumes. *Ing.-Arch.*, **8**, 229–245.

Shindo, Y., 1981: Normal compression waves scattering at a flat annular crack in an infinite elastic solid. *Quart. Appl. Math.*, **39**, 305–315.

Sidhu, R. S., 1971: SH waves from torsional sources in semi-infinite heterogeneous media. *Pure Appl. Geophys.*, **87**, 55–65.

Singh, B. M., J. Rokne, and R. S. Dhaliwal, 1983: Diffraction of torsional wave by a circular rigid disc at the interface of two bonded dissimilar elastic solids. *Acta Mech.*, **49**, 139–146.

Singh, S. J., 1966: Point source in a layer overlying a semi-infinite solid with special reference to diffracted waves. *Geophys. J. R. Astr. Soc.*, **11**, 433–452.

Wadhwa, S. K., 1971: Reflected and refracted waves from a linear transition layer. *Pure Appl. Geophys.*, **89**, 45–66.

Thermoelasticity

Watts, R. G., 1969: Temperature distributions in solid and hollow cylinders due to a moving circumferential ring heat source. *J. Heat Transfer*, **91**, 465–470.

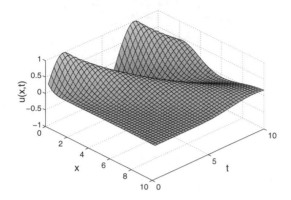

Chapter 4
Methods Involving
Multivalued Laplace Transforms

In the previous two chapters, we showed how we can use the techniques from an undergraduate complex variable course to solve partial differential equations. When transients are present, multivalued functions such as $\sqrt{z}$ appear in the transform and certain special efforts are necessary. The inversion of Laplace transforms that contain a multivalued function is shown in Section 4.1. Then, with these new techniques, we solve the wave and heat equations where these multivalued functions arise in Section 4.3 and Section 4.4.

4.1 INVERSION OF LAPLACE TRANSFORMS BY CONTOUR INTEGRATION

In Section 2.1 we showed that we can always use Bromwich's inversion integral to invert Laplace transforms. There, we restricted ourselves to transforms that are only single-valued. In many applications the transform will have a multivalued function in it. In this section, we show how we may apply Bromwich's integral in the inversion of these transforms.

307

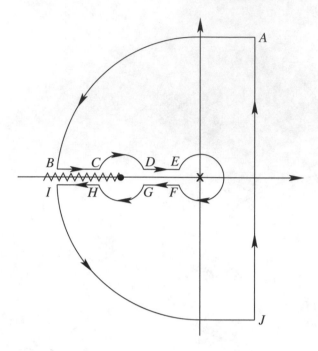

Figure 4.1.1: The contour used to invert the Laplace transform given by Equation 4.1.1.

• **Example 4.1.1**

Let us find the inverse[1] of

$$F(s) = \frac{\exp(-\beta y)}{\beta s}, \qquad 0 < x, y, \qquad (4.1.1)$$

where $\beta = \sqrt{\omega^2 + s/x}$, $\mathrm{Re}(\beta) > 0$, by evaluating the Bromwich integral

$$f(t) = \frac{1}{2\pi i} \int_{c-\infty i}^{c+\infty i} e^{-\beta y + tz} \frac{dz}{\beta z}. \qquad (4.1.2)$$

The branch cut runs along the negative real axis of the z-plane from the branch point $z = -x\omega^2$ out to $-\infty$. Consequently, we can close the contour as shown in Figure 4.1.1 where the point $z = 0$ is a simple pole.

Because the arcs AB and IJ vanish as the semicircle expands to infinity and the integral along FG cancels the integral along DE,

---

[1] Adapted from Koizumi, T., 1970: Transient thermal stresses in a semi-infinite body heated axi-symmetrically. *Z. Angew. Math. Mech.*, **50**, 747–757. See also Koizumi, T., 1970: Thermal stresses in a semi-infinite body with an instantaneous heat source on its surface. *Bull. JSME*, **13**, 26–33.

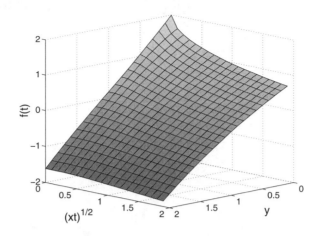

Figure 4.1.2: The inverse of the Laplace transform given by Equation 4.1.1 when $\omega = 0.5$.

$$f(t) = -\frac{1}{2\pi i} \left(\int_{BC} F(z)e^{tz}\,dz + \int_{CD} F(z)e^{tz}\,dz + \int_{EF} F(z)e^{tz}\,dz \right.$$
$$\left. + \int_{GH} F(z)e^{tz}\,dz + \int_{HI} F(z)e^{tz}\,dz \right) \tag{4.1.3}$$

$$= \frac{1}{\pi} \int_{\infty}^{0} \frac{d\lambda}{\omega^2 + \lambda^2} \exp[-i\lambda y - x\left(\omega^2 + \lambda^2\right)t]$$

$$- \frac{\sqrt{x}}{2\pi} \lim_{r \to 0} \int_{\pi}^{0} \frac{\sqrt{r}e^{\theta i/2}}{re^{\theta i} - x\omega^2} \exp\left[-y\sqrt{\frac{r}{x}}e^{\theta i/2}\left(re^{\theta i} - x\omega^2\right)t\right]d\theta$$

$$+ \frac{e^{-\beta y}}{\beta}\bigg|_{z=0} - \frac{\sqrt{x}}{2\pi} \lim_{r \to 0} \int_{0}^{-\pi} \frac{\sqrt{r}e^{\theta i/2}}{re^{\theta i} - x\omega^2} \exp\left[-y\sqrt{\frac{r}{x}}e^{\theta i/2} + \left(re^{\theta i} - x\omega^2\right)t\right]d\theta$$

$$- \frac{1}{\pi} \int_{0}^{\infty} \frac{d\lambda}{\omega^2 + \lambda^2} \exp\left[i\lambda y - x\left(\omega^2 + \lambda^2\right)t\right] \tag{4.1.4}$$

$$= \frac{e^{-\omega y}}{\omega} - \frac{2}{\pi} \int_{0}^{\infty} \frac{\cos(\lambda y)}{\omega^2 + \lambda^2} \exp[-x\left(\omega^2 + \lambda^2\right)t]\,d\lambda \tag{4.1.5}$$

$$= \frac{e^{-\omega y}}{\omega} - \frac{1}{2\omega}\left[e^{-\omega y}\operatorname{erf}\left(\omega\sqrt{xt} - \frac{y}{2\sqrt{xt}}\right) + e^{\omega y}\operatorname{erf}\left(\omega\sqrt{xt} + \frac{y}{2\sqrt{xt}}\right)\right]. \tag{4.1.6}$$

This inverse is illustrated in Figure 4.1.2.

• **Example 4.1.2**

In this example, let us find the inverse of the Laplace transform

$$F(s) = \frac{K_1(sr)}{\alpha s K_0(s) + K_1(s)} \tag{4.1.7}$$

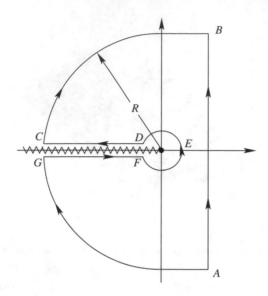

Figure 4.1.3: The contour used to invert the Laplace transform given by Equation 4.1.7.

by contour integration where α and r are constants. Because we can express $K_0(\cdot)$ and $K_1(\cdot)$ in terms of infinite power series,[2] they are tabulated functions, like sine and cosine. For the present, however, we need only note that the power series for $K_n(z)$ contains a logarithm. Consequently, $K_0(z)$ and $K_1(z)$ are multivalued functions with branch points at $z = 0$ and infinity. Figure 4.1.3 shows the contour used in the inversion integral with the branch cut along the negative real axis of the s-plane.

We begin by noting that for large z, $K_n(z)$ behaves[3] as

$$K_n(z) \sim \left(\frac{\pi}{2z}\right)^{1/2} e^{-z}, \qquad (4.1.8)$$

so that

$$e^{tz} F(z) \sim \frac{r^{-1/2}}{1 + \alpha z} \exp[z(t - r + 1)] \qquad (4.1.9)$$

in the limit $|z| \to \infty$. Consequently, the contribution along the arcs BC and GA vanishes as $R \to \infty$ if $t > r - 1$. For $t < r - 1$, $f(t) = 0$.

Turning to the nature of $F(z)$, there is a conjugate pair $z_{1,2} = -x \pm iy$ of simple poles if $\alpha > 0$. Figure 4.1.4 gives the values of x and y. For the

---

[2] Watson, G. N., 1966: *A Treatise on the Theory of Bessel Functions*. Cambridge University Press, p. 80.

[3] *Ibid.*, p. 202.

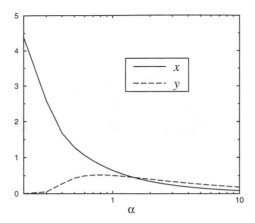

Figure 4.1.4: The values of x and y that give the position of the simple poles $z_{1,2} = -x \pm iy$ for the transform given by Equation 4.1.7.

branch cut integrals, we have along FG $z = \eta e^{-\pi i}$ and $dz = d\eta\, e^{-\pi i}$ so that

$$\int_{FG} F(z)e^{tz}dz = \int_{\infty}^{0} \frac{K_1(\eta) - i\pi I_1(\eta)}{-\alpha\eta[K_0(\eta) + i\pi I_0(\eta)] - K_1(\eta) + i\pi I_1(\eta)}e^{-t\eta}\,d\eta \tag{4.1.10}$$

and along CD, $z = \eta e^{\pi i}$ and $dz = d\eta\, e^{\pi i}$ so that

$$\int_{CD} F(z)e^{tz}dz = \int_{0}^{\infty} \frac{K_1(\eta) + i\pi I_1(\eta)}{-\alpha\eta[K_0(\eta) - i\pi I_0(\eta)] - K_1(\eta) - i\pi I_1(\eta)}e^{-t\eta}\,d\eta, \tag{4.1.11}$$

because[4]

$$K_\nu\big(ze^{m\pi i}\big) - e^{-m\nu\pi i}K_\nu(z) - i\pi\frac{\sin(m\nu\pi)}{\sin(\nu\pi)}I_\nu(z). \tag{4.1.12}$$

The integral for the small circle DEF vanishes. After a little algebra, we combine Equation 4.1.10 and Equation 4.1.11 together and find

$$f(t) = 2\alpha e^{-xt}\frac{[\alpha^2(x^2 + y^2) + 2\alpha - 1][x\cos(yt) - y\sin(yt)]}{[\alpha^2(x^2 - y^2) + 2\alpha - 1]^2 + 4\alpha^4 x^2 y^2}H(\alpha) \tag{4.1.13}$$
$$- \int_{0}^{\infty} \frac{I_1(r\eta)[\alpha\eta K_0(\eta) + K_1(\eta)] + K_1(r\eta)[\alpha\eta I_0(\eta) - I_1(\eta)]}{[\alpha\eta K_0(\eta) + K_1(\eta)]^2 + \pi^2[\alpha\eta I_0(\eta) - I_1(\eta)]^2}e^{-\eta t}\,d\eta,$$

if $t > r - 1$. The first term in Equation 4.1.13 arises from the sum of the residues. This inverse is illustrated in Figure 4.1.5.

---

[4] *Ibid.*, p. 80.

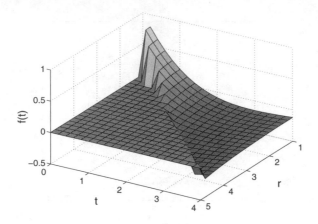

Figure 4.1.5: The inverse of the Laplace transform given by Equation 4.1.7 when $\alpha = 1$.

• **Example 4.1.3**

Recently, Bobylev and Cercignani[5] developed a theorem concerning the inversion of multivalued transforms that are analytic everywhere in the s-plane except along the negative real axis. Their theorem is as follows:

Theorem: *Let $f(t)$ denote a real-valued function whose Laplace transform $F(s)$ exists. Let $F(s)$ satisfy the following criteria:*

• *$F(s)$ is a multivalued function which has no singularities in the cut s-plane. The branch cut lies along the negative real axis $(-\infty, 0]$.*

• *$F^*(s) = F(s^*)$, where the star denotes the complex conjugate.*

• *$F^\pm(\eta) = \lim_{\varphi \to \pi^-} F\left(\eta e^{\pm \varphi i}\right)$ and $F^+(\eta) = [F^-(\eta)]^*$.*

• *$F(s) = o(1)$ as $|s| \to \infty$ and $F(s) = o(1/|s|)$ as $|s| \to 0$, uniformly in any sector $|\arg(s)| < \pi - \eta$, $0 < \eta < \pi$.*

• *There exists $\epsilon > 0$, such that for every $\pi - \epsilon < \varphi \leq \pi$, $F(re^{\pm \varphi i})/(1+r) \in L^1(R_+)$ and $|F(re^{\pm \varphi i})| < a(r)$, where $a(r)$ does not depend on φ and $a(r)e^{-\delta r} \in L^1(R_+)$ for any $\delta > 0$.*

Then

$$f(t) = \frac{1}{\pi} \int_0^\infty \text{Im}[F^-(\eta)] \, e^{-t\eta} \, d\eta. \tag{4.1.14}$$

To illustrate this theorem, consider the function $f(t) = t^\alpha$, $0 < \alpha < 1$. Its Laplace transform is $F(s) = \Gamma(\alpha)/s^\alpha$, where $\Gamma(\cdot)$ is the gamma function.

[5] Reprinted from *Appl. Math. Lett.*, **15**, A. V. Bobylev and C. Cercignani, The inverse Laplace transform of some analytic functions with an application to the eternal solutions of the Boltzmann equation, 807–813, ©2002, with permission from Elsevier.

Therefore,

$$F^-(\eta) = \lim_{\varphi \to \pi^-} F\left(\eta e^{-\varphi i}\right) = \lim_{\varphi \to \pi^-} \frac{\Gamma(\alpha)}{[\eta e^{-\varphi i}]^\alpha} = \frac{\Gamma(\alpha)}{\eta^\alpha} e^{\pi \alpha i} \qquad (4.1.15)$$

and

$$\mathrm{Im}\left[F^-(\eta)\right] = \sin(\pi\alpha)\frac{\Gamma(\alpha)}{\eta^\alpha}. \qquad (4.1.16)$$

Substituting these results into Equation 4.1.14,

$$f(t) = \sin(\pi\alpha)\frac{\Gamma(\alpha)}{\pi} \int_0^\infty \frac{e^{-t\eta}}{\eta^\alpha}\, d\eta = \frac{\sin(\pi\alpha)}{\pi}\Gamma(\alpha)\Gamma(1-\alpha)t^{\alpha-1} = t^{\alpha-1}, \qquad (4.1.17)$$

since

$$\Gamma(\alpha)\Gamma(1-\alpha) = \frac{\pi}{\sin(\alpha\pi)}. \qquad (4.1.18)$$

For another illustration, consider the Laplace transform

$$F(s) = \frac{K_0(r\sqrt{s})}{\sqrt{s}\,K_1(a\sqrt{s}) + hK_0(a\sqrt{s})}, \qquad 0 < a < r, \quad 0 < h. \qquad (4.1.19)$$

It satisfies all of the conditions listed in the theorem. Then,

$$F^-(x) = \frac{K_0(-ir\sqrt{x})}{-i\sqrt{x}\,K_1(-ia\sqrt{x}) + hK_0(-ia\sqrt{x})} \qquad (4.1.20)$$

$$= \frac{J_0(r\sqrt{x}) + iY_0(r\sqrt{x})}{g_J(\sqrt{x}) + ig_Y(\sqrt{x})} \qquad (4.1.21)$$

$$= \frac{J_0(r\sqrt{x})\,g_J(\sqrt{x}) + Y_0(r\sqrt{x})\,g_Y(\sqrt{x})}{g_J^2(\sqrt{x}) + g_Y^2(\sqrt{x})}$$

$$+ i\frac{Y_0(r\sqrt{x})\,g_J(\sqrt{x}) - J_0(r\sqrt{x})\,g_Y(\sqrt{x})}{g_J^2(\sqrt{x}) + g_Y^2(\sqrt{x})}, \qquad (4.1.22)$$

since $K_1(-\eta i) = -\pi\left[J_1(\eta) + iY_1(\eta)\right]/2$ and $K_0(-\eta i) = -\pi i\left[J_0(\eta) + iY_0(\eta)\right]$ $/2$ for $\eta > 0$. Here, $g_J(\eta) = \eta J_1(a\eta) + hJ_0(a\eta)$ and $g_Y(\eta) = \eta Y_1(a\eta) + hY_0(a\eta)$. Therefore,

$$\mathrm{Im}\left[F^-(x)\right] = \frac{Y_0(r\sqrt{x})\,g_J(\sqrt{x}) - J_0(r\sqrt{x})\,g_Y(\sqrt{x})}{g_J^2(\sqrt{x}) + g_Y^2(\sqrt{x})}. \qquad (4.1.23)$$

Substituting into Equation 4.1.14,

$$f(t) = \frac{1}{\pi}\int_0^\infty \frac{Y_0(r\sqrt{x})\,g_J(\sqrt{x}) - J_0(r\sqrt{x})\,g_Y(\sqrt{x})}{g_J^2(\sqrt{x}) + g_Y^2(\sqrt{x})} e^{-tx}\, dx \qquad (4.1.24)$$

$$= \frac{2}{\pi}\int_0^\infty \frac{Y_0(r\eta)\,g_J(\eta) - J_0(r\eta)\,g_Y(\eta)}{g_J^2(\eta) + g_Y^2(\eta)} e^{-t\eta^2}\eta\, d\eta, \qquad (4.1.25)$$

where $x = \eta^2$.

Consider now $F(s)/s$. By the convolution theorem,

$$\mathcal{L}\left[\frac{F(s)}{s}\right] = \int_0^t f(\tau)\, d\tau = \frac{2}{\pi} \int_0^\infty \frac{Y_0(r\eta)\, g_J(\eta) - J_0(r\eta)\, g_Y(\eta)}{g_J^2(\eta) + g_Y^2(\eta)} \left(1 - e^{-t\eta^2}\right) \frac{d\eta}{\eta}. \tag{4.1.26}$$

As $t \to \infty$,

$$\frac{2}{\pi} \int_0^\infty \frac{Y_0(r\eta)\, g_J(\eta) - J_0(r\eta)\, g_Y(\eta)}{g_J^2(\eta) + g_Y^2(\eta)} \frac{d\eta}{\eta} = F(0) = \frac{1}{h} \tag{4.1.27}$$

by the final-value theorem. Consequently,

$$\int_0^\infty \frac{Y_0(r\eta)\, g_J(\eta) - J_0(r\eta)\, g_Y(\eta)}{g_J^2(\eta) + g_Y^2(\eta)} \frac{d\eta}{\eta} = \frac{\pi}{2h}. \tag{4.1.28}$$

In the previous examples, we found the inverse of the Laplace transform when the inversion integral had branch points and cuts that lay along the negative real axis. In those cases when the multivalued functions have branch points and cuts that are more complicated in their structure, a more formal bookkeeping system is necessary to keep track of the phase of these functions. In the following examples we handle these more difficult situations.

● **Example 4.1.4**

Let us find the inverse of the Laplace transform

$$F(s) = \frac{1}{s} \exp\left(-\frac{xs}{\sqrt{s^2 + b^2}}\right), \qquad 0 < x. \tag{4.1.29}$$

From the inversion integral

$$f(t) = \frac{1}{2\pi i} \int_{c-\infty i}^{c+\infty i} \exp\left(-\frac{z}{\sqrt{z^2 + b^2}} x + tz\right) \frac{dz}{z} \tag{4.1.30}$$

$$= \frac{1}{2\pi i} \int_C \exp\left(-\frac{z}{\sqrt{z^2 + b^2}} x + tz\right) \frac{dz}{z}, \tag{4.1.31}$$

where the line integral C comprises the line integrals $C_1, C_2, \ldots, C_8$ plus the integrations around the infinitesimally small circles at $z = 0$, $z = ib$ and $z = -ib$. See Figure 4.1.6.

As in previous examples, the integrals along C_1 and C_8 vanish as the semicircle expands without bound. Furthermore, the integral along C_2 cancels that along C_7.

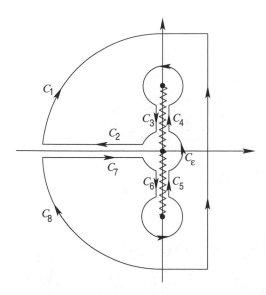

Figure 4.1.6: The contour used to invert the Laplace transform given by Equation 4.1.29.

Turning to the integrals along C_3, C_4, C_5 and C_6, we introduce the following notation:

$$z = \eta e^{\theta i}, \qquad\qquad 0 < \theta < 2\pi, \qquad (4.1.32)$$

$$z - ib = r_1 e^{\theta_1 i}, \qquad -\pi/2 < \theta_1 < 3\pi/2, \qquad (4.1.33)$$

$$z + ib = r_2 e^{\theta_2 i}, \qquad -\pi/2 < \theta_2 < 3\pi/2. \qquad (4.1.34)$$

Therefore, $\sqrt{z^2 + b^2} = \sqrt{r_1 r_2}\, e^{\psi i}$, where $\psi = (\theta_1 + \theta_2)/2$ and $-\pi/2 < \psi < 3\pi/2$. With these definitions of the phase, we have from purely geometric considerations:

Along C_3, $z = \eta e^{\pi i/2}$, $z - ib = (b - \eta)e^{3\pi i/2}$, $z + ib = (b + \eta)e^{\pi i/2}$. (**4.1.35**)

Along C_4, $z = \eta e^{\pi i/2}$, $z - ib = (b - \eta)e^{-\pi i/2}$, $z + ib = (b + \eta)e^{\pi i/2}$. (**4.1.36**)

Along C_5, $z = \eta e^{-\pi i/2}$, $z - ib = (b + \eta)e^{-\pi i/2}$, $z + ib = (b - \eta)e^{\pi i/2}$. (**4.1.37**)

Along C_6, $z = \eta e^{-\pi i/2}$, $z - ib = (b + \eta)e^{3\pi i/2}$, $z + ib = (b - \eta)e^{\pi i/2}$ (**4.1.38**)

with $0 \le \eta \le b$. Upon substituting these relationships into the integrals and simplifying,

$$\int_{C_3} F(z)e^{tz}\, dz = \int_b^0 \exp\!\left(\frac{ix\eta}{\sqrt{b^2 - \eta^2}} + it\eta\right) \frac{d\eta}{\eta}, \qquad (4.1.39)$$

$$\int_{C_4} F(z)e^{tz}\, dz = \int_0^b \exp\!\left(-\frac{ix\eta}{\sqrt{b^2 - \eta^2}} + it\eta\right) \frac{d\eta}{\eta}, \qquad (4.1.40)$$

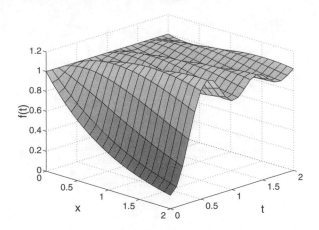

Figure 4.1.7: The inverse of the Laplace transform given by Equation 4.1.29 when $b = 10$.

$$\int_{C_5} F(z)e^{tz}\, dz = \int_b^0 \exp\left(\frac{ix\eta}{\sqrt{b^2 - \eta^2}} - it\eta\right) \frac{d\eta}{\eta}, \tag{4.1.41}$$

and

$$\int_{C_6} F(z)e^{tz}\, dz = \int_0^b \exp\left(-\frac{ix\eta}{\sqrt{b^2 - \eta^2}} - it\eta\right) \frac{d\eta}{\eta}. \tag{4.1.42}$$

Turning to the integration around the three infinitesimal circles, the integrations around $z = ib$ and $z = -ib$ are zero. However, because of the singularity at the origin, the integration around $z = 0$ is nonzero. When we combine the contributions from the four quarter circles, the value of the integral equals

$$\int_{C_\epsilon} \exp\left(-\frac{xz}{\sqrt{z^2 + b^2}} + tz\right) \frac{dz}{z} = 2\pi i. \tag{4.1.43}$$

Combining our results,

$$f(t) = 1 - \frac{2}{\pi} \int_0^b \cos(\eta t) \sin\left(\frac{x\eta}{\sqrt{b^2 - \eta^2}}\right) \frac{d\eta}{\eta}. \tag{4.1.44}$$

This inverse is illustrated in Figure 4.1.7.

● **Example 4.1.5**

In this example we find the inverse[6] of

$$F(s) = \frac{\sqrt{s^2 + a^2}}{s^2} \tag{4.1.45}$$

---

[6] A more complicated version appeared in Dimaggio, F. L., 1956: Effect of an acoustic medium on the dynamic buckling of plates. *J. Appl. Mech.*, **23**, 201–206.

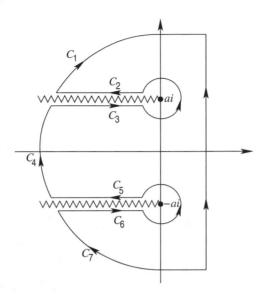

Figure 4.1.8: The contour used to invert the Laplace transform given by Equation 4.1.45.

by contour integration. The inverse is

$$f(t) = \frac{1}{2\pi i} \int_{c-\infty i}^{c+\infty i} \frac{\sqrt{z^2 + a^2}}{z^2} e^{tz}\, dz = \frac{1}{2\pi i} \oint_{\Gamma} \frac{\sqrt{z^2 + a^2}}{z^2} e^{tz}\, dz, \qquad (4.1.46)$$

where Γ is the closed contour shown in Figure 4.1.8. As in the previous problem, there are branch points associated with the square root at $z = \pm ai$. However, we have chosen the branch cuts to run from each of the branch points out to infinity along paths that are parallel to the real axis rather than along a segment of the imaginary axis.

Within the contour Γ, there is a second-order pole at $z = 0$. Consequently, the inverse equals

$$f(t) = \frac{d}{dz}\left(e^{tz}\sqrt{z^2 + a^2}\right)\Big|_{z=0} - \frac{1}{2\pi i} \int_{C} \frac{\sqrt{z^2 + a^2}}{z^2} e^{tz}\, dz, \qquad (4.1.47)$$

where C is the contour comprising the line integrals over $C_1, \ldots, C_7$ plus the integrations around the infinitesimal circles at $z = \pm ai$.

We only need to evaluate C_2, C_3, C_5 and C_6. The integrals over C_1, C_4 and C_7 vanish as the semicircle expands without bound by Jordan's lemma. Similarly, the integrals around the circles at $z = \pm ai$ vanish as their radii tend to zero.

Let us consider the integral along C_2. From purely geometric considerations,

$$z = -\eta + ia, \quad z - ia = \eta e^{\pi i} \quad \text{and} \quad z + ia = \sqrt{\eta^2 + 4a^2}\, e^{\theta i}, \qquad (4.1.48)$$

where $\theta = \arctan(-2a/\eta)$ and $0 < \eta < \infty$. Now,

$$\sqrt{z^2 + a^2} = \sqrt{(z - ia)(z + ia)} = \eta^{1/2}(\eta^2 + 4a^2)^{1/4}e^{\pi i/2}e^{\theta i/2} \qquad (4.1.49)$$

$$= i\eta^{1/2}(\eta^2 + 4a^2)^{1/4}\left[\cos(\theta/2) + i\sin(\theta/2)\right] \qquad (4.1.50)$$

$$= i\eta^{1/2}(\eta^2 + 4a^2)^{1/4}\left\{\left[\frac{1 + \cos(\theta)}{2}\right]^{1/2} + i\left[\frac{1 - \cos(\theta)}{2}\right]^{1/2}\right\}$$
$$(4.1.51)$$

$$= i\eta^{1/2}\left[\left(\sqrt{a^2 + \eta^2/4} - \eta/2\right)^{1/2} + i\left(\sqrt{a^2 + \eta^2/4} + \eta/2\right)^{1/2}\right]$$
$$(4.1.52)$$

because $\cos(\theta) = -\eta/\sqrt{4a^2 + \eta^2}$.

Substituting these results into the line integral and simplifying,

$$\int_{C_3} F(z)e^{tz}\,dz = \int_{C_2} F(z)e^{tz}\,dz = -\int_0^\infty i\eta^{1/2}\frac{f_1 + if_2}{\eta^2 - a^2 - 2ia}e^{-t\eta}e^{iat}\,d\eta, \qquad (4.1.53)$$

where

$$f_1 = \left(\sqrt{a^2 + \eta^2/4} - \eta/2\right)^{1/2} \qquad (4.1.54)$$

and

$$f_2 = \left(\sqrt{a^2 + \eta^2/4} + \eta/2\right)^{1/2}. \qquad (4.1.55)$$

From similar considerations,

$$\int_{C_6} F(z)e^{tz}\,dz = \int_{C_5} F(z)e^{tz}\,dz = -\int_0^\infty i\eta^{1/2}\frac{f_1 - if_2}{\eta^2 - a^2 + 2ia}e^{-t\eta}e^{-iat}\,d\eta. \qquad (4.1.56)$$

Combining these results, the inverse equals

$$f(t) = at - \frac{2}{\pi}\int_0^\infty \eta^{1/2}e^{-t\eta}\,d\eta$$
$$\times \frac{(\eta^2 - a^2)[f_1\cos(at) - f_2\sin(at)] - 2a\eta[f_1\sin(at) + f_2\cos(at)]}{(\eta^2 + a^2)^2}.$$
$$(4.1.57)$$

This inverse is illustrated in Figure 4.1.9.

• Example 4.1.6

Our next example arises in the analysis of wave solutions in beams. We shall find the inverse of the Laplace transform[7]

$$F(s) = \frac{\left(s - N\sqrt{s^2 - a^2}\right)^{1/2}}{\sqrt{s(s^2 - a^2)}}, \qquad 0 < N < 1, \qquad (4.1.58)$$

---

[7] A similar problem occurred in Boley, B. A., and C. C. Chao, 1955: Some solutions to the Timoshenko beam equation. *J. Appl. Mech.*, **22**, 579–586.

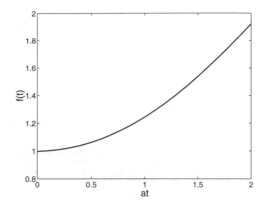

Figure 4.1.9: The inverse of the Laplace transform given by Equation 4.1.45.

by the inversion integral

$$f(t) = \frac{1}{2\pi i} \int_{c-\infty i}^{c+\infty i} \frac{\left(z - N\sqrt{z^2 - a^2}\right)^{1/2}}{\sqrt{z(z^2 - a^2)}} \, e^{tz} \, dz, \qquad (\mathbf{4.1.59})$$

which we transform into the contour integral

$$f(t) = \frac{1}{2\pi i} \int_{C} \frac{\left(z - N\sqrt{z^2 - a^2}\right)^{1/2}}{\sqrt{z(z^2 - a^2)}} \, e^{tz} \, dz, \qquad (\mathbf{4.1.60})$$

where Figure 4.1.10 shows C. The integrand of Equation 4.1.60 has several multivalued functions, so we must introduce several branch cuts. The function $\sqrt{z}$ results in a branch cut along the negative real axis from the branch point $z = 0$ out to $-\infty$. Similarly, the function $\sqrt{z^2 - a^2}$ results in a branch cut along the real axis between the branch points $z = \pm a$.

Let us now turn our attention to $\left(z - N\sqrt{z^2 - a^2}\right)^{1/2}$. We note that

$$\left(z - N\sqrt{z^2 - a^2}\right)^{1/2} = \frac{(1 - N^2)\sqrt{z^2 + b^2}}{(z + N\sqrt{z^2 - a^2})^{1/2}}, \qquad (\mathbf{4.1.61})$$

where $b = aN/\sqrt{1 - N^2}$. The numerator dictates a branch cut that runs from $z = -bi$ to $z = bi$. It only remains to show that the denominator of Equation 4.1.61 does not introduce any further branch points. This would occur if $z + N\sqrt{z^2 + a^2}$ vanished or $z^2 = N^2(z^2 - a^2)$ or $z = \pm bi$. Consider the case $z = ib$. From our definition of the branch cuts so far, $\sqrt{z^2 - a^2} = (a^2 + b^2)i$ and both z and $\sqrt{z^2 - a^2}$ have imaginary parts greater than zero. Therefore, their sum cannot be zero and we need no further branch cuts. The phase of $z + ib$ and $z - ib$ lies between $-\pi/2$ and $3\pi/2$.

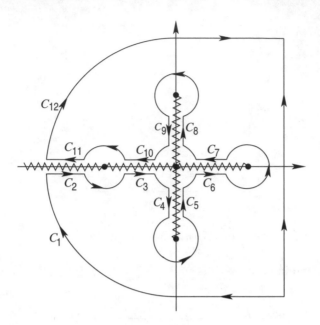

Figure 4.1.10: The contour used to invert the Laplace transform given by Equation 4.1.58.

With our branch cuts now clearly defined, we begin the evaluation of the various line integrals. From Jordan's lemma, the contribution from paths C_1 and C_{12} are zero. Now, along C_2 we have for $a < \eta < \infty$,

$$z = \eta e^{-\pi i}, \quad z - a = (\eta + a)e^{-\pi i}, \quad z + a = (\eta - a)e^{-\pi i}, \qquad (4.1.62)$$

$$\sqrt{z^2 - a^2} = \sqrt{\eta^2 - a^2}\, e^{-\pi i}, \qquad (4.1.63)$$

$$\left(z - N\sqrt{z^2 - a^2}\right)^{1/2} = \left(\eta - N\sqrt{\eta^2 - a^2}\right)^{1/2} e^{-\pi i/2} \qquad (4.1.64)$$

so that

$$\int_{C_2} \frac{\left(z - N\sqrt{z^2 - a^2}\right)^{1/2}}{\sqrt{z(z^2 - a^2)}}\, e^{tz}\, dz = \int_{\infty}^{a} \frac{\left(\eta - N\sqrt{\eta^2 - a^2}\right)^{1/2}}{\sqrt{\eta(\eta^2 - a^2)}}\, e^{-t\eta}\, d\eta,$$
$$(4.1.65)$$

while along C_{11} for $a < \eta < \infty$,

$$z = \eta e^{\pi i}, \quad z - a = (\eta + a)e^{\pi i}, \quad z + a = (\eta - a)e^{\pi i}, \qquad (4.1.66)$$

$$\sqrt{z^2 - a^2} = \sqrt{\eta^2 - a^2}\, e^{\pi i}, \qquad (4.1.67)$$

$$\left(z - N\sqrt{z^2 - a^2}\right)^{1/2} = \left(\eta - N\sqrt{\eta^2 - a^2}\right)^{1/2} e^{\pi i/2} \qquad (4.1.68)$$

so that

$$\int_{C_6} \frac{\left(z - N\sqrt{z^2 - a^2}\right)^{1/2}}{\sqrt{z(z^2 - a^2)}}\, e^{tz}\, dz = \int_a^\infty \frac{\left(\eta - N\sqrt{\eta^2 - a^2}\right)^{1/2}}{\sqrt{\eta(\eta^2 - a^2)}}\, e^{-t\eta}\, d\eta.$$

(4.1.69)

Consequently, the contribution from C_{11} cancels that from C_2 and we will discard them at this point.

Along C_3 we have for $0 < \eta < a$,

$$z = \eta e^{-\pi i}, \quad z - a = (\eta + a)e^{-\pi i}, \quad z + a = (a - \eta), \qquad (4.1.70)$$

$$\sqrt{z^2 - a^2} = \sqrt{a^2 - \eta^2}\, e^{-\pi i/2}, \qquad (4.1.71)$$

and

$$z - N\sqrt{z^2 - a^2} = -\eta + iN\sqrt{a^2 - \eta^2} \qquad (4.1.72)$$

$$= -\eta + i\sqrt{r^2 - \eta^2}, \qquad (4.1.73)$$

where $r = \sqrt{(1 - N^2)(\eta^2 + b^2)}$. Consequently,

$$\left(z - N\sqrt{z^2 - a^2}\right)^{1/2} = \left(\frac{r - \eta}{2}\right)^{1/2} + i\left(\frac{r + \eta}{2}\right)^{1/2}. \qquad (4.1.74)$$

Similarly, along C_{10} for $0 < \eta < a$,

$$z = \eta e^{\pi i}, \quad z - a = (a + \eta)e^{\pi i}, \quad z + a = (a - \eta), \qquad (4.1.75)$$

$$\sqrt{z^2 - a^2} = \sqrt{a^2 - \eta^2}\, e^{\pi i/2}, \qquad (4.1.76)$$

and

$$z - N\sqrt{z^2 - a^2} = -\eta + iN\sqrt{a^2 - \eta^2} \qquad (4.1.77)$$

so that

$$\left(z - N\sqrt{z^2 - a^2}\right)^{1/2} = \left(\frac{r - \eta}{2}\right)^{1/2} - i\left(\frac{r + \eta}{2}\right)^{1/2}. \qquad (4.1.78)$$

Along C_6 with $0 < \eta < a$,

$$z = \eta, \quad z - a = (a - \eta)e^{-\pi i}, \quad z + a = (a + \eta), \qquad (4.1.79)$$

$$\sqrt{z^2 - a^2} = \sqrt{a^2 - \eta^2}\, e^{-\pi i/2}, \qquad (4.1.80)$$

and

$$z - N\sqrt{z^2 - a^2} = \eta + iN\sqrt{a^2 - \eta^2} \qquad (4.1.81)$$

so that

$$\left(z - N\sqrt{z^2 - a^2}\right)^{1/2} = \left(\frac{r+\eta}{2}\right)^{1/2} + i\left(\frac{r-\eta}{2}\right)^{1/2}. \qquad (4.1.82)$$

Along C_7 with $0 < \eta < a$,

$$z = \eta, \quad z - a = (a-\eta)e^{\pi i}, \quad z + a = (a+\eta), \qquad (4.1.83)$$

$$\sqrt{z^2 - a^2} = \sqrt{a^2 - \eta^2}\, e^{\pi i/2}, \qquad (4.1.84)$$

and

$$z - N\sqrt{z^2 - a^2} = \eta - iN\sqrt{a^2 - \eta^2} \qquad (4.1.85)$$

so that

$$\left(z - N\sqrt{z^2 - a^2}\right)^{1/2} = \left(\frac{r+\eta}{2}\right)^{1/2} - i\left(\frac{r-\eta}{2}\right)^{1/2}. \qquad (4.1.86)$$

Consequently, direct substitution leads to

$$\frac{1}{2\pi i}\int_{C_3 \cup C_6 \cup C_7 \cup C_{10}} \frac{\left(z - N\sqrt{z^2 - a^2}\right)^{1/2}}{\sqrt{z(z^2 - a^2)}}\, e^{tz}\, dz$$
$$= \frac{2}{\pi}\int_0^a \left(\frac{r+\eta}{2}\right)^{1/2} \frac{\cosh(t\eta)}{\sqrt{\eta(a^2 - \eta^2)}}\, d\eta. \qquad (4.1.87)$$

Finally, turning to the contours along the imaginary axis, we have for C_4 with $0 < \eta < b$,

$$z = \eta e^{-\pi i/2}, \quad z + a = \sqrt{a^2 + \eta^2}\, e^{-\theta i}, \quad z - a = \sqrt{a^2 + \eta^2}\, e^{-\pi i + \theta i}, \qquad (4.1.88)$$

where $\theta = \arctan(\eta/a)$. Therefore,

$$\sqrt{z^2 - a^2} = \sqrt{a^2 + \eta^2}e^{-\pi i/2}. \qquad (4.1.89)$$

For $z - N\sqrt{z^2 - a^2}$,

$$z + N\sqrt{z^2 - a^2} = \left(\eta + N\sqrt{\eta^2 + a^2}\right)e^{-\pi i/2}. \qquad (4.1.90)$$

Now

$$z - ib = (b+\eta)e^{3\pi i/2}, \qquad (4.1.91)$$

and

$$z + ib = (b-\eta)e^{\pi i/2} \qquad (4.1.92)$$

so that

$$z - N\sqrt{z^2 - a^2} = \frac{b^2 - \eta^2}{\eta + N\sqrt{a^2 + \eta^2}} e^{5\pi i/2}. \tag{4.1.93}$$

Along C_5 with $0 < \eta < b$,

$$z = \eta e^{-\pi i/2}, \quad \sqrt{z^2 - a^2} = \sqrt{a^2 + \eta^2}\, e^{-\pi i/2}, \tag{4.1.94}$$

$$z + ib = (b - \eta)e^{\pi i/2}, \quad z - ib = (b + \eta)e^{-\pi i/2}, \tag{4.1.95}$$

and

$$z - N(z^2 - a^2) = \frac{b^2 - \eta^2}{\eta + N\sqrt{a^2 + \eta^2}} e^{\pi i/2}. \tag{4.1.96}$$

Along C_8 with $0 < \eta < b$,

$$z = \eta e^{\pi i/2}, \quad z + a = \sqrt{\eta^2 + a^2}\, e^{\theta i}, \quad z - a = \sqrt{\eta^2 + a^2}\, e^{\pi i - \theta i}, \tag{4.1.97}$$

$$z + ib = (b + \eta)e^{\pi i/2}, \quad z - ib = (b - \eta)e^{-\pi i/2}, \tag{4.1.98}$$

and

$$z - N\sqrt{z^2 - a^2} = \frac{b^2 - \eta^2}{\eta + N\sqrt{a^2 + \eta^2}} e^{-\pi i/2}. \tag{4.1.99}$$

Along C_9 with $0 < \eta < b$,

$$z = \eta e^{\pi i/2}, \quad \sqrt{z^2 - a^2} = \sqrt{\eta^2 + a^2}\, e^{\pi i/2}, \tag{4.1.100}$$

$$z + ib = (b + \eta)e^{\pi i/2}, \quad z - ib = (b - \eta)e^{3\pi i/2}, \tag{4.1.101}$$

and

$$z - N\sqrt{z^2 - a^2} = \frac{b^2 - \eta^2}{\eta + N\sqrt{a^2 + \eta^2}} e^{3\pi i/2}. \tag{4.1.102}$$

Substitution of these results into the line integrals and simplification results in

$$\frac{1}{2\pi i} \int_{C_4 \cup C_5 \cup C_8 \cup C_9} \frac{\left(z - N\sqrt{z^2 - a^2}\right)^{1/2}}{\sqrt{z(z^2 - a^2)}} e^{tz}\, dz \tag{4.1.103}$$

$$= \frac{2}{\pi} \int_0^b \left(\frac{b^2 - \eta^2}{\eta + N\sqrt{a^2 + \eta^2}}\right)^{1/2} \frac{\cos(t\eta)}{\sqrt{\eta(a^2 + \eta^2)}}\, d\eta.$$

We omit the demonstration that the small circles around the branch points do not contribute to the inverse. The inverse is the sum of Equation 4.1.87 and Equation 4.1.103 and is illustrated in Figure 4.1.11.

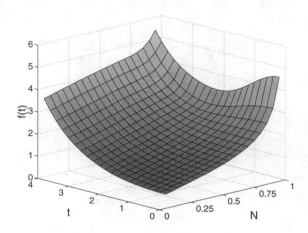

Figure 4.1.11: The inverse of the Laplace transform given by Equation 4.1.58 when $a = 1$.

• **Example 4.1.7**

We now employ the inversion formula on the transform[8]

$$F(s) = \frac{I_0(\zeta)}{\zeta I_0'(\zeta)} = \frac{I_0(\zeta)}{\zeta I_1(\zeta)}, \qquad (4.1.104)$$

where

$$\zeta = \sqrt{M^2(s + ik)^2 - s^2}, \qquad (4.1.105)$$

or

$$s = \frac{-ikM^2 \pm \sqrt{(M^2 - 1)\zeta^2 - M^2 k^2}}{M^2 - 1}, \qquad (4.1.106)$$

with $M > 1$ and $k \neq 0$. Upon converting the transform into a contour integration,

$$f(t) = \frac{1}{2\pi i} \oint_C \frac{I_0(\zeta)}{\zeta I_1(\zeta)} e^{tz} \, dz, \qquad (4.1.107)$$

where Figure 4.1.12 shows the contour C and ζ is now a complex variable. From the definition of ζ, the branch points are located at $z = s_1 i$ and $z = s_2 i$, where $s_1 = -kM/(M + 1)$ and $s_2 = -kM/(M - 1)$, with the branch cut running along the imaginary axis between these two points. Because $I_0'(z) = I_1(z) = -iJ_1(iz)$, an infinite number of simple poles are located at

$$z_{n1} = \frac{-ikM^2 + i\sqrt{(M^2 - 1)\alpha_n^2 + M^2 k^2}}{M^2 - 1}, \qquad n = 1, 2, 3, \ldots \qquad (4.1.108)$$

---

[8] Adapted from Widnall, S. E., and E. H. Dowell, 1967: Aerodynamic forces on an oscillating cylindrical duct with an internal flow. *J. Sound Vib.*, **6**, 71–85. Published by Academic Press Ltd., London, UK.

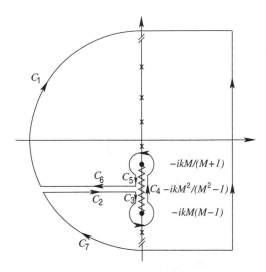

Figure 4.1.12: The contour used to invert the Laplace transform given by Equation 4.1.104.

above the branch point $z = s_1 i$, and

$$z_{n2} = \frac{-ikM^2 - i\sqrt{(M^2 - 1)\alpha_n^2 + M^2 k^2}}{M^2 - 1}, \quad n = 1, 2, 3, \ldots \qquad (4.1.109)$$

below the branch point $z = s_2 i$ with $J_1(\alpha_n) = 0$ except for $\alpha_0 = 0$.

From the asymptotic expansion for $I_0(z)$, the contribution from the line integrals on C_1 and C_7 vanish as we allow the semicircle to approach infinity. Furthermore, the integration along C_2 cancels the integration along C_6. Now, along C_4,

$$z = (s_1 - s_0 + s_0\eta)e^{\pi i/2}, \qquad (4.1.110)$$

$$z - is_1 = s_0(1 - \eta)e^{-\pi i/2}, \qquad (4.1.111)$$

$$z - is_2 = s_0(1 + \eta)e^{\pi i/2} \qquad (4.1.112)$$

and

$$\zeta = s_0\sqrt{(M^2 - 1)(1 - \eta^2)} = x, \qquad (4.1.113)$$

where $s_0 = kM/(M^2 - 1)$ and $-1 < \eta < 1$. The argument of $z - is_1$ and $z - is_2$ lies between $-\pi/2$ and $3\pi/2$. Therefore,

$$\int_{C_4} F(z)e^{tz}\, dz = s_0 i \int_{-1}^{1} \frac{I_0(x)\exp[i(s_1 - s_0 + s_0\eta)]}{xI_1(x)}\, d\eta. \qquad (4.1.114)$$

Along both C_3 and C_5,

$$z = (s_1 - s_0 + s_0\eta)e^{\pi i/2}, \qquad (4.1.115)$$

$$z - is_1 = s_0(1 - \eta)e^{3\pi i/2}, \tag{4.1.116}$$

$$z - is_2 = s_0(1 + \eta)e^{\pi i/2} \tag{4.1.117}$$

and

$$\zeta = -s_0\sqrt{(M^2 - 1)(1 - \eta^2)} = -x \tag{4.1.118}$$

so that

$$\int_{C_3 \cup C_5} F(z)e^{tz}\, dz = s_0 i \int_1^{-1} \frac{I_0(-x)\exp[i(s_1 - s_0 + s_0\eta)]}{(-x)I_1(-x)}\, d\eta \tag{4.1.119}$$

$$= -s_0 i \int_{-1}^1 \frac{I_0(x)\exp[i(s_1 - s_0 + s_0\eta)]}{xI_1(x)}\, d\eta. \tag{4.1.120}$$

Consequently,

$$\int_{C_3 \cup C_4 \cup C_5} F(z)e^{tz}\, dz = 0. \tag{4.1.121}$$

Finally, we must find the contribution from integrating around the branch points $z = is_1$ and $z = is_2$. For the integration around s_1,

$$z = is_1 + \epsilon e^{\theta i} \tag{4.1.122}$$

and

$$\zeta^2 = 2ikM\epsilon e^{\theta i} + (M^2 - 1)\epsilon^2 e^{2\theta i}, \tag{4.1.123}$$

where $-\pi/2 \le \theta \le 3\pi/2$. Consequently, the integral around the branch cut is

$$\int_{C_{\epsilon_1}} F(z)e^{tz}\, dz = \lim_{\epsilon \to 0} 2 \int_{-\pi/2}^{3\pi/2} \frac{\exp\left[-ikMt/(M+1) + \epsilon e^{\theta i}t\right]}{2ikM\epsilon e^{\theta i} + (M^2 - 1)\epsilon^2 e^{2\theta i}} i\epsilon e^{\theta i}\, d\theta \tag{4.1.124}$$

$$= \frac{\exp[-ikMt/(M+1)]}{kM} \int_{-\pi/2}^{3\pi/2} d\theta \tag{4.1.125}$$

$$= 2\pi \frac{\exp[-ikMt/(M+1)]}{kM}, \tag{4.1.126}$$

because

$$\frac{I_0(\zeta)}{\zeta I_1(\zeta)} \simeq \frac{2}{\zeta^2} \tag{4.1.127}$$

as $\zeta \to 0$.

At the other branch point,

$$z = is_2 + \epsilon e^{\theta i} \tag{4.1.128}$$

and

$$\zeta^2 = -2ikM\epsilon e^{\theta i} + (M^2 - 1)\epsilon^2 e^{2\theta i} \tag{4.1.129}$$

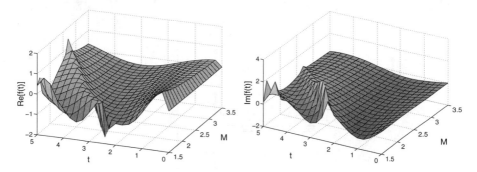

Figure 4.1.13: The inverse of the Laplace transform given by Equation 4.1.104.

with $-3\pi/2 < \theta < \pi/2$. Therefore, the integral around this point is

$$\int_{C_{\epsilon_2}} F(z)e^{tz}\, dz = \lim_{\epsilon \to 0} \int_{-3\pi/2}^{\pi/2} \frac{2\exp[-ikMt/(M-1) + \epsilon e^{\theta i}t]}{-2ikM\epsilon e^{\theta i} + (M^2-1)\epsilon^2 e^{2\theta i}} i\epsilon e^{\theta i}\, d\theta \tag{4.1.130}$$

$$= -2\pi \frac{\exp[-ikMt/(M-1)]}{kM}. \tag{4.1.131}$$

Applying the residue theorem,

$$f(t) = \sum_{n=1}^{\infty} \frac{I_0(\zeta_{n1})\exp(z_{n1}t)}{\frac{d}{d\zeta}[\zeta I_1(\zeta)]\frac{d\zeta}{dz}|_{\zeta=\zeta_{n1},z=z_{n1}}} + \sum_{n=1}^{\infty} \frac{I_0(\zeta_{n2})\exp(z_{n2}t)}{\frac{d}{d\zeta}[\zeta I_1(\zeta)]\frac{d\zeta}{dz}|_{\zeta=\zeta_{n2},z=z_{n2}}}$$
$$- \frac{i}{kM}\left[\exp\left(\frac{-ikMt}{M+1}\right) - \exp\left(\frac{-ikMt}{M-1}\right)\right], \tag{4.1.132}$$

or

$$f(t) = 2\exp[-ikM^2t/(M^2-1)] \sum_{n=1}^{\infty} \frac{\sin\left[t\sqrt{(M^2-1)\alpha_n^2 + M^2k^2}/(M^2-1)\right]}{\sqrt{(M^2-1)\alpha_n^2 + M^2k^2}}$$
$$- \frac{i}{kM}\left[\exp\left(\frac{-ikMt}{M+1}\right) - \exp\left(\frac{-ikMt}{M-1}\right)\right], \tag{4.1.133}$$

because $I_1'(\zeta_{ni}) = [I_0(\zeta_{ni}) + I_2(\zeta_{ni})]/2 = I_0(\zeta_{ni})$ since $I_0(\zeta_{ni}) - I_2(\zeta_{ni}) = 2I_1(\zeta_{ni})/\zeta_{ni} = 0$. This inverse is illustrated in Figure 4.1.13.

• **Example 4.1.8**

In this example, we invert the transform[9]

$$F(s) = \frac{s}{s^2 + k^2} e^{-\mu(s,k)z}, \tag{4.1.134}$$

[9] Taken from Cole, J. D., and C. Greifinger, 1969: Acoustic-gravity waves from an energy source at the ground in an isothermal atmosphere. *J. Geophys. Res.*, **74**, 3693–3703. ©1969 American Geophysical Union. Reproduced/modified by permission of American Geophysical Union.

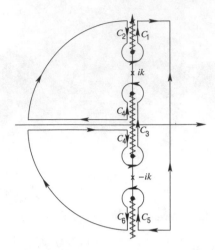

Figure 4.1.14: The contour used to invert the Laplace transform given by Equation 4.1.134.

where

$$\mu(s,k) = \frac{1}{s}\left[s^4 + \left(\frac{1}{4} + k^2\right)s^2 + \beta^2 k^2 \right]^{1/2}, \qquad 0 < \mathrm{Re}(\mu), \qquad (4.1.135)$$

and k and β are real, $0 < \beta < 1/2$, and $0 < z$. For convenience, we can rewrite $\mu(s,k)$ as

$$\mu(s,k) = \frac{1}{s}\left\{ \left[s^2 + \omega_1^2(k) \right]\left[s^2 + \omega_2^2(k) \right] \right\}^{1/2}, \qquad (4.1.136)$$

where

$$\omega_{1,2}(k) = \frac{1}{2}\left[\left(\tfrac{1}{4} + k^2 + 2\beta k\right)^{1/2} \mp \left(\tfrac{1}{4} + k^2 - 2\beta k\right)^{1/2} \right]. \qquad (4.1.137)$$

The point of interest in this example is the presence of an essential singularity at $s = 0$.

Figure 4.1.14 illustrates the contour integration used to invert the Laplace transform given by Equation 4.1.134. Note that this contour excludes the essential singularity and, therefore, avoids the necessity of computing its residue. The arguments for the various functions are

$$s + i\omega_2 = re^{\theta i}, \qquad -\pi/2 < \theta < 3\pi/2, \qquad (4.1.138)$$
$$s - i\omega_2 = re^{\theta i}, \qquad -3\pi/2 < \theta < \pi/2, \qquad (4.1.139)$$
$$s + i\omega_1 = re^{\theta i}, \qquad -3\pi/2 < \theta < \pi/2, \qquad (4.1.140)$$
$$s - i\omega_1 = re^{\theta i}, \qquad -\pi/2 < \theta < 3\pi/2. \qquad (4.1.141)$$

The contribution from the poles at $z = \pm ik$ is

$$\text{Res}\left[F(z)e^{tz}; \pm ik\right] = \tfrac{1}{2}e^{-z\sqrt{\frac{1}{4}-\beta^2}}e^{\pm ikt}, \tag{4.1.142}$$

or

$$\text{Res}\left[F(z)e^{tz}; ik\right] + \text{Res}\left[F(z)e^{tz}; -ik\right] = e^{-z\sqrt{\frac{1}{4}-\beta^2}}\cos(kt). \tag{4.1.143}$$

From the branch cut integrals,

$$\int_{C_1} F(z)e^{tz}\,dz = \frac{1}{2\pi i}\int_{\omega_2}^{\infty} e^{i(\omega t - \mu_1 z)}\frac{\omega}{\omega^2 - k^2}\,d\omega, \tag{4.1.144}$$

and

$$\int_{C_2} F(z)e^{tz}\,dz = \frac{1}{2\pi i}\int_{\infty}^{\omega_2} e^{i(\omega t + \mu_1 z)}\frac{\omega}{\omega^2 - k^2}\,d\omega, \tag{4.1.145}$$

where

$$\mu_1 = \frac{1}{\omega}\sqrt{\omega^4 - \left(\frac{1}{4} + k^2\right)\omega^2 + \beta^2 k^2}. \tag{4.1.146}$$

Next,

$$\int_{C_5} F(z)e^{tz}\,dz = \frac{1}{2\pi i}\int_{-\infty}^{-\omega_2} e^{i(\omega t + \mu_5 z)}\frac{\omega}{\omega^2 - k^2}\,d\omega, \tag{4.1.147}$$

and

$$\int_{C_6} F(z)e^{tz}\,dz = \frac{1}{2\pi i}\int_{-\omega_2}^{-\infty} e^{i(\omega t - \mu_5 z)}\frac{\omega}{\omega^2 - k^2}\,d\omega, \tag{4.1.148}$$

where $\mu_5 = -\mu_1$. Therefore,

$$\int_{C_1 \cup C_2 \cup C_5 \cup C_6} F(z)e^{tz}\,dz = \frac{1}{2\pi i}\int_{-\infty}^{\infty} \frac{\omega}{\omega^2 - k^2}\left[e^{i(\omega t - \mu_1 z)} - e^{i(\omega t + \mu_1 z)}\right]d\omega, \tag{4.1.149}$$

where we only include the frequencies $|\omega| > \omega_2$. In a similar manner,

$$\int_{C_3 \cup C_4} F(z)e^{tz}\,dz = \frac{1}{2\pi i}\int_{-\omega_1}^{\omega_1} \frac{\omega}{k^2 - \omega^2}\left[e^{i(\omega t - \mu_1 z)} - e^{i(\omega t + \mu_1 z)}\right]d\omega. \tag{4.1.150}$$

Therefore,

$$\int_{C_1 \cup \ldots \cup C_6} F(z)e^{tz}\,dz = \frac{1}{2\pi i}\int_{-\infty}^{\infty} \frac{\omega}{|\omega^2 - k^2|}\left[e^{i(\omega t - \mu_1 z)} - e^{i(\omega t + \mu_1 z)}\right]d\omega, \tag{4.1.151}$$

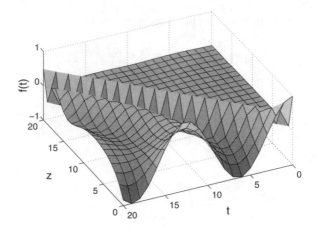

z

t

Figure 4.1.15: The inverse of the Laplace transform given by Equation 4.1.134.

where we only include those values of ω such that $\omega_2 < |\omega|$ or $|\omega| < \omega_1$. The inverse equals the sum of Equation 4.1.143 and Equation 4.1.151 and is illustrated in Figure 4.1.15 when $k = 0.5$ and $\beta = 0.25$.

- **Example 4.1.9**

Let us now invert the Laplace transform[10]

$$F(s) = \frac{\gamma_3 e^{-\gamma_3 x} - \gamma_4 e^{-\gamma_4 x}}{(\gamma_3 - \gamma_4)[c_3^2(\gamma_3^2 + \gamma_4^2 + \gamma_3\gamma_4) - s^2]}, \qquad 0 < x, \qquad (4.1.152)$$

where

$$\gamma_{3,4} = \frac{1}{c_3 c_4 \sqrt{2}} \sqrt{s^2(c_3^2 + c_4^2) + c_5^2 \nu^2 \pm A}, \qquad (4.1.153)$$

$$A^2 = [s^2(c_3^2 + c_4^2) + c_5^2 \nu^2]^2 - 4c_3^2 c_4^2 s^2 (s^2 + \nu^2) \qquad (4.1.154)$$

$$= \left[s^2(c_3^2 - c_4^2) + \frac{\nu^2(c_5^2 c_3^2 + c_5^2 c_4^2 - 2c_3^2 c_4^2)}{c_3^2 - c_4^2} \right]^2 + \frac{4\nu^4 c_3^2 c_4^2 (c_3^2 - c_5^2)(c_5^2 - c_4^2)}{(c_3^2 - c_4^2)^2},$$

$$(4.1.155)$$

ν is a real, γ_3 and γ_4 correspond to the upper and lower signs, respectively, $c_4 < c_5 < c_3$ and we use the positive square root of A^2 for A. Under these conditions, $\gamma_{3,4}$ have branch points at $s = \pm i\nu$ and $s = R[\pm\cos(\varphi) \pm i\sin(\varphi)]$, where

$$R = \frac{\nu c_5}{\sqrt{c_3^2 - c_4^2}} \qquad (4.1.156)$$

---

[10] Taken from Eason, G., 1970: The displacement produced in a semi-infinite linear Cosserat continuum by an impulsive force. *Proc. Vib. Prob.*, **11**, 199–220.

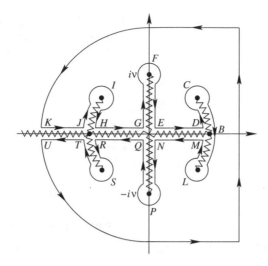

Figure 4.1.16: The contour used to evaluate the Laplace transform given by Equation 4.1.152.

and

$$\varphi = \arctan\left(\frac{c_3\sqrt{c_5^2 - c_4^2}}{c_4\sqrt{c_3^2 - c_5^2}}\right). \tag{4.1.157}$$

Figure 4.1.16 shows our choices for the branch cuts.

In addition to the branch points, there may be poles, if the denominator on the right side of Equation 4.1.152 vanishes. One case would be $\gamma_3 = \gamma_4$; however, further analysis shows that this does not occur. The other possibility would be

$$G(s) = c_3^2(\gamma_3^2 + \gamma_4^2 + \gamma_3\gamma_4) - s^2 \tag{4.1.158}$$

$$= \left(s^2 c_3^2 + \nu^2 c_5^2 + c_3 c_4 s\sqrt{s^2 + \nu^2}\right)/c_4^2 = 0. \tag{4.1.159}$$

With the branch cut for $\sqrt{s^2 + \nu^2}$ lying along the imaginary axis, $G(s)$ has no zeros on either the real or imaginary axes. Finally, because $\text{Im}[G(s)] > 0$ in the first and third quadrants and < 0 in the second and fourth quadrants, $G(s)$ has no zeros at all.

To find $f(t)$, we write the inversion as the contour integral

$$f(t) = \frac{1}{2\pi i}\int_{c-\infty i}^{c+\infty i} \frac{\gamma_3 e^{ts - x\gamma_3} - \gamma_4 e^{ts - x\gamma_4}}{(\gamma_3 - \gamma_4)[c_3^2(\gamma_3^2 + \gamma_4^2 + \gamma_3\gamma_4) - s^2]}\, ds. \tag{4.1.160}$$

To close the contour in the left half-plane, we must examine γ_3 and γ_4 for large s. Because $c_3 > c_4$, γ_4 behaves as s/c_3 for large s. We must close the $e^{st - \gamma_4 x}$ portion of the line integral in Equation 4.1.160 in the right half-plane when $c_3 t < x$ and in the left half-plane when $c_3 t > x$ according to Jordan's lemma.

Because there are no singularities in the right half-plane, the contribution from this term is zero until $c_3 t > x$. Similar considerations lead to the conclusion that the contribution from $e^{st-\gamma_3 x}$ term will be zero until $t > x/c_4$. When we close the contour in the left half-plane, the original Bromwich contour equals the negative of integral along the contour $KJIHGFEDCBLMNPQRSTU$. Table 4.1.1 gives the various values of s, γ_3 and γ_4 along various segments of Figure 4.1.16.

As stated earlier, prior to the arrival of the first wave at time $t = x/c_3$, $f(t) = 0$. For $x/c_3 < t < x/c_4$, there are contributions from cuts CL, FP and IS. Performing the integrations along each of the contours gives

$$
\begin{aligned}
f(t) = I_2 + \int_0^\varphi \frac{R\chi_2}{E^2 + F^2} &\Big\{ \cos[Rt\sin(\theta) - x\chi_2\sin(\Theta_2)] \\
&\times \sinh[Rt\cos(\theta) - x\chi_2\cos(\Theta_2)][E\cos(\Theta_2 + \theta) + F\sin(\Theta_2 + \theta)] \\
&- \sin[Rt\sin(\theta) - x\chi_2\sin(\Theta_2)] \\
&\times \cosh[Rt\cos(\theta) - x\chi_2\cos(\Theta_2)][E\sin(\Theta_2 + \theta) - F\cos(\Theta_2 + \theta)] \Big\}\, d\theta \\
+ \int_0^\varphi \frac{R\chi_1}{E^2 + F^2} &\Big\{ \cos[Rt\sin(\theta) - x\chi_1\sin(\Theta_1)] \\
&\times \sinh[Rt\cos(\theta) - x\chi_1\cos(\Theta_1)][E\cos(\Theta_1 + \theta) + F\sin(\Theta_1 + \theta)] \\
&- \sin[Rt\sin(\theta) - x\chi_1\sin(\Theta_1)] \\
&\times \cosh[Rt\cos(\theta) - x\chi_1\cos(\Theta_1)][E\sin(\Theta_1 + \theta) - F\cos(\Theta_1 + \theta)] \Big\}\, d\theta,
\end{aligned}
$$

$$(4.1.161)$$

where

$$
I_2 = -\int_0^\nu \frac{\delta_1\delta_2(r^2 - c_3^2\delta_1^2)\sin(rt)\cosh(\delta_2 x)}{\delta_2^2(r^2 + c_3^2\delta_2^2)^2 + \delta_1^2(r^2 - c_3^2\delta_1^2)^2}\, dr
$$
$$
- \int_0^\nu \frac{\delta_2^2(r^2 - c_3^2\delta_2^2)\cos(rt)\sinh(\delta_2 x)}{\delta_2^2(r^2 + c_3^2\delta_2^2)^2 + \delta_1^2(r^2 - c_3^2\delta_1^2)^2}\, dr, \qquad (4.1.162)
$$

$$
E = -c_3^2\chi_1^3\cos(3\Theta_1) + c_3^2\chi_2^3\cos(3\Theta_2) + R^2\chi_1\cos(2\theta + \Theta_1) - R^2\chi_2\cos(2\theta + \Theta_2) \tag{4.1.163}
$$

and

$$
F = -c_3^2\chi_1^3\sin(3\Theta_1) + c_3^2\chi_2^3\sin(3\Theta_2) + R^2\chi_1\sin(2\theta + \Theta_1) - R^2\chi_2\sin(2\theta + \Theta_2). \tag{4.1.164}
$$

For $t > x/c_4$, there are contributions from both the $e^{\gamma_3 x}$ and $e^{\gamma_4 x}$ terms. Because of symmetries present in the values of γ_3 and γ_4, the contribution from cuts CL and IL cancel out and there is only the contribution from cut FP. This yields

$$
f(t) = I_2 + \int_0^\nu \frac{\delta_1\delta_2(r^2 + c_3^2\delta_2^2)\sin(rt - \delta_1 x)}{\delta_2^2(r^2 + c_3^2\delta_2^2)^2 + \delta_1^2(r^2 - c_3^2\delta_1^2)^2}\, dr. \tag{4.1.165}
$$

Table 4.1.1: The Values of γ_3 and γ_4 Along Various Segments of Figure 4.1.16, Where $0 < \theta < \varphi$.

Segment	s	γ_3	γ_4
AC	r	η_1	η_2
BC	$Re^{\theta i}$	$\chi_1 e^{\Theta_1 i}$	$\chi_2 e^{\Theta_2 i}$
CD	$Re^{\theta i}$	$\chi_2 e^{\Theta_2 i}$	$\chi_1 e^{\Theta_1 i}$
BL	$Re^{-\theta i}$	$\chi_1 e^{-\Theta_1 i}$	$\chi_2 e^{-\Theta_2 i}$
LM	$Re^{-\theta i}$	$\chi_2 e^{-\Theta_2 i}$	$\chi_1 e^{-\Theta_1 i}$
DE and MN	$s = r$	η_2	η_1
EF	ri	$\delta_1 i$	δ_2
FG	ri	$\delta_1 i$	$-\delta_2$
NP	$-ri$	$-\delta_1 i$	δ_2
PQ	$-ri$	$-\delta_1 i$	$-\delta_2$
GH and QR	$-r$	$-\eta_2$	$-\eta_1$
HI	$-Re^{-\theta i}$	$-\chi_2 e^{-\Theta_2 i}$	$-\chi_1 e^{-\Theta_1 i}$
IJ	$-Re^{-\theta i}$	$-\chi_1 e^{-\Theta_1 i}$	$-\chi_2 e^{-\Theta_2 i}$
RS	$-Re^{\theta i}$	$-\chi_2 e^{\Theta_1 i}$	$-\chi_1 e^{\Theta_2 i}$
ST	$-Re^{\theta i}$	$-\chi_1 e^{\Theta_1 i}$	$-\chi_2 e^{\Theta_2 i}$
JK and TU	$s = -r$	$-\eta_1$	$-\eta_2$

$$\delta_1 = \frac{1}{c_3 c_4 \sqrt{2}} \sqrt{r^2(c_3^2 + c_4^2) - \nu^2 c_5^2 + |c_3^2 - c_4^2|D},$$

$$\delta_2 = \frac{1}{c_3 c_4 \sqrt{2}} \sqrt{|c_3^2 - c_4^2|D - r^2(c_3^2 + c_4^2) + \nu^2 c_5^2},$$

$$\eta_{1,2} = \frac{1}{c_3 c_4 \sqrt{2}} \sqrt{J \pm (c_3^2 - c_4^2)\sqrt{r^4 + R^4 - 2R^2 r^2 \cos(2\varphi)}},$$

$$\chi_{1,2} = \frac{1}{c_3 c_4 \sqrt{2}} \left\{ \Omega_1^2 + [(c_3^2 + c_4^2)R^2 \sin(2\theta) \mp 2R^2(c_3^2 - c_4^2)B]^2 \right\}^{1/4},$$

$$\Omega_{1,2} = \nu^2 c_5^2 + (c_3^2 + c_4^2)R^2 \cos(2\theta) \pm 2R^2(c_3^2 - c_4^2)C,$$

$$\Theta_i = \arctan\left[\sqrt{\frac{2c_3^2 c_4^2 \chi_i^2 - \Omega_i}{2c_3^2 c_4^2 \chi_i^2 + \Omega_i}} \right], \qquad i = 1, 2,$$

$$B = \sin(\theta)\sqrt{\cos^2(\theta) - \cos^2(\varphi)}, \qquad C = \cos(\theta)\sqrt{\cos^2(\theta) - \cos^2(\varphi)},$$

$$D = \sqrt{4r^2 k_1^2 + (k_1^2 - k_2^2 - r^2)^2}, \qquad J = \nu^2 c_5^2 + r^2(c_3^2 + c_4^2),$$

$$k_1 = \frac{\nu c_4 \sqrt{c_3^2 - c_5^2}}{|c_3^2 - c_4^2|}, \qquad k_2 = \frac{\nu c_3 \sqrt{c_4^2 - c_5^2}}{|c_3^2 - c_4^2|}$$

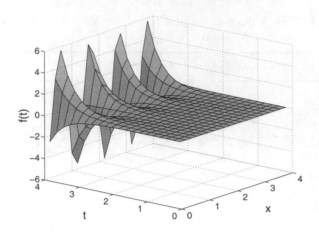

Figure 4.1.17: The inverse of the Laplace transform given by Equation 4.1.152 when $c_4 = 1$, $c_5 = 2$, $c_3 = 3$ and $\nu = 1$.

This inverse is illustrated in Figure 4.1.17.

• **Example 4.1.10**

During a study of solute transport from an injection well into an aquifer, C.-S. Chen[11] employed a contour integration to invert the Laplace transform

$$F(s) = \frac{\text{Ai}(\lambda)}{s[\text{Ai}(\lambda_0)/2 - s^{1/3}\text{Ai}'(\lambda_0)]}, \qquad a < r < \infty, \qquad (4.1.166)$$

where $\lambda = (1 + 4rs)/(4s^{2/3})$, $\lambda_0 = (1 + 4as)/(4s^{2/3})$, and $\text{Ai}(\cdot)$ is the Airy function of the first kind.

We begin by closing the Bromwich integral as shown in Figure 4.1.3 where the branch cut for $s^{1/3}$ lies along the negative real axis of the s-plane. By Jordan's lemma, the contribution from the arcs at infinity are zero. Consequently, the inversion consists of two possible parts. First, we have terms arising from the poles of $F(s)$. If we denote each pole by s_n, then s_n is a solution of

$$\text{Ai}(\lambda_n)/2 - s_n^{1/3}\text{Ai}'(\lambda_n) = 0, \qquad (4.1.167)$$

where $\lambda_n = (1 + 4as_n)/(4s_n^{2/3})$. Note that in the present case, $s = 0$ is outside the contour because it is a branch point of $s^{1/3}$. Upon differentiating Equation 4.1.167 with respect to a and using the differential equation that defines the Airy function, we can rewrite Equation 4.1.167

$$(1/4 - s_n^{2/3}\lambda_n)\text{Ai}(\lambda_n) = 0. \qquad (4.1.168)$$

[11] Taken from Chen, C.-S., 1987: Analytical solutions for radial dispersion with Cauchy boundary at injection well. *Water Resour. Res.*, **23**, 1217–1224. ©1987 American Geophysical Union. Reproduced/modified by permission of American Geophysical Union.

Because $\text{Ai}(\lambda_n)$ does not contain any zeros within the closed contour,

$$\frac{1}{4} = s_n^{2/3}\lambda_n = \frac{1 + 4as_n}{4}, \quad \text{or} \quad s_n = 0, \tag{4.1.169}$$

which is outside of our closed contour. Consequently, the only contribution to the inversion will come from the branch cut integral.

Turning to these integrals, we have along CD that $s = \eta^2 e^{\pi i}$. Thus,

$$\frac{1}{2\pi i}\int_{CD} F(z)e^{tz}\,dz \tag{4.1.170}$$

$$= \frac{1}{\pi i}\int_\infty^0 \frac{e^{-t\eta^2}}{\eta}\,d\eta\left\{\frac{\text{Ai}[\chi(\eta)e^{-2\pi i/3}]}{\text{Ai}[\chi_0(\eta)e^{-2\pi i/3}] - \eta^{2/3}e^{\pi i/3}\,\text{Ai}'[\chi_0(\eta)e^{-2\pi i/3}]}\right\},$$

where $\chi_0(\eta) = (1-4a\eta^2)/(4\eta^{4/3})$ and $\chi(\eta) = (1-4r\eta^2)/(4\eta^{4/3})$. Substituting the relationships that

$$\text{Ai}(\eta e^{\pm 2\pi i/3}) = e^{\pm\pi i/3}[\text{Ai}(\eta) \mp i\,\text{Bi}(\eta)]/2 \tag{4.1.171}$$

and

$$\text{Ai}'(\eta e^{\pm 2\pi i/3}) = e^{\mp\pi i/3}[\text{Ai}'(\eta) \mp i\,\text{Bi}'(\eta)]/2 \tag{4.1.172}$$

into Equation 4.1.170 and simplifying,

$$\frac{1}{2\pi i}\int_{CD} F(z)e^{tz}\,dz \tag{4.1.173}$$

$$= -\frac{1}{\pi i}\int_0^\infty e^{-t\eta^2}\frac{d\eta}{\eta}\left\{\frac{\text{Ai}(\chi) + i\,\text{Bi}(\chi)}{[\text{Ai}(\chi_0) + i\,\text{Bi}(\chi_0)]/2 + \eta^{2/3}[\text{Ai}'(\chi_0) + i\,\text{Bi}'(\chi_0)]}\right\},$$

where $\text{Bi}(\cdot)$ denotes the Airy function of the second kind. Similarly, along FG, $s = \eta^2 e^{-\pi i}$ and

$$\frac{1}{2\pi i}\int_{FG} F(z)e^{tz}\,dz \tag{4.1.174}$$

$$= \frac{1}{\pi i}\int_0^\infty e^{-t\eta^2}\frac{d\eta}{\eta}\left\{\frac{\text{Ai}(\chi) - i\,\text{Bi}(\chi)}{[\text{Ai}(\chi_0) - i\,\text{Bi}(\chi_0)]/2 + \eta^{2/3}[\text{Ai}'(\chi_0) - i\,\text{Bi}'(\chi_0)]}\right\}.$$

Thus, the integrals along the branch cuts yield

$$\frac{1}{2\pi i}\int_{CD\cup FG} F(z)e^{tz}\,dz = \frac{4}{\pi}\int_0^\infty e^{-t\eta^2}\left[\frac{\text{Ai}(\chi)f_1 - \text{Bi}(\chi)f_2}{f_1^2 + f_2^2}\right]\frac{d\eta}{\eta}, \tag{4.1.175}$$

where

$$f_1 = \text{Bi}(\chi_0) + 2\eta^{2/3}\text{Bi}'(\chi_0) \tag{4.1.176}$$

and

$$f_2 = \text{Ai}(\chi_0) + 2\eta^{2/3}\text{Ai}'(\chi_0). \tag{4.1.177}$$

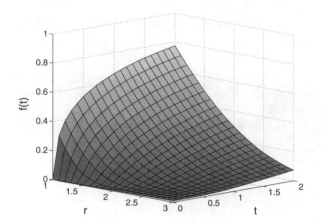

Figure 4.1.18: The inverse of the Laplace transform given by Equation 4.1.166 when $a = 1$.

Finally, we must deal with the integration around the branch point $z = 0$:

$$\frac{1}{2\pi i} \int_{DEF} F(z)e^{tz} \, dz = \frac{1}{2\pi i} \int_{\pi}^{-\pi} \left\{ \frac{\text{Ai}(\xi)}{\text{Ai}(\xi_0)/2 - [\epsilon \exp(\theta i)]^{1/3}\text{Ai}'(\xi_0)} \right\} i \, d\theta, \tag{4.1.178}$$

$$= \frac{1}{2\pi i} \int_{\pi}^{-\pi} e^{(a-r)/2} i \, d\theta = -e^{(a-r)/2}, \tag{4.1.179}$$

where $\xi = (1 + 4r\epsilon e^{\theta i})/[4(\epsilon e^{\theta i})^{2/3}]$ and $\xi_0 = (1 + 4a\epsilon e^{\theta i})/[4(\epsilon e^{\theta i})^{2/3}]$. Therefore,

$$f(t) = e^{(a-r)/2} - \frac{4}{\pi} \int_0^{\infty} e^{-t\eta^2} \left[\frac{\text{Ai}(\chi)f_1 - \text{Bi}(\chi)f_2}{f_1^2 + f_2^2} \right] \frac{d\eta}{\eta}. \tag{4.1.180}$$

This inverse is illustrated in Figure 4.1.18 when $a = 1$.

- **Example 4.1.11**

Let us invert the Laplace transform[12]

$$F(s) = \frac{\exp\left(-x\sqrt{k^2 + s^2/c^2}\right)}{k^2 + s^2/c^2}, \qquad 0 < c, k, x, \tag{4.1.181}$$

by contour integration. If $t > x/c$, Jordan's lemma allows us to close Bromwich's contour as shown in Figure 4.1.19. Thus, if C denotes the contour $ABCDEF$ shown in Figure 4.1.19, then

$$f(t) = \frac{c^2}{2\pi i} \int_C \frac{\exp\left[tz - (x/c)\sqrt{z^2 + k^2 c^2}\right]}{z^2 + k^2 c^2} \, dz, \tag{4.1.182}$$

---

[12] Taken from Watanabe, K., 1981: Transient response of an inhomogeneous elastic half space to a torsional load. *Bull. JSME*, **24**, 1537–1542.

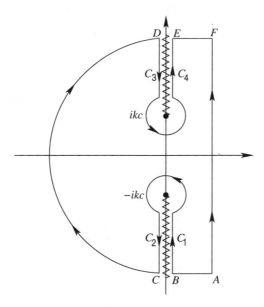

Figure 4.1.19: Contour used in the integration of Equation 4.1.182.

if $t > x/c$; if $t < x/c$, $f(t) = 0$.

Our first task is to compute the phases of the various functions along the four legs that comprise the branch cut integration; the contribution from the integrations along the arcs at infinity vanish. Along C_1,

$$z - kci = kc(u+1)e^{-\pi i/2}, \ z + kci = kc(u-1)e^{-\pi i/2}, \ 1 < u < \infty, \quad (\textbf{4.1.183})$$

so that

$$\sqrt{z^2 + k^2c^2} = kc\sqrt{u^2 - 1} \ e^{-\pi i/2}. \quad (\textbf{4.1.184})$$

Along C_2,

$$z - kci = kc(u-1)e^{-\pi i/2}, \ z + kci = kc(u+1)e^{3\pi i/2}, \ 1 < u < \infty, \quad (\textbf{4.1.185})$$

and

$$\sqrt{z^2 + k^2c^2} = kc\sqrt{u^2 - 1} \ e^{\pi i/2}. \quad (\textbf{4.1.186})$$

Next, along C_3,

$$z - kci = kc(u-1)e^{-3\pi i/2}, \ z + kci = kc(u+1)e^{\pi i/2}, \ 1 < u < \infty, \quad (\textbf{4.1.187})$$

and

$$\sqrt{z^2 + k^2c^2} = kc\sqrt{u^2 - 1} \ e^{-\pi i/2}. \quad (\textbf{4.1.188})$$

Finally, along C_4,

$$z - kci = kc(u-1)e^{\pi i/2}, \ z + kci = kc(u+1)e^{\pi i/2}, \ 1 < u < \infty, \quad (\textbf{4.1.189})$$

and

$$\sqrt{z^2 + k^2 c^2} = kc\sqrt{u^2 - 1}\; e^{\pi i/2}. \qquad (4.1.190)$$

Substituting these definitions into the contour integration,

$$f(t) = \frac{c}{k}\sin(ckt) + \frac{c}{2\pi ki}\int_\infty^1 \frac{\exp\left(-cktui + ikx\sqrt{u^2-1}\right)}{1-u^2}(-i\,du)$$

$$+ \frac{c}{2\pi ki}\int_1^\infty \frac{\exp\left(-cktui - ikx\sqrt{u^2-1}\right)}{1-u^2}(-i\,du)$$

$$+ \frac{c}{2\pi ki}\int_\infty^1 \frac{\exp\left(cktui + ikx\sqrt{u^2-1}\right)}{1-u^2}(i\,du)$$

$$+ \frac{c}{2\pi ki}\int_1^\infty \frac{\exp\left(cktui - ikx\sqrt{u^2-1}\right)}{1-u^2}(i\,du), \qquad (4.1.191)$$

where the first term in Equation 4.1.191 arises from the integration around the branch points $z = \pm kci + \epsilon e^{\theta i}$. Combining the exponentials,

$$f(t) = \frac{c}{k}\sin(ckt) - \frac{2c}{k\pi}\int_1^\infty \frac{\sin\left(kx\sqrt{u^2-1}\right)}{u^2-1}\sin(cktu)\,du \qquad (4.1.192)$$

$$= \frac{c}{k}\sin(ckt) - \frac{2c}{k\pi}\int_0^\infty \frac{\sin\left(ckt\sqrt{v^2+1}\right)}{v\sqrt{v^2+1}}\sin(xkv)\,dv \qquad (4.1.193)$$

$$= \frac{c}{k}\sin(ckt) - \frac{2c}{\pi}\int_0^x\left[\int_0^\infty \frac{\sin\left(ckt\sqrt{v^2+1}\right)}{v\sqrt{v^2+1}}\cos(\eta kv)\,dv\right]d\eta \qquad (4.1.194)$$

if $u^2 = v^2 + 1$. To obtain Equation 4.1.194 we used the relationship that

$$\sin(xkv) = kv\int_0^x \cos(kv\eta)\,d\eta. \qquad (4.1.195)$$

We have done this because we can evaluate the integral inside the square brackets exactly. From a table of integrals,

$$\int_0^\infty \frac{\sin(b\sqrt{a^2+x^2}\,)\cos(xy)}{\sqrt{a^2+x^2}}\,dx = \frac{\pi}{2}H(b-y)J_0\left(a\sqrt{b^2-y^2}\right), \qquad (4.1.196)$$

so that

$$f(t) = \frac{c}{k}\sin(kct) - c\int_0^x J_0\left(k\sqrt{c^2t^2-\eta^2}\right)\,d\eta. \qquad (4.1.197)$$

Finally, because

$$\sin(x) = \int_0^x J_0\left(\sqrt{x^2-\eta^2}\right)\,d\eta, \qquad (4.1.198)$$

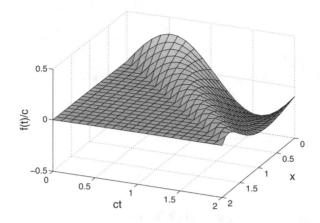

Figure 4.1.20: The inverse of the Laplace transform given by Equation 4.1.181 when $k = 3$.

our final result is

$$f(t) = c\,H(t - x/c) \int_x^{ct} J_0\Big(k\sqrt{c^2t^2 - \eta^2}\,\Big)\, d\eta. \qquad (\mathbf{4.1.199})$$

This inverse is illustrated in Figure 4.1.20 when $k = 3$.

- **Example 4.1.12**

Let us invert[13]

$$F(s) = \frac{e^{-k\sqrt{s}}}{\sqrt{s} - \zeta}, \qquad 0 < k, \qquad (\mathbf{4.1.200})$$

where ζ is a complex number with a nonpositive real part.

As before, we use Bromwich's integral to invert the transform, or

$$f(t) = \frac{1}{2\pi i} \int_{c-\infty i}^{c+\infty i} \frac{e^{tz - k\sqrt{z}}}{\sqrt{z} - \zeta}\, dz. \qquad (\mathbf{4.1.201})$$

Cutting the z-plane along the negative real axis as shown in Figure 4.1.3, we can rewrite Equation 4.1.201 as

$$f(t) = \frac{1}{\pi\sqrt{t}} \int_{-\infty}^{\infty} \frac{e^{-2y\eta i - \eta^2}}{\eta + ai}\, d\eta \qquad (\mathbf{4.1.202})$$

$$= \frac{e^{-y^2}}{\pi\sqrt{t}} \int_{-\infty}^{\infty} \frac{\eta}{\eta + ai}\, e^{-(\eta + iy)^2}\, d\eta \qquad (\mathbf{4.1.203})$$

$$= \frac{e^{-y^2}}{\pi\sqrt{t}} \int_{-\infty + yi}^{\infty + yi} \frac{\chi - yi}{\chi - b}\, e^{-\chi^2}\, d\chi, \qquad (\mathbf{4.1.204})$$

[13] Taken from Menikoff, R., R. C. Mjolsness, D. H. Sharp, C. Zemach, and B. J. Doyle, 1978: Initial value problem for Rayleigh-Taylor instability of viscous fluids. *Phys. Fluids*, **21**, 1674–1687.

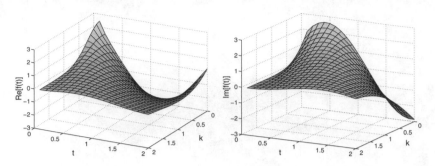

Figure 4.1.21: The inverse of the Laplace transform given by Equation 4.1.200 when $\zeta = 1 + i$.

where $zt = -\eta^2$, $\sqrt{zt} = \eta i$, $k = 2y\sqrt{t}$, $a = \zeta\sqrt{t}$, $\chi = \eta + yi$, and $b = (y - a)i$ with $\text{Re}(a) < 0$. Here, $-\infty < \eta < 0$ along GF and $0 < \eta < \infty$ along DC.

Because $\text{Im}(b) = y - \text{Re}(a) < y$, we can deform the contour in Equation 4.1.204 to the real axis, or

$$f(t) = \frac{e^{-y^2}}{\pi\sqrt{t}} \int_{-\infty}^{\infty} e^{-\chi^2} \, d\chi - \frac{aie^{-y^2}}{\pi\sqrt{t}} \int_{-\infty}^{\infty} \frac{e^{-\chi^2}}{\chi - b} \, d\chi \qquad \textbf{(4.1.205)}$$

$$= \frac{e^{-y^2}}{\sqrt{\pi t}} - \frac{aie^{-y^2}}{\pi\sqrt{t}} \int_{-\infty}^{\infty} \frac{\chi + b}{\chi^2 + (y - a)^2} e^{-\chi^2} \, d\chi \qquad \textbf{(4.1.206)}$$

$$= \frac{e^{-y^2}}{\sqrt{\pi t}} - \frac{aie^{-y^2}}{\pi\sqrt{t}} \int_{-\infty}^{\infty} \frac{\chi e^{-\chi^2}}{\chi^2 + (y - a)^2} \, d\chi$$

$$- \frac{2iabe^{-y^2}}{\pi\sqrt{t}} \int_{0}^{\infty} \frac{e^{-\chi^2}}{\chi^2 + (y - a)^2} \, d\chi \qquad \textbf{(4.1.207)}$$

$$= \frac{e^{-k^2/(4t)}}{\sqrt{\pi t}} + \zeta \exp(\zeta^2 t - k\zeta) \, \text{erfc}\left(\frac{k}{2\sqrt{t}} - \zeta\sqrt{t} \right), \quad \textbf{(4.1.208)}$$

since the second integral in Equation 4.1.207 vanishes. We used tables[14] to evaluate the last integral in Equation 4.1.207. This inverse is illustrated in Figure 4.1.21.

● **Example 4.1.13**

In Example 4.1.2, we inverted a Laplace transform that was multivalued due to the presence of modified Bessel functions of the second kind. Let us consider the additional complication of introducing a square root. In particular, let us invert

$$F(s) = \frac{K_1(a\sqrt{s})}{\sqrt{s}}, \qquad\qquad 0 < a. \qquad \textbf{(4.1.209)}$$

---

[14] Gradshteyn, I. S., and I. M. Ryzhik, 1965: *Table of Integrals, Series and Products.* Academic Press, Section 8.252, Formula 4.

As before, we use Bromwich's integral to invert the transform, or

$$f(t) = \frac{1}{2\pi i} \int_{c-\infty i}^{c+\infty i} \frac{K_1(a\sqrt{z})}{\sqrt{z}} e^{tz}\, dz. \qquad (4.1.210)$$

If we introduce the branch cuts for both the square root and the modified Bessel function along the negative real axis, we can replace Bromwich's contour with the contours shown in Figure 4.1.3. Because[15]

$$\frac{K_1(a\sqrt{s})}{\sqrt{s}} \approx \frac{1}{as}, \qquad \text{as} \qquad s \to 0, \qquad (4.1.211)$$

we have a simple pole at $z = 0$ and

$$\frac{1}{2\pi i} \int_{FED} \frac{K_1(a\sqrt{z})}{\sqrt{z}} e^{tz}\, dz = \frac{1}{a}. \qquad (4.1.212)$$

Along GH, $z = \eta^2 e^{-\pi i}$ and $\sqrt{z} = \eta e^{-\pi i/2}$. Therefore, since[16]

$$K_\nu(z) = \frac{\pi i}{2} e^{\nu\pi i/2} H_\nu^{(1)}(z e^{\pi i/2}), \qquad -\pi < \arg(z) < \frac{\pi}{2}, \qquad (4.1.213)$$

and

$$H_\nu^{(1)}(z) = J_\nu(z) + iY_\nu(z), \qquad (4.1.214)$$

we find that

$$K_1(-a\eta i) = -\pi\left[J_1(a\eta) + iY_1(a\eta)\right]/2, \qquad (4.1.215)$$

and

$$\frac{1}{2\pi i} \int_{GH} \frac{K_1(a\sqrt{z})}{\sqrt{z}} e^{tz}\, dz = -\frac{1}{2} \int_0^\infty \left[J_1(a\eta) + iY_1(a\eta)\right] e^{-t\eta^2}\, d\eta. \quad (4.1.216)$$

Similarly, along DC, $z = \eta^2 e^{\pi i}$, $\sqrt{z} = \eta e^{\pi i/2}$,

$$K_\nu(z) = -\frac{\pi i}{2} e^{-\nu\pi i/2} H_\nu^{(2)}(z e^{-\pi i/2}), \qquad -\frac{\pi}{2} < \arg(z) < \pi, \qquad (4.1.217)$$

and

$$H_\nu^{(2)}(z) = J_\nu(z) - iY_\nu(z), \qquad (4.1.218)$$

so that

$$K_1(a\eta i) = -\pi\left[J_1(a\eta) - iY_1(a\eta)\right]/2, \qquad (4.1.219)$$

---

[15] See Lebedev, N. N., 1972: *Special Functions and Their Applications.* Dover Publications, Inc., Equation 5.7.12.

[16] *Ibid.*, Equations 5.6.1 and 5.7.5.

and

$$\frac{1}{2\pi i}\int_{GH}\frac{K_1(a\sqrt{z})}{\sqrt{z}}\,e^{tz}\,dz = -\frac{1}{2}\int_0^\infty [J_1(a\eta) - iY_1(a\eta)]\,e^{-t\eta^2}\,d\eta. \quad (4.1.220)$$

Therefore,

$$f(t) = \frac{1}{a} - \int_0^\infty J_1(a\eta)\,e^{-t\eta^2}\,d\eta. \quad (4.1.221)$$

Evaluating the integral,[17] Equation 4.1.221 simplifies to

$$f(t) = \frac{1}{a} - \sqrt{\frac{\pi}{4t}}\exp\left(-\frac{a^2}{8t}\right)I_{\frac{1}{2}}\left(\frac{a^2}{8t}\right) = \frac{1}{a}\exp\left(-\frac{a^2}{4t}\right). \quad (4.1.222)$$

• Example 4.1.14: Asymptotic Expansions

In Example 2.2.9 we showed how we can obtain approximate expressions for a Laplace transform $F(s)$ when t is large. Here we expand these ideas when the Laplace transform contains both isolated singularities as well as branch points and cuts.

In constructing asymptotic expansions, some care is needed. Some[18] have tried to apply the final-value theorem to this problem, stating that "large t corresponds to small s." Unfortunately the final-value theorem yields only the limit value of the inverse and says nothing about the rate at which this limit is approached. The correct method is as follows.

Consider a Laplace transform $F(s)$ that contains isolated singularities such as poles as well as branch points. Let us now deform the initial Bromwich contour by translating it to the left, as shown in Figure 4.1.22. The contribution from the vertical portions of the new contour go to zero as they move farther to left due to the presence of e^{tz} in the integrand. Consequently, the inverse equals

$$f(t) = \text{Res}[F(z)e^{tz};\alpha_1] + \text{Res}[F(z)e^{tz};\alpha_2]$$
$$+ \frac{1}{2\pi i}\int_{C_1}F(z)e^{tz}\,dz + \frac{1}{2\pi i}\int_{C_2}F(z)e^{tz}\,dz. \quad (4.1.223)$$

The contribution from the isolated singularities is given by the residues, just as it has been before. The interesting point here is how we evaluate the branch cut integrals. One method would be to use the method of steepest descent to

[17] Gradshteyn and I. M. Ryzhik, *op. cit.*, Formula 6.618.1.

[18] See the discussion in Mathias, S. A., and R. W. Zimmerman, 2003: Laplace transform inversion for late-time behavior of groundwater flow problems. *Water Resour. Res.*, **39**, Art. No. 1283.

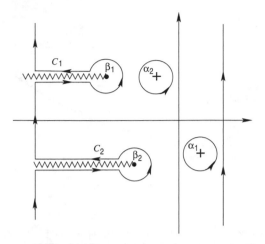

Figure 4.1.22: The deformation of the original Bromwich contour associated with the Laplace transform $F(s)$ to the one used in the asymptotic analysis.

find the asymptotic form of Bromwich's integral.[19] As an alternative, suppose that we can express $F(s)$ as

$$F(s) = \sum_{n=0}^{\infty} b_n (s - \beta)^{n+\nu} \tag{4.1.224}$$

about the branch point $s = \beta$. Then it can be proven[20] that the contribution to the inverse from that branch point is

$$f(t; \beta) \sim e^{\beta t} \sum_{n=0}^{\infty} \frac{b_n}{\Gamma(-n - \nu)} t^{-n-\nu-1} \tag{4.1.225}$$

$$\sim e^{\beta t} \frac{\sin(\nu\pi)}{\pi} \sum_{n=0}^{\infty} (-1)^{n+1} b_n \Gamma(1 + n + \nu) t^{-n-\nu-1}, \tag{4.1.226}$$

where $\Gamma(\cdot)$ is the gamma function. We can then construct an expansion for each of the transform's branch points.

Let us illustrate our analysis by considering the Laplace transform

$$F(s) = \frac{1}{\sqrt{s}\,(\sqrt{s} - b)} = \frac{\sqrt{s} + b}{\sqrt{s}\,(s - b^2)} = \frac{1}{s - b^2} + \frac{b}{\sqrt{s}\,(s - b^2)}. \tag{4.1.227}$$

[19] See, for example, Willemsen, J., and R. Burridge, 1991: Diffusion and decay in two dissimilar half-spaces in contact. *Phys. Review, Ser. A*, **43**, 735–748.

[20] See Watson, E. J., 1981: *Laplace Transforms and Applications*. Van Nostrand Reinhold Co., Section 3.4.

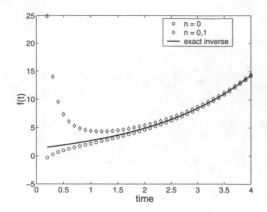

Figure 4.1.23: Comparison of the asymptotic expansion given by Equation 4.1.230 with the exact inverse given by Equation 4.1.228 with various truncations of the asymptotic expansion when $b = 0.5$.

In Problem 1 you will show that its inverse is

$$f(t) = 2e^{b^2 t} - \frac{2b}{\pi} \int_0^\infty \frac{e^{-t\eta^2}}{b^2 + \eta^2} \, d\eta. \tag{4.1.228}$$

The inverse of the first term on the right side of Equation 4.1.227 is $e^{b^2 t}$. The second term has two singularities: a simple pole at $s = b^2$ and a branch point at $s = 0$. The contribution from the simple pole is $e^{b^2 t}$. Near $s = 0$, we have

$$\frac{b}{\sqrt{s}\,(s - b^2)} = -\frac{1}{b\sqrt{s}} \left(1 + \frac{s}{b^2} + \frac{s^2}{b^4} + \cdots \right) = -\sum_{n=0}^{\infty} \frac{s^{n-\frac{1}{2}}}{b^{2n+1}}. \tag{4.1.229}$$

Consequently, the contribution from this branch point is

$$-\sum_{n=0}^{\infty} \frac{t^{-n-\frac{1}{2}}}{b^{2n+1}\Gamma(\frac{1}{2} - n)}$$

and the asymptotic expansion for $F(s)$ is

$$f(t) \sim 2e^{b^2 t} - \sum_{n=0}^{\infty} \frac{t^{-n-\frac{1}{2}}}{b^{2n+1}\Gamma(\frac{1}{2} - n)}. \tag{4.1.230}$$

Figure 4.1.23 illustrates this asymptotic expression for various truncations of the series given by Equation 4.1.230. Note how Equation 4.1.230 diverges from the correct value for fixed t as an additional term is added. This is a common property of asymptotic expansions.

Problems

1. Consider the Laplace transform

$$F(s) = \frac{1}{\sqrt{s}\,\left(\sqrt{s} - b\right)}, \qquad 0 < b.$$

Taking the branch cut associated with $\sqrt{s}$ along the negative real axis of the s-plane, show that

$$f(t) = 2e^{b^2 t} - \frac{2b}{\pi} \int_0^\infty \frac{e^{-t\eta^2}}{b^2 + \eta^2}\, d\eta.$$

This inverse $f(t)$ is illustrated in the figure labeled Problem 1.

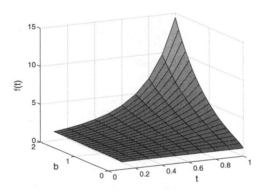

Problem 1

2. Consider the Laplace transform

$$F(s) = \frac{1}{1 + (as)^p}, \qquad 0 < p < 1.$$

Step 1: Taking the branch cut associated with s^p along the negative real axis of the s-plane, show that[21]

$$f(t) = \frac{\sin(p\pi)}{\pi} \int_0^\infty \frac{e^{-t\eta}}{a^p \eta^p + a^{-p} \eta^{-p} + 2\cos(p\pi)}\, d\eta.$$

---

[21] Taken from Meshkov, S. I., 1970: The integral representation of fractionally exponential functions and their application to dynamic problems of linear visco-elasticity. *J. Appl. Mech. Tech. Phys.*, **11**, 100–107.

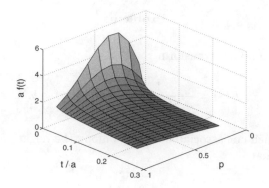

Problem 2

This inverse $f(t)$ is illustrated in the figure labeled Problem 2.

Step 2: Using the convolution theorem, show that

$$\mathcal{L}^{-1}\left[\frac{F(s)}{s}\right] = \frac{\sin(p\pi)}{\pi}\int_0^\infty \frac{1 - e^{-t\eta}}{a^p\eta^p + a^{-p}\eta^{-p} + 2\cos(p\pi)}\frac{d\eta}{\eta}.$$

Step 3: Again, taking the branch cut associated with s^p along the negative real axis of the s-plane, use Bromwich's integral to show that

$$\mathcal{L}^{-1}\left[\frac{F(s)}{s}\right] = 1 - \frac{\sin(p\pi)}{\pi}\int_0^\infty \frac{e^{-t\eta}}{a^p\eta^p + a^{-p}\eta^{-p} + 2\cos(p\pi)}\frac{d\eta}{\eta}.$$

Step 4: Because the results from Step 2 and Step 3 must be equivalent, show that

$$\int_0^\infty \frac{d\eta}{\eta\left[a^p\eta^p + a^{-p}\eta^{-p} + 2\cos(p\pi)\right]} = \frac{\pi}{\sin(p\pi)}, \qquad 0 < p < 1.$$

3. Consider the Laplace transform

$$F(s) = \frac{a}{a + s^{1/n}}, \qquad 0 < a, \quad 1 < n.$$

Step 1: Taking the branch cut associated with $s^{1/n}$ along the negative real axis of the s-plane, show that

$$f(t) = \frac{n}{\pi}a^n \sin\left(\frac{\pi}{n}\right)\int_0^\infty \frac{\eta^n \exp(-a^n t\eta^n)}{\eta^2 + 2\eta\cos(\pi/n) + 1}\, d\eta.$$

This inverse $f(t)$ is illustrated in the figure labeled Problem 3.

Step 2: Using the convolution theorem, show that

$$\mathcal{L}^{-1}\left[\frac{F(s)}{s}\right] = \frac{n}{\pi}\sin\left(\frac{\pi}{n}\right)\int_0^\infty \frac{1 - \exp(-a^n t\eta^n)}{\eta^2 + 2\eta\cos(\pi/n) + 1}\, d\eta.$$

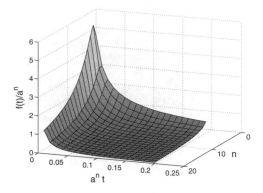

Problem 3

Step 3: Again, taking the branch cut associated with $s^{1/n}$ along the negative real axis of the s-plane, use Bromwich's integral to show[22] that

$$f(t) = 1 - \frac{n}{\pi} \sin\left(\frac{\pi}{n}\right) \int_0^\infty \frac{\exp(-a^n t \eta^n)}{\eta^2 + 2\eta \cos(\pi/n) + 1} \, d\eta.$$

Step 4: Because the results from Step 2 and Step 3 must be equivalent, show that

$$\int_0^\infty \frac{d\eta}{\eta^2 + 2\eta \cos(\pi/n) + 1} = \frac{\pi}{n \sin(\pi/n)}.$$

4. Consider the Laplace transform

$$F(s) = \frac{1}{a + s + b\sqrt{s}}, \qquad 0 < a, b.$$

Step 1: Taking the branch cut associated with $\sqrt{s}$ along the negative real axis of the s-plane, show that

$$f(t) = \frac{b}{\pi} \int_0^\infty \frac{\sqrt{\eta}\, e^{-t\eta}}{(a - \eta)^2 + b^2 \eta} \, d\eta.$$

This inverse $f(t)$ is illustrated in the figure labeled Problem 4.

Step 2: Using the convolution theorem, show that

$$\mathcal{L}^{-1}\left[\frac{F(s)}{s}\right] = \frac{b}{\pi} \int_0^\infty \frac{1 - e^{-t\eta}}{(a - \eta)^2 + b^2 \eta} \frac{d\eta}{\sqrt{\eta}}.$$

[22] Reprinted from *Int. J. Heat Mass Transfer*, **21**, R. Ghez, Mass transport and surface reactions in Lévêque's approximation, pp. 745–750, ©1978, with kind permission from Pergamon Press Ltd., Headington Hill Hall, Oxford OX3 0BW, UK.

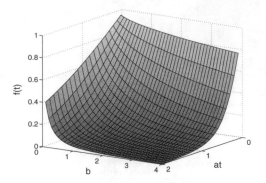

Problem 4

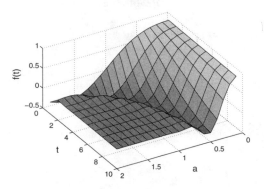

Problem 5

Step 3: Again, taking the branch cut associated with $\sqrt{s}$ along the negative real axis of the s-plane, use Bromwich's integral to show that

$$\mathcal{L}^{-1}\left[\frac{F(s)}{s}\right] = \frac{1}{a} - \frac{b}{\pi}\int_0^\infty \frac{e^{-t\eta}}{(a-\eta)^2 + b^2\eta}\frac{d\eta}{\sqrt{\eta}}.$$

Step 4: Because the results from Step 2 and Step 3 must be equivalent, show that

$$\int_0^\infty \frac{d\eta}{(a-\eta^2)^2 + b^2\eta^2} = \frac{\pi}{2ab}.$$

5. Show that the inverse of the Laplace transform

$$F(s) = \frac{\sqrt{s}}{s\sqrt{s} + a^3}, \qquad\qquad 0 < a,$$

is

$$f(t) = \frac{4}{3}e^{-a^2t/2}\cos\left(\frac{\sqrt{3}}{2}a^2t\right) - \frac{2a^3}{\pi}\int_0^\infty e^{-t\eta^2}\frac{\eta^2}{\eta^6 + a^6}\,d\eta.$$

Take the branch cut associated with the square root along the negative real axis of the s-plane. This inverse $f(t)$ is illustrated in the figure labeled Problem 5 for various values of a.

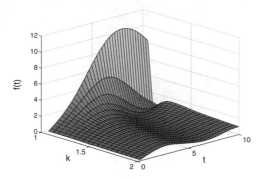

Problem 6

6. Find the inverse[23] of the Laplace transform

$$F(s) = \frac{s\sqrt{1 + s^2/k^2}}{(s^2 + 2)^2 - 4\sqrt{(1 + s^2/k^2)(s^2 + 1)}}, \qquad 1 < k.$$

The transform $F(s)$ has a simple pole at $s = 0$ because $F(s) \sim k^2/[2s(k^2 - 1)]$ as $s \to 0$, simple poles at $s = \pm i/\gamma$ where $\gamma > 1$, and branch points at $s = \pm i$ and $s = \pm ki$. If we take the branch cuts associated with $\sqrt{s + i}$, $\sqrt{s - i}$, $\sqrt{s + ki}$ and $\sqrt{s - ki}$ along the imaginary axis of the s-plane in a manner similar to that shown in Figure 4.1.14, show that

$$f(t) = \frac{k^2}{2(k^2 - 1)} + \frac{2\sqrt{1 - 1/(\gamma k)^2}}{\gamma[-ig'(i/\gamma)]} \cos(t/\gamma)$$
$$- \frac{8}{\pi} \int_1^k \frac{(1 - \eta^2/k^2)\eta\sqrt{\eta^2 - 1}}{(2 - \eta^2)^4 + 16(1 - \eta^2/k^2)(\eta^2 - 1)} \cos(t\eta) \, d\eta$$
$$- \frac{2}{\pi} \int_k^\infty \frac{\eta\sqrt{\eta^2/k^2 - 1}}{(2 - \eta^2)^2 + 4\sqrt{(\eta^2/k^2 - 1)(\eta^2 - 1)}} \cos(t\eta) \, d\eta,$$

where $g(s)$ is the denominator of the Laplace transform. This inverse $f(t)$ is illustrated in the figure labeled Problem 6.

7. Find the inverse[24] of the Laplace transform

$$F(s) = \frac{1}{s}\sqrt{\frac{v}{c}\left(s + \frac{\beta}{2}\right) + \sqrt{s(s + \beta) + \frac{v^2\beta^2}{4c^2}}}, \qquad 0 < v/c < 1, \quad 0 < \beta.$$

[23] Reprinted from *Int. J. Solids Struct.*, **7**, Y. M. Tsai, Dynamic contact stresses produced by the impact of an axisymmetrical projectile on an elastic half-space, pp. 543–558, ©1971, with kind permission from Pergamon Press Ltd., Headington Hill Hall, Oxford OX3 0BW, UK.

[24] Taken from Mondal, S. C., and M. L. Ghosh, 1990: Moving punch on a viscoelastic semi-infinite medium. *Indian J. Pure Appl. Math.*, **21**, 847–864.

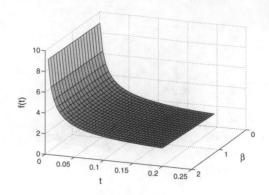

Problem 7

The transform has three branch points at $s = -\beta$ and

$$s = \alpha_{1,2} = \frac{\beta}{2}\left(-1 \pm \sqrt{1 - \frac{v^2}{c^2}}\right),$$

so that $-\beta < \alpha_2 < \alpha_1$. Taking the branch cut associated with each of these branch points along the negative real axis of the s-plane, show that the inverse is

$$f(t) = \sqrt{\frac{v\beta}{c}} - \int_0^{\beta\sqrt{1-(v/c)^2}} \frac{e^{t(\alpha_1-\eta)}}{\alpha_1 - \eta}\sqrt{\frac{r-x}{2}}\, d\eta$$

$$- \int_{\beta\sqrt{1-(v/c)^2}}^{\infty} \frac{e^{t(\alpha_1-\eta)}}{\alpha_1 - \eta}\, d\eta$$

$$\times \sqrt{\sqrt{\frac{v}{c}\left[\eta - \frac{\beta}{2}\sqrt{1-(v/c)^2}\right]} + \sqrt{\eta[\eta - \beta\sqrt{1-(v/c)^2}]}},$$

where

$$r = \sqrt{\eta[\beta\sqrt{1-(v/c)^2} - \eta] + \frac{v^2}{c^2}\left[\frac{\beta}{2}\sqrt{1-(v/c)^2} - \eta\right]^2}$$

and

$$x = \frac{v}{c}\left[\frac{\beta}{2}\sqrt{1-(v/c)^2} - \eta\right].$$

This inverse $f(t)$ is illustrated in the figure labeled Problem 7 when $v/c = 0.8$.

8. Show that the inverse[25] of the Laplace transform

$$F(s) = \frac{1}{s[a + b\log(1 + c/s)]}, \qquad 0 < a, b, c,$$

---

[25] Reprinted from *Int. J. Solids Struct.*, **29**, J. Lubliner and V. P. Panoskaltsis, The modified Kuhn model of linear viscoelasticity, pp. 3099–3112, ©1992, with kind permission from Pergamon Press Ltd., Headington Hill Hall, Oxford OX3 0BW, UK.

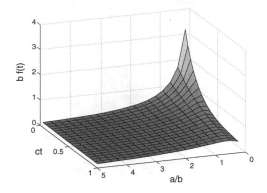

Problem 8

is

$$f(t) = \frac{\exp[-ct/(1 - e^{-a/b})]}{b(e^{a/b} - 1)} + b\int_0^1 \frac{\exp(-ct\eta)}{\eta\{[a + b\ln(1/\eta - 1)]^2 + b^2\pi^2\}}\, d\eta,$$

if we take the branch cut along the negative real axis of the s-plane between its branch points at $s = -c$ and $s = 0$. This inverse $f(t)$ is illustrated in the figure labeled Problem 8.

9. Using the contour shown in Figure 4.1.19 with the branch points located at $z = \pm ai$, show that the inverse[26] of the Laplace transform

$$F(s) = \frac{1}{\ln(s^2 + a^2)}, \qquad 0 < a,$$

is

$$f(t) = \frac{\sinh(t\sqrt{1 - a^2}\,)}{\sqrt{1 - a^2}} + 2\int_a^\infty \frac{\sin(t\eta)}{\ln^2(\eta^2 - a^2) + \pi^2}\, d\eta,$$

if $0 < 1 < a$,

$$f(t) = t + 2\int_1^\infty \frac{\sin(t\eta)}{\ln^2(\eta^2 - 1) + \pi^2}\, d\eta,$$

if $a = 1$, and

$$f(t) = \frac{\sin(t\sqrt{a^2 - 1}\,)}{\sqrt{a^2 - 1}} + 2\int_a^\infty \frac{\sin(t\eta)}{\ln^2(\eta^2 - a^2) + \pi^2}\, d\eta,$$

---

[26] Taken from Apelblat, A., and N. Kravitsky, 1998: Improper integrals associated with the inverse Laplace transforms of $[\ln(s^2 + a^2)]^{-n}, n = 1, 2$. *Z. Angew. Math. Mech.*, **78**, 565–571.

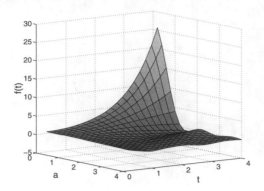

Problem 9

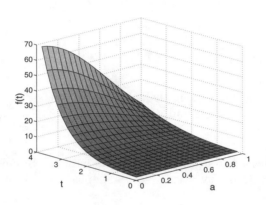

Problem 10

if $1 < a < \infty$. This inverse $f(t)$ is illustrated in the figure labeled Problem 9.

10. Using the contour shown in Figure 4.1.19 with branch points at $z = \pm ai$, show that the inverse[27] of the Laplace transform

$$F(s) = \frac{1}{[\ln(s^2 + a^2)]^2}, \qquad 0 < a < 1,$$

is

$$f(t) = \frac{(1 - 2a^2)\sinh\left(t\sqrt{1-a^2}\,\right)}{2(1-a^2)^{3/2}} + \frac{t\cosh\left(t\sqrt{1-a^2}\,\right)}{2(1-a^2)}$$

$$+ 4\int_a^\infty \frac{\ln(\eta^2 - a^2)\sin(t\eta)}{\left[\ln^2(\eta^2 - a^2) + \pi^2\right]^2}\, d\eta.$$

This inverse $f(t)$ is illustrated in the figure labeled Problem 10.

---

[27] *Ibid.*

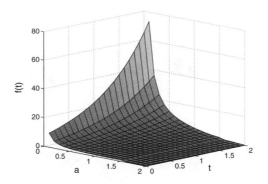

Problem 11

11. Show that the inverse of the Laplace transform

$$F(s) = \frac{1}{a - \arctan(a/s)}, \qquad 0 < a < \infty,$$

where

$$\arctan\left(\frac{a}{s}\right) = \frac{1}{2i} \log\left(\frac{s + ai}{s - ai}\right),$$

is

$$f(t) = \frac{a}{\sin^2(a)} \exp\left[\frac{at}{\tan(a)}\right] H\left(\frac{\pi}{2} - |a|\right)$$

$$+ \frac{1}{2\pi} \int_{-1}^{1} \frac{e^{iat\eta}}{\left(a - \frac{\pi}{2}\right) + \frac{i}{2} \ln\left(\frac{1+\eta}{1-\eta}\right)} \, d\eta - \frac{1}{2\pi} \int_{-1}^{1} \frac{e^{iat\eta}}{\left(a + \frac{\pi}{2}\right) + \frac{i}{2} \ln\left(\frac{1+\eta}{1-\eta}\right)} \, d\eta.$$

Hint: Show that $F(s)$ has the simple pole $s_0 = a/\tan(a)$ only when $|a| < \pi/2$ with $0 \le s_0 \le 1$. Take the branch cut of the logarithm to lie along the imaginary axis in the s-plane between $s = \pm ai$. This inverse $f(t)$ is illustrated in the figure labeled Problem 11.

12. Find the inverse of the Laplace transform

$$F(s) = \frac{1}{s} \exp\left(-x\sqrt{\frac{s}{s + a}}\right), \qquad 0 < a, x.$$

If we take the branch cuts associated with $\sqrt{s}$ and $\sqrt{s + a}$ along the negative real axis of the s-plane, show that

$$f(t) = 1 - \frac{2}{\pi} \int_0^\infty \exp\left(-\frac{at\eta^2}{1 + \eta^2}\right) \frac{\sin(x\eta)}{\eta(1 + \eta^2)} \, d\eta.$$

The inverse $f(t)$ is plotted as a function of x and at in the figure labeled Problem 8 at the end of Section 4.4.

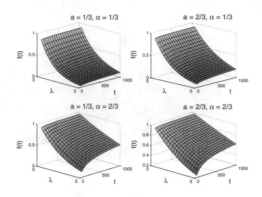

Problem 13

13. Find the inverse[28] of the Laplace transform

$$F(s) = \frac{1}{s} \exp\left(-\lambda\sqrt{\frac{s^\alpha}{s^\alpha + a}}\right), \qquad 0 < a, \alpha < 1, \quad 0 \le \lambda.$$

If the branch cut associated with s^α lies along the negative real axis of the s-plane, show that the inverse is

$$f(t) = 1 - \frac{1}{\pi} \int_0^\infty e^{-t\eta} \exp\left[-\lambda\sqrt{\frac{\eta^\alpha}{\rho}} \cos\left(\frac{\alpha\pi - \varphi}{2}\right)\right]$$

$$\times \sin\left[\lambda\sqrt{\frac{\eta^\alpha}{\rho}} \sin\left(\frac{\alpha\pi - \varphi}{2}\right)\right] \frac{d\eta}{\eta},$$

where

$$\rho = \sqrt{\eta^{2\alpha} + a^2 + 2a\eta^\alpha \cos(\alpha\pi)}$$

and

$$\varphi = \arctan\left[\frac{\eta^\alpha \sin(\alpha\pi)}{\eta^\alpha \cos(\alpha\pi) + a}\right], \qquad 0 < \varphi < \pi.$$

This inverse $f(t)$ is illustrated in the figure labeled Problem 13.

14. Find the inverse[29] of the Laplace transform

$$F(s) = \frac{1}{s} \exp\left(-x\sqrt{s\frac{1+s}{1+as}}\right), \qquad 0 < a, x < \infty,$$

---

[28] Reprinted from *Int. J. Solids Struct.*, **25**, D. C. Lagoudas, C.-Y. Hui and S. L. Phoenix, Time evolution of overstress profiles near broken fibers in a composite with a viscoelastic matrix, pp. 45–66, ©1989, with kind permission from Pergamon Press Ltd., Headington Hill Hall, Oxford OX3 0BW, UK.

[29] Taken from Tanner, R. I., 1962: Note on the Rayleigh problem for a visco-elastic fluid. *Z. Angew. Math. Phys.*, **13**, 573–580. Published by Birkhäuser Verlag, Basel, Switzerland. Tsamopoulos and Borkar [Tsamopoulos, J., and A. Borkar, 1994: Transient rotational flow of an Oldroyd-B fluid over a disk. *Phys. Fluid*, **6**, 1144–1157.] have generalized this problem by introducing a parameter b so that the argument of the square root reads $s(s+b)/(1+as)$.

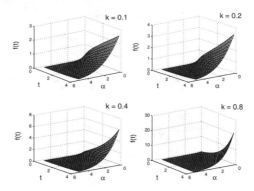

Problem 15

by taking the branch cuts associated with the branch points $s = 0$, $s = -1$ and $s = -1/a$ along the negative real axis of the s-plane. Then we may deform the Bromwich contour along the imaginary axis of the s-plane except for a small semicircle around $s = 0$. Using this contour, you should find that

$$f(t) = \frac{1}{2} + \frac{1}{\pi} \int_0^\infty \exp\left\{-x\sqrt{\frac{\eta}{2}} \, M[\cos(\theta) - \sin(\theta)]\right\} \frac{d\eta}{\eta}$$
$$\times \sin\left\{t\eta - r\sqrt{\frac{\eta}{2}} \, M[\cos(\theta) + \sin(\theta)]\right\},$$

where

$$M = \left(\frac{1 + \eta^2}{1 + a^2\eta^2}\right)^{1/4} \qquad \text{and} \qquad 2\theta = \arctan(\eta) - \arctan(a\eta).$$

The inverse $f(t)$ is plotted as a function of x and t for various values of a in the figure labeled Problem 2 at the end of Section 4.3.

15. Find the inverse[30] of the Laplace transform

$$F(s) = \frac{\exp(-\alpha\sqrt{s})}{(s - k)\sqrt{s}}, \qquad 0 < \alpha, k.$$

If we take the branch cut associated with $\sqrt{s}$ along the negative real axis of the s-plane, you should find that

$$f(t) = \frac{\exp\left(kt - \alpha\sqrt{k}\right)}{\sqrt{k}} - \frac{2}{\pi} \int_0^\infty \frac{\cos(\alpha\eta)}{\eta^2 + k} e^{-t\eta^2} \, d\eta.$$

This inverse $f(t)$ is illustrated in the figure labeled Problem 15.

[30] Taken from Duffy, P., 1989: The acceleration of cometary ions by Alfvén waves. *J. Plasma Phys.*, **42**, 13–25. Reprinted with the permission of Cambridge University Press.

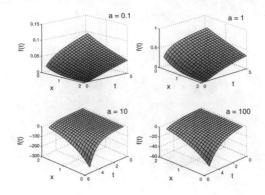

Problem 16

16. Find the inverse[31] of the Laplace transform

$$F(s) = \frac{a\exp(-x\sqrt{s})}{s\left[a + (2-a)\sqrt{\pi s/2}\right]}, \qquad 0 < a, x.$$

If we take the branch cut associated with $\sqrt{s}$ along the negative real axis of the s-plane, show that the inverse is

$$f(t) = 1 - 2\exp\left[\frac{2t}{\pi}\left(\frac{a}{a-2}\right)^2 - \sqrt{\frac{2}{\pi}}\frac{ax}{a-2}\right]H(a-2)$$

$$- \frac{2ac}{\pi}\int_0^\infty \frac{\cos(x\eta)}{a^2 + c^2\eta^2}e^{-t\eta^2}\,d\eta - \frac{2a^2}{\pi}\int_0^\infty \frac{\sin(x\eta)}{\eta(a^2 + c^2\eta^2)}e^{-t\eta^2}\,d\eta,$$

where $c^2 = \pi(2-a)^2/2$. This inverse $f(t)$ is illustrated in the figure labeled Problem 16.

17. Find the inverse of the Laplace transform

$$F(s) = \frac{\exp\left[-x\sqrt{s(s+b)}\right]}{s + a\sqrt{s(s+b)}}, \qquad 0 < a, b, x.$$

If we take the branch cuts associated with $\sqrt{s}$ and $\sqrt{s+b}$ along the negative real axis of the s-plane, show that the inverse is

$$f(t) = \frac{2}{\pi b}\int_0^1 \frac{a\sqrt{1-\eta^2}\cos\left(bx\eta\sqrt{1-\eta^2}\right) - \eta\sin\left(bx\eta\sqrt{1-\eta^2}\right)}{a^2 + (1-a^2)\eta^2}e^{-bt\eta^2}\,d\eta$$

---

[31] Originally solved by Kuznetsov [Kuznetsov, M. M., 1975: Unsteady-state slip of a gas near an infinite plane with diffusion-mirror reflection of molecules. *J. Appl. Mech. Tech. Phys.*, **16**, 853–858.] for $0 < \sigma < 1$.

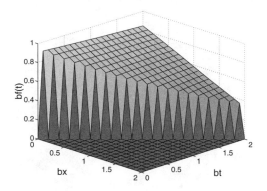

Problem 17

if $t > x$; $f(t) = 0$ otherwise. This inverse $f(t)$ is illustrated in the figure labeled Problem 17 when $a = 0.1$.

18. Find the inverse[32] of the Laplace transform

$$F(s) = \frac{1 + r\sqrt{s}}{s\left(1 + a\sqrt{s}\right)} e^{-(r-a)\sqrt{s}}, \qquad 0 < a < r.$$

If we take the branch cut associated with the square root along the negative real axis of the s-plane, show that

$$f(t) = 1 - \frac{2}{\pi} \int_0^\infty \frac{1 + ar\eta^2}{1 + a^2\eta^2} \sin[(r-a)\eta] e^{-t\eta^2} \frac{d\eta}{\eta}$$
$$+ \frac{2}{\pi}(r-a) \int_0^\infty \frac{1}{1 + a^2\eta^2} \cos[(r-a)\eta] e^{-t\eta^2} \, d\eta.$$

The inverse $a^3 f(t)/r^{3/2}$ is plotted as a function of r/a and t/a^2 in the figure labeled Problem 6 at the end of Section 4.4.

19. Find the inverse of the Laplace transform

$$F(s) = \frac{1}{s} \exp\left(-a\sqrt{s}\, \frac{1 - ce^{-b\sqrt{s}}}{1 + ce^{-b\sqrt{s}}}\right), \qquad 0 < a, b, \quad |c| < 1.$$

If we take the branch cut associated with the square root along the negative real axis of the s-plane, then

$$f(t) = 1 - \frac{1}{\pi} \int_0^\infty e^{-t\eta + a\sqrt{\eta}\, u(\eta)} \sin[a\sqrt{\eta}\, v(\eta)] \frac{d\eta}{\eta},$$

---

[32] Taken from Seth, S. S., 1977: Unsteady motion of viscoelastic fluid due to rotation of a sphere. *Indian J. Pure Appl. Math.*, **8**, 302–308.

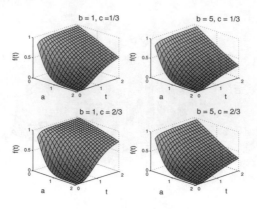

Problem 19

where

$$u(\eta) = \frac{2c\sin(b\sqrt{\eta}\,)}{1 + c^2 + 2c\cos(b\sqrt{\eta}\,)} \qquad \text{and} \qquad v(\eta) = \frac{1 - c^2}{1 + c^2 + 2c\cos(b\sqrt{\eta}\,)}.$$

This inverse $f(t)$ is illustrated in the figure labeled Problem 19.

20. Find the inverse[33] of the Laplace transform

$$F(s) = \frac{\exp(-|x|\sqrt{s^2 + 1}\,)}{s^2 - a^2}, \qquad 0 < a, \quad -\infty < x < \infty.$$

If we take the branch cut associated with the square root along the imaginary axis of the s-plane between $s = -i$ and $s = i$, show that

$$f(t) = \left[\frac{\sinh(at - |x|\sqrt{1 + a^2}\,)}{a} + \frac{2}{\pi} \int_0^1 \frac{\sinh\!\left(|x|\sqrt{1 - \eta^2}\,\right)\cos(t\eta)}{a^2 + \eta^2}\, d\eta \right] H(t - |x|).$$

This inverse $f(t)$ is illustrated in the figure labeled Problem 20 when $a = 0.25$.

21. Find the inverse[34]

$$F(s) = \frac{s\exp(-|x|\sqrt{s^2 + 1}\,)}{(s^2 - a^2)\sqrt{s^2 + 1}}, \qquad 0 < a, \quad -\infty < x < \infty.$$

---

[33] Reprinted from *Wave Motion*, **24**, A. R. M. Wolfert, A. V. Metrikine and H. A. Dieterman, Wave radiation in a one-dimensional system due to a non-uniformly moving constant load, 185–196, ©1996, with permission from Elsevier.

[34] *Ibid.*

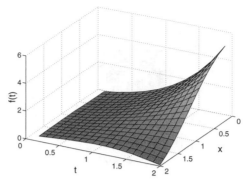

Problem 20

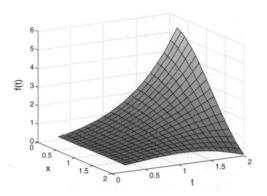

Problem 21

If we take the branch cut associated with the square root along the imaginary axis of the s-plane between $s = -i$ and $s = i$, show that

$$f(t) = \left[\frac{\sinh\left(at - |x|\sqrt{1+a^2}\right)}{\sqrt{1+a^2}} + \frac{2}{\pi} \int_0^1 \frac{\cosh(x\eta)\sin\left(t\sqrt{1-\eta^2}\right)}{1+a^2-\eta^2} \, d\eta \right] H(t - |x|).$$

This inverse $f(t)$ is illustrated in the figure labeled Problem 21 when $a = 0.25$.

22. Find the inverse[35] of the Laplace transform

$$F(s) = \frac{a\sqrt{s}\exp\left[-x\sqrt{s(s+1)}\right]}{s\left[a\sqrt{s+1} + (2-a)\sqrt{\pi s/2}\right]}, \qquad 0 < a, x.$$

---

[35] Kuznetsov, *op. cit.*

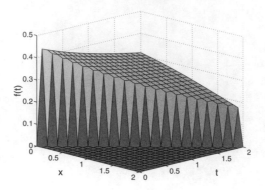

Problem 22

If we take the branch cuts associated with $\sqrt{s}$ and $\sqrt{s+1}$ along the negative real axis of the s-plane, show that the inverse is

$$
f(t) = \left[\frac{2a^2}{\pi} \int_0^1 \frac{\sqrt{1-\eta^2}}{a^2 + b^2\eta^2} \cos\!\left(x\eta\sqrt{1-\eta^2} \right) e^{-t\eta^2}\, d\eta \right.
$$
$$
\left. - a(2-a)\sqrt{\frac{2}{\pi}} \int_0^1 \frac{\eta}{a^2 + b^2\eta^2} \sin\!\left(x\eta\sqrt{1-\eta^2} \right) e^{-t\eta^2}\, d\eta \right] H(t-x),
$$

where $b^2 = \pi(2-a)^2/2 - a^2$. This inverse $f(t)$ is illustrated in the figure labeled Problem 22 when $a = 1$.

23. Show that the Laplace transform

$$
F(s) = \frac{\omega}{s^2 + \omega^2} e^{-x\sqrt{s/\kappa}}, \qquad 0 < x, \kappa, \omega,
$$

has the inverse

$$
f(t) = e^{-x\sqrt{\omega/(2\kappa)}} \sin\!\left(\omega t - x\sqrt{\frac{\omega}{2\kappa}} \right) + \frac{2\kappa\omega}{\pi} \int_0^\infty \frac{\eta\, e^{-\kappa t\eta^2}}{\omega^2 + \kappa^2\eta^4} \sin(x\eta)\, d\eta.
$$

Take the branch cut for $\sqrt{s}$ along the negative real axis of the s-plane. This inverse $f(t)$ is illustrated in the figure labeled Problem 23.

24. Find the inverse of Laplace transform

$$
F(s) = \frac{1}{s^2 + 1} \exp\!\left(-\frac{sx}{\sqrt{s+1}} \right), \qquad 0 < x,
$$

by taking the branch cut associated with the branch point $s = -1$ along the negative real axis of the s-plane and using Bromwich's integral. You should find that

$$
f(t) = \exp\!\left[-x\sin(\pi/8)/\sqrt[4]{2} \right] \sin\!\left[t - x\cos(\pi/8)/\sqrt[4]{2} \right]
$$
$$
- \frac{e^{-t}}{\pi} \int_0^\infty \frac{\exp(-t\eta)}{\eta^2 + 2\eta + 2} \sin\!\left[\frac{(1+\eta)x}{\sqrt{\eta}} \right] d\eta.
$$

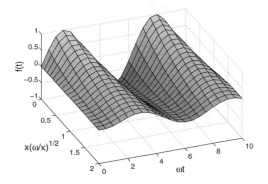

Problem 23

The inverse $f(t)$ is plotted as a function of x and t in the figure labeled Problem 1 at the end of Section 4.3.

25. Find the inverse[36] of the Laplace transform

$$F(s) = \frac{2\pi}{s^2 + 4\pi^2} \exp\left(-x\sqrt{s\frac{1 + as}{1 + bs}}\right), \qquad 0 < a < b.$$

If we take the branch cuts from the three branch points $s = 0$, $s = -1/b$ and $s = -1/a$ along the negative real axis of the s-plane, you should find that

$$f(t) = \sin[2\pi t - x\Delta\sin(\psi/2)]\exp[-x\Delta\cos(\psi/2)]$$
$$+ 2\int_0^{1/b} F(x, t, \eta)\, d\eta + 2\int_{1/a}^{\infty} F(x, t, \eta)\, d\eta,$$

where

$$F(x, t, \eta) = \frac{\exp(-t\eta)}{\eta^2 + 4\pi^2}\sin\left[x\sqrt{\frac{\eta(1 - a\eta)}{1 - b\eta}}\right], \qquad \Delta = [(d_1^2 + d_2^2)d_3^2]^{1/4},$$

$$d_1 = 1 + 4\pi^2 ab, \quad d_2 = 2\pi(b - a), \quad d_3 = \frac{2\pi}{1 + 4\pi^2 b^2} \quad \text{and} \quad \psi = \arctan(d_1/d_2).$$

A plot of the inverse $f(t)$ is presented with Problem 3 in Section 4.3.

26. Find the inverse of the Laplace transform

$$F(s) = (1 + i)e^{-(1+i)a\sqrt[4]{s}}/s + (1 - i)e^{-(1-i)a\sqrt[4]{s}}/s, \qquad 0 < a.$$

---

[36] Taken from Zheltov, Yu. P., and V. S. Kutlyarov, 1965: Transient motion of a liquid in a fissured porous stratum subject to periodic pressure variation at the boundary. *J. Appl. Mech. Tech. Phys.*, **6, No. 6**, 45–49.

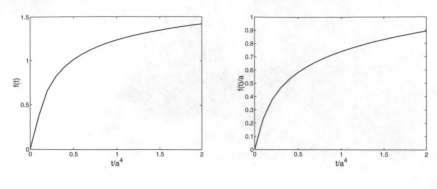

Problem 26   **Problem 27**

If we take the branch cut associated with $\sqrt[4]{s}$ along the negative real axis of the s-plane, show that

$$f(t) = 2 + \frac{4}{\pi} \int_0^\infty \left[\cos\left(\sqrt{2}a\eta\right) - \sin\left(\sqrt{2}a\eta\right) - e^{-\sqrt{2}a\eta} \right] e^{-t\eta^4} \frac{d\eta}{\eta}.$$

This inverse $f(t)$ is illustrated in the figure labeled Problem 26.

27. Find the inverse of the Laplace transform

$$F(s) = \frac{i}{s\sqrt[4]{s}} \left[e^{-(1+i)a\sqrt[4]{s}} - e^{-(1-i)a\sqrt[4]{s}} \right], \qquad 0 < a.$$

If we take the branch cut associated with $\sqrt[4]{s}$ along the negative real axis of the s-plane, show that

$$f(t) = 2a + \frac{2\sqrt{2}}{\pi} \int_0^\infty \left[\cos\left(\sqrt{2}a\eta\right) - \sin\left(\sqrt{2}a\eta\right) - e^{-\sqrt{2}a\eta} \right] e^{-t\eta^4} \frac{d\eta}{\eta^2}.$$

This inverse $f(t)$ is illustrated in the figure labeled Problem 27.

28. Find the inverse of the Laplace transform

$$F(s) = \frac{K_1(as)}{s\, K_1(bs)}, \qquad 0 < b < a.$$

If we take the branch cut associated with the Bessel function along the negative real axis of the s-plane, show that

$$f(t) = \frac{b}{a} + \int_0^\infty \left[\frac{I_1(a\eta)K_1(b\eta) - K_1(a\eta)I_1(b\eta)}{K_1^2(b\eta) + \pi^2 I_1^2(b\eta)} \right] \frac{e^{-t\eta}}{\eta}\, d\eta,$$

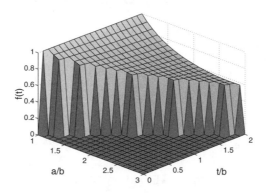

Problem 28

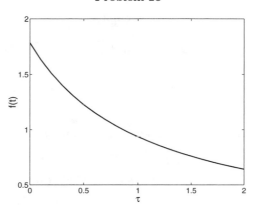

Problem 29

if $t > a - b$; $f(t) = 0$ otherwise. This inverse $f(t)$ is illustrated in the figure labeled Problem 28.

29. Find the inverse[37] of the Laplace transform

$$F(s) = \frac{\exp(-sr/c)}{s \, K_0(sa/c)}, \qquad 0 < a < r, \quad 0 < c.$$

If we take the branch cut associated with $K_0(\cdot)$ along the negative real axis of the s-plane, you should find that

$$f(t) = H(\tau + 1) \int_0^\infty \frac{I_0(\eta)}{K_0^2(\eta) + \pi^2 I_0^2(\eta)} e^{-\tau \eta} \frac{d\eta}{\eta},$$

where $\tau = (ct - r)/a$. This inverse $f(t)$ is illustrated in the figure labeled Problem 29.

---

[37] Taken from Pine, Z. L., and F. M. Tesche, 1973: Calculation of the early time radiated electric field from a linear antenna with a finite source gap. *IEEE Trans. Antennas Propag.*, **AP-21**, 740–743. ©1973 IEEE.

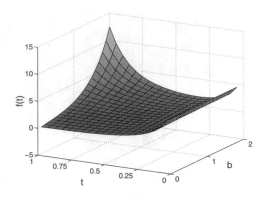

Problem 30

30. Find the inverse of the Laplace transform

$$F(s) = \frac{K_0(a\sqrt{s})}{\sqrt{s} - b}, \qquad 0 < a, b.$$

If we take the branch cut associated with $\sqrt{s}$ and $K_0(\cdot)$ along the negative real axis of the s-plane, show that

$$f(t) = 2bK_0(ab)e^{b^2 t} - \int_0^\infty \frac{bJ_0(a\eta) + \eta Y_0(a\eta)}{b^2 + \eta^2}\, \eta\, e^{-t\eta^2}\, d\eta.$$

The solution is shown in the figure labeled Problem 30 when $a = 1$.

31. Invert the Laplace transform

$$F(s) = I_\nu(as)K_\nu(bs), \qquad 0 < a < b, \quad 0 \le \nu.$$

Step 1: Using the asymptotic expansions for modified Bessel functions, show that $f(t) = 0$ if $t < b - a$.

Step 2: For $t > b - a$, you might be tempted to deform the Bromwich contour into the one shown in Figure 4.1.3. However, from the asymptotic expansion for $I_\nu(\cdot)$, it *grows* exponentially in the left side of the complex s-plane. Therefore, this deformation is permissible only if $t > a + b$. In that case, show that

$$f(t) = \cos(\nu\pi) \int_0^\infty I_\nu(a\eta)I_\nu(b\eta)e^{-t\eta}\, d\eta.$$

For $b - a < t < b + a$, we must use numerical techniques. The numerical inversion is shown in the figure labeled Problem 31 when $a = 1$ and $\nu = 0$.

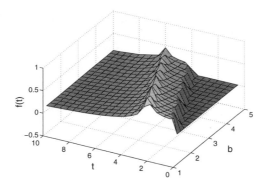

Problem 31

32. Find the inverse[38] of the Laplace transform

$$F(s) = I_{-\nu}\left(a\sqrt{s}\right) K_\nu\left(b\sqrt{s}\right), \qquad 0 < a, b, \quad 0 < \nu < 1.$$

If we take the branch cuts associated with the modified Bessel functions and $\sqrt{s}$ along the negative real axis of the s-plane, show that

$$f(t) = \frac{1}{2t} \exp\left(-\frac{a^2 + b^2}{4t}\right) I_{-\nu}\left(\frac{ab}{2t}\right).$$

Hint: You will need to use properties given by Equation 9.1.5, Equation 9.6.2 and Equation 9.6.3 in Abramowitz and Stegun's *Handbook of Mathematical Functions*[39] and the integral[40]

$$\int_0^\infty J_{-\nu}(a\sqrt{\eta})\, J_{-\nu}(b\sqrt{\eta})\, e^{-t\eta}\, d\eta = \frac{1}{t} \exp\left(-\frac{a^2 + b^2}{4t}\right) I_{-\nu}\left(\frac{ab}{2t}\right).$$

This inverse is shown in the figure labeled Problem 32 when $b = 2$ and $a = 1$.

33. Find the inverse[41] of the Laplace transform

$$F(s) = \frac{K_0(r\sqrt{s})}{s^{3/2}\, K_1(a\sqrt{s})}, \qquad 0 < a < r,$$

[38] Taken from Peyraud, J., and J. Coste, 1976: Some diffusion equations in plasma physics. *Phys. Fluids*, **19**, 388–393.

[39] Abramowitz, M., and I. A. Stegun, 1968: *Handbook of Mathematical Functions with Formulas, Graphs, and Mathematical Tables.* Dover Publ., Inc., 1046 pp.

[40] Gradshteyn and Ryzhik, *op. cit.*, Formula 6.615.

[41] Taken from Goldstein, R. J., and D. G. Briggs, 1964: Transient free convection about vertical plates and circular cylinders. *J. Heat Transfer*, **86**, 490–500.

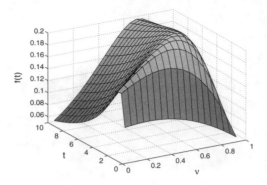

Problem 32

by first showing that

$$\mathcal{L}^{-1}\left[\frac{K_0(r\sqrt{s})}{\sqrt{s}\,K_1(a\sqrt{s})}\right] = -\frac{2}{\pi}\int_0^\infty e^{-\eta^2 t}\frac{J_0(\eta r)Y_1(\eta a) - Y_0(\eta r)J_1(\eta a)}{J_1^2(\eta a) + Y_1^2(\eta a)}\,d\eta.$$

Take the branch cuts associated with $K_0(\cdot)$, $K_1(\cdot)$ and $\sqrt{s}$ along the negative real axis of the s-plane. Then complete the problem by showing that

$$f(t) = \int_0^t \mathcal{L}^{-1}\left[\frac{K_0(r\sqrt{s})}{\sqrt{s}\,K_1(a\sqrt{s})}\right]\,d\tau$$

$$= -\frac{2}{\pi}\int_0^\infty \left(1 - e^{-t\eta^2}\right)\frac{J_0(\eta r)Y_1(\eta a) - Y_0(\eta r)J_1(\eta a)}{J_1^2(\eta a) + Y_1^2(\eta a)}\frac{d\eta}{\eta^2}.$$

The inverse $f(t)$ is plotted as a function of r/a and t in the figure labeled Problem 3 at the end of Section 4.4.

34. Find the inverse of the Laplace transform

$$F(s) = \frac{K_n(\sqrt{s})}{\sqrt{s}\,K_{n-1}(\sqrt{s})}, \qquad 1 \le n.$$

If you take the branch cuts associated with $\sqrt{s}$ and $K_n(\cdot)$ along the negative real axis of the s-plane, you should find that

$$f(t) = 2(n-1)H(n-3/2) + \frac{4}{\pi^2}\int_0^\infty \frac{e^{-t\eta^2}}{J_{n-1}^2(\eta) + Y_{n-1}^2(\eta)}\frac{d\eta}{\eta}.$$

Hint:

$$J_\nu(z)Y_{\nu+1}(z) - J_{\nu+1}(z)Y_\nu(z) = -\frac{2}{\pi z}.$$

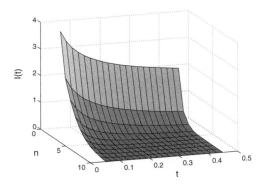

Problem 34

The integral portion of the inverse is illustrated in the figure labeled Problem 34.

35. Find the inverse of the Laplace transform

$$F(s) = \frac{K_0(r\sqrt{s+1})}{s\,K_0(\sqrt{s+1})}, \qquad 1 < r.$$

If you take the branch cut associated with $\sqrt{s+1}$ and $K_0(\cdot)$ along the negative real axis of the s-plane, you should find that

$$f(t) = \frac{K_0(r)}{K_0(1)} - \frac{2}{\pi}e^{-t}\int_0^\infty \frac{\eta}{1+\eta^2}\frac{J_0(\eta)Y_0(r\eta) - J_0(r\eta)Y_0(\eta)}{J_0^2(\eta) + Y_0^2(\eta)}e^{-t\eta^2}\,d\eta.$$

The inverse $f(t)$ is plotted as a function of r and t in the figure labeled Problem 4 at the end of Section 4.4.

36. Show that the inverse of the Laplace transform

$$F(s) = \frac{K_0\!\left(r\sqrt{s/\nu + 1/b^2}\right)}{s\,K_0\!\left(a\sqrt{s/\nu + 1/b^2}\right)}, \qquad 0 < a < r, \quad 0 < b, \nu,$$

is

$$f(t) = \frac{K_0(r/b)}{K_0(a/b)} + \frac{2e^{-\nu t/b^2}}{\pi}\int_0^\infty \frac{\eta\,e^{-\nu t\eta^2/a^2}}{\eta^2 + (a/b)^2}\frac{J_0(r\eta/a)Y_0(\eta) - Y_0(r\eta/a)J_0(\eta)}{J_0^2(\eta) + Y_0^2(\eta)}\,d\eta.$$

Take the branch cut along the negative real axis of the s-plane from its branch point at $s = -\nu/b^2$. The inverse $f(t)$ is plotted as a function of r/a and $\nu t/b^2$ when $a/b = 1$ in the figure labeled Problem 2 at the end of Section 4.4.

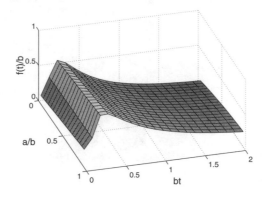

Problem 37

37. Find the inverse[42] of the Laplace transform

$$F(s) = K_0\left[r\sqrt{\frac{s(s+b)}{c(s+a)}}\,\right], \qquad 0 < a < b,$$

where c and r are real and positive. If we take all of the branch cuts along the negative real axis of the s-plane, then

$$f(t) = \frac{1}{2}\int_0^a J_0\left[r\sqrt{\frac{\eta(b-\eta)}{c(a-\eta)}}\,\right]e^{-t\eta}\,d\eta + \frac{1}{2}\int_b^\infty J_0\left[r\sqrt{\frac{\eta(\eta-b)}{c(\eta-a)}}\,\right]e^{-t\eta}\,d\eta.$$

This inverse $f(t)$ is illustrated in the figure labeled Problem 37 with $r\sqrt{b/c} = 1$.

38. Find the inverse[43] of the Laplace transform

$$F(s) = \frac{\omega}{s^2 + \omega^2}\frac{K_1(r\sqrt{s}\,)}{K_1(\sqrt{s}\,)}, \qquad 1 < r, \omega.$$

If we take the branch cut associated with $K_1(\cdot)$ along the negative real axis of the s-plane, show that

$$f(t) = \frac{N_1(r\sqrt{\omega}\,)}{N_1(\sqrt{\omega}\,)}\sin[\omega t + \phi_1(r\sqrt{\omega}\,) - \phi_1(\sqrt{\omega}\,)]$$
$$- \frac{2\omega}{\pi}\int_0^\infty \frac{\eta}{\eta^4 + \omega^2}e^{-\eta^2 t}\frac{J_1(r\eta)Y_1(\eta) - Y_1(r\eta)J_1(\eta)}{J_1^2(\eta) + Y_1^2(\eta)}\,d\eta,$$

[42] Taken from Cooley, R. L., and C. M. Case, 1973: Effect of a water table aquitard on drawdown in an underlying pumped aquifer. *Water Resour. Res.*, **9**, 434–447. ©1973 American Geophysical Union. Reproduced/modified by permission of American Geophysical Union.

[43] Taken from Schwarz, W. H., 1963: The unsteady motion of an infinite oscillating cylinder in an incompressible Newtonian fluid at rest. *Appl. Sci. Res.*, Ser. A, **11**, 115–124. Reprinted by permission of Kluwer Academic Publishers.

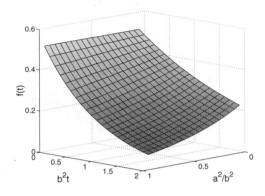

Problem 39

where

$$N_1(z) = \sqrt{\ker_1^2(z) + \kei_1^2(z)}, \qquad \phi_1(z) = \arctan[\kei_1(z)/\ker_1(z)],$$

and $\ker_1(\cdot)$ and $\kei_1(\cdot)$ are Kelvin's ker and kei functions, respectively. The inverse $f(t)$ is plotted as a function of r and t when $\omega = 4$ in the figure labeled Problem 5 at the end of Section 4.4.

39. Find the inverse[44] of the Laplace transform

$$F(s) = \frac{1}{s + b^2} K_0\left(rc\sqrt{\frac{s + a^2}{s + b^2}} \right), \qquad 0 < a < b,$$

where c and r are real and positive. If we take the branch cuts associated with $\sqrt{s + a^2}$, $\sqrt{s + b^2}$ and $K_0(\cdot)$ along the negative real axis of the s-plane, show that

$$f(t) = \frac{1}{2} \int_{a^2}^{b^2} \frac{1}{b^2 - \eta} J_0\left(rc\sqrt{\frac{\eta - a^2}{b^2 - \eta}} \right) e^{-t\eta} \, d\eta.$$

This inverse $f(t)$ is illustrated in the figure labeled Problem 39 with $rc = 1$.

40. Find the inverse[45] of the Laplace transform

$$F(s) = \frac{K_1(a\sqrt{s})}{s \left[bK_1(\sqrt{s}) + \sqrt{s}\, K_0(\sqrt{s}) \right]}, \qquad 1 < a, \quad 0 < b.$$

[44] Taken from Raichenko, L. M., 1976: Flow of liquid to an incomplete well in a bed of fissile-porous rocks. *Sov. Appl. Mech.*, **12**, 1196–1199.

[45] A more complicated version occurred in Das, A., A. Bello, and R. A. Jishi, 1990: Theoretical considerations relating to the characteristic curves of the silver chalcogenide glass inorganic photoresists. *J. Appl. Phys.*, **68**, 3957–3963.

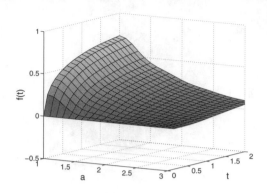

Problem 40

If you take the branch cuts associated with $K_0(\cdot)$, $K_1(\cdot)$ and $\sqrt{s}$ along the negative real axis of the s-plane, you should find that

$$f(t) = \frac{2}{\pi} \int_0^\infty \left(1 - e^{-t\eta^2}\right) \frac{d\eta}{\eta}$$
$$\times \frac{J_1(a\eta)[\eta Y_0(\eta) - bY_1(\eta)] - Y_1(a\eta)[\eta J_0(\eta) - bJ_1(\eta)]}{[\eta J_0(\eta) - bJ_1(\eta)]^2 + [\eta Y_0(\eta) - bY_1(\eta)]^2}.$$

This inverse $f(t)$ is illustrated in the figure labeled Problem 40 with $b = 1$.

41. Find the inverse[46] of the Laplace transform

$$F(s) = \frac{aM}{s} \frac{K_1(r\sqrt{s})}{(Is + 2\pi\mu a^2)K_1(a\sqrt{s}) - 2\pi\mu a^3 \partial K_1(r\sqrt{s})/\partial r|_{r=a}}, \quad 0 < a < r,$$

where M, I and μ are positive constants. If we take the branch cut associated with the Bessel function along the negative real axis of the s-plane, show that

$$f(t) = \frac{M}{4\pi\mu r} \left[1 + \frac{4\chi r}{a\pi} \int_0^\infty \frac{P(\eta)}{\eta^2 Q(\eta)} e^{-t\eta^2/a^2} \, d\eta\right],$$

where

$$P(\eta) = Y_1(r\eta/a)[\eta J_1(\eta) - \chi J_2(\eta)] - J_1(r\eta/a)[\eta Y_1(\eta) - \chi Y_2(\eta)],$$

$$Q(\eta) = [\eta J_1(\eta) - \chi J_2(\eta)]^2 + [\eta Y_1(\eta) - \chi Y_2(\eta)]^2,$$

and $\chi = 2\pi\mu a^4/I$. The inverse $f(t)$ is plotted as a function of r and t in the figure labeled Problem 7 at the end of Section 4.4.

---

[46] Taken from Sennitskii, V. L., 1980: Unsteady rotation of a cylinder in a viscous fluid. *J. Appl. Mech. Tech. Phys.*, **21**, 347–349.

42. Find the inverse[47] of the Laplace transform

$$F(s) = \frac{\exp[-(h-\delta)s/a]}{\sqrt{s}\,[s+a/(2\delta)]\sinh\{[s+a/(2\delta)](R\pi/c)\}}, \qquad 0 < a, c, R, \delta.$$

Step 1: Use the properties of geometric series to show that

$$\frac{1}{\sinh\{[s+a/(2\delta)](R\pi/c)\}} = \frac{2\exp\{-[s+a/(2\delta)](R\pi/c)\}}{1-\exp\{-2[s+a/(2\delta)](R\pi/c)\}}$$

$$= 2\sum_{n=0}^{\infty}\exp\left[-(2n+1)\left(\frac{aR\pi}{2\delta c}\right)\right]\exp\left[-s(2n+1)\frac{R\pi}{c}\right]$$

so that

$$F(s) = 2\sum_{n=0}^{\infty}\exp\left[-(2n+1)\left(\frac{aR\pi}{2\delta c}\right)\right]f_n(s),$$

where

$$f_n(s) = \frac{\exp\{-s[(h-\delta)/a+(2n+1)(R\pi/c)]\}}{\sqrt{s}\,[s+a/(2\delta)]}.$$

Step 2: Use the convolution theorem to show that

$$\mathcal{L}^{-1}\left\{\frac{1}{\sqrt{s}\,[s+a/(2\delta)]}\right\} = \frac{\exp[-at/(2\delta)]}{\sqrt{\pi}}\int_0^t \frac{\exp[a\eta/(2\delta)]}{\sqrt{\eta}}\,d\eta = 2\sqrt{\frac{2\delta}{\pi a}}\,D\left(\sqrt{\frac{at}{2\delta}}\right),$$

where

$$D(x) = e^{-x^2}\int_0^x e^{\eta^2}\,d\eta$$

is Dawson's function.[48]

Step 3: Use the second shifting theorem to show that

$$f(t) = 4\sqrt{\frac{2\delta}{\pi a}}\sum_{n=0}^{\infty}\exp\left[-(2n+1)\left(\frac{aR\pi}{2\delta c}\right)\right]$$

$$\times D\left\{\sqrt{\frac{a}{2\delta}}\left[\tau-(2n+1)\left(\frac{\pi R}{c}\right)\right]\right\}H\left[\tau-(2n+1)\left(\frac{\pi R}{c}\right)\right],$$

[47] Taken from Kahalas, S. L., 1965: Excitation of extremely low frequency electromagnetic waves in the earth-ionosphere cavity by high-altitude nuclear detonations. *J. Geophys. Res.*, **70**, 3587–3594. ©1965 American Geophysical Union. Reproduced/modified by permission of American Geophysical Union.

[48] Dawson, H. G., 1898: On the numerical value of $\int_0^h e^{x^2}\,dx$. *Proc. London Math. Soc., Ser. 1*, **29**, 519–522.

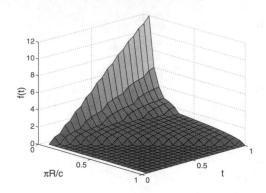

Problem 42

where $\tau = t - (h - \delta)/a$. This inverse $f(t)$ is illustrated in the figure labeled Problem 42.

4.2 NUMERICAL INVERSION OF LAPLACE TRANSFORMS

Due to the importance of Laplace transforms in science and engineering, several methods have been developed for their numerical inversion. Most methods evaluate Bromwich's integral using Fourier series. These are the techniques that will be highlighted here. Another method involves expansions using Laguerre polynomials.[49] Despite work by Weideman,[50] this method is infrequently used. In the case when the transform is not given by a closed analytic function, the inversion process becomes ill-posed.[51] Recently, several authors[52] have sought to use regularization theory to construct a stable inversion scheme. These methods are still in their infancy.

[49] Weeks, W. T., 1966: Numerical inversion of Laplace transforms using Laguerre functions. *J. ACM*, **13**, 419–426; Piessens, R., and M. Branders, 1971: Numerical inversion of the Laplace transform using generalised Laguerre polynomials. *Proc. IEE*, **118**, 1517–1522; Lyness, J. N., and G. Giunta, 1986: A modification of the Weeks method for numerical inversion of the Laplace transform. *Math. Comput.*, **47**, 313–333; Gabutti, B., and P. Lepora, 1987: The numerical performance of Tricomi's formula for inverting the Laplace transform. *Numer. Math.*, **51**, 369–380; Giunta, G., G. Laccetti, and M. R. Rizzardi, 1988: More on the Weeks method for the numerical inversion of the Laplace transform. *Numer. Math.*, **54**, 193–200; Garbow, B. S., G. Giunta, and J. N. Lyness, 1988: Software for an implementation of Weeks' method for the inverse Laplace transform problem. *ACM Trans. Math. Softw.*, **14**, 163–170.

[50] Weideman, J. A. C., 1999: Algorithms for parameter selection in the Weeks method for inverting the Laplace transform. *SIAM J. Scient. Comput.*, **21**, 111–128.

[51] Craig, I. J. D., 1994: Why Laplace transforms are difficult to invert numerically. *Comput. Phys.*, **8**, 648–654.

[52] Kryzhniy, V. V., 2003: Regularized inversion of integral transformations of Mellin convolution type. *Inverse Prob.*, **19**, 1227–1240; Schuster, T., 2003: A stable inversion scheme for the Laplace transform using arbitrarily distributed data scanning points. *J. Inverse Ill-Posed Prob.*, **11**, 263–288.

Having surveyed the various techniques, let us turn to the simplest of all of the inversion schemes: using Fourier series to invert the Laplace transform. Re-expressing Bromwich's integral as a Fourier integral:

$$f(t) = \frac{2e^{ct}}{\pi} \int_0^\infty \mathrm{Re}[F(c + i\eta)] \cos(t\eta)\, d\eta \qquad (4.2.1)$$

$$= -\frac{2e^{ct}}{\pi} \int_0^\infty \mathrm{Im}[F(c + i\eta)] \sin(t\eta)\, d\eta, \qquad (4.2.2)$$

where c denotes any real number whose value is larger than the real part of any singularity of $F(s)$. Applying the trapezoidal rule to Equation 4.2.1, we obtain

$$f(t) = \frac{\Delta\eta}{\pi} e^{ct} \mathrm{Re}[F(c)] + \frac{2\Delta\eta}{\pi} e^{ct} \sum_{n=1}^\infty \mathrm{Re}[F(c + in\Delta\eta)] \cos(n\Delta t). \qquad (4.2.3)$$

A version of Equation 4.2.3 due to Dubner and Abate[53] is

$$f(t) = \frac{e^{A/2}}{2t} \mathrm{Re}\left[F\left(\frac{A}{2t}\right)\right] + \frac{e^{A/2}}{t} \sum_{n=1}^N (-1)^n \, \mathrm{Re}\left[F\left(\frac{A + 2n\pi i}{2t}\right)\right], \qquad (4.2.4)$$

where we have set $\Delta\eta = \pi/(2t)$ and $A = 2ct$. As an alternative to Equation 4.2.4, Hosono[54] applied the midpoint rule to Equation 4.2.2 and showed that

$$f(t) = \frac{e^{A/2}}{t} \sum_{n=1}^N (-1)^n \, \mathrm{Im}\left\{F\left[\frac{A + (2n-1)\pi i}{2t}\right]\right\}. \qquad (4.2.5)$$

To illustrate Equation 4.2.4 and Equation 4.2.5, consider the Laplace transform in Example 4.1.4, namely,

$$F(s) = \frac{1}{s} \exp\left(-\frac{xs}{\sqrt{s^2 + b^2}}\right), \qquad 0 < x. \qquad (4.2.6)$$

As shown there, this transform contains both an isolated singularity (simple pole at $s = 0$) as well as branch points (at $s = \pm bi$). Its inverse is

$$f(t) = 1 - \frac{2}{\pi} \int_0^b \cos(t\eta) \sin\left(\frac{x\eta}{\sqrt{b^2 - \eta^2}}\right) \frac{d\eta}{\eta}. \qquad (4.2.7)$$

[53] Dubner, H., and J. Abate, 1968: Numerical inversion of Laplace transforms by relating them to the finite Fourier cosine transform. *J. ACM*, **15**, 115–123.

[54] Hosono, T., 1979: Numerical inversion of Laplace transform (in Japanese). *Denki Gakkai Ronbunshi (Trans. Inst. Electr. Eng. Japan)*, Ser. A, **99**, 494–500; Hosono, T., 1981: Numerical inversion of Laplace transform and some applications to wave optics. *Radio Sci.*, **16**, 1015–1019.

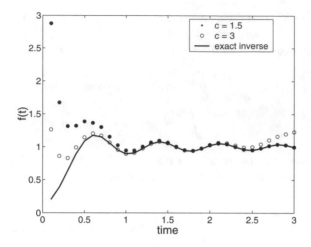

Figure 4.2.1: The numerical inversion of the Laplace transform given by Equation 4.2.6 when $b = 10$ and $x = 2$ using the Dubner–Abate scheme.

Figure 4.2.1 illustrates the numerical inversion of the Laplace transform given by Equation 4.2.6 by the Dubner and Abate technique, Equation 4.2.4, when $N = 42$ for $x = 2$ and $b = 10$ using the MATLAB code

```
clear; A = 2 * c * time; sign = -1; arg = A / (2 * time);
sum = 0.5 * exp(-arg * x / sqrt(arg*arg+b*b)) / arg;
for n = 1:N
arg = arg + pi * i / time;
sum = sum + sign * exp(-arg * x / sqrt(arg*arg+b*b)) / arg;
sign = - sign;
end
f_DB = exp(0.5*A) * real(sum) / time;
```

We see that for a smaller c, the results are poorest at small times and are quite good for large t. When we increases c, the results become better for smaller t and poorer at large times. Why do we obtain better results at small t but lose accuracy at large t as c increases?

Recall that the singularities occur at $s = 0$ and $s = \pm bi$. Therefore, a large c yields a Bromwich contour that passes these singularities at a greater distance and the series gives a better representation of the integrand. On the other hand, the exponential grows as ct and the discretization errors increase. Therefore, if we were to use this method, it would be best to continually decrease c as t increases.

Figure 4.2.2 illustrates the numerical inversion of the Laplace transform given by Equation 4.2.6 using the Hosono scheme when $N = 42$. This figures was created using the MATLAB code

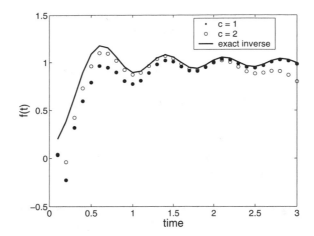

Figure 4.2.2: Same as Figure 4.2.1 except that the Hosono scheme was used.

```
A = 2 * c * time; sign = -1; sum = 0;
arg = (A + i * pi) / (2 * time);
for n = 1:N
sum = sum + sign * exp(-arg * x / sqrt(arg*arg+b*b)) / arg;
sign = - sign;
arg = arg + pi * i / time;
end
f_Hosono = exp(0.5*A) * imag(sum) / time;
```

It is not surprising that the scheme has the same strengths and weaknesses as the Dubner–Abate scheme.

As Figure 4.2.1 and Figure 4.2.2 show, the Dubner–Abate and Hosono schemes, as currently presented, need improvement. Abate and Whitt[55] noted that Equation 4.2.4 and Equation 4.2.5 form alternating series due to the presence of $(-1)^n$. Consequently, they suggested using Euler summation to improve the Dubner–Abate scheme. If we rewrite Equation 4.2.4 as the partial sum $S_n = \sum_{k=0}^{n}(-1)^k a_k(t)$, then the inverse is given by

$$f_{AW}(t) = 2^{-M} \sum_{m=0}^{M} C(m, M)S_{N+m}(t), \qquad C(m, M) = \binom{M}{m}. \qquad (4.2.8)$$

Let us test this scheme using the following MATLAB code. We begin by constructing $C(m, M)$:

```
C(1) = 1;
```

[55] Abate, J., and W. Whitt, 1992: The Fourier-series method for inverting transforms of probability distributions. *Queueing Syst.*, **10**, 5–88.

```
for k = 2:M+1
C(k) = (M+2-k) * C(k-1) / (k-1); end
```

Note that $C(k)$ could also be computed by calling the MATLAB function nchoosek(k,M).

Next, we compute S_N, excluding the e^{ct}/t coefficient:

```
h = pi / time; arg = c;
sum = 0.5 * exp(-arg * x / sqrt(arg*arg+b*b)) / arg;
arg = arg + i * h; sign = -1;
for n = 1:N
sum = sum + sign * exp(-arg * x / sqrt(arg*arg+b*b)) / arg;
sign = - sign;
arg = arg + i * h;
end
S_N(1) = real(sum);
```

Having compute S_N, we compute the additional sums S_{N+m}, $m = 1, 2, \ldots, M$:

```
for k = 1:M
S_N(k+1) = S_N(k) + sign ...
        * real(exp(-arg * x / sqrt(arg*arg+b*b)) / arg);
sign = - sign;
arg = arg + i * h;
end
```

Finally, we compute the inverse using Equation 4.2.8 and include the e^{ct}/t factor:

```
f_AW = 0;
for k = 1:M+1
f_AW = f_AW + C(k) * S_N(k);
end
f_AW = exp(c * time) * f_AW / (2^M * time);
```

Figure 4.2.3 illustrates the numerical inversion of the Laplace transform given by Equation 4.2.6 by the Abate and Whitt technique given by Equation 4.2.8 when $N = 30$ and $M = 11$ for $x = 2$ and $b = 10$. This figure shows that this scheme has two advantages over the Dubner–Abate and Hosono schemes. First, the results are excellent at large t, even when a large c is used. Second, because a large c can be used, the results at small time are better because we can pass the singularities at a larger distance.

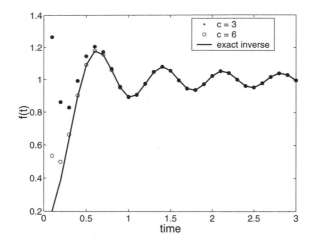

Figure 4.2.3: The numerical inversion of the Laplace transform given by Equation 4.2.6 when $b = 10$ and $x = 2$ using the Abate–Whitt scheme.

As an alternative to the Euler method, De Hoog, Knight and Stokes[56] use the so-called epsilon algorithm to accelerate convergence. Once again, we begin with a Fourier series representation of Bromwich's integral which we write

$$f_N(t) = \frac{e^{ct}}{T} \sum_{n=0}^{N} \mathrm{Re}\left(a_n z^n\right), \qquad (4.2.9)$$

where

$$a_0 = \tfrac{1}{2} F(c), \qquad a_n = F(c + in\pi/T), \qquad z = \exp(in\pi t/T), \qquad (4.2.10)$$

and T is twice the largest t to be found.

As it stands, Equation 4.2.9 could be used to find the inverse. De Hoog et al.'s contribution was to accelerate the convergence of this series by computing a diagonal Padé approximation to this power series. To this end, they developed a quotient-difference algorithm to calculate a trigonometric rational approximation to Equation 4.2.10 in the form of continued fractions:

$$u(z, N) = \sum_{n=0}^{2N} a_n z^n, \quad v(z, N) = d_0/(1 + d_1 z/(1 + \cdots + d_{2N} z)). \qquad (4.2.11)$$

Then, the inverse equals $e^{ct} \mathrm{Re}[v(z, N)]/T$.

[56] De Hoog, F. R., J. H. Knight, and A. N. Stokes, 1982: An improved method for numerical inversion of Laplace transforms. *SIAM J. Sci. Stat. Comput.*, **3**, 357–366.

The coefficients d_n are computed by the following method. Introducing the sequences $e_r^{(k)}$ and $q_r^{(k)}$, we compute their value by

$$e_0^{(k)} = 0, \quad k = 0, 1, \ldots, 2N; \qquad q_1^{(k)} = a_{k+1}/a_k, \quad k = 0, 1, \ldots, 2N - 1;$$
$$(4.2.12)$$

$$e_r^{(k)} = q_r^{(k+1)} - q_r^{(k)} + e_{r-1}^{(k+1)}, \qquad k = 0, 1, \ldots, 2N - 2r, \quad r = 1, 2, \ldots, N,$$
$$(4.2.13)$$

and

$$q_r^{(k)} = q_{r-1}^{(k+1)} e_{r-1}^{(k+1)} / e_{r-1}^{(k)}, \qquad k = 0, 1, \ldots, 2N - 2r - 1, \quad r = 2, 3, \ldots, N.$$
$$(4.2.14)$$

Then,

$$d_0 = a_0, \quad d_{2n-1} = -q_n^{(0)}, \quad d_{2n} = -e_n^{(0)}, \qquad n = 1, 2, \ldots, N. \qquad (4.2.15)$$

For the Laplace transform given by Equation 4.2.6, the MATLAB script that realizes Equation 4.2.10 through Equation 4.2.15 reads

```
arg = c1; TIME = 2*T(length(T));
a(1) = 0.5 * exp(-arg * x / sqrt(arg*arg+b*b)) / arg;
for n = 1:2*N
arg = arg + i * pi / TIME;
a(n+1) = exp(-arg * x / sqrt(arg*arg+b*b)) / arg;
end
% set up e_0^(k) and q_1^(k)
for k = 0:2*N; e(k+1,1) = 0; end
for k = 0:2*N-1; q(k+1,2) = a(k+2) / a(k+1); end
% compute the remaining e_r^(k)
for r = 1:N
k_max = 2 * (N-r);
for k = 0:k_max
e(k+1,r+1) = q(k+2,r+1) - q(k+1,r+1) + e(k+2,r);
end
% compute the remaining q_r^(k)
if (r < N)
for k = 0:k_max-1
q(k+1,r+2) = q(k+2,r+1) * e(k+2,r+1) / e(k+1,r+1);
end; end; end;
% now compute the sequence d_n
d(1) = a(1);
for n = 1:N
d(2*n) = - q(1,n+1);
d(2*n+1) = - e(1,n+1);
end
```

It is important to note here that all of the previous computations are independent of time. For each time t and the corresponding value of z, we find the inverse by first computing

$$A_n = A_{n-1} + d_n z A_{n-2}, \quad B_n = B_{n-1} + d_n z B_{n-2}, \qquad n = 1, 2, \ldots, 2N,$$
$$(4.2.16)$$

with the initial values $A_{-1} = 0$, $B_{-1} = 1$, $A_0 = d_0$ and $B_0 = 1$. The inverse then equals $e^{ct} \text{Re}(A_{2N}/B_{2N})/T$.

So far we have only accelerated the power series representing the inverse. We can also apply the acceleration procedure to the continued fraction itself. De Hoog et al. showed that this occurs if we replace the last evaluation of the recurrence relationship $d_{2N} z$ with

$$R_{2N} = -h_{2N}\left(1 - \sqrt{1 + d_{2N}z/h_{2N}^2}\right) \quad \text{and} \quad h_{2N} = \tfrac{1}{2}\left[1 + (d_{2N-1} - d_{2N})z\right].$$
$$(4.2.17)$$

In this case, the inverse equals $e^{ct} \text{Re}(A'_{2N}/B'_{2N})/T$, where

$$A'_{2N} = A_{2N-1} + R_{2N} A_{2N-2} \quad \text{and} \quad B'_{2N} = B_{2N-1} + R_{2N} B_{2N-2}. \quad (4.2.18)$$

The corresponding MATLAB script code to compute the inverse $f(t)$ at time t is

```
% compute A_n and B_n
z = exp(i*pi*time/TIME);
A(1) = 0; B(1) = 1; A(2) = d(1); B(2) = 1;
for n = 3:2*N+2
term = d(n-1) * z;
A(n) = A(n-1) + term * A(n-2);
B(n) = B(n-1) + term * B(n-2);
end
h_2M = (d(2*N) - d(2*N+1)) * z;
h_2M = 0.5 * (1 + h_2M);
R_2M = 1 + d(2*N+1) * z / (h_2M * h_2M);
R_2M = - h_2M * (1 - sqrt(R_2M));
A_2M = A(2*N+1) + R_2M * A(2*N);
B_2M = B(2*N+1) + R_2M * B(2*N);
inverse = exp(c1*time) * real(A_2M / B_2M) / TIME;
```

Figure 4.2.4 illustrates the DeHoog–Knight–Stokes algorithm for the Laplace transform given by Equation 4.2.6 when $N = 12$.

Albrecht and Honig[57] have analyzed the discretization and truncation errors that arise from the use of Fourier series approximations to Bromwich's

---

[57] Albrecht, P., and G. Honig, 1977: Numerische Inversion der Laplace-Transformierten. *Angew. Info.*, **19**, 336–345; Honig, G., and U. Hirdes, 1984: A method for the numerical inversion of Laplace transforms. *J. Comput. Appl. Math.*, **10**, 113–132.

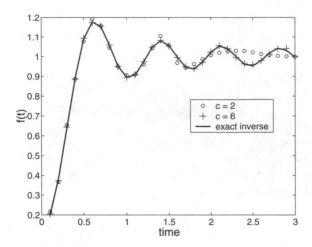

Figure 4.2.4: The numerical inversion of the Laplace transform given by Equation 4.2.6 when $b = 10$ and $x = 2$ using the DeHoog–Knight–Stokes scheme.

integral. They noted that if you chose the parameters so that the discretization becomes arbitrarily small, the truncation error then grows to infinity and vice versa. By analyzing the discretization error, they showed that the Fourier series approximation given by Equation 4.2.10 can be improved by subtracting the leading term of the discretization error, $e^{-2cT} f_N(t + 2T)$. Thus, we can choose a cT that yields both small discretization and truncation errors.

Let us implement Albrecht and Honig's algorithm using the Laplace transform given by Equation 4.2.6. We begin by choosing a value of cT which we denote by `con`. Then, by referring to the uncorrected inversion by `iter = 1` and the correction term by `iter = 2`, we set our parameters for computing f_N as follows:

```
if (iter == 1)
t = time;
TT = TIME;
c1 = con / TT;
else
t = time + 2 * TIME;
TT = 6 * TIME;
c1 = con / TT;
end
```

Here the time at which the inversion is desired is called `time` while `TIME` equals twice the largest value of `time`. Regardless of value of `iter`, f_N is computed as follows:

```
arg = c1; z = exp(i*pi*t/TT); factor = z;
```

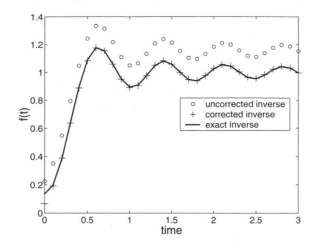

Figure 4.2.5: The numerical inversion of the Laplace transform given by Equation 4.2.6 when $b = 10$ and $x = 2$ using the Albrecht–Honig method.

```
sum = 0.5 * exp(-arg * x / sqrt(arg*arg+b*b)) / arg;
for n = 1:N
arg = arg + i * pi / TT;
a_n = exp(-arg * x / sqrt(arg*arg+b*b)) / arg;
sum = sum + a_n * factor;
factor = factor * z;
end
```

Our final step is to compute the inverse. For graphical purposes, we store the values of the inverse without correction and the inversion with correction:

```
if (iter == 1)
inverse_original=exp(c1*t)*real(sum)/TT;
else
inverse_corrected=inverse_original-exp(c1*t-2*con)*real(sum)/TT;
end
```

Figure 4.2.5 illustrates the Albrecht–Honig method when $N = 20$ and `con = 1`. This method can also be combined with the acceleration methods that we explored earlier.

So far, all of the algorithms have performed an integration along the Bromwich contour. In 1979, Talbot[58] developed an inversion scheme that deforms the Bromwich contour into one whose end points lie deep in the half-plane $\mathrm{Re}(s) < 0$. This new contour eliminates the truncation errors.

---

[58] Talbot, A., 1979: The accurate numerical inversion of Laplace transforms. *J. Inst. Math. Appl.*, **23**, 97–120.

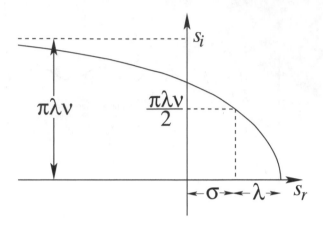

Figure 4.2.6: Talbot's contour.

The scheme consists of two parts. In the first part, the deformed contour is constructed from the equation

$$s(\theta) = \sigma + \lambda\left[\theta\cot(\theta) + i\nu\theta\right], \qquad \theta \in (-\pi, \pi), \tag{4.2.19}$$

and is illustrated in Figure 4.2.6. Of course, for this deformation to be permissible, all of the singularities s_0 must remain to the left of the new contour. Talbot does this by finding the "dominant" singularity – essentially the singularity that lies farthest from the origin of the s-plane – of the num_sing singularities which have the real and imaginary parts s_r and s_i, respectively. A MATLAB realization of his method is as follows:

```
%%%%%%%%%%%%%%%%%%%%%%%%%%%%%%%%%%%%%%%%%%%%%%%%%%%%%%%%%%%%%%%%%%%%
% find the maximum value of s_r of all of the singularities
%%%%%%%%%%%%%%%%%%%%%%%%%%%%%%%%%%%%%%%%%%%%%%%%%%%%%%%%%%%%%%%%%%%%
sigma_0 = 0; p_max = s_r(1);
if (num_sing > 1)
for n = 2:num_sing
if (s_r(n) > p_max) p_max = s_r(n); end % Talbot's eq (58)
end; end
if (p_max > sigma_0) sigma_0 = p_max; end % Talbot's eq (59)
%%%%%%%%%%%%%%%%%%%%%%%%%%%%%%%%%%%%%%%%%%%%%%%%%%%%%%%%%%%%%%%%%%%%
% find ''dominant'' singularity
%%%%%%%%%%%%%%%%%%%%%%%%%%%%%%%%%%%%%%%%%%%%%%%%%%%%%%%%%%%%%%%%%%%%
% assume that all of the singularities are real
r_d = 0; q_d = 0; theta_d = pi;
% now check and see if that is true
for n = 1:num_sing
if (s_i(n) > 0)
```

```
% it's not true, compute new values
eta_n = atan2(sigma_0-s_r(n),s_i(n));
theta_n = eta_n + 0.5 * pi;
r_n = s_i(n) / theta_n;
% see if this is the dominant singularity.
% if so, store its values
if (r_d < r_n)
r_d = r_n; theta_d = theta_n; q_d = s_i(n); % Talbot's eq (61)
end; end
end
```

Because of symmetry we must only consider the singularities in the upper half-plane $\text{Im}(s) \geq 0$. For example, for the transform given by Equation 4.2.6, we have that s_r(1) = 0, s_i(1) = 0 and s_r(2) = 0, s_i(2) = b with num_sing = 2.

Having found the dominant singularity, we can now compute λ, σ and ν so that Talbot's contour passes to the right of all of the singularities. A critical parameter in this calculation is a variable that Talbot calls omega. Its value is based on a numerical analysis of round-off error so that the inversion scheme gives an inverse that possesses cc significant digits. Once omega is computed, λ, σ and ν follow directly:

```
v = q_d * time; % Talbot's eq (66)
omega = min(0.4*(cc+1)+0.5*v,2*(cc+1)/3); % Talbot's eq (77)
if (1.8 * v <= omega * theta_d)
% λ, σ and ν for Talbot's Case 1
lambda = omega / time; % Talbot's eq (68)
sigma = sigma_0; % Talbot's eq (68)
nu = 1; % Talbot's eq (68)
else
kappa = 1.6 + 12 / (v + 25); % Talbot's eq (75)
phi = 1.05 + 1050 / max(553., 800.-v); % Talbot's eq (75)
mu = ( omega / time + sigma_0 - p_max ) ...
    / ( kappa/phi - cot(phi) ); % Talbot's eq (73)
lambda = kappa * mu / phi; % Talbot's eq (74)
sigma = p_max - mu / tan(phi); % Talbot's eq (74)
nu = q_d / mu; % Talbot's eq (74)
end
```

Having computed λ, σ and ν, we are ready to perform the numerical integration. Talbot used trapezoidal integration. Subsequently, Murli and Rizzardi[59] replaced this scheme with a Chebyshev–Clenshaw sum as follows:

---

[59] Murli, A., and M. Rizzardi, 1990: Algorithm 682: Talbot's method for the Laplace inversion problem. *ACM Trans. Math. Softw.*, **16**, 158–168.

```
pi_over_N = pi / N; tau = lambda * time; % Talbot's eq (29)
psi = pi_over_N * tau * nu; c = cos(psi);
b_r = 0; b_i = 0; db_r = 0; db_i = 0;
% compute the first term of the summation
s = lambda + sigma;
F = exp(-s * x / sqrt(s*s+b*b)) / s;
f(j) = 0.5 * nu * exp(tau) * real(F); % n = 0 term
if (N > 1)
%%%%%%%%%%%%%%%%%%%%%%%%%%%%%%%%%%%%%%%%%%%%%%%%%%%%%%%%%%%%%%
% do the case cos(psi) <= 0
%%%%%%%%%%%%%%%%%%%%%%%%%%%%%%%%%%%%%%%%%%%%%%%%%%%%%%%%%%%%%%
if (c <= 0)
u = 4 * cos(0.5 * psi)^2;
for k = N-1:-1:1
theta = k * pi_over_N;
alpha = theta * cot(theta); % Talbot's eq (26)
beta(k) = theta + alpha * (alpha-1) / theta; % Talbot's eq (27)
ss_r = lambda * alpha + sigma; ss_i = lambda * nu * theta;
eat(k) = exp(alpha*tau);
s = ss_r + i * ss_i;
temp = sqrt(s*s+b*b);
if (imag(temp) < 0) temp = - temp; end
F = exp(-s * x / temp) / s;
g(k) = real(F); h(k) = imag(F);
end
for k = N-1:-1:1
b_r = db_r - b_r; b_i = db_i - b_i;
db_r = u * b_r - db_r + eat(k) * (g(k) * nu - h(k) * beta(k));
db_i = u * b_i - db_i + eat(k) * (h(k) * nu + g(k) * beta(k));
end
b_r = db_r - b_r; b_i = db_i - b_i; db_r = u * b_r - db_r;
else
%%%%%%%%%%%%%%%%%%%%%%%%%%%%%%%%%%%%%%%%%%%%%%%%%%%%%%%%%%%%%%
% do the case cos(psi) > 0
%%%%%%%%%%%%%%%%%%%%%%%%%%%%%%%%%%%%%%%%%%%%%%%%%%%%%%%%%%%%%%
u = -4 * sin(0.5*psi)^2;
for k = N-1:-1:1
theta = k * pi_over_N;
alpha = theta * cot(theta); % Talbot's eq (26)
beta(k) = theta + alpha * (alpha-1) / theta; % Talbot's eq (27)
ss_r = lambda * alpha + sigma; ss_i = lambda * nu * theta;
eat(k) = exp(alpha*tau);
s = ss_r + i * ss_i;
temp = sqrt(s*s+b*b);
if (imag(temp) < 0) temp = - temp; end
```

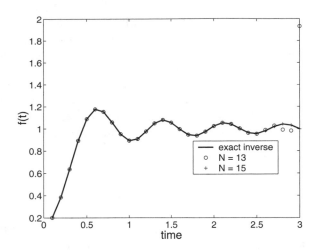

Figure 4.2.7: The numerical inversion of the Laplace transform given by Equation 4.2.6 when $b = 10$ and $x = 2$ using Talbot's method.

```
F = exp(-s * x / temp) / s;
g(k) = real(F); h(k) = imag(F);
end
for k = N-1:-1:1
b_r = db_r + b_r; b_i = db_i + b_i;
db_r = u * b_r + db_r + eat(k) * (g(k) * nu - h(k) * beta(k));
db_i = u * b_i + db_i + eat(k) * (h(k) * nu + g(k) * beta(k));
end
b_r = db_r + b_r; b_i = db_i + b_i; db_r = u * b_r + db_r;
end
%%%%%%%%%%%%%%%%%%%%%%%%%%%%%%%%%%%%%%%%%%%%%%%%%%%%%%%%%%%%%%%%%%%
```

$$\% \text{ compute} \qquad \sum_{k=1}^{N-1} e^{\lambda t s_\nu(\theta_k)} F[\sigma + \lambda s_\nu(\theta_k)] s_\nu'(\theta_k)/i$$

```
%%%%%%%%%%%%%%%%%%%%%%%%%%%%%%%%%%%%%%%%%%%%%%%%%%%%%%%%%%%%%%%%%%%
f(j) = f(j) + db_r - 0.5 * b_r * u - b_i * sin(psi);
end
% Murli and Rizzardi's eq (2.6)
f(j) = lambda * f(j) * exp(sigma * time) / N;
```

Figure 4.2.7 illustrates Talbot's method for the Laplace transform given by Equation 4.2.6. For a given time, the figure illustrates how sensitive the results depend upon N. Indeed, Talbot has developed an algorithm to determine the optimal N for a desired accuracy.

This example also provides a cautionary tale concerning Talbot's method when applied to multivalued Laplace transforms. When the code was initially tested, it used `sqrt(s*s+b*b)` in the evaluation of the Laplace transform.

The results were poor and became worse with increasing N. A closer examination showed that MATLAB's square root is defined so that the real part is always positive (a common definition). Unfortunately, this definition introduces branch cuts along the imaginary axis running from $-\infty i$ to $-bi$ and from bi to ∞i. Consequently, when the integration is taken along Talbot's contour, we cross a branch cut and compute incorrect values of $F(s)$. To eliminate this problem, the square root is modified so that its imaginary part is alway positive; this yields the branch cut that runs from $-bi$ to bi. Once this was done, excellent results were obtained. Similar results occurred when testing $F(s) = 1/\sqrt{s^2 + 1}$.

4.3 THE WAVE EQUATION

In Section 2.2 and Section 2.3 we showed how we may successfully employ Laplace transforms in solving linear wave and heat equations. However, at that time we restricted ourselves to situations where the Laplace transform was single-valued. In this section, we extend our technique to multivalued transforms.

• Example 4.3.1

For our first example, we find the sound waves that propagate away from an exponentially flared loud speaker.[60] To find the pressure field $p(x,t)$ at a distance x away from the throat of the speaker, we must first find the velocity potential $u(x,t)$ because $p(x,t) = \rho_0 u_t(x,t)$, where ρ_0 is the density of air. The wave equation

$$\frac{\partial^2 u}{\partial x^2} + \beta \frac{\partial u}{\partial x} = \frac{1}{c^2}\frac{\partial^2 u}{\partial t^2} \tag{4.3.1}$$

governs the velocity potential where c is the speed of sound and β is the flaring index. At time $t = 0$, we subject the air at the throat ($x = 0$) to a forcing

$$\left.\frac{\partial u}{\partial x}\right|_{x=0} = -\frac{d\xi}{dt} = \omega \xi_0 \sin(\omega t), \tag{4.3.2}$$

where $\omega > \beta c/2$. If $\omega < \beta c/2$, there are no wave solutions. Initially the air is at rest.

We begin our analysis by first taking the Laplace transform of Equation 4.3.1 and find that

$$\frac{d^2 U(x,s)}{dx^2} + \beta \frac{dU(x,s)}{dx} - \frac{s^2}{c^2} U(x,s) = 0, \tag{4.3.3}$$

[60] Taken from McLachlan, N. W., and A. T. McKay, 1936: Transient oscillations in a loud-speaker horn. *Proc. Cambridge Philos. Soc.*, **32**, 265–275. Reprinted with the permission of Cambridge University Press.

where $U(x, s)$ denotes the Laplace transform of $u(x, t)$. The general solution to Equation 4.3.3 is

$$U(x, s) = A(s)e^{-\beta x/2}e^{-x\sqrt{\beta^2 + 4s^2/c^2}/2}, \qquad (4.3.4)$$

where we have discarded the exponentially growing solution because the solution must be finite in the limit of $x \to \infty$.

To evaluate $A(s)$, we take the Laplace transform of Equation 4.3.2 which gives

$$-\frac{dU(x, s)}{dx}\bigg|_{x=0} = \tfrac{1}{2}A(s)\left(\beta + \sqrt{\beta^2 + 4s^2/c^2}\right) = -\frac{\omega^2 \xi_0}{s^2 + \omega^2}, \qquad (4.3.5)$$

or

$$A(s) = \frac{\omega^2 c^2 \xi_0 \left(\beta - \sqrt{\beta^2 + 4s^2/c^2}\right)}{2s^2(s^2 + \omega^2)}. \qquad (4.3.6)$$

Thus, we may write the Laplace transform of the velocity potential

$$U(x, s) = \tfrac{1}{2}\omega^2 c^2 \xi_0 e^{-\beta x/2}\left(\beta + 2\frac{\partial}{\partial x}\right)\left[\frac{e^{-x\sqrt{\beta^2 + 4s^2/c^2}/2}}{s^2(s^2 + \omega^2)}\right]. \qquad (4.3.7)$$

Because $P(x, s) = \rho_0 s U(x, s)$,

$$p(x, t) = \tfrac{1}{2}\rho_0 \omega^2 c^2 \xi_0 e^{-\beta x/2}\left(\beta + 2\frac{\partial}{\partial x}\right)\left[\frac{1}{2\pi i}\int_{c-\infty i}^{c+\infty i}\frac{e^{tz - x\sqrt{\beta^2 + 4z^2/c^2}/2}}{z(z^2 + \omega^2)}\,dz\right]. \qquad (4.3.8)$$

We have written $p(x, t)$ in this form because it is easier to invert the transform inside the brackets rather than in the original expression.

We may evaluate Bromwich's integral in Equation 4.3.8 using complex variables. The integrand has poles at $z = 0$ and $z = \pm\omega i$ and branch points at $z = \pm\beta ci/2 = \pm ai$. For $t < x/c$, Jordan's lemma dictates that we close the contour in the right half-plane and $p(x, t) = 0$ because there are no singularities. However, when $t > x/c$, we must close the contour in the left half-plane as shown in Figure 4.3.1.

Turning our attention first to the poles,

$$\text{Res}\left[\frac{e^{tz - x\sqrt{\beta^2 + 4z^2/c^2}/2}}{z(z^2 + \omega^2)}; 0\right] = \lim_{z \to 0}\frac{e^{tz - x\sqrt{\beta^2 + 4z^2/c^2}/2}}{z^2 + \omega^2} = \frac{e^{-\beta x/2}}{\omega^2}, \qquad (4.3.9)$$

$$\text{Res}\left[\frac{e^{tz - x\sqrt{\beta^2 + 4z^2/c^2}/2}}{z(z^2 + \omega^2)}; \omega i\right] = \lim_{z \to \omega i}\frac{e^{tz - x\sqrt{\beta^2 + 4z^2/c^2}/2}}{z(z + \omega i)} = -\frac{e^{i(\omega t - mx)}}{2\omega^2}, \qquad (4.3.10)$$

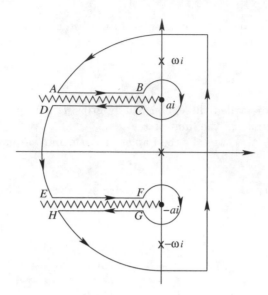

Figure 4.3.1: The contour used to evaluate the Laplace transform in Equation 4.3.7.

and

$$\text{Res}\left[\frac{e^{tz-x\sqrt{\beta^2+4z^2/c^2}/2}}{z(z^2+\omega^2)};-\omega i\right] = \lim_{z\to-\omega i}\frac{e^{tz-x\sqrt{\beta^2+4z^2/c^2}/2}}{z(z-\omega i)} = -\frac{e^{-i(\omega t-mx)}}{2\omega^2},$$

$$(4.3.11)$$

where $m = \sqrt{\omega^2/c^2 - \beta^2/4}$. Thus, the total contribution from the poles is $[e^{-\beta x/2} - \cos(\omega t - mx)]/\omega^2$ and the pressure contribution is

$$p_1(x,t) = -\tfrac{1}{2}\rho_0 c^2 \xi_0 e^{-\beta x/2} H\!\left(t - \frac{x}{c}\right)[\beta\cos(\omega t - mx) + 2m\sin(\omega t - mx)].$$

$$(4.3.12)$$

We now turn to the branch cut integrals. The contributions from the integrations around the branch points $z = \pm ai$ vanish. Patterning our analysis along the lines in Example 4.1.5, we have along AB

$$z = -\eta + ai, \quad \sqrt{z^2 + a^2} = -f_2 + f_1 i, \quad z(z^2 + \omega^2) = -f_3 + f_4 i, \quad (4.3.13)$$

where

$$f_1 = \eta^{1/2}\left(\sqrt{a^2 + \eta^2/4} - \eta/2\right)^{1/2}, \qquad (4.3.14)$$

$$f_2 = \eta^{1/2}\left(\sqrt{a^2 + \eta^2/4} + \eta/2\right)^{1/2}, \qquad (4.3.15)$$

$$f_3 = \eta(\eta^2 + \omega^2 - 3a^2), \qquad (4.3.16)$$

$$f_4 = a(3\eta^2 + \omega^2 - a^2), \qquad (4.3.17)$$

and $0 < \eta < \infty$. Along CD, the only change is $\sqrt{z^2 + a^2} = f_2 - f_1 i$. Along the other branch cut,

$$z = -\eta - ai, \quad \sqrt{z^2 + a^2} = f_2 + f_1 i, \quad z(z^2 + \omega^2) = -f_3 - f_4 i \quad \textbf{(4.3.18)}$$

for EF. Finally, the only difference between EF and GH is

$$\sqrt{z^2 + a^2} = -f_2 - f_1 i. \quad \textbf{(4.3.19)}$$

Then,

$$\int_{AB} F(z)e^{tz}\, dz = -\int_{\infty}^{0} \frac{e^{-t\eta}e^{iat}e^{-if_1 x/c}e^{f_2 x/c}}{-f_3 + f_4 i}\, d\eta, \quad \textbf{(4.3.20)}$$

$$\int_{CD} F(z)e^{tz}\, dz = -\int_{0}^{\infty} \frac{e^{-t\eta}e^{iat}e^{if_1 x/c}e^{-f_2 x/c}}{-f_3 + f_4 i}\, d\eta, \quad \textbf{(4.3.21)}$$

$$\int_{EF} F(z)e^{tz}\, dz = -\int_{\infty}^{0} \frac{e^{-t\eta}e^{-iat}e^{-if_1 x/c}e^{-f_2 x/c}}{-f_3 - f_4 i}\, d\eta, \quad \textbf{(4.3.22)}$$

and

$$\int_{GH} F(z)e^{tz}\, dz = -\int_{0}^{\infty} \frac{e^{-t\eta}e^{-iat}e^{if_1 x/c}e^{f_2 x/c}}{-f_3 - f_4 i}\, d\eta. \quad \textbf{(4.3.23)}$$

We may combine Equation 4.3.20 through Equation 4.3.23 together, which yields

$$\frac{1}{2\pi i} \int_{ABUCDUEFUGH} \frac{e^{tz - (x/c)\sqrt{z^2 + a^2}}}{z(z^2 + \omega^2)}\, dz$$

$$= -\frac{1}{\pi} \int_{0}^{\infty} \frac{f_3 e^{-t\eta}}{f_3^2 + f_4^2}\, d\eta$$

$$\times \left[e^{f_2 x/c} \sin(at - f_1 x/c) - e^{-f_2 x/c} \sin(at + f_1 x/c) \right]$$

$$- \frac{1}{\pi} \int_{0}^{\infty} \frac{f_4 e^{-t\eta}}{f_3^2 + f_4^2}\, d\eta$$

$$\times \left[e^{f_2 x/c} \cos(at - f_1 x/c) - e^{-f_2 x/c} \cos(at + f_1 x/c) \right],$$

$$\textbf{(4.3.24)}$$

and

$$p_2(x, t) = \frac{\rho_0 \omega^2 c^2 \xi_0}{2\pi} e^{-\beta x/2} H\left(t - \frac{x}{c}\right) \int_{0}^{\infty} \frac{f_3 e^{-t\eta}}{f_3^2 + f_4^2}\, d\eta$$

$$\times \left[\beta e^{f_2 x/c} \sin(at - f_1 x/c) - \beta e^{-f_2 x/c} \sin(at + f_1 x/c) \right.$$

$$\left. + 2f_2 e^{f_2 x/c} \sin(at - f_1 x/c)/c + 2f_2 e^{-f_2 x/c} \sin(at + f_1 x/c)/c \right.$$

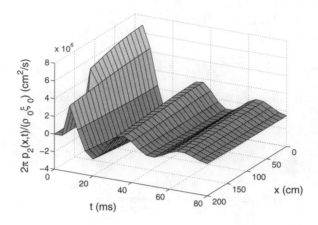

Figure 4.3.2: The evolution of the transient pressure field $2\pi p_2(x,t)/(\rho_0\xi_0)$ generated by the sounding of a loud speaker for parameters stated in the text.

$$
\begin{aligned}
&\left. - 2f_1 e^{f_2 x/c} \cos(at - f_1 x/c)/c - 2f_1 e^{-f_2 x/c} \cos(at + f_1 x/c)/c \right] \\
&+ \frac{\rho_0 \omega^2 c^2 \xi_0}{2\pi} e^{-\beta x/2} H\left(t - \frac{x}{c}\right) \int_0^\infty \frac{f_4 e^{-t\eta}}{f_3^2 + f_4^2}\, d\eta \\
&\times \left[\beta e^{f_2 x/c} \cos(at - f_1 x/c) - \beta e^{-f_2 x/c} \cos(at + f_1 x/c) \right. \\
&+ 2f_2 e^{f_2 x/c} \cos(at - f_1 x/c)/c + 2f_2 e^{-f_2 x/c} \cos(at + f_1 x/c)/c \\
&\left. + 2f_1 e^{f_2 x/c} \sin(at - f_1 x/c)/c + 2f_1 e^{-f_2 x/c} \sin(at + f_1 x/c)/c \right].
\end{aligned}
$$

$$(4.3.25)$$

The sum of $p_1(x,t)$ and $p_2(x,t)$ gives the total radiation field. Figure 4.3.2 presents $p_2(x,t)$ for $c = 3.43 \times 10^4$ cm s^{-1}, $\beta = 0.0119$ cm^{-1} and $\omega = 256$ Hz.

Problems

1. Solve the partial differential equation[61]

$$
\frac{\partial^2 u}{\partial t^2} = \frac{\partial^2 u}{\partial x^2} + \frac{\partial^3 u}{\partial t \partial x^2}, \qquad 0 < x < \infty, \quad 0 < t,
$$

with the boundary conditions $u(0,t) = \sin(t)$ and $\lim_{x\to\infty} |u(x,t)| < \infty$, and the initial conditions $u(x,0) = 0$ and $u_t(x,0) = u_{xx}(x,0)$.

---

[61] For a more complicated version of this problem, see Jordan, P. M., M. R. Meyer, and A. Puri, 2000: Causal implications of viscous damping in compressible fluid flow. *Phys. Review E*, **62**, 7918–7926.

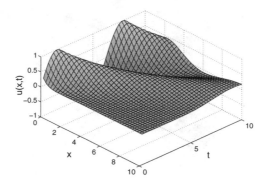

Problem 1

Step 1: Show that the Laplace transform of the partial differential equation yields the boundary-value problem

$$\frac{d^2 U(x,s)}{dx^2} - \frac{s^2}{s+1} U(x,s) = 0,$$

with

$$U(0,s) = \frac{1}{s^2 + 1} \qquad \text{and} \qquad \lim_{x \to \infty} |U(x,s)| < \infty.$$

Step 2: Show that the solution to the boundary-value problem is

$$U(x,s) = \frac{1}{s^2 + 1} \exp\left(-\frac{sx}{\sqrt{1+s}} \right).$$

Step 3: Show that the singularities in the Laplace transform consist of simple poles at $s = \pm i$ and a branch point at $s = -1$.

Step 4: By defining the branch cut along the negative real axis of the s-plane, show that we may express the solution as a sum of the residues at the poles plus an integral along the branch cut:

$$u(x,t) = \exp\left[-x \sin(\pi/8) / \sqrt[4]{2} \right] \sin\left[t - x \cos(\pi/8) / \sqrt[4]{2} \right]$$

$$- \frac{e^{-t}}{\pi} \int_0^\infty \frac{\exp(-t\eta)}{\eta^2 + 2\eta + 2} \sin\left[\frac{(1+\eta)x}{\sqrt{\eta}} \right] d\eta.$$

This result shows that the solution consists of two parts: The first term is the steady-state solution while the integral gives the transients that dies away exponentially fast. Unfortunately, this solution is of little value in any numerical computation due to the behavior of the sine term in the integrand. The figure labeled Problem 1 was constructed by numerically inverting the Laplace transform using Talbot's method.

2. Solve the partial differential equation[62]

$$\frac{\partial^2 u}{\partial t^2} + \frac{\partial u}{\partial t} = \frac{\partial^2 u}{\partial x^2} + a \frac{\partial^3 u}{\partial t \partial x^2}, \qquad 0 < x < \infty, \quad 0 < a, t,$$

---

[62] Tanner, *op. cit.*

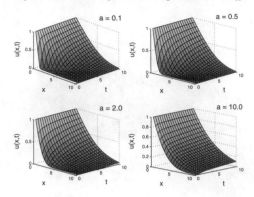

Problem 2

with the boundary conditions $u(0, t) = 1$ and $\lim_{x \to \infty} |u(x, t)| < \infty$, and the initial conditions $u(x, 0) = 0$ and $u_t(x, 0) = au_{xx}(x, 0)$.

Step 1: Show that the Laplace transform of the partial differential equation and boundary conditions yields the boundary-value problem

$$(1 + as)\frac{d^2U(x, s)}{dx^2} - s(1 + s)U(x, s) = 0,$$

with

$$U(0, s) = \frac{1}{s} \qquad \text{and} \qquad \lim_{x \to \infty} |U(x, s)| < \infty.$$

Step 2: Show that the solution to the boundary-value problem is

$$U(x, s) = \frac{1}{s}\exp\left[-x\sqrt{\frac{s(1 + s)}{1 + as}}\right].$$

Step 3: Show that the singularities in the Laplace transform consist of a simple pole at $s = 0$ and branch points at $s = 0$, $s = -1$ and $s = -1/a$.

Step 4: By defining all of the branch cuts along the negative real axis of the s-plane and deforming the Bromwich contour along the imaginary axis of the s-plane except for a small semicircle around $s = 0$, show that

$$u(x, t) = \frac{1}{2} + \frac{1}{\pi}\int_0^\infty \exp\left\{-x\sqrt{\frac{\eta}{2}}\, M[\cos(\theta) - \sin(\theta)]\right\}\frac{d\eta}{\eta}$$

$$\times \sin\left\{t\eta - x\sqrt{\frac{\eta}{2}}\, M[\cos(\theta) + \sin(\theta)]\right\},$$

where

$$M = \left(\frac{1 + \eta^2}{1 + a^2\eta^2}\right)^{1/4} \qquad \text{and} \qquad 2\theta = \arctan(\eta) - \arctan(a\eta).$$

The figure labeled Problem 2 illustrates the solution for various values of a when $t > 0$.

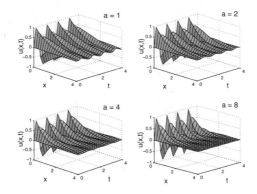

Problem 3

3. Solve the partial differential equation[63]

$$\frac{\partial u}{\partial t} + a\frac{\partial^2 u}{\partial t^2} - b\frac{\partial^3 u}{\partial x^2 \partial t} - \frac{\partial^2 u}{\partial x^2} = 0, \qquad 0 < a < b, \quad 0 < x < \infty, \quad 0 < t,$$

with the boundary conditions that $u(0,t) = \sin(2\pi t)$ and $\lim_{x\to\infty} |u(x,t)| < \infty$, and the initial conditions that $u(x,0) = u_t(x,0) = 0$.

Step 1: By taking the Laplace transform of the partial differential equation and boundary conditions, show that the partial differential equation reduces to the boundary-value problem

$$\frac{d^2 U(x,s)}{dx^2} - \frac{s(1+as)}{1+bs}U(x,s) = 0,$$

with $U(0,s) = 2\pi/(s^2 + 4\pi^2)$ and $\lim_{x\to\infty} |U(x,s)| < \infty$.

Step 2: Show that the solution to the boundary-value problem is

$$U(x,s) = \frac{2\pi}{s^2 + 4\pi^2} \exp\left[-x\sqrt{\frac{s(1+as)}{1+bs}}\right].$$

Step 3: Show that the Laplace transform in Step 2 has simple poles at $s = \pm 2\pi i$ and branch points at $s = 0$, $s = -1/b$ and $s = -1/a$.

Step 4: By taking the branch cuts associated with each branch point along the negative real axis of the s-plane, use Bromwich's integral and show that

$$u(x,t) = \sin[2\pi t - x\Delta\sin(\psi/2)]\exp[-x\Delta\cos(\psi/2)] + 2\int_0^{1/b} F(x,t,\eta)\,d\eta + 2\int_{1/a}^\infty F(x,t,\eta)\,d\eta,$$

where

$$F(x,t,\eta) = \frac{\exp(-t\eta)}{\eta^2 + 4\pi^2}\sin\left[x\sqrt{\frac{\eta(1-a\eta)}{1-b\eta}}\right],$$

$$\Delta = [(d_1^2 + d_2^2)d_3^2]^{1/4}, \quad d_1 = 1 + 4\pi^2 ab, \quad d_2 = 2\pi(b-a),$$

$$d_3 = \frac{2\pi}{1+4\pi^2 b^2}, \qquad \text{and} \qquad \psi = \arctan(d_1/d_2).$$

The figure labeled Problem 3 illustrates the solution for various values of a when $b = 10$.

---

[63] Zheltov and Kutlyarov, *op. cit.*

4.4 THE HEAT EQUATION

In Chapter 2, we showed how we may successfully employ Laplace transforms in solving linear partial differential equations. However, at that time we restricted ourselves to situations where the Laplace transform was single-valued. In this section we extend our technique to multivalued transforms.

• Example 4.4.1

In their study of the seepage of a homogeneous fluid in fissured rocks, Barenblatt et al.[64] solved an equation similar to

$$\frac{\partial u}{\partial t} = \frac{\partial^2 u}{\partial x^2} + \frac{\partial^3 u}{\partial t \partial x^2}, \qquad 0 < x < \infty, \quad 0 < t, \tag{4.4.1}$$

with the boundary conditions

$$u(0, t) = 1 - e^{-t} \qquad \text{and} \qquad \lim_{x \to \infty} |u(x, t)| < \infty, \qquad 0 < t, \tag{4.4.2}$$

and the initial condition

$$u(x, 0) = 0, \qquad 0 < x < \infty. \tag{4.4.3}$$

Upon applying the Laplace transform to Equation 4.4.1 and using the boundary and initial conditions, it becomes

$$\frac{d^2 U(x, s)}{dx^2} - \frac{s}{s+1} U(x, s) = 0 \tag{4.4.4}$$

with

$$U(0, s) = \frac{1}{s(s+1)} \qquad \text{and} \qquad \lim_{x \to \infty} |U(x, s)| < \infty. \tag{4.4.5}$$

The general solution to Equation 4.4.4 that satisfies the boundary conditions is

$$U(x, s) = \frac{1}{s(s+1)} \exp\left(-x\sqrt{\frac{s}{s+1}}\right). \tag{4.4.6}$$

The inversion of Equation 4.4.7 is

$$u(x, t) = \frac{1}{2\pi i} \int_{c-\infty i}^{c+\infty i} \frac{e^{ts}}{s(s+1)} \exp\left(-x\sqrt{\frac{s}{s+1}}\right) ds, \tag{4.4.7}$$

---

[64] Reprinted from *J. Appl. Math. Mech.*, **24**, G. I. Barenblatt, Iu. P. Zheltov and I. N. Kochina, Basic concepts in the theory of seepage of homogeneous liquids in fissured rocks (strata), 1286–1303, ©1960, with kind permission from Pergamon Press Ltd., Headington Hill Hall, Oxford OX3 0BX, UK.

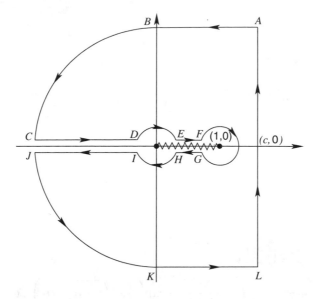

Figure 4.4.1: The contour used to evaluate Equation 4.4.8.

where $c > 0$. We facilitate the evaluation of the integral in Equation 4.4.7 by the change of variables $s = z - 1$, yielding

$$u(x,t) = \frac{e^{-t}}{2\pi i} \int_{c-\infty i}^{c+\infty i} \frac{e^{tz}}{z(z-1)} \exp\left(-x\sqrt{\frac{z-1}{z}}\right) dz, \qquad (4.4.8)$$

where $c > 1$.

Because of the square root in the exponential in Equation 4.4.8, the integrand is a multivalued function. A quick check shows that $z = 0$ and $z = 1$ are the branch points of the integrand. A convenient choice for the branch cut is the line segment lying along the real axis running between $z = 0$ and $z = 1$.

At this point we convert the line integral running from $c-\infty i$ to $c+\infty i$ into a closed contour so that we can apply the residue theorem. Figure 4.4.1 shows the contour that we will use. The contribution from the arcs ABC and JKL are negligibly small by Jordan's lemma (see Section 1.4). The contribution from the line segment CD cancels the contribution from IJ. Consequently, the dumbbell-shaped contour integral shown in Figure 4.4.2 is equivalent to the contour $ABCDEFGHIJKL$ shown in Figure 4.4.1. Because there are no singularities inside the closed contour, the value given by the contour integral shown in Figure 4.4.2 must equal the negative of the integral from $c - \infty i$ to $c + \infty i$.

Along C_1, $z = \epsilon e^{\theta i}$ and $dz = i\epsilon e^{\theta i} d\theta$ so that

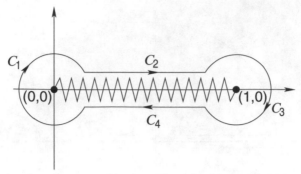

Figure 4.4.2: Another contour used in the evaluation of Equation 4.4.8.

$$\frac{e^{-t}}{2\pi i} \int_{C_1} \frac{e^{tz}}{z(z-1)} \exp\left(-x\sqrt{\frac{z-1}{z}}\right) dz$$

$$= \frac{e^{-t}}{2\pi} \lim_{\epsilon \to 0} \int_{2\pi}^{0} \frac{\exp(\epsilon t e^{\theta i})}{-1 + \epsilon e^{\theta i}} \exp\left(-x\sqrt{1 - e^{-\theta i}/\epsilon}\right) d\theta \qquad (4.4.9)$$

$$= -\frac{e^{-t}}{2\pi} \lim_{\epsilon \to 0} \int_{2\pi}^{0} \exp\left(-x\frac{e^{-\theta i/2 + i\pi/2}}{\sqrt{\epsilon}}\right) d\theta = 0. \qquad (4.4.10)$$

Along C_3, $z = 1 + \epsilon e^{\theta i}$, $dz = i\epsilon e^{\theta i} d\theta$ and

$$\frac{e^{-t}}{2\pi i} \int_{C_3} \frac{e^{tz}}{z(z-1)} \exp\left(-x\sqrt{\frac{z-1}{z}}\right) dz$$

$$= \frac{e^{-t}}{2\pi} \lim_{\epsilon \to 0} \int_{\pi}^{-\pi} \frac{\exp(\epsilon t e^{\theta i})}{1 + \epsilon e^{\theta i}} \exp\left(-x\sqrt{\frac{\epsilon e^{\theta i}}{1 + \epsilon e^{\theta i}}}\right) d\theta = -1. \quad (4.4.11)$$

Along C_2, $z = \sigma e^{0i}$, $dz = d\sigma$, $z - 1 = (1-\sigma)e^{\pi i}$ and

$$\frac{e^{-t}}{2\pi i} \int_{C_2} \frac{e^{tz}}{z(z-1)} \exp\left(-x\sqrt{\frac{z-1}{z}}\right) dz$$

$$= \frac{e^{-t}}{2\pi i} \int_0^1 \frac{e^{\sigma t}}{\sigma(\sigma - 1)} \exp\left(-ix\sqrt{\frac{1-\sigma}{\sigma}}\right) d\sigma. \quad (4.4.12)$$

Along C_4, $z = \sigma e^{0i}$, $dz = d\sigma$, $z - 1 = (1-\sigma)e^{-\pi i}$ and

$$\frac{e^{-t}}{2\pi i} \int_{C_4} \frac{e^{tz}}{z(z-1)} \exp\left(-x\sqrt{\frac{z-1}{z}}\right) dz$$

$$= \frac{e^{-t}}{2\pi i} \int_1^0 \frac{e^{\sigma t}}{\sigma(\sigma - 1)} \exp\left(ix\sqrt{\frac{1-\sigma}{\sigma}}\right) d\sigma. \quad (4.4.13)$$

Therefore,

$$u(x,t) = 1 + \frac{1}{\pi} \int_0^1 \frac{e^{t(\sigma-1)}}{\sigma(\sigma-1)} \sin\left(x\sqrt{\frac{1-\sigma}{\sigma}}\right) d\sigma \qquad (4.4.14)$$

$$= 1 - \frac{1}{\pi} \int_0^1 \frac{e^{-t\eta}}{\eta(1-\eta)} \sin\left(x\sqrt{\frac{\eta}{1-\eta}}\right) d\eta \qquad (4.4.15)$$

$$= 1 - \frac{2}{\pi} \int_0^\infty \frac{\sin(x\nu)}{\nu} \exp\left(-\frac{t\nu^2}{1+\nu^2}\right) d\nu, \qquad (4.4.16)$$

where we introduced the new variables $\eta = 1 - \sigma$ and $\nu^2 = \eta/(1-\eta)$. Figure 4.4.3 illustrates Equation 4.4.16.

• **Example 4.4.2**

Next, let us find the heat flow in a two-layer earth,[65] where

$$\frac{\partial u_1}{\partial t} = a_1 \frac{\partial^2 u_1}{\partial x^2}, \qquad 0 \le x \le h, \quad 0 < t, \qquad (4.4.17)$$

and

$$\frac{\partial u_2}{\partial t} = a_2 \frac{\partial^2 u_2}{\partial x^2}, \qquad h \le x, \quad 0 < t, \qquad (4.4.18)$$

a_i denotes the diffusivity of the layer, u_i denotes the temperature and x denotes the distance down from the surface. At the earth's surface, there is a diurnal cycle $u_1(0,t) = T_0 \sin(\omega t)$, while at the interface between the two layers,

$$u_1(h,t) = u_2(h,t) \qquad \text{and} \qquad k_1 \frac{\partial u_1(h,t)}{\partial x} = k_2 \frac{\partial u_2(h,t)}{\partial x}, \qquad (4.4.19)$$

where k_i denotes the thermal conductivity. We also require that $\lim_{x \to \infty} |u_2(x,t)| < \infty$. The initial conditions are $u_1(x,0) = u_2(x,0) = 0$.

Taking the Laplace transform of Equation 4.4.17 and Equation 4.4.18,

$$\frac{d^2 U_i(x,s)}{dx^2} - \frac{s}{a_i} U_i(x,s) = 0, \qquad i = 1,2, \qquad (4.4.20)$$

while the boundary conditions are

$$U_1(0,s) = \frac{T_0\omega}{s^2 + \omega^2}, \qquad U_1(h,s) = U_2(h,s), \qquad (4.4.21)$$

---

[65] The steady-state solution was found by Parasnis, D. S., 1976: Effect of a uniform overburden on the passage of a thermal wave and the temperatures in the underlying rock. *Geophys. J. R. Astr. Soc.*, **46**, 189–192.

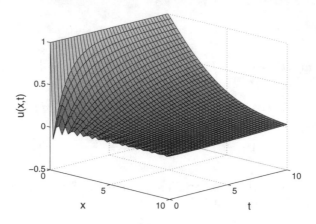

Figure 4.4.3: The seepage of a homogeneous fluid in fissured rocks as given by Equation 4.4.16 as a function of distance x and time t.

and

$$k_1 \frac{dU_1(h,s)}{dx} = k_2 \frac{dU_2(h,s)}{dx}, \qquad \lim_{x \to \infty} |U_2(x,s)| < \infty. \qquad (4.4.22)$$

The solutions to Equation 4.4.20 through Equation 4.4.22 are

$$U_1(x,s) = \frac{T_0 \omega}{s^2 + \omega^2} \exp\!\left(x\sqrt{s/a_1}\right) - \frac{T_0 \omega}{s^2 + \omega^2} \frac{\exp\!\left(x\sqrt{s/a_1}\right)}{1 + \beta \exp\!\left(-2h\sqrt{s/a_1}\right)}$$

$$+ \frac{T_0 \omega}{s^2 + \omega^2} \frac{\exp\!\left(-x\sqrt{s/a_1}\right)}{1 + \beta \exp\!\left(-2h\sqrt{s/a_1}\right)} \qquad (4.4.23)$$

and

$$U_2(x,s) = \frac{T_0 \omega (1+\beta)}{s^2 + \omega^2} \frac{\exp\!\left[h\!\left(\sqrt{s/a_2} - \sqrt{s/a_1}\right) - x\sqrt{s/a_2}\right]}{1 + \beta \exp\!\left(-2h\sqrt{s/a_1}\right)}, \qquad (4.4.24)$$

where $\beta = \left(k_1\sqrt{a_2} - k_2\sqrt{a_1}\right) / \left(k_1\sqrt{a_2} + k_2\sqrt{a_1}\right)$. To invert Equation 4.4.23 and Equation 4.4.24, we note that we can express $\left[1 + \beta \exp\!\left(-2h\sqrt{s/a_1}\right)\right]^{-1}$ as the geometric series $\sum_{n=0}^{\infty}(-1)^n \beta^n e^{-2nh\sqrt{s/a_1}}$, if $\left|\beta e^{-2h\sqrt{s/a_1}}\right| < 1$. Upon expanding the denominator of these equations, we find

$$U_1(x,s) = \frac{T_0 \omega}{s^2 + \omega^2} \exp\!\left(x\sqrt{s/a_1}\right)$$

$$- \frac{T_0 \omega}{s^2 + \omega^2} \sum_{n=0}^{\infty} (-1)^n \beta^n \exp\!\left[-(2nh - x)\sqrt{s/a_1}\right]$$

$$+ \frac{T_0 \omega}{s^2 + \omega^2} \sum_{n=0}^{\infty} (-1)^n \beta^n \exp\!\left[-(2nh + x)\sqrt{s/a_1}\right] \qquad (4.4.25)$$

and

$$U_2(x, s) = \frac{T_0 \omega (1 + \beta)}{s^2 + \omega^2} \tag{4.4.26}$$

$$\times \sum_{n=0}^{\infty} (-1)^n \beta^n \exp\left\{ - \left[(2n + 1)h + (x - h)\sqrt{a_1/a_2} \right] \sqrt{s/a_1} \right\}.$$

Using Bromwich's integral, we can show (see Problem 23 at the end of Section 4.1) that the inverse of

$$F(x, s) = \frac{\omega}{s^2 + \omega^2} \exp\left(x \sqrt{s/\kappa} \right) \tag{4.4.27}$$

is

$$f(x, t) = \exp\left(x\sqrt{\omega/2\kappa} \right) \sin\left(\omega t + x\sqrt{\omega/2\kappa} \right)$$

$$+ \frac{2\kappa}{\pi} \int_0^{\infty} \frac{\omega}{\omega^2 + \kappa^2 \eta^4} \exp(-\kappa t \eta^2) \sin(x\eta)\, \eta\, d\eta. \tag{4.4.28}$$

At this point, we note that we can express both Equation 4.4.25 and Equation 4.4.26 in terms of $f(x, t)$. Inverting these equations term by term, the final solution is

$$u_1(x, t) = T_0 f(x, t) - T_0 \sum_{n=0}^{\infty} (-1)^n \beta^n f(x - 2nh, t)$$

$$+ T_0 \sum_{n=0}^{\infty} (-1)^n \beta^n f(-x - 2nh, t) \tag{4.4.29}$$

and

$$u_2(x, t) = (1 + \beta) T_0 \sum_{n=0}^{\infty} (-1)^n \beta^n f\left[(h - x)\sqrt{a_1/a_2} - (2n + 1)h, t \right]. \tag{4.4.30}$$

• Example 4.4.3

For our final example,[66] we solve the partial differential equation

$$\frac{\partial u}{\partial t} = \frac{\partial}{\partial x} \left[(1 + x^2) \frac{\partial u}{\partial x} \right], \qquad 0 < x < \infty, \quad 0 < t, \tag{4.4.31}$$

---

[66] Taken from Moshinskii, A. I., 1987: Effective diffusion of a dynamically passive impurity in narrow channels. *J. Appl. Mech. Tech. Phys.*, **28**, 374–382.

subject to the boundary conditions $\lim_{x\to\infty}|u_x(x,t)| < \infty$ and $u(0,t) = 0$, and the initial condition $u(x,0) = 1$.

If the Laplace transform of $u(x,t)$ is

$$U(x,s) = \int_0^\infty u(x,t)e^{-st}\,dt, \qquad (4.4.32)$$

the partial differential equation and boundary conditions reduce to the boundary-value problem

$$\frac{d}{dx}\left[(1+x^2)\frac{dU(x,s)}{dx}\right] - sU(x,s) = -1, \qquad 0 < x < \infty, \qquad (4.4.33)$$

subject to the boundary conditions $\lim_{x\to\infty}|U'(x,s)| < \infty$ and $U(0,s) = 0$. The particular solution to Equation 4.4.33 is $1/s$. If we introduce $s = \nu(\nu+1)$ into Equation 4.4.33, the left side becomes Legendre's equation. Consequently, the homogeneous solution is

$$U(x,s) = AP_\nu(xi) + BQ_\nu(xi), \qquad (4.4.34)$$

where $\nu = -\frac{1}{2} + \sqrt{\frac{1}{4} + s}$, $P_\nu(\cdot)$ and $Q_\nu(\cdot)$ are Legendre functions of order ν of the first and second kind,[67] respectively. Because the power series for $P_\nu(z)$ becomes unbounded as $|z| \to \infty$ while $Q_\nu(z) \to 0$, we must discard the $P_\nu(z)$ solutions. Consequently, the solution to Equation 4.4.33 is

$$U(x,s) = \frac{1}{s} - \frac{Q_\nu(xi)}{sQ_\nu(0^+i)}. \qquad (4.4.35)$$

We have written 0^+i for the argument of $Q_\nu(\cdot)$ because it is a multivalued function and we must cut the x-plane with a branch cut along the real axis from $-\infty$ to 1. Consequently, 0^+i denotes a point lying on the imaginary axis just above the branch cut.

From Bromwich's integral, the inverse of $U(x,s)$ is

$$u(x,t) = \frac{1}{2\pi i}\int_{c-\infty i}^{c+\infty i}\left[\frac{1}{z} - \frac{Q_\nu(xi)}{zQ_\nu(0^+i)}\right]e^{tz}\,dz. \qquad (4.4.36)$$

We close the line integral in Equation 4.4.36 with an infinite semicircle on the left side of the complex z-plane excluding the branch cut along the negative real axis from $z = -\infty$ to $z = -\frac{1}{4}$ which is associated with the multivalued ν. We also have a simple pole at $z = 0$. The intriguing aspect of this problem is the presence of a multivalued function in the order of the Legendre function.

---

[67] See Lebedev, *op. cit.*, Chapter 7.

Computing first the residue at $z = 0$,

$$\text{Res}\left\{\left[\frac{1}{z} - \frac{Q_\nu(xi)}{zQ_\nu(0^+i)}\right]e^{tz}; 0\right\} = \left[1 - \frac{Q_0(xi)}{Q_0(0^+i)}\right] = 1 + \frac{2}{\pi}\arctan(x), \quad (4.4.37)$$

because $Q_0(z) = \tanh(z) = \frac{1}{2}\log[(z+1)/(z-1)]$. On the other hand, along the branch cut, we have $z + \frac{1}{4} = \eta^2 e^{\pm\pi i}$ so that along the top of the branch cut,

$$\int_{\substack{top \\ branch\ cut}}\left[\frac{1}{z} - \frac{Q_\nu(xi)}{zQ_\nu(0^+i)}\right]e^{tz}\,dz$$

$$= -2e^{-t/4}\int_0^\infty \frac{Q_{-\frac{1}{2}+\eta i}(xi)}{(\frac{1}{4}+\eta^2)Q_{-\frac{1}{2}+\eta i}(0^+i)}e^{-t\eta^2}\eta\,d\eta, \quad (4.4.38)$$

while just below the branch cut,

$$\int_{\substack{bottom \\ branch\ cut}}\left[\frac{1}{z} - \frac{Q_\nu(xi)}{zQ_\nu(0^+i)}\right]e^{tz}\,dz$$

$$= 2e^{-t/4}\int_0^\infty \frac{Q_{-\frac{1}{2}-\eta i}(xi)}{(\frac{1}{4}-\eta^2)Q_{-\frac{1}{2}-\eta i}(0^+i)}e^{-t\eta^2}\eta\,d\eta. \quad (4.4.39)$$

Combining Equation 4.4.38 and Equation 4.4.39 yields

$$\int_{\substack{branch \\ cut}}\left[\frac{1}{z} - \frac{Q_\nu(xi)}{zQ_\nu(0^+i)}\right]e^{tz}\,dz = 2e^{-t/4}\int_0^\infty\left[\frac{Q_{-\frac{1}{2}-\eta i}(xi)}{Q_{-\frac{1}{2}-\eta i}(0^+i)} - \frac{Q_{-\frac{1}{2}+\eta i}(xi)}{Q_{-\frac{1}{2}+\eta i}(0^+i)}\right]$$

$$\times\, e^{-\eta^2 t}\frac{\eta}{\frac{1}{4}+\eta^2}\,d\eta. \quad (4.4.40)$$

Let us introduce the relationships[68]

$$Q_\nu(z) - Q_{-\nu-1}(z) = \pi\cot(\nu\pi)P_\nu(z), \quad (4.4.41)$$

$$\frac{2}{\pi}\sin(\nu\pi)Q_\nu(z) = P_\nu(z)e^{-\nu\pi i} - P_\nu(-z), \quad \text{Im}(z) > 0, \quad (4.4.42)$$

$$Q_\nu(0^+i) = Q_\nu(0) - \pi i P_\nu(0)/2, \quad (4.4.43)$$

$$P_\nu(0) = \frac{1}{\sqrt{\pi}}\cos\left(\frac{\pi\nu}{2}\right)\frac{\Gamma(\frac{1}{2}+\nu/2)}{\Gamma(1+\nu/2)} \quad (4.4.44)$$

and

$$Q_\nu(0) = -\frac{\sqrt{\pi}}{2}\sin\left(\frac{\pi\nu}{2}\right)\frac{\Gamma(\frac{1}{2}+\nu/2)}{\Gamma(1+\nu/2)}, \quad (4.4.45)$$

---

[68] Taken from Erdélyi, A., W. Magnus, F. Oberhettinger, and F. G. Tricomi, 1953: *Higher Transcendental Functions.* McGraw-Hill Book Co., Inc., Chapter 3.

where $\Gamma(\cdot)$ is the gamma function. Combining Equation 4.4.42 through Equation 4.4.45 yields

$$iQ_{\eta i-\frac{1}{2}}(0^+i) = \frac{\sqrt{\pi}}{2}\frac{\Gamma(\frac{1}{4}+\eta i/2)}{\Gamma(\frac{3}{4}+\eta i/2)}e^{\pi i/4}e^{\eta\pi/2}. \tag{4.4.46}$$

Because $\Gamma(z)\Gamma(1-z) = \pi/\sin(\pi z)$,

$$iQ_{\eta i-\frac{1}{2}}(0^+i) = \frac{1}{2\sqrt{\pi}}\Gamma(\tfrac{1}{4}+\eta i/2)\Gamma(\tfrac{1}{4}-\eta i/2)s\cos\left(\frac{\eta\pi i}{2}+\frac{\pi}{4}\right)e^{\pi i/4}e^{\eta\pi/2}. \tag{4.4.47}$$

From Equation 4.4.41 and Equation 4.4.42,

$$\frac{Q_{-\frac{1}{2}-\eta i}(xi)}{Q_{-\frac{1}{2}-\eta i}(0^+i)} - \frac{Q_{-\frac{1}{2}+\eta i}(xi)}{Q_{-\frac{1}{2}+\eta i}(0^+i)}$$

$$= \pi i\tanh(\eta\pi)\frac{P_{\eta i-\frac{1}{2}}(0)Q_{\eta i-\frac{1}{2}}(xi) - Q_{\eta i-\frac{1}{2}}(0^+i)P_{\eta i-\frac{1}{2}}(xi)}{Q_{\eta i-\frac{1}{2}}(0^+i)Q_{-\eta i-\frac{1}{2}}(0^+i)} \tag{4.4.48}$$

$$= -\frac{2\pi\sqrt{\pi}\tanh(\eta\pi)[P_{\eta i-\frac{1}{2}}(xi) - P_{\eta i-\frac{1}{2}}(-xi)]}{\Gamma(\frac{1}{4}+\eta i/2)\Gamma(\frac{1}{4}-\eta i/2)\cosh(\eta\pi)}. \tag{4.4.49}$$

Consequently, the solution equals the residue minus the branch cut integral after we have divided the sum by $2\pi i$, or

$$u(x,t) = 1 + \frac{2}{\pi}\arctan(x)$$

$$- 4\sqrt{\pi}\,e^{-t/4}\int_0^\infty \frac{\eta\tanh(\eta\pi)}{\cosh(\eta\pi)\Gamma(\frac{1}{4}+\eta i/2)\Gamma(\frac{1}{4}-\eta i/2)(\frac{1}{4}+\eta^2)}$$

$$\times e^{-\eta^2 t}\frac{P_{\eta i-\frac{1}{2}}(xi) - P_{\eta i-\frac{1}{2}}(-xi)}{2i}\,d\eta. \tag{4.4.50}$$

Problems

1. Solve the partial differential equation

$$\frac{\partial u}{\partial t} = a^2\frac{\partial^2 u}{\partial x^2}, \qquad 0 < x < \infty, \quad 0 < t.$$

This partial differential equation is subject to the boundary conditions $u(0,t) = 1$ and $\lim_{x\to\infty}|u(x,t)| < \infty$, and the initial condition $u(x,0) = e^{-kx}$ with $0 < k$.

Step 1: By taking the Laplace transform of the partial differential equation and boundary conditions, show that it reduces to the boundary-value problem

$$\frac{d^2 U(x,s)}{dx^2} - \frac{s}{a^2}U(x,s) = -\frac{e^{-kx}}{a^2}$$

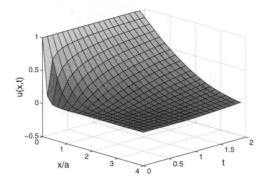

Problem 1

with $U(0, s) = 1/s$ and $\lim_{x \to \infty} |U(x, s)| < \infty$.

Step 2: Show that the solution to the boundary-value problem is

$$U(x, s) = \frac{e^{-x\sqrt{s}/a}}{s} + \frac{e^{-kx}}{s - a^2 k^2} - \frac{e^{-x\sqrt{s}/a}}{s - a^2 k^2}.$$

Step 3: Show that the singularities of the transform in Step 2 are a simple pole at $s = 0$, a removable pole at $s = a^2 k^2$, and a branch point located at $s = 0$.

Step 4: By taking the branch cut along the negative real axis of the s-plane from $s = 0$ to $s = -\infty$, use Bromwich's integral to invert $U(x, s)$ and show that

$$u(x, t) = \operatorname{erfc}\left(\frac{x}{\sqrt{4a^2 t}}\right) + \frac{2}{\pi} \int_0^\infty \frac{\eta \exp(-t\eta^2)}{\eta^2 + a^2 k^2} \sin\left(\frac{x\eta}{a}\right) d\eta.$$

This solution $u(x, t)$ is illustrated in the figure labeled Problem 1 as a function of distance x and time t with $ak = 10$.

2. Solve the partial differential equation

$$\frac{\partial u}{\partial t} = \nu \left(\frac{\partial^2 u}{\partial r^2} + \frac{1}{r} \frac{\partial u}{\partial r} - \frac{u}{b^2}\right), \qquad a < r < \infty, \quad 0 < a, b, t, \nu.$$

This partial differential equation is subject to the boundary conditions $u(a, t) = 1$ and $\lim_{r \to \infty} |u(r, t)| < \infty$, and the initial condition $u(r, 0) = 0$.

Step 1: By taking the Laplace transform of the partial differential equation and boundary conditions, show that it reduces to the boundary-value problem

$$\frac{d^2 U(r, s)}{dr^2} + \frac{1}{r} \frac{dU(r, s)}{dr} - \left(\frac{1}{b^2} + \frac{s}{\nu}\right) U(r, s) = 0$$

with $U(a, s) = 1/s$ and $\lim_{r \to \infty} |U(r, s)| < \infty$.

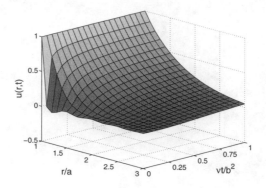

Problem 2

Step 2: Show that the solution to Step 1 is

$$U(r,s) = \frac{K_0\left(r\sqrt{s/\nu + 1/b^2}\right)}{s\, K_0\left(a\sqrt{s/\nu + 1/b^2}\right)}.$$

Step 3: Show that the singularities of the transform in Step 2 are a simple pole at $s = 0$ and branch point located at $s = -\nu/b^2$.

Step 4: By taking the branch cut along the negative real axis of the s-plane from $s = -\nu/b^2$ to $s = -\infty$, use Bromwich's integral to invert $U(r,s)$ and show that

$$u(r,t) = \frac{K_0(r/b)}{K_0(a/b)} + \frac{2}{\pi}e^{-\nu t/b^2}\int_0^\infty \frac{\exp(-\nu t\eta^2/a^2)}{\eta^2 + (a/b)^2}\frac{J_0(r\eta/a)Y_0(\eta) - Y_0(r\eta/a)J_0(\eta)}{J_0^2(\eta) + Y_0^2(\eta)}\eta\, d\eta.$$

This solution $u(r,t)$ is illustrated in the figure labeled Problem 2 as a function of distance r and time t when $a/b = 1$.

3. Solve the partial differential equation[69]

$$\frac{\partial u}{\partial t} = \frac{\partial^2 u}{\partial r^2} + \frac{1}{r}\frac{\partial u}{\partial r}, \qquad a < r < \infty, \quad 0 < t,$$

with the boundary conditions $\lim_{r\to\infty}|u(r,t)| < \infty$ and $u_r(a,t) = -1$, and the initial condition $u(r,0) = 0$.

Step 1: By taking the Laplace transform of the partial differential equation and boundary conditions, show that they reduce to the boundary-value problem

$$\frac{d^2 U(r,s)}{dr^2} + \frac{1}{r}\frac{dU(r,s)}{dr} - sU(r,s) = 0, \qquad a < r < \infty,$$

with $\lim_{r\to\infty}|U(r,s)| < \infty$ and $U'(a,s) = -1/s$.

---

[69]　Goldstein and Briggs, *op. cit.*

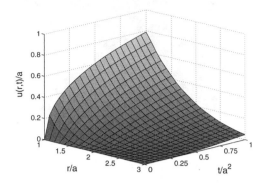

Problem 3

Step 2: Show that the solution to the boundary-value problem is

$$U(r,s) = \frac{K_0\left(r\sqrt{s}\right)}{s^{3/2}\, K_1\left(a\sqrt{s}\right)}.$$

Step 3: Show that we can invert the Laplace transform in Step 2 as follows:

$$u(r,t) = \int_0^t \mathcal{L}^{-1}\left[\frac{K_0\left(r\sqrt{s}\right)}{\sqrt{s}\, K_1\left(a\sqrt{s}\right)}\right]\, d\tau.$$

Step 4: Show that

$$\mathcal{L}^{-1}\left[\frac{K_0(r\sqrt{s})}{\sqrt{s}K_1(a\sqrt{s})}\right] = -\frac{2}{\pi}\int_0^\infty e^{-t\eta^2}\frac{J_0(\eta r)Y_1(\eta a) - Y_0(\eta r)J_1(\eta a)}{J_1^2(\eta a) + Y_1^2(\eta a)}\, d\eta$$

so that

$$u(r,t) = -\frac{2}{\pi}\int_0^\infty \left(1 - e^{-t\eta^2}\right)\frac{J_0(\eta r)Y_1(\eta a) - Y_0(\eta r)J_1(\eta a)}{J_1^2(\eta a) + Y_1^2(\eta a)}\frac{d\eta}{\eta^2}.$$

Take the branch cuts associated with the square root and $K_0(\cdot)$ along the negative real axis of the s-plane. This solution $u(r,t)$ is illustrated in the figure labeled Problem 3 as a function of distance r and time t.

4. Solve the partial differential equation

$$\frac{\partial u}{\partial t} = \frac{\partial^2 u}{\partial r^2} + \frac{1}{r}\frac{\partial u}{\partial r} - u, \qquad 1 < r < \infty, \quad 0 < t,$$

with the boundary conditions $\lim_{r\to\infty} |u(r,t)| < \infty$ and $u(1,t) = 1$, and the initial condition $u(r,0) = 0$.

Step 1: By taking the Laplace transform of the partial differential equation and boundary conditions, show that they reduce to the boundary-value problem

$$\frac{d^2 U(r,s)}{dr^2} + \frac{1}{r}\frac{dU(r,s)}{dr} - (s+1)U(r,s) = 0, \qquad 1 < r < \infty,$$

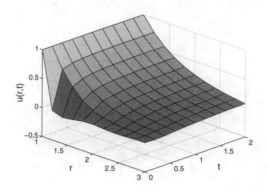

Problem 4

with $\lim_{r \to \infty} |U(r, s)| < \infty$ and $U(1, s) = 1/s$.

Step 2: Show that the solution to Step 1 is

$$U(r, s) = \frac{K_0\left(r\sqrt{s+1}\right)}{s \, K_0\left(\sqrt{s+1}\right)}.$$

Step 3: Show that the Laplace transform in Step 2 has a simple pole at $s = 0$ and a branch point at $s = -1$.

Step 4: Use Bromwich's integral to invert $U(r, s)$. If you take the branch cut associated with $K_0(\cdot)$ along the negative real axis of the s-plane, you should find that

$$u(r, t) = \frac{K_0(r)}{K_0(1)} - \frac{2}{\pi} e^{-t} \int_0^\infty \frac{\eta}{1 + \eta^2} e^{-t\eta^2} \frac{J_0(\eta)Y_0(r\eta) - J_0(r\eta)Y_0(\eta)}{J_0^2(\eta) + Y_0^2(\eta)} \, d\eta.$$

This solution $u(r, t)$ is illustrated in the figure labeled Problem 4 as a function of distance r and time t.

5. Solve the partial differential equation[70]

$$\frac{\partial u}{\partial t} = \frac{\partial^2 u}{\partial r^2} + \frac{1}{r} \frac{\partial u}{\partial r} - \frac{u}{r^2}, \qquad 1 \le r < \infty, \quad 0 < t,$$

with the boundary conditions that $\lim_{r \to \infty} |u(r, t)| < \infty$ and $u(1, t) = \sin(\omega t)$, and the initial condition that $u(r, 0) = 0$.

Step 1: By taking the Laplace transform of the partial differential equation and boundary conditions, show that they reduce to the boundary-value problem

$$\frac{d^2 U(r, s)}{dr^2} + \frac{1}{r} \frac{dU(r, s)}{dr} - \left(s + \frac{1}{r^2}\right) U(r, s) = 0$$

---

[70] Schwarz, *op. cit.*

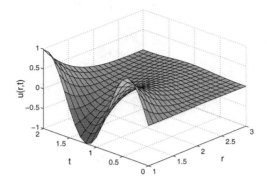

Problem 5

with $\lim_{r\to\infty} |U(r,s)| < \infty$ and $U(1,s) = \omega/(s^2 + \omega^2)$.

Step 2: Show that the solution to the boundary-value problem is

$$U(r,s) = \frac{\omega}{s^2 + \omega^2} \frac{K_1\left(r\sqrt{s}\right)}{K_1\left(\sqrt{s}\right)}.$$

Step 3: Show that the Laplace transform in Step 2 has simple poles at $s = \pm\omega i$ and a branch point at $s = 0$.

Step 4: By taking the branch cut associated with $K_1(\cdot)$ along the negative real axis of the s-plane, use Bromwich's integral and show that

$$u(r,t) = \frac{N_1\left(r\sqrt{\omega}\right)}{N_1\left(\sqrt{\omega}\right)} \sin\left[\omega t + \phi_1\left(r\sqrt{\omega}\right) - \phi_1\left(\sqrt{\omega}\right)\right]$$
$$- \frac{2\omega}{\pi} \int_0^\infty \frac{\eta}{\eta^4 + \omega^2} e^{-t\eta^2} \frac{J_1(r\eta)Y_1(\eta) - Y_1(r\eta)J_1(\eta)}{J_1^2(\eta) + Y_1^2(\eta)} \, d\eta,$$

where

$$N_1(z) = \sqrt{\ker_1^2(z) + \kei_1^2(z)}, \qquad \phi_1(z) = \arctan[\kei_1(z)/\ker_1(z)],$$

and $\ker_1(\cdot)$ and $\kei_1(\cdot)$ are Kelvin's ker and kei functions, respectively. This solution $u(r,t)$ is illustrated in the figure labeled Problem 5 as a function of distance r and time t when $\omega = 4$.

6. Solve the partial differential equation[71]

$$\frac{\partial u}{\partial t} = \frac{\partial^2 u}{\partial r^2} + \frac{1}{r}\frac{\partial u}{\partial r} - \frac{9u}{4r^2}, \qquad a < r < \infty, \quad 0 < t,$$

---

[71] Seth, *op. cit.*

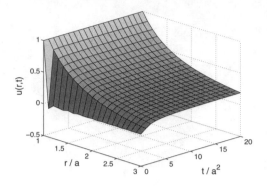

Problem 6

with the boundary conditions that $u(a, t) = a^{3/2}$ and $\lim_{r \to \infty} |u(r, t)| < \infty$, and the initial condition that $u(r, 0) = 0$.

Step 1: By taking the Laplace transform of the partial differential equation and boundary conditions, show that they reduce to the boundary-value problem

$$\frac{d^2 U(r, s)}{dr^2} + \frac{1}{r} \frac{dU(r, s)}{dr} - \left(s + \frac{9}{4r^2} \right) U(r, s) = 0$$

with $\lim_{r \to \infty} |U(r, s)| < \infty$ and $U(a, s) = a^{3/2}/s$.

Step 2: Show that the solution to the boundary-value problem is

$$U(r, s) = \frac{a^{3/2} K_{3/2}\left(r\sqrt{s} \right)}{s\, K_{3/2}\left(a\sqrt{s} \right)} = \frac{a^3 \left(1 + r\sqrt{s} \right)}{s r^{3/2} \left(1 + a\sqrt{s} \right)} e^{-(r-a)\sqrt{s}}.$$

Step 3: Show that the Laplace transform in Step 2 has a branch point at $s = 0$.

Step 4: By taking the branch cut associated with the square root along the negative real axis of the s-plane, use Bromwich's integral and show that

$$u(r, t) = \frac{a^3}{r^{3/2}} \left\{ 1 - \frac{2}{\pi} \int_0^\infty \frac{1 + ar\eta^2}{1 + a^2\eta^2} \sin[(r - a)\eta] e^{-t\eta^2} \frac{d\eta}{\eta} \right.$$
$$\left. + \frac{2}{\pi}(r - a) \int_0^\infty \frac{1}{1 + a^2\eta^2} \cos[(r - a)\eta] e^{-t\eta^2}\, d\eta \right\}.$$

This solution $u(r, t)/a^{3/2}$ is illustrated in the figure labeled Problem 6.

7. Solve the partial differential equation[72]

$$\frac{\partial u}{\partial t} = \frac{\partial^2 u}{\partial r^2} + \frac{1}{r} \frac{\partial u}{\partial r} - \frac{u}{r^2}, \qquad a < r < \infty, \quad 0 < t,$$

---

[72] Sennitskii, *op. cit.*

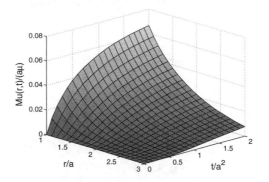

Problem 7

with the boundary conditions that $\lim_{r\to\infty} |u(r,t)| < \infty$,

$$u(a,t) = a\Omega(t), \qquad \text{and} \qquad \frac{d\Omega(t)}{dt} = M + 2\pi\mu a^2\left[\frac{\partial u(a,t)}{\partial r} - \Omega(t)\right],$$

where I, M and μ are constants. Initially all systems are at rest so that $u(r,0) = 0$ and $\Omega(0) = 0$.

Step 1: By taking the Laplace transform of the partial differential equation and boundary conditions, show that they reduce to the boundary-value problem

$$\frac{d^2U(r,s)}{dr^2} + \frac{1}{r}\frac{dU(r,s)}{dr} - \left(s + \frac{1}{r^2}\right)U(r,s) = 0$$

with $\lim_{r\to\infty} |U(r,s)| < \infty$ and $(Is + 2\pi\mu a^2)U(a,s) - 2\pi\mu a^3 U'(a,s) = aM/s$.

Step 2: Show that the solution to the boundary-value problem is

$$U(r,s) = \frac{aM}{s}\frac{K_1\left(r\sqrt{s}\right)}{(Is + 2\pi\mu a^2)K_1\left(a\sqrt{s}\right) - 2\pi\mu a^3 \partial K_1\left(r\sqrt{s}\right)/\partial r|_{r=a}}.$$

Step 3: Show that the Laplace transform in Step 2 has a branch point at $s = 0$.

Step 4: By taking the branch cut associated with the square root along the negative real axis of the s-plane, use Bromwich's integral and show that

$$u(r,t) = \frac{M}{4\pi\mu r}\left[1 + \frac{4\chi r}{a\pi}\int_0^\infty \frac{P(\eta)}{\eta^2 Q(\eta)}e^{-t\eta^2/a^2}\,d\eta\right],$$

where

$$P(\eta) = Y_1(r\eta/a)[\eta J_1(\eta) - \chi J_2(\eta)] - J_1(r\eta/a)[\eta Y_1(\eta) - \chi Y_2(\eta)],$$

$$Q(\eta) = [\eta J_1(\eta) - \chi J_2(\eta)]^2 + [\eta Y_1(\eta) - \chi Y_2(\eta)]^2,$$

and $\chi = 2\pi\mu a^4/I$. This solution, $Mu(r,t)/(a\mu)$, is illustrated in the figure labeled Problem 7 when $\chi = 1$.

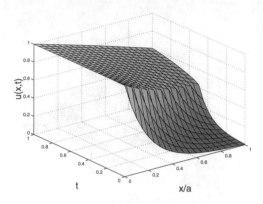

Problem 8

8. Solve the partial differential equation

$$\frac{\partial u}{\partial t} = a\frac{\partial^2 u}{\partial x \partial t} + b\frac{\partial^2 u}{\partial x^2}, \qquad 0 < x < \infty, \quad 0 < a, b, t,$$

with the boundary conditions $u(0,t) = 1$, $\lim_{x\to\infty}|u(x,t)| < \infty$, $0 < t$, and the initial condition $u(x,0) = e^{-kx}$, $0 < x, k < \infty$.

Step 1: Show that the Laplace transform of the partial differential equation and boundary conditions yields the boundary-value problem

$$b\frac{d^2 U(x,s)}{dx^2} + as\frac{dU(x,s)}{dx} - sU(x,s) = -(1+ak)e^{-kx}, \qquad 0 < x < \infty,$$

with $U(0,s) = 1/s$ and $\lim_{x\to\infty}|U(x,s)| < \infty$.

Step 2: Show that the solution to boundary-value problem is

$$U(x,s) = \frac{e^{-kx}}{s-\lambda} - \frac{\lambda}{s(s-\lambda)}\exp\left[-\frac{axs}{2b} - \frac{ax}{2b}\sqrt{s\left(s + \frac{4b}{a^2}\right)}\right].$$

Step 3: Show that the Laplace transform in Step 2 has simple poles at $s = 0$ and $s = \lambda$ and branch points at $s = 0$ and $s = -4b/a^2$.

Step 4: By taking the branch cut between the two branch points $s = -4b/a^2$ and $s = 0$ along the negative real axis of the s-plane, use Bromwich's integral and show that

$$u(x,t) = e^{\lambda t - kx} - H\left(t - \frac{ax}{b}\right)\left\{e^{\lambda t - m(\lambda)x} - 1\right.$$

$$\left. + \frac{\lambda}{\pi}\int_0^{4b/a^2}\exp\left(\frac{ax\eta}{2b} - t\eta\right)\sin\left[\frac{ax}{2b}\sqrt{\eta\left(\frac{4b}{a^2} - \eta\right)}\right]\frac{d\eta}{\eta(\eta+\lambda)}\right\},$$

where

$$m(s) = \frac{as}{2b} + \frac{a}{2b}\sqrt{s\left(s + \frac{4b}{a^2}\right)}.$$

The solution $u(x,t)$ is shown in the plot labeled Problem 8 with $ak = 5$ and $b/a^2 = 2$.

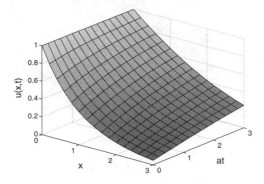

Problem 9

9. Solve the partial differential equation

$$\frac{\partial u}{\partial t} = \left(a + \frac{\partial}{\partial t}\right)\frac{\partial^2 u}{\partial x^2}, \qquad 0 < x < \infty, \quad 0 < a, t,$$

with the boundary conditions $u(0, t) = 1$ and $\lim_{x \to \infty} |u(x, t)| < \infty$, and the initial condition $u(x, 0) = u_{xx}(x, 0)$. Note that we do *not* know the value of $u(x, 0)$.

Step 1: By taking the Laplace transform of the partial differential equation and boundary conditions, show that they reduce to the boundary-value problem

$$\frac{d^2 U(x, s)}{dx^2} - \frac{s}{s + a} U(x, s) = 0$$

with $U(0, s) = 1/s$ and $\lim_{x \to \infty} |U(x, s)| < \infty$.

Step 2: Show that the solution to the boundary-value problem is

$$U(x, s) = \frac{1}{s} \exp\left(-x\sqrt{\frac{s}{s + a}}\right).$$

Step 3: Show that the Laplace transform in Step 2 has branch points at $s = 0$ and $s = -a$.

Step 4: By taking the branch cuts associated with the square roots $\sqrt{s}$ and $\sqrt{s + a}$ along the negative real axis of the s-plane, use Bromwich's integral and show that

$$u(x, t) = 1 - \frac{2}{\pi} \int_0^\infty \exp\left(-\frac{a t \eta^2}{1 + \eta^2}\right) \frac{\sin(x\eta)}{\eta(1 + \eta^2)} \, d\eta.$$

Step 5: Show that $u(x, 0) = e^{-x}$. The solution $u(x, t)$ is shown in the plot labeled Problem 9.

10. Solve the partial differential equation

$$a\frac{\partial u}{\partial t} + 2b\frac{\partial u}{\partial x} + c^2 u = \frac{\partial^2 u}{\partial x^2}, \qquad 0 < x < \infty, \quad 0 < a, t,$$

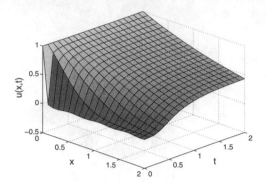

Problem 10

with the boundary conditions $u(0,t) = 1$ and $\lim_{x\to\infty} |u(x,t)| < \infty$, and the initial condition $u(x,0) = 0$.

Step 1: By taking the Laplace transform of the partial differential equation and boundary conditions, show that they reduce to the boundary-value problem

$$\frac{d^2U(x,s)}{dx^2} - 2b\frac{dU(x,s)}{dx} - (as + c^2)U(x,s) = 0,$$

with $U(0,s) = 1/s$ and $\lim_{x\to\infty} |U(x,s)| < \infty$.

Step 2: Show that the solution to Step 1 is

$$U(x,s) = \frac{e^{bx}}{s} \exp\!\left(-x\sqrt{b^2 + c^2 + as}\,\right).$$

Step 3: Show that the Laplace transform in Step 2 has a simple pole at $s = 0$ and a branch point at $s = -(b^2 + c^2)/a$.

Step 4: By taking the branch cut associated with $\sqrt{s + (b^2 + c^2)/a}$ along the negative real axis of the s-plane, use Bromwich's integral and show that

$$u(x,t) = e^{x\left(b-\sqrt{b^2+c^2}\right)} - \frac{1}{\pi} \exp\!\left(bx - t\sqrt{b^2 + c^2}/a\right) \int_0^\infty \frac{e^{-t\eta}\sin\!\left(x\sqrt{a\eta}\,\right)}{\eta + (b^2 + c^2)/a}\,d\eta.$$

The solution $u(x,t)$ is shown in the plot labeled Problem 10 when $a = b = c = 1$.

11. Solve the partial differential equation[73]

$$\frac{\partial u}{\partial t} + bu = \frac{\partial^2 u}{\partial x^2} + a\frac{\partial^3 u}{\partial x^2 \partial t}, \qquad 0 < x < \infty, \quad 0 < a,b,t,$$

---

[73] Reprinted from *Int. J. Non-linear Mech.*, **38**, P. M. Jordan and P. Puri, Stokes' first problem for a Rivlin-Ericksen fluid of second grade in a porous half-space, 1019–1025, ©2002, with permission from Elsevier.

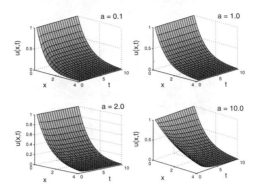

Problem 11

with the boundary conditions $u(0,t) = 1$ and $\lim_{x\to\infty} |u(x,t)| < \infty$, and the initial condition $u(x,0) = au_{xx}(x,0)$. Note that we do *not* know the value of $u(x,0)$.

Step 1: By taking the Laplace transform of the partial differential equation and boundary conditions, show that they reduce to the boundary-value problem

$$(1 + as)\frac{d^2 U(x,s)}{dx^2} - (s+b)U(x,s) = 0,$$

with $U(0,s) = 1/s$ and $\lim_{x\to\infty} |U(x,s)| < \infty$.

Step 2: Show that the solution to the boundary-value problem is

$$U(x,s) = \frac{1}{s}\exp\left(-x\sqrt{\frac{s+b}{1+as}}\right).$$

Step 3: Show that the Laplace transform in Step 2 has a simple pole at $s = 0$ and branch points at $s = -1/a$ and $s = -b$.

Step 4: By taking the branch cuts associated with the square roots $\sqrt{s+1/a}$ and $\sqrt{s+b}$ along the negative real axis of the s-plane, use Bromwich's integral and show that

$$u(x,t) = e^{-bx} + \frac{1}{\pi}\int_{1/a}^{b} e^{-t\eta}\sin\left(x\sqrt{\frac{b-\eta}{a\eta-1}}\right)\frac{d\eta}{\eta},$$

if $ab > 1$;

$$u(x,t) = e^{-ax},$$

if $ab = 1$; and

$$u(x,t) = e^{-bx} - \frac{1}{\pi}\int_{b}^{1/a} e^{-t\eta}\sin\left(x\sqrt{\frac{\eta-b}{1-a\eta}}\right)\frac{d\eta}{\eta},$$

if $1 > ab$. The solution $u(x,t)$ is shown in the plot labeled Problem 11 for various values of a with $b = 1$.

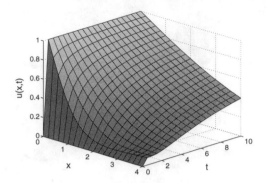

Problem 12

12. Solve the partial differential equation[74]

$$\frac{\partial u}{\partial t} = \frac{\partial^2 u}{\partial x^2} - (a - b)\left[u - be^{-bt}\int_0^t u(x, \eta)e^{b\eta}\, d\eta\right], \qquad 0 < x < \infty, \quad 0 < t,$$

with the boundary conditions $u(0, t) = 1$ and $\lim_{x\to\infty}|u(x, t)| < \infty$, and the initial condition $u(x, 0) = 0$. Here, $b < a$.

Step 1: By taking the Laplace transform of the partial differential equation and boundary conditions, show that they reduce to the boundary-value problem

$$(s + b)\frac{d^2 U(x, s)}{dx^2} - s(s + a)U(x, s) = 0,$$

with $U(0, s) = 1/s$ and $\lim_{x\to\infty}|U(x, s)| < \infty$.

Step 2: Show that the solution to boundary-value problem is

$$U(x, s) = \frac{1}{s}\exp\left(-x\sqrt{s\frac{s + a}{s + b}}\right).$$

Step 3: Show that the Laplace transform in Step 2 has a simple pole at $s = 0$ and branch points at $s = 0$, $s = -a$ and $s = -b$.

Step 4: By taking the branch cuts associated with the square roots $\sqrt{s}$, $\sqrt{s + a}$ and $\sqrt{s + b}$ along the negative real axis of the s-plane, use Bromwich's integral and show that

$$u(x, t) = 1 - \frac{1}{\pi}\int_0^b e^{-t\eta}\sin\left(x\sqrt{\eta\frac{a - \eta}{b - \eta}}\right)\frac{d\eta}{\eta} - \frac{1}{\pi}\int_a^\infty e^{-t\eta}\sin\left(x\sqrt{\eta\frac{\eta - a}{\eta - b}}\right)\frac{d\eta}{\eta}.$$

The solution $u(x, t)$ is shown in the plot labeled Problem 12 when $a = 10$ and $b = 8$.

---

[74] Reprinted from *Astronaut. Acta*, **17**, J. Val. Healy and H. T. Yang, The Stokes problems for a suspension of particles, 851–856, ©1972, with permission from Elsevier.

13. Solve the partial differential equation[75]

$$\frac{\partial u}{\partial t} - \chi \frac{\partial}{\partial t}\left(\frac{\partial^2 u}{\partial r^2} + \frac{1}{r}\frac{\partial u}{\partial r} + \frac{\partial^2 u}{\partial z^2}\right) = \kappa\left(\frac{\partial^2 u}{\partial r^2} + \frac{1}{r}\frac{\partial u}{\partial r} + \frac{\partial^2 u}{\partial z^2}\right),$$

where $0 < r < \infty$, $0 < z < h$, $0 < t$, and χ and κ are real and positive, subject to the boundary conditions

$$\lim_{r\to\infty}|u(r,z,t)| < \infty, \qquad \frac{\partial u(r,0,t)}{\partial z} = \frac{\partial u(r,h,t)}{\partial z} = 0,$$

and

$$\lim_{r\to 0} r\frac{\partial u(r,z,t)}{\partial r} = \begin{cases} e^{-\kappa t/\chi} - 1, & \text{if} \quad 0 \le z \le \ell, \\ 0, & \text{if} \quad \ell < z \le h, \end{cases}$$

and the initial condition $u(r,z,0) = 0$.

Step 1: By taking the Laplace transform of the partial differential equation and boundary conditions, show that they reduce to the partial differential equation

$$sU(r,z,s) - (\kappa + \chi s)\left[\frac{\partial^2 U(r,z,s)}{\partial r^2} + \frac{1}{r}\frac{\partial U(r,z,s)}{\partial r} + \frac{\partial^2 U(r,z,s)}{\partial z^2}\right] = 0$$

with the boundary conditions

$$\lim_{r\to\infty}|U(r,z,s)| < \infty, \qquad \frac{\partial U(r,0,s)}{\partial z} = \frac{\partial U(r,h,s)}{\partial z} = 0,$$

and

$$\lim_{r\to 0} r\frac{\partial U(r,z,s)}{\partial r} = \begin{cases} -\kappa/[s(\kappa + \chi s)], & \text{if} \quad 0 \le z \le \ell, \\ 0, & \text{if} \quad \ell < z \le h. \end{cases}$$

Step 2: Using the method of separation of variables, solve the partial differential equation in Step 1 and show that

$$U(r,z,s) = \frac{\kappa\ell}{h}\frac{K_0(r\sigma_0)}{s(\kappa + \chi s)} + \sum_{n=1}^{\infty}\frac{2\kappa K_0(r\sigma_n)}{n\pi s(\kappa + \chi s)}\sin(\lambda_n\ell)\cos(\lambda_n z)$$

$$= \frac{\kappa\ell}{h}\frac{c_0^2}{s(s+b^2)}K_0\left(rc_0\sqrt{\frac{s+a_0^2}{s+b^2}}\right)$$

$$+ \frac{2\kappa}{s\pi}\sum_{n=1}^{\infty}\frac{c_0^2}{n(s+b^2)}K_0\left(rc_n\sqrt{\frac{s+a_n^2}{s+b^2}}\right)\sin(\lambda_n\ell)\cos(\lambda_n z),$$

where $a_n^2 = b^2\lambda_n^2/c_n^2$, $b^2 = \kappa/\chi$, $c_n^2 = \lambda_n^2 + 1/\chi$, $\lambda_n = n\pi/h$ and $\sigma_n^2 = s/(\kappa + \chi s) + \lambda^2$. Note that $b > a_n$.

---

[75] Raichenko, *op. cit.*

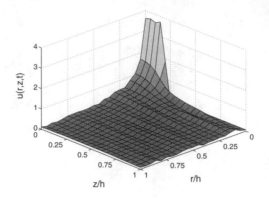

Problem 13

Step 3: Taking the branch cuts associated with $\sqrt{s + a_n^2}$ and $\sqrt{s + b^2}$ along the negative real axis of the s-plane, show that

$$\mathcal{L}^{-1}\left[\frac{c_0^2}{s + b^2} K_0\left(rc_n\sqrt{\frac{s + a_n^2}{s + b^2}}\right)\right] = \int_{a_n^2}^{b^2} \frac{c_0^2}{b^2 - \eta} J_0\left(rc_n\sqrt{\frac{\eta - a_n^2}{b^2 - \eta}}\right) e^{-t\eta} \, d\eta.$$

Step 4: Because $\mathcal{L}^{-1}[F(s)/s] = \int_0^t f(\tau) \, d\tau$, show that the final solution is

$$u(r, z, t) = \frac{\ell\kappa}{2h}\int_0^{b^2} \frac{1 - e^{-t\eta}}{\eta(\kappa - \chi\eta)} J_0\left(r\sqrt{\frac{\eta}{\kappa - \chi\eta}}\right) \, d\eta$$

$$+ \frac{\kappa}{\pi}\sum_{n=1}^{\infty}\frac{\sin(\lambda_n\ell)\cos(\lambda_n z)}{n}\int_{a_n^2}^{b^2}\frac{1 - e^{-t\eta}}{\eta(\kappa - \chi\eta)} J_0\left(r\sqrt{\frac{\eta}{\kappa - \chi\eta} - \lambda_n^2}\right) \, d\eta.$$

The solution $u(r, z, t)$ is shown in the plot labeled Problem 13 when $t = 2$, $\ell/h = 0.2$ and $h^2/\chi = 2$.

Papers Using Laplace Transforms
to Solve Partial Differential Equations

Aquifers, Reservoirs and Porous Media

Chen, C.-S., 1987: Analytical solutions for radial dispersion with Cauchy boundary at injection well. *Water Resour. Res.*, **23**, 1217–1224.

Chen, C.-S., and G. D. Woodside, 1988: Analytical solution for aquifer decontamination by pumping. *Water Resour. Res.*, **24**, 1329–1338.

Closmann, P. J., 1968: Steam zone growth in a preheated reservoir. *Soc. Pet. Eng. J.*, **8**, 313–320.

De Wiest, J. M. R., 1963: Flow to an eccentric well in a leaky circular aquifer with varied lateral replenishment. *Geofis. Pura Appl.*, **54**, 87–102.

Hantush, M. S., 1959: Nonsteady flow to flowing wells in leaky aquifers. *J. Geophys. Res.*, **64**, 1043–1052.

Karasaki, K., J. C. S. Long, and P. A. Witherspoon, 1988: A new analytical model for fracture-dominated reservoirs. *SPE Form. Eval.*, **3**, 242–250.

Löfqvist, T., and G. Rehbinder, 1993: Transient flow towards a well in an aquifer including the effect of fluid inertia. *Appl. Sci. Res.*, **51**, 611–623.

Loucks, T. L., and E. T. Guerrero, 1961: Pressure drop in a composite reservoir. *Soc. Pet. Eng. J.*, **1**, 170–176.

Oliver, D. S., 1990: The averaging process in permeability estimation from well-test data. *SPE Form. Eval.*, **5**, 319–324.

Papadopulos, I. S., and H. H. Cooper, 1967: Drawdown in a well of large diameter. *Water Resour. Res.*, **3**, 241–244.

Raichenko, L. M., 1973: The problem of the influx of liquid to a perfect well in a layer of fissured-porous rock in the presence of an end-face zone. *Sov. Appl. Mech.*, **9**, 1225–1228.

Raichenko, L. M., 1976: Flow of liquid to an incomplete well in a bed of fissile-porous rocks. *Sov. Appl. Mech.*, **12**, 1196–1200.

Rehbinder, G., and N. Apazidis, 1994: Oscillation of the water level in a well connected to a confined aquifer. *Appl. Sci. Res.*, **52**, 51–65.

Seth, M. S., and K. E. Gray, 1968: Transient stresses and displacement around a wellbore due to fluid flow in transversely isotropic, porous media: I. Infinite reservoirs. *Soc. Pet. Eng. J.*, **8**, 63–78.

Shvidler, M. I., 1965: Sorption in a plane-radial filtration flow. *J. Appl. Mech. Tech. Phys.*, **6, No. 3**, 77–79.

Tang, D. H. E., and D. W. Peaceman, 1987: New analytical and numerical solutions for the radial convection-dispersion problem. *SPE Reservoir Eng.*, **2**, 343–359.

Wikramaratna, R. S., 1984: An analytical solution for the effects of abstraction from a multi-layered confined aquifer with no cross flow. *Water Resour. Res.*, **20**, 1067–1074.

Wu, Y.-S., and K. Pruess, 1990: An analytical solution for wellbore heat transmission in layered formations. *SPE Reservoir Eng.*, **5**, 531–538.

Yates, S. R., 1990: An analytical solution for one-dimensional transport in heterogeneous porous media. *Water Resour. Res.*, **26**, 2331–2338.

Yeh, H.-D., S.-Y. Yang, and H.-Y. Peng, 2003: A new closed-form solution for a radial two-layer drawdown equation for groundwater under constant-flux pumping in a finite-radius well. *Adv. Water Resour.*, **26**, 747–757.

Yew, C. H., and P. N. Jogi, 1976: Study of wave motions in fluid-saturated porous rocks. *J. Acoust. Soc. Am.*, **60**, 2–8.

Zheltov, Yu. P., and V. S. Kutlyarov, 1965: Transient motion of a liquid in a fissured porous stratum subject to periodic pressure variation at the boundary. *J. Appl. Mech. Tech. Phys.*, **6, No. 6**, 69–76.

Diffusion

Akkaş, N., and F. Erdoğan, 1989: The residual variable method applied to the diffusion equation in cylindrical coordinates. *Acta Mech.*, **79**, 207–219.

Berlyand, O. S., 1961: A closed solution of the equation of turbulent diffusion. *Dokl. Acad. Sci. USSR, Earth Sci. Section*, **130**, 111–113.

Masamune, S., and J. M. Smith, 1965: Adsorption rate studies – Interaction of diffusion and surface processes. *AICHE J.*, **11**, 34–40.

Moshinskii, A. I., 1987: Effective diffusion of a dynamically passive impurity in narrow channels. *J. Appl. Mech. Tech. Phys.*, **28**, 374–382.

Price, W. S., A. V. Barzykin, K. Hayamizu, and M. Tachiya, 1998: A model for diffusive transport through a spherical interface probed by pulsed-field gradient NMR. *Biophys. J.*, **74**, 2259–2271.

Rasmuson, A., and I. Neretnieks, 1980: Exact solution of a model for diffusion in particles and longitudinal dispersion in packed beds. *AICHE J.*, **26**, 686–690.

Rosen, J. B., 1952: Kinetics of a fixed bed system for solid diffusion into spherical particles. *J. Chem. Phys.*, **20**, 387–394.

Volkov, F. G., and A. M. Golovin, 1970: Thermal and diffusive relaxation of an evaporating drop with internal heat liberation. *J. Appl. Mech. Tech. Phys.*, **11**, 76–85.

Electromagnetism

Belluigi, A., 1959: Aspetto elettrogeosmotico bidimensionale variabile col tempo. *Geof. Pura Appl.*, **43**, 182–194.

Boerner, W. M., and Y. M. Antar, 1972: Aspects of electromagnetic pulse scattering from a grounded dielectric slab. *Arch. für Electronik Übertragungstechnik*, **26**, 14–21.

Ciric, I. R., and J. Ma, 1991: Transient currents on a cylinder excited by a parallel line-current. *Can. J. Phys.*, **69**, 1242–1248.

Ma, J., and I. R. Ciric, 1992: Transient response of a circular cylinder to an electromagnetic pulse. *Radio Sci.*, **27**, 561–567.

Pal, S. C., M. L. Ghosh, and P. K. Chowdhuri, 1985: Spectral representation of a certain class of self-adjoint differential operators and its application to axisymmetric boundary value problems in electrodynamics. *J. Tech. Phys.*, **26**, 97–115.

Fluid Dynamics

Ehlers, F. E., and T. Strand, 1958: The flow of a supersonic jet in a supersonic stream at an angle of attack. *J. Aerospace Sci.*, **25**, 497–506.

Engevik, L., 1971: A note on a stability problem in hydrodynamics. *Acta Mech.*, **12**, 143–153.

Jordan, P. M., and P. Puri, 1999: Exact solutions for the flow of a dipolar fluid on a suddenly accelerated flat plate. *Acta Mech.*, **137**, 183–194.

Healy, J. Val., and H. T. Yang, 1972: The Stokes problems for a suspension of particles. *Astronaut. Acta*, **17**, 851–856.

Iben, H., 1974: Ebene, kreissymmetrische Rand-Anfangswertprobleme zäher Flüssigkeiten. *Z. Angew. Math. Mech.*, **54**, 213–224.

Johri, A. K., and R. K. Singhal, 1980: Unsteady flow of visco-elastic fluid. *Indian J. Theoret. Phys.*, **28**, 197–203.

Jordan, P. M., M. R. Meyer, and A. Puri, 2000: Causal implications of viscous damping in compressible fluid flow. *Phys. Review E*, **62**, 7918–7926.

Jordan, P. M., and P. Puri, 2002: Stokes' first problem for a Rivlin-Ericksen fluid of second grade in a porous half-space. *Int. J. Non-linear Mech.*, **38**, 1019–1025.

Kildal, A., 1970: On the motion generated by a plate vibrating in a stratified fluid. *Acta Mech.*, **9**, 78–104.

Majumdar, S. R., 1969: On the motion of an infinite circular cylinder in a rotating viscous liquid. *Rev. Roum. Sci. Tech., Ser. Mec. Appl.*, **14**, 701–719.

Mallick, D. D., 1957: Nonuniform rotation of an infinite circular cylinder in an infinite viscous liquid. *Z. Angew. Math. Mech.*, **37**, 385–392.

Miles, J. W., 1951: On virtual mass and transient motion in subsonic compressible flow. *Quart. J. Mech. Appl. Math.*, **4**, 388–400.

Mittal, M. L., 1963: Unsteady hydrodynamic viscous flow in an annular channel. *Appl. Sci. Res., Ser. B*, **10**, 86–90.

Nanda, R. S., 1960: Unsteady circulatory flow about a circular cylinder with suction. *Appl. Sci. Res., Ser. A*, **9**, 85–92.

Pack, D. C., 1956: The oscillations of a supersonic gas jet embedded in a supersonic stream. *J. Aeronaut. Sci.*, **23**, 747–753.

Schwarz, W. H., 1963: The unsteady motion of an infinite oscillating cylinder in an incompressible Newtonian fluid at rest. *Appl. Sci. Res., Ser. A*, **11**, 115–124.

Sennitskii, V. L., 1981: Unsteady rotation of a cylinder in a viscous fluid. *J. Appl. Phys. Tech. Phys.*, **21**, 347–349.

Seth, S. S., 1977: Unsteady motion of viscoelastic fluid due to rotation of a sphere. *Indian J. Pure Appl. Math.*, **8**, 302–308.

Standing, R. G., 1971: The Rayleigh problem for a slightly diffusive density-stratified fluid. *J. Fluid Mech.*, **48**, 673–688.

Tanner, R. I., 1962: Note on the Rayleigh problem for a visco-elastic fluid. *Z. Angew. Math. Phys.*, **13**, 573–580.

Tsamopoulos, J., and A. Borkar, 1994: Transient rotational flow of an Oldroyd-B fluid over a disk. *Phys. Fluid*, **6**, 1144–1157.

Vimala, C. S., and G. Nath, 1976: Unsteady motion of a slightly rarefied gas about an infinite circular cylinder. *Rev. Roum. Sci. Tech., Ser. Mec. Appl.*, **21**, 219–227.

Widnall, S. E., and E. H. Dowell, 1967: Aerodynamic forces on an oscillating cylindrical duct with an internal flow. *J. Sound Vib.*, **6**, 71–85.

Wilks, G., 1969: The flow around a semi-infinite oscillating plate and the skin friction on arbitrarily cross-sectioned infinite cylinders oscillating parallel to their length. *Proc. Cambridge Philos. Soc.*, **66**, 163–187.

General

Carslaw, H. S., and J. C. Jaeger, 1963: *Operational Methods in Applied Mathematics*. Dover Publications, Inc., Chapters 5–10.

McLachlan, N. W., 1939: *Complex Variables & Operational Calculus with Technical Applications*. Cambridge University Press, Chapters 13–15.

Puri, P., and P. K. Kythe, 1988: Some inverse Laplace transforms of exponential form. *ZAMP*, **39**, 150–156.

Thomson, W. T., 1950: *Laplace Transforms*. Prentice-Hall, Inc., Chapter 8.

Geophysical Sciences

Haldar, K., 1986: Study of storm surge problem on a continential shelf with sloping bottom. *Mausam*, **37**, 169–172.

Knighting, E., 1950: A note on nocturnal cooling. *Quart. J. R. Meteor. Soc.*, **76**, 173–181.

Mæland, E., 1983: On the response of a wind-driven current over a continental shelf. *J. Geophys. Res.*, **88**, 4534–4538.

Mareschal, J. C., and A. F. Gangi, 1977: A linear approximation to the solution of a one-dimensional Stefan problem and its geophysical implications. *Geophys. J. R. Astr. Soc.*, **49**, 443–458.

Smith, F. B., 1957: The diffusion of smoke from a continuous elevated point-source into a turbulent atmosphere. *J. Fluid Mech.*, **2**, 49–76.

Toptygin, I. N., 1972: Time dependence of cosmic-ray intensity at the anisotropic stage of solar flares. *Geomagnet. Aeronomy*, **12**, 860–865.

Yamashita, T., 1977: Dependence of source time function on tectonic field. *J. Phys. Earth*, **25**, 419–445.

Heat Conduction

Balasubramaniam, T. A., and H. F. Bowman, 1974: Temperature field due to a time dependent heat source of spherical geometry in an infinite medium. *J. Heat Transfer*, **96**, 296–299.

Barletta, A., 1996: Hyperbolic propagation of an axisymmetric thermal signal in an infinite solid medium. *Int. J. Heat Mass Transfer*, **39**, 3261–3271.

Buller, F. H., 1951: Thermal transients on buried cables. *AIEE Trans., Part 1*, **70**, 45–52.

Carslaw, H. S., and J. C. Jaeger, 1938: Some problems in mathematical theory of the conduction of heat. *Philos. Mag., Ser. 7*, **26**, 473–495.

Chester W., R. Bobone, and E. Brocher, 1984: Transient conduction through a two-layer medium. *Int. J. Heat Mass Transfer*, **27**, 2167–2170.

Goldenberg, H., 1952: Heat flow in an infinite medium heated by a sphere. *Brit. J. Appl. Phys.*, **3**, 296–298.

Goldstein, R. J., and D. G. Briggs, 1964: Transient free convection about vertical plates and circular cylinders. *J. Heat Transfer*, **86**, 490–500.

Hudson, J. L., and S. G. Bankoff, 1964: An exact solution of unsteady heat transfer to a shear flow. *Chem. Eng. Sci.*, **19**, 591–598.

Ioffe, I. A., 1959: Plane nonstationary heat conduction problem for a semi-infinite body with an internal isothermal cylindrical source of heat. *Sov. Phys. Tech. Phys.*, **4**, 369–374.

Jaeger, J. C., 1941: Heat conduction in composite circular cylinders. *Philos. Mag., Ser. 7*, **32**, 324–335.

Jaeger, J. C., 1944: Some problems involving line sources in conduction of heat. *Philos. Mag., Ser. 7*, **35**, 169–179.

Jain, R. K., and P. M. Gullino, 1980: Analysis of transient temperature distributions in a perfused medium due to a spherical heat source with application to heat transfer in tumors: Homogeneous and infinite medium. *Chem. Eng. Commun.*, **4**, 95–118.

Pugh, H. Ll. D., 1964: The temperature distribution in a semi-infinite solid whose surface is maintained at an arbitrary temperature. *Proc. Edinburgh Math. Soc., Ser. 2*, **14**, 143–148.

Schubert, G. U., 1950: Über eine in der Theorie der elektrischen Schmelzsicherungen auftretende Lösung der Wärmeleitungsgleichung. *Z. Angew. Phys.*, **2**, 174–179.

Watson, K., 1973: Periodic heating of a layer over a semi-infinite solid. *J. Geophys. Res.*, **78**, 5904–5910.

Magnetohydrodynamics

Ghosh, A. K., and L. Debnath, 1986: Hydromagnetic Stokes flow in a rotating fluid with suspended small particles. *Appl. Sci. Res.*, **43**, 165–192.

Walker, J. S., 1978: Solitary fluid transients in rectangular ducts with transverse magnetic fields. *Z. Angew. Math. Phys.*, **29**, 35–53.

Yano, T., and Y. Inoue, 1993: Weakly nonlinear waves radiated by pulsations of a cylinder. *J. Acoust. Soc. Am.*, **93**, 132–141.

Other

Blackwell, J. H., 1954: A transient-flow method for determination of thermal constants of insulating material in bulk. *J. Appl. Phys.*, **25**, 137–144.

Duffy, P., 1989: The acceleration of cometary ions by Alfvén waves. *J. Plasma Phys.*, **42**, 13–25.

Huang, J. C., D. Rothstein, and R. Madey, 1984: Analytical solution for a first-order reaction in a packed bed with diffusion. *AICHE J.*, **30**, 660–662.

Jauho, P., and A. Virjo, 1968: Neutron thermalization in heavy gas with nonuniform temperature distribution. *Nucl. Sci. Eng.*, **31**, 102–109.

Kuznetsov, M. M., 1975: Unsteady-state slip of a gas near an infinite plane with diffusion-mirror reflection of molecules. *J. Appl. Mech. Tech. Phys.*, **16**, 853–858.

Lenau, C. W., and A. T. Hjelmfelt, 1992: River bed degradation due to abrupt outfall lowering. *J. Hydraul. Eng. Div. (Am. Soc. Civ. Eng.)*, **118**, 918–933.

McLachlan, N. W., and A. T. McKay, 1936: Transient oscillations in a loud-speaker horn. *Proc. Cambridge Philos. Soc.*, **32**, 265–275.

Rasmuson, A., 1982: Transport processes and conversion in an isothermal fixed-bed catalytic reactor. *Chem. Eng. Sci.*, **37**, 411–415.

Rehbinder, G., 1993: Damped pitching of a slender body partially immersed in a liquid. *Appl. Sci. Res.*, **50**, 69–81.

Rushchitskii, Ya. Ya., and B. B. Érgashev, 1987: Nonsteady effects in a massive sample of fiber composite under the short-term action of a harmonic pulse. *Sov. Appl. Mech.*, **23**, 1177–1181.

Shams, K., 2001: Sorption dynamics of a fixed-bed system of thin-film-coated monodisperse spherical particles/hollow spheres. *Chem. Eng. Sci.*, **56**, 5383–5390.

Silvestroni, P., and L. Rampazzo, 1964: Current-time curves on the dropping mercury electrode calculated for interactions between the absorbed substance and the depolarizer. *J. Electroanal. Chem.*, **7**, 73–80.

Snider, A. D., D. L. Akins, and R. L. Birke, 1977: Transient analysis of electrolytically initiated polymerization. *J. Electroanal. Chem.*, **79**, 31–47.

Tzou, D. Y., and Y. S. Zhang, 1995: An analytical study on the fast-transient process in small scales. *Int. J. Eng. Sci.*, **33**, 1449–1463.

Zaitsev, A. S., 1974: Dynamics of a rod with an elastic shock absorber. *Sov. Appl. Mech.*, **10**, 1003–1008.

Solid Mechanics

Barenblatt, G. I., Iu. P. Zheltov, and I. N. Kochina, 1960: Basic concepts in the theory of seepage of homogeneous liquids in fissured rocks [strata]. *J. Appl. Math. Mech.*, **24**, 1286–1303.

Baron, M. L., and A. T. Matthews, 1961: Diffraction of a pressure wave by a cylindrical cavity in an elastic medium. *J. Appl. Mech.*, **28**, 347–354.

Baron, M. L., and R. Parnes, 1962: Displacements and velocities produced by the diffraction of a pressure wave by a cylindrical cavity in an elastic medium. *J. Appl. Mech.*, **29**, 385–395.

Boley, B. A., and C. C. Chao, 1955: Some solutions of the Timoshenko beam equation. *J. Appl. Mech.*, **22**, 579–586.

Broberg, K. B., 1995: Intersonic mode. II. Crack expansion. *Arch. Mech.*, **47**, 859–871.

Dimaggio, F. L., 1956: Effect of an acoustic medium on the dynamic buckling of plates. *J. Appl. Mech.*, **23**, 201–206.

Eason, G., 1969: Wave propagation in inhomogeneous elastic media, solution in terms of Bessel functions. *Acta Mech.*, **7**, 137–160.

Eason, G., 1970: The displacement produced in a semi-infinite linear Cosserat continuum by an impulsive force. *Proc. Vib. Prob.*, **11**, 199–221.

Florence, A. L., 1965: Traveling force on a Timoshenko beam. *J. Appl. Mech.*, **32**, 351–358.

Forrestal, M. J., and W. E. Alzheimer, 1968: Transient motion of a rigid cylinder produced by elastic and acoustic waves. *J. Appl. Mech.*, **55**, 134–138.

Forrestal, M. J., 1968: Response of an elastic cylindrical shell to a transverse, acoustic pulse. *J. Appl. Mech.*, **35**, 614–616.

Forrestal, M. J., 1968: Transient response at the boundary of a cylindrical cavity in an elastic medium. *Int. J. Solids Struct.*, **4**, 391–395.

Forrestal, M. J., and M. J. Sagartz, 1971: Radiated pressure in an acoustic medium produced by pulsed cylindrical and spherical shells. *J. Appl. Mech.*, **38**, 1057–1060.

Forrestal, M. J., G. E. Sliter, and M. J. Sagartz, 1972: Stresses emanating from the supports of a cylindrical shell produced by a lateral pressure pulse. *J. Appl. Mech.*, **39**, 124–128.

Guan, F., and M. Novak, 1994: Transient response of an elastic homogeneous half-space to suddenly applied rectangular loading. *J. Appl. Mech.*, **61**, 256–264.

Guan, F., I. D. Moore, and C. C. Spyrakos, 1998: Two dimensional transient fundamental solution due to suddenly applied load in a half-space. *Soil Dyn. Earthq. Eng.*, **17**, 269–277.

Geers, T. L., 1969: Excitation of an elastic cylindrical shell by a transient acoustic wave. *J. Appl. Mech.*, **36**, 459–469.

Gonsovskii, V. L., S. I. Meshkov, and Yu. A. Rossikhin, 1972: Impact of a viscoelastic rod onto a rigid target. *Sov. Appl. Mech.*, **8**, 1109–1113.

Gonsovskii, V. L., and Yu. A. Rossikhin, 1973: Stress waves in a viscoelastic medium with a singular hereditary kernel. *J. Appl. Mech. Tech. Phys.*, **14**, 595–597.

Holt, M., and S. L. Strack, 1961: Supersonic panel flutter of a cylindrical shell of finite length. *J. Aerospace Sci.*, **28**, 197–208.

Kramer, B. M., and N. P. Suh, 1973: Plane strain pulse propagation in a semi-infinite viscoelastic Maxwell solid. *J. Sound Vib.*, **29**, 435–442.

Lee, I.-M., 1996: Transient groundmotion in an elastic homogeneous halfspace to blasting load. *Soil Dyn. Earthq. Eng.*, **15**, 151–159.

Lou, Y. K., and J. M. Klosner, 1973: Transient response of a point-excited submerged spherical shell. *J. Appl. Mech.*, **40**, 1078–1084.

Lubliner, J., and V. P. Panoskaltsis, 1992: The modified Kuhn model of linear viscoelasticity. *Int. J. Solids Struct.*, **29**, 3099–3112.

Matsumoto, H., E. Tsuchida, S. Miyao, and N. Tsunadu, 1976: Torsional stress wave propagation in a semi-infinite conical bar. *Bull. JSME*, **19**, 8–14.

Miklowitz, J., 1960: Plane-stress unloading waves emanating from a suddenly punched hole in a stretched elastic plate. *J. Appl. Mech.*, **27**, 165–171.

Miklowitz, J., 1960: Flexural stress waves in an infinite elastic plate due to a suddenly applied concentrated transverse load. *J. Appl. Mech.*, **27**, 681–689.

Misiak, P., A. Papliński, and E. Włodarczyk, 1975: Oblique stress biwaves generated in elastic medium by a moving axisymmetrical concentrated load. *J. Tech. Phys.*, **16**, 303–314.

Movsisyan, L. A., 1966: Vibration of a semi-infinite beam of variable length. *Mech. Solids*, **1, No. 1**, 123–124.

Odaka, T., and I. Nakahara, 1967: Stresses in an infinite beam impacted by an elastic bar. *Bull. JSME*, **10**, 863–872.

Parnes, R., 1969: Response of an infinite elastic medium to traveling loads in a cylindrical bore. *J. Appl. Mech.*, **36**, 51–58.

Parnes, R., 1980: Progressing torsional loads along a bore in an elastic medium. *Int. J. Solids Struct.*, **16**, 653–670.

Parnes, R., and L. Banks-Sills, 1983: Transient response of an elastic medium to torsional loads on a cylindrical cavity. *J. Appl. Mech.*, **50**, 397–404.

Sagartz, M. J., and M. J. Forrestal, 1969: Transient stresses at a clamped support of a circular cylindrical shell. *J. Appl. Mech.*, **36**, 367–369.

Selberg, H. L., 1952: Transient compression waves from spherical and cylindrical cavities. *Ark. Fys.*, **5**, 97–108.

Shibuya, T., I. Nakahara, T. Koizumi, and K. Kaibara, 1975: Impact stress analysis of a semi-infinite plate by the finite difference method. *Bull. JSME*, **18**, 649–655.

Tsai, Y. M., 1971: Dynamic contact stresses produced by the impact of an axisymmetrical projectile on an elastic half-space. *Int. J. Solids Struct.*, **7**, 543–558.

Uflyand, Ya. S., 1948: The propagation of waves in the transverse vibration of bars and plates (in Russian). *Prikl. Mat. Mek.*, **12**, 287–300.

Usuki, T., and A. Maki, 2003: Behavior of beams under transverse impact according to higher-order beam theory. *Int. J. Solids Struct.*, **40**, 3737–3785.

Valathur, M., 1972: Wave propagation in a truncated conical shell. *Int. J. Solids Struct.*, **8**, 1223–1233.

Wolfert, A. R. M., A. V. Metrikine, and H. A. Dieterman, 1996: Wave radiation in a one-dimensional system due to a non-uniformly moving constant load. *Wave Motion*, **24**, 185–196.

Yao, S., and T. Nogami, 1994: Lateral cyclic response of pile in viscoelastic Winkler subgrade. *J. Eng. Mech. Div. (Am. Soc. Civ. Eng.)*, **120**, 758–775.

Zakout, U., and N. Akkas, 1997: Transient response of a cylindrical cavity with and without a bonded shell in an infinite elastic medium. *Int. J. Eng. Sci.*, **35**, 1203–1220.

Thermoelasticity and Magnetoelasticity

Chaudhuri, B. R., 1971: Dynamical problems of thermo-magnetoelasticity for cylindrical regions. *Indian J. Pure Appl. Math.*, **2**, 631–656.

Dhaliwal, R. S., and K. L. Chowdhury, 1968: Dynamical problems of thermoelasticity for cylindrical regions. *Arch. Mech. Stosow.*, **20**, 46–66.

Dillon, O. W., 1967: Coupled thermoelasticity of bars. *J. Appl. Mech.*, **34**, 137–145.

Ghosn, A. H., and M. Sabbaghian, 1982: Quasi-static coupled problems of thermoelasticity for cylindrical regions. *J. Therm. Stresses*, **5**, 299–313.

Koizamu, T., 1970: Transient thermal stresses in a semi-infinite body heated axi-symmetrically. *Z. Angew. Math. Mech.*, **50**, 747–757.

Sherief, H. H., and H. A. Saleh, 1998: A problem for an infinite thermoelastic body with a spherical cavity. *Int. J. Eng. Sci.*, **36**, 473–487.

Tanaka, K., and T. Kurokawa, 1972: Thermoelastic effect caused by longitudinal collision of bars. *Bull. JSME*, **15**, 816–821.

Wilms, E. V., 1966: Temperature induced in a medium due to a suddenly applied pressure inside a spherical cavity. *J. Appl. Mech.*, **33**, 941–943.

Wave Equation

Barakat, R., 1965: Diffraction of a plane step pulse by a perfectly conducting circular cylinder. *J. Opt. Soc. Am.*, **55**, 998–1002.

Baron, M. L., 1961: Response of nonlinearly supported cylindrical boundaries to shock waves. *J. Appl. Mech.*, **28**, 135–136.

Bhattacharyya, S., 1972: Note on diffraction of waves incident obliquely on a half-plane. *Rev. Roum. Phys.*, **17**, 723–728.

Blackstock, D. T., 1967: Transient solution for sound radiated into a viscous fluid. *J. Acoust. Soc. Am.*, **41**, 1312–1319.

Chen, Y. M., 1964: The transient behavior of diffraction of plane pulse by a circular cylinder. *Int. J. Eng. Sci.*, **2**, 417–429.

Daimaruya, M., M. Naitoh, and K. Hamada, 1983: Propagation of elastic waves in a finite length bar with a variable cross section (in Japanese). *Nihon Kikai Gakkai Rombunshu (Trans. Japan Soc. Mech. Eng.), Ser. A*, **49**, 1119–1125.

Minster, J. B., 1978: Transient and impulsive responses of a one-dimensional linearly attenuating medium – I. Analytical results. *Geophys. J. R. Astr. Soc.*, **52**, 479–501.

Yano, T., and Y. Inoue, 1993: Weakly nonlinear waves radiated by pulsations of a cylinder. *J. Acoust. Soc. Am.*, **93**, 132–141.

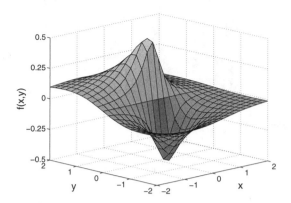

Chapter 5
Methods Involving
Multivalued Fourier Transforms

In the previous chapter we showed that some of the Laplace transforms involved in solving the heat or wave equation can be multivalued. In the case of Fourier transforms, similar considerations hold and we explore them in this chapter.

5.1 INVERSION OF FOURIER TRANSFORMS BY CONTOUR INTEGRATION

In Section 3.1 we showed that we may invert single-valued Fourier transforms by using the inversion integral and complex variables. Here we show how we may apply the techniques of Section 1.5 and Section 1.6 to the inversion of Fourier transforms that possess branch points.

• **Example 5.1.1**

Consider the bilateral Fourier transform

$$F(\omega) = -i\frac{\exp\left(-\sqrt{a^2 + \omega^2}\right)}{\omega\sqrt{a - i\omega}}, \qquad 0 < a. \tag{5.1.1}$$

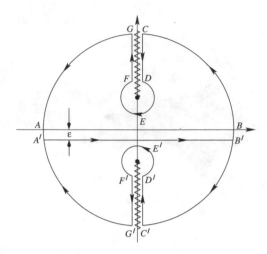

Figure 5.1.1: The contour used to invert the Fourier transform given by Equation 5.1.1.

Let us invert[1] it by contour integration.

We know that this is a bilateral Fourier transform because of the singularity on the real axis at $\omega = 0$. Taking the contour of integration just below the real axis, Figure 5.1.1 shows the closed contour. From the residue theorem,

$$\oint_C \frac{\exp\left(itz - \sqrt{a^2 + z^2}\right)}{z\sqrt{a - iz}}\, dz = \frac{2\pi i}{\sqrt{a}} e^{-a}. \qquad (5.1.2)$$

From Figure 5.1.1, the closed contour for $t > 0$ may be broken down into several line integrations

$$2\pi f(t) + \int_{CD} F(z)e^{itz}\, dz + \int_{DEF} F(z)e^{itz}\, dz + \int_{FG} F(z)e^{itz}\, dz = \frac{2\pi}{\sqrt{a}} e^{-a}, \qquad (5.1.3)$$

because the arcs BC and GA vanish as the radius of the semicircle becomes infinite.

Turning to the remaining integrals, the contribution from the infinitesimally small circle DEF vanishes. Along CD,

$$z = \chi e^{\pi i/2}, \quad z - ia = (\chi - a)e^{\pi i/2}, \quad z + ia = (\chi + a)e^{\pi i/2}, \qquad (5.1.4)$$

so that

$$\int_{CD} F(z)e^{itz}\, dz = -i \int_\infty^a e^{-t\chi}\, \frac{\exp\left(-i\sqrt{\chi^2 - a^2}\right)}{\chi\sqrt{\chi + a}}\, d\chi \qquad (5.1.5)$$

---

[1] A similar Fourier transform arose in Azpeitia, A. G., and G. F. Newell, 1958: Theory of oscillation type viscometers III: A thin disk. *Z. Angew. Math. Phys.*, **9a**, 98–118. Published by Birkhäuser Verlag, Basel, Switzerland.

$$\int_{CD} F(z)e^{itz}\,dz = \frac{i}{\sqrt{a}}\int_1^\infty e^{-at\eta}\,\frac{\exp\!\left(-ai\sqrt{\eta^2-1}\right)}{\eta\sqrt{\eta+1}}\,d\eta. \qquad (5.1.6)$$

Alternatively, along FG,

$$z = \chi e^{\pi i/2}, \quad z - ia = (\chi - a)e^{-3\pi i/2}, \quad z + ia = (\chi + a)e^{\pi i/2}, \qquad (5.1.7)$$

so that

$$\int_{FG} F(z)e^{itz}\,dz = -i\int_a^\infty e^{-t\chi}\,\frac{\exp\!\left(i\sqrt{\chi^2-a^2}\right)}{\chi\sqrt{\chi+a}}\,d\chi \qquad (5.1.8)$$

$$= -\frac{i}{\sqrt{a}}\int_1^\infty e^{-at\eta}\,\frac{\exp\!\left(ai\sqrt{\eta^2-1}\right)}{\eta\sqrt{\eta+1}}\,d\eta. \qquad (5.1.9)$$

Substituting these relations into Equation 5.1.3,

$$f(t) = \frac{e^{-a}}{\sqrt{a}} - \frac{1}{\pi\sqrt{a}}\int_1^\infty e^{-at\eta}\,\frac{\sin\!\left(a\sqrt{\eta^2-1}\right)}{\eta\sqrt{\eta+1}}\,d\eta. \qquad (5.1.10)$$

For the case $t < 0$, we must use the contour in the lower half-plane. Because the contour does not contain any singularities, the inverse of the Fourier transform is

$$2\pi f(t) + \int_{C'D'} F(z)e^{itz}\,dz + \int_{D'E'F'} F(z)e^{itz}\,dz + \int_{F'G'} F(z)e^{itz}\,dz = 0;$$
$$(5.1.11)$$

the contributions from the arcs $B'C'$ and $G'A'$ vanish if the semicircle is of infinite radius.

The contribution from the infinitesimal circle $D'E'F'$ vanishes. For $C'D'$,

$$z = \chi e^{-\pi i/2}, \quad z - ia = (\chi + a)e^{-\pi i/2}, \quad z + ia = (\chi - a)e^{-\pi i/2}, \qquad (5.1.12)$$

so that

$$\int_{C'D'} F(z)e^{itz}\,dz = \int_\infty^a e^{t\chi}\,\frac{\exp\!\left(i\sqrt{\chi^2-a^2}\right)}{\chi\sqrt{\chi-a}}\,d\chi \qquad (5.1.13)$$

$$= -\frac{1}{\sqrt{a}}\int_1^\infty e^{at\eta}\,\frac{\exp\!\left(ai\sqrt{\eta^2-1}\right)}{\eta\sqrt{\eta-1}}\,d\eta. \qquad (5.1.14)$$

Alternatively, along $F'G'$,

$$z = \chi e^{-\pi i/2}, \quad z - ia = (\chi + a)e^{-\pi i/2}, \quad z + ia = (\chi - a)e^{3\pi i/2}, \qquad (5.1.15)$$

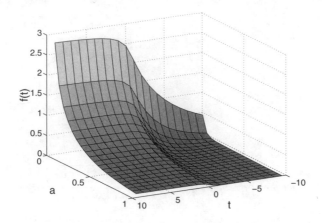

Figure 5.1.2: A plot of the inverse $f(t)$ of the Fourier transform given by Equation 5.1.1.

so that

$$\int_{F'G'} F(z)e^{itz}\,dz = -\int_a^\infty e^{t\chi}\frac{\exp\left(-i\sqrt{\chi^2-a^2}\right)}{\chi\sqrt{\chi-a}}\,d\chi \qquad (5.1.16)$$

$$= -\frac{1}{\sqrt{a}}\int_1^\infty e^{at\eta}\frac{\exp\left(-ai\sqrt{\eta^2-1}\right)}{\eta\sqrt{\eta-1}}\,d\eta. \qquad (5.1.17)$$

Consequently, the inverse of the Fourier transform given by Equation 5.1.1 for $t < 0$ is

$$f(t) = \frac{1}{\pi\sqrt{a}}\int_1^\infty e^{at\eta}\frac{\cos\left(a\sqrt{\eta^2-1}\right)}{\eta\sqrt{\eta-1}}\,d\eta. \qquad (5.1.18)$$

Figure 5.1.2 illustrates $f(t)$.

● **Example 5.1.2**

We now find the inverse of the bilateral Fourier transform[2]

$$F(\omega) = \frac{\exp\left(-\omega^2 + \omega\sqrt{\omega^2-1}\right)}{i\omega}, \qquad (5.1.19)$$

or

$$f(t) = \frac{1}{2\pi}\int_{-\infty-\epsilon i}^{\infty-\epsilon i}\exp\left(-\omega^2 + \omega\sqrt{\omega^2-1} + it\omega\right)\frac{d\omega}{i\omega}, \qquad 0 < \epsilon \ll 1.$$
$$(5.1.20)$$

---

[2] Taken from Menkes, J., 1972: The propagation of sound in the ionosphere. *J. Sound Vib.*, **20**, 311–319. Published by Academic Press Ltd., London, UK.

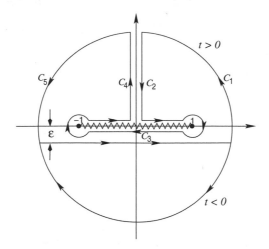

Figure 5.1.3: The contour used to invert the Fourier transform given by Equation 5.1.19.

Because the singularities at $\omega = 0$ and $\omega = \pm 1$ lie above the contour, $f(t) = 0$ for $t < 0$. For $t > 0$, we convert the line integral into a closed contour integral in the upper half-plane shown in Figure 5.1.3.

If our branch cut lies on the real axis from $z = -1$ to $z = 1$, the singularity at $z = 0$ also lies on the branch cut. This poses no difficulty, but we can simplify the integration by a trick. Note that

$$\frac{df}{dt} = \frac{1}{2\pi} \int_{-\infty-\epsilon i}^{\infty-\epsilon i} \exp\left(-z^2 + z\sqrt{z^2-1} + itz\right) dz \tag{5.1.21}$$

only has the branch points at $z = \pm 1$; there is no singularity at $z = 0$. If we introduce the closed contour shown in Figure 5.1.3 with $C = C_1 \cup C_2 \cup C_3 \cup C_4 \cup C_5$,

$$\frac{df}{dt} + \frac{1}{2\pi} \int_C \exp\left(-z^2 + z\sqrt{z^2-1} + itz\right) dz = 0, \tag{5.1.22}$$

because the function is analytic in the interior. As we expand the semicircle without limit, the integrations along C_1 and C_5 tend to zero if we choose the positive square root. Next, the line integral C_2 cancels the line integral C_4. This leaves only the path integral C_3. Along the top, to the right of the ordinate,

$$z = \eta e^{0i}, \quad z - 1 = (1-\eta)e^{\pi i}, \quad z + 1 = (1+\eta)e^{0i}, \tag{5.1.23}$$

$$\sqrt{z^2-1} = \sqrt{1-\eta^2}\, e^{\pi i/2} \tag{5.1.24}$$

with $0 < \eta < 1$. To the left of the ordinate,

$$z = \eta e^{\pi i}, \quad z - 1 = (1+\eta)e^{\pi i}, \quad z + 1 = (1-\eta)e^{0i}, \tag{5.1.25}$$

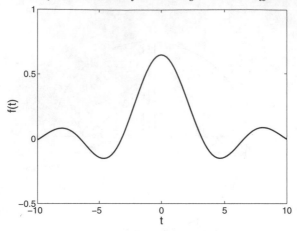

Figure 5.1.4: A plot of the inverse $f(t)$ of the Fourier transform given by Equation 5.1.19 with $C = 0$.

$$\sqrt{z^2 - 1} = \sqrt{1 - \eta^2}\, e^{\pi i/2}. \tag{5.1.26}$$

Below the branch cut, to the right of the ordinate, with $0 < \eta < 1$,

$$z = \eta e^{2\pi i}, \quad z - 1 = (1 - \eta)e^{\pi i}, \quad z + 1 = (1 + \eta)e^{2\pi i}, \tag{5.1.27}$$

and

$$\sqrt{z^2 - 1} = \sqrt{1 - \eta^2}\, e^{3\pi i/2} \tag{5.1.28}$$

and, to the left of the ordinate,

$$z = \eta e^{\pi i}, \quad z - 1 = (1 - \eta)e^{\pi i}, \quad z + 1 = (1 + \eta)e^{2\pi i}, \tag{5.1.29}$$

so that

$$\sqrt{z^2 - 1} = \sqrt{1 - \eta^2}\, e^{3\pi i/2}. \tag{5.1.30}$$

Upon substitution,

$$\frac{df}{dt} = -\frac{2}{\pi} \int_0^1 e^{-\eta^2} \sin\left(\eta\sqrt{1 - \eta^2}\right) \sin(t\eta)\, d\eta. \tag{5.1.31}$$

We obtain the final result by integrating Equation 5.1.31,

$$f(t) = \frac{2}{\pi} \int_0^1 e^{-\eta^2} \sin\left(\eta\sqrt{1 - \eta^2}\right) \cos(t\eta)\, \frac{d\eta}{\eta} + C. \tag{5.1.32}$$

We specify the constant of integration from the initial value of $f(t)$. Figure 5.1.4 illustrates $f(t)$.

• Example 5.1.3

For our next example, we will find the inverse[3] of

$$F(\omega) = 2\frac{\tanh(\omega b)}{\omega b}\cos\left(t\sqrt{\omega^2 + \frac{n^2\pi^2}{b^2}}\right), \tag{5.1.33}$$

or

$$f(x,t) = \frac{1}{\pi}\int_{-\infty}^{\infty}\frac{\tanh(\omega b)}{\omega b}\cos\left(t\sqrt{\omega^2 + \frac{n^2\pi^2}{b^2}}\right)e^{ix\omega}\,d\omega, \tag{5.1.34}$$

where $1 \leq n$, $0 < b$ and $0 < x < t$. For the case $t < x$, we may invert Equation 5.1.34 using the techniques shown earlier in this section. The case of $0 < x < t$ is much more complicated and we will now do it in detail.

First, we replace the cosine with its complex definition and obtain

$$f(x,t) = \frac{1}{2\pi}\int_{-\infty}^{\infty}\frac{\sinh(\omega b)}{\omega b\cosh(\omega b)}\left(e^{ix\omega+it\sqrt{\omega^2+N_n^2}} + e^{ix\omega-it\sqrt{\omega^2+N_n^2}}\right)d\omega, \tag{5.1.35}$$

where $N_n = n\pi/b$. Next, we distort the original contour along the real axis into contours that include arcs with infinite radii. See Figure 5.1.5. For $t > x$, the behavior of $\sqrt{\omega^2 + N_n^2}$ determines the choice of the contour rather than ω as $|\omega| \to \infty$. Let us define the phase of $\omega - iN_n$ as lying between $-3\pi/2$ and $\pi/2$, and the phase of $\omega + iN_n$ as lying between $-\pi/2$ and $3\pi/2$. Consequently, $e^{it\sqrt{\omega^2+N_n^2}} \to 0$ as $|\omega| \to \infty$ in the first and third quadrants while $e^{-it\sqrt{\omega^2+N_n^2}} \to 0$ as $|\omega| \to \infty$ in the second and fourth quadrant. With these considerations, we may distort the line integration along the real axis as shown in Figure 5.1.5. We use the Γ_i contours with $e^{it\sqrt{\omega^2+N_n^2}}$ while we employ the C_i contours with $e^{-it\sqrt{\omega^2+N_n^2}}$.

Turning now to the actual evaluation of the various contour integrals, those over C_1, C_7, Γ_1 and Γ_7 vanish as $|\omega| \to \infty$. For the other contours,

$$\int_{C_2} F(z)e^{itz}\,dz = \frac{1}{2\pi}\int_{\infty}^{N_n}\frac{\sin(\eta b)}{\eta b\cos(\eta b)}e^{-x\eta-t\sqrt{\eta^2-N_n^2}}\,i\,d\eta$$
$$+ \frac{1}{2b^2}\sum_{m=n}^{\infty}\frac{e^{-M_mx-t\sqrt{M_m^2-N_n^2}}}{M_m}, \tag{5.1.36}$$

$$\int_{C_6} F(z)e^{itz}\,dz = \frac{1}{2\pi}\int_{N_n}^{\infty}\frac{\sin(\eta b)}{\eta b\cos(\eta b)}e^{x\eta-t\sqrt{\eta^2-N_n^2}}(-i\,d\eta)$$
$$+ \frac{1}{2b^2}\sum_{m=n}^{\infty}\frac{e^{M_mx-t\sqrt{M_m^2-N_n^2}}}{M_m}, \tag{5.1.37}$$

---

[3] A more complicated version appeared in Walker, J. S., 1978: Solitary fluid transients in rectangular ducts with transverse magnetic fields. *Z. Angew. Math. Phys.*, **A29**, 35–53. Published by Birkhäuser Verlag, Basel, Switzerland.

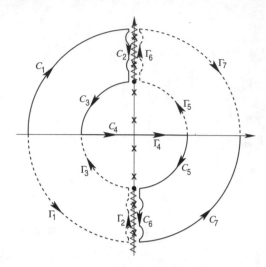

Figure 5.1.5: The contour used in the evaluation of the Fourier transform given by Equation 5.1.33.

$$\int_{\Gamma_2} F(z)e^{itz}\,dz = \frac{1}{2\pi}\int_{\infty}^{N_n}\frac{\sin(\eta b)}{\eta b\cos(\eta b)}e^{x\eta - t\sqrt{\eta^2 - N_n^2}}(-i\,d\eta)$$

$$+\frac{1}{2b^2}\sum_{m=n}^{\infty}\frac{e^{M_m x - t\sqrt{M_m^2 - N_n^2}}}{M_m},\qquad (5.1.38)$$

$$\int_{\Gamma_6} F(z)e^{itz}\,dz = \frac{1}{2\pi}\int_{N_n}^{\infty}\frac{\sin(\eta b)}{\eta b\cos(\eta b)}e^{-x\eta - t\sqrt{\eta^2 - N_n^2}}\,i\,d\eta$$

$$+\frac{1}{2b^2}\sum_{m=n}^{\infty}\frac{e^{-M_m x - t\sqrt{M_m^2 - N_n^2}}}{M_m},\qquad (5.1.39)$$

$$\int_{C_4} F(z)e^{itz}\,dz = \frac{1}{2\pi}\int_{-N_n}^{N_n}\frac{\sinh(\eta b)}{\eta b\cosh(\eta b)}e^{ix\eta - it\sqrt{\eta^2 + N_n^2}}\,d\eta,\qquad (5.1.40)$$

and

$$\int_{\Gamma_4} F(z)e^{itz}\,dz = \frac{1}{2\pi}\int_{-N_n}^{N_n}\frac{\sinh(\eta b)}{\eta b\cosh(\eta b)}e^{ix\eta + it\sqrt{\eta^2 + N_n^2}}\,d\eta,\qquad (5.1.41)$$

where $M_n = (n + \frac{1}{2})\pi/b$.

The remaining contour integrals are more involved. Along C_3, from purely geometric considerations,

$$\omega = -N_n e^{-\theta i},\quad \omega - iN_n = N_n\sqrt{2[1 - \sin(\theta)]}\,e^{-3\pi i/4 - \theta i/2},\qquad (5.1.42)$$

$$\omega + iN_n = N_n\sqrt{2[1 + \sin(\theta)]}\,e^{3\pi i/4 - \theta i/2},\qquad (5.1.43)$$

$$\sqrt{\omega^2 + N_n^2} = N_n \sqrt{2\cos(\theta)}\, e^{-\theta i/2}, \tag{5.1.44}$$

where θ is clockwise from the negative real axis, and

$$\int_{C_3} F(z)e^{itz}\, dz = \frac{1}{2\pi b} \int_{\pi/2}^{0} \frac{\sinh(-N_n b e^{-\theta i})}{-N_n e^{-\theta i}\cosh(-N_n b e^{-\theta i})} iN_n e^{-\theta i}\, d\theta$$

$$\times \exp\!\left[-iN_n x e^{-\theta i} - itN_n\sqrt{2\cos(\theta)}\, e^{-\theta i/2}\right] \tag{5.1.45}$$

$$= -\frac{i}{2\pi b} \int_{0}^{\pi/2} \frac{\exp[-N_n x \sin(\theta) - \xi \sin(\theta/2)]}{\cosh(\kappa) + \cos(\lambda)}\, d\theta$$

$$\times \Big\{ \cos[\zeta(-x)]\sinh(\kappa) - \sin(\lambda)\sin[\zeta(-x)] \tag{5.1.46}$$

$$- i\sin(\lambda)\cos[\zeta(-x)] - i\sinh(\kappa)\sin[\zeta(-x)] \Big\},$$

where $\zeta(x) = \xi\cos(\theta/2) - N_n x \cos(\theta)$, $\xi = N_n t\sqrt{2\cos(\theta)}$, $\kappa = 2n\pi\cos(\theta)$ and $\lambda = 2n\pi\sin(\theta)$. Similarly, along C_5,

$$\omega = N_n e^{-\theta i}, \quad \omega - iN_n = N_n\sqrt{2[1 + \sin(\theta)]}\, e^{-\pi i/4 - \theta i/2}, \tag{5.1.47}$$

$$\omega + iN_n = N_n\sqrt{2[1 - \sin(\theta)]}\, e^{\pi i/4 - \theta i/2}, \tag{5.1.48}$$

$$\sqrt{\omega^2 + N_n^2} = N_n\sqrt{2\cos(\theta)}\, e^{-\theta i/2}, \tag{5.1.49}$$

where θ is clockwise from the positive real axis, and

$$\int_{C_5} F(z)e^{itz}\, dz = \frac{1}{2\pi b} \int_{0}^{\pi/2} \frac{\sinh(N_n b e^{-\theta i})}{N_n e^{-\theta i}\cosh(N_n b e^{-\theta i})} (-iN_n e^{-\theta i})\, d\theta$$

$$\times \exp\!\left[iN_n x e^{-\theta i} - itN_n\sqrt{2\cos(\theta)}\, e^{-\theta i/2}\right] \tag{5.1.50}$$

$$= -\frac{i}{2\pi b} \int_{0}^{\pi/2} \frac{\exp[N_n x \sin(\theta) - \xi \sin(\theta/2)]}{\cosh(\kappa) + \cos(\lambda)}\, d\theta$$

$$\times \Big\{ \cos[\zeta(x)]\sinh(\kappa) - \sin(\lambda)\sin[\zeta(x)] \tag{5.1.51}$$

$$- i\sin(\lambda)\cos[\zeta(x)] - i\sinh(\kappa)\sin[\zeta(x)] \Big\}.$$

Along Γ_3,

$$\omega = -N_n e^{-\theta i}, \quad \omega - iN_n = N_n\sqrt{2[1 + \sin(\theta)]}\, e^{-3\pi i/4 + \theta i/2}, \tag{5.1.52}$$

$$\omega + iN_n = N_n\sqrt{2[1 - \sin(\theta)]}\, e^{3\pi i/4 + \theta i/2}, \tag{5.1.53}$$

$$\sqrt{\omega^2 + N_n^2} = N_n\sqrt{2\cos(\theta)}\, e^{\theta i/2} \tag{5.1.54}$$

where θ measures counterclockwise from the negative axis. Therefore,

$$\int_{\Gamma_3} F(z)e^{itz}\,dz = \frac{1}{2\pi b}\int_{\pi/2}^{0} \frac{\sinh(-N_n b e^{\theta i})}{-N_n e^{\theta i}\cosh(-N_n b e^{\theta i})}(-iN_n e^{\theta i})\,d\theta$$

$$\times \exp\!\left[-iN_n x e^{\theta i} + itN_n\sqrt{2\cos(\theta)}\,e^{\theta i/2}\right] \quad \textbf{(5.1.55)}$$

$$= \frac{i}{2\pi b}\int_{0}^{\pi/2} \frac{\exp[N_n x \sin(\theta) - \xi\sin(\theta/2)]}{\cosh(\kappa) + \cos(\lambda)}\,d\theta$$

$$\times \Big\{\cos[\zeta(x)]\sinh(\kappa) - \sin(\lambda)\sin[\zeta(x)] \quad \textbf{(5.1.56)}$$

$$+\, i\sin(\lambda)\cos[\zeta(x)] + i\sinh(\kappa)\sin[\zeta(x)]\Big\}.$$

Finally, along Γ_5,

$$\omega = N_n e^{\theta i}, \quad \omega - iN_n = N_n\sqrt{2[1-\sin(\theta)]}\,e^{-\pi i/4 + \theta i/2}, \quad \textbf{(5.1.57)}$$

$$\omega + iN_n = N_n\sqrt{2[1+\sin(\theta)]}\,e^{\pi i/4 + \theta i/2}, \quad \textbf{(5.1.58)}$$

$$\sqrt{\omega^2 + N_n^2} = N_n\sqrt{2\cos(\theta)}\,e^{\theta i/2}, \quad \textbf{(5.1.59)}$$

where θ increases counterclockwise from the positive real axis. Therefore,

$$\int_{\Gamma_5} F(z)e^{itz}\,dz = \frac{1}{2\pi b}\int_{0}^{\pi/2} \frac{\sinh(N_n b e^{\theta i})}{N_n e^{\theta i}\cosh(N_n b e^{\theta i})}(iN_n e^{\theta i})\,d\theta$$

$$\times \exp\!\left[iN_n x e^{\theta i} + itN_n\sqrt{2\cos(\theta)}\,e^{\theta i/2}\right] \quad \textbf{(5.1.60)}$$

$$= \frac{i}{2\pi b}\int_{0}^{\pi/2} \frac{\exp[-N_n x \sin(\theta) - \xi\sin(\theta/2)]}{\cosh(\kappa) + \cos(\lambda)}\,d\theta$$

$$\times \Big\{\cos[\zeta(-x)]\sinh(\kappa) - \sin(\lambda)\sin[\zeta(-x)] \quad \textbf{(5.1.61)}$$

$$+\, i\sin(\lambda)\cos[\zeta(-x)] + i\sinh(\kappa)\sin[\zeta(-x)]\Big\}.$$

Summing the results from Equation 5.1.36 to Equation 5.1.41, Equation 5.1.46, Equation 5.1.51, Equation 5.1.56 and Equation 5.1.61, we obtain the final result that

$$f(x,t) = \frac{2}{b^2}\sum_{m=n}^{\infty} \frac{\cosh(M_n x)}{M_n} e^{-t\sqrt{M_n^2 - N_n^2}}$$

$$+ \frac{2}{\pi b}\int_{0}^{N_n} \frac{\sinh(\eta b)}{\eta\cosh(\eta b)}\cos[t\sqrt{\eta^2 + N_n^2}]\,d\eta$$

$$- \frac{1}{\pi b}\int_{0}^{\pi/2} \frac{\exp[-N_n x \sin(\theta) - \xi\sin(\theta/2)]}{\cosh(\kappa) + \cos(\lambda)}\,d\theta$$

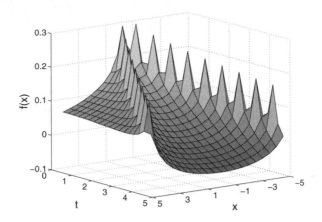

Figure 5.1.6: A plot of the inverse $f(x, t)$ of the Fourier transform given by Equation 5.1.33 for $b = 10$ and $n = 1$.

$$\times \left\{ \sin(\lambda) \cos[\zeta(x)] + \sinh(\kappa) \sin[\zeta(x)] \right\}$$

$$- \frac{1}{\pi b} \int_0^{\pi/2} \frac{\exp[N_n x \sin(\theta) - \xi \sin(\theta/2)]}{\cosh(\kappa) + \cos(\lambda)} \, d\theta$$

$$\times \left\{ \sin(\lambda) \cos[\zeta(-x)] + \sinh(\kappa) \sin[\zeta(-x)] \right\}. \quad (\textbf{5.1.62})$$

Figure 5.1.6 illustrates $f(x, t)$.

- **Example 5.1.4**

For our final example, let us find the inverse[4]

$$f(x, y) = \frac{1}{2\pi} \int_{-\infty}^{\infty} \exp\left(ixk - y\sqrt{k^2 - \frac{ikV}{c}} \right) dk, \quad (\textbf{5.1.63})$$

where the square root must have a positive real part. To recast Equation 5.1.63 into a more symmetric form, we introduce $\alpha = k - iV/2c$ so that

$$f(x, y) = \frac{e^{-Vx/2c}}{2\pi} \int_{-\infty - iV/2c}^{\infty - iV/2c} \exp\left[ix\alpha - y\sqrt{\alpha^2 + \left(\frac{V}{2c}\right)^2} \right] d\alpha. \quad (\textbf{5.1.64})$$

In order that the radical has a positive real part, we define the function $\alpha - iV/2c$ with a phase between $-3\pi/2$ to $\pi/2$ and $\alpha + iV/2c$ with a phase

---

[4] Patterned after Rudnicki, J. W., and E. A. Roeloffs, 1990: Plane-strain shear disloca-tions moving steadily in linear elastic diffusive solids. *J. Appl. Mech.*, **57**, 32–38. See also Chowdhury, K. L., and P. G. Glockner, 1980: On a boundary value problem for an elastic dielectric half-space. *Acta Mech.*, **37**, 65–74.

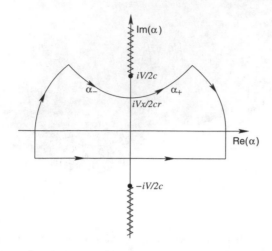

Figure 5.1.7: The contour used to invert the Fourier transform given by Equation 5.1.64.

between $-\pi/2$ to $3\pi/2$. See Figure 5.1.7. We will give the details for $x > 0$; the case $x < 0$ follows by analog.

One method of integrating Equation 5.1.64 would be to deform the contour so that it consists of arcs at infinity in the first and second quadrants of the α-plane plus integrations along the branch cuts. Here we will use an alternative method that will be explored in greater detail in Section 6.3 where we will use it to invert Laplace transforms.

Let us define

$$ix\alpha - y\sqrt{\alpha^2 + \left(\frac{V}{2c}\right)^2} = -s. \qquad (5.1.65)$$

A little algebra shows that we must define α by

$$\alpha_{\pm} = \frac{ixs}{r^2} \pm \frac{y}{r^2}\sqrt{s^2 - \left(\frac{rV}{2c}\right)^2}, \qquad \frac{rV}{2c} \leq s \leq \infty, \qquad (5.1.66)$$

if we introduce the polar coordinates $x = r\cos(\theta)$ and $y = r\sin(\theta)$. We now deform our original contour so that it conforms to α_+ in the first quadrant and α_- in the second quadrant of the α-plane. This deformation is permissible because we do not cross any singularities and the integration along the arcs at infinity vanishes by Jordan's lemma. Our motivation lies in the simplicity that this transformation introduces in the exponential.

Upon introducing this new contour into Equation 5.1.64,

$$f(x,y) = e^{-Vx/2c}\frac{y}{\pi r^2}\int_{rV/2c}^{\infty}\frac{se^{-s}}{\sqrt{s^2 - r^2V^2/(4c^2)}}\,ds \qquad (5.1.67)$$

$$= e^{-Vx/2c}\frac{Vy}{2\pi cr}\int_{1}^{\infty}\frac{\xi e^{-rV\xi/2c}}{\sqrt{\xi^2 - 1}}\,d\xi \qquad (5.1.68)$$

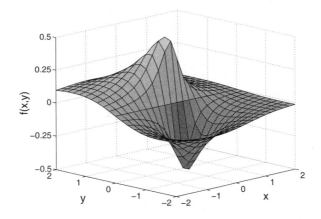

Figure 5.1.8: A plot of the solution $f(x,y)$ given by Equation 5.1.69 as a function of x and y when $V/c = 1$.

$$f(x,y) = e^{-Vx/2c}\frac{y}{\pi}\left(\frac{V}{2c}\right)^2\int_1^\infty e^{-rV\xi/2c}\sqrt{\xi^2-1}\,d\xi \qquad (5.1.69)$$

after an integration by parts and introducing $s = rV\xi/2c$. Figure 5.1.8 illustrates $f(x,y)$ when $V/c = 1$; the numerical quadrature was done with Simpson's rule.

Problems

1. Show that the inverse Fourier transform of $F(\omega) = 1/\sqrt{1+i\omega}$ is $f(t) = e^{-t}H(t)/\sqrt{\pi t}$. Take the branch cut associated with $\sqrt{\omega - i}$ along the imaginary axis of the ω-plane from $[i, \infty i)$.

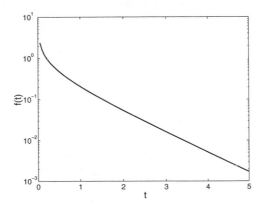

Problem 1

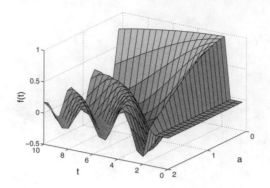

Problem 2

2. Find the Fourier inverse[5] of the bilateral Fourier transform

$$F(\omega) = -i\frac{\exp(-it_0\sqrt{\omega^2 - a^2})}{\sqrt{\omega^2 - a^2}},$$

where a and t_0 are real. If you take the branch cuts along the real axis from $(-\infty, -a]$ and $[a, \infty)$, show that

$$f(t) = -\frac{i}{2\pi}\int_{-\infty-\epsilon i}^{\infty-\epsilon i} \frac{\exp(it\omega - it_0\sqrt{\omega^2 - a^2})}{\sqrt{\omega^2 - a^2}}\, d\omega = J_0\left(a\sqrt{t^2 - t_0^2}\right)H(t - t_0),$$

where $0 < \epsilon \ll 1$. This inverse is illustrated in the figure labeled Problem 2 with $t_0 = 1$.

3. Find the inverse[6] of the Fourier transform

$$F(\omega) = \frac{\exp\left[-i\omega\sqrt{r^2 + (z + a/\omega i)^2}\right]}{\sqrt{r^2 + (z + a/\omega i)^2}}, \qquad 0 < a, r, z.$$

Define the square root such that $\text{Im}\left[\omega\sqrt{r^2 + (z + a/\omega i)^2}\right] \leq 0$. Show that

$$f(t) = \frac{2}{\pi r_i}\frac{d}{dt}\left[e^{-\alpha t}J_0\left(\beta\sqrt{t^2 - r_i^2}\right)H(t - r_i)\right],$$

---

[5] For a problem which uses this inverse, see Row, R. V., 1967: Acoustic-gravity waves in the upper atmosphere due to a nuclear detonation and an earthquake. *J. Geophys. Res.*, **72**, 1599–1610. ©1967 American Geophysical Union. Reproduced/modified by permission of American Geophysical Union.

[6] Taken from Nikoskinen, K. I., and I. V. Lindell, 1990: Time-domain analysis of the Sommerfeld VMD problem based on the exact image theory. *IEEE Trans. Antennas Propag.*, **AP-38**, 241–250. ©1990 IEEE.

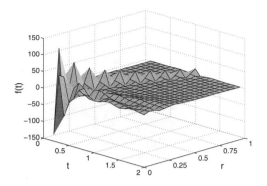

Problem 3

where $r_i = \sqrt{r^2 + z^2}$, $\alpha = az/r_i^2$ and $\beta = ar/r_i^2$. This inverse is illustrated in the figure labeled Problem 3 with $a = 1$ and $z = 0.01$.

4. One method[7] for dealing with Fourier transforms with singularities on the real axis which also have symmetry properties in t is to compute the inverse using the formula

$$f(t) = \frac{1}{2\pi}\left[\int_{-\infty-\epsilon i}^{\infty-\epsilon i} F(\omega)e^{it\omega}\,d\omega + \int_{-\infty+\epsilon i}^{\infty+\epsilon i} F(\omega)e^{it\omega}\,d\omega\right], \qquad 0 < \epsilon \ll 1.$$

Using this formula, find the inverse of

$$F(\omega) = \frac{e^{x\sqrt{\omega^2+k^2}}}{k - \sqrt{\omega^2 + k^2}}, \qquad 0 < k, \quad x < 0.$$

If we take the branch cuts associated with $\sqrt{\omega^2 + k^2}$ along the imaginary axis from $(-\infty i, -ki]$ and $[ki, \infty i)$, show that the inverse is

$$f(t) = k|t|e^{kx} - \frac{1}{\pi}\int_k^\infty \frac{e^{-|t|\eta}}{\eta^2}\,d\eta$$
$$\times \left[k\sin\left(x\sqrt{\eta^2 - k^2}\right) + \sqrt{\eta^2 - k^2}\cos\left(x\sqrt{\eta^2 - k^2}\right)\right].$$

This inverse is illustrated in the figure labeled Problem 4.

5. Show that the inverse[8] of

$$F(\omega) = -\frac{|\omega|\exp(-s|\omega|)}{|\omega| - a_1 + i\epsilon a_2\mathrm{sgn}(\omega)}, \qquad 0 < a_1, a_2, s,$$

[7] Taken from Haren, P., and C. C. Mei, 1981: Head-sea diffraction by a slender raft with application to wave-power absorption. *J. Fluid Mech.*, **104**, 505–526. Reprinted with the permission of Cambridge University Press.

[8] Taken from Savage, M. D., 1967: Stationary waves at a plasma-magnetic field interface. *J. Plasma Phys.*, **1**, 229–239. Reprinted with the permission of Cambridge University Press.

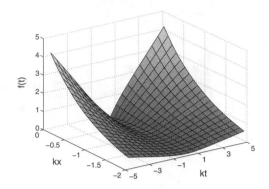

Problem 4

where $\epsilon \to 0$. By deforming the path of integration along the real ω axis so that it passes over $\omega = \pm a_1$ and taking the branch cuts associated with $|\omega|$ along the imaginary axis from $(-\infty i, 0^- i]$ and $[0^+ i, \infty i)$, show that

$$f(t) = -2a_1 e^{-a_1 s} \sin(a_1 t) H(-t)$$

$$- \frac{1}{\pi} \int_0^\infty \frac{\eta e^{-|t|\eta}}{\eta^2 + a_1^2} [a_1 \cos(s\eta) + \eta \sin(s\eta)] \, d\eta.$$

[Hint: Treat the absolute value as $|z| = \sqrt{(z - 0^+ i)(z - 0^- i)}$, where the real part of the square root must be always positive.] The inverse $f(t)$ is plotted as a function of $s = a - y_0$, $t = x$ and $f(t) = \eta(x)/a_1$ in the figure labeled Problem 3 at the end of Section 5.3.

6. Find the inverse of the Fourier transform

$$F(\omega) = \frac{1}{\omega^2 + a^2} \log\left(\frac{b + i\omega}{b - i\omega}\right), \qquad 0 < a < b.$$

Take the branch cuts associated with the logarithm along the imaginary axis from $(-\infty i, -bi]$ and $[bi, \infty i)$. Show that the inverse is

$$f(t) = \text{sgn}(t) \frac{e^{-a|t|}}{2a} \ln\left(\frac{b - a}{b + a}\right) - \text{sgn}(t) \int_b^\infty \frac{e^{-|t|\eta}}{a^2 - \eta^2} \, d\eta.$$

This inverse is illustrated in the figure labeled Problem 6 for $b = 2$.

7. Find the inverse[9] of the Fourier transform

$$F(\omega) = \frac{V(\omega)i}{\omega[1 - cV(\omega)]} - \frac{V(\omega)\omega i}{(\omega^2 + \alpha^2)[1 - cV(\omega)]}, \qquad 0 < \alpha, \omega_0 < 1, \quad c \neq 1,$$

---

[9] Taken from Williams, M. M. R., 1965: Neutron transport in differentially heated media. *Brit. J. Appl. Phys.*, **16**, 1727–1732. See also Pearlstein, L. D., and G. W. Stuart, 1961: Effect of collisional energy loss on ionization growth in H_2. *Phys. Fluids*, **4**, 1293–1297.

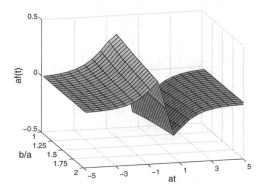

Problem 6

where

$$V(\omega) = \frac{1}{2i\omega} \log\left(\frac{1+\omega i}{1-i\omega}\right) = \frac{\arctan(\omega)}{\omega},$$

and $cV(i\omega_0) = 1$. Taking the branch cuts associated with the logarithms along the imaginary axis from $(-\infty i, -i]$ and $[i, \infty i)$, show that

$$f(t) = \frac{V(\alpha i)e^{-\alpha t}}{2[1 - cV(\alpha i)]} - \frac{1}{1-c} - \frac{\alpha^2(1-\omega_0^2)e^{-\omega_0 t}}{c(\alpha^2 - \omega_0^2)(\omega_0^2 + c - 1)}$$

$$- \frac{\alpha^2}{2} \int_0^1 \frac{\eta^2}{1-\alpha^2\eta^2} e^{-t/\eta} \left\{ \left[1 + \frac{c\eta}{2}\ln\left(\frac{1-\eta}{1+\eta}\right)\right]^2 + \frac{\pi^2 c^2 \eta^2}{4} \right\}^{-1} d\eta$$

if $t > 0$. This inverse is illustrated in the figure labeled Problem 7 for $\alpha = 0.5$ and $c = 2$.

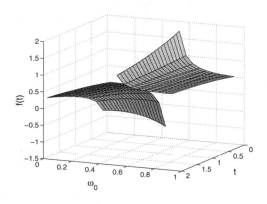

Problem 7

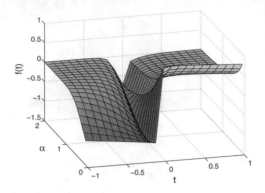

Problem 8

8. Find the inverse of the Fourier transform

$$F(\omega) = \frac{1}{\omega i + 2\alpha\sqrt{k^2 + \omega i}}, \qquad 0 < k, \alpha.$$

Take the branch cuts associated with the square root along the imaginary axis from $-k^2 i$ to ∞i. Show that

$$f(t) = \frac{\left(\sqrt{\alpha^2 + k^2} - \alpha\right)\exp\left(2\alpha^2 t - 2\alpha t\sqrt{\alpha^2 + k^2}\right)}{\sqrt{\alpha^2 + k^2}}$$

$$- \frac{4\alpha}{\pi}\int_0^\infty \frac{\eta^2 e^{-(k^2 + \eta^2)t}}{(k^2 + \eta^2)^2 + 4\alpha^2\eta^2}\, d\eta$$

for $t > 0$, and

$$f(t) = -\frac{\left(\sqrt{\alpha^2 + k^2} + \alpha\right)\exp\left(2\alpha^2 t + 2\alpha t\sqrt{\alpha^2 + k^2}\right)}{\sqrt{\alpha^2 + k^2}}$$

for $t < 0$. This inverse is illustrated in the figure labeled Problem 8 for $k = 1$.

9. Find the inverse of the generalized Fourier transform

$$F(\omega) = \frac{\log(i\omega\tau)}{\left(\lambda + \sqrt{\lambda^2 + \omega i}\right)^2}, \qquad 0 < \lambda, \tau.$$

If we take the branch cut associated with the square root along the positive imaginary axis of the ω-plane, from $\omega = \lambda^2 i$ to ∞i, and the branch cut of $\log(z)$ lies along the positive imaginary axis, then show that

$$f(t) = \frac{1}{\pi}\int_1^\infty e^{-\lambda^2 t\eta}\left[2\sqrt{\eta - 1}\ln(\lambda^2\tau\eta) - \pi(2 - \eta)\right]\frac{d\eta}{\eta^2} - \int_0^1 \frac{e^{-\lambda^2 t\eta}}{\left(1 + \sqrt{1 - \eta}\right)^2}\, d\eta$$

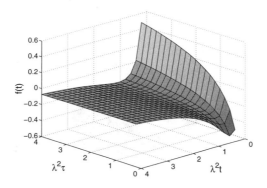

Problem 9

if $t > 0$; $f(t) = 0$ if $t < 0$. This inverse $f(t)$ is illustrated in the figure labeled Problem 9.

10. Show that the inverse[10] of

$$F(\omega, y) = e^{-|k|y} \qquad \text{is} \qquad f(x, y) = \pi \frac{\partial[\ln(r)]}{\partial y}.$$

Step 1: Use polar coordinates to show that

$$f(x, y) = \frac{1}{2\pi} \int_{-\infty}^{\infty} e^{r[ik\cos(\theta) - |k|\sin(\theta)]} \, dk.$$

Step 2: Introduce the transformation $\eta = -ik\cos(\theta) + |k|\sin(\theta) = ke^{i(-\pi/2\pm\theta)}$, $0 < \eta < \infty$, where we take the positive sign for $0 < k$ and the minus sign for $k < 0$. Then $k = \eta e^{i(\pi/2\mp\theta)}$ and $dk/d\eta = e^{i(\pi/2\mp\theta)}$. By breaking the integral in Step 1 into two parts, from $-\infty$ to 0 and 0 to ∞, show that

$$f(x, y) = \pi \sin(\theta) \int_{0}^{\infty} e^{-r\eta} \, d\eta = \frac{\pi \sin(\theta)}{r}.$$

11. Rework Example 5.1.4 with the transform

$$F(k, y) = \frac{\exp\left(-y\sqrt{k^2 - ikV/c}\right)}{\sqrt{k^2 - ikV/c}}, \qquad 0 < y,$$

and show that

$$f(x, y) = \frac{\exp[-Vx/(2c)]}{\pi} \int_{1}^{\infty} \frac{\exp[-rV\eta/(2c)]}{\sqrt{\eta^2 - 1}} \, d\eta,$$

---

[10] Taken from Chowdhury, K. L., and P. G. Glockner, 1980: On a boundary value problem for an elastic dielectric half-space. *Acta Mech.*, **37**, 65–74.

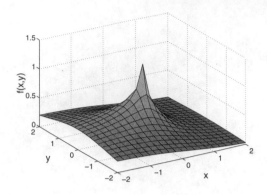

Problem 11

where $r = \sqrt{x^2 + y^2}$. This inverse, $f(x, y)$, is illustrated in the figure labeled Problem 11 for $V/c = 1$.

5.2 NUMERICAL INVERSION OF FOURIER TRANSFORMS

So far in this book we have focused on analytic techniques to invert the Fourier transform. In many instances these techniques fail and we must resort to numerical techniques. The purpose of this section is to present several numerical options for inversion.

In its simplest form, the numerical inversion of a Fourier transform is a line integration of a complex integrand along the real axis. The first step in its numerical inversion replaces the integration from $-\infty$ to ∞ with

$$f(t) \approx \frac{1}{2\pi} \int_{-L}^{L} F(\omega)e^{it\omega} \, d\omega, \tag{5.2.1}$$

where L is a large number that will be specified later. The next step approximates the integral in Equation 5.2.1. The simplest method is the first-order extended rectangular rule, or

$$f(t) \approx \frac{\Delta\omega}{2\pi} \sum_{m=-M}^{M-1} F(\omega_m)e^{it\omega_m}, \tag{5.2.2}$$

where $\omega_m = m\Delta\omega$ and $\Delta\omega = L/M$. Although we could perform the summation in Equation 5.2.2 for a given time t, let us examine the case when $t = t_n = \pi n/L$,

$$f_n = f(t_n) \approx \frac{\Delta\omega}{2\pi} \sum_{m=-M}^{M-1} F(\omega_m)e^{\pi imn/M}, \tag{5.2.3}$$

or

$$f_n \approx \frac{\Delta\omega}{2\pi} \sum_{m=-M}^{-1} F(\omega_m) e^{\pi imn/M} + \frac{\Delta\omega}{2\pi} \sum_{m=0}^{M-1} F(\omega_m) e^{\pi imn/M}. \qquad (5.2.4)$$

Invoking the periodicity condition $F(\omega_m) = F(\omega_{m+2M})$,

$$f_n \approx \frac{L}{\pi} \left[\frac{1}{2M} \sum_{m=0}^{2M-1} F(\omega_m) e^{\pi imn/M} \right]. \qquad (5.2.5)$$

The quantity within the square brackets is the inverse of the discrete Fourier transform. We see now why we choose to evaluate Equation 5.2.2 at $t = t_n$; it can be rapidly computed. Equation 5.2.5 then yields the inverse for times $0 \leq t_n \leq (2M-1)\pi/L$. For other times outside of this range, the periodicity condition gives $f_n = f_{n\pm 2M}$.

To illustrate this method, consider the Fourier transform

$$F(\omega) = \frac{1}{\omega^2 + a^2} \log\left(\frac{b + i\omega}{b - i\omega}\right), \qquad 0 < a < b. \qquad (5.2.6)$$

It provides a good test because it has both isolated singularities at $\omega = \pm ai$ and branch points at $\omega = \pm bi$. As shown in Problem 6 of the previous section, its inverse is

$$f(t) = \mathrm{sgn}(t) \frac{e^{-a|t|}}{2a} \ln\left(\frac{b-a}{b+a}\right) - \mathrm{sgn}(t) \int_b^\infty \frac{e^{-|t|\eta}}{a^2 - \eta^2} \, d\eta. \qquad (5.2.7)$$

If we program Equation 5.2.5 using MATLAB, our inversion scheme is

```
% d_omega = Δω
clear; a = 0.1; b = 0.2; L = 100; M = 500; d_omega = L / M;
for m = 1:2*M
omega = -L + (m-1) * d_omega; % compute ωm
arg = (b + i * omega) / (b - i * omega);
bot = omega * omega + a * a;
if (omega < 0)
F(m+M) = log(arg) / bot;
else
F(m-M) = log(arg) / bot;
end; end

f = ifft(F,2*M); % compute inverse of discrete Fourier transform
```

We have special code for F(m) so that F(m) runs from $m = 1$ to $m = 2M$. To correctly order the computed values of f_n so that they run from $t = -M\pi/L$ to $(M-1)\pi/L$, we need

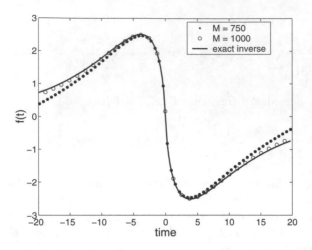

Figure 5.2.1: The numerical inversion of the Fourier transform $F(\omega)$ with $a = 0.1$ and $b = 0.2$ using Equation 5.2.5.

```
for m = 1:2*M
time(m) = pi * (m-1-M) / L;
if (time(m) >= 0)
f_n(m) = L * real(f(m-M)) / pi;
else
f_n(m) = L * real(f(m+M)) / pi;
end; end
```

Figure 5.2.1 shows the results when $L = 100$ for two different values of M. From this figure, the results are rather good even when the resolution is $\Delta\omega = 0.13333$.

This method suffers the drawback that its resolution π/L is determined only by the integration range. Therefore, the range of the inverse in t can only be increased by increasing M. This could be disadvantageous in two ways. First, if a large M was needed to reduce the truncation error, we might compute the inverse integral a superfluous number of times. Second, if we desire the inverse over a large temporal range, we take such a large M that we have inordinate levels of accuracy.

In 1994, Bailey and Swarztrauber[11] derived an algorithm using circular convolution and fractional Fourier transforms to decouple the temporal and spatial resolution. Recently, Inverarity[12] obtained Bailey and Swarztrauber's

[11] Bailey, D. H., and P. N. Swarztrauber, 1994: A fast method for the numerical evaluation of continuous Fourier and Laplace transforms. *SIAM J. Sci. Comput.*, **15**, 1105–1110.

[12] Inverarity, G. W., 2002: Fast computation of multidimensional Fourier integrals. *SIAM J. Sci. Comput.*, **24**, 645–651.

algorithm without using fractional Fourier transforms. His derivation is as follows:

Let us introduce a $\omega_m = -L + m\Delta\omega$ and $t_n = -T + n\Delta t$, where $\Delta\omega = L/M$, $\Delta t = T/M$ and $n, m = 0, 1, 2, \ldots, 2M - 1$. Then,

$$f_n = \frac{\Delta\omega}{2\pi} \sum_{m=0}^{2M-1} F(\omega_m) e^{i\omega_m t_n}. \tag{5.2.9}$$

Because

$$i\omega_m t_n = -iLt_n + i\delta n^2 - im\Delta\omega T + i\delta m^2 - i\delta(n - m)^2, \tag{5.2.10}$$

where $\delta = \Delta\omega\Delta t/2$,

$$f_n = \frac{\Delta\omega}{2\pi} e^{-iLt_n + i\delta n^2} \sum_{m=0}^{2M-1} \left[F(\omega_m) e^{-im\Delta\omega T + i\delta m^2} \right] e^{-i\delta(n-m)^2}. \tag{5.2.11}$$

The summation in Equation 5.2.11 is very similar to circular convolution from the theory of discrete Fourier transforms. If we introduce the sequences:

$$y_m = \begin{cases} F(\omega_m) e^{-im\Delta\omega T + i\delta m^2}, & 0 \leq m \leq 2M - 1, \\ 0, & 2M \leq m \leq 4M - 1, \end{cases} \tag{5.2.12}$$

and

$$z_m = \begin{cases} e^{-i\delta m^2}, & 0 \leq m \leq 2M - 1, \\ e^{-i\delta(m-4M)^2}, & 2M \leq m \leq 4M - 1, \end{cases} \tag{5.2.13}$$

then $f_n \propto \sum_{m=0}^{4M-1} y_m z_{n-m}$. We may then use the circular convolution theorem to compute the sum. Coding this in MATLAB, the script reads

```
clear; a = 0.1; b = 0.2; L = 100; T = 20;
M = 500; M2 = 2 * M; d_omega = L / M; dt = T / M;
delta = 0.5 * dt * d_omega;
%%%%%%%%%%%%%%%%%%%%%%%%%%%%%%%%%%%%%%%%%%%%%%%%%%%%%%%%%%%%%%%%%
%                    create sequence z_m
%%%%%%%%%%%%%%%%%%%%%%%%%%%%%%%%%%%%%%%%%%%%%%%%%%%%%%%%%%%%%%%%%
for m = 0:2*M2-1
y(m+1) = 0;
if (m < M2 )
z(m+1) = exp(-i*delta*m*m);
else
z(m+1) = exp(-i*delta*(m-2*M2)*(m-2*M2));
end; end
%                take the FFT of the sequence z_m
Z = fft(z,2*M2);
```

```
%%%%%%%%%%%%%%%%%%%%%%%%%%%%%%%%%%%%%%%%%%%%%%%%%%%%%%%%%%%%%%%
%                        create sequence y_m
%%%%%%%%%%%%%%%%%%%%%%%%%%%%%%%%%%%%%%%%%%%%%%%%%%%%%%%%%%%%%%%
for m = 0:M2-1
omega = -L + m * d_omega; % eqn(1)
arg = (b + i * omega) / (b - i * omega);
                   % compute the Fourier transform
y(m+1) = log(arg) / (omega*omega+a*a);
y(m+1) = y(m+1) * exp(i*delta*m*m-i*m*d_omega*T);
end
%                  take the FFT of the sequence y_m
Y = fft(y,2*M2);
%%%%%%%%%%%%%%%%%%%%%%%%%%%%%%%%%%%%%%%%%%%%%%%%%%%%%%%%%%%%%%%
% use circular convolution theorem to compute sum in Eq 5.2.11
%%%%%%%%%%%%%%%%%%%%%%%%%%%%%%%%%%%%%%%%%%%%%%%%%%%%%%%%%%%%%%%
for m = 1:2*M2
PROD(m) = Y(m) .* Z(m);
end
prod = ifft(PROD,2*M2);
%%%%%%%%%%%%%%%%%%%%%%%%%%%%%%%%%%%%%%%%%%%%%%%%%%%%%%%%%%%%%%%
%                     compute the inverse f_n
%%%%%%%%%%%%%%%%%%%%%%%%%%%%%%%%%%%%%%%%%%%%%%%%%%%%%%%%%%%%%%%
k = 0;
for m = 0:M2-1
time(m+1) = -T + m * dt;
f_n(m+1) = d_omega * real(exp(i*delta*m*m-i*L*time(m+1)) ...
        * prod(m+1)) / (2*pi); % eqn(2)
end
```

Figure 5.2.2 shows a sample of the results when $L = 100$ and $T = 20$ for two different values of M. Excellent results were obtained using smaller values of M than shown in Figure 5.2.1.

As Inverarity[13] has shown, a simple modification of our code, namely replacing `eqn(1)` with `omega=-L+(m+0.5)*d_omega` and replacing `eqn(2)` with `f_n(m+1) = d_omega*real(exp(i*delta*m*m + i*(-L+0.5*d_omega)*time (m+1))*prod(m+1))/(2*pi)`, yields a second-order extended midpoint quadrature rule. The advantage of this approach is that it can be implemented adaptively using Richardson extrapolation. This method yields great accuracy even for large $|t|$ with a surprisingly low number of interval subdivisions. See Inverarity[14] for further details.

---

[13] *Ibid.*

[14] Inverarity, G. W., 2003: Numerically inverting a class of singular Fourier transforms: Theory and application to mountain waves. *Proc. R. Soc. London, Ser. A*, **459**, 1153–1170.

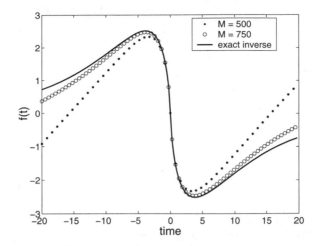

Figure 5.2.2: Same as Figure 5.2.1, but the resolution is different for the temporal and spectral resolutions.

As Figures 5.2.1 and 5.2.2 show, the worst results occur at large $|t|$. This occurs because of the oscillatory nature of the integrand $F(\omega)e^{it\omega}$. This problem was first addressed by Filon[15] who modified Simpson's rule so that its accuracy depends only on the smoothness of $F(\omega)$. Applying his scheme to Equation 5.2.1,

$$f(t) = \frac{\Delta\omega}{2\pi}\left\{i\alpha(\Delta\omega t)\left[F(\omega_0)e^{i\omega_0 t} - F(\omega_{2M})e^{i\omega_{2M}t}\right]\right.$$

$$\left. + \beta(\Delta\omega t)S_{2M} + \gamma(\Delta\omega t)S_{2M-1}\right], \qquad \textbf{(5.2.14)}$$

where

$$S_{2M} = \sum_{j=0}^{M} F(\omega_{2j})e^{i\omega_{2j}t} - \tfrac{1}{2}\left[F(\omega_0)e^{i\omega_0 t} + F(\omega_{2M})e^{i\omega_{2M}t}\right], \qquad \textbf{(5.2.15)}$$

$$S_{2M-1} = \sum_{j=1}^{M} F(\omega_{2j-1})e^{i\omega_{2j-1}t}, \qquad \textbf{(5.2.16)}$$

$$\alpha(\tau) = \frac{1}{\tau} + \frac{\sin(2\tau)}{2\tau^2} - \frac{2\sin^2(\tau)}{\tau^3}, \qquad \textbf{(5.2.17)}$$

$$\beta(\tau) = 2\left[\frac{1+\cos^2(\tau)}{\tau^2} - \frac{\sin(2\tau)}{\tau^3}\right], \qquad \textbf{(5.2.18)}$$

[15] Filon, L. N. G., 1928-29: On a quadrature formula for trigonometric integrals. *Proc. R. Soc. Edinburgh*, **49**, 38–47. See Section 2.10.2 in Davis, P. J., and P. Rabinowitz, 1975: *Methods of Numerical Integration*. Academic Press, 459 pp.

$$\gamma(\tau) = 4 \left[\frac{\sin(\tau)}{\tau^3} - \frac{\cos(\tau)}{\tau^2} \right], \qquad \textbf{(5.2.19)}$$

$\omega_j = -L + j\Delta\omega$ and $\Delta\omega = L/M$. The MATLAB script for executing this method is

```
d_omega = LL/M; tau = d_omega * time
%%%%%%%%%%%%%%%%%%%%%%%%%%%%%%%%%%%%%%%%%%%%%%%%%%%%%%%%%%%%%%%
%              Compute the coefficients alpha, beta, and gamma
%%%%%%%%%%%%%%%%%%%%%%%%%%%%%%%%%%%%%%%%%%%%%%%%%%%%%%%%%%%%%%%
alpha = 1/tau + sin(2*tau) / (2*tau*tau) ...
      - 2 * sin(tau) * sin(tau) / (tau * tau * tau);
alpha = d_omega * alpha;
beta = (1 + cos(tau) * cos(tau)) / (tau * tau) ...
     - sin(2*tau) / (tau * tau * tau);
beta = 2 * beta * d_omega;
gamma = sin(tau) / (tau * tau * tau) - cos(tau) / (tau * tau);
gamma = 4 * gamma * d_omega;
%%%%%%%%%%%%%%%%%%%%%%%%%%%%%%%%%%%%%%%%%%%%%%%%%%%%%%%%%%%%%%%
%              Compute S_2M and S_2M-1
%%%%%%%%%%%%%%%%%%%%%%%%%%%%%%%%%%%%%%%%%%%%%%%%%%%%%%%%%%%%%%%
sum_2nm1 = 0;
omega = -LL;
arg = (bb + i * omega) / (bb - i * omega);
temp = log(arg) * exp(i*omega*time) / (omega*omega+aa*aa);
sum_2n = 0.5 * temp;
endpt = i * temp;
for m = 1:M-1
omega = omega + d_omega;
arg = (bb + i * omega) / (bb - i * omega);
sum_2nm1 = sum_2nm1 ...
         + log(arg) * exp(i*omega*time) / (omega*omega+aa*aa);
omega = omega + d_omega;
arg = (bb + i * omega) / (bb - i * omega);
sum_2n = sum_2n ...
       + log(arg) * exp(i*omega*time) / (omega*omega+aa*aa);
end
omega = LL;
arg = (bb + i * omega) / (bb - i * omega);
temp = log(arg) * exp(i*omega*time) / (omega*omega+aa*aa);
sum_2n = sum_2n + 0.5 * temp;
endpt = endpt - i * temp;
%%%%%%%%%%%%%%%%%%%%%%%%%%%%%%%%%%%%%%%%%%%%%%%%%%%%%%%%%%%%%%%
%              Compute the inverse
%%%%%%%%%%%%%%%%%%%%%%%%%%%%%%%%%%%%%%%%%%%%%%%%%%%%%%%%%%%%%%%
```

```
f = real(alpha * endpt + beta * sum_2n ...
         + gamma * sum_2nm1) / (2*pi);
```

Recently, Evans and Webster[16] took another line of attack, developing a quadrature method of the form

$$\int_{-1}^{1} f(x)e^{i\tau x}\,dx \approx \sum_{j=0}^{N} w_j f(x_j). \tag{5.2.20}$$

To determine the weights w_j, they required that Equation 5.2.20 becomes exact for the set of functions $f(x) = i\tau p_k(x) + p'_k(x)$. Presently, both $p_k(x_j)$ and x_j are undetermined. The motivation behind this choice is the fact that we can directly evaluate the left side of Equation 5.2.20 to obtain

$$\sum_{j=0}^{N} \left[i\tau p_k(x_j) + p'_k(x_j) \right] w_j = \left. p_k(x)e^{i\tau x} \right|_{-1}^{1}, \tag{5.2.21}$$

where $k = 0, 1, \ldots, N$. They then tested

$$x_j = -1 + 2j/N \quad \text{and} \quad p_k(x) = x^k, \tag{5.2.22}$$

and

$$x_j = \cos(j\pi/N) \quad \text{and} \quad p_k(x) = T_k(x), \tag{5.2.23}$$

where $T_k(x)$ denotes the kth Chebyshev polynomial, as possible candidates for x_j and p_k. They found that Equation 5.2.23 is particularly effective. The weights w_j are found using simple linear algebra.

Having found the coefficients w_j, we must now rewrite Equation 5.2.1 in terms of Equation 5.2.20. We do this by dividing the interval $[-L, L]$ into $2M + 1$ subdivisions of half-thickness $\Delta\omega = L/(2M + 1)$. Within each subdivision, $\omega = \omega_c + \Delta\omega\, x$ with $\omega_c = 2m\Delta\omega$, $x \in [-1, 1]$ and $m = -M, -M + 1, \ldots, M$. From these definitions, Evans and Webster's method yields

$$\int_{-L}^{L} F(\omega)e^{it\omega}\,d\omega = \Delta\omega \sum_{m=-M}^{M} e^{it\omega_c} \left[\int_{-1}^{1} F(\omega)e^{it\Delta\omega x}\,dx \right]. \tag{5.2.24}$$

Let us now convert this routine into a MATLAB script. First, we must find Chebyshev polynomials and x_j. This does not depend on t and must be computed only once.

%%%

---

[16] Evans, G. A., and J. R. Webster, 1997: A high order, progressive method of the evaluation of irregular oscillatory integrals. *Appl. Numer. Math.*, **23**, 205–218.

```
%                Compute the Chebyshev polynomials at x_j
%%%%%%%%%%%%%%%%%%%%%%%%%%%%%%%%%%%%%%%%%%%%%%%%%%%%%%%%%%%%%%%%%
clear; aa = 0.1; bb = 0.2; N = 16; M = 100; LL = 100; % LL = L
d_omega = LL / (2*M+1); % d_omega = Δω
tau = time * d_omega
for j = 0:n
x(j+1) = cos(pi*j/N);
% *** compute T_n(x_j) and T'_n(x_j) for n = 0,1
t(j+1,1) = 1; t(j+1,2) = x(j+1);
tp(j+1,1) = 0; tp(j+1,2) = 1;
% *** compute T_n(1) and T'_n(1)
if (j == 0)
for n = 2:N
t(1,n+1) = 2 * x(1) * t(1,n) - t(1,n-1);
tp(1,n+1) = n * n;
end; end
% *** compute T_n(-1) and T'_n(-1)
if (j == N)
for n = 2:N
t(i+1,n+1) = 2 * x(i+1) * t(i+1,n) - t(i+1,n-1);
tp(i+1,n+1) = n * n;
tp(i+1,n+1) = (-1)^(n-1) * tp(i+1,n+1);
end; end
% *** compute T_n(x_j) and T'_n(x_j) elsewhere
if ((j > 0) & (j < N))
for n = 2:N
t(j+1,n+1) = 2 * x(j+1) * t(j+1,n) - t(j+1,n-1);
tp(j+1,n+1) = - n * x(j+1) * t(j+1,n+1) + n * t(j+1,n);
tp(j+1,n+1) = tp(j+1,n+1) / (1 - x(j+1) * x(j+1));
end; end
end
```

Having found the Chebyshev polynomials and x_j, we now find the weights w_j by inverting Equation 5.2.21.

```
%%%%%%%%%%%%%%%%%%%%%%%%%%%%%%%%%%%%%%%%%%%%%%%%%%%%%%%%%%%%%%%%%
% Set up the elements in the matrix given by Equation 5.2.21
%%%%%%%%%%%%%%%%%%%%%%%%%%%%%%%%%%%%%%%%%%%%%%%%%%%%%%%%%%%%%%%%%
tau = time * d_omega
for j = 0:N
for m = 0:N
A(m+1,j+1) = tp(j+1,m+1) + i * tau * t(j+1,m+1);
end; end
for m = 0:N
b(m+1,1) = t(1,m+1) * exp(i*tau) - t(n+1,m+1) * exp(-i*tau);
```

```
end
%%%%%%%%%%%%%%%%%%%%%%%%%%%%%%%%%%%%%%%%%%%%%%%%%%%%%%%%%%%%%%%%%%
%            Invert Equation 5.2.21 to find coefficient w_j
%%%%%%%%%%%%%%%%%%%%%%%%%%%%%%%%%%%%%%%%%%%%%%%%%%%%%%%%%%%%%%%%%%
[L,U] = lu(A);
coeff = U\(L\b);
```

Having found the coefficients w_j, we finally employ Equation 5.2.24 for the inversion.

```
inverse = 0;
for m = -M:M
omega_c = 2 * d_omega * m; coef = exp(i*time*omega_c);
%%%%%%%%%%%%%%%%%%%%%%%%%%%%%%%%%%%%%%%%%%%%%%%%%%%%%%%%%%%%%%%%%%
%    For each subdivision m, use the Evans-Webster method to
%            compute the contribution from that subdivision
%%%%%%%%%%%%%%%%%%%%%%%%%%%%%%%%%%%%%%%%%%%%%%%%%%%%%%%%%%%%%%%%%%
sum = 0;
for j = 0:N
omega = omega_c + d_omega * x(N+1-j);
arg = (bb + i * omega) / (bb - i * omega);
f_transform = log(arg) / (omega*omega+aa*aa);
sum = sum + coeff(N+1-j) * f_transform;
end
%%%%%%%%%%%%%%%%%%%%%%%%%%%%%%%%%%%%%%%%%%%%%%%%%%%%%%%%%%%%%%%%%%
%    Now add this contribution on the value of the inverse
%%%%%%%%%%%%%%%%%%%%%%%%%%%%%%%%%%%%%%%%%%%%%%%%%%%%%%%%%%%%%%%%%%
inverse = inverse + d_omega * real(coef * sum) / (2 * pi);
end
```

Figure 5.2.3 illustrates both Filon's and Evans and Webster's schemes. One difficulty in comparing results is the question of what is meant by $\Delta\omega$. For Filon's method, $\Delta\omega_{Filon} = L/M_{Filon}$ with $\omega_m = -L + m\Delta\omega_{Filon}$, where $m = 0, 1, 2, \ldots, 2M_{Filon}$. In the case of Evans and Webster's calculation, $\Delta\omega_{EW} = L/(2M_{EW}+1)$ with each panel having the width of $2\Delta\omega_{EW}$. For the present calculation with $n = 4$, $M_{EW} = 300$, and $L = 100$, $\Delta\omega_{EW} = 0.1664$. To ensure similar resolution using the Filon scheme, M_{Filon} was set equal to $4M_{EW} + 2$ which yields $\Delta\omega_{Filon} = 0.0832$. As Figure 5.2.3 shows, for the present example and using MATLAB, Filon's method is slightly better than the case when $n = 4$. An important difference between these two schemes is that the method by Evans and Webster requires a matrix inversion for each tau. No effort was made here to optimize this inversion process.

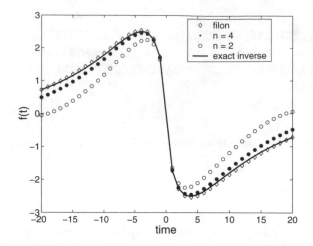

Figure 5.2.3: Same as Figure 5.2.1 except that Filon's and Evans and Webster's methods have been used.

5.3 SOLUTION OF PARTIAL DIFFERENTIAL EQUATIONS

Having shown how to invert a Fourier transform which contains a branch point, we are ready to apply our knowledge to some specific problems.

• Example 5.3.1

For our first example, we solve the partial differential equation

$$\frac{\partial^2 u}{\partial t^2} = \frac{\partial^2 u}{\partial x^2} - a^2 u, \qquad -\infty < x < \infty, \quad 0 < t, \quad 0 \le b < a, \qquad (\mathbf{5.3.1})$$

with the boundary conditions

$$\lim_{|x|\to\infty} u(x,t) \to 0, \qquad\qquad 0 < t, \qquad\qquad (\mathbf{5.3.2})$$

and the initial conditions

$$u(x,0) = e^{-b|x|}, \qquad u_t(x,0) = 0, \qquad -\infty < x < \infty, \qquad (\mathbf{5.3.3})$$

where $0 \le b < a$.

We begin by introducing the Fourier transform

$$U(k,t) = \int_{-\infty}^{\infty} u(x,t)e^{-ikx}\,dx, \qquad\qquad (\mathbf{5.3.4})$$

so that we may rewrite Equation 5.3.1 to Equation 5.3.3 as

$$\frac{d^2 U(k,t)}{dt^2} + (k^2 + a^2)U(k,t) = 0 \qquad\qquad (\mathbf{5.3.5})$$

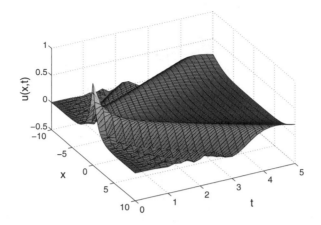

Figure 5.3.1: A plot of the solution $u(x,t)$ to Equation 5.1.1 through Equation 5.1.3 as a function of x and t when $a = 2$ and $b = 1$.

with

$$U(k,0) = \frac{2b}{k^2 + b^2} \quad \text{and} \quad U'(k,0) = 0. \qquad (\mathbf{5.3.6})$$

The solution that satisfies Equation 5.3.5 and Equation 5.3.6 is

$$U(k,t) = \frac{b}{k^2 + b^2}\left(e^{it\sqrt{k^2+a^2}} + e^{-it\sqrt{k^2+a^2}}\right), \qquad (\mathbf{5.3.7})$$

or

$$u(x,t) = \frac{2b}{\pi}\int_0^\infty \frac{\cos\left(t\sqrt{k^2+a^2}\right)}{k^2 + b^2}\cos(kx)\,dk. \qquad (\mathbf{5.3.8})$$

Equation 5.3.8 can be evaluated by numerical quadrature. Figure 5.3.1 shows the solution when $a = 2$ and $b = 1$. On the other hand, we can find alternative expressions for the solution via complex integration on the k-plane.

In Figure 5.3.2 we introduce several contours that are useful in finding an alternative representation to Equation 5.3.8. If $x > t > 0$, consider the contour consisting of contours C_1 to C_5. Then,

$$\frac{b}{2\pi}\int_{-\infty}^{\infty} \frac{e^{it\sqrt{k^2+a^2}} + e^{-it\sqrt{k^2+a^2}}}{k^2 + b^2}e^{ikx}\,dk$$

$$+ \frac{b}{2\pi}\int_{C_1\cup\cdots\cup C_5} \frac{e^{it\sqrt{z^2+a^2}} + e^{-it\sqrt{z^2+a^2}}}{z^2 + b^2}e^{izx}\,dz$$

$$= bi\operatorname{Res}\left(\frac{e^{it\sqrt{z^2+a^2}} + e^{-it\sqrt{z^2+a^2}}}{z^2 + b^2}e^{izx}; bi\right). \qquad (\mathbf{5.3.9})$$

A quick check shows that the contributions from contours C_1 and C_5 vanish as $|z| \to 0$. The contribution from the integration around the branch point

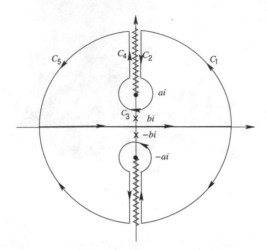

Figure 5.3.2: The contour used to invert the Fourier transform given by Equation 5.3.7.

$z = ai$ also vanishes. Along C_2,

$$z = \eta e^{\pi i/2}, \quad z - ai = (\eta - a)e^{\pi i/2}, \quad z + ai = (\eta + a)e^{\pi i/2}, \quad a < \eta < \infty, \tag{5.3.10}$$

while along C_4,

$$z = \eta e^{\pi i/2}, \quad z - ai = (\eta - a)e^{-3\pi i/2}, \quad z + ai = (\eta + a)e^{\pi i/2}, \quad a < \eta < \infty. \tag{5.3.11}$$

Substituting these values into Equation 5.3.9 and simplifying, we find that the branch cut integrals along C_2 and C_4 sum to zero. Therefore, for $x > t > 0$,

$$u(x,t) = e^{-bx} \cos\!\left(t\sqrt{a^2 - b^2}\,\right). \tag{5.3.12}$$

For $x < -t < 0$, a similar analysis yields

$$u(x,t) = e^{bx} \cos\!\left(t\sqrt{a^2 - b^2}\,\right). \tag{5.3.13}$$

Therefore, for $|x| > t > 0$,

$$u(x,t) = e^{-b|x|} \cos\!\left(t\sqrt{a^2 - b^2}\,\right). \tag{5.3.14}$$

What about $0 < |x| < t$? An analysis similar to that given in Example 5.1.3 fails to yield an expression simpler than Equation 5.3.8. In Problem 3 at the end of Section 6.1, we show how this problem can be solved with a joint application of Fourier and Laplace transforms.

• Example 5.3.2

Let us find the velocity potential[17] for an incompressible, irrotational, three-dimensional fluid of infinite depth. At the free surface, $z = 0$, we apply a specified pressure distribution. The governing equations are

$$\frac{\partial^2 \phi}{\partial y^2} + \frac{\partial^2 \phi}{\partial z^2} - \nu^2 \phi = 0, \qquad -\infty < z < 0, \qquad (5.3.15)$$

and

$$-\kappa\phi + \frac{\partial \phi}{\partial z} = p(y), \qquad z = 0, \qquad (5.3.16)$$

where we have factored out $\exp[i(\kappa x - \omega t)]$ and the usual dispersion relationship $\omega^2 = g\kappa$ holds. The relationship between κ and ν is $\nu = \kappa \cos(\chi)$, where χ is the heading angle. We assume that the pressure field is symmetric and given by

$$p(y) = p_0 e^{-\mu|y| - i|y|\sqrt{\gamma^2 - \nu^2}}. \qquad (5.3.17)$$

The presence of the term $e^{-\mu|y|}$ plays a very important role in our analysis. Mathematically, it ensures that the Fourier transform for $p(y)$ exists. Physically, it introduces some friction into the system. In Section 3.2 we showed that a certain ambiguity exists in nondissipative systems because incoming and outgoing waves are both present. We resolved this problem by requiring that energy radiate outward, away from the source of the disturbance, the so-called "Sommerfeld radiation condition." The addition of a small amount of friction is another method of imposing the same condition, because any energy propagating inward from infinity would be dissipated before it reaches any areas of interest.

Let us now define the Fourier transform and inverse for the velocity potential as

$$\Phi(k, z) = \int_{-\infty}^{\infty} \phi(y, z) e^{-iky} \, dy \qquad (5.3.18)$$

and

$$\phi(y, z) = \frac{1}{2\pi} \int_{-\infty}^{\infty} \Phi(k, z) e^{iky} \, dk. \qquad (5.3.19)$$

Upon substituting Equation 5.3.19 into Equation 5.3.15, we find

$$\Phi(k, z) = C(k) e^{z\sqrt{\nu^2 + k^2}}, \qquad (5.3.20)$$

where $\mathrm{Re}\left(\sqrt{\nu^2 + k^2}\right) > 0$ so that the solution does not grow exponentially for $z < 0$. From the boundary condition at $z = 0$, Equation 5.3.16, we obtain

$$-\kappa C(k) + C(k)\sqrt{\nu^2 + k^2} = P(k), \qquad (5.3.21)$$

---

[17] Taken from Troesch, A. W., 1979: The diffraction forces for a ship moving in oblique seas. *J. Ship Res.*, **23**, 127–139.

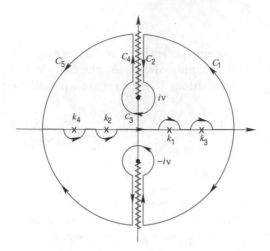

Figure 5.3.3: The contour used to invert the Fourier transform given by Equation 5.3.25.

where $P(k)$ is the Fourier transform of $p(y)$. From the definition of the Fourier transform, we find that

$$P(k) = p_0 \int_{-\infty}^{\infty} e^{-\mu|y| - i|y|\sqrt{\gamma^2 - \nu^2}} e^{-iky} \, dy \qquad (5.3.22)$$

$$= -p_0 \left[\frac{1}{-\mu + i\left(k - \sqrt{\gamma^2 - \nu^2}\right)} + \frac{1}{-\mu - i\left(k + \sqrt{\gamma^2 - \nu^2}\right)} \right], \qquad (5.3.23)$$

and

$$C(k) = p_0 \frac{2\mu + 2i\sqrt{\gamma^2 - \nu^2}}{[i(k - k_1)][-i(k - k_2)]\left(-\kappa + \sqrt{\nu^2 + k^2}\right)}, \qquad (5.3.24)$$

where $k_1 = \sqrt{\gamma^2 - \nu^2} - i\mu$ and $k_2 = -\sqrt{\gamma^2 - \nu^2} + i\mu$. Then, we may write $\phi(y, z)$ as

$$\phi(y, z) = \frac{p_0}{2\pi} \int_{-\infty}^{\infty} e^{i\zeta y + z\sqrt{\nu^2 + \zeta^2}} \frac{2\mu + 2i\sqrt{\gamma^2 - \nu^2}}{(\zeta - k_1)(\zeta - k_2)\left(-\kappa + \sqrt{\nu^2 + \zeta^2}\right)} \, d\zeta. \qquad (5.3.25)$$

To invert Equation 5.3.25, we employ the residue theorem where ζ is a complex variable. The integrand has branch points at $\zeta = \pm i\nu$ and simple poles at $\zeta = k_1$, $\zeta = k_2$, $\zeta = k_3 = \sqrt{\kappa^2 - \nu^2}$ and $\zeta = k_4 = -\sqrt{\kappa^2 - \nu^2}$. In Figure 5.3.3 we show that we have taken the branch cuts to lie along the imaginary axis. Note the important role that friction plays here; if it were not present, we would have singularities along the path of integration for the inverse.

For $y > 0$, we close the contour with an infinite semicircle in the upper half ζ-plane as dictated by Jordan's lemma. Therefore, the contribution from contours C_1 and C_5 are zero. Further analysis shows that the contribution near the branch point $\zeta = i\nu$ is also zero. However, the integration along the branch cuts yields

$$\int_{C_2} e^{iy\zeta + z\sqrt{\nu^2 + \zeta^2}} \frac{2\mu + 2i\sqrt{\gamma^2 - \nu^2}}{(\zeta - k_1)(\zeta - k_2)\left(-\kappa + \sqrt{\zeta^2 + \nu^2}\right)} d\zeta$$

$$= \int_{\infty}^{\nu} e^{-ky + iz\sqrt{k^2 - \nu^2}} \frac{2i\sqrt{\gamma^2 - \nu^2}\,(i\,dk)}{(ik - k_1)(ik - k_2)\left(-\kappa + i\sqrt{k^2 - \nu^2}\right)} \qquad (5.3.26)$$

and

$$\int_{C_4} e^{iy\zeta + z\sqrt{\nu^2 + \zeta^2}} \frac{2\mu + 2i\sqrt{\gamma^2 - \nu^2}}{(\zeta - k_1)(\zeta - k_2)\left(-\kappa + \sqrt{\zeta^2 + \nu^2}\right)} d\zeta$$

$$= \int_{\nu}^{\infty} e^{-ky - iz\sqrt{k^2 - \nu^2}} \frac{2i\sqrt{\gamma^2 - \nu^2}\,(i\,dk)}{(ik - k_1)(ik - k_2)\left(-\kappa - i\sqrt{k^2 - \nu^2}\right)}. \qquad (5.3.27)$$

The residue at k_2 equals

$$\mathrm{Res}\left[e^{iy\zeta + z\sqrt{\nu^2 + \zeta^2}} \frac{2\mu + 2i\sqrt{\gamma^2 - \nu^2}}{(\zeta - k_1)(\zeta - k_2)\left(-\kappa + \sqrt{\zeta^2 + \nu^2}\right)} ; k_2 \right]$$

$$= e^{ik_2 y + z\sqrt{\nu^2 + k_2^2}} \frac{2i\sqrt{\gamma^2 - \nu^2}}{(k_2 - k_1)\left(-\kappa + \sqrt{\nu^2 + k_2^2}\right)} \qquad (5.3.28)$$

$$= e^{-iy\sqrt{\gamma^2 - \nu^2} + \gamma z} \frac{2i\sqrt{\gamma^2 - \nu^2}}{\left(-2\sqrt{\gamma^2 - \nu^2}\right)(\gamma - \kappa)}. \qquad (5.3.29)$$

An interesting case occurs as $\gamma \to \kappa$. The forcing produces a resonant solution. Expanding $\sqrt{\gamma^2 - \nu^2}$ with $\gamma = \kappa - \delta$ and $\delta/\kappa \ll 1$,

$$\sqrt{\gamma^2 - \nu^2} = \sqrt{\kappa^2 - \nu^2} - \frac{\kappa\delta}{\sqrt{\kappa^2 - \nu^2}} + O(\delta^2) \qquad (5.3.30)$$

so that

$$\mathrm{Res}\left[e^{iy\zeta + z\sqrt{\nu^2 + \zeta^2}} \frac{2\mu + 2i\sqrt{\gamma^2 - \nu^2}}{(\zeta - k_1)(\zeta - k_2)\left(-\kappa + \sqrt{\zeta^2 + \nu^2}\right)} ; k_2 \right]$$

$$= \lim_{\delta \to 0} i\left(\frac{1}{\delta} - z + \frac{i\kappa y}{\sqrt{\kappa^2 - \nu^2}} \right) e^{-iy\sqrt{\kappa^2 - \nu^2} + \kappa z} \qquad (5.3.31)$$

while at the other residue $k = k_4$,

$$\operatorname{Res}\left[e^{iy\zeta + z\sqrt{\nu^2 + \zeta^2}}\frac{2\mu + 2i\sqrt{\gamma^2 - \nu^2}}{(\zeta - k_1)(\zeta - k_2)\left(-\kappa + \sqrt{\zeta^2 + \nu^2}\right)}; k_4\right]$$

$$= \lim_{\delta \to 0}\frac{-ie^{-iy\sqrt{\kappa^2 - \nu^2} + \kappa z}}{\delta}. \tag{5.3.32}$$

Consequently, the sum of the two residues is

$$\operatorname{Res}\left[e^{iy\zeta + z\sqrt{\nu^2 + \zeta^2}}\frac{2\mu + 2i\sqrt{\gamma^2 - \nu^2}}{(\zeta - k_1)(\zeta - k_2)\left(-\kappa + \sqrt{\zeta^2 + \nu^2}\right)}; k_2\right]$$

$$+ \operatorname{Res}\left[e^{iy\zeta + z\sqrt{\nu^2 + \zeta^2}}\frac{2\mu + 2i\sqrt{\gamma^2 - \nu^2}}{(\zeta - k_1)(\zeta - k_2)\left(-\kappa + \sqrt{\zeta^2 + \nu^2}\right)}; k_4\right]$$

$$= -ie^{-iy\sqrt{\kappa^2 - \nu^2} + \kappa z}\left(z - \frac{i\kappa y}{\sqrt{\kappa^2 - \nu^2}}\right), \tag{5.3.33}$$

where the singular nature of k_2 and k_4 cancel out.

For $y < 0$, a similar analysis follows with the exception that we close the contour by introducing an infinite semicircle in the lower half of the k-plane. Combining our results yields the final result for the special case $\kappa = \gamma$:

$$\phi(y, z) = p_0 e^{-i|y|\sqrt{\kappa^2 - \nu^2} + \kappa z}\left(z - \frac{i\kappa|y|}{\sqrt{\kappa^2 - \nu^2}}\right) \tag{5.3.34}$$

$$- \frac{p_0\sqrt{\kappa^2 - \nu^2}}{\pi}\int_\nu^\infty \frac{e^{-k|y|}}{\left(ik - \sqrt{\kappa^2 - \nu^2}\right)\left(ik + \sqrt{\kappa^2 - \nu^2}\right)}$$

$$\times \left(\frac{e^{iz\sqrt{\kappa^2 - \nu^2}}}{-\kappa + i\sqrt{\kappa^2 - \nu^2}} + \frac{e^{-iz\sqrt{\nu^2 + k^2}}}{\kappa + i\sqrt{\nu^2 + k^2}}\right) dk.$$

The numerical inversion of Equation 5.3.25 is illustrated in Figure 5.3.4.

• Example 5.3.3

For our second example, we solve a classic problem from seismology. In 1904, Lamb[18] modeled the seismic waves generated by an earthquake as the response of an infinite, elastic half-space to an oscillating, impulsive load. If we denote the y-component of the displacement by u, then the governing equation is the Helmholtz equation

$$\frac{\partial^2 u}{\partial x^2} + \frac{\partial^2 u}{\partial z^2} + \frac{\omega^2}{V_S^2}u = 0, \qquad 0 < z < \infty, \tag{5.3.35}$$

---

[18] Lamb, H., 1904: The propagation of tremors over the surface of an elastic solid. *Philos. Trans. R. Soc. London, Ser. A*, **203**, 1–42.

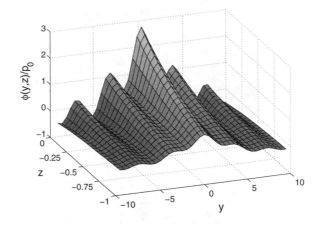

Figure 5.3.4: A plot of the solution $\phi(y,z)$ to Equation 5.3.15 and Equation 5.3.16 as a function of y and z when $\kappa = 1 + 0.1i$, $\gamma = 2$, $\mu = 0.1$ and $\nu = 1$.

where the forcing has a frequency ω and the phase speed of the shear waves is V_S. Assuming that the Fourier transform exists, we have

$$u(x, z, \omega) = \frac{1}{2\pi} \int_{-\infty}^{\infty} U(k, z, \omega) e^{-ikx}\, dk, \qquad (5.3.36)$$

and the transformed equation is

$$\frac{d^2 U(k, z, \omega)}{dz^2} - \left(k^2 - \frac{\omega^2}{V_S^2}\right) U(k, z, \omega) = 0. \qquad (5.3.37)$$

The general solution to Equation 5.3.37 is

$$U(k, z, \omega) = A(k, \omega) e^{-\nu z} + B(k, \omega) e^{\nu z}, \qquad (5.3.38)$$

where $\nu = \sqrt{k^2 - \omega^2/V_S^2}$. At the surface of the half-plane, $z = 0$, the boundary condition relates the surface stress to the impulsive point load,

$$\mu \frac{\partial u}{\partial z} = -P\delta(x), \qquad (5.3.39)$$

where μ is one of Lamé's constants. The Fourier transformed version of Equation 5.3.39 is

$$\mu \frac{dU}{dz} = -P. \qquad (5.3.40)$$

Upon substituting Equation 5.3.38 into Equation 5.3.40, $A(k, \omega) = P/(\mu\nu)$, or

$$u(x, z, \omega) = \frac{P}{2\pi\mu} \int_{-\infty}^{\infty} \frac{e^{-\nu z - ikx}}{\nu}\, dk. \qquad (5.3.41)$$

Figure 5.3.5: Graduating from Cambridge as second wrangler in 1872, Horace Lamb (1849–1934) spent his entire life in academia, first at the University of Adelaide, Australia (1875–1885) and then at Owens College in Manchester, England (1885–1920). In common with Stokes and Maxwell, Lamb's work exhibits not only great mathematical skill but also keen physical insight on such diverse fields as seismology to hydrodynamics. (Portrait courtesy of ©The Royal Society.)

In the derivation of Equation 5.3.41, we have assumed that $\mathrm{Re}(\nu) \geq 0$. Therefore, $B(k, \omega) = 0$ so that the solution remains finite as $z \to \infty$.

At this point, we introduce polar coordinates $x = r\cos(\theta)$ and $y = r\sin(\theta)$ along with $k = \omega\cos(\zeta)/V_S$ and $\nu = i\omega\sin(\zeta)/V_S$. Equation 5.3.41 then becomes

$$u(x, z, \omega) = \frac{P}{2\pi\mu i} \int_{0-i\infty}^{\pi+i\infty} e^{-i\omega r \sin(\zeta+\theta)/V_S} \, d\zeta. \qquad (5.3.42)$$

Before reducing Equation 5.3.42 to its final form, we must consider integrals of the form

$$I(\rho) = \frac{1}{\pi} \int_W e^{i\rho\cos(w)} e^{in(w-\pi/2)} \, dw. \qquad (5.3.43)$$

Sommerfeld[19] showed that for a wide class of contour integrals in the complex w-plane, $I(\rho)$ equals either the first or second Hankel functions,[20] $H_n^{(1)}(\rho)$ and $H_n^{(2)}(\rho)$. Figure 5.3.6 illustrates these so-called "*Sommerfeld contours.*" We

---

[19] Sommerfeld, A., 1949: *Partial Differential Equations in Physics*. Academic Press, Inc., Section 19.

[20] Watson, G. N., 1966: *A Treatise on the Theory of Bessel Functions*. Cambridge University Press, Section 3.6.

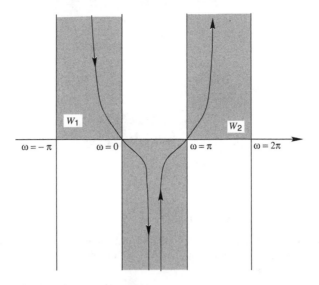

Figure 5.3.6: The Sommerfeld contours associated with Equation 5.3.43. Integration along W_1 gives $H_n^{(1)}(\rho)$, while integration along W_2 gives $H_n^{(2)}(\rho)$. The contours W_1 and W_2 may lie anywhere in the shaded area.

transform our integral into a contour integral similar to W_2 by defining $n = 0$, $w = \zeta + \theta + \pi/2$ because $-\pi/2 < \theta < \pi/2$ so that

$$u(x, z, \omega) = \frac{P}{2i\mu} H_0^{(2)}(\omega r/V_S). \qquad (5.3.44)$$

For large $\omega r/V_S$, we can use the asymptotic approximation for the Hankel functions[21] and find that

$$u(x, z, \omega) = \frac{P}{2i\mu} \sqrt{\frac{2V_S}{\pi \omega r}} e^{-i\omega r/V_S + \pi i/4}. \qquad (5.3.45)$$

Problems

1. Solve

$$\frac{\partial^2 u}{\partial x^2} + \frac{\partial^2 u}{\partial y^2} - a^2 u = 0, \qquad -\infty < x < \infty, \quad 0 < y < \infty, \quad 0 < b \le a,$$

with the boundary conditions $\lim_{|x| \to \infty} u(x, y) \to 0, 0 < y < \infty$, and $u(x, 0) = e^{-b|x|}$, $\lim_{y \to \infty} u(x, y) \to 0$, $-\infty < x < \infty$.

---

[21] *Ibid.*, p. 196.

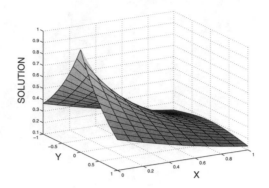

Problem 1

Step 1: Taking the Fourier transforms of the partial differential equation and boundary conditions, show that you obtain the boundary-value problem

$$\frac{d^2 U(k,y)}{dy^2} + (k^2 + a^2)U(k,y) = 0, \qquad U(k,0) = \frac{2b}{k^2 + b^2}, \qquad \lim_{y \to \infty} U(k,y) \to 0.$$

Step 2: Show that the solution to the boundary-value problem is

$$U(k,y) = \frac{2b}{k^2 + b^2} e^{-y\sqrt{k^2 + a^2}}.$$

Step 3: Invert the Fourier transform $U(k,y)$ and show that

$$u(x,y) = \frac{2b}{\pi} \int_0^\infty \frac{\cos(kx)}{k^2 + b^2} e^{-y\sqrt{k^2 + a^2}} \, dk.$$

Step 4: Using Figure 5.3.2, show that

$$u(x,y) = e^{-bx - y\sqrt{a^2 - b^2}} - \frac{2b}{\pi} \int_a^\infty \frac{e^{-x\eta}}{\eta^2 - b^2} \sin\left(y\sqrt{\eta^2 - a^2}\right) d\eta.$$

This solution is illustrated in the figure labeled Problem 1.

2. Solve[22]

$$\frac{\partial^2 u}{\partial x^2} + \frac{\partial^2 u}{\partial y^2} - 2a \frac{\partial u}{\partial y} = 0, \qquad 0 < x < \infty, \qquad -\infty < y < \infty, \qquad 0 < a,$$

---

[22] First solved by King, L. V., 1914: On the convection of heat from small cylinders in a stream of fluid: Determination of the convection constants of small platinum wires with applications to hot-wire anemometry. *Philos. Trans. R. Soc. London, Ser. A*, **214**, 373–432.

with the boundary conditions $u_x(0, y) = s(y) = f(y)[H(y) - H(y - b)]$, $\lim_{x \to \infty} u(x, y) \to 0$, $-\infty < y < \infty$, and $\lim_{|y| \to \infty} u(x, y) \to 0$, $0 < x < \infty$, where $b > 0$.

Step 1: Taking the Fourier transforms of the partial differential equation and boundary conditions, show that you obtain the boundary-value problem

$$\frac{d^2 U(x, \ell)}{dx^2} - (\ell^2 + 2ai\ell)U(x, \ell) = 0,$$

with

$$U'(0, \ell) = S(\ell) = \int_0^b f(y)e^{-i\ell y} \, dy \quad \text{and} \quad \lim_{x \to \infty} U(x, \ell) \to 0,$$

where

$$U(x, \ell) = \int_{-\infty}^{\infty} u(x, y)e^{-i\ell y} \, dy.$$

Step 2: Show that the solution to the boundary-value problem is

$$U(x, \ell) = -S(\ell)\frac{\exp\!\left(-x\sqrt{\ell^2 + 2ai\ell}\right)}{\sqrt{\ell^2 + 2ai\ell}}.$$

Step 3: If the branch cut for $\sqrt{\ell + 2ai}$ is taken along the imaginary axis from $-2ai$ to $-\infty i$, then the branch cut for $\sqrt{\ell}$ must be taken along the imaginary axis from $0i$ to ∞i so that $\sqrt{\ell^2 + 2ai\ell} \to \infty$ as $|\ell| \to \infty$. Using this choice of branch cuts with a cut ℓ-plane similar to that shown in Figure 5.1.1, verify

$$\mathcal{F}^{-1}\!\left[\frac{\exp\!\left(-x\sqrt{\ell^2 + 2ai\ell}\right)}{\sqrt{\ell^2 + 2ai\ell}}\right] = \frac{1}{2\pi i} \int_{-\infty - \epsilon i}^{\infty - \epsilon i} \frac{\exp\!\left(iy\ell - x\sqrt{\ell^2 + 2ai\ell}\right)}{\sqrt{\ell^2 + 2ai\ell}} \, d\ell$$

$$= \frac{1}{\pi} \int_0^{\infty} \frac{e^{(2a+\eta)y} \cos\!\left(x\sqrt{\eta^2 + 2a\eta}\right)}{\sqrt{\eta^2 + 2a\eta}} \, d\eta$$

$$= \frac{e^{ay}}{\pi} \int_0^{\infty} \frac{e^{y\sqrt{a^2 + \tau^2}} \cos(x\tau)}{\sqrt{a^2 + \tau^2}} \, d\tau$$

$$= \frac{e^{ay}}{\pi} K_0\!\left(a\sqrt{x^2 + y^2}\right),$$

if $y < 0$, where $0 < \epsilon < 2a$.

Step 4: In a similar manner, show that

$$\mathcal{F}^{-1}\!\left[\frac{\exp\!\left(-x\sqrt{\ell^2 + 2ai\ell}\right)}{\sqrt{\ell^2 + 2ai\ell}}\right] = \frac{e^{ay}}{\pi} K_0\!\left(a\sqrt{x^2 + y^2}\right),$$

if $y > 0$.

Step 5: Use the convolution theorem and show that

$$u(x, y) = -\frac{1}{\pi} \int_0^b f(\tau)e^{a(y-\tau)} K_0\!\left[a\sqrt{x^2 + (y - \tau)^2}\right] d\tau.$$

3. Find $\eta(x)$ defined by

$$\eta(x) = \frac{\partial u(x, y_0)}{\partial x}, \qquad -\infty < x < \infty,$$

where the partial differential equation[23]

$$\frac{\partial^2 u}{\partial x^2} + \frac{\partial^2 u}{\partial y^2} = \delta'(x)\delta(y - a), \qquad -\infty < x < \infty, \quad y_0 < y < \infty,$$

gives $u(x, y)$ with the boundary conditions that $\lim_{y \to \infty} u(x, y) \to 0$ and

$$\frac{\partial^2 u(x, y_0)}{\partial x^2} - a_1 \frac{\partial u(x, y_0)}{\partial y} - \epsilon a_2 \frac{\partial u(x, y_0)}{\partial x} = 0$$

with $0 < y_0 < a$, $0 < a_1, a_2$ and $\epsilon \to 0$.

Step 1: If

$$u(x, y) = \frac{1}{2\pi} \int_{-\infty}^{\infty} U(k, y)e^{ikx} \, dk \quad \text{and} \quad \eta(x) = \frac{1}{2\pi} \int_{-\infty}^{\infty} H(k)e^{ikx} \, dk,$$

show that the partial differential equations reduce to the ordinary differential equations

$$H(k) = ikU(k, y_0) \qquad \text{and} \qquad \frac{d^2 U(k, y)}{dy^2} - k^2 U(k, y) = ik\delta(y - a)$$

with the boundary conditions that $\lim_{y \to \infty} U(k, y) \to 0$ and

$$a_1 \frac{dU(k, y_0)}{dy} + (k^2 + ik\epsilon a_2)U(k, y_0) = 0.$$

Step 2: By integrating the second differential equation in Step 1 over the interval $[a^-, a^+]$, where a^+ and a^- are points just above and below a, respectively, show that we can replace the differential equation in Step 1 by

$$\frac{d^2 U(k, y)}{dy^2} - k^2 U(k, y) = 0, \qquad \text{if} \qquad y \neq a,$$

with

$$\lim_{y \to \infty} U(k, y) \to 0, \qquad U(k, a^+) = U(k, a^-), \qquad \frac{dU(k, a^+)}{dy} - \frac{dU(k, a^-)}{dy} = ik,$$

and

$$a_1 \frac{dU(k, y_0)}{dy} + (k^2 + ik\epsilon a_2)U(k, y_0) = 0.$$

---

[23] Savage, *op. cit.*

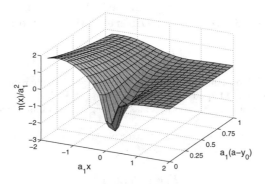

Problem 3

Step 3: Show that the solution to Step 2 is

$$U(k,y) = \frac{i}{2}\operatorname{sgn}(k)\left[\frac{|k| + a_1 + i\epsilon a_2 \operatorname{sgn}(k)}{|k| - a_1 + i\epsilon a_2 \operatorname{sgn}(k)}\right] e^{(2y_0 - a - y)|k|} - \frac{i}{2}\operatorname{sgn}(k)e^{|k|(a-y)}$$

if $a < y$, or

$$U(k,y) = \frac{i}{2}\operatorname{sgn}(k)\left[\frac{|k| + a_1 + i\epsilon a_2 \operatorname{sgn}(k)}{|k| - a_1 + i\epsilon a_2 \operatorname{sgn}(k)}\right] e^{(2y_0 - a - y)|k|} - \frac{i}{2}\operatorname{sgn}(k)e^{|k|(y-a)}$$

if $y_0 < y < a$, and

$$H(k) = -\frac{a_1|k|\exp[-(a - y_0)|k|]}{|k| - a_1 + i\epsilon a_2 \operatorname{sgn}(k)}.$$

Step 4: Show that the inverse of $H(k)$ is

$$\eta(x) = -2a_1^2 e^{-a_1(a-y_0)}\sin(a_1 x)H(-x)$$

$$- \frac{a_1}{\pi}\int_0^\infty \frac{\eta e^{-x\eta}}{\eta^2 + a_1^2}\{a_1\cos[(a - y_0)\eta] + \eta\sin[(a - y_0)\eta]\}\,d\eta$$

by deforming the path of integration along the real k-axis so that it passes over the $k = \pm a_1$ and taking the branch cuts associated with $|k|$ along the imaginary axis from $(-\infty i, 0^- i]$ and $[0^+ i, \infty i)$. [Hint: Treat the absolute value as $|z| = \sqrt{(z - 0^+ i)(z - 0^- i)}$, where the real part of the square root is always positive.] This inverse $\eta(x)$ is illustrated in the figure labeled Problem 3.

Papers Using Fourier Transforms
to Solve Partial Differential Equations

Electromagnetism

Lee, T. J., 1980/81: Transient electromagnetic response of a sphere in a layered medium. *Pure Appl. Geophys.*, **119**, 309–338.

Savage, M. D., 1967: Stationary waves at a plasma-magnetic field interface. *J. Plasma Phys.*, **1**, 229–239.

Singh, S. K., 1973: Electromagnetic transient response of a conducting sphere embedded in a conductive medium. *Geophysics*, **38**, 864–893.

Fluid Dynamics

Azpeitia, A. G., and G. F. Newell, 1958: Theory of oscillatory type viscometers. III: A thin disk. *Z. Angew. Math. Phys.*, **9a**, 97–118.

Carrier, G. F., and R. C. DiPrima, 1957: On the unsteady motion of a viscous fluid past a semi-infinite flat plate. *J. Math. Phys. (Cambridge, MA)*, **35**, 359–383.

Goldstein, M. E., 1975: Cascade with subsonic leading-edge locus. *AIAA J.*, **13**, 1117–1119.

Haren, P., and C. C. Mei, 1981: Head-sea diffraction by a slender raft with application to wave-power absorption. *J. Fluid Mech.*, **104**, 505–526.

Geophysical Sciences

Menkes, J., 1972: The propagation of sound in the ionosphere. *J. Sound Vib.*, **20**, 311–319.

Sezawa, K., and K. Kanai, 1932: Possibility of free oscillations of strata excited by seismic waves. Part III. *Bull. Earthq. Res. Inst.*, **10**, 1–19.

Weaver, A. J., L. A. Mysak, and A. F. Bennett, 1988: The steady state response of the atmosphere to midlatitude heating with various zonal structures. *Geophys. Astrophys. Fluid Dyn.*, **41**, 1–44.

Magnetohydrodynamics

Bhutani, O. M., and K. D. Nanda, 1968: A general theory of thin airfoils in nonequilibrium magnetogasdynamics. Part I: Aligned magnetic field. *AIAA J.*, **6**, 1757–1762.

Other

Beauwens, R., and J. Devooght, 1968: The study of one-speed multiregion transport problems in plain geometry by the method of boundary sources. *Nucl. Sci. Eng.*, **32**, 249–261.

Williams, M. M. R., 1965: Neutron transport in differentially heated media. *Brit. J. Appl. Phys.*, **16**, 1727–1732.

Solid Mechanics

Chowdhury, K. L., and P. G. Glockner, 1980: On a boundary value problem for an elastic dielectric half-plane. *Acta Mech.*, **37**, 65–74.

Dieterman, H. A., and A. V. Kononov, 1997: Uniform motion of a constant load along a string on an elastically supported membrane. *J. Sound Vib.*, **208**, 575–586.

Karasudhi, P., L. M. Keer, and S. L. Lee, 1968: Vibration motion of a body on an elastic half space. *J. Appl. Mech.*, **35**, 697–705.

Nagaya, K., and Y. Hirano, 1976: Responses of an infinite medium with cavities to an impact load at one of the cavities. *Bull. JSME*, **19**, 1430–1434.

Nakano, H., 1925: On Rayleigh waves. *Jap. J. Astron. Geophys.*, **2**, 233–326.

Norwood, F. R., and J. Miklowitz, 1967: Diffraction of transient elastic waves by a spherical cavity. *J. Appl. Mech.*, **34**, 735–744.

Papliński, A., and E. Włodarczyk, 1980: Response of elastic medium to a traveling line load applied in a cylindrical bore. *J. Tech. Phys.*, **21**, 313–335.

Rudnicki, J. W., and E. A. Roeloffs, 1990: Plane-strain shear dislocations moving steadily in linear elastic diffusive solids. *J. Appl. Mech.*, **57**, 32–39.

Skalak, R., and M. B. Friedman, 1958: Reflection of an acoustic step wave from an elastic cylinder. *J. Appl. Mech.*, **25**, 103–108.

Wijeyewickrema, A. C., and L. M. Keer, 1986: Antiplane transient response of embedded cylinder. *J. Eng. Mech. Div. (Am. Soc. Civ. Eng.)*, **112**, 536–549.

Thermoelasticity

Chian, C. T., and F. C. Moon, 1981: Magnetically induced cylindrical stress waves in a thermoelastic conductor. *Int. J. Solids Struct.*, **17**, 1021–1035.

Wave Equation

Goswami, S. K., 1982: A note on the problem of scattering of surface waves by a submerged fixed vertical barrier. *Z. Angew. Math. Mech.*, **62**, 637–639.

Marcus, S. W., 1991: A generalized impedance method for application of the parabolic approximation to underwater acoustics. *J. Acoust. Soc. Am.*, **90**, 391–398.

Papliński, A., and E. Włodarczyk, 1977: Propagation in acoustic medium of two-dimensional, cylindrical pressure waves excited by a moving load. *J. Tech. Phys.*, **18**, 81–97.

Wimp, J., C. Rorres, and R. F. Wayland, 1992: Acoustic impulse responses for nonuniform media. *J. Comput. Appl. Math.*, **42**, 89–107.

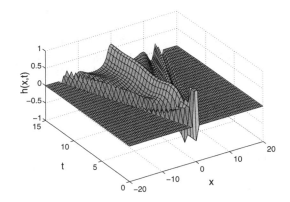

Chapter 6

The Joint Transform Method

So far we have used Laplace and Fourier transforms separately to solve partial differential equations. In Section 6.1 and Section 6.2 we shall show how we can apply them together to the same end. In general, this requires a double inversion of the transformed solution — often a formidable task. However, over the last sixty years, several analytic techniques, in particular, the Cagniard–De Hoop method, have been developed for the joint inversion problem. We will explore these methods in Section 6.3 and Section 6.4.

6.1 THE WAVE EQUATION

The joint transform method is a popular method for solving linear wave equations in an infinite or semi-infinite spatial domain and with specified initial conditions. The general procedure is as follows: We use the Laplace transform to eliminate the temporal dependence while we apply a Fourier or Hankel transform in the spatial dimension. This results in an algebraic or ordinary differential equation which we solve to obtain the joint transform. We then compute the inverses. Whether we invert the Laplace or the spatial transform first is usually dictated by the nature of the joint transform.

• Example 6.1.1

To illustrate this technique, we use it to find the shallow-water gravity waves excited on an infinite, one-dimensional, flat earth. Here, we neglect variations of the Coriolis parameter with latitude. The nondimensional x-momentum, y-momentum and continuity equations are

$$\frac{\partial u}{\partial t} + \frac{\partial h}{\partial x} - v = 0, \tag{6.1.1}$$

$$\frac{\partial v}{\partial t} + u = 0, \tag{6.1.2}$$

and

$$\frac{\partial h}{\partial t} + c^2 \frac{\partial u}{\partial x} = 0, \tag{6.1.3}$$

where c is the nondimensional phase speed of the shallow water waves.

To solve Equation 6.1.1 through Equation 6.1.3, we assume that a Fourier transform exists for each of the dependent variables. For example,

$$u(x,t) = \frac{1}{2\pi} \int_{-\infty}^{\infty} U(k,t) e^{ikx} \, dk, \tag{6.1.4}$$

with similar expressions for $v(x,t)$ and $h(x,t)$. If the perturbations vanish at infinity, Equation 6.1.1 through Equation 6.1.3 reduce to the ordinary differential equations

$$\frac{dU(k,t)}{dt} + ik\Theta(k,t) - V(k,t) = 0, \tag{6.1.5}$$

$$\frac{dV(k,t)}{dt} + U(k,t) = 0, \tag{6.1.6}$$

and

$$\frac{d\Theta(k,t)}{dt} + ikc^2 U(k,t) = 0. \tag{6.1.7}$$

To solve the system, Equation 6.1.5 through Equation 6.1.7, we take their Laplace transform by defining

$$\overline{U}(k,s) = \int_0^{\infty} U(k,t) e^{-st} \, dt \tag{6.1.8}$$

with similar expressions for $\overline{V}(k,s)$ and $\overline{\Theta}(k,s)$.

At this point we must specify the initial conditions. In this problem, we choose to find the solution if the initial height field is $\theta(x,0) = H(x+a) - H(x-a)$. Taking the Laplace transform of Equation 6.1.5 through Equation 6.1.7,

$$s\overline{U}(k,s) + ik\overline{\Theta}(k,s) - \overline{V}(k,s) = 0, \tag{6.1.9}$$

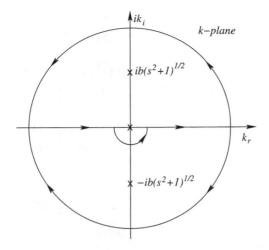

Figure 6.1.1: Contours used in the inversion of the joint transform given by Equation 6.1.15.

$$sV(k,s) + \overline{U}(k,s) = 0, \tag{6.1.10}$$

and

$$s\overline{\Theta}(k,s) + ikc^2\overline{U}(k,s) = \Theta(k,0). \tag{6.1.11}$$

Because $\Theta(k,0) = \sin(ka)/k$, we obtain the following transformed solutions

$$\overline{U}(k,s) = \frac{-ik}{c^2k^2 + s^2 + 1} \frac{\sin(ka)}{k}, \tag{6.1.12}$$

$$\overline{V}(k,s) = \frac{i}{c^2k^2 + s^2 + 1} \frac{\sin(ka)}{s}, \tag{6.1.13}$$

and

$$\overline{\Theta}(k,s) = \frac{s^2 + 1}{c^2k^2 + s^2 + 1} \frac{\sin(ka)}{ks}. \tag{6.1.14}$$

Since the transform $\overline{V}(k,s)$ is the most involved, we present a detailed analysis for this case; the remaining transforms follow by analogy.

If we rewrite $\sin(ka)$ in terms of complex exponentials, then the inverse Fourier transform is

$$V(x,s) = \frac{b^2}{2\pi s} \int_{-\infty}^{\infty} \frac{e^{ik(x+a)} - e^{-ik(x-a)}}{k^2 + b^2 s^2 + b^2} \, dk, \tag{6.1.15}$$

where $b = 1/c$. To evaluate Equation 6.1.15, we introduce an infinite semicircle in the upper or lower half-plane (see Figure 6.1.1) as dictated by Jordan's lemma and evaluate the closed contour by Cauchy's residue theorem. The

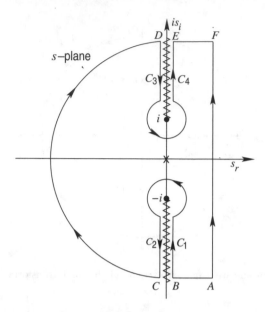

Figure 6.1.2: The contours used in the inversion of the Laplace transform given by Equation 6.1.16.

integrand possesses simple poles at $k = \pm ib(s^2 + 1)^{1/2}$ and an application of the residue theorem yields

$$V(x, s) = \frac{b}{2} \frac{\exp\left(-b|x + a|\sqrt{s^2 + 1}\right) - \exp\left(-b|x - a|\sqrt{s^2 + 1}\right)}{s\sqrt{s^2 + 1}}. \qquad (6.1.16)$$

We invert the Laplace transform $V(x, s)$ by applying Bromwich's integral on the complex plane, i.e.,

$$v(x, t) = \frac{b}{4\pi i} \int_C \frac{1}{s\sqrt{s^2 + 1}} \, ds \qquad (6.1.17)$$
$$\times \left[\exp\left(st - b|x + a|\sqrt{s^2 + 1}\right) - \exp\left(st - b|x - a|\sqrt{s^2 + 1}\right)\right],$$

where the contour C runs to the right of any singularities. Treating each term in Equation 6.1.17 separately, we convert each line integral into a closed contour and then use the residue theorem. Because the analysis is essentially identical for both terms, we shall only give the details for one of them.

For the first term, Jordan's lemma requires that we close the contour with an infinite semicircle in the right half-plane for $t < b|x + a|$. Because there are no singularities within the contour, the integral vanishes, as we would expect, from causality. When $t > b|x + a|$, we close the contour with an infinite semicircle in the left half-plane. However, we must be careful because the integrand is multivalued with branch points at $s = \pm i$. There are several possible choices and Figure 6.1.2 shows the one that we shall use.

Because the argument of the square root must be positive for large positive s, we define the amplitude and phase along each contour as follows:

$$C_1: \; s = \rho e^{-\pi i/2}, \quad s - i = (\rho + 1)e^{-\pi i/2}, \quad s + i = (\rho - 1)e^{-\pi i/2}, \quad \textbf{(6.1.18)}$$

$$C_2: \; s = \rho e^{-i\pi/2}, \quad s - i = (\rho + 1)e^{-\pi i/2}, \quad s + i = (\rho - 1)e^{3\pi i/2}, \quad \textbf{(6.1.19)}$$

$$C_3: \; s = \rho e^{\pi i/2}, \quad s - i = (\rho - 1)e^{-3\pi i/2}, \quad s + i = (\rho + 1)e^{\pi i/2}, \quad \textbf{(6.1.20)}$$

and

$$C_4: \; s = \rho e^{\pi i/2}, \quad s - i = (\rho - 1)e^{\pi i/2}, \quad s + i = (\rho + 1)e^{\pi i/2}, \quad \textbf{(6.1.21)}$$

where $1 \le \rho \le \infty$. The contribution from the arcs AB, CD and EF at infinity vanish.

In addition to line integrals, contributions come from the simple pole at $s = 0$, and integrations around the branch points at $s = \pm i$. We compute the contribution from the simple pole $s = 0$ in the usual manner. For the branch points, we introduce the variable $s = \pm i + \epsilon e^{\theta i}$ and perform an integration around the infinitesimally small circles at the branch points as $\epsilon \to 0$.

Thus,

$$
\begin{aligned}
\frac{b}{4\pi i} &\int_C \frac{\exp\left(st - b|x + a|\sqrt{s^2 + 1}\right)}{s\sqrt{s^2 + 1}} \, ds \\
&= \frac{b}{2} \exp(-b|x + a|) \\
&+ \frac{b}{4\pi i} \int_\infty^1 \frac{e^{-i\rho t}(i\, d\rho)}{(-i\rho)[-i(\rho^2 - 1)^{1/2}]} \exp\left[-b|x + a|(-i)\sqrt{\rho^2 - 1}\right] \\
&+ \frac{b}{4\pi i} \int_1^\infty \frac{e^{-i\rho t}(-i\, d\rho)}{(-i\rho)[i(\rho^2 - 1)^{1/2}]} \exp\left(-b|x + a|i\sqrt{\rho^2 - 1}\right) \\
&+ \frac{b}{4\pi i} \int_\infty^1 \frac{e^{i\rho t}i\, d\rho}{i\rho[-i(\rho^2 - 1)^{1/2}]} \exp\left[-b|x + a|(-i)\sqrt{\rho^2 - 1}\right] \\
&+ \frac{b}{4\pi i} \int_1^\infty \frac{e^{i\rho t}i\, d\rho}{i\rho[i(\rho^2 - 1)^{1/2}]} \exp\left(-b|x + a|i\sqrt{\rho^2 - 1}\right). \quad \textbf{(6.1.22)}
\end{aligned}
$$

In this case, the contribution from the branch points vanishes although this is not always the case. We may further simplify Equation 6.1.22 by introducing the transformation $\eta^2 = \rho^2 - 1$. When we carry out similar analyses for Equation 6.1.12 and Equation 6.1.14, we find that

$$
\begin{aligned}
u(x, t) = {}&\frac{b}{2} J_0\left(\sqrt{t^2 - b^2|x - a|^2}\right) H(t - b|x - a|) \\
&- \frac{b}{2} J_0\left(\sqrt{t^2 - b^2|x + a|^2}\right) H(t - b|x + a|), \quad \textbf{(6.1.23)}
\end{aligned}
$$

LIVERPOOL JOHN MOORES UNIVERSITY
LEARNING SERVICES

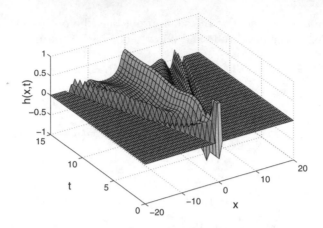

Figure 6.1.3: A plot of $h(x,t)$ given by Equation 6.1.25 as a function of x and t when $a = b = 1$.

$$v(x,t) = H(t - b|x + a|) \left[\frac{b}{2} e^{-b|x+a|} \right.$$
$$\left. - \frac{b}{\pi} \int_0^\infty \cos\left(t\sqrt{1+\eta^2}\right) \cos(b|x+a|\eta) \frac{d\eta}{1+\eta^2} \right]$$
$$- H(t - b|x - a|) \left[\frac{b}{2} e^{-b|x-a|} \right.$$
$$\left. - \frac{b}{\pi} \int_0^\infty \cos\left(t\sqrt{1+\eta^2}\right) \cos(b|x-a|\eta) \frac{d\eta}{1+\eta^2} \right] \qquad \mathbf{(6.1.24)}$$

and

$$h(x,t) = H(a - |x|) \qquad\qquad\qquad\qquad\qquad \mathbf{(6.1.25)}$$
$$- \operatorname{sgn}(x + a) H(t - b|x + a|)$$
$$\times \left[\tfrac{1}{2} e^{-b|x+a|} - \frac{1}{\pi} \int_0^\infty \frac{\eta}{1+\eta^2} \cos\left(t\sqrt{1+\eta^2}\right) \sin(b|x+a|\eta)\, d\eta \right]$$
$$+ \operatorname{sgn}(x - a) H(t - b|x - a|)$$
$$\times \left[\tfrac{1}{2} e^{-b|x-a|} - \frac{1}{\pi} \int_0^\infty \frac{\eta}{1+\eta^2} \cos\left(t\sqrt{1+\eta^2}\right) \sin(b|x-a|\eta)\, d\eta \right].$$

Figure 6.1.3 illustrates Equation 6.1.25 as a function of x and t.

• Example 6.1.2

When there is axial symmetry, we replace the Fourier transform by the Hankel transform. To illustrate this, let us find the response[1] $u(r, z, t)$ of a

---

[1] Taken from Duffy, D. G., 1992: On the generation of oceanic surface waves by underwater volcanic explosions. *J. Volcanol. Geotherm. Res.*, **50**, 323–344.

LIVERPOOL JOHN MOORES UNIVERSITY
LEARNING SERVICES

quiescent, compressible ocean when we force it impulsively at $r = 0$ and $z = d$. The corresponding governing equation is

$$\frac{1}{V_0^2}\frac{\partial^2 u}{\partial t^2} - \frac{\partial^2 u}{\partial z^2} - \frac{\partial^2 u}{\partial r^2} - \frac{1}{r}\frac{\partial u}{\partial r} = 4\pi\delta(r)\delta(z-d)H(t), \qquad \textbf{(6.1.26)}$$

where V_0 is the (nondimensional) speed of sound in the ocean and $0 < d < 1$. The boundary condition at the free surface is

$$\frac{\partial^2 u(r,0,t)}{\partial t^2} = \frac{\partial u(r,0,t)}{\partial z}, \qquad \textbf{(6.1.27)}$$

while at the bottom,

$$\frac{\partial u(r,1,t)}{\partial z} = 0. \qquad \textbf{(6.1.28)}$$

We solve Equation 6.1.26 through Equation 6.1.28 by a joint application of Laplace and Hankel transforms,

$$u(r,z,t) = \frac{1}{2\pi i}\int_{c-\infty i}^{c+\infty i} e^{ts}\int_0^\infty \overline{U}(k,z,s)J_0(kr)\, k\, dk\, ds. \qquad \textbf{(6.1.29)}$$

The application of these transforms yields the system of equations

$$\frac{d^2\overline{U}(k,z,s)}{dz^2} - m^2\overline{U}(k,z,s) = 0, \qquad 0 \le z \le 1, \qquad \textbf{(6.1.30)}$$

$$s^2\overline{U}(r,0,s) = \frac{d\overline{U}(k,0,s)}{dz}, \qquad \textbf{(6.1.31)}$$

$$\overline{U}(k,d^+,s) = \overline{U}(k,d^-,s), \qquad \textbf{(6.1.32)}$$

$$\frac{d\overline{U}(k,d^+,s)}{dz} - \frac{d\overline{U}(k,d^-,s)}{dz} = -\frac{2}{s}, \qquad \textbf{(6.1.33)}$$

and

$$\frac{d\overline{U}(k,1,s)}{dz} = 0, \qquad \textbf{(6.1.34)}$$

where $m^2 = k^2 + s^2/V_0^2$, and d^+ and d^- are points slightly greater than and less than $z = d$, respectively. We obtain Equation 6.1.33 by integrating the transformed form of Equation 6.1.26 across a narrow strip between d^+ and d^-.

The solution which satisfies Equation 6.1.30 with Equation 6.1.31 and Equation 6.1.34 is

$$\overline{U}(k,z,s) = \frac{2\cosh[m(1-d)]\cosh(mz)}{s[s^2\cosh(m) + m\sinh(m)]} + \frac{2s\cosh[m(1-d)]\sinh(mz)}{m[s^2\cosh(m) + m\sinh(m)]}$$
$$\textbf{(6.1.35)}$$

for $0 \le z \le d$, and

$$\overline{U}(k, z, s) = \frac{2m \cosh(md) + 2s^2 \sinh(md)}{sm[s^2 \cosh(m) + m \sinh(m)]} \cosh[m(1 - z)] \qquad (6.1.36)$$

for $d \le z \le 1$. Because $m = \sqrt{k^2 + s^2/V_0^2}$, it appears that we must deal with a branch cut to invert the joint transform given by Equation 6.1.35 and Equation 6.1.36. However, when we expand the hyperbolic functions, the transforms are purely even functions of m and thus no branch points or cuts exist. Unlike our previous example, we first invert the Laplace transform by converting the line integral into a closed contour by introducing an infinite semicircle in the left side of the complex s-plane. By Jordan's lemma, the contribution from the semicircle at infinity is zero. For this reason, the inversion follows as a straightforward application of the residue theorem. This transform has singularities at $s = 0$ and

$$m \tanh(m) + s^2 = 0. \qquad (6.1.37)$$

The solutions of this transcendental equation given by Equation 6.1.37 lie along the imaginary axis. Although we must compute the precise solutions numerically, an asymptotic expansion in powers of the small quantity $1/V_0^2$ gives

$$s_0 = i\omega_0 = i\sqrt{k \tanh(k)}\left\{1 - \frac{1}{4V_0^2}\left[\operatorname{sech}^2(k) + \frac{\tanh(k)}{k}\right] + \cdots\right\} \qquad (6.1.38)$$

and

$$s_n = i\omega_n = iV_0\sqrt{(2n-1)^2\pi^2/4 + k^2}\left\{1 + \frac{(2n-1)^2\pi^2/4}{V_0^2[(2n-1)^2\pi^2/4 + k^2]^2}\right.$$
$$\left. + \frac{4\pi^2(2n-1)^2[12k^2 - 7(2n-1)^2\pi^2]}{V_0^4[(2n-1)^2\pi^2/4 + k^2]^4} + \cdots\right\}, \qquad (6.1.39)$$

where $n = 1, 2, 3, \ldots$. We observe that Equation 6.1.38 gives the external surface gravity wave with a correction due to the compressibility of the water. The solutions expressed by Equation 6.1.39 correspond to sound waves within the ocean.

Using these poles we can invert the Laplace transform and then the Hankel transform. The solution to Equation 6.1.26 through Equation 6.1.28, $u(r, z, t)$, is

$$u(r, z, t) = \int_0^\infty J_0(kr)k\,dk$$
$$\times \left\{\frac{\cosh[k(1 - z - d)] + \cosh[k(1 - |z - d|)]}{k\sinh(k)}\right.$$

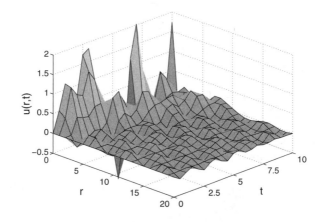

Figure 6.1.4: A plot of sea surface elevation $\eta(r,t)$ given by Equation 6.1.41 as a function of r and t when $b = 2$ and $V_0 = 5$.

$$-2V_0^2 \frac{m_0\{\cosh[m_0(1-z-d)] + \cosh[m_0(1+z-d)]\}\cos(\omega_0 t)}{\omega_0^2\{m_0(1+2V_0^2)\cosh(m_0) + (1-\omega_0^2)\sinh(m_0)\}}$$

$$-2V_0^2 \frac{\{\sinh[m_0(1-z-d)] - \sinh[m_0(1-|z-d|)]\}\cos(\omega_0 t)}{m_0(1+2V_0^2)\cosh(m_0) + (1-\omega_0^2)\sinh(m_0)}$$

$$-2V_0^2 \sum_{n=1}^{\infty} \frac{m_n\{\cos[m_n(1-z-d)] + \cos[m_n(1+z-d)]\}\cos(\omega_n t)}{\omega_n^2\{m_n(1+2V_0^2)\cos(m_n) + (1-\omega_n^2)\sin(m_n)\}}$$

$$\left. -2V_0^2 \sum_{n=1}^{\infty} \frac{\{\sin[m_n(1-z-d)] - \sin[m_n(1-|z-d|)]\}\cos(\omega_n t)}{m_n(1+2V_0^2)\cos(m_n) + (1-\omega_n^2)\sin(m_n)} \right\},$$

$$(6.1.40)$$

where $m_0 = \sqrt{k^2 - \omega_0^2/V_0^2}$ and $m_n = \sqrt{\omega_n^2/V_0^2 - k^2}$.

A useful way to visualize our results is with a plot of the elevation $\eta(r,t)$ of the sea surface caused by a forcing, $S(r,t) = [H(r) - H(r-b)]H(t)$. If we denote the joint transform of $\eta(r,t)$ by $\overline{H}(k,s)$, then the joint transform equals $s^2\overline{S}(k,s)\overline{U}(k,0,s)$. Inverting the joint transform, we find

$$\eta(r,t) = 4bV_0^2 \int_0^{\infty} J_1(kb)J_0(kr)\,dk$$

$$\times \left\{ \frac{m_0\cosh[m_0(1-d)]\sin(\omega_0 t)}{\omega_0[m_0(1+2V_0^2)\cosh(m_0) + (1-\omega_0^2)\sinh(m_0)]} \right.$$

$$\left. + \sum_{n=1}^{\infty} \frac{m_n\cos[m_n(1-d)]\sin(\omega_n t)}{\omega_n[m_n(1+2V_0^2)\cos(m_n) + (1-\omega_n^2)\sin(m_n)]} \right\}.$$

$$(6.1.41)$$

We have plotted the surface elevation $\eta(r,t)$ in Figure 6.1.4 when $b = 2$ and $V_0 = 5$.

• Example 6.1.3

For our next example[2] with Hankel transforms, let us solve the wave equation in cylindrical coordinates:

$$\frac{\partial^2 u}{\partial r^2} + \frac{1}{r}\frac{\partial u}{\partial r} + \frac{\partial^2 u}{\partial z^2} - \frac{u}{r^2} = \frac{1}{c^2}\frac{\partial^2 u}{\partial t^2}, \qquad 0 < r < \infty, \quad 0 < z < \infty, \quad 0 < t.$$
$$(6.1.42)$$

Assuming that the system is initially at rest, $u(r, z, 0) = u_t(r, z, 0) = 0$, we have the following boundary conditions:

$$|u(0, z, t)| < \infty, \qquad \lim_{r \to \infty} u(r, z, t) \to 0, \qquad \lim_{z \to \infty} u(r, z, t) \to 0, \quad (6.1.43)$$

and

$$\mu\frac{\partial u(r, 0, t)}{\partial z} = \begin{cases} 0, & \text{if} \quad 0 < r < a, \\[2mm] \dfrac{Pa\delta(t)}{r\sqrt{r^2 - a^2}}, & \text{if} \quad a < r. \end{cases} \qquad (6.1.44)$$

We begin by defining the Laplace transform,

$$U(r, z, s) = \int_0^\infty u(r, z, t)e^{-st}\, dt \qquad (6.1.45)$$

so that Equation 6.1.42 becomes

$$\frac{\partial^2 U(r, z, s)}{\partial r^2} + \frac{1}{r}\frac{\partial U(r, z, s)}{\partial r} + \frac{\partial^2 U(r, z, s)}{\partial z^2} - \frac{U(r, z, s)}{r^2} - \frac{s^2}{c^2}U(r, z, s) = 0. \qquad (6.1.46)$$

Next, we introduce the Hankel transform,

$$\overline{U}(k, z, s) = \int_0^\infty U(r, z, s)\, J_1(kr)\, r\, dr, \qquad (6.1.47)$$

so that Equation 6.1.46 becomes

$$\frac{d^2\overline{U}(k, z, s)}{dz^2} - \left(k^2 + \frac{s^2}{c^2}\right)\overline{U}(k, z, s) = 0, \qquad (6.1.48)$$

which has the solution

$$\overline{U}(k, z, s) = A(k, s)\exp\left[-z\sqrt{k^2 + (s/c)^2}\right]. \qquad (6.1.49)$$

We have assumed that $\mathrm{Re}[\sqrt{k^2 + (s/c)^2}] \geq 0$ and also discarded the exponentially growing solution.

---

[2] Taken from Sarkar, G., 1966: Wave motion due to impulsive twist on the surface. *Pure Appl. Geophys.*, **65**, 43–47. Published by Birkhäuser Verlag, Basel, Switzerland.

To compute $A(k, s)$, we take the Hankel and Laplace transforms of the boundary condition given by Equation 6.1.44 and find that

$$\mu \frac{d\overline{U}(k, 0, s)}{dz} = P \frac{\sin(ka)}{k} \qquad (6.1.50)$$

so that

$$A(k, s) = -\frac{P}{\mu} \frac{\sin(ka)}{\sqrt{k^2 + (s/c)^2}}. \qquad (6.1.51)$$

Therefore, taking the inverse of the Hankel transform,

$$U(r, z, s) = -\frac{P}{\mu} \int_0^\infty \sin(ka) \frac{\exp\left[-z\sqrt{k^2 + (s/c)^2}\right]}{\sqrt{k^2 + (s/c)^2}} J_1(kr)\, dk. \qquad (6.1.52)$$

Consider now the integral

$$\int_{C_1} H_1^{(1)}(\zeta r) \sin(\zeta a) \frac{\exp\left[-z\sqrt{\zeta^2 + (s/c)^2}\right]}{\sqrt{\zeta^2 + (s/c)^2}}\, d\zeta. \qquad (6.1.53)$$

The contour C_1 is pie-shaped and includes the positive real axis ($\zeta = k$), the positive imaginary axis ($\zeta = ki$) and an arc at infinity in the first quadrant of the ζ-plane. By Jordan's lemma, the contribution from the arc vanishes. Therefore,

$$\int_0^\infty H_1^{(1)}(kr) \sin(ka) \frac{\exp\left[-z\sqrt{k^2 + (s/c)^2}\right]}{\sqrt{k^2 + (s/c)^2}}\, dk$$

$$= \int_0^\infty H_1^{(1)}(ikr) \sinh(ka) \frac{\exp\left[-z\sqrt{(s/c)^2 - k^2}\right]}{\sqrt{(s/c)^2 - k^2}}\, dk. \qquad (6.1.54)$$

Similarly, consider the integral

$$\int_{C_2} H_1^{(2)}(\zeta r) \sin(\zeta a) \frac{\exp\left[-z\sqrt{\zeta^2 + (s/c)^2}\right]}{\sqrt{\zeta^2 + (s/c)^2}}\, d\zeta. \qquad (6.1.55)$$

The contour C_2 is pie-shaped and includes the positive real axis ($\zeta = k$), the negative imaginary axis ($\zeta = -ki$) and an arc at infinity in the fourth quadrant of the ζ-plane. By Jordan's lemma, the contribution from the arc vanishes. Therefore,

$$\int_0^\infty H_1^{(2)}(kr) \sin(ka) \frac{\exp\left[-z\sqrt{k^2 + (s/c)^2}\right]}{\sqrt{k^2 + (s/c)^2}}\, dk$$

$$= \int_0^\infty H_1^{(2)}(-ikr) \frac{\sinh(ka) \exp\left[-z\sqrt{(s/c)^2 - k^2}\right]}{\sqrt{(s/c)^2 - k^2}}\, dk. \qquad (6.1.56)$$

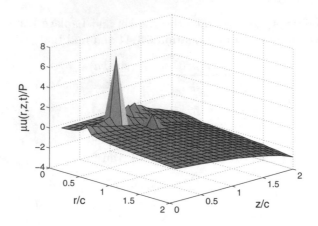

Figure 6.1.5: A plot of $u(r,z,t)$ given by Equation 6.1.59 when $t = 2.5$ and $a/c = 2$.

Combining Equation 6.1.54 and Equation 6.1.56,

$$U(r,z,s) = \frac{2P}{\mu\pi} \int_0^\infty K_1(kr) \frac{\sinh(ka) \exp\left[-z\sqrt{(s/c)^2 - k^2}\right]}{\sqrt{(s/c)^2 - k^2}} \, dk, \quad \textbf{(6.1.57)}$$

because $K_1(z) = -\pi H_1^{(1)}(iz)/2$ if $-\pi < \arg(z) < \pi/2$, $H_1^{(2)}(-z) = H_1^{(1)}(z)$ and $J_1(x) = [H_1^{(1)}(x) + H_1^{(2)}(x)]/2$. Replacing the hyperbolic sine with its exponential equivalent,

$$U(r,z,s) = \frac{P}{\mu\pi} \int_0^1 K_1\left(\frac{srp}{c}\right) \left(e^{pas/c} - e^{-pas/c}\right) \frac{\exp\left(-sz\sqrt{1 - p^2}/c\right)}{\sqrt{1 - p^2}} \, dp. \tag{6.1.58}$$

Using the second shifting theorem, we can take the inverse Laplace transform of Equation 6.1.58 and obtain the final solution

$$u(r,z,t) = -\frac{Pc}{\mu\pi} \int_0^1 \left[\frac{\tau_1 H(\tau_1 - rp/c)}{\sqrt{\tau_1^2 - (rp/c)^2}} - \frac{\tau_2 H(\tau_2 - rp/c)}{\sqrt{\tau_2^2 - (rp/c)^2}}\right] \frac{dp}{rp\sqrt{1 - p^2}}, \tag{6.1.59}$$

where $\tau_1 = t - z\sqrt{1 - p^2}/c + ap/c$ and $\tau_2 = t - z\sqrt{1 - p^2}/c - ap/c$. Figure 6.1.5 illustrates Equation 6.1.59 when $t = 2.5$ and $a/c = 2$.

• **Example 6.1.4**

For our final problem, we find the Green's function for the wave equation in cylindrical coordinates:

$$\frac{1}{r}\frac{\partial}{\partial r}\left(r\frac{\partial u}{\partial r}\right) + \frac{\partial^2 u}{\partial z^2} - \frac{1}{c^2}\frac{\partial^2 u}{\partial t^2} = -4\pi\delta(r)\delta(z - h)\delta(t), \tag{6.1.60}$$

subject to the boundary conditions $|u(0, z, t)| < \infty$, $\lim_{r \to \infty} u(r, z, t) \to 0$, and $u_{rz}(r, 0, t) = u_{rz}(r, D, t) = 0$, and the initial conditions $u(r, z, 0) = u_t(r, z, 0) = 0$, where $0 < z, h < D$ and $0 < r, t < \infty$.

We begin by taking the Laplace transform

$$U(r, z, s) = \int_0^\infty u(r, z, t) e^{-st} \, dt \tag{6.1.61}$$

and the Hankel transform

$$\overline{U}(k, z, s) = \int_0^\infty U(r, z, s) J_0(kr) \, r \, dr \tag{6.1.62}$$

of Equation 6.1.60 and the boundary conditions. This results in the boundary-value problem

$$\frac{d^2 \overline{U}(k, z, s)}{dz^2} - m^2 \overline{U}(k, z, s) = -2\delta(z - h) \tag{6.1.63}$$

with

$$\overline{U}'(k, 0, s) = \overline{U}'(k, D, s) = 0, \tag{6.1.64}$$

where $m^2 = k^2 + s^2/c^2$. We construct solutions to Equation 6.1.63 and Equation 6.1.64 in the regions $0 < z < h$ and $h < z < D$ and then use the conditions that $\overline{U}(k, h^+, s) = \overline{U}(k, h^-, s)$ and $\overline{U}'(k, h^+, s) - \overline{U}'(k, h^-, s) = -2$ to piece the two parts together. After some algebra,

$$U(x, z, s) = 2 \int_0^\infty \frac{\cosh(ma)\cosh(mb)}{m \sinh(mD)} J_0(kr) \, k \, dk, \tag{6.1.65}$$

where we have taken the inverse Hankel transform, $a = z$ and $b = D - h$ for $0 \le z \le h$, and $a = h$ and $b = D - z$ for $h \le z \le D$.

At this point, we replace $J_0(kr)$ with its representation in terms of Hankel functions of the first and second kind, $2J_0(kr) = H_0^{(1)}(kr) + H_0^{(2)}(kr)$. Therefore, we can write Equation 6.1.65 as

$$U(x, z, s) = \int_0^\infty \frac{\cosh(ma)\cosh(mb)}{m \sinh(mD)} H_0^{(1)}(kr) \, k \, dk$$

$$+ \int_0^\infty \frac{\cosh(ma)\cosh(mb)}{m \sinh(mD)} H_0^{(2)}(kr) \, k \, dk. \tag{6.1.66}$$

Using the residue theorem, it follows that

$$\oint_{C_1} z \frac{\cosh(ma)\cosh(mb)}{m \sinh(mD)} H_0^{(1)}(rz) \, dz$$

$$= 2\pi i \sum_{n=0}^\infty \text{Res}\left[z \frac{\cosh(ma)\cosh(mb)}{m \sinh(mD)} H_0^{(1)}(rz); iZ_n \right] \tag{6.1.67}$$

and

$$\oint_{C_2} z \frac{\cosh(ma)\cosh(mb)}{m\sinh(mD)} H_0^{(2)}(rz)\, dz$$

$$= -2\pi i \sum_{n=0}^{\infty} \text{Res}\left[z \frac{\cosh(ma)\cosh(mb)}{m\sinh(mD)} D_0^{(1)}(rz); -iZ_n \right], (6.1.68)$$

where $Z_n = \sqrt{s^2/c^2 + n^2\pi^2/D^2}$. C_1 is a pie-shaped closed contour in the first quadrant of the z-plane and consists of the positive real and imaginary axes and a circular arc at infinity that connects them. C_2 is the mirror image of C_1 in the fourth quadrant. Because of the behavior of $H_0^{(1)}$ and $H_0^{(2)}$, the contribution from the arcs at infinity vanish. Furthermore, the integration along the positive imaginary axis in Equation 6.1.67 gives the negative of the integration along the negative imaginary axis in Equation 6.1.68 because $H_0^{(1)}(ikr) = -H_0^{(2)}(-ikr)$. Consequently,

$$U(x,z,s) = 2\pi i \left(\sum_{n=0}^{\infty} \left\{ \text{Res}\left[z \frac{\cosh(ma)\cosh(mb)}{m\sinh(mD)} H_0^{(1)}(rz); iZ_n \right] \right. \right. \quad (6.1.69)$$

$$\left. \left. - \text{Res}\left[z \frac{\cosh(ma)\cosh(mb)}{m\sinh(mD)} H_0^{(1)}(rz); -iZ_n \right] \right\} \right).$$

We now compute the residues and they equal

$$\text{Res}\left[z \frac{\cosh(ma)\cosh(mb)}{m\sinh(mD)} H_0^{(1)}(rz); \frac{is}{c} \right] = \frac{1}{D\pi i} K_0\left(\frac{rs}{c}\right), \quad (6.1.70)$$

$$\text{Res}\left[z \frac{\cosh(ma)\cosh(mb)}{m\sinh(mD)} H_0^{(1)}(rz); -\frac{is}{c} \right] = -\frac{1}{D\pi i} K_0\left(\frac{rs}{c}\right), \quad (6.1.71)$$

$$\text{Res}\left[z \frac{\cosh(ma)\cosh(mb)}{m\sinh(mD)} H_0^{(1)}(rz); iZ_n \right]$$

$$= \frac{2(-1)^n}{D\pi i} \cos\left(\frac{n\pi a}{D}\right) \cos\left(\frac{n\pi b}{D}\right) K_0(rZ_n), \quad (6.1.72)$$

and

$$\text{Res}\left[z \frac{\cosh(ma)\cosh(mb)}{m\sinh(mD)} H_0^{(1)}(rz); -iZ_n \right]$$

$$= -\frac{2(-1)^n}{D\pi i} \cos\left(\frac{n\pi a}{D}\right) \cos\left(\frac{n\pi b}{D}\right) K_0(rZ_n), \quad (6.1.73)$$

because $H_0^{(1)}(ri) = 2K_0(r)/(\pi i)$ and $H_0^{(2)}(ri) = -2K_0(r)/(\pi i)$. A summation of the residues gives

$$U(x,z,s) = \frac{2}{D}\left[K_0\left(\frac{rs}{c}\right) + 2\sum_{n=1}^{\infty}(-1)^n \cos\left(\frac{n\pi a}{D}\right)\cos\left(\frac{n\pi b}{D}\right) K_0(rZ_n) \right].$$

$$(6.1.74)$$

Finally, taking the inverse of the Laplace transform, we obtain the desired solution

$$u(r, z, t) = \frac{2}{D\sqrt{t^2 - r^2/c^2}} \left[1 + 2\sum_{n=1}^{\infty} (-1)^n \cos\left(\frac{n\pi a}{D}\right) \cos\left(\frac{n\pi b}{D}\right) \right.$$

$$\left. \times \cos\left(\frac{n\pi c}{D}\sqrt{t^2 - r^2/c^2}\right) \right] \qquad \textbf{(6.1.75)}$$

if $t > r/c$; otherwise, $u(r, z, t) = 0$.

Problems

1. Solve the partial differential equation[3]

$$\frac{\partial^2 u}{\partial t^2} = \frac{\partial^2 u}{\partial x^2} + \frac{\partial^2 u}{\partial z^2}, \qquad -\infty < x < \infty, \quad 0 \le z < \infty, \quad 0 < t,$$

with the boundary conditions $\lim_{|x|\to\infty} u(x, z, t) \to 0$, $\lim_{z\to\infty} u(x, z, t) \to 0$ and $u(x, 0, t) = H(x + t) - H(x - t)$ with the initial conditions $u(x, z, 0) = u_t(x, z, 0) = 0$.

Step 1: Introducing the Laplace transform

$$U(x, z, s) = \int_0^{\infty} u(x, z, t)e^{-st}\, dt,$$

and the Fourier transform

$$\overline{U}(k, z, s) = \int_{-\infty}^{\infty} U(x, z, s)e^{-ikx}\, dx,$$

show that the partial differential equation and boundary conditions reduce to the boundary-value problem

$$\frac{d^2\overline{U}(k, z, s)}{dz^2} - (s^2 + k^2)\overline{U}(k, z, s) = 0, \qquad 0 \le z < \infty,$$

with $\lim_{z\to\infty} \overline{U}(k, z, s) \to 0$ and $\overline{U}(k, 0, s) = 2/(s^2 + k^2)$.

Step 2: Show that the solution to boundary-value problem is

$$\overline{U}(k, z, s) = 2\frac{\exp\left(-z\sqrt{s^2 + k^2}\right)}{s^2 + k^2}.$$

---

[3] Reprinted from *Int. J. Mech. Sci.*, **11**, W. R. Spillers and A. Callegari, Impact of two elastic cylinders: Short-time solution, pp. 846–851, ©1969, with kind permission from Pergamon Press Ltd., Headington Hill Hall, Oxford OX3 0BW, UK.

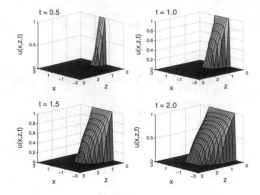

Problem 1

Step 3: Using the relationship[4] that

$$\mathcal{L}^{-1}\left[\frac{F\left(\sqrt{s^2 + k^2}\right)}{\sqrt{s^2 + k^2}}\right] = \int_0^t J_0\left(k\sqrt{t^2 - \eta^2}\right) f(\eta)\, d\eta$$

if $\mathcal{L}[f(t)] = F(s)$, show that

$$\mathcal{F}^{-1}\left\{\mathcal{L}^{-1}\left[\frac{F\left(\sqrt{s^2 + k^2}\right)}{\sqrt{s^2 + k^2}}\right]\right\} = \frac{1}{\pi}\int_0^{\sqrt{t^2 - x^2}}\frac{f(\eta)}{\sqrt{t^2 - x^2 - \eta^2}}\, d\eta.$$

Step 4: If we choose $F(s) = 2e^{-zs}/s$, show that

$$u(x, z, t) = \frac{2}{\pi}\int_0^{\sqrt{t^2 - x^2}}\frac{H(t - z)}{\sqrt{t^2 - x^2 - \eta^2}}\, d\eta = \frac{2}{\pi}\left[\frac{\pi}{2} - \arcsin\left(\frac{z}{\sqrt{t^2 - x^2}}\right)\right] H(t^2 - x^2 - z^2).$$

This solution is illustrated in the figure labeled Problem 1.

2. Solve the partial differential equation[5]

$$\frac{\partial^2 u}{\partial t^2} = \frac{\partial^2 u}{\partial z^2} + \frac{\partial^2 u}{\partial r^2} + \frac{1}{r}\frac{\partial u}{\partial r}, \qquad 0 < r, z < \infty, \quad 0 < t,$$

---

[4] From Erdélyi, A., W. Magnus, F. Oberhettinger, and F. G. Tricomi, 1954: *Table of Integral Transforms. Volume 1.* McGraw-Hill Book Co., Inc., p. 133, we have

$$\mathcal{L}\left[\int_0^t \left(\frac{t - \eta}{t + \eta}\right)^\nu J_{2\nu}\left(\sqrt{t^2 - \eta^2}\right) f(\eta)\, d\eta\right] = \frac{\left(\sqrt{s^2 + 1} - s\right)^{-2\nu} F\left(\sqrt{s^2 + 1}\right)}{\sqrt{s^2 + 1}}.$$

[5] Taken from Miles, J. W., 1953: Transient loading of a baffled piston. *J. Acoust. Soc. Am.*, **25**, 200–203.

with the boundary conditions $|u(0, z, t)| < \infty$, $\lim_{r \to \infty} u(r, z, t) \to 0$, $\lim_{z \to \infty} u(r, z, t) \to 0$, and $u_z(r, 0, t) = H(1 - r)H(t)$ with the initial conditions $u(r, z, 0) = u_t(r, z, 0) = 0$.

Step 1: By defining the Laplace transform

$$U(r, z, s) = \int_0^\infty u(r, z, t)e^{-st}\, dt,$$

and the Hankel transform

$$\overline{U}(k, z, s) = \int_0^\infty U(r, z, s)\, J_0(kr)\, r\, dr,$$

show that the partial differential equation and boundary conditions reduce to

$$\frac{d^2 \overline{U}(k, z, s)}{dz^2} - (s^2 + k^2)\overline{U}(k, z, s) = 0, \qquad 0 < z < \infty,$$

with $\lim_{z \to \infty} \overline{U}(k, z, s) \to 0$ and $\overline{U}'(k, 0, s) = J_1(k)/(ks)$.

Step 2: Show that the solution to Step 1 is

$$\overline{U}(k, z, s) = -J_1(k)\frac{\exp\left(-z\sqrt{k^2 + s^2}\right)}{sk\sqrt{k^2 + s^2}}.$$

Step 3: Take the inverse Laplace transform and show that

$$\overline{u}_t(k, z, t) = -\frac{J_1(k)}{k} J_0\left(k\sqrt{t^2 - z^2}\right) H(t - z).$$

Step 4: Take the inverse Hankel transform and show that

$$u_t(r, z, t) = -H(t - z)\int_0^\infty J_1(k) J_0\left(k\sqrt{t^2 - z^2}\right) J_0(kr)\, dk.$$

Step 5: Finally, use the relationship[6]

$$\int_0^\infty J_\mu(at) J_\nu(bt) J_\nu(ct) t^{1-\mu}\, dt = \begin{cases} 0, & \text{if } a^2 < (b+c)^2, a^2 < (b-c)^2, \\ \arccos\left(\dfrac{b^2 + c^2 - a^2}{2bc}\right), & \text{if } (b-c)^2 < a^2 < (b+c)^2, \\ \pi, & \text{if } (b+c)^2 < a^2, (b-c)^2 < a^2, \end{cases}$$

and show that

$$u_t(r, z, t) = \begin{cases} 0, & \text{if } \quad 1 < |R_-| \text{ or } t < z, \\ -\dfrac{1}{\pi} \arccos\left(\dfrac{t^2 + r^2 - z^2 - 1}{2r\sqrt{t^2 - z^2}}\right), & \text{if } \quad |R_-| < 1 < R_+, z < t, \\ -1, & \text{if } \quad 1 < R_+, z < t, \end{cases}$$

[6] Taken from Watson, G. N., 1966: *A Treatise on the Theory of Bessel Functions.* Cambridge University Press, p. 411.

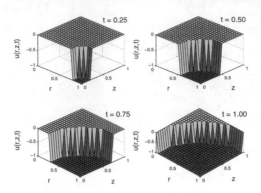

Problem 2

where $R_- = r - \sqrt{t^2 - z^2}$ and $R_+ = r + \sqrt{t^2 - z^2}$. This solution is illustrated in the figure labeled Problem 2.

3. Solve the Klein–Gordon equation

$$\frac{\partial^2 u}{\partial t^2} = \frac{\partial^2 u}{\partial x^2} - a^2 u, \qquad -\infty < x < \infty, \quad 0 < t,$$

subject to the boundary conditions $\lim_{|x| \to \infty} u(x,t) \to 0$, $0 < t$, and the initial conditions $u(x,0) = e^{-b|x|}$, $u_t(x,0) = 0$, $-\infty < x < \infty$, where $0 < b \le a$.

Step 1: Take the Laplace transform of the partial differential equation and show that

$$\frac{d^2 U(x,s)}{dx^2} - (s^2 + a^2) U(x,s) = -s e^{-b|x|}, \qquad -\infty < x < \infty,$$

with the boundary conditions $\lim_{|x| \to \infty} U(x,s) \to 0$.

Step 2: Using Fourier transforms, show that the solution to the ordinary differential equation in Step 1 is

$$U(x,s) = \frac{s}{s^2 + a^2 - b^2} e^{-b|x|} - \frac{bs}{s^2 + a^2 - b^2} \frac{e^{-|x|\sqrt{s^2 + a^2}}}{\sqrt{s^2 + a^2}}.$$

Step 3: Using tables to invert the first term in Step 2 and integrating on the complex s-plane to invert the second term, show that

$$u(x,t) = \cos\left(t\sqrt{a^2 - b^2}\right) e^{-b|x|} - \frac{2b}{\pi} H(t - |x|) \int_a^\infty \frac{\eta \cos(t\eta) \cos\left(x\sqrt{\eta^2 - a^2}\right)}{(\eta^2 + b^2 - a^2)\sqrt{\eta^2 - a^2}} \, d\eta$$

$$= \cos\left(t\sqrt{a^2 - b^2}\right) e^{-b|x|} - \frac{2b}{\pi} H(t - |x|) \int_0^\infty \frac{\cos\left(t\sqrt{\eta^2 + a^2}\right) \cos(x\eta)}{\eta^2 + b^2} \, d\eta.$$

This solution is illustrated in Figure 5.3.1.

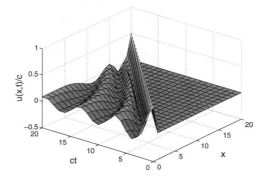

Problem 4

4. Find the Green's function for the Klein–Gordon equation

$$\frac{\partial^2 u}{\partial x^2} - \frac{1}{c^2}\frac{\partial^2 u}{\partial t^2} - \beta^2 u = -\delta(x)\delta(t), \qquad -\infty < x < \infty, \quad 0 < t,$$

subject to the boundary condition $\lim_{|x|\to\infty}|u(x,t)| < \infty$, $0 < t$, and the initial conditions $u(x,0) = u_t(x,0) = 0$, $-\infty < x < \infty$.

Step 1: Take the Laplace transform of the partial differential equation and boundary conditions and show that

$$\frac{d^2 U(x,s)}{dx^2} - \left(\frac{s^2}{c^2} + \beta^2\right) U(x,s) = -\delta(x), \qquad -\infty < x < \infty,$$

with the boundary condition $\lim_{|x|\to\infty}|U(x,s)| < \infty$.

Step 2: Using Fourier transforms, show that the solution to the ordinary differential equation in Step 1 is

$$U(x,s) = \frac{\exp\left(-|x|\sqrt{s^2/c^2 + \beta^2}\right)}{2\sqrt{s^2/c^2 + \beta^2}}.$$

Step 3: Using tables, show that the Green's function is

$$u(x,t) = \frac{c}{2} J_0\left(\beta\sqrt{c^2 t^2 - x^2}\right) H(ct - |x|).$$

This solution is illustrated in the figure labeled Problem 4 when $\beta = 1$.

5. Solve the partial differential equation[7]

$$\frac{\partial^2 u}{\partial r^2} + \frac{1}{r}\frac{\partial u}{\partial r} - \frac{\partial^2 u}{\partial t^2} - 2\epsilon\frac{\partial u}{\partial t} = -2\pi\delta(r)\delta(t), \qquad 0 < \epsilon, \quad 0 < r < \infty, \quad 0 < t,$$

---

[7] This result was derived in a different manner by Sezginer, A., and W. C. Chew, 1984: Closed form expression of the Green's function for the time-domain wave equation for a lossy two-dimensional medium. *IEEE Trans. Antennas Propag.*, **AP-32**, 527–528. ©1984 IEEE.

with the boundary conditions $|u(0,t)| < \infty$ and $\lim_{r \to \infty} u(r,t) \to 0$, and the initial conditions $u(r,0) = u_t(r,0) = 0$.

Step 1: By defining the Laplace transform

$$U(r,s) = \int_0^\infty u(r,t)e^{-st}\,dt,$$

and the Hankel transform

$$\overline{U}(k,s) = \int_0^\infty U(r,s)\,J_0(kr)\,r\,dr,$$

show that the partial differential equation and boundary conditions reduce to the algebraic equation $\overline{U}(k,s) = 1/[k^2 + s(s + 2\epsilon)]$.

Step 2: Using a table of integrals,[8] invert the Hankel transform and show that

$$\mathcal{L}\big[e^{\epsilon t}u(r,t)\big] = \int_0^\infty \frac{J_0(kr)}{k^2 + s^2 - \epsilon^2}\,k\,dk = K_0\left(r\sqrt{s^2 - \epsilon^2}\right),$$

where $\mathrm{Re}\left(\sqrt{s^2 - \epsilon^2}\right) > 0$.

Step 3: By using a table of Laplace transform[9] or by direct evaluation using Bromwich's integral, invert the Laplace transform in Step 2 and show that

$$u(r,t) = e^{-\epsilon t}\frac{\cosh\left(\epsilon\sqrt{t^2 - r^2}\right)}{\sqrt{t^2 - r^2}}H(t-r).$$

6. Solve the partial differential equation[10]

$$EI\frac{\partial^4 u}{\partial x^4} + ku + m\frac{\partial^2 u}{\partial t^2} = P_0\delta(x)\delta(t), \qquad -\infty < x < \infty, \quad 0 < t,$$

with the initial conditions $u(x,0) = u_t(x,0) = 0$ by the joint transform method.

Step 1: Applying the Fourier transform

$$U(k,t) = \int_{-\infty}^\infty u(x,t)e^{-ikx}\,dx$$

---

[8] For example, Gradshteyn, I. S., and I. M. Ryzhik, 1965: *Table of Integrals, Series, and Products*. Academic Press, 1086 pp.

[9] For example, Erdélyi, A., W. Magnus, F. Oberhettinger, and F. G. Tricomi, 1954: *Table of Integral Transforms. Vol. 1.* McGraw-Hill Book Co., Inc., 391 pp.

[10] Taken from Stadler, W., and R. W. Shreever, 1970: The transient and steady-state response of an infinite Bernoulli-Euler beam with damping and an elastic foundation. *Quart. J. Mech. Appl. Math.*, **23**, 197–208. By permission of Oxford University Press.

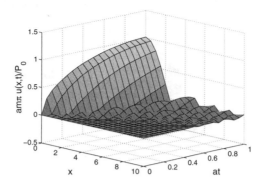

Problem 6

and the Laplace transform

$$\overline{U}(k, s) = \int_0^\infty U(k, t)e^{-st}\, dt$$

to the partial differential equation, show that the joint transform is

$$\overline{U}(k, s) = \frac{P_0/m}{s^2 + a^2 k^4 + \omega^2}$$

with $a^2 = EI/m$ and $\omega^2 = k/m$.

Step 2: Find the inverse Laplace transform of $\overline{U}(k, s)$ and show that

$$U(k, t) = \frac{P_0}{m} \frac{\sin\!\left(t\sqrt{a^2 k^4 + \omega^2}\right)}{\sqrt{a^2 k^4 + \omega^2}}.$$

Step 3: Use the fact that $U(k, t)$ is an even function in k to show that the inverse of the Fourier transform is

$$u(x, t) = \frac{P_0}{m\pi} \int_0^\infty \frac{\sin\!\left(t\sqrt{a^2 k^4 + \omega^2}\right)}{\sqrt{a^2 k^4 + \omega^2}} \cos(kx)\, dk.$$

This solution is illustrated in the figure labeled Problem 6 when $\omega/a = 1$.

7. Solve the partial differential equation[11]

$$\frac{\partial^2 u}{\partial x^2} - \frac{1}{c^2}\frac{\partial^2 u}{\partial t^2} = P\cos(\omega t)\delta[x - X(t)], \qquad -\infty < x < \infty, \quad 0 < t,$$

with the boundary conditions that $\lim_{|x|\to\infty} u(x, t) \to 0$ and the initial conditions $u(x, 0) = u_t(x, 0) = 0$ by the joint transform method.

---

[11] Taken from Knowles, J. K., 1968: Propagation of one-dimensional waves from a source in random motion. *J. Acoust. Soc. Am.*, **43**, 948–957.

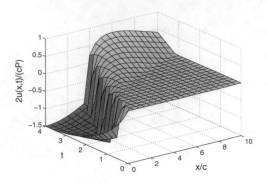

Problem 7

Step 1: By taking the Laplace transform in t and the Fourier transform in x, show that the partial differential equation and boundary conditions reduce to the algebraic equation

$$\overline{U}(k, s) = -\frac{c^2 P F(s)}{k^2 c^2 + s^2},$$

where

$$F(s) = \mathcal{L}\left[\cos(\omega t)e^{-ikX(t)}\right].$$

Step 2: Use the convolution theorem as it applies to Laplace transforms and show that

$$U(k, t) = -cP \int_0^t \frac{\sin[kc(t - \eta)]}{k} \cos(\omega \eta)e^{-ikX(\eta)} \, d\eta$$

and

$$u(x, t) = -\frac{cP}{2\pi} \int_{-\infty}^{\infty} \int_0^t \frac{\sin[kc(t - \eta)]}{k} \cos(\omega \eta)e^{ik[x - X(\eta)]} \, d\eta \, dk.$$

Step 3: By reversing the order of integration, show that

$$u(x, t) = -\frac{cP}{2} \int_0^t H\left[t - \eta - \frac{|X(\eta) - x|}{c}\right] \cos(\omega \eta) \, d\eta.$$

This solution is illustrated in the figure labeled Problem 7 when $\omega = 1$ and $X(t)/c = t^2$.

8. Solve the partial differential equations[12]

$$\frac{\partial^2 u}{\partial z^2} + \frac{\partial^2 u}{\partial x^2} = \frac{1}{c^2}\frac{\partial^2 u}{\partial t^2}, \qquad -\infty < x < \infty, \quad 0 < z < \infty, \quad 0 < t,$$

and

$$\frac{\partial^2 u}{\partial z^2} + \frac{\partial^2 u}{\partial x^2} = 0, \qquad -\infty < x < \infty, \quad -\infty < z < 0, \quad 0 < t,$$

---

[12] Taken from Lee, K. H., G. Liu, and H. F. Morrison, 1989: A new approach to modeling the electromagnetic response of conductive media. *Geophysics*, **54**, 1180–1192.

with the boundary conditions $\lim_{|z|\to\infty} u(x,z,t) \to 0$, $u(x,0^+,t) = u(x,0^-,t)$ and $u_z(x,0^+,t) - u_z(x,0^-,t) = -\delta(x)\delta(t)$. Assume that the initial conditions are $u(x,z,0) = u_t(x,z,0) = 0$.

Step 1: By taking the Laplace and Fourier transforms of the partial differential equations and boundary conditions, show that

$$\frac{d^2\overline{U}(k,z,s)}{dz^2} - \left(k^2 + \frac{s^2}{c^2}\right)\overline{U}(k,z,s) = 0, \qquad 0 < z,$$

and

$$\frac{d^2\overline{U}(k,z,s)}{dz^2} - k^2\overline{U}(k,z,s) = 0, \qquad z < 0,$$

where $\overline{U}(k,z,s)$ is the joint transform of $u(x,z,t)$. The boundary conditions become $\lim_{|z|\to\infty}\overline{U}(k,z,s) \to 0$, $\overline{U}(k,0^+,s) = \overline{U}(k,0^-,s)$ and $\overline{U}'(k,0^+,s) - \overline{U}'(k,0^-,s) = -1$.

Step 2: Show that the solution to the transform equations is

$$\overline{U}(k,z,s) = \frac{c^2(|k|+m)}{s^2}e^{|k|z}, \qquad z < 0,$$

and

$$\overline{U}(k,z,s) = \frac{c^2(|k|+m)}{s^2}e^{-mz}, \qquad 0 < z,$$

where $m = \sqrt{k^2 + s^2/c^2}$.

Step 3: For $z > 0$, show that we may write the inverse of the joint transform

$$u(x,z,t) = I_1 + I_2,$$

where

$$I_1 = \frac{c^2}{4\pi^2 i}\int_{c-\infty i}^{c+\infty i} e^{st}\left(\int_{-\infty}^{\infty}\frac{|k|}{s^2}e^{-mz}e^{ikx}\,dk\right)ds$$

and

$$I_2 = \frac{c^2}{4\pi^2 i}\int_{c-\infty i}^{c+\infty i} e^{st}\left(\int_{-\infty}^{\infty}\frac{m}{s^2}e^{-mz}e^{ikx}\,dk\right)ds.$$

Step 4: Using tables of Laplace transforms and integrals, verify the following evaluations of I_1 and I_2:

$$I_1 = \frac{c^2}{\pi} \frac{\partial^2}{\partial x \partial z} \int_0^\infty \mathcal{L}^{-1}\left(\frac{1}{s^2}\right) * \mathcal{L}^{-1}\left[\frac{\exp\left(-z\sqrt{k^2 + s^2/c^2}\right)}{\sqrt{k^2 + s^2/c^2}}\right] \sin(kx)\, dk$$

$$= \frac{c^3}{\pi}\begin{cases} 0, & 0 < t < z/c, \\ \dfrac{\partial^2}{\partial x \partial z} \displaystyle\int_0^\infty t * J_0\left(ck\sqrt{t^2 - z^2/c^2}\right) \sin(kx)\, dk, & z/c < t, \end{cases}$$

$$= \frac{c^2}{\pi}\begin{cases} 0, & 0 < t < z/c, \\ \dfrac{\partial^2}{\partial x \partial z} \displaystyle\int_{z/c}^t \dfrac{t-y}{\sqrt{r^2/c^2 - y^2}}\, dy, & z/c < t < r/c, \\ \dfrac{\partial^2}{\partial x \partial z} \displaystyle\int_{z/c}^{r/c} \dfrac{t-y}{\sqrt{r^2/c^2 - y^2}}\, dy, & r/c < t, \end{cases}$$

$$= \frac{c^2}{\pi r^4}\begin{cases} 0, & 0 < t < z/c, \\ \dfrac{(x^2 - z^2)t + |x|z(2t^2 - r^2/c^2)}{\sqrt{r^2/c^2 - t^2}}, & z/c < t < r/c, \\ (x^2 - z^2)t, & r/c < t, \end{cases}$$

where $r^2 = x^2 + z^2$, and

$$I_2 = \frac{c^2}{2\pi i} \frac{\partial^2}{\partial z^2}\left\{\int_{c-\infty i}^{c+\infty i} \frac{e^{st}}{s^2}\left[\frac{1}{\pi}\int_0^\infty \frac{e^{-mz}}{m}\cos(kx)\, dk\right] ds\right\}$$

$$= \frac{c^2}{2\pi i} \frac{\partial^2}{\partial z^2}\left[\int_{c-\infty i}^{c+\infty i} K_0(rs/c)\frac{e^{st}}{s^2}\, ds\right]$$

$$= \frac{c^2}{\pi}\begin{cases} 0, & 0 < t < r/c, \\ \dfrac{\partial^2}{\partial z^2} \displaystyle\int_{r/c}^t \dfrac{t-y}{\sqrt{y^2 - r^2/c^2}}\, dy, & r/c < t, \end{cases}$$

$$= \frac{c^2}{\pi r^4}\begin{cases} 0, & 0 < t < r/c, \\ -x^2\sqrt{t^2 - r^2/c^2} + \dfrac{z^2 t^2}{\sqrt{t^2 - r^2/c^2}}, & r/c < t. \end{cases}$$

Thus, the solution for $z > 0$ is

$$u(x,z,t) = \frac{c^2}{\pi r^4}\begin{cases} 0, & 0 < t < z/c, \\ \dfrac{(x^2 - z^2)t + |x|z(2t^2 - r^2/c^2)}{\sqrt{r^2/c^2 - t^2}}, & z/c < t < r/c, \\ (x^2 - z^2)t - x^2\sqrt{t^2 - r^2/c^2} + \dfrac{z^2 t^2}{\sqrt{t^2 - r^2/c^2}}, & r/c < t. \end{cases}$$

This solution is illustrated in the figure labeled Problem 8 when $t = 5$.

9. Solve the partial differential equation[13]

$$\frac{\partial^2 u}{\partial t^2} + \frac{\partial u}{\partial t} = \frac{1}{r}\frac{\partial}{\partial r}\left(r\frac{\partial u}{\partial r}\right) + \frac{\partial^2 u}{\partial z^2}, \qquad 0 \le r < \infty, \quad 0 < t, z,$$

---

[13] Suggested by a similar problem in Vedavarz, A., K. Mitra, and S. Kumar, 1994: Hyperbolic temperature profiles for laser surface interactions. *J. Appl. Phys.*, **76**, 5014–5021.

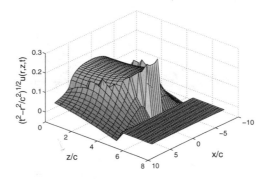

Problem 8

with the boundary conditions

$$\lim_{r\to 0} |u(r,z,t)| < \infty, \quad \lim_{r\to\infty} |u(r,z,t)| < \infty, \quad 0 < t, z,$$

$$u_z(r,0,t) = -e^{-r^2/2}\delta(t), \quad \lim_{z\to\infty} |u(r,z,t)| < \infty, \quad 0 \le r < \infty, \quad 0 < t,$$

and the initial conditions $u(r,z,0) = u_t(r,z,0) = 0$, $0 \le r < \infty$, $0 < z$.

Step 1: By taking the Laplace and Hankel transforms of the partial differential equation and boundary conditions, show that you obtain the boundary-value problem

$$\frac{d^2\overline{U}(k,z,s)}{dz^2} - (k^2 + s^2 + s)\overline{U}(k,z,s) = 0, \quad 0 < z < \infty,$$

with

$$\overline{U}'(k,0,s) = -e^{-k^2/2} \quad \text{and} \quad \lim_{z\to\infty} |\overline{U}(k,z,s)| < \infty,$$

where $\overline{U}(k,z,s)$ is the joint transform of $u(x,z,t)$.

Step 2: Show that the solution to the boundary-value problem is

$$\overline{U}(k,z,s) = e^{-k^2/2} \frac{\exp\left[-z\sqrt{\left(s+\frac{1}{2}\right)^2 + k^2 - \frac{1}{4}}\right]}{\sqrt{\left(s+\frac{1}{2}\right)^2 + k^2 - \frac{1}{4}}}.$$

Step 3: Take the inverse Laplace transform and show that

$$\overline{u}(k,z,t) = e^{-k^2/2-t/2} \begin{cases} I_0\left(\sqrt{\frac{1}{4} - k^2}\sqrt{t^2 - z^2}\right) H(t-z), & 0 < k < \frac{1}{2}, \\ J_0\left(\sqrt{k^2 - \frac{1}{4}}\sqrt{t^2 - z^2}\right) H(t-z), & \frac{1}{2} < k < \infty. \end{cases}$$

Step 4: Take the inverse Hankel transform and show that

$$u(k,z,t) = e^{-t/2}H(t-z)\int_0^{1/2} e^{-k^2/2}I_0\left(\sqrt{\frac{1}{4} - k^2}\sqrt{t^2 - z^2}\right) J_0(kr)\, k\, dk$$

$$+ e^{-t/2}H(t-z)\int_{1/2}^\infty e^{-k^2/2}J_0\left(\sqrt{k^2 - \frac{1}{4}}\sqrt{t^2 - z^2}\right) J_0(kr)\, k\, dk.$$

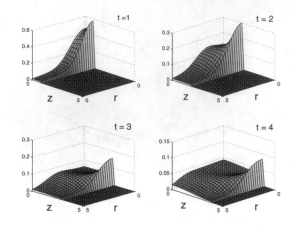

Problem 9

This solution is illustrated in the figure labeled Problem 9.

6.2 THE HEAT AND OTHER PARTIAL DIFFERENTIAL EQUATIONS

The joint transform method can also be applied to other types of linear partial differential equations in addition to the wave equation. The following examples provide a sample of its use.

• Example 6.2.1

In Example 2.2.1, we solved the heat conduction problem

$$\frac{\partial u}{\partial t} = a^2 \frac{\partial^2 u}{\partial x^2}, \qquad 0 < x < \infty, \quad 0 < t, \qquad (6.2.1)$$

subject to the boundary conditions

$$u_x(0,t) = 1, \qquad \lim_{x \to \infty} |u(x,t)| < \infty, \qquad 0 < t, \qquad (6.2.2)$$

and the initial condition

$$u(x,0) = 1, \qquad 0 < x < \infty. \qquad (6.2.3)$$

In that example, we applied Laplace transforms to eliminate the temporal dependence. Here, we will do that again. In the joint transform method, we also apply Laplace transforms to eliminate the x-dependence.

As usual, we begin by taking the Laplace transform with respect to time. This yields

$$\frac{d^2 U(x,s)}{dx^2} - \frac{s}{a^2} U(x,s) = -\frac{1}{a^2}, \qquad (6.2.4)$$

with

$$U'(0, s) = \frac{1}{s} \quad \text{and} \quad \lim_{x \to \infty} |U(x, s)| < \infty. \tag{6.2.5}$$

Here, the prime denotes differentiation with respect to x.

In Example 2.2.1 we solved Equation 6.2.4 using the standard techniques from a differential equations course. Alternatively we could take the Laplace transform of Equation 6.2.4 with respect to x. This gives

$$k^2 \overline{U}(k, s) - U'(0, s) - kU(0, s) - \frac{s}{a^2} \overline{U}(k, s) = -\frac{1}{ka^2}, \tag{6.2.6}$$

where

$$\overline{U}(k, s) = \int_0^\infty U(x, s) e^{-kx} \, dx. \tag{6.2.7}$$

Although Equation 6.2.5 provides the value for $U'(0, s)$, we do not know the value of $U(0, s)$. Let us denote it by A. Then,

$$\left(k^2 - \frac{s}{a^2}\right) \overline{U}(k, s) = \frac{1}{s} - kA - \frac{1}{ka^2}, \tag{6.2.8}$$

or

$$\overline{U}(k, s) = \frac{1}{s(k^2 - s/a^2)} - \frac{kA}{k^2 - s/a^2} - \frac{1}{ka^2(k^2 - s/a^2)} \tag{6.2.9}$$

$$= \frac{1}{sk} + \left(\frac{a}{2s^{3/2}} - \frac{A}{2} - \frac{1}{2s}\right) \frac{1}{k - \sqrt{s}/a}$$

$$- \left(\frac{a}{2s^{3/2}} + \frac{A}{2} + \frac{1}{2s}\right) \frac{1}{k + \sqrt{s}/a}. \tag{6.2.10}$$

Because the second term in Equation 6.2.10 will produce terms such as $e^{x\sqrt{s}/a}$, we must choose A so that this term vanishes. Therefore,

$$A = \frac{a}{s^{3/2}} - \frac{1}{s}, \tag{6.2.11}$$

so that

$$\overline{U}(k, s) = \frac{1}{sk} - \frac{a}{s^{3/2}} \frac{1}{k + \sqrt{s}/a}, \tag{6.2.12}$$

or

$$U(x, s) = \frac{1}{s} + \frac{a}{s^{3/2}} e^{-x\sqrt{s}/a}. \tag{6.2.13}$$

Taking the inverse Laplace transform of Equation 6.2.13 yields the final answer,

$$u(x, t) = 1 - 2a\sqrt{\frac{t}{\pi}} \exp\left(-\frac{x^2}{4a^2 t}\right) - x \operatorname{erfc}\left(\frac{x}{2a\sqrt{t}}\right). \tag{6.2.14}$$

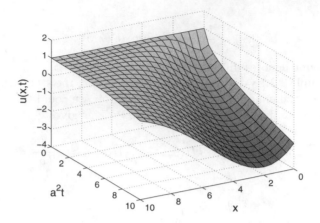

Figure 6.2.1: A plot of $u(x,t)$ given by Equation 6.2.14 as a function of x and a^2t.

Figure 6.2.1 illustrates Equation 6.2.14 as a function of x and a^2t.

● **Example 6.2.2**

Just as we applied Fourier and Laplace transforms to solve the wave equation, we can use the same technique to solve the heat equation. For example, let us solve[14]

$$\frac{\partial u}{\partial t} = \frac{\partial^2 u}{\partial x^2} + \frac{\partial^2 u}{\partial y^2}, \qquad 0 < x, y < \infty, \quad 0 < t, \qquad (6.2.15)$$

subject to the boundary conditions

$$u(0, y, t) - u_x(0, y, t) = 1, \qquad \lim_{x \to \infty} |u(x, y, t)| < \infty, \qquad 0 < y, t, \quad (6.2.16)$$

and

$$u(x, 0, t) = 0, \qquad \lim_{y \to \infty} u(x, y, t) \to 0, \qquad 0 < x, t, \qquad (6.2.17)$$

and the initial condition

$$u(x, y, 0) = 0, \qquad 0 < x, y < \infty. \qquad (6.2.18)$$

In the previous example, we applied Laplace transforms to eliminate both the x and t dependence. We could do this here, too. However, an alternative would be to apply Fourier sine transforms. Let us see how this works.

---

[14] Reprinted from *Int. J. Heat Mass Transfer*, **21**, D. A. Spence and D. B. R. Kenning, Unsteady heat conduction in a quarter plane, with an application to bubble growth models, 719–724, ©1972, with permission of Elsevier.

As usual, we begin by taking the Laplace transform with respect to time. This yields

$$\frac{\partial^2 U(x,y,s)}{\partial x^2} + \frac{\partial^2 U(x,y,s)}{\partial y^2} - sU(x,y,s) = 0, \tag{6.2.19}$$

subject to the boundary conditions

$$U(0,y,s) - U_x(0,y,s) = \frac{1}{s}, \qquad \lim_{x \to \infty} |U(x,y,s)| < \infty, \tag{6.2.20}$$

and

$$U(x,0,s) = 0, \qquad \lim_{y \to \infty} U(x,y,s) \to 0. \tag{6.2.21}$$

To eliminate the y-dependence, we use the Fourier sine transform

$$U(x,y,s) = \frac{2}{\pi} \int_0^\infty \overline{U}(x,\ell,s) \sin(\ell y) \, d\ell. \tag{6.2.22}$$

We have chosen this transform because it automatically satisfies the boundary condition along $y = 0$. Taking the Fourier sine transform of Equation 6.2.19 and Equation 6.2.20 yields

$$\frac{d^2 \overline{U}(x,\ell,s)}{dx^2} - (s + \ell^2)\overline{U}(x,\ell,s) = 0, \tag{6.2.23}$$

with the boundary conditions

$$\overline{U}(0,\ell,s) - \overline{U}'(0,\ell,s) = \frac{1}{s\ell} \quad \text{and} \quad \lim_{x \to \infty} |\overline{U}(x,\ell,s)| < \infty. \tag{6.2.24}$$

Here, the prime denotes differentiation with respect to x.

The solution to Equation 6.2.23 that satisfies the boundary conditions given by Equation 6.2.24 is

$$\overline{U}(x,\ell,s) = \frac{\exp\left(-x\sqrt{s + \ell^2}\right)}{s\ell\left(1 + \sqrt{s + \ell^2}\right)}. \tag{6.2.25}$$

Applying the results from Example 4.1.12 and the convolution theorem, we can invert Equation 6.2.25 with respect to s. This gives

$$\overline{u}(x,\ell,t) = \int_0^t \frac{e^{-\ell^2\tau}}{\ell} \left\{ \frac{\exp[-x^2/(4\tau)]}{\sqrt{\pi t}} - e^{x+\tau} \operatorname{erfc}\left(\frac{x}{2\sqrt{\tau}} + \sqrt{\tau}\right) \right\} d\tau. \tag{6.2.26}$$

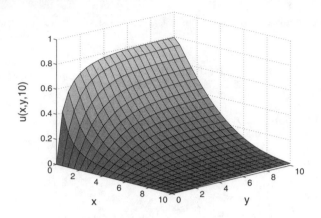

Figure 6.2.2: A plot of $u(x, y, t)$ given by Equation 6.2.28 as a function of x and y when $t = 10$.

Finally, we take the inverse of the Fourier sine transform and find that

$$u(x, y, t) = \frac{2}{\pi} \int_0^t \left[\int_0^\infty \frac{e^{-\ell^2 \tau}}{\ell} \sin(\ell y) \, d\ell \right] \tag{6.2.27}$$

$$\times \left\{ \frac{\exp[-x^2/(4\tau)]}{\sqrt{\pi \tau}} - e^{x+\tau} \mathrm{erfc}\left(\frac{x}{2\sqrt{\tau}} + \sqrt{\tau} \right) \right\} d\tau$$

$$= \int_0^t \mathrm{erf}\left(\frac{y}{2\sqrt{\tau}} \right) \left\{ \frac{\exp[-x^2/(4\tau)]}{\sqrt{\pi \tau}} - e^{x+\tau} \mathrm{erfc}\left(\frac{x}{2\sqrt{\tau}} + \sqrt{\tau} \right) \right\} d\tau. \tag{6.2.28}$$

Figure 6.2.2 illustrates Equation 6.2.28 as a function of x and y when $t = 10$.

Problems

1. Solve the heat equation[15]

$$\frac{\partial u}{\partial t} = \frac{\partial^2 u}{\partial r^2} + \frac{1}{r} \frac{\partial u}{\partial r} - \lambda u + S(r), \qquad 0 \le r < \infty, \quad 0 < t,$$

subject to the boundary conditions

$$\lim_{r \to 0} |u(r, t)| < \infty, \qquad \lim_{r \to \infty} |u(r, t)| < \infty, \qquad 0 < t,$$

and the initial condition

$$u(r, 0) = u_0(r), \qquad 0 \le r < \infty.$$

---

[15] Taken from Hassan, M. H. A., 1988: Ion distribution functions during ion cyclotron resonance heating at the fundamental frequency. *Phys. Fluids*, **31**, 596–601.

Step 1: By defining the Laplace transform

$$U(r, s) = \int_0^\infty u(r, t)e^{-st}\, dt$$

and the Hankel transform

$$\overline{U}(k, s) = \int_0^\infty U(r, s)J_0(kr)\, r\, dr,$$

show that the partial differential equation and boundary conditions reduce to

$$(s + k^2 + \lambda)\overline{U}(k, s) = \overline{u}_0(k) + \overline{S}(k)/s,$$

where

$$\overline{u}_0(k) = \int_0^\infty u_0(r)J_0(kr)\, r\, dr \quad \text{and} \quad \overline{S}(k) = \int_0^\infty S(r)J_0(kr)\, r\, dr.$$

Step 2: Show that the solution to Step 1 is

$$\overline{U}(k, s) = \frac{\overline{u}_0(k)}{s + k^2 + \lambda} + \frac{\overline{S}(k)}{(k^2 + \lambda)s} - \frac{\overline{S}(k)}{(k^2 + \lambda)(s + k^2 + \lambda)}.$$

Step 3: Inverting the Laplace transform, show that

$$\overline{u}(k, t) = \overline{u}_0(k)e^{-(k^2 + \lambda)t} + \frac{\overline{S}(k)}{k^2 + \lambda}\left[1 - e^{-(k^2 + \lambda)t}\right].$$

Step 4: Complete the problem by inverting the Hankel transform and show that

$$u(r, t) = \int_0^\infty \eta\, u_0(\eta) \int_0^\infty e^{-(k^2 + \lambda)t} J_0(k\eta)J_0(kr)\, k\, dk\, d\eta$$

$$+ \int_0^\infty \eta\, S(\eta) \int_0^\infty \left[1 - e^{-(k^2 + \lambda)t}\right] \frac{J_0(k\eta)J_0(kr)}{k^2 + \lambda}\, k\, dk\, d\eta$$

$$= \frac{e^{-r^2/(4t) - \lambda t}}{2t} \int_0^\infty \eta\, u_0(\eta)\, e^{-\eta^2/(4t)} I_0\!\left(\frac{r\eta}{2t}\right) d\eta$$

$$+ \int_0^\infty \eta\, S(\eta) \int_0^\infty \left[1 - e^{-(k^2 + \lambda)t}\right] \frac{J_0(k\eta)J_0(kr)}{k^2 + \lambda}\, k\, dk\, d\eta.$$

2. Solve the partial differential equation

$$\frac{\partial u}{\partial t} = \frac{\partial^2 u}{\partial r^2} + \frac{1}{r}\frac{\partial u}{\partial r} + \frac{\partial^2 u}{\partial z^2}, \qquad 0 \le r, z < \infty, \quad 0 < t,$$

subject to the boundary conditions

$$\frac{\partial u(r, 0, t)}{\partial z} = \begin{cases} -\dfrac{1}{\sqrt{a^2 - r^2}}, & \text{if} \quad 0 \le r < a, \\[2mm] 0, & \text{if} \quad a < r, \end{cases}$$

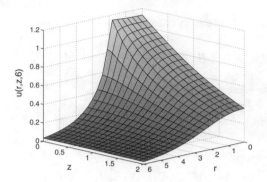

Problem 2

$$\frac{\partial u(0, z, t)}{\partial r} = 0, \qquad 0 \le z < \infty,$$

$$\lim_{r \to \infty} u(r, z, t) \to 0, \qquad \text{and} \qquad \lim_{z \to \infty} u(r, z, t) \to 0,$$

and the initial condition $u(r, z, 0) = 0$.

Step 1: By defining the Laplace transform

$$U(r, z, s) = \int_0^\infty u(r, z, t) e^{-st}\, dt$$

and the Hankel transform

$$\overline{U}(k, z, s) = \int_0^\infty U(r, z, s) J_0(kr)\, r\, dr,$$

show that the partial differential equation and boundary conditions reduce to the boundary-value problem

$$\frac{d^2 \overline{U}(k, z, s)}{dz^2} - (k^2 + s) \overline{U}(k, z, s) = 0$$

with $\lim_{z \to \infty} \overline{U}(k, z, s) \to 0$ and $\overline{U}'(k, 0, s) = -\sin(ka)/(ks)$.

Step 2: Show that the solution to boundary-value problem is

$$\overline{U}(k, z, s) = \frac{\sin(ka)}{ks\sqrt{k^2 + s}} e^{-z\sqrt{k^2 + s}} = \frac{\sin(ka)}{k(s' - k^2)\sqrt{s'}} e^{-z\sqrt{s'}}$$

$$= \frac{\sin(ka)}{2k^2\sqrt{s'}} e^{-z\sqrt{s'}} \left(\frac{1}{\sqrt{s'} - k} - \frac{1}{\sqrt{s'} + k} \right),$$

where $s' = s + k^2$.

Step 3: Using the first shifting theorem and tables, invert the Laplace transform and show that

$$\overline{u}(k, z, t) = \frac{\sin(ka)}{2k^2} \left[e^{-kz} \operatorname{erfc}\left(\frac{z}{2\sqrt{t}} - k\sqrt{t} \right) - e^{kz} \operatorname{erfc}\left(\frac{z}{2\sqrt{t}} + k\sqrt{t} \right) \right].$$

Step 4: Complete the problem by inverting the Hankel transform and showing that

$$u(r, z, t) = \int_0^\infty \frac{\sin(ka)}{2k} \left[e^{-kz} \operatorname{erfc}\left(\frac{z}{2\sqrt{t}} - k\sqrt{t} \right) - e^{kz} \operatorname{erfc}\left(\frac{z}{2\sqrt{t}} + k\sqrt{t} \right) \right] J_0(kr)\, dk.$$

This solution $u(r, z, t)$ is illustrated in the figure labeled Problem 2 when $t = 6$ and $a = 2$.

3. Solve the partial differential equation[16]

$$\frac{\partial u}{\partial t} - a^2 \frac{\partial^2 u}{\partial r^2} - \frac{a^2}{r} \frac{\partial u}{\partial r} - a^2 \frac{\partial^2 u}{\partial z^2} = (1 - r^2) \frac{\cos(t)}{\sqrt{4\pi a^2 t}} e^{-z^2/(4a^2 t)},$$

with $0 \le r < 1$, $-\infty < z < \infty$, and $0 < t$, subject to the boundary conditions

$$\lim_{r \to 0} |u(r, z, t)| < \infty, \qquad u_r(1, z, t) = 0, \qquad -\infty < z < \infty, \quad 0 < t,$$

and

$$\lim_{|z| \to \infty} u(r, z, t) \to 0, \qquad 0 \le r < 1, \quad 0 < t,$$

and the initial condition $u(r, z, 0) = 0$, $0 \le r < 1$, $-\infty < z < \infty$.

Step 1: Taking the Fourier transform of the partial differential equation in z, show that the governing equation and boundary conditions become

$$\frac{\partial \overline{u}}{\partial t} - a^2 \frac{\partial^2 \overline{u}}{\partial r^2} - \frac{a^2}{r} \frac{\partial \overline{u}}{\partial r} + a^2 m^2 \overline{u} = (1 - r^2) \frac{\cos(t)}{\sqrt{2\pi}} e^{-a^2 m^2 t}, \qquad 0 \le r < 1, \quad 0 < t,$$

subject to the boundary conditions

$$\lim_{r \to 0} |\overline{u}(r, m, t)| < \infty, \qquad \overline{u}_r(1, m, t) = 0, \qquad 0 < t,$$

and the initial condition $\overline{u}(r, m, 0) = 0$, $0 \le r < 1$, where

$$\overline{u}(r, m, t) = \int_{-\infty}^{\infty} u(r, z, t) e^{-imz} \, dz.$$

Step 2: Taking the Laplace transform of the partial differential equation in Step 1, show that it reduces to the boundary-value problem

$$s\overline{U} - a^2 \frac{d^2 \overline{U}}{dr^2} - \frac{a^2}{r} \frac{d\overline{U}}{dr} + a^2 m^2 \overline{U} = \frac{(1 - r^2)}{\sqrt{2\pi}} \frac{s + a^2 m^2}{(s + a^2 m^2)^2 + 1}, \qquad 0 \le r < 1,$$

with

$$\lim_{r \to 0} |\overline{U}(r, m, s)| < \infty, \qquad \overline{U}'(1, m, s) = 0,$$

where

$$\overline{U}(r, m, s) = \int_0^{\infty} \overline{u}(r, m, t) e^{-st} \, dt.$$

Step 3: Show that

$$1 - r^2 = \frac{1}{2} - 4 \sum_{n=1}^{\infty} \frac{J_0(k_n r)}{k_n^2 J_0(k_n)}, \qquad 0 \le r < 1,$$

---

[16] Solved in a slightly different way by Purtell, L. P., 1981: Molecular diffusion in oscillating laminar flow in a pipe. *Phys. Fluids*, **24**, 789–793.

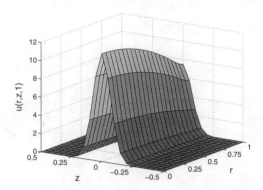

Problem 3

where k_n is the nth root of $J_1(k) = 0$.

Step 4: Show that the solution to the boundary-value problem in Step 2 is

$$\overline{U}(r, m, s) = \frac{1}{2\sqrt{2\pi}} \frac{1}{(s + a^2 m^2)^2 + 1} - \frac{4}{\sqrt{2\pi}} \sum_{n=1}^{\infty} \frac{J_0(k_n r)}{k_n^2 (1 + a^4 k_n^4) J_0(k_n)}$$

$$\times \left[\frac{1 + a^2 k_n^2 (s + a^2 m^2)}{(s + a^2 m^2)^2 + 1} - \frac{a^2 k_n^2}{s + a^2 m^2 + a^2 k_n^2} \right].$$

Step 5: Invert the Laplace transform and show that

$$\overline{u}(r, m, t) = \frac{1}{2\sqrt{2\pi}} \sin(t) e^{-a^2 m^2 t} - \frac{4 e^{-a^2 m^2 t}}{\sqrt{2\pi}} \sum_{n=1}^{\infty} \frac{J_0(k_n r)}{k_n^2 (1 + a^4 k_n^4) J_0(k_n)}$$

$$\times \left[\sin(t) + a^2 k_n^2 \cos(t) - a^2 k_n^2 e^{-a^2 k_n^2 t} \right].$$

Step 6: Invert the Fourier transform and show that

$$u(r, z, t) = \frac{\exp[-z^2/(4a^2 t)]}{\sqrt{\pi a^2 t}} \left(\frac{\sin(t)}{4} \right.$$

$$\left. - 2 \sum_{n=1}^{\infty} \frac{J_0(k_n r)}{k_n^2 (1 + a^4 k_n^4) J_0(k_n)} \left\{ \sin(t) + a^2 k_n^2 \left[\cos(t) - e^{-a^2 k_n^2 t} \right] \right\} \right).$$

This solution $u(r, z, t)$ is illustrated in the figure labeled Problem 3 when $t = 1$ and $a = 0.05$.

4. Solve the partial differential equation[17]

$$x \frac{\partial u}{\partial t} + \frac{\partial u}{\partial x} = f(x), \qquad -\infty < x < \infty, \quad 0 < t,$$

---

[17] Suggested by a considerably more complicated problem given by Shankar, P. N., 1981: On the evolution of disturbances at an inviscid interface. *J. Fluid Mech.*, **108**, 159–170. Reprinted with the permission of Cambridge University Press.

with the boundary conditions $\lim_{|x|\to\infty} u(x,t) \to 0$, and the initial condition $u(x,0) = 0$ by the joint transform method. The function $f(x)$ is an odd function.

Step 1: Defining the Laplace transform of $u(x,t)$ by

$$U(x,s) = \int_0^\infty u(x,t)e^{-st}\,dt,$$

show that the Laplace transform of the partial differential equation is

$$\frac{dU(x,s)}{dx} + sxU(x,s) = \frac{f(x)}{s}.$$

Step 2: Defining the Fourier transform of $U(x,s)$ by

$$\overline{U}(k,s) = \int_{-\infty}^\infty U(x,s)e^{-ikx}\,dx$$

and assuming that $U(x,s)$ goes to zero sufficiently rapid as $|x| \to \infty$, show that the joint transform of the partial differential equation is

$$\frac{d\overline{U}(k,s)}{dk} + \frac{k}{s}\overline{U}(k,s) = \frac{F(k)}{is^2}.$$

Step 3: Show that the solution to Step 2 is

$$\overline{U}(k,s) = \int_0^k F(\xi)\frac{\exp\left[-(k^2 - \xi^2)/(2s)\right]}{s^2 i}\,d\xi.$$

Step 4: Find the inverse Laplace transform of $\overline{U}(k,s)$ and show that

$$\overline{u}(k,t) = \sqrt{2t}\int_0^k F(\xi)\frac{J_1\left[\sqrt{2(k^2 - \xi^2)t}\right]}{i\sqrt{k^2 - \xi^2}}\,d\xi.$$

Step 5: Find the inverse Fourier transform of $\overline{u}(k,t)$ and show that

$$u(x,t) = \frac{\sqrt{2t}}{\pi}\int_0^\infty \left\{\int_0^k F(\xi)\frac{J_1\left[\sqrt{2(k^2 - \xi^2)t}\right]}{\sqrt{k^2 - \xi^2}}\,d\xi\right\}\sin(kx)\,dk.$$

5. Solve the partial differential equation[18]

$$\frac{\partial^2 u}{\partial x^2} + \frac{\partial^2 u}{\partial z^2} = 0, \qquad -h < z < 0, \quad -\infty < x < \infty, \quad 0 < t,$$

---

[18] Modeled after a problem that occurred in Hammack, J. L., 1973: A note on tsunamis: Their generation and propagation in an ocean of uniform depth. *J. Fluid Mech.*, **60**, 769–799.

with the boundary conditions

$$\frac{\partial u(x, -h, t)}{\partial z} = \xi_0(1 - e^{-\alpha t})\, H(b^2 - x^2)\, H(t)$$

and

$$\frac{\partial^2 u(x, 0, t)}{\partial t^2} + g\frac{\partial u(x, 0, t)}{\partial z} = 0,$$

and the initial conditions $u(x, z, 0) = u_t(x, z, 0) = 0$ by the joint transform method.

Step 1: If we define the Fourier transform by

$$U(k, z, t) = \int_{-\infty}^{\infty} u(x, z, t)e^{-ikx}\, dx$$

and the Laplace transform by

$$\overline{U}(k, z, s) = \int_{0}^{\infty} U(k, z, t)e^{-st}\, dt,$$

show that the joint transform of the partial differential equation is

$$\frac{d^2\overline{U}(k, z, s)}{dz^2} - k^2\overline{U}(k, z, s) = 0, \qquad -h < z < 0,$$

with

$$\frac{d\overline{U}(k, -h, s)}{dz} = \frac{2\xi_0\alpha\sin(kb)}{sk(s + \alpha)}$$

and

$$\frac{d\overline{U}(k, 0, s)}{dz} + \frac{s^2}{g}\overline{U}(k, 0, s) = 0.$$

Step 2: Show that the solution to Step 1 is

$$\overline{U}(k, z, s) = -\frac{2g\alpha\xi_0\sin(kb)}{ks(s + \alpha)(s^2 + \omega^2)\cosh(kh)}\left[\cosh(kz) - \frac{s^2}{gk}\sinh(kz)\right],$$

where $\omega^2 = gk\tanh(kh)$.

Step 3: Find the inverse Laplace transform of $\overline{U}(k, z, s)$ and show that

$$U(k, z, t) = \frac{2g\xi_0\sin(kb)}{k\cosh(kh)}\left[\frac{\sinh(kz)}{gk}\left\{\frac{\alpha^2}{\omega^2 + \alpha^2}\cos(\omega t) - \frac{\alpha^2\, e^{-\alpha t}}{\omega^2 + \alpha^2} + \frac{\alpha\omega}{\omega^2 + \alpha^2}\sin(\omega t)\right\}\right.$$

$$\left. - \cosh(kz)\left\{\frac{1 - \cos(\omega t)}{\omega^2} - \frac{1}{\omega^2 + \alpha^2}\left[e^{-\alpha t} - \cos(\omega t) + \frac{\alpha}{\omega}\sin(\omega t)\right]\right\}\right].$$

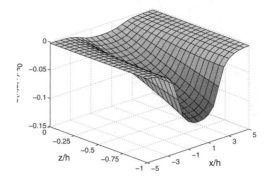

Problem 5

Step 4: Use the fact that $U(k, z, t)$ is an even function in k to show that the inverse of the Fourier transform is

$$u(x, z, t) = \frac{2g\xi_0}{\pi} \int_0^\infty \frac{\sin(kb)\cos(kx)}{k\cosh(kh)}\, dk \left[\frac{\sinh(kz)}{gk} \left\{ \frac{\alpha^2}{\omega^2 + \alpha^2} \cos(\omega t) \right. \right.$$

$$\left. - \frac{\alpha^2}{\omega^2 + \alpha^2} e^{-\alpha t} + \frac{\alpha\omega}{\omega^2 + \alpha^2}\sin(\omega t) \right\}$$

$$\left. - \cosh(kz) \left\{ \frac{1 - \cos(\omega t)}{\omega^2} - \frac{1}{\omega^2 + \alpha^2} \left[e^{-\alpha t} - \cos(\omega t) + \frac{\alpha}{\omega}\sin(\omega t) \right] \right\} \right].$$

This solution $u(x, z, t)/\xi_0$ is illustrated in the figure labeled Problem 5 when $b/h = 2$, $\omega t = 1$ and $\alpha/\omega = 0.1$.

6. Solve the partial differential equation[19]

$$\frac{\partial^2 u}{\partial r^2} + \frac{1}{r}\frac{\partial u}{\partial r} + \frac{\partial^2 u}{\partial z^2} + \nu\frac{\partial u}{\partial z} = -2I\rho_0\delta(r)\delta(z), \qquad 0 < r, z < \infty,$$

with the boundary conditions $\lim_{r\to\infty} u(r, z) \to 0$, $\lim_{z\to\infty} u(r, z) \to 0$, $|u(0, z)| < \infty$ and $u_z(r, 0) = -I\rho_0\delta(r)$.

Step 1: By defining the Hankel transform

$$U(k, z) = \int_0^\infty u(r, z)J_0(kr)\, r\, dr,$$

show that the partial differential equation becomes the boundary-value problem

$$\frac{d^2 U}{dz^2} + \nu\frac{dU}{dz} - k^2 U = -\frac{I\rho_0}{\pi}\delta(z)$$

---

[19] Taken from Paul, M. K., and B. Banerjee, 1970: Electrical potentials due to a point source upon models of continuously varying conductivity. *Pure Appl. Geophys.*, **80**, 218–237. Published by Birkhäuser Verlag, Basel, Switzerland.

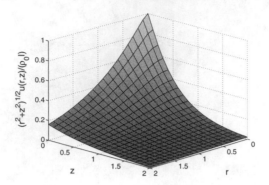

Problem 6

with $\lim_{z \to \infty} U(k, 0) \to 0$ and $U'(k, 0) = -I\rho_0/(2\pi)$.

Step 2: By taking the Fourier transform of the ordinary differential equation in Step 1, show that the particular solution reduces to the algebraic equation

$$\overline{U}_p(k, n) = \frac{I\rho_0}{\pi} \frac{1}{n^2 - i\nu n + k^2},$$

where

$$\overline{U}(k, n) = \int_{-\infty}^{\infty} U(k, z) e^{-inz} \, dz.$$

Step 3: Use the residue theorem to invert $\overline{U}_p(k, n)$ and show that

$$U_p(k, z) = \frac{I\rho_0}{\pi} \frac{\exp\left[-\left(\nu + \sqrt{\nu^2 + 4k^2}\right) z/2\right]}{\sqrt{\nu^2 + 4k^2}},$$

while the general solution is

$$U(k, z) = \frac{I\rho_0}{\pi} \frac{\exp\left[-\left(\nu + \sqrt{\nu^2 + 4k^2}\right) z/2\right]}{\sqrt{\nu^2 + 4k^2}} + A(k) \exp\left[-\left(\nu + \sqrt{\nu^2 + 4k^2}\right) z/2\right],$$

because $\lim_{z \to \infty} U(k, z) \to 0$.

Step 4: Using the boundary condition $U'(k, 0) = -I\rho_0/2\pi$, show that

$$u(r, z) = \frac{I\rho_0}{\pi} \int_0^\infty \left\{ \frac{\exp\left[-\left(\nu + \sqrt{\nu^2 + 4k^2}\right) z/2\right]}{\sqrt{\nu^2 + 4k^2}} \right.$$
$$\left. - \frac{\nu \exp\left[-\left(\nu + \sqrt{\nu^2 + 4k^2}\right) z/2\right]}{\sqrt{\nu^2 + 4k^2} \left(\nu + \sqrt{\nu^2 + 4k^2}\right)} \right\} J_0(kr) \, k \, dk,$$

or

$$u(r, z) = \frac{I\rho_0}{2\pi} \left\{ \frac{\exp\left[-\nu \left(z + \sqrt{r^2 + z^2}\right)/2\right]}{\sqrt{r^2 + z^2}} \right.$$
$$\left. - 2\nu \int_0^\infty \frac{\exp\left[-\left(\nu + \sqrt{\nu^2 + 4k^2}\right) z/2\right]}{\sqrt{\nu^2 + 4k^2} \left(\nu + \sqrt{\nu^2 + 4k^2}\right)} J_0(kr) \, k \, dk \right\}.$$

This solution $u(r, z)$ is illustrated in the figure labeled Problem 6 when $\nu = 2$.

Figure 6.3.1: Today, Louis Paul Emile Cagniard (1900–1971) is best known for his pioneering work on telluric currents – currents that flow in the ground and result from natural causes, such as the earth's magnetic field or auroral activity. He discovered his inversion method during his thesis work in mathematics (1938), a second doctorate after his first in physics (1928).

6.3 INVERSION OF THE JOINT TRANSFORM BY CAGNIARD'S METHOD

Although the vast majority of joint transform problems are solved along the lines given in Section 6.1 and Section 6.2, this direct assault is by no means the only method. In 1939, Cagniard published a book[20] in which he suggested that by using a series of transformations he could convert the joint transform into the integral definition of the Laplace transform. In this new form, he could then extract the inversion by inspection.

- **Example 6.3.1**

Consider the anisotropic wave equation[21]

$$\rho\frac{\partial^2 u}{\partial t^2} = N\frac{\partial^2 u}{\partial x^2} + L\frac{\partial^2 u}{\partial z^2}, \quad -\infty < x, z < \infty, \ 0 < t, \tag{6.3.1}$$

[20] The most accessible version is Cagniard, L., E. A. Flinn, and C. H. Dix, 1962: *Reflection and Refraction of Progressive Seismic Waves*. McGraw-Hill, 282 pp.

[21] Taken from Sakai, Y., and I. Kawasaki, 1990: Analytic waveforms for a line source in a transversely isotropic medium. *J. Geophys. Res.*, **95**, 11333–11344. ©1990 American Geophysical Union. Reproduced/modified by permission of American Geophysical Union.

where N, L and ρ are constants. Assuming that the system is initially quiescent, the Laplace transform of Equation 6.3.1 gives

$$N\frac{\partial^2 U}{\partial x^2} + L\frac{\partial^2 U}{\partial z^2} = \rho s^2 U, \qquad -\infty < x, z < \infty, \tag{6.3.2}$$

while the Fourier transform of Equation 6.3.2 yields

$$\frac{d^2 \overline{U}(k, z, s)}{dz^2} - \left(\frac{\rho s^2 + N k^2}{L}\right)\overline{U}(k, z, s) = 0, \tag{6.3.3}$$

where $\overline{U}(k, z, s)$ is the joint Fourier–Laplace transform of $u(x, z, t)$. The solution of Equation 6.3.3 is

$$\overline{U}(k, z, s) = \begin{cases} a_+ e^{-\beta z}, & \text{if} \quad z > 0, \\ a_- e^{+\beta z}, & \text{if} \quad z < 0, \end{cases} \tag{6.3.4}$$

where

$$\beta = \sqrt{\frac{\rho s^2 + N k^2}{L}}, \qquad \mathrm{Re}(\beta) \geq 0. \tag{6.3.5}$$

Let us now assume that there is an interface located at $z = 0$. The boundary conditions across this interface are

$$u(x, 0^+, t) = u(x, 0^-, t) \tag{6.3.6}$$

and

$$L\frac{\partial u(x, 0^+, t)}{\partial z} - L\frac{\partial u(x, 0^-, t)}{\partial z} = M\delta(t)\delta(x). \tag{6.3.7}$$

The transformed form of Equation 6.3.6 and Equation 6.3.7 is then

$$\overline{U}(k, 0^+, s) = \overline{U}(k, 0^-, s) \tag{6.3.8}$$

and

$$L\frac{d\overline{U}(k, 0^+, s)}{dz} - L\frac{d\overline{U}(k, 0^-, s)}{dz} = M. \tag{6.3.9}$$

From these boundary conditions, $a_\pm = -M/(2L\beta)$, and

$$U(x, z, s) = -\frac{M}{4\pi L}\int_{-\infty}^{\infty}\frac{e^{-\beta z + ikx}}{\beta}\, dk, \qquad 0 < z, \tag{6.3.10}$$

and

$$U(x, z, s) = -\frac{M}{4\pi L}\int_{-\infty}^{\infty}\frac{e^{\beta z + ikx}}{\beta}\, dk, \qquad z < 0. \tag{6.3.11}$$

From this point forward, we will only treat the case $z > 0$; the case of $z < 0$ follows by analogy.

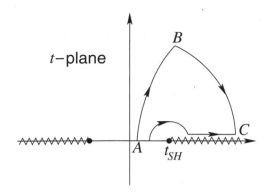

Figure 6.3.2: Contours used in the integration of the Laplace transform given by Equation 6.3.17.

Let us now introduce $k = -sp/V_{SH}$ which results in $\beta = s\nu/V_{SV}$, where $\nu = \sqrt{1+p^2}$, $V_{SH} = \sqrt{N/\rho}$ and $V_{SV} = \sqrt{L/\rho}$. Then, Equation 6.3.10 becomes

$$U(x,z,s) = -\frac{M\zeta}{4\pi L} \int_{-\infty}^{\infty} \frac{\exp[-s(\nu z + i\zeta px)/V_{SV}]}{\nu} \, dp, \qquad (6.3.12)$$

where $\zeta = \sqrt{L/N}$. Because the integrand in the region $p < 0$ is the complex conjugate to that in $p > 0$,

$$U(x,z,s) = -\frac{M\zeta}{2\pi L} \text{Re}\left\{ \int_{0}^{\infty} \frac{\exp[-s(\nu z + i\zeta px)/V_{SV}]}{\nu} \, dp \right\}. \qquad (6.3.13)$$

We now transform the variable of integration from p to a *complex* variable t defined by

$$t = \frac{\nu z + i\zeta px}{V_{SV}}. \qquad (6.3.14)$$

Solving for p,

$$p = \frac{-i\zeta V_{SV} xt \pm z\sqrt{V_{SV}^2 t^2 - \zeta^2 x^2 - z^2}}{\zeta^2 x^2 + z^2}. \qquad (6.3.15)$$

One of the roots given by Equation 6.3.15 corresponds to an upward-propagating wave in the upper half-space, while the other solution is spurious. A check of Equation 6.3.10 shows that we should choose the top sign. With that choice,

$$\frac{1}{\nu}\frac{\partial p}{\partial t} = -\frac{1}{\sqrt{t^2 - t_{SH}^2}}, \qquad (6.3.16)$$

where $t_{SH}^2 = (\zeta^2 x^2 + z^2)/V_{SV}^2$. Consider now the change of limits. As p increases from 0 to ∞, t changes from z/V_{SV} to $\infty + \infty i$ along a contour AB in the first quadrant of the complex t-plane shown in Figure 6.3.2. Then,

$$U(x,z,s) = -\frac{M\zeta}{2\pi L}\text{Re}\left(\int_{z/V_{SV}}^{\infty+\infty i} \frac{1}{\nu}\frac{\partial p}{\partial t} e^{-st} \, dt \right). \qquad (6.3.17)$$

Finally, because there is no singularity inside the region enclosed by the loop ABC, we may deform the contour AB to AC provided we pass above the branch point at $t = t_{SH}$. Therefore,

$$U(x, z, s) = -\frac{M\zeta}{2\pi L}\text{Re}\left(\int_{z/V_{SV}}^{\infty} \frac{1}{\nu}\frac{\partial p}{\partial t}e^{-st}\,dt\right). \qquad (6.3.18)$$

Because the integrand is purely imaginary for $t < t_{SH}$, we can modify Equation 6.3.18 so that

$$U(x, z, s) = \int_0^{\infty} u(x, z, t)e^{-st}\,dt = -\frac{M\zeta}{2\pi L}\text{Re}\left(\int_0^{\infty} \frac{1}{\nu}\frac{\partial p}{\partial t}e^{-st}\,dt\right). \qquad (6.3.19)$$

Consequently, by inspection,

$$u(x, z, t) = \frac{M\zeta}{2\pi L}\frac{H(t - t_{SH})}{\sqrt{t^2 - t_{SH}^2}}. \qquad (6.3.20)$$

• Example 6.3.2

Our next example is much more complicated. We find the velocity potential $u(x, z, t)$ that results from an explosion[22] located at $z = h$ within a compressible ocean having the speed of sound c. The governing equation is

$$\frac{\partial^2 u}{\partial t^2} - c^2\left(\frac{\partial^2 u}{\partial x^2} + \frac{\partial^2 u}{\partial z^2}\right) - g\frac{\partial u}{\partial z} = -c^2\delta(x)\delta(z - h)\delta(t), \qquad (6.3.21)$$

where g denotes the gravitational acceleration. The boundary condition at the free surface $z = 0$ is

$$\frac{\partial^2 u}{\partial t^2} = g\frac{\partial u}{\partial z}. \qquad (6.3.22)$$

Taking the Laplace transform of Equation 6.3.21,

$$c^2\left[\frac{\partial^2 U(x, z, s)}{\partial x^2} + \frac{\partial^2 U(x, z, s)}{\partial z^2}\right] + g\frac{\partial U(x, z, s)}{\partial z} - s^2 U(x, z, s) = c^2\delta(x)\delta(z - h),$$
$$(6.3.23)$$

assuming that the fluid is initially at rest. If we now take the Fourier transform of Equation 6.3.23, we find that

$$c^2\frac{d^2\overline{U}(k, z, s)}{dz^2} + g\frac{d\overline{U}(k, z, s)}{dz} - (k^2c^2 + s^2)\overline{U}(k, z, s) = c^2\delta(z - h), \qquad (6.3.24)$$

---

[22] Taken from Ross, R. A., 1961: The effect of an explosion in a compressible fluid under gravity. *Can. J. Phys.*, **39**, 1330–1346.

where $\overline{U}(k, z, s)$ denotes the joint Fourier–Laplace transform.

The solution of Equation 6.3.24 is

$$\overline{U}(k, z, s) = \frac{1}{2} \exp\left[-\frac{g(z - h)}{2c^2}\right] \frac{e^{-|z-h|\lambda}}{\lambda}, \tag{6.3.25}$$

where

$$\lambda = \sqrt{k^2 + \frac{s^2}{c^2} + \frac{g^2}{4c^4}}, \tag{6.3.26}$$

and the inverse Fourier transform is

$$U(x, z, s) = \frac{1}{4\pi} \exp\left[-\frac{g(z - h)}{2c^2}\right] \int_{-\infty}^{\infty} \frac{e^{ikx - |z-h|\lambda}}{\lambda} \, dk. \tag{6.3.27}$$

The parameter λ is a multivalued function with branch points at

$$k = \pm i\sqrt{\frac{s^2}{c^2} + \frac{g^2}{4c^4}}. \tag{6.3.28}$$

In order for Equation 6.3.27 to be finite as $|z - h| \to \infty$, $\text{Re}(\lambda) \geq 0$. We ensure this if one branch cut lies between $i\sqrt{s^2/c^2 + g^2/(4c^4)}$ and ∞i, while the other lies between $-i\sqrt{s^2/c^2 + g^2/(4c^4)}$ and $-\infty i$.

Equation 6.3.27 is the particular solution of Equation 6.3.23. Consequently, the most general solution is this particular solution plus the homogeneous solution, or

$$\begin{aligned} U(x, z, s) = {} & \frac{1}{4\pi} \exp\left[-\frac{g(z - h)}{2c^2}\right] \int_{-\infty}^{\infty} \frac{e^{ikx - |z-h|\lambda}}{\lambda} \, dk \\ & + \frac{1}{4\pi} \exp\left(-\frac{gz}{2c^2}\right) \int_{-\infty}^{\infty} A(k) e^{ikx - \lambda z} \, dk, \end{aligned} \tag{6.3.29}$$

where $A(k)$ is an arbitrary function of k. We determine this function by satisfying the Laplace transformed Equation 6.3.22,

$$g\frac{\partial U(x, 0, s)}{\partial z} = s^2 U(x, 0, s). \tag{6.3.30}$$

Upon substituting Equation 6.3.29 into Equation 6.3.30, we find that

$$\begin{aligned} U(x, z, s) = {} & \frac{1}{2\pi} \exp\left[-\frac{g(z - h)}{2c^2}\right] \text{Re}\left(\int_0^{\infty} \frac{e^{ikx - |z-h|\lambda}}{\lambda} \, dk\right) \\ & - \frac{1}{2\pi} \exp\left[-\frac{g(z - h)}{2c^2}\right] \text{Re}\left(\int_0^{\infty} \frac{e^{ikx - |z+h|\lambda}}{\lambda} \, dk\right) \\ & + \frac{1}{\pi} \exp\left[-\frac{g(z - h)}{2c^2}\right] \text{Re}\left[\int_0^{\infty} \frac{e^{ikx - (z+h)\lambda}}{s^2/g + g/(2c^2) + \lambda} \, dk\right]. \tag{6.3.31} \end{aligned}$$

Our next step is to write the right side of Equation 6.3.31 as the forward Laplace transform

$$U(x, z, s) = \int_0^\infty u(x, z, t)e^{-st}\, dt. \tag{6.3.32}$$

To this end we deform the path of integration in the second and third integrals of Equation 6.3.31 into a new path given by

$$k = \frac{[gx/(2c^2) - (z+h)s/c]\sqrt{c^2p^2 - x^2 - (z+h)^2}}{x^2 + (z+h)^2} + \frac{ip[g(z+h)/(2c) + xs]}{x^2 + (z+h)^2}, \tag{6.3.33}$$

where p is a real parameter varying from 0 to ∞. We deform the first integral into the path given by Equation 6.3.33 where we replace $z + h$ with $|z - h|$. These deformations are permissible because there are no contributions from poles (Cauchy's theorem) and the contribution from the arcs at infinity vanish (Jordan's lemma).

Because $p = 0$ corresponds to

$$k = i\frac{gx/(2c^2) - s(z+h)/c}{\sqrt{x^2 + (z+h)^2}}, \tag{6.3.34}$$

a portion of the new path lies along the negative imaginary axis. However, because the integrands are purely imaginary along this segment, there is no contribution from this portion of the contour integration. Consequently, the first two integrals in Equation 6.3.31 become

$$U_1(x, z, s) = \frac{1}{2\pi} \exp\left[-\frac{g(z-h)}{2c^2}\right] \mathrm{Re}\left[\int_0^\infty \frac{e^{-sp+gi\sqrt{p^2 - k_1^2}/(2c)}}{\sqrt{p^2 - k_1^2}}\, dp\right]$$
$$- \frac{1}{2\pi} \exp\left[-\frac{g(z-h)}{2c^2}\right] \mathrm{Re}\left[\int_0^\infty \frac{e^{-sp+gi\sqrt{p^2 - k_2^2}/(2c)}}{\sqrt{p^2 - k_2^2}}\, dp\right], \tag{6.3.35}$$

where

$$k_1 = \sqrt{x^2 + (z+h)^2}/c \tag{6.3.36}$$

and

$$k_2 = \sqrt{x^2 + (z-h)^2}/c. \tag{6.3.37}$$

Because the first term in Equation 6.3.35 vanishes if $0 < p < k_1$ (because the integral is imaginary) and the second one vanishes if $0 < p < k_2$, its contribution to $u(x, z, t)$ is

$$u_1(x, z, t) = \frac{1}{2\pi} \exp\left[-\frac{g(z-h)}{2c^2}\right] \frac{\cos[g\sqrt{t^2 - k_1^2}/(2c)]}{\sqrt{t^2 - k_1^2}} H(t - k_1)$$
$$- \frac{1}{2\pi} \exp\left[-\frac{g(z-h)}{2c^2}\right] \frac{\cos[g\sqrt{t^2 - k_2^2}/(2c)]}{\sqrt{t^2 - k_2^2}} H(t - k_2). \tag{6.3.38}$$

The third integral in Equation 6.3.31 becomes

$$U_2(x, z, s) = \frac{1}{2\pi} \exp\left[-\frac{g(z-h)}{2c^2}\right]$$

$$\times \mathrm{Re}\left\{\int_{k_1}^{\infty} \frac{F_1(p) \exp\left[-sp - gi\sqrt{p^2 - k_1^2}/(2c)\right]}{[s + G_1(p)]\sqrt{p^2 - k_1^2}} \, dp\right\}$$

$$-\frac{1}{2\pi} \exp\left[-\frac{g(z-h)}{2c^2}\right]$$

$$\times \mathrm{Re}\left\{\int_{k_1}^{\infty} \frac{F_2(p) \exp\left[-sp - gi\sqrt{p^2 - k_1^2}/(2c)\right]}{[s + G_2(p)]\sqrt{p^2 - k_1^2}} \, dp\right\}, \quad (6.3.39)$$

where

$$F_{1,2}(p) = \frac{1}{2}\left[\sqrt{\omega^2 + \frac{1}{c^2}} \pm \frac{\omega^2 + 1/c^2 - i\omega/c}{\omega - i/c}\right], \quad (6.3.40)$$

$$G_{1,2}(p) = \frac{g}{2}\left[\sqrt{\omega^2 + \frac{1}{c^2}} \pm \left(\omega - \frac{i}{c}\right)\right], \quad (6.3.41)$$

and

$$\omega = \frac{ixp - (z+h)\sqrt{p^2 - k_1^2}}{c^2 k_1^2}. \quad (6.3.42)$$

We now note that

$$\int_p^{\infty} e^{-G_{1,2}(p)(t-p)} e^{-st} \, dt = \frac{e^{-sp}}{s + G_{1,2}(p)}. \quad (6.3.43)$$

Upon substituting this into Equation 6.3.39 and interchanging the order of integration, we obtain a forward Laplace transform. Thus, the complete expression for $u(x, z, t)$ is

$$u(x, z, t) = \frac{1}{2\pi} \exp\left[-\frac{g(z-h)}{2c^2}\right] \frac{\cos[g\sqrt{t^2 - k_1^2}/(2c)]}{\sqrt{t^2 - k_1^2}} H(t - k_1)$$

$$-\frac{1}{2\pi} \exp\left[-\frac{g(z-h)}{2c^2}\right] \frac{\cos[g\sqrt{t^2 - k_2^2}/(2c)]}{\sqrt{t^2 - k_2^2}} H(t - k_2)$$

$$-\frac{g}{\pi} \exp\left[-\frac{g(z-h)}{2c^2}\right] H(t - k_1) \quad (6.3.44)$$

$$\times \mathrm{Re}\left\{\int_{k_1}^{t} \frac{F_1(p) \exp\left[gi\sqrt{p^2 - k_1^2}/(2c) - (t-p)G_1(p)\right]}{\sqrt{p^2 - k_1^2}} \, dp\right\}$$

$$-\frac{g}{\pi} \exp\left[-\frac{g(z-h)}{2c^2}\right] H(t - k_1)$$

$$\times \mathrm{Re}\left\{\int_{k_1}^{t} \frac{F_2(p) \exp\left[gi\sqrt{p^2 - k_1^2}/(2c) - (t-p)G_2(p)\right]}{\sqrt{p^2 - k_1^2}} \, dp\right\}.$$

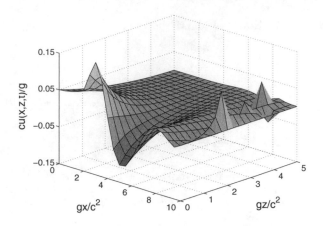

Figure 6.3.3: A plot of $u(x, z, t)$ given by Equation 6.3.44 as a function of x and z when $gt/c = 10$.

The first two terms represent the source and image, while the integrals are the surface effects. Figure 6.3.3 illustrates Equation 6.3.44 as a function of x and z when $gt/c = 10$.

• **Example 6.3.3**

Finally, let us solve the wave equation[23]

$$\frac{\partial^2 u}{\partial x^2} + \frac{\partial^2 u}{\partial y^2} = \frac{1}{c^2}\frac{\partial^2 u}{\partial t^2}, \qquad -\infty < x, y < \infty, \quad 0 < t, \qquad (6.3.45)$$

where c denotes the phase speed. The medium is initially at rest so that $u(x, y, 0) = u_t(x, y, 0) = 0$. At infinity, the disturbances die away,

$$\lim_{|y| \to \infty} u(x, y, t) \to 0, \qquad\qquad 0 < t, \qquad (6.3.46)$$

while at $y = 0$,

$$u(x, 0^+, t) - u(x, 0^-, t) = \begin{cases} vt - |x|, & \text{if} \quad 0 < |x| < vt, \\ 0, & \text{if} \quad vt < |x|, \end{cases} \qquad (6.3.47)$$

and

$$\frac{\partial u(x, 0^+, t)}{\partial y} = \frac{\partial u(x, 0^-, t)}{\partial y}, \qquad\qquad 0 < t. \qquad (6.3.48)$$

We begin by taking the Laplace transform of Equation 6.3.45, followed by applying the Fourier transform. If

$$U(x, y, s) = \int_0^\infty u(x, y, t)e^{-st}\, dt \qquad (6.3.49)$$

---

[23] Taken from Bakhshi, V. S., 1965: Propagation of a fracture in an infinite medium. *Pure Appl. Geophys.*, **62**, 23–34. Published by Birkhäuser Verlag, Basel, Switzerland.

and

$$\overline{U}(k, y, s) = \int_{-\infty}^{\infty} U(x, y, s)e^{-ikx}\, dx, \tag{6.3.50}$$

then Equation 6.3.45 becomes

$$\frac{d^2\overline{U}(k, y, s)}{dy^2} - \left(k^2 + \frac{s^2}{c^2}\right)\overline{U}(k, y, s) = 0 \tag{6.3.51}$$

with $\lim_{|y|\to\infty} \overline{U}(k, y, s) \to 0$,

$$\overline{U}(k, 0^+, s) - \overline{U}(k, 0^-, s) = \frac{2}{s}\left[\frac{1}{k^2 + (s/v)^2}\right], \tag{6.3.52}$$

and

$$\frac{d\overline{U}(k, 0^+, s)}{dy} = \frac{d\overline{U}(k, 0^-, s)}{dy}. \tag{6.3.53}$$

The solution to Equation 6.3.51 through Equation 6.3.53 is

$$\overline{U}(k, y, s) = \operatorname{sgn}(y)\frac{v^2}{s(s^2 + v^2k^2)}e^{-|y|\sqrt{k^2 + s^2/c^2}}. \tag{6.3.54}$$

Therefore, taking the inverse Fourier transform,

$$U(x, y, s) = \operatorname{sgn}(y)\frac{1}{2\pi s}\int_{-\infty}^{\infty}\frac{v^2}{s^2 + v^2k^2}e^{ikx - |y|\sqrt{k^2 + s^2/c^2}}\, dk \tag{6.3.55}$$

$$= \operatorname{sgn}(y)\frac{1}{\pi s}\int_0^{\infty}\frac{v^2}{s^2 + v^2k^2}\cos(kx)e^{-|y|\sqrt{k^2 + s^2/c^2}}\, dk \tag{6.3.56}$$

$$= \operatorname{sgn}(y)\frac{1}{\pi s^2}\operatorname{Re}\left[\int_0^{\infty}\frac{e^{-s\left(iwx + |y|\sqrt{w^2 + 1/c^2}\right)}}{w^2 + 1/v^2}\, dw\right]. \tag{6.3.57}$$

We now invert $U(x, y, s)$ by Cagniard's method for the most pratical case $c > v$. Consider the case of $y > 0$; the case for $y < 0$ follows by analog. Let us introduce the transformation $t = iwx + y\sqrt{w^2 + 1/c^2}$, where $\operatorname{Re}\left(\sqrt{w^2 + 1/c^2}\right) \geq 0$. We define the multivalued square root by introducing branch points at $w = \pm i/c$ and a branch cut along the imaginary axis of the w-plane. Our definition of the phases is

$$w - i/c = re^{\theta i}, \qquad -\pi/2 \leq \theta < 3\pi/2, \tag{6.3.58}$$

and

$$w + i/c = re^{\theta i}, \qquad -\pi/2 \leq \theta < 3\pi/2. \tag{6.3.59}$$

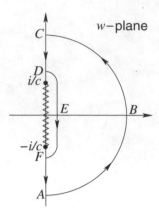

Figure 6.3.4: Contour used in the integration of Equation 6.3.57.

Table 6.3.1. Values of Various Functions Along the Contour $CDEFA$

w	ℓ	$\sqrt{w^2 + 1/c^2}$	t
C to D	∞ to $1/c$	$i\sqrt{\ell^2 - 1/c^2}$	$-\ell x + iy\sqrt{\ell^2 - 1/c^2}$
D	$1/c$	0	$-x/c$
D to E	$1/c$ to 0	$\sqrt{1/c^2 - \ell^2}$	$-\ell x + y\sqrt{1/c^2 - \ell^2}$
E	0	$1/c$	y/c
E to F	0 to $-1/c$	$\sqrt{1/c^2 - \ell^2}$	$-\ell x + y\sqrt{1/c^2 - \ell^2}$
F	$-1/c$	0	x/c
F to A	$-1/c$ to $-\infty$	$-i\sqrt{\ell^2 - 1/c^2}$	$-\ell x - iy\sqrt{\ell^2 - 1/c^2}$

Consider the path $ABCDEFA$ shown in Figure 6.3.4. The semicircle ABC is of infinite radius. Along this semicircle, $t = iwx + yx + O(1/w)$. Along the imaginary axis, $CDEFA$, we set $w = \ell i$ and give the values of t in Table 6.3.1. Figure 6.3.5 shows the closed contour in the t-plane with corresponding points marked with primes. Note that as we move from E to F, the corresponding curve in the t-plane moves along the real axis until $t(\ell_0)$, marked by T', and then back-tracks to F'. To maintain uniqueness, we introduce a cut along the real axis of the t-plane from $E'T'$ and then take that k in the definition of t whose modulus is smaller.

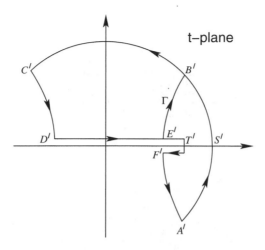

Figure 6.3.5: New contour used in the integration of Equation 6.3.56.

Carrying out the change of variables,

$$U(x, y, s) = \frac{1}{\pi s^2} \mathrm{Re} \left[\int_\Gamma f(x, y, t) \frac{\partial w}{\partial t} e^{-st} \, dt \right], \qquad (6.3.60)$$

where $f(x, y, t)$ is the expression for $[w^2 + 1/v^2]^{-1}$ after we transform it to t. The path of integration Γ starts at the point y/c and extends out to infinity in the first quadrant along the curve shown in Figure 6.3.5.

We would now like to deform Γ to an integration along the real axis so that the analogy with the forward Laplace transform is complete. This is permissible because the contribution from the arc $S'B'$ vanishes as the radius of the semicircle $A'B'C'$ becomes infinite. Since $c > v$, the branch points are of no consequence and

$$U(x, y, s) = \frac{1}{\pi s^2} \mathrm{Re} \left[\int_{E'S'} f(x, y, t) \frac{\partial w}{\partial t} e^{-st} \, dt \right] \qquad (6.3.61)$$

$$= \frac{1}{\pi s^2} \mathrm{Re} \left[\int_{y/c}^{\infty} f(x, y, t) \frac{\partial w}{\partial t} e^{-st} \, dt \right] \qquad (6.3.62)$$

$$= \frac{1}{\pi s^2} \mathrm{Re} \left[\int_0^{\infty} f(x, y, t) \frac{\partial w}{\partial t} e^{-st} H(t - y/c) \, dt \right]. \qquad (6.3.63)$$

Because the value of w along the real axis is $w = (-ixt + k)/r^2$, where $k = y\sqrt{t^2 - p^2 r^2}$, $r^2 = x^2 + y^2$ and $p = 1/c$, we can write down the inverse of Equation 6.3.63 as a convolution

$$u(x, y, t) = \frac{t}{\pi} * \frac{r^2 [2kx^2 t + y^2 t (L^2 r^4 + k^2 - x^2 t^2)/k]}{(L^2 r^4 + k^2 - x^2 t^2)^2 + 4x^2 k^2 t^2} H(t - r/c), \qquad (6.3.64)$$

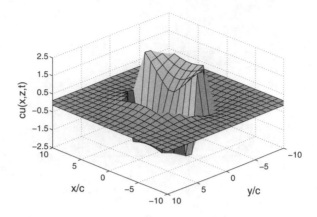

Figure 6.3.6: A plot of $u(x, y, t)$ given by Equation 6.3.66 as a function of x and y when $t = 10$.

where $L^2 = 1/v^2$, or

$$u(x,y,t) = \frac{r^2}{\pi} H(t - r/c) \int_{r/c}^t \frac{[2kx^2\eta + y^2\eta(L^2r^4 + k^2 - x^2\eta^2)/k](t-\eta)}{(L^2r^4 + k^2 - x^2\eta^2)^2 + 4x^2k^2\eta^2} \, d\eta.$$

$$(6.3.65)$$

Evaluating the integral in Equation 6.3.65,

$$u(x,y,t) = H(t - r/c)\left[\frac{vt}{2\pi} \arctan\left(\frac{2vy\sqrt{t^2 - r^2/c^2}}{r^2 + v^2y^2/c^2 - v^2t^2} \right) \right.$$ $$(6.3.66)$$

$$- \frac{y}{\pi} \ln\left(\frac{ct}{r} + \frac{\sqrt{t^2 - r^2/c^2}}{r/c} \right)$$

$$- \frac{x + y\sqrt{(v/c)^2 - 1}}{2\pi} \arctan\left(\frac{v^2xy/c^2 + r^2\sqrt{(v/c)^2 - 1}}{r^2 - v^2x^2/c^2} \frac{\sqrt{t^2 - r^2/c^2}}{t} \right)$$

$$\left. - \frac{x - y\sqrt{(v/c)^2 - 1}}{2\pi} \arctan\left(\frac{v^2xy/c^2 - r^2\sqrt{(v/c)^2 - 1}}{r^2 - v^2x^2/c^2} \frac{\sqrt{t^2 - r^2/c^2}}{t} \right) \right].$$

Figure 6.3.6 illustrates Equation 6.3.66 as a function of x and y when $t = 10$.

Problems

1. Given the Laplace transform

$$F(s) = \int_0^\infty e^{-a\sqrt{s^2+k^2}} \frac{\left[1 + \delta\sqrt{(s^2 + k^2)/(\gamma s^2 + k^2)}\right]^n s^{2n-1}}{(s^2 + \alpha^2 k^2)^n (s^2 + k^2)^{(n+1)/2}} \, k \, dk,$$

where a, α, δ and γ are real and positive and $n > 1$, use the Cagniard technique to find the inverse $f(t)$.

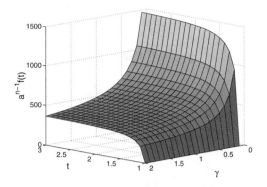

Problem 1

Step 1: Introducing $k = sz$, show that we can rewrite $F(s)$

$$F(s) = \int_0^\infty z \frac{\left[1 + \delta\sqrt{(1+z^2)/(\gamma+z^2)}\right]^n}{s^n(1+\alpha^2 z^2)^n(1+z^2)^{(n+1)/2}} e^{-as\sqrt{z^2+1}}\, dz.$$

Step 2: Introducing $a\sqrt{z^2+1} = t$, show that we can rewrite $F(s)$

$$F(s) = \frac{1}{as^n} \int_0^\infty \frac{\left(a/t + \delta a/\sqrt{\gamma a^2 - a^2 + t^2}\right)^n}{[1 + \alpha^2(t^2/a^2 - 1)]^n} H(t-a)e^{-st}\, dt.$$

Step 3: Using the convolution theorem, show that

$$f(t) = \frac{H(t-a)}{a(n-1)!} \int_a^t \frac{(t-\tau)^{n-1} \left(a/\tau + \delta a/\sqrt{\gamma a^2 - a^2 + \tau^2}\right)^n}{[1 + \alpha^2(\tau^2/a^2 - 1)]^n}\, d\tau.$$

This inverse $f(t)$ is illustrated in the figure labeled Problem 2 when $n = 2$, $\alpha = 3$ and $\delta = 1$.

2. Solve the ideal string problem[24]

$$\frac{\partial^2 u}{\partial t^2} = c^2 \frac{\partial^2 u}{\partial x^2} + \frac{F}{\rho}\delta[t - f(x)]H(x)$$

by Cagniard's method, where $u(x,t)$ is the displacement, c is the phase speed, F is the amplitude of the load, ρ is the density of the string and $f(0) = 0$. Assume $u(x,0) = u_t(x,0) = 0$, and $\lim_{|x| \to \infty} u(x,t) \to 0$.

---

[24] Reprinted from *Int. J. Solids Struct.*, **4**, F. T. Flaherty, Transient resonance of an ideal string under a load moving with varying speed, pp. 1221–1231, ©1968, with kind permission from Pergamon Press Ltd., Headington Hill Hall, Oxford OX3 0BW, UK.

Step 1: Show that the Laplace transform of the partial differential equation is

$$s^2 U(x, s) = c^2 \frac{d^2 U(x, s)}{dx^2} + \frac{F}{\rho} e^{-sf(x)} H(x).$$

Step 2: Take the Fourier transform of the ordinary differential equation in Step 1 and show that the joint transform is

$$\rho \overline{U}(k, s) = \frac{F}{s^2 + c^2 k^2} \int_0^\infty e^{-sf(\xi) - ik\xi} \, d\xi.$$

Step 3: Show that we can write Fourier inverse (after reversing the order of integration) of Step 2 as

$$\rho s U(x, s) = -\frac{F}{2\pi i} \int_0^\infty \int_{-\infty i}^{\infty i} \frac{e^{-s[f(\xi) + p(\xi - x)]}}{c^2 (p - 1/c)(p + 1/c)} \, dp \, d\xi$$

after setting $p = ik/s$.

Step 4: Use the residue theorem to evaluate the p integration by closing the contour on the left or right side of the p-plane, as dictated by Jordan's lemma. Keeping the sign of $\xi - x$ in mind, show that

$$2cps U(x, s) = F \int_0^x e^{-s[f(\xi) - (\xi - x)/c]} \, d\xi + F \int_x^\infty e^{-s[f(\xi) + (\xi - x)/c]} \, d\xi, \qquad 0 < x,$$

and

$$2cps U(x, s) = F \int_0^\infty e^{-s[f(\xi) + (\xi - x)/c]} \, d\xi, \qquad x < 0.$$

Step 5: Define the Cagniard contours $f(\xi) \pm (\xi - x)/c = t$, so that you can write the integrals in Step 4 as

$$2cps U(x, s) = F \int_{x/c}^{f(x)} \xi_1'(t) e^{-st} \, dt + F \int_{f(x)}^\infty \xi_2'(t) e^{-st} \, dt, \qquad 0 < x,$$

and

$$2cps U(x, s) = F \int_{-x/c}^\infty \xi_2'(t) e^{-st} \, dt, \qquad x < 0,$$

where ξ_1 and ξ_2 are given by $t = f(\xi_1) - (\xi_1 - x)/c$ and $t = f(\xi_2) + (\xi_2 - x)/c$, respectively.

Step 6: Obtain the inverse of $sU(x, s)$ by using the definition of the forward Laplace transform. Show that for $x > 0$,

$$2c\rho u_t(x, t) = F \begin{cases} \xi_1'(t), & \text{if} \quad x/c < t < f(x), \\ \xi_2'(t), & \text{if} \quad f(x) < t, \\ 0, & \text{otherwise,} \end{cases}$$

while for $x < 0$,

$$2c\rho u_t(x, t) = F \begin{cases} \xi_2'(t), & \text{if} \quad -x/c < t, \\ 0, & \text{otherwise.} \end{cases}$$

6.4 THE MODIFICATION OF CAGNIARD'S METHOD BY De HOOP

Although the Cagniard technique is very clever, its use in more complicated problems becomes increasingly difficult because we must use a series of transformations.[25] In 1960, De Hoop[26] suggested a modification that simplified matters by playing the Fourier transform off against the Laplace transform. It is this modification that has had the greatest acceptance.

• Example 6.4.1

To illustrate this technique,[27] consider the two-dimensional wave equation

$$\frac{\partial^2 u}{\partial x^2} + \frac{\partial^2 u}{\partial y^2} = \frac{1}{c^2}\frac{\partial^2 u}{\partial t^2} - M\delta(x)\delta(y)\delta(t), \tag{6.4.1}$$

where $-\infty < x, y < \infty$ and $0 < t$. If all of the initial conditions are zero, the transformed form of Equation 6.4.1 is

$$\frac{d^2\overline{U}(\alpha, y, s)}{dy^2} - s^2\left(\alpha^2 + \frac{1}{c^2}\right)\overline{U}(\alpha, y, s) = -M\delta(y), \tag{6.4.2}$$

where we have taken the Laplace transform with respect to time and the Fourier transform with respect to x. We have also replaced the Fourier transform parameter k with αs. The solution of Equation 6.4.2 that tends to zero as $|y| \to \infty$ is

$$\overline{U}(\alpha, y, s) = \frac{M}{2s\beta}e^{-s\beta|y|}, \tag{6.4.3}$$

where

$$\beta(\alpha) = \left(\alpha^2 + \frac{1}{c^2}\right)^{1/2} \quad \text{with} \quad \text{Re}(\beta) \geq 0. \tag{6.4.4}$$

We first invert the Fourier transform. This yields

$$U(x, y, s) = \frac{M}{2\pi}\int_{-\infty}^{\infty}\frac{e^{-s(\beta|y|-i\alpha x)}}{2\beta}\,d\alpha, \tag{6.4.5}$$

where $-\infty < x, y < \infty$. Introducing $\alpha = iw$, Equation 6.4.5 becomes

$$U(x, y, s) = \frac{M}{2\pi i}\int_{-\infty i}^{\infty i}\frac{e^{-s(wx+\beta|y|)}}{2\beta}\,dw \tag{6.4.6}$$

---

[25] Dix, C. H., 1954: The method of Cagniard in seismic pulse problems. *Geophysics*, **19**, 722–738.

[26] De Hoop, A. T., 1960: A modification of Cagniard's method for solving seismic pulse problems. *Appl. Sci. Res., Ser. B*, **8**, 349–356.

[27] Patterned after Gopalsamy, K., and B. D. Aggarwala, 1972: Propagation of disturbances from randomly moving sources. *Z. Angew. Math. Mech.*, **52**, 31–35.

with

$$\beta = \left(\frac{1}{c^2} - w^2 \right)^{1/2} \quad \text{with} \quad \text{Re}(\beta) \geq 0. \tag{6.4.7}$$

As before, we note that choosing a path of integration so that

$$wx + \beta|y| = t \tag{6.4.8}$$

is real and positive results in a forward Laplace transform. In this case, however, we must deform our integral to the new Cagniard contour with great care. There are branch points at $w = 1/c$ and $w = -1/c$. To ensure that $\text{Re}(\beta) \geq 0$, the branch cuts are taken along the real axis from $1/c$ to ∞ and from $-1/c$ to $-\infty$.

From Equation 6.4.8, a little algebra gives

$$w = \frac{xt}{r^2} \pm \frac{i|y|}{r^2} \sqrt{t^2 - \frac{r^2}{c^2}}, \qquad \frac{r}{c} < t < \infty, \tag{6.4.9}$$

where $r^2 = x^2 + y^2$. Furthermore, along this hyperbola,

$$\beta = \frac{|y|t}{r^2} \mp \frac{xi}{r^2} \sqrt{t^2 - \frac{r^2}{c^2}} \tag{6.4.10}$$

and

$$\frac{\partial w}{\partial t} = \pm \frac{i\beta}{\sqrt{t^2 - r^2/c^2}}. \tag{6.4.11}$$

Upon using the symmetry of the path of integration, transforming Equation 6.4.6 to an integration in t yields

$$U(x, y, s) = \frac{M}{2\pi} \int_{r/c}^{\infty} \frac{e^{-st}}{\sqrt{t^2 - r^2/c^2}} \, dt. \tag{6.4.12}$$

Consequently, by inspection,

$$u(x, y, t) = \frac{M}{2\pi} \frac{H(t - r/c)}{\sqrt{t^2 - r^2/c^2}}. \tag{6.4.13}$$

Equation 6.4.13 is the Green's function for the two-dimensional wave equation.

● **Example 6.4.2**

There are several ways that we can generalize the previous example. One[28] of them is to break the homogeneous domain into two dielectric half-spaces where the speed of light is c/n in Region 1 and c in Region 2. (See

---

[28] Taken from De Hoop, A. T., 1979: Pulsed electromagnetic radiation from a line source in a two-media configuration. *Radio Sci.*, **14**, 253–268. ©1979 American Geophysical Union. Reproduced/modified by permission of American Geophysical Union. See also Shirai, H., 1995: Transient scattering responses from a plane interface between dielectric half-spaces. *Electron. Comm. Japan, Part 2*, **78**, No. 9, 52–62.

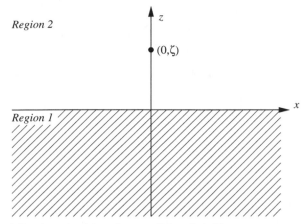

Figure 6.4.1: A line impulse lying above a dielectric half-space.

Figure 6.4.1.) We locate the impulse forcing at $x = 0$ and $z = \zeta$. The Green's functions then satisfy the equations

$$\frac{\partial^2 u_1}{\partial x^2} + \frac{\partial^2 u_1}{\partial z^2} - \frac{n^2}{c^2}\frac{\partial^2 u_1}{\partial t^2} = 0, \qquad z < 0, \tag{6.4.14}$$

and

$$\frac{\partial^2 u_2}{\partial x^2} + \frac{\partial^2 u_2}{\partial z^2} - \frac{1}{c^2}\frac{\partial^2 u_2}{\partial t^2} = -\delta(z - \zeta)\delta(x)\delta(t), \qquad 0 < z, \tag{6.4.15}$$

where x is the horizontal distance. We require at the interface $z = 0$ that

$$u_1(x, 0, t) = u_2(x, 0, t), \tag{6.4.16}$$

and

$$\frac{\partial u_1(x, 0, t)}{\partial z} = \frac{\partial u_2(x, 0, t)}{\partial z}. \tag{6.4.17}$$

At infinity, $\lim_{|z| \to \infty} |u_j(x, z, t)| < \infty$.

We solve Equation 6.4.14 through Equation 6.4.17 by the joint application of Laplace and Fourier transforms. We denote this joint transform of $u_j(x, z, t)$ by $\overline{U}_j(k, z, s)$ with $j = 1, 2$, where s and k are the transform variables of the Laplace and Fourier transforms, respectively. Assuming that the system is initially at rest,

$$\frac{d^2\overline{U}_1}{dz^2} - \nu_1^2\overline{U}_1 = 0, \qquad z < 0, \tag{6.4.18}$$

and

$$\frac{d^2\overline{U}_2}{dz^2} - \nu_2^2\overline{U}_2 = -\delta(z - \zeta), \qquad 0 < z, \tag{6.4.19}$$

with the boundary conditions

$$\overline{U}_1(k, 0, s) = \overline{U}_2(k, 0, s), \tag{6.4.20}$$

and

$$\overline{U}_1'(k, 0, s) = \overline{U}_2'(k, 0, s), \tag{6.4.21}$$

where $\nu_1 = \sqrt{k^2 + n^2 s^2/c^2}$ and $\nu_2 = \sqrt{k^2 + s^2/c^2}$.

Solutions that satisfy Equation 6.4.18 through Equation 6.4.21 are

$$\overline{U}_1(k, z, s) = \frac{2\nu_2}{\nu_1 + \nu_2} \frac{e^{-\nu_2\zeta + \nu_1 z}}{2\nu_2} \tag{6.4.22}$$

and

$$\overline{U}_2(k, z, s) = \frac{e^{-\nu_2|z-\zeta|}}{2\nu_2} + \frac{\nu_2 - \nu_1}{\nu_1 + \nu_2} \frac{e^{-\nu_2(z+\zeta)}}{2\nu_2}. \tag{6.4.23}$$

The physical interpretation of Equation 6.4.22 and Equation 6.4.23 is straightforward. Equation 6.4.22 gives the wave that has been transmitted into Region 1 from Region 2. On the other hand, Equation 6.4.23 gives the direct wave emanating from the source (the first term) and a reflected wave (the second term).

(a) Direct Wave

Let us now invert Equation 6.4.22 and Equation 6.4.23. We begin with the first term in Equation 6.4.23. If we denote this term by $\overline{U}_2^{(d)}(k, z, s)$, then the inverse Fourier transform gives

$$U_2^{(d)}(x, z, s) = \frac{1}{4\pi} \int_{-\infty}^{\infty} \frac{e^{ikx - \nu_2|z-\zeta|}}{\nu_2} \, dk. \tag{6.4.24}$$

Setting $x = r\sin(\varphi)$, $|z - \zeta| = r\cos(\varphi)$, where $r^2 = x^2 + (z - \zeta)^2$ and $k = s\sinh(w)/c$, then Equation 6.4.24 becomes

$$U_2^{(d)}(x, z, s) = \frac{1}{4\pi} \int_{-\infty}^{\infty} e^{-sr\cosh(w-\varphi i)/c} \, dw, \tag{6.4.25}$$

with the requirement that $\text{Re}[\cosh(w - \varphi i)] > 0$. Introducing $w - \varphi i = \beta$,

$$U_2^{(d)}(x, z, s) = \frac{1}{4\pi} \int_{-\infty-\varphi i}^{\infty-\varphi i} e^{-sr\cosh(\beta)/c} \, d\beta \tag{6.4.26}$$

$$= \frac{1}{4\pi} \int_{-\infty}^{\infty} e^{-sr\cosh(\beta)/c} \, d\beta \tag{6.4.27}$$

$$= \frac{1}{2\pi} \int_{0}^{\infty} e^{-sr\cosh(\beta)/c} \, d\beta \tag{6.4.28}$$

$$= \frac{1}{2\pi} \int_{r/c}^{\infty} \frac{e^{-st}}{\sqrt{t^2 - r^2/c^2}} \, dt, \tag{6.4.29}$$

where $t = r\cosh(\beta)/c$. We can deform the contour in Equation 6.4.26 to the one in Equation 6.4.27 because the integrand has no singularities.

Examining the integrand in Equation 6.4.29 closely, we note that it is the Laplace transform of the function $H(t-r/c)/\sqrt{t^2 - r^2/c^2}$. Consequently, we immediately have

$$u_2^{(d)}(x, z, t) = \frac{H(t - r/c)}{2\pi\sqrt{t^2 - r^2/c^2}}. \tag{6.4.30}$$

Equation 6.4.30 agrees with the two-dimensional, free-space Green's function that we found for a line source, Equation 6.4.13.

(b) Reflected Wave

We now turn our attention to the waves that are reflected as the result of the interface at $z = 0$. If we denote the Laplace transform of that portion of the Green's function by $U^{(r)}(x, z, s)$, it is governed by the equation

$$U_2^{(r)}(x, z, s) = \frac{1}{4\pi} \int_{-\infty}^{\infty} \left(\frac{\nu_2 - \nu_1}{\nu_2 + \nu_1}\right) \frac{e^{ikx - \nu_2(z+\zeta)}}{\nu_2} \, dk. \tag{6.4.31}$$

Using the same substitution that we used to evaluate the direct wave, we have

$$U_2^{(r)}(x, z, s) = \frac{1}{4\pi} \int_{-\infty}^{\infty} \left[\frac{\cosh(w) - \sqrt{n^2 + \sinh^2(w)}}{\cosh(w) + \sqrt{n^2 + \sinh^2(w)}}\right] e^{-s\rho \cosh(w - \varphi_r i)/c} \, dw, \tag{6.4.32}$$

where $\rho^2 = x^2 + (z+\zeta)^2$, $x = \rho\sin(\varphi_r)$ and $z + \zeta = \rho\cos(\varphi_r)$. Because the integrand of Equation 6.4.32 has branch points at $\pm\arcsin(n)i$, two branch cuts must be introduced. We take one to run from $\arcsin(n)i$ out to ∞i while the other runs from $-\arcsin(n)i$ out to $-\infty i$. We will do the analysis for $n < 1$.

When $n < 1$, we introduce $w = \beta + \varphi_r i$ so that

$$U_2^{(r)}(x, z, s) = \frac{1}{4\pi} \int_{-\infty - \varphi_r i}^{\infty - \varphi_r i} e^{-s\rho \cosh(\beta)/c} \, d\beta \tag{6.4.33}$$

$$\times \left[\frac{\cosh(\beta + \varphi_r i) - \sqrt{n^2 + \sinh^2(\beta + \varphi_r i)}}{\cosh(\beta + \varphi_r i) + \sqrt{n^2 + \sinh^2(\beta + \varphi_r i)}}\right].$$

Turning to the case $\arcsin(n) > |\varphi_r|$ first, we can still deform the contour $\beta - \varphi_r i$ to the real axis because we will not cross any branch points. Then,

$$U_2^{(r)}(x, z, s) = \frac{1}{4\pi} \int_{-\infty}^{\infty} F(\beta) e^{-s\rho \cosh(\beta)/c} \, d\beta \tag{6.4.34}$$

$$U_2^{(r)}(x, z, s) = \frac{1}{4\pi} \int_0^\infty [F(\beta) + F(-\beta)] e^{-s\rho \cosh(\beta)/c} \, d\beta \qquad (6.4.35)$$

$$= \frac{1}{4\pi} \int_{\rho/c}^\infty \{F[\cosh^{-1}(ct/\rho)] + F[-\cosh^{-1}(ct/\rho)]\}$$

$$\times \frac{e^{-st}}{\sqrt{t^2 - \rho^2/c^2}} \, dt, \qquad (6.4.36)$$

where

$$F(\beta) = \frac{\cosh(\beta + \varphi_r i) - \sqrt{n^2 + \sinh^2(\beta + \varphi_r i)}}{\cosh(\beta + \varphi_r i) + \sqrt{n^2 + \sinh^2(\beta + \varphi_r i)}}, \qquad (6.4.37)$$

$$\rho^2 = x^2 + (z + \zeta)^2, \qquad \varphi_r = \arctan\left(\frac{x}{z + \zeta}\right), \qquad (6.4.38)$$

$t = \rho \cosh(\beta)/c$, and the real part of the radical must be positive.

Because we recognize Equation 6.4.36 as the definition of the Laplace transform, we can immediately write down

$$u_2^{(r)}(x, z, t) = \frac{\{F[\cosh^{-1}(ct/\rho)] + F[-\cosh^{-1}(ct/\rho)]\}}{4\pi\sqrt{t^2 - \rho^2/c^2}} H(t - \rho/c) \quad (6.4.39)$$

$$= \text{Re}\left\{F\left[\cosh^{-1}\left(\frac{ct}{\rho}\right)\right]\right\} \frac{H(t - \rho/c)}{2\pi\sqrt{t^2 - \rho^2/c^2}}, \qquad (6.4.40)$$

since $F(-\beta) = F^*(\beta)$.

Consider now the case when $|\varphi_r| > w_b = \arcsin(n)$. Let us take the branch cut associated with the branch points $\pm(|\varphi_r| - w_b)i$ to run along the imaginary axis between the two branch points. Consequently, when we deform the original contour from $\beta - \varphi_r i$ to the real axis, we have additional contour integrals arising from integrations along the branch cut. For example, if $\varphi_r > 0$, then the branch cut integrals would be: (1) a line integral running from the real axis (and just to the left of the branch cut) down to the branch point, (2) an integration around the branch point, and (3) a line integral from the branch point (and just to the right of the branch cut) running up to the real axis. See Figure 6.4.2. Since the integration around the branch point equals zero, we have

$$u_2^{(r)}(x, z, s) = \frac{1}{4\pi} \int_{-\infty}^0 F(\beta) e^{-s\rho \cosh(\beta)/c} \, d\beta$$

$$- \frac{1}{4\pi} \int_0^{|\varphi_r| - w_b} \frac{\cos(|\varphi_r| - y) - \sqrt{n^2 - \sin^2(|\varphi_r| - y)}}{\cos(|\varphi_r| - y) + \sqrt{n^2 - \sin^2(|\varphi_r| - y)}}$$

$$\times e^{-s\rho \cos(y)/c} \, dy$$

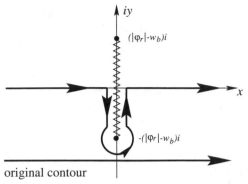

Figure 6.4.2: Schematic of the deformation of the contour used in Equation 6.4.32 to the one used in Equation 6.4.42.

$$-\frac{1}{4\pi}\int_{|\varphi_r|-w_b}^{0}\frac{\cos(|\varphi_r|-y)+\sqrt{n^2-\sin^2(|\varphi_r|-y)}}{\cos(|\varphi_r|-y)-\sqrt{n^2-\sin^2(|\varphi_r|-y)}}$$

$$\times\,e^{-s\rho\cos(y)/c}\,dy$$

$$+\frac{1}{4\pi}\int_0^\infty F(\beta)e^{-s\rho\cosh(\beta)/c}\,d\beta \qquad (6.4.41)$$

$$=\frac{1}{4\pi}\int_0^\infty [F(\beta)+F(-\beta)]e^{-s\rho\cosh(\beta)/c}\,d\beta$$

$$+\frac{1}{4\pi}\int_0^{|\varphi_n|-w_b}\frac{\cos(|\varphi_r|-y)\sqrt{n^2-\sin^2(|\varphi_r|-y)}}{1-n^2}e^{-s\rho\cos(y)/c}\,dy$$

$$(6.4.42)$$

$$=\frac{1}{4\pi}\int_{\rho/c}^\infty\left\{F[\cosh^{-1}(ct/\rho)]+F[-\cosh^{-1}(ct/\rho)]\right\}\frac{e^{-st}}{\sqrt{t^2-\rho^2/c^2}}\,dt$$

$$+\frac{1}{\pi}\int_{\rho/c}^{\rho\cos(|\varphi_r|-w_b)/c}G[|\varphi_r|-\arccos(ct/\rho)]\frac{e^{-st}}{\sqrt{\rho^2/c^2-t^2}}\,dt,$$

$$(6.4.43)$$

where

$$G(\beta)=\frac{\cos(\beta)|n^2-\sin^2(\beta)|^{1/2}}{1-n^2}. \qquad (6.4.44)$$

In particular, we recognize Equation 6.4.43 as the definition of the Laplace transform, in which case we can immediately write down

$$u_2^{(r)}(x,z,t)=\mathrm{Re}\left\{F\left[\cosh^{-1}\left(\frac{ct}{\rho}\right)\right]\right\}\frac{H(t-\rho/c)}{2\pi\sqrt{t^2-\rho^2/c^2}}$$

$$+\frac{G[|\varphi_r|-\arccos(ct/\rho)]H(|\varphi_r|-w_b)}{\pi\sqrt{\rho^2/c^2-t^2}}$$

$$\times\left\{H[t-\rho\cos(w_b-|\varphi_r|)/c]-H(t-\rho/c)\right\}. \qquad (6.4.45)$$

Let us examine Equation 6.4.45 more closely. The first term is merely the reflected wave from the interface that is felt at any point in Region 2 after the time $t = \rho/c$. The second term is new and radically different. First, it exists *before* and *disappears when* the conventional reflected wave arrives at those points where $|\varphi_r| > \arcsin(n)$; it is a *precursor* to the arrival of the reflected wave. Second, this wave occurs only for sufficiently large $|\varphi_r|$; this corresponds to large $|x|$. Consequently, these waves occur near the lateral sides of the expanding reflected wave field. For this reason, they are called *lateral waves*. In geophysics, similar waves occur due to different elastic properties within the solid earth. Because these lateral waves are recorded at the beginning or head of a seismographic record, they are often called *head waves*.

Why do these lateral or head waves exist? Recall that the case $n < 1$ corresponds to light traveling faster in Region 1 than in Region 2. Consequently, some of the wave energy that enters Region 1 near $x = 0$ can outrun its counterpart in Region 2 and then reemerge into Region 2 where it is observed as head waves. Thus, lateral waves can reach an observer sooner than the direct wave, much as a car may reach a destination sooner by taking an indirect route involving an expressway rather than a direct route that consists of residential streets: the higher speeds of the highway more than compensate for the increased distance traveled.

(c) Transmitted Wave

We complete our analysis by finding the Green's function for the waves that pass through the interface into Region 1. If $U^{(t)}(x, y, s)$ denotes the Laplace transform of the Green's function, it is found by evaluating

$$U_1^{(t)}(x, z, s) = \frac{1}{2\pi} \int_{-\infty}^{\infty} \left(\frac{\nu_2}{\nu_2 + \nu_1} \right) \frac{e^{ikx - \nu_2 \zeta + \nu_1 z}}{\nu_2} \, dk. \qquad (6.4.46)$$

At this point, we introduce the new variable p such that $k = isp$ and

$$U_1^{(t)}(x, z, s) = \frac{1}{2\pi i} \int_{-\infty i}^{\infty i} \frac{e^{-s\left(px + \zeta\sqrt{1/c^2 - p^2} - z\sqrt{n^2/c^2 - p^2}\right)}}{\sqrt{n^2/c^2 - p^2} + \sqrt{1/c^2 - p^2}} \, dp. \qquad (6.4.47)$$

We next change to the variable

$$t = px + \zeta\sqrt{\frac{1}{c^2} - p^2} - z\sqrt{\frac{n^2}{c^2} - p^2}. \qquad (6.4.48)$$

Just as we have done twice before, we would like to deform the original contour (along the imaginary axis) into a contour C along which t is real so that

$$U_1^{(t)}(x, z, s) = \frac{1}{2\pi i} \int_C \frac{e^{-st}}{\sqrt{n^2/c^2 - p^2} + \sqrt{1/c^2 - p^2}} \, dp \qquad (6.4.49)$$

$$= \frac{1}{\pi} \int_{t_{min}}^{\infty} \text{Im}\left(\frac{1}{\sqrt{n^2/c^2 - p^2} + \sqrt{1/c^2 - p^2}} \frac{dp}{dt} \right) e^{-st} \, dt, \qquad (6.4.50)$$

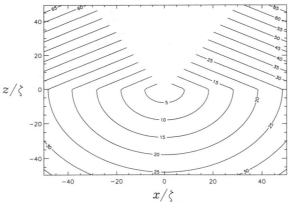

$$x/\zeta$$

Figure 6.4.3: The minimum arrival times ct_{min}/ζ for the laterally reflected [$z > 0$ and $|\varphi_r| > \arcsin(n)$] and the transmitted ($z < 0$) waves as a function of position when $n = 0.5$.

and

$$\frac{dp}{dt} = \frac{1}{x - \zeta pc/\sqrt{1 - p^2c^2} + zpc/\sqrt{n^2 - p^2c^2}}. \qquad (6.4.51)$$

What is the nature of C? From a detailed analysis of Equation 6.4.49, we discover the following properties:

• C must remain on the same Riemann surface as the original contour. We take the branch cuts along the real axis from the branch points $(\pm 1/c, 0)$ and $(\pm n/c, 0)$ out to infinity.

• For $x > 0$, the contour C lies in the first and fourth quadrants of the complex p-plane.

• For $x > 0$, the contour C crosses the real axis with $0 < p < 1/c$ or n/c, depending upon which is smaller. At this crossing, t assumes its smallest value, t_{min}, which corresponds to the minimum travel time for a transmitted wave to arrive at a point in Region 1. At that time, dp/dt is of infinite magnitude and

$$x - \frac{\zeta pc}{\sqrt{1 - p^2c^2}} + \frac{zpc}{\sqrt{n^2 - p^2c^2}} = 0. \qquad (6.4.52)$$

Combining this result with Equation 6.4.48, we have

$$\frac{ct_{min}}{\zeta} = \frac{1}{\sqrt{1 - p^2c^2}} - \frac{n^2(z/\zeta)}{\sqrt{n^2 - p^2c^2}}. \qquad (6.4.53)$$

Figure 6.4.3 gives this minimum arrival time for the reflected lateral and transmitted waves when $n = 0.5$. Because head waves have their origin in the leaking of energy from Region 1 into Region 2, the arrival times must match at the interface.

Having found t_{min}, we now find the transmitted wave solution for any time $t > t_{min}$, given x, z, ζ, c and n. To do this, we must find the corresponding value of p that is the (complex) zero of the analytic function given

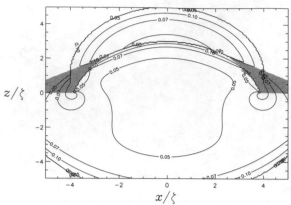

Figure 6.4.4: The Green's function $\zeta u_j(x, z, t)/c$ for various values of x/ζ and z/ζ when $ct/\zeta = 4$. The shaded region denotes that region where head waves are present.

in Equation 6.4.52. Once the value of p is found, then

$$u_1^{(t)}(x, z, t) = \frac{1}{\pi}\text{Im}\left(\frac{c}{\sqrt{n^2 - p^2 c^2} + \sqrt{1 - p^2 c^2}}\frac{dp}{dt}\right) H\left(t - t_{min}\right). \quad (\mathbf{6.4.54})$$

In Figure 6.4.4, the Green's function $\zeta u_j(x, z, t)/c$ has been plotted for various values of x and z when $ct/\zeta = 4$. The shaded region of the wave field denotes that region where head waves are present. We clearly see how these wave are at the lateral sides of the reflected wave field.

● **Example 6.4.3**

In the previous example, we explored the wave solutions created by an impulse forcing at $z = \zeta$ in a domain consisting of two half-spaces. See Figure 6.4.1. Similar considerations[29] hold with the heat equation. Here the governing equations are

$$\frac{\partial u_1}{\partial t} - a_1^2\left(\frac{\partial^2 u_1}{\partial x^2} + \frac{\partial^2 u_1}{\partial z^2}\right) = 0, \qquad z < 0, \qquad (\mathbf{6.4.55})$$

and

$$\frac{\partial u_2}{\partial t} - a_2^2\left(\frac{\partial^2 u_2}{\partial x^2} + \frac{\partial^2 u_2}{\partial z^2}\right) = \delta(z - \zeta)\delta(x)\delta(t), \qquad 0 < z, \qquad (\mathbf{6.4.56})$$

---

[29] Taken from Shendeleva, M. L., 2001: Reflection and refraction of a transient temperature field at a plane interface using Cagniard–De Hoop approach. *Phys. Review, Ser. E*, **64**, Art. no. 036612. For the case when the source lies at the interface, see Shendeleva, M. L., 2002: Temperature fields generated by impulsive interfacial heat sources. *J. Appl. Phys.*, **91**, 3444–3451.

where x is the horizontal distance, z is the distance from the interface, t is time and a_j^2 is the thermal diffusivity of layer j. We require at the interface $z = 0$ that

$$u_1(x, 0, t) = u_2(x, 0, t) \qquad (6.4.57)$$

and

$$\kappa_1 \frac{\partial u_1(x, 0, t)}{\partial z} = \kappa_2 \frac{\partial u_2(x, 0, t)}{\partial z}, \qquad (6.4.58)$$

where κ_j is the thermal conductivity of layer j. At infinity, $\lim_{|z| \to \infty} |u_j(x, z, t)| < \infty$. The initial condition is

$$u_1(x, z, 0) = u_2(x, z, 0) = 0. \qquad (6.4.59)$$

We solve Equation 6.4.55 through Equation 6.4.58 by the joint application of Laplace and Fourier transforms. We denote this joint transform of $u_j(x, z, t)$ by $\overline{U}_j(k, z, s)$ with $j = 1, 2$, where s and k are the transform variables of the Laplace and Fourier transforms, respectively. Applying the initial conditions given by Equation 6.4.59,

$$\frac{d^2 \overline{U}_1}{dz^2} - \left(k^2 + \frac{s}{a_1^2} \right) \overline{U}_1 = 0, \qquad z < 0, \qquad (6.4.60)$$

and

$$\frac{d^2 \overline{U}_2}{dz^2} - \left(k^2 + \frac{s}{a_2^2} \right) \overline{U}_2 = -\delta(z - \zeta), \qquad 0 < z, \qquad (6.4.61)$$

with the boundary conditions

$$\overline{U}_1(k, 0, s) = \overline{U}_2(k, 0, s), \qquad (6.4.62)$$

and

$$\kappa_1 \overline{U}_1'(k, 0, s) = \kappa_2 \overline{U}_2'(k, 0, s). \qquad (6.4.63)$$

Solutions that satisfy Equation 6.4.60 through Equation 6.4.63 may be written

$$U_1(x, z, s) = U_P(x, z, s) \quad \text{and} \quad U_2(x, z, s) = U_S(x, z, s) + U_R(x, z, s) \qquad (6.4.64)$$

where

$$U_P(x, z, s) = \frac{1}{4\pi a_2^2} \int_{-\infty}^{\infty} \exp\left[\sqrt{\frac{s}{a_2^2}} \left(ikx + z\sqrt{k^2 + n^2} - \zeta\sqrt{k^2 + 1} \right) \right]$$
$$\times \frac{P(k)}{\sqrt{k^2 + n^2}} \, dk, \qquad (6.4.65)$$

$$U_R(x, z, s) = \frac{1}{4\pi a_2^2} \int_{-\infty}^{\infty} \exp\left\{ \sqrt{\frac{s}{a_2^2}} \left[ikx - (z + \zeta)\sqrt{k^2 + 1} \right] \right\} \frac{R(k)}{\sqrt{k^2 + 1}} \, dk, \qquad (6.4.66)$$

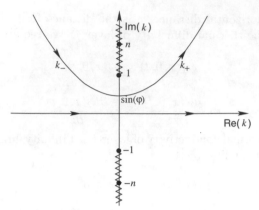

Figure 6.4.5: The Cagniard contour used to invert $U_S(x, z, s)$ and $U_R(x, z, s)$ when $n > \sin(\varphi)$.

$$U_S(x, z, s) = \frac{1}{4\pi a_2^2} \int_{-\infty}^{\infty} \exp\left[\sqrt{\frac{s}{a_2^2}}\left(ikx - |z - \varsigma|\sqrt{k^2 + 1}\right)\right] \frac{dk}{\sqrt{k^2 + 1}},$$

(6.4.67)

$$P(k) = \frac{2\sqrt{k^2 + n^2}}{\sqrt{k^2 + 1} + \chi\sqrt{k^2 + n^2}}, \qquad R(k) = \frac{\sqrt{k^2 + 1} - \chi\sqrt{k^2 + n^2}}{\sqrt{k^2 + 1} + \chi\sqrt{k^2 + n^2}},$$

(6.4.68)

$n = a_2/a_1$ and $\chi = \kappa_1/\kappa_2$. The physical interpretation of Equation 6.4.64 through Equation 6.4.68 is straightforward. The Laplace transform $U_1(x, y, s)$ gives that portion of the solution that is transmitted into Region 1 from Region 2. On the other hand, $U_S(x, z, s)$ gives the solution that would occur if there was no boundary while $U_R(x, z, s)$ yields that portion of the solution which is reflected back into Region 2.

(a) Free-Space Solution

Let us now invert Equation 6.4.65 through Equation 6.4.67. We begin with $U_S(x, z, s)$ and introduce the real parameter β such that

$$-ik|x| + |z - \varsigma|\sqrt{k^2 + 1} = r\sqrt{1 + \beta^2},$$

(6.4.69)

where $r^2 = x^2 + (z - \varsigma)^2$. Solving for k,

$$k_\pm = i\sqrt{1 + \beta^2}\sin(\varphi) \pm \beta\cos(\varphi),$$

(6.4.70)

where φ is defined by $\sin(\varphi) = |x|/r$. As β varies from 0 to ∞, $k_\pm$ describes a hyperbola in the complex k-plane. The hyperbola intersects the imaginary k-axis at the point $k = i\sin(\varphi)$ when $\beta = 0$. See Figure 6.4.5.

If we define the branch cuts associated with $\sqrt{k^2 + 1}$ as lying along the imaginary axis from $(-\infty i, -i]$ and $[i, \infty i)$, then the hyperbola does not cross any branch cuts. Furthermore, the integrand in Equation 6.4.67 has no poles.

Therefore, we can deform the contour along the real k-axis into the Cagniard hyperbola $k_\pm$. Since

$$\frac{\partial k_\pm}{\partial \beta} = \pm \frac{\sqrt{1 + k_\pm^2}}{\sqrt{1 + \beta^2}} \qquad \text{and} \qquad k_- = -k_+^*, \tag{6.4.71}$$

$$u_S(x, z, t) = \frac{1}{4\pi^2 i a_2^2} \int_{c-\infty i}^{c+\infty i} e^{tz} \left\{ \int_0^\infty \exp\left[-\frac{r}{a_2} \sqrt{z(1 + \beta^2)} \right] \frac{d\beta}{\sqrt{1 + \beta^2}} \right\} dz. \tag{6.4.72}$$

Interchanging the order of integration,

$$u_S(x, z, t) = \frac{r}{4a_2^3(\pi t)^{3/2}} \int_0^\infty \exp\left[-\frac{r^2(1 + \beta^2)}{4a_2^2 t} \right] d\beta = \frac{\exp[-r^2/(4a_2^2 t)]}{4\pi a_2^2 t}. \tag{6.4.73}$$

Equation 6.4.73 gives the free-space Green's function for the two-dimensional heat equation. It describes the diffusion in an unbounded medium of an impulse that is introduced at $(0, \zeta)$ when $t = 0$.

(b) Reflection

We now turn our attention to $U_R(x, z, s)$. Once again, we introduce a Cagniard contour, namely,

$$-ik|x| + (z + \zeta)\sqrt{k^2 + 1} = r\sqrt{1 + \beta^2}, \tag{6.4.74}$$

or

$$k_\pm = i\sqrt{1 + \beta^2} \sin(\varphi) \pm \beta \cos(\varphi), \tag{6.4.75}$$

where r^2 now equals $x^2 + (z + \zeta)^2$ and φ is defined by $\sin(\varphi) = |x|/r$. The hyperbola intersects the imaginary axis at the point $\sin(\varphi)$ when $\beta = 0$.

The integrand in Equation 6.4.66 has two pairs of branch points: $k = \pm i$ and $k = \pm in$. The most interesting case is when $n < 1$ and the heat source is located in the medium with the lower thermal diffusivity. As before, we wish to deform the integration along the real k-axis to the Cagniard contour. If $n > \sin(\varphi)$, this is permissible because we do not cross any poles or branch points. See Figure 6.4.5. Carrying out deformation,

$$u_R(x, z, t) = \frac{1}{4\pi^2 i a_2^2} \int_{c-\infty i}^{c+\infty i} e^{tz} \left\{ \int_0^\infty \exp\left[-\frac{r}{a_2} \sqrt{z(1 + \beta^2)} \right] \frac{R(k_+)}{\sqrt{1 + \beta^2}} d\beta \right.$$

$$\left. - \int_\infty^0 \exp\left[-\frac{r}{a_2} \sqrt{z(1 + \beta^2)} \right] \frac{R(k_-)}{\sqrt{1 + \beta^2}} d\beta \right\} dz, \tag{6.4.76}$$

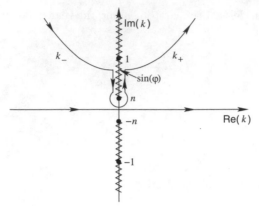

Figure 6.4.6: The Cagniard contour used to invert $U_R(x,z,s)$ when $n < \sin(\varphi) < 1$.

where $k_\pm$ is given by Equation 6.4.75. Interchanging the order of integration,

$$u_R(x,z,t) = \frac{r}{4a_2^3(\pi t)^{3/2}} \int_0^\infty \exp\left[-\frac{r^2(1+\beta^2)}{4a_2^2 t}\right] \mathrm{Re}[R(k_+)]\, d\beta. \qquad (6.4.77)$$

When $n < \sin(\varphi) < 1$, then there will be a contribution along the branch cuts from in to $i\sin(\varphi)$. See Figure 6.4.6. For this portion of the integration, we introduce $\gamma = -i\beta$. In the case of $n < 1$, Equation 6.4.77 becomes

$$u_R(x,z,t) = \frac{r}{4a_2^3(\pi t)^{3/2}} \int_0^\infty \exp\left[-\frac{r^2(1+\beta^2)}{4a_2^2 t}\right] \mathrm{Re}[R(k)]\, d\beta$$
$$- \frac{r}{4a_2^3(\pi t)^{3/2}} \int_0^{\sin(\varphi - \alpha)} \exp\left[-\frac{r^2(1-\gamma^2)}{4a_2^2 t}\right] \mathrm{Im}[R(k_-)]\, d\gamma,$$

$$(6.4.78)$$

where $k_- = i\left[\sin(\varphi)\sqrt{1-\gamma^2} - \gamma\cos(\varphi)\right]$ and $\alpha = \arcsin(n)$. The second term in Equation 6.4.78 is the contribution from the integration along the imaginary axis between in and i. Figure 6.4.7 illustrates this second term when $n = 0.5$, $\chi = 1$ and $a_2^2 t/\zeta^2 = 3$.

(c) Transmission

We complete our analysis by finding $u_P(x,z,t)$. Here, we introduce the Cagniard contour

$$-ik|x| - z\sqrt{k^2 + n^2} + \zeta\sqrt{k^2 + 1} = \tau. \qquad (6.4.79)$$

From a detailed analysis of Equation 6.4.79, we find that k is purely imaginary or complex for real τ. The smallest value of τ, $\tau_{\min}$, occurs at $k = i$ or $k = ni$

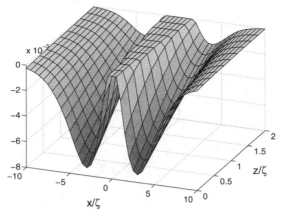

Figure 6.4.7: The value of the second term in Equation 6.4.78 for various values of x/ζ and z/ζ when $n = 0.5$, $\chi = 1$ and $a_2^2 t/\zeta^2 = 3$.

if $n < 1$. For $\tau > \tau_{min}$, k increases in magnitude as the contour moves along the imaginary axis. Eventually $\partial k/\partial \tau$ becomes infinite when

$$k = ip, \qquad |x| + \frac{pz}{\sqrt{n^2 - p^2}} - \frac{\zeta p}{\sqrt{1 - p^2}} = 0. \tag{6.4.80}$$

The value of τ at $k = ip$ is τ_0. For $\tau > \tau_0$, the value of k along the Cagniard contour becomes complex. Since this contour never crosses the branch cuts along the imaginary axis, we can deform the integration along the real k-axis to the contour involving τ. If we define $\beta = \sqrt{\tau^2/\tau_0^2 - 1}$, then

$$u_P(x, z, t) = \frac{\tau_0^2}{4a_2^3 (\pi t)^{3/2}} \int_0^\infty \beta \exp\left[-\frac{\tau_0^2(1 + \beta^2)}{4a_2^2 t}\right] d\beta \tag{6.4.81}$$

$$\times \operatorname{Re}\left[\frac{\sqrt{k^2 + 1}\, P(k)}{-i|x|\sqrt{k^2 + 1}\sqrt{k^2 + n^2} - zk\sqrt{k^2 + 1} + \zeta k \sqrt{k^2 + n^2}}\right],$$

where k is computed from Equation 6.4.79.

Figure 6.4.8 illustrates the total solution $u(r, z, t)$ when $n = 0.5$, $\chi = 1$ and $a_2^2 t/\zeta^2 = 3$.

- **Example 6.4.4**

For our next example, we find the (SH) wave motion[30] in a two layer, semi-infinite elastic medium that we subject to a stress-discontinuity of magnitude P that grows in areal coverage $a < x < a + Vt$ and occurs at the depth h from the free surface. In both layers, $i = 1$ or 2, the governing equation is

$$\frac{\partial^2 u_i}{\partial x^2} + \frac{\partial^2 u_i}{\partial z^2} = \frac{1}{V_s^2}\frac{\partial^2 u_i}{\partial t^2} \tag{6.4.82}$$

---

[30] Taken from Nag, K. R., 1962: Disturbance due to shearing-stress discontinuity in a semi-infinite elastic medium. *Geophys. J.*, **6**, 468–478.

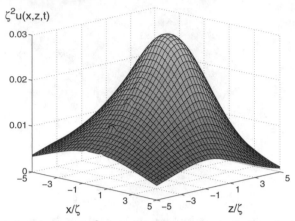

Figure 6.4.8: The total solution $\zeta^2 u(r, z, t)$ for various values of x/ζ and z/ζ when $n = 0.5$, $\chi = 1$ and $a_2^2 t/\zeta^2 = 3$.

in the domain $-\infty < x < \infty$, $-h < z < \infty$ and $0 < t$. At the interfaces, the wave solution must satisfy the boundary conditions

$$u_1(x, 0, t) = u_2(x, 0, t), \tag{6.4.83}$$

$$\frac{\partial u_1(x, -h, t)}{\partial z} = 0, \tag{6.4.84}$$

and

$$\frac{\partial u_1(x, 0, t)}{\partial z} - \frac{\partial u_2(x, 0, t)}{\partial z} = \begin{cases} P, & a \leq x \leq a + Vt, \\ 0, & \text{otherwise,} \end{cases} \tag{6.4.85}$$

for $-\infty < x < \infty$ and $0 < t$. We add the restriction that $V < V_s$.

The Laplace transform of Equation 6.4.82 yields

$$\frac{\partial^2 U_i(x, z, s)}{\partial x^2} + \frac{\partial^2 U_i(x, z, s)}{\partial z^2} - \frac{s^2}{V_s^2} U_i(x, z, s) = 0, \tag{6.4.86}$$

if the medium is initially at rest while the Fourier transform of Equation 6.4.86 gives

$$\frac{d^2 \overline{U}_i(k, z, s)}{dz^2} - \nu^2 \overline{U}_i(k, z, s) = 0, \qquad \nu = \sqrt{k^2 + s^2/V_s^2}. \tag{6.4.87}$$

The solution to Equation 6.4.87 for the two regions is

$$\overline{U}_1(k, z, s) = A_1(k, s) \cosh(\nu z) + B_1(k, s) \sinh(\nu z), \qquad -h < z < 0, \tag{6.4.88}$$

and

$$\overline{U}_2(k, z, s) = A_2(k, s) e^{-\nu z}, \qquad 0 < z, \tag{6.4.89}$$

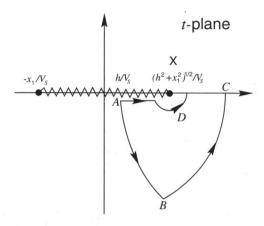

Figure 6.4.9: Contours used in the inversion of Equation 6.4.93.

with $\text{Re}(\nu) \geq 0$.

If we substitute Equation 6.4.88 and Equation 6.4.89 into the boundary conditions given by Equation 6.4.83 through Equation 6.4.85 after we have taken their Laplace and Fourier transforms, then

$$U_1(x, -h, s) = \frac{P}{2\pi s} \int_{-\infty}^{\infty} \frac{e^{\nu h - ika + ikx}}{\nu(ik + s/V)} \, dk \qquad (6.4.90)$$

$$= \frac{P}{\pi s} \int_0^{\infty} \frac{e^{-\nu h}}{\nu(ik + s/V)} \cos(kx_1) \, dk \qquad (6.4.91)$$

$$= \frac{P}{\pi s} \text{Re}\left[\int_0^{\infty} \frac{e^{ikx_1 - \nu h}}{\nu(ik + s/V)} \, dk \right], \qquad (6.4.92)$$

where $x_1 = x - a$.

Let us change the integration variable from k to $p = kV_s/s$. Then,

$$U_1(x, -h, s) = \frac{PV_s}{\pi s^2} \text{Re}\left[\int_0^{\infty} \frac{e^{-st}}{(V_s/V + ip)\sqrt{p^2 + 1}} \, dp \right], \qquad (6.4.93)$$

where $t = \left(-ipx_1 + h\sqrt{p^2 + 1} \right)/V_s$, or

$$p(t) = \frac{V_s}{x_1^2 + h^2}\left[itx_1 + h\sqrt{t^2 - (x_1^2 + h^2)/V_s^2} \right]. \qquad (6.4.94)$$

The integrand of Equation 6.4.94 has branch points at $p = \pm i$ and a simple pole at $p = iV_s/V$.

From the definition of t, we see that as p varies from 0 to ∞ in the p-plane, t starts at h/V_s, moves into the fourth quadrant of the t-plane and approaches the limit $t \to phe^{-\theta i}/V_s$ as $p \to \infty$, where $\theta = \arctan(x_1/h)$. Figure 6.4.9 illustrates this new contour by the curve AB. Because there are no poles in the fourth quadrant, we may deform the contour from AB to the real axis. Hence,

$$\int_0^\infty (\ ) \, dp = \int_A^C (\ ) \frac{dp}{dt} \, dt = \int_{h/V_s}^\infty (\ ) \frac{dp}{dt} \, dt, \qquad (6.4.95)$$

or

$$U_1(x, -h, s) = \frac{PV_s}{\pi s^2} \operatorname{Re} \left\{ \int_{h/V_s}^\infty F_1[p(t)] \frac{dp}{dt} e^{-st} \, dt \right\}, \qquad (6.4.96)$$

where

$$F_1[p(t)] = \frac{1}{[ip(t) + V_s/V] \sqrt{p(t)^2 + 1}}. \qquad (6.4.97)$$

Since $F_1[p(t)]$ is real and dp/dt is imaginary for $h/V_s < t < \sqrt{x_1^2 + h^2}/V_s$, we can replace the lower limit of h/V_s by $\sqrt{x_1^2 + h^2}/V_s$ so that

$$U_1(x, -h, s) = \frac{PV_s}{\pi s^2} \operatorname{Re} \left\{ \int_{\sqrt{x_1^2 + h^2}/V_s}^\infty F_1[p(t)] \frac{dp}{dt} e^{-st} \, dt \right\}. \qquad (6.4.98)$$

We recognize that $u_1(x, -h, t)$ equals the convolution of $tH(t)$ with a function $g(x, t)$ whose Laplace transform $G(x, s)$ is

$$G(x, s) = \operatorname{Re} \left\{ \int_{\sqrt{x_1^2 + h^2}/V_s}^\infty F_1[p(t)] \frac{dp}{dt} e^{-st} \, dt \right\}. \qquad (6.4.99)$$

Because Equation 6.4.99 is in the form of a forward Laplace transform, we immediately have

$$g[x, p(t)] = \operatorname{Re} \left\{ F_1[p(t)] \frac{dp}{dt} \right\} H\left(t - \sqrt{x_1^2 + h^2}/V_s \right) \qquad (6.4.100)$$

$$= \frac{(x_1^2 + h^2)(h^2 + x_1^2 - Vtx_1)}{\sqrt{t^2 - (x_1^2 + h^2)/V_s^2}} \qquad (6.4.101)$$

$$\times \frac{H\left(t - \sqrt{x_1^2 + h^2}/V_s \right)}{\{(h^2 + x_1^2 - Vtx_1)^2 + V^2 h^2 [t^2 - (x_1^2 + h^2)/V_s^2]\}}$$

and

$$u_1(x, -h, t) = \frac{PV}{\pi} \int_{\sqrt{x_1^2 + h^2}/V_s}^t (t - \lambda) g[x, p(\lambda)] \, d\lambda. \qquad (6.4.102)$$

Figure 6.4.10 illustrates Equation 6.4.102 when $h/V_s = 2$ and $V/V_s = 0.4$.

Problems

1. Using Sommerfeld's integral given by Equation 3.2.35, Wiggins and Helmberger[31] showed that

$$U(x, z, s) = \frac{e^{-sr/c}}{r} = \int_0^\infty J_0(kx) \frac{\exp\left(-z\sqrt{k^2 + s^2/c^2}\right)}{\sqrt{k^2 + s^2/c^2}} \, k \, dk,$$

---

[31] Wiggins, R. A., and D. V. Helmberger, 1974: Synthetic seismogram computation by expansion in generalized rays. *Geophys. J. R. Astr. Soc.*, **37**, 73–90.

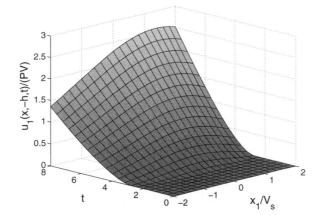

Figure 6.4.10: A plot of Equation 6.4.102 when $h/V_s = 2$ and $V/V_s = 0.4$.

where $r^2 = x^2 + z^2$. Let us use the Cagniard–De Hoop method to find a representation of $u(x, z, t) = \delta(t - r/c)$ for large positive x and $z > 0$.

Step 1: By changing variables via $k = -isp$, convert the original transform to

$$U(x, z, s) = \frac{2s}{\pi} \, \text{Im}\left[\int_0^{\infty i} \frac{K_0(spx)e^{-s\eta z}}{\eta} p \, dp \right],$$

where $\eta = \sqrt{1/c^2 - p^2}$. Figure 6.4.11 shows the branch cuts for η.

Step 2: Using the leading term from the asymptotic expansion for $K_0(\cdot)$, show that for $x \gg 1$

$$U(x, z, s) = \left(\frac{2s}{\pi x} \right)^{1/2} \text{Im}\left[\int_0^{\infty i} \frac{\sqrt{p}}{\eta} e^{-s(px + \tau)} \, dp \right],$$

where $\tau = z\eta$.

Step 3: By setting $t = px + \tau$, we can deform the original p contour into one where t is real, as shown in Figure 6.4.11. Show then that

$$U(x, z, s) = \left(\frac{2s}{\pi x} \right)^{1/2} \int_0^{\infty} \text{Im}\left[\frac{\sqrt{p(t)}}{\eta(t)} \frac{dp}{dt} \right] e^{-st} \, dt,$$

or

$$u(x, z, t) = \delta(t - r/c) = \left(\frac{2}{\pi x} \right)^{1/2} \frac{d}{dt}\left[\frac{H(t)}{\sqrt{t}} \right] * f(t),$$

where

$$f(t) = \text{Im}\left[\frac{\sqrt{p(t)}}{\eta(t)} \frac{dp}{dt} \right], \qquad p(t) = \begin{cases} \dfrac{xt}{r^2} - \dfrac{z\sqrt{r^2/c^2 - t^2}}{r^2}, & \text{if} \quad t \leq r/c, \\[4mm] \dfrac{xt}{r^2} + \dfrac{iz\sqrt{t^2 - r^2/c^2}}{r^2}, & \text{if} \quad t \geq r/c. \end{cases}$$

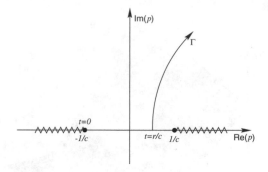

Figure 6.4.11: Contour arising in the use of Cagniard–De Hoop to solve Problem 1.

2. Consider the following system of diffusion equations:[32]

$$\frac{\partial u_1}{\partial t} = \frac{\partial^2 u_1}{\partial x^2} + \frac{\partial^2 u_1}{\partial z^2}, \qquad -\infty < x < \infty, \quad 0 < z < \infty, \quad 0 < t,$$

and

$$\frac{\partial u_2}{\partial t} = \frac{\partial^2 u_2}{\partial x^2} + \frac{\partial^2 u_2}{\partial z^2}, \qquad -\infty < x < \infty, \quad -\infty < z < 0, \quad 0 < t,$$

with the boundary conditions

$$\lim_{z \to \infty} u_1(x, z, t) \to 0, \qquad \lim_{z \to -\infty} u_2(x, z, t) \to 0,$$

$$u_1(x, 0^+, t) = u_2(x, 0^-, t), \quad \text{and} \quad \frac{\partial u_1(x, 0^+, t)}{\partial z} - \frac{\partial u_2(x, 0^-, t)}{\partial z} = \delta(x)\delta(t),$$

and the initial conditions $u_1(x, z, 0) = u_2(x, z, 0) = 0$. Assume $a_2 > a_1$.

Step 1: By defining the Laplace and Fourier transforms as follows:

$$U_i(x, z, s) = \int_0^\infty u_i(x, z, t)e^{-st}\, dt \quad \text{and} \quad \overline{U}_i(k, z, s) = \int_{-\infty}^\infty U_i(x, z, s)e^{-i\sqrt{s}kx}\, dx,$$

show that the partial differential equations and boundary conditions become

$$\frac{d^2 \overline{U}_1(k, z, s)}{dz^2} - s\left(k^2 - \frac{1}{a_1^2}\right)\overline{U}_1(k, z, s) = 0, \qquad 0 < z < \infty,$$

---

[32] Patterned after De Hoop, A. T., and M. L. Oristaglio, 1988: Application of the modified Cagniard technique to transient electromagnetic diffusion problems. *Geophys. J.*, **94**, 387–397.

and

$$\frac{d^2\overline{U}_2(k,z,s)}{dz^2} - s\left(k^2 - \frac{1}{a_2^2}\right)\overline{U}_2(k,z,s) = 0, \qquad -\infty < z < 0,$$

with

$$\lim_{z\to\infty}\overline{U}_1(k,z,s) \to 0, \qquad \lim_{z\to-\infty}\overline{U}_2(k,z,s) \to 0,$$

$$U_1(k,0^+,s) = U_2(k,0^-,s), \quad \text{and} \quad U_1'(k,0^+,s) - U_2'(k,0^-,s) = 1.$$

Step 2: Show that the solutions to Step 1 are

$$\overline{U}_1(k,z,s) = -\frac{\exp\left(-\sqrt{s}\gamma_1 z\right)}{\sqrt{s}\,(\gamma_1 + \gamma_2)}, \qquad 0 < z < \infty,$$

and

$$\overline{U}_2(k,z,s) = -\frac{\exp\left(\sqrt{s}\gamma_2 z\right)}{\sqrt{s}\,(\gamma_1 + \gamma_2)}, \qquad -\infty < z < 0,$$

where $\gamma_i = \sqrt{k^2 + 1/a_i^2}$ with $\text{Re}(\gamma_i) > 0$.

Step 3: Show that

$$U_1(x,y,s) = \frac{\sqrt{s}}{2\pi}\int_{-\infty}^{\infty}\overline{U}_1(k,z,s)e^{i\sqrt{s}kx}\,dk = -\frac{1}{2\pi}\int_{-\infty}^{\infty}\frac{\exp\left[-\sqrt{s}(-ikx + \gamma_1 z)\right]}{\gamma_1 + \gamma_2}\,dk$$

$$= -\frac{1}{2\pi i}\int_{-\infty i}^{\infty i}\frac{\exp\left[-\sqrt{s}(px + \overline{\gamma}_1 z)\right]}{\overline{\gamma}_1 + \overline{\gamma}_2}\,dp,$$

where $\overline{\gamma}_i = \sqrt{1/a_i^2 - p^2}$.

Step 4: Let us choose the branch cut of $\overline{\gamma}_i$ to lie along the real axis of the p-plane between the branch points $(-\infty, -1/a_i]$ and $[1/a_i, \infty)$. Then $\text{Re}(\gamma_i) > 0$. Next, introduce the real, positive variable $\kappa = px + \overline{\gamma}_1|z|$. For $z > 0$,

$$p_1 = \frac{\kappa x}{r^2} + \frac{iz}{r^2}\sqrt{\kappa^2 - \kappa_1^2}, \qquad \kappa_1 < \kappa < \infty,$$

where $r = \sqrt{x^2 + z^2}$ and $\kappa_1 = r/a_1$. Show that we can deform the integration in Step 3 to this new path, yielding

$$U_1(x,z,s) = -\frac{1}{\pi}\int_{\kappa_1}^{\infty}\text{Im}\left(\frac{\partial_\kappa p_1}{\overline{\gamma}_1 + \overline{\gamma}_2}\right)e^{-\sqrt{s}\kappa}\,d\kappa.$$

See Example 5.1.4.

Step 5: Finish the problem by noting that the only quantity in the integral in Step 4 that contains s is $e^{-\kappa\sqrt{s}}$. Therefore, by inspection,

$$u_1(x,z,t) = -\frac{1}{\pi}\int_{\kappa_1}^{\infty}\text{Im}\left(\frac{\partial_\kappa p_1}{\overline{\gamma}_1 + \overline{\gamma}_2}\right)g(t,\kappa)\,d\kappa,$$

where

$$\partial_\kappa p_1 = \frac{i\overline{\gamma}_1}{\sqrt{\kappa^2 - \kappa_1^2}} \quad \text{and} \quad g(t,\kappa) = \frac{\kappa}{\sqrt{4\pi t^3}}e^{-\kappa^2/(4t)}H(t).$$

This solution is illustrated in the figure labeled Problem 2 when $t = 2$, $a_1 = 1$ and $a_2 = 2$.

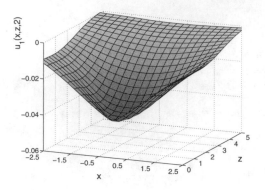

Problem 2

Papers Using the Joint Transform Techniques

Aquifers, Reservoirs and Porous Media

Boulton, N. S., and T. D. Streltsova, 1978: Unsteady flow to a pumped well in a fissured aquifer with a free surface level maintained constant. *Water Resour. Res.*, **14**, 527–532.

Chrysikopoulos, C. V., P. K. Kitanidis, and P. V. Roberts, 1990: Analysis of one-dimensional solute transport through porous media with spatially variable retardation factor. *Water Resour. Res.*, **26**, 437–446.

Chrysikopoulos, C. V., E. A. Voudrias, and M. M. Fyrillas, 1994: Modeling of contaminant transport resulting from dissolution of nonaqueous phase liquid pools in saturated porous media. *Transport Porous Media*, **16**, 125–145.

Cooley, R. L., and C. M. Case, 1973: Effect of a water table aquitard on drawdown in an underlying pumped aquifer. *Water Resour. Res.*, **9**, 434–447.

Durrant, A. J., and R. K. M. Thambynayagam, 1986: Wellbore heat transmission and pressure drop for steam/water injection and geothermal production: A simple solution technique. *SPE Reservoir Eng.*, **1**, 148–162.

Goode, P. A., and R. K. M. Thambynayagam, 1987: Pressure drawdown and buildup analysis of horizontal wells in anisotropic media. *SPE Form. Eval.*, **2**, 683–697.

Halepaska, J. C., 1972: Drawdown distribution around wells partially penetrating thick leaky artesian aquifers. *Water Resour. Res.*, **8**, 1332–1337.

Hantush, M. S., 1967: Flow of groundwater in relatively thick leaky aquifers. *Water Resour. Res.*, **3**, 583–590.

Hantush, M. S., 1967: Flow of wells in aquifers separated by a semipervious layer. *J. Geophys. Res.*, **72**, 1709–1720.

Javandel, I., and P. A. Witherspoon, 1983: Analytical solution of a partially penetrating well in a two-layer aquifer. *Water Resour. Res.*, **19**, 567–578.

Leij, F. J., T. H. Skaggs, and M. Th. van Genuchten, 1991: Analytical solutions for solute transport in three-dimensional semi-infinite porous media. *Water Resour. Res.*, **27**, 2719–2733.

Mohamed, F. A., R. B. Guenther, R. T. Hudspeth, and W. G. McDougal, 1992: A damped wave equation for groundwater flow. *Adv. Water Resour.*, **15**, 117–123.

Neuman, S. P., and P. A. Witherspoon, 1969: Theory of flow in a confined two aquifer system. *Water Resour. Res.*, **5**, 803–816.

Neuman, S. P., 1972: Theory of flow in unconfined aquifers considering delayed response of the water table. *Water Resour. Res.*, **8**, 1031–1045.

Neuman, S. P., 1974: Effect of partial penetration in flow in unconfined aquifers considering delayed gravity response. *Water Resour. Res.*, **10**, 303–312.

Russell, D. G., and M. Prats, 1962: Performance of layered reservoirs with crossflow – Single-compressible-fluid case. *Soc. Pet. Eng. J.*, **2**, 53–67.

Sim, Y., and C. V. Chrysikopoulos, 1999: Analytical solutions for solute transport in saturated porous media with semi-infinite or finite thickness. *Adv. Water Resour.*, **22**, 507–519.

Yaxley, L. M., 1987: Effect of a partially communicating fault on transient pressure behavior. *SPE Form. Eval.*, **2**, 590–598.

Diffusion

Han, C. D., 1970: The effect of radial diffusion on the performance of a liquid-liquid displacement process. *Appl. Sci. Res.*, **22**, 223–238.

Roulet, C., 1972: Solution de l'équation de Fick avec source appliquée à la diffusion d'adatomes d'argent sur le cuivre. *Z. Angew. Math. Phys.*, **23**, 412–419.

Electromagnetism

Chen, K. C., 1987: Transient response of an infinite cylindrical antenna is a dissipative medium. *IEEE Trans. Antennas Propag.*, **AP-35**, 562–573.

Geronimi, C., F. Bouchut, M. R. Feix, H. Ghalila, M. Valentini, and J. M. Buzzi, 1997: Transient electromagnetic field. *Eur. J. Phys.*, **18**, 102–107.

Pekeris, C. L., and Z. Alterman, 1957: Radiation resulting from an impulsive current in a vertical antenna placed on a dielectric ground. *J. Appl. Phys.*, **28**, 1317–1323.

Poritsky, H., 1955: Propagation of transient fields from dipoles near the ground. *Brit. J. Appl. Phys.*, **6**, 421–426.

Fluid Dynamics

Ashpis, D. E., and E. Reshotko, 1990: The vibrating ribbon problem revisited. *J. Fluid Mech.*, **213**, 531–547.

Bhattacharyya, P., 1968: Note on the rotatory motion set up in a semi-infinite viscous fluid by a certain impulsive velocity prescribed over a circular area on the surface. *Rev. Roum. Sci. Tech., Ser. Mec. Appl.*, **13**, 259–263.

Bukreev, V. I., A. V. Gusev, and I. V. Sturova, 1983: Unsteady motion of a circular cylinder in a two-layer liquid. *J. Appl. Mech. Tech. Phys.*, **24**, 856–861.

Carrier, G. F., and H. P. Greenspan, 1958: Water waves of finite amplitude on a sloping beach. *J. Fluid Mech.*, **4**, 97–109.

Chaudhuri, K., 1975: Effect of viscosity on the surface waves produced by an explosion above a liquid. *Indian J. Pure Appl. Math.*, **6**, 49–68.

Das, D. K., 1977: Motion of a viscous incompressible liquid due to a surface pressure. *Rev. Roum. Sci. Tech., Ser. Mec. Appl.*, **22**, 819–828.

Greenspan, H. P., 1956: The generation of edge waves by moving pressure distributions. *J. Fluid Mech.*, **1**, 574–592.

Gupta, S. C., 1979: Unsteady flow of a dusty gas in a channel whose cross-section is an annular section. *Indian J. Pure Appl. Math.*, **10**, 704–714.

Hultgren, L. S., and L. H. Gustavsson, 1981: Algebraic growth of disturbances in a laminar boundary layer. *Phys. Fluids*, **24**, 1000–1004.

Imberger, J., and C. Fandry, 1975: Withdrawal of a stratified fluid from a vertical two-dimensional duct. *J. Fluid Mech.*, **70**, 321–332.

Jones, J. R., and T. S. Walters, 1966: Oscillatory motion of an elastico-viscous liquid contained in a cylindrical cup. I. Theoretical. *Brit. J. Appl. Phys.*, **17**, 937–944.

Joshi, B. K., and R. S. Sharma, 1976: Unsteady flow of visco-elastic fluid through long circular ducts. *Indian J. Pure Appl. Math.*, **7**, 1081–1090.

Kanwal, R. P., 1959: Impulsive rotatory motion of a circular disk in a viscous fluid. *Z. Angew. Math. Phys.*, **10**, 552–557.

Kumar, R., 1961: Propagation of small disturbances in viscous liquid contained between two infinite co-axial cylinders. *Proc. Natl. Inst. Sci. India*, **A27**, 178–184.

MacDonald, D. A., 1966: The Rayleigh problem for two-layer fluid. *Quart. J. Mech. Appl. Math.*, **19**, 198–215.

Menikoff, R., R. C. Mjolsness, D. H. Sharp, and C. Zemach, 1978: Initial value problem for Rayleigh-Taylor instability of viscous fluids. *Phys. Fluids*, **21**, 1674–1687.

Miles, J. W., 1958: On the disturbed motion of a plane vortex sheet. *J. Fluid Mech.*, **4**, 538–552.

Miles, J. W., 1968: The Cauchy-Poisson problem for a viscous liquid. *J. Fluid Mech.*, **34**, 359–370.

Milgram, J. H., 1969: The motion of a fluid in a cylindrical container with a free surface following vertical impact. *J. Fluid Mech.*, **37**, 435–448.

Pramanik, A. K., 1972: The effect of viscosity on the waves due to a moving oscillating surface pressure. *Z. Angew. Math. Phys.*, **23**, 85–95.

Raichenko, L. M., 1977: Fluid inflow to an imperfect borehole in an inhomogeneous medium. *Sov. Appl. Mech.*, **13**, 937–942.

Rawat, M. L., 1970: Flow of viscous incompressible fluids through a tube with sector of a circle as cross-section. *Bull. Calcutta Math. Soc.*, **62**, 149–156.

Roy, P. C., 1978: Unsteady flow of a Maxwell fluid past a flat plate. *Indian J. Pure Appl. Math.*, **9**, 157–166.

Trogdon, S. A., and M. T. Farmer, 1991: Unsteady axisymmetric creeping flow from an orifice. *Acta Mech.*, **88**, 61–75.

General

Estrin, T. A., and T. J. Higgins, 1951: The solution of boundary value problems by multiple Laplace transformations. *J. Franklin Inst.*, **252**, 153–167.

Jaeger, J. C., 1940: The solution of boundary value problems by a double Laplace transformation. *Bull. Am. Math. Soc.*, **46**, 687–693.

Geophysical Sciences

Cherkesov, L. V., 1966: The influence of viscosity on the propagation of tsunami type waves. *Izv. Acad. Sci. USSR, Atmos. Ocean. Phys.*, **2**, 793–797.

Cherkesov, L. V., 1970: Influence of viscosity on tsunami waves in a basin of variable depth. *Izv. Acad. Sci. USSR, Atmos. Ocean. Phys.*, **6**, 46–48.

Cole, J. D., and C. Greifinger, 1969: Acoustic-gravity waves from an energy source at the ground in an isothermal atmosphere. *J. Geophys. Res.*, **74**, 3693–3703.

Das Gupta, S. C., 1959: On coda waves of earthquakes. *Geofis. Pura Appl.*, **43**, 45–74.

Duffy, D. G., 1990: Geostrophic adjustment in a baroclinic atmosphere. *J. Atmos. Sci.*, **47**, 457–473.

Duffy, D. G., 1992: On the generation of oceanic surface waves by underwater volcanic explosions. *J. Volcanol. Geotherm. Res.*, **50**, 323–344.

Gilbert, F., and L. Knopoff, 1960: Seismic scattering from topographic irregularities. *J. Geophys. Res.*, **65**, 3437–3444.

Grigor'ev, G. I., N. G. Denisov, and O. N. Savina, 1987: Emission of acoustic-gravity waves and a Lamb surface wave in an isothermal atmosphere. *Radiophys. Quantum Electron.*, **30**, 207–212.

Grigor'ev, G. I., N. G. Denisov, and O. N. Savina, 1989: Transition radiation of acoustic waves by sources moving in the atmosphere above the earth's surface. *Radiophys. Quantum Electron.*, **32**, 110–115.

Hibiya, T., 1986: Generation mechanism of internal waves by tidal flow over a sill. *J. Geophys. Res.*, **91**, 7697–7708.

Kamenkovich, V. M., 1989: Development of Rossby waves generated by localized effects. *Oceanology*, **29**, 1–11.

Lowndes, J. S., 1957: A transient magnetic dipole source above a two-layer earth. *Quart. J. Mech. Appl. Math.*, **10**, 79–89.

Miyoshi, H., 1954: Generation of the tsunami in compressible water. *J. Oceanogr. Soc. Japan*, **10**, 1–9.

Miyoshi, H., 1955: Generation of the tsunami in compressible water (Part II). *Rec. Oceanogr. Works Japan*, **2**, 49–56.

Moore, D. W., R. C. Kloosterziel, and W. S. Kessler, 1998: Evolution of mixed Rossby gravity waves. *J. Geophys. Res.*, **103**, 5331–5346.

Oi, M., 1976: Some theoretical considerations upon the transient topographical perturbation of a zonal current. *J. Geophys. Res.*, **81**, 1084–1094.

Pavlov, V. A., 1979: Effect of earthquakes and volcanic eruptions on the ionospheric plasma. *Radiophys. Quantum Electron.*, **22**, 10–23.

Pekeris, C. L., 1955: The seismic surface pulse. *Proc. Natl. Acad. Sci.*, **41**, 469–480.

Pekeris, C. L., 1955: The seismic buried pulse. *Proc. Natl. Acad. Sci.*, **41**, 629–639.

Phinney, R. A., 1961: Leaky modes in the crustal waveguide. Part 1. The oceanic PL waves. *J. Geophys. Res.*, **66**, 1445–1469.

Thomson, R. E., 1970: On the generation of Kelvin-type waves by atmospheric disturbances. *J. Fluid Mech.*, **42**, 657–670.

Voyt, S. S., 1964: Integration of equations for the tides in one case of unstabilized motion. *Dokl. Acad. Sci. USSR, Earth Sci. Section*, **144**, 1–3.

Heat Conduction

Burku, A. L., 1968: Unsteady radiative heating of a cylindrical body. *J. Appl. Mech. Tech. Phys.*, **9**, 184–186.

Chu, S. C., and S. G. Bankoff, 1965: Unsteady heat transfer to slug flows: Effect of axial conduction. *AICHE J.*, **11**, 607–612.

Ganguly, M., 1986: Propagation of disturbances in a thin infinite piezoelectric rod due to a randomly moving heat source acting for a finite interval. *Proc. Indian Natl. Sci. Acad., Part A*, **52**, 1218–1236.

Gupta, P. M., and V. P. Saxena, 1971: Heat conduction in a moving anisotropic rectangular slab. *Indian J. Pure Appl. Math.*, **2**, 290–296.

Hector, L. G., W.-S. Kim, and M. N. Özisik, 1992: Propagation and reflection of thermal waves in a finite medium due to axisymmetric surface sources. *Int. J. Heat Mass Transfer*, **35**, 897–912.

Kartashov, E. M., G. M. Bartenev, and A. B. Sidorov, 1967: A certain problem of the thermal conductivity of a system of cylindrical tubes. *Sov. Phys. J.*, **10, No. 12**, 16–22.

Ölçer, N. Y., 1965: On the solar heating of rotating space vehicles. *Acta Mech.*, **1**, 148–170.

Sabherwal, K. C., 1965: An inverse problem in transient heat conduction. *Indian J. Pure Appl. Phys.*, **3**, 397–398.

Spence, D. A., and D. B. R. Kenning, 1978: Unsteady heat conduction in a quarter plate, with an application to bubble growth models. *Int. J. Heat Mass Transfer*, **21**, 719–724.

Vedavarz, A., K. Mitra, and S. Kumar, 1994: Hyperbolic temperature profiles for laser surface interactions. *J. Appl. Phys.*, **76**, 5014–5021.

Magnetohydrodynamics

Chee-Seng, L., 1974: Unsteady current-induced perturbation of a magnetically contained magnetohydrodynamic flow. *J. Fluid Mech.*, **63**, 273–299.

Das, K. P., 1971: Hydromagnetic disturbance produced by a momentary current in a semi-infinite fluid of finite electrical conductivity. *Indian J. Theoret. Phys.*, **19**, 131–139.

Eraslan, A. H., 1967: Oscillatory characteristics of unsteady MHD channel flows under Heaviside-type applied magnetic fields. *J. Appl. Mech.*, **34**, 854–859.

Gupta, P. C., 1983: Unsteady magneto-hydrodynamic flow through porous media in a channel whose cross-section is a circular section. *Bull. Calcutta Math. Soc.*, **75**, 264–270.

Jaggi, R. K., 1967: Intensification of magnetic field in plasmas. *Phys. Fluids*, **10**, 648–651.

Mohandis, M. G. S. el, 1969: Magnetohydrodynamic disturbances in the earth's core. I, II, III. *Pure Appl. Geophys.*, **72**, 155–192.

Mohandis, M. G. S. el,1969: Magnetohydrodynamic disturbances in the earth's core. IV. *Pure Appl. Geophys.*, **74**, 45–56.

Sanyal, D. C., and S. K. Samanta, 1989: Unsteady motion of a semi-infinite and conducting liquid by a suddenly applied velocity on its surface. *Indian J. Pure Appl. Math.*, **20**, 1146–1156.

Sengupta, P. R., and J. Raymahrpatra, 1975: On the motion set up in a semi-infinite viscous conducting fluid. *Bull. Calcutta Math. Soc.*, **67**, 29–35.

Other

Aoki, K., 1990: Theory of current distribution at a conical electrode under diffusion control with time dependence. *J. Electroanal. Chem.*, **281**, 29–40.

Comstock, W. D., and E. M. Williams, 1959: Current distribution in the cylindrical source plane-electrode configuration. *AIEE Trans., Part 1*, **78**, 252–256.

Solid Mechanics

Aggarwal, H. R., and C. M. Ablow, 1965: Disturbance from a circularly symmetric load spreading over an acoustic half-space. *Bull. Seism. Soc. Am.*, **55**, 673–691.

Ben-Menahem, A., and A. G. Sena, 1990: Seismic source theory in stratified anisotropic media. *J. Geophys. Res.*, **95**, 15395–15427.

Bo, J., 1998: Elastic halfspace under impulsive, distributed, vertical loading at the surface: Exact solution at the center for a punch-like distribution. *Soil Dynam. Earthq. Eng.*, **17**, 311–315.

Chakraborty, M., 1985: Disturbance of SH type due to body forces and due to shearing stress-discontinuity in a pre-stressed semi-infinite viscoelastic medium. *Indian J. Pure Appl. Math.*, **16**, 309–322.

Chao, C. C., 1960: Dynamical response of an elastic half-space to tangential surface loadings. *J. Appl. Mech.*, **27**, 559–567.

Chao, C. C., H. H. Bleich, and J. Sackman, 1961: Surface waves in an elastic half space. *J. Appl. Mech.*, **28**, 300–301.

Derski, W., 1960: On transient thermal stresses in an infinite thin plate. *Proc. Edinburgh Math. Soc., Ser. 2*, **12**, 69–73.

Duffy, D. G., 1990: The response of an infinite railroad track to a moving, vibrating mass. *J. Appl. Mech.*, **57**, 66–73.

Eason, G., J. Fulton, and I. N. Sneddon, 1956: The generation of waves in an infinite elastic solid by variable body forces. *Philos. Trans. R. Soc. London, Ser. A*, **248**, 575–607.

Eason, G., 1964: On the torsional impulsive loading of an elastic half space. *Quart. J. Mech. Appl. Math.*, **17**, 279–292.

Eason, G., 1965: The stresses produced in a semi-infinite solid by a moving surface force. *Int. J. Eng. Sci.*, **2**, 581–609.

Eason, G., 1966: The displacements produced in an elastic half-space by a suddenly applied surface force. *J. Inst. Math. Appl.*, **2**, 299–326.

Filippov, I. G., and K. N. Smamutov, 1975: Axisymmetric nonstationary problem for an anisotropic inhomogeneous half-space and a layer. *Sov. Appl. Mech.*, **11**, 614–618.

Flinn, E. A., 1961: Exact transient solution to some elementary problems of elastic wave propagation. *J. Acoust. Soc. Am.*, **33**, 623–627.

Gerdes, K. F., and R. L. Bell, 1986: Transport parameters in two-dimensional dispersive-convective systems from temporal moments of pulsed tracer experiments. *Chem. Eng. Sci.*, **41**, 1699–1709.

Ghosh, M., and M. L. Ghosh, 1985: Harmonic rocking of a rigid strip on a semi-infinite elastic medium. *Indian J. Pure Appl. Math.*, **16**, 938–955.

Ghosh, S. S., 1969: On the disturbances in a thin elastic circular plate resting on a viscoelastic foundation of Pasternak type. *Pure Appl. Geophys.*, **75**, 88–92.

Gilbert, F., and L. Knopoff, 1959: Scattering of impulsive elastic waves by a rigid cylinder. *J. Acoust. Soc. Am.*, **31**, 1169–1175.

Gilbert, F., 1960: Scattering of impulsive elastic waves by a smooth convex cylinder. *J. Acoust. Soc. Am.*, **32**, 841–857.

Guan, F., and M. Novak, 1994: Transient response of an elastic homogeneous half-space to suddenly applied rectangular loading. *J. Appl. Mech.*, **61**, 256–263.

Guan, F., I. D. Moore, and C. C. Spyrakos, 1998: Two dimensional transient fundamental solution due to suddenly applied load in a half-space. *Soil Dynam. Earthq. Eng.*, **17**, 269–277.

Hudson, J. A., 1963: SH waves in a wedge-shaped medium. *Geophys. J. Astr. R. Soc.*, **7**, 517–546.

Jain, N. K., 1967: Torsional damped vibrations of an elastic half-space subjected to an impulsive radial stress distribution. *Indian J. Pure Appl. Phys.*, **5**, 599–601.

Johnson, E. R., and R. Parnes, 1973: Propagation of axisymmetric waves in an elastic half-space containing a cylindrical inclusion. Part I: Formulation and general integral solution. *Quart. J. Mech. Appl. Math.*, **30**, 235–253.

Kobayashi, N., and H. Takeuchi, 1955: Propagation of tremors over the sphere of an elastic solid. *J. Phys. Earth*, **3**, 17–22.

Kobayashi, N., and H. Takeuchi, 1957: Wave generation from line sources within the ground. *J. Phys. Earth*, **5**, 25–32.

Kubenko, V. D., and Yu. B. Moscenkov, 1987: Interaction of unsteady acoustic waves with plates and hollow shells in a fluid. *Sov. Appl. Mech.*, **23**, 951–957.

Lehner, F. K., V. C. Li, and J. R. Rice, 1981: Stress diffusion along rupturing plate boundaries. *J. Geophys. Res.*, **86**, 6155–6169.

Liu, T., 1996: Dynamic response of a half-space to radial shear surface loadings. *Arch. Appl. Mech.*, **66**, 137–148.

Matsumoto, H., I. Nakahara, and Y. Matsuoko, 1978: The impact of an elastic cylinder on an elastic solid. *Bull. JSME*, **21**, 579–586.

Mitra, A., 1966: Mixed boundary value problem in an elastic quarter-space. *Pure Appl. Geophys.*, **63**, 60–67.

Mitra, M., 1958: Disturbance produced in an elastic half-space by an impulsive twisting moment applied to an attached rigid circular disc. *Z. Angew. Math. Mech.*, **38**, 40–43.

Mucichescu, D., 1973: Response of infinite, mass-attached string on elastic foundation to an impact point load. *Rev. Roum. Sci. Tech., Ser. Mec. Appl.*, **18**, 651–662.

Muller, P., 1976: Propagation d'ondes de flexion dans une tige micropolaire élastique tendue. *J. Mécanique*, **15**, 493–520.

Mukhopadhyay, A., 1966: Disturbances produced by variable body forces in a layer of finite thickness resting on a rigid foundation. *Pure Appl. Geophys.*, **65**, 29–36.

Mukhopadhyay, A., 1967: Disturbances produced by a visco-elastic medium by transient torsional body forces. *Pure Appl. Geophys.*, **67**, 43–53.

Parnes, R., 1978: Dynamic surface stresses on an elastic half-space containing a cylindrical inclusion. *J. Sound Vib.*, **58**, 167–178.

Parton, V. Z., 1972: The axially symmetric temperature problem for the space with a disk-shaped crack. *J. Appl. Math. Mech.*, **36**, 104–111.

Patil, S. P., 1988: Response of infinite railroad track to vibrating mass. *J. Eng. Mech. Div. (Am. Soc. Civ. Eng.)*, **114**, 688–703.

Payton, R. G., 1962: Initial bending stresses in elastic shells impacting into compressible fluids. *Quart. J. Mech. Appl. Math.*, **15**, 77–90.

Payton, R. G., 1965: Dynamic bond stress in a composite structure subjected to a sudden pressure rise. *J. Appl. Mech.*, **32**, 643–650.

Payton, R. G., 1968: Epicenter motion of an elastic half-space due to buried stationary and moving line sources. *Int. J. Solids Struct.*, **4**, 287–300.

Payton, R. G., 1969: Two-dimensional pulse propagation in a two-parameter anisotropic elastic solid. *Q. Appl. Math.*, **27**, 147–160.

Peck, J. C., and J. Miklowitz, 1969: Shadow-zone response in the diffraction of a plane compressional pulse by a circular cavity. *Int. J. Solids Struct.*, **5**, 437–454.

Pekeris, C. L., and H. Lifson, 1957: Motion of the surface of a uniform elastic half-space produced by a buried pulse. *J. Acoust. Soc. Am.*, **29**, 1233–1238.

Pekeris, C. L., and I. M. Longman, 1958: The motion of the surface of a uniform elastic half-space produced by a buried torque-pulse. *Geophys. J. R. Astr. Soc.*, **1**, 146–153.

Pekeris, C. L., and I. M. Longman, 1958: Ray-theory solution of the problem of propagation of explosive sound in a layered liquid. *J. Acoust. Soc. Am.*, **30**, 323–328.

Pekeris, C. L., Z. Alterman, and F. Abramovici, 1963: Propagation of an SH-torque pulse in a layered solid. *Bull. Seism. Soc. Am.*, **53**, 39–57.

Pekeris, C. L., Z. Alterman, F. Abramovici, and A. Jarosch, 1965: Propagation of a compressional pulse in a layered solid. *Rev. Geophys.*, **3**, 25–47.

Peralta, L. A., and S. Raynor, 1964: Initial response of a fluid-filled, elastic, circular, cylindrical shell to a shock wave in acoustic medium. *J. Acoust. Soc. Am.*, **36**, 476–488.

Pinney, E., 1954: Surface motion due to a point source in a semi-infinite elastic medium. *Bull. Seism. Soc. Am.*, **44**, 571–596.

Rahman, M., 1995: Some fundamental axisymmetric singular solutions of elastodynamics. *Quart. J. Mech. Appl. Math.*, **48**, 329–342.

Rosenbaum, J. H., 1960: The long-time response of a layered elastic medium to explosive sound. *J. Geophys. Res.*, **65**, 1577–1613.

Roy, A. B., 1964: Disturbances in a viscoelastic medium due to a twist of finite duration on the surface of a spherical cavity. *Pure Appl. Geophys.*, **59**, 21–26.

Sarkar, G., 1966: Wave motion due to impulsive twist on the surface. *Pure Appl. Geophys.*, **65**, 43–47.

Sarkar, G., 1966: Displacement due to impulsive loadings on the surface of elastic half-space. *Pure Appl. Geophys.*, **66**, 37–47.

Sarker, A. K., 1963: On SH type of motion due to body forces and due to stress-discontinuity in a semi-infinite viscoelastic medium. *Geofis. Pura Appl.*, **55**, 42–52.

Scavuzzo, R. J., J. L. Bailey, and D. D. Raftopoulos, 1971: Lateral structure interaction with seismic waves. *J. Appl. Mech.*, **38**, 125–134.

Schimmerl, J., 1965: Setzung einer Tonschicht endlicher Mächtigkeit bei zylindersymmetrischer Belastung. *Z. Angew. Math. Mech.*, **45**, 553–562.

Sheehan, J. P., and L. Debnath, 1972: On the dynamic response of an infinite Bernoulli-Euler beam. *Pure Appl. Geophys.*, **97**, 100–110.

Sherwood, J. W. C., 1958: Elastic wave propagation in a semi-infinite solid medium. *Proc. Phys. Soc. London*, **71**, 207–219.

Shibuya, T., H. Matsumoto, and I. Nakahara, 1968: The semi-infinite plate subjected to impact load on the free boundary. *Bull. JSME*, **11**, 203–210.

Shibuya, T., and I. Nakahara, 1968: The semi-infinite body subjected to a concentrated impact load on the surface. *Bull. JSME*, **11**, 983–992.

Shindo, Y., 1981: Sudden twisting of a flat annular crack. *Int. J. Solids Struct.*, **17**, 1103–1112.

Shindo, Y., H. Nozaki, and H. Higaki, 1986: Impact response of a finite crack in an orthotropic strip. *Acta Mech.*, **62**, 87–104.

Shindo, Y., 1988: Impact response of a crack in a semi-infinite body with a surface layer under longitudinal shear. *Acta Mech.*, **73**, 147–162.

Sidhu, R. S., 1972: Disturbances in semi-infinite heterogeneous media generated by torsional sources. I. *Bull. Seism. Soc. Am.*, **62**, 541–550.

Spillers, W. R., and A. Callegari, 1969: Impact of two elastic cylinders: Short-time solution. *Int. J. Mech. Sci.*, **11**, 846–851.

Srivastava, K. N., R. M. Palaiya, and D. S. Karaulia, 1981: Diffraction of SH waves by two coplanar Griffith cracks at the interface of two bonded dissimilar elastic half-spaces. *Indian J. Pure Appl. Math.*, **12**, 242–252.

Stadler, W., and R. W. Shreeves, 1970: The transient and steady-state response of the infinite Bernoulli-Euler beam with damping and an elastic foundation. *Quart. J. Mech. Appl. Math.*, **23**, 197–208.

Stick, E., 1959: Propagation of elastic wave motion from an impulsive source along a fluid/solid interface. Part II. Theoretical pressure response. *Philos. Trans. R. Soc. London, Ser. A*, **251**, 465–488.

Stick, E., 1959: Propagation of elastic wave motion from an impulsive source along a fluid/solid interface. Part III. The pseudo-Rayleigh wave. *Philos. Trans. R. Soc. London, Ser. A*, **251**, 488–523.

Sve, C., and J. Miklowitz, 1973: Thermally induced stress waves in an elastic layer. *J. Appl. Mech.*, **40**, 161–167.

Takemiya, H., and F. Guan, 1993: Transient Lamb's solution for surface strip impulses. *J. Eng. Mech. Div. (Am. Soc. Civ. Eng.)*, **119**, 2385–2403.

Takeuchi, H., and N. Kobayashi, 1955: Wave generations from line sources within the ground. *J. Phys. Earth*, **3**, 7–15.

Takeuchi, H., and N. Kobayachi, 1956: Wave generations in a superficial layer resting on a semi-infinite lower layer. *J. Phys. Earth*, **4**, 21–30.

Tanabe, Y., I. Maekawa, S. Handa, and T. Hara, 1989: The propagation of shear wave in a viscoelastic rod (in Japanese). *Nihon Kikai Gakkai Rombunshu (Trans. Japan Soc. Mech. Eng.), Ser. A*, **55**, 2452–2457.

Tawari, G., 1971: Effect of couple-stresses in a semi-infinite elastic medium due to impulsive twist over the surface. *Pure Appl. Geophys.*, **91**, 71–75.

Vasudevaiah, M., and K. V. S. N. Prasad, 1990: Viscous impulsive rotation of an annular disk. *Indian J. Pure Appl. Math.*, **21**, 1125–1136.

Watanabe, K., 1981: Transient response of an inhomogeneous elastic half space to a torsional load (constant SH-wave velocity) (in Japanese). *Nihon Kikai Gakkai Rombunshu (Trans. Japan Soc. Mech. Eng.), Ser. A*, **47**, 92–100.

Watanabe, K., 1981: Transient response of an inhomogeneous elastic solid to an impulsive SH-source (variable SH-wave velocity) (in Japanese). *Nihon Kikai Gakkai Rombunshu (Trans. Japan Soc. Mech. Eng.), Ser. A*, **47**, 740–746.

Watanabe, K., 1981: Scattering of SH-wave by a cylindrical discontinuity in inhomogeneous elastic media (varying SH-wave velocity) (in Japanese). *Nihon Kikai Gakkai Rombunshu (Trans. Japan Soc. Mech. Eng.), Ser. A*, **47**, 1222–1228.

Watanabe, K., 1981: Response of an elastic plate on a Pasternak foundation to a moving load. *Bull. JSME*, **24**, 775–780.

Watanabe, K., 1981: Transient response of an inhomogeneous elastic half space to a torsional load. *Bull. JSME*, **24**, 1537–1542.

Watanabe, K., 1982: Transient response of an inhomogeneous elastic solid to an impulse SH-source (Variable SH-wave velocity). *Bull. JSME*, **25**, 315–320.

Watanabe, K., 1984: Transient response of an inhomogeneous elastic solid to a moving torsional load in a cylindrical bore. *Int. J. Solids Struct.*, **20**, 359–376.

Yen, D. H. Y., 1968: Dynamic response of an infinite plate subjected to a steadily moving transverse force. *Z. Angew. Math. Phys.*, **19**, 257–270.

Yen, D. H. Y., and C. C. Chou, 1974: An addendum to the paper 'Dynamic response of an infinite plate subjected to a steadily moving transverse force'. *Z. Angew. Math. Phys.*, **25**, 463–485.

Żórawski, M., 1961: Moving dynamic heat sources in a visco-elastic space and corresponding basic solutions for moving sources. *Arch. Mech. Stosow.*, **13**, 257–274.

Thermoelasticity and Magnetoelasticity

Achari, R. M., 1975: A quasi-static thermoelastic problem for a semi-space. *Z. Angew. Math. Mech.*, **55**, 688–692.

Alexandrov, V. M., 2003: High-speed motion of an extended thermal-force source in elastic medium. *Mech. Solids*, **38, No. 3**, 58–64.

Chakravarti, S., 1977: One dimensional thermo-elastic wave in a non-simple medium. *Bull. Calcutta Math. Soc.*, **64**, 129–135.

Chandrasekharaiah, D. S., and K. S. Srinath, 1998: Thermoelastic interactions without energy dissipation due to a line heat source. *Acta Mech.*, **128**, 243–251.

Goshima, T., and K. Miyao, 1981: Transient thermal stresses in circular disk due to a moving heat source (in Japanese). *Nihon Kikai Gakkai Rombunshu (Trans. Japan Soc. Mech. Eng.), Ser. A*, **47**, 619–625.

Goshima, T., and K. Miyao, 1991: Transient thermal stresses in a circular disk with a hole due to a rotating heat source (in Japanese). *Nihon Kikai Gakkai Rombunshu (Trans. Japan Soc. Mech. Eng.), Ser. A*, **57**, 53–58.

Kill, I. D., 1966: Thermoelastic stresses inside a half-space. *Mech. Solids*, **1, No. 1**, 97–99.

Koizumi, T., 1970: Thermal stresses in a semi-infinite body with an instantaneous heat source on its surface. *Bull. JSME*, **13**, 26–33.

Kolyano, Yu. M., and E. A. Pakala, 1967: Temperature stresses in a heat-source heated plate strip with heat transfer. *Sov. Appl. Mech.*, **3, No. 3**, 48–54.

Kolyano, Yu. M., 1967: Temperature stresses in an orthotropic plate strip with heat outflow. *Sov. Appl. Mech.*, **3, No. 6**, 23–27.

Kolyano, Yu. M., and E. A. Pakula, 1970: Two-dimensional dynamic problem of thermoelasticity for heated thin plates. *Sov. Appl. Mech.*, **6**, 174–179.

Misra, J. C., and R. M. Achari, 1980: Temperature stresses in an infinite disk having a heat source in the vicinity. *J. Therm. Stresses*, **3**, 57–66.

Noda, T., 1988: On a certain inverse problem of coupled thermal stress fields in a thick plate. *Z. Angew. Math. Mech.*, **68**, 411–415.

Podstrigach, Ya. S., and B. I. Kolodii, 1970: Two-dimensional temperature and stress field in induction heating of an elastic half-space. *Sov. Appl. Mech.*, **6**, 1329–1333.

Rafalski, P., 1965: Dynamic thermal stresses in viscoelastic slab. *Arch. Mech. Stosow.*, **17**, 617–631.

Rożnowski, T., 1969: The plane problem of thermoelasticity with a moving boundary condition. *Arch. Mech. Stosow.*, **21**, 657–677.

Takeuti, Y., Y. Tanigawa, and Y. Ohtao, 1981: Three-dimensional unsteady thermal stress problems in a finite circular composite cylinder under periodic heating (in Japanese). *Nihon Kikai Gakkai Rombunshu (Trans. Japan Soc. Mech. Eng.), Ser. A*, **47**, 425–432.

Tanigawa, Y., Y. Ootao, and N. Takada, 1991: Transient thermal stresses and thermal deformations of a semi-infinite solid cylinder with heated moving end surface (Effect of heat transfer on the end surface) (in Japanese). *Nihon Kikai Gakkai Rombunshu (Trans. Japan Soc. Mech. Eng.), Ser. A*, **57**, 852–857.

Youngdahl, C. K., and E. Sternberg, 1961: Transient thermal stresses in a circular cylinder. *J. Appl. Mech.*, **28**, 25–34.

Wave Equation

Akulenko, L. D., S. A. Mikhailov, and S. V. Nesterov, 1990: The oscillations of an oscillator near the interface between two liquids. *J. Appl. Mech. Math.*, **54**, 30–38.

Chadwick, P., and G. E. Tupholme, 1967: Generation of an acoustic pulse by a baffled circular piston. *Proc. Edinburgh Math. Soc., Ser. 2*, **15**, 263–277.

Leij, F. J., and J. H. Dane, 1990: Analytical solutions of the one-dimensional advection equation and two- or three-dimensional dispersion equation. *Water Resour. Res.*, **26**, 1475–1482.

Levine, H., 1973: A note on problems of wave generation in semi-infinite media by surface forces. *Appl. Sci. Res.*, **28**, 207–222.

Liu, D. T., 1959: Wave propagation in a liquid layer. *Geophysics*, **24**, 658–666.

Mishra, S. K., 1964: Propagation of sound pulses in a semi-infinite stratified medium. *Proc. Indian Acad. Sci.*, **A59**, 21–48.

Morioka, S., and G. Matsui, 1975: Pressure-wave propagation through a separated gas-liquid layer in a duct. *J. Fluid Mech.*, **70**, 721–731.

Norwood, F. R., 1968: Propagation of transient sound signals into a viscous fluid. *J. Acoust. Soc. Am.*, **44**, 450–457.

Okeke, E. O., 1977: The acoustic vibrations induced in a homogeneous fluid by gravity waves. *Geophys. Astrophys. Fluid Dyn.*, **8**, 155–161.

Pekeris, C. L., 1956: Solution of an integral equation occurring in impulsive wave propagation problems. *Proc. Natl. Acad. Sci.*, **42**, 439–443.

Peterson, E. W., 1974: Acoustic wave propagation along a fluid-filled cylinder. *J. Appl. Phys.*, **45**, 3340–3350.

Rosenbaum, J. II., 1959: A note on the propagation of a sound pulse in a two-layer liquid medium. *J. Geophys. Res.*, **64**, 95–102.

Stronge, W. J., 1970: A load accelerating on the surface of an acoustic half space. *J. Appl. Mech.*, **37**, 1077–1082.

Papers Using the Cagniard Technique

Electromagnetism

Vlaar, N. J., 1964: The transient electromagnetic field from an antenna near the plane boundary between two dielectric halfspaces. II. A closer investigation of the field. *Appl. Sci. Res., Ser. B*, **11**, 49–66.

Fluid Dynamics

Bykovtsev, A. S., and D. B. Kramarovskii, 1987: The propagation of a complex fracture area. The exact three-dimensional solution. *J. Appl. Math. Mech.*, **51**, 89–98.

Ross, R. A., 1961: The effect of an explosion in a compressible fluid under gravity. *Can. J. Phys.*, **39**, 1330–1346.

Geophysical Sciences

Ghosh, S. K., 1965: The disturbance due to oceanic explosion of finite extent. *Proc. Nat. Inst. Sci. India*, **A31**, 423–433.

Magnetoelasticity

Abubakar, I., 1964: Magneto-elastic SH-type of motion. *Pure Appl. Geophys.*, **59**, 10–20.

Solid Mechanics

Abramovici, F., 1996: Vertical and near-vertical incidence of P waves in a layered solid. *Bull. Seismol. Soc. Am.*, **86**, 406–415.

Abubakar, I., 1962: Disturbance due to a line source in a semi-infinite transversely isotropic elastic medium. *Geophys. J.*, **6**, 337–359.

Ang, D. D., 1960: Elastic waves generated by a force moving along a crack. *J. Math. Phys. (Cambridge, MA)*, **38**, 246–256.

Bakhski, V. S., 1965: Propagation of a fracture in an infinite medium. *Pure Appl. Geophys.*, **62**, 23–34.

Ben-Menahem, A., and M. Vered, 1973: Extension and interpretation of the Cagniard–Pekeris method for dislocation sources. *Bull. Seism. Soc. Am.*, **63**, 1611–1636.

Bhattacharyya, S., 1973: SH waves in a semi-infinite elastic space due to torsional disturbance. *Pure Appl. Geophys.*, **111**, 2216–2222.

Blowers, R. M., 1969: On the response of an elastic solid to droplet impact. *J. Inst. Math. Appl.*, **5**, 167–193.

Bortfeld, R., 1962: Exact solution of the reflection and refraction of arbitrary spherical compressional waves at liquid-liquid interfaces and at solid-solid interfaces with equal shear velocities and equal densities. *Geophys. Prospect.*, **10**, 35–67.

Bortfeld, R., 1962: Reflection and refraction of spherical compressional waves at arbitrary plane interfaces. *Geophys. Prospect.*, **10**, 517–538.

Broberg, K. B., 1960: The propagation of a brittle crack. *Ark. Fyk.*, **18**, 159–192.

Chandra, U., 1967: Propagation of an SH-torque pulse in a three layered solid half-space. *Pure Appl. Geophys.*, **67**, 54–64.

Chen, P.-L., 1988: The investigation of transient wave propagation in a dipping layered medium overlaying a half space. *J. Chinese Inst. Eng.*, **11**, 455–465.

Chen, P.-L., 1989: Transient SH waves propagation in a double dipping layered medium overlaying half-space. *J. Chinese Inst. Eng.*, **12**, 9–21.

Chen, P.-L., and Y.-S. Jeng, 1989: The amplification effect of soil media due to SH wave. *J. Chinese Inst. Eng.*, **12**, 567–575.

Chen, P.-L., 1989: On the reponses of seismic wave propagating in a single dipping layered medium due to shear fault. *J. Chinese Inst. Eng.*, **12**, 671–687.

Chwalczyk, F., J. Rafa, and E. Włodarczyk, 1972: Propagation of two-dimensional non-stationary stress waves in a semi-infinite viscoelastic body, produced by a normal load moving over the surface with subseismic velocity. *Proc. Vib. Prob.*, **13**, 241–257.

Chwalczyk, F., J. Rafa, and E. Włodarczyk, 1984: Propagation of non-stationary elastic waves in an anisotropic half-space. Part I. Analytic solution. *J. Tech. Phys.*, **25**, 23–34.

Crosson, R. S., and C.-C. Chao, 1970: Radiation of Rayleigh waves from a vertical directional charge. *Geophysics*, **35**, 45–56.

Eason, G., and R. R. M. Wilson, 1971: The displacements produced in a composite infinite solid by an impulsive torsional body force. *Quart. J. Mech. Appl. Math.*, **24**, 169–185.

Emura, K., 1960: Propagation of the disturbances in the medium consisting of semi-infinite liquid and solid. *Sci. Rep. Tohoku Univ., Ser. 5 Geophys.*, **12**, 63–100.

Ghosh, M. L., 1964/65: Disturbance in an elastic half space due to an impulsive twisting moment applied to an attached rigid circular disc. *Appl. Sci. Res., Ser. A*, **14**, 31–42.

Ghosh, S. C., 1970: Disturbance produced in an elastic half-space by impulsive normal pressure. *Pure Appl. Geophys.*, **80**, 71–83.

Ghosh, S. K., 1973: The transient disturbance produced in an elastic layer by a buried spherical source. *Pure Appl. Geophys.*, **105**, 781–801.

Jingu, T., H. Matsumoto, and K. Nezu, 1985: Transient stress in an elastic half-space subjected to an uniform impulsive load on a rectangular region of its surface (in Japanese). *Nihon Kikai Gakkai Rombunshu (Trans. Japan Soc. Mech. Eng.), Ser. A*, **51**, 1131–1140.

Jingu, T., H. Matsumoto, and K. Nezu, 1985: The transient stress in an elastic half-space excited by impulsive loading over one quarter of its surface (in Japanese). *Nihon Kikai Gakkai Rombunshu (Trans. Japan Soc. Mech. Eng.), Ser. A*, **51**, 1401–1408.

Jingu, T., H. Matsumoto, and K. Nezu, 1985: The transient stress in an elastic half-space subjected to the semi-infinite line load varying as the unit step function on its surface (in Japanese). *Nihon Kikai Gakkai Rombunshu (Trans. Japan Soc. Mech. Eng.), Ser. A*, **51**, 1410–1418.

Jingu, T., H. Matsumoto, and K. Nezu, 1985: Transient stress of an elastic half space subjected to a uniform impulsive load in a rectangular region of its surface. *Bull. JSME*, **28**, 2881–2889.

Kawasaki, I., Y. Suzuki, and R. Sato, 1973: Seismic waves due to a shear fault in a semi-infinite medium. Part I: Point source. *J. Phys. Earth*, **21**, 251–284.

Knopoff, L., 1958: The surface motions of a thick plate. *J. Appl. Phys.*, **29**, 661–670.

Maiti, N. C., and L. Debnath, 1990: Transient wave motions due to an asymmetric shear stress discontinuity in a layered elastic medium. *Z. Angew. Math. Mech.*, **70**, 35–40.

Mandal, S. B., 1972: Diffraction of elastic waves by a rigid half-space. *Proc. Vib. Prob.*, **13**, 331–343.

Mandal, S. B., 1974: Diffraction of elastic pulses by a rigid half-plane. *Indian J. Pure Appl. Math.*, **5**, 594–600.

Mencher, A. G., 1953: Epicentral displacement caused by elastic waves in an elastic slab. *J. Appl. Phys.*, **24**, 1240–1246.

Miklowitz, J., 1982: Wavefront analysis in the nonseparable elastodynamic quarter-plate problems. Part 1: The general method. *J. Appl. Mech.*, **49**, 797–807.

Miklowitz, J., 1982: Wavefront analysis in the nonseparable elastodynamic quarter-plate problems. Part 2: Wavefront events in the edge uniform pressure problem. *J. Appl. Mech.*, **49**, 808–815.

Mitra, M., 1958: Exact transient solution of the buried line source problem for an asymmetric source. *Z. Angew. Math. Phys.*, **9a**, 322–331.

Mitra, M., 1960: On the application of Cagniard's method to dynamical problems of elasticity. *Gerland's Beit. Geophys.*, **69**, 73–86.

Mitra, M., 1963: The disturbance produced by axially-symmetric, time-dependent body forces in an elastic half-space. *Proc. Nat. Inst. Sci. India*, **A29**, 271–281.

Mitra, M., 1970: On a finite SH type source in a layered half-space. II. *Pure Appl. Geophys.*, **80**, 147–151.

Mittal, J. P., and R. S. Sidhu, 1982: Generation of SH waves from a nonuniformly moving stress discontinuity in a layered half space. *Indian J. Pure Appl. Math.*, **13**, 682–695.

Müller, G., 1967: Theoretical seismograms for some types of point-sources in layered media. Part I: Theory. *Z. Geophys.*, **33**, 15–35.

Nag, K. R., 1961: On "SH" type of motion due to body forces in a semi-infinite elastic medium. *Gerland's Beitr. Geophys.*, **70**, 221–232.

Nag, K. R., 1962: Disturbance due to shearing-stress discontinuity in a semi-infinite elastic medium. *Geophys. J.*, **6**, 468–478.

Nag, K. R., 1963: Generation of Love waves due to a point source in a layered medium. *Indian J. Theoret. Phys.*, **11**, 105–118.

Nag, K. R., 1963: Generation of 'SH' type of waves due to stress-discontinuity in a layered medium. *Quart. J. Mech. Appl. Math.*, **16**, 293–303.

Nag, K. R., 1963: Transient disturbance due to tangential stress applied on the surface of an elastic half-space. *Proc. Nat. Inst. Sci. India*, **A29**, 359–369.

Nag, K. R., 1965: Disturbance due to a point source in a transversely isotropic half-space. *Bull. Calcutta Math. Soc.*, **57**, 16–24.

Nag, K. R., 1972: Disturbance due to a point source in a layered half-space. *Pure Appl. Geophys.*, **98**, 72–86.

Niazy, A., 1973: Elastic displacements caused by a propagating crack in an infinite medium: An exact solution. *Bull. Seism. Soc. Am.*, **63**, 357–379.

Niazy, A., 1975: An exact solution for a finite, two-dimensional moving dislocation in an elastic half-space with application to San Fernando earthquake of 1971. *Bull. Seism. Soc. Am.*, **65**, 1797–1826.

Norwood, F. R., 1973: Similarity solutions in plane elastodynamics. *Int. J. Solids Struct.*, **9**, 789–803.

Norwood, F. R., 1975: Transient response of an elastic plate to loads with finite characteristic dimensions. *Int. J. Solids Struct.*, **11**, 33–51.

Pal, P. C., 1983: On the disturbance produced by an impulsive shearing stress on the surface of a semi-infinite poro-elastic medium. *J. Acoust. Soc. Am.*, **74**, 586–590.

Pao, Y.-H., and F. Ziegler, 1982: Transient SH-waves in a wedge-shaped layer. *Geophys. J. R. Astr. Soc.*, **71**, 57–77.

Paul, S., 1976: On the displacements produced in a porous elastic half-space by an impulsive line load. (Non-dissipative case.) *Pure Appl. Geophys.*, **114**, 605–614.

Paul, S., 1976: On the disturbance produced in a semi-infinite poroelastic medium by a surface load. *Pure Appl. Geophys.*, **114**, 615–627.

Paul, S., 1976: Lamb's line load problem for a porous elastic half-space: Non-dissipative case. *Indian J. Pure Appl. Math.*, **7**, 854–867.

Paul, S., 1978: On the disturbance in a poro-elastic medium by an expanding ring and disk load. *Indian J. Pure Appl. Math.*, **9**, 324–331.

Payton, R. G., 1988: Stresses in a constrained transversely isotropic elastic solid caused by a moving dislocation. *Acta Mech.*, **74**, 35–49.

Ping, L. X., and L. C. Tu, 1995: Elastodynamic stress-intensity factors for a semi-infinite crack under 3-D combined mode loading. *Int. J. Fract.*, **69**, 319–339.

Rajhans, B. K., and P. Kesari, 1988: Diffraction of elastic spherical P waves by a cylindrical cavity. *Acta Mech.*, **72**, 309–325.

Ravera, R. J., and G. C. Sih, 1970: Transient analysis of stress waves around cracks under antiplane strain. *J. Acoust. Soc. Am.*, **47**, 875–881.

Roy, A., 1974: Surface displacements in an elastic half space due to a buried moving point source. *Geophys. J. R. Astr. Soc.*, **40**, 289–304.

Roy, A., 1974: Exact transient response of an elastic half space to a non-uniformly expanding circular ring. *Indian J. Pure Appl. Math.*, **5**, 1063–1080.

Sato, R., 1972: Seismic waves in the mean field. *J. Phys. Earth*, **20**, 357–375.

Sur, S. P., 1968: Transient motion of a line load on the surface of a transversely isotropic half-space. *Proc. Natl. Inst. Sci. India*, **A34**, 86–100.

Tada, M., K. Watanabe, and Y. Hirano, 1989: Nonsteady response for longitudinal shear force on inhomogeneous elastic halfspace (in Japanese). *Nihon Kikai Gakkai Rombunshu (Trans. Japan Soc. Mech. Eng.)*, Ser. A, **55**, 246–250.

Thiruvenkatachar, V. R., 1960: Stress-wave propagation induced in an infinite slab by an impulse over a circular area of one face – I. *Proc. Natl. Inst. Sci. India, Suppl. II*, **A26**, 31-47.

Vlaar, N. J., 1964: The seismic pulse in a semi-infinite medium. *Appl. Sci. Res., Ser. B*, **11**, 67–83.

Watanabe, K., 1977: Transient response of a layered elastic half space subjected to a reciprocating anti-plane shear load. *Int. J. Solids Struct.*, **13**, 63–74.

Watanabe, K., 1980: Transient response of an elastic half-space to moving loads (in Japanese). *Nihon Kikai Gakkai Rombunshu (Trans. Japan Soc. Mech. Eng.), Ser. A*, **46**, 1121–1128.

Watanabe, K., 1980: Transient response of an infinite elastic solid to a moving load (in Japanese). *Nihon Kikai Gakkai Rombunshu (Trans. Japan Soc. Mech. Eng.), Ser. A*, **46**, 1238–1248.

Watanabe, K., 1981: Transient response of an inhomogeneous elastic half space to an impulsive anti-plane shear load (constant velocity of SH-wave) (in Japanese). *Nihon Kikai Gakkai Rombunshu (Trans. Japan Soc. Mech. Eng.), Ser. A*, **47**, 203–210.

Wright, T. W., 1969: Impact on an elastic quarter space. *J. Acoust. Soc. Am.*, **45**, 935–943.

Wave Equation

Chwalczyk, F., J. Rafa, and E. Włodarczyk, 1973: Propagation of two-dimensional non-stationary pressure waves in a layer of perfect compressible liquid. *Proc. Vib. Prob.*, **14**, 245–256.

Donato, R. J., 1976: Spherical-wave reflection from a boundary of reactive impedance using a modification of Cagniard's method. *J. Acoust. Soc. Am.*, **60**, 999–1002.

Flaherty, F. T., 1968: Transient resonance of an ideal string under a load moving with varying speed. *Int. J. Solids Struct.*, **4**, 1221–1231.

Garvin, W. W., 1956: Exact transient solution of the buried line source problem. *Proc. R. Soc. London, Ser. A*, **234**, 528–541.

Gopalsamy, K., 1974: Response of an acoustic half-space excited by a randomly moving pressure pulse on the surface. *Pure Appl. Geophys.*, **112**, 240–252.

Knopoff, L., F. Gilbert, and W. L. Pilant, 1960: Wave propagation in a medium with a single layer. *J. Geophys. Res.*, **65**, 265–278.

Longman, I. M., 1961: Solution of an integral equation occurring in the study of certain wave-propagation problems in layered media. *J. Acoust. Soc. Am.*, **33**, 954–958.

Mitra, M., 1963: Exact solution of the source problem for a finite threedimensional source. *Geofis. Pura Appl.*, **56**, 31–38.

Mitra, M., 1964: Propagation of explosive sound in a layered liquid. *J. Acoust. Soc. Am.*, **36**, 1145–1149.

Papadopoulos, M., 1963: The reflexion and refraction of point-source fields. *Proc. R. Soc. London, Ser. A*, **273**, 198–221.

Papadopoulos, M., 1963: The use of singular integrals in wave propagation problems; with application to the point source in a semi-infinite elastic medium. *Proc. R. Soc. London, Ser. A*, **276**, 204–237.

Rajhans, B. K., and P. Kesari, 1986: Scattering of compressional waves by a cylindrical cavity. *J. Math. Phys. Sci.*, **20**, 429–444.

Rajhans, B. K., and S. K. Samal, 1992: Diffraction of compressional waves by a fluid cylinder in a homogeneous medium. *Indian J. Pure Appl. Math.*, **23**, 603–616.

Towne, D. H., 1968: Pulse shapes of spherical waves reflected and refracted at a plane interface separating two homogeneous fluids. *J. Acoust. Soc. Am.*, **44**, 65–76.

Papers Using the Cagniard–De Hoop Technique

Aquifers, Reservoirs and Porous Media

Oliver, D., 1994: Application of a wave transform to pressure transient testing in porous media. *Transport Porous Media*, **16**, 209–236.

Electromagnetism

Dai, R., and C. T. Young, 1997: Transient fields of a horizontal electric dipole on a multi-layered dielectric medium. *IEEE Trans. Antenn. Propag.*, **45**, 1023–1031.

De Hoop, A. T., and H. J. Frankena, 1960: Radiation of pulses generated by a vertical electric dipole above a plane, non-conducting, earth. *Appl. Sci. Res., Ser. B*, **8**, 369–377.

De Hoop, A. T., 1979: Pulsed electromagnetic radiation from a line source in a two-media configuration. *Radio Sci.*, **14**, 253–268.

De Hoop, A. T., and M. L. Oristaglio, 1988: Application of the modified Cagniard technique to transient electromagnetic diffusion problems. *Geophys. J.*, **94**, 387–397.

Du Cloux, R., 1984: Pulsed electromagnetic radiation from a line source in the presence of a semi-infinite screen in the plane interface of two different media. *Wave Motion*, **6**, 459–476.

Ezzeddine, A., J. A. Kong, and L. Tsang, 1981: Transient fields of a vertical electric dipole over a two-layer nondispersive dielectric. *J. Appl. Phys.*, **52**, 1202–1208.

Finkler, H., and K. J. Langenberg, 1975: Das Einschwingverhalten der elektrischen Feldstärke in einem atmosphärischen Bodenwellenleiter bei beliebiger Empfängerentfernung in Abhängigkeit von der Trägerfrequenz. *Arch. Electronik Übertragung.*, **29**, 37–45.

Frankena, H. J., 1960: Transient phenomena associated with Sommerfeld's horizontal dipole problem. *Appl. Sci. Res., Ser. B*, **8**, 357–368.

Kooij, B. J., 1990: Transient electromagnetic field of a vertical magnetic dipole above a plane conducting Earth. *Radio Sci.*, **25**, 349–356.

Kooij, B. J., 1996: The electromagnetic field emitted by a pulsed current traveling along finite, straight wires above the interface of a two-media configuration. *Radio Sci.*, **31**, 81–93.

Kooij, B. J., 1996: The electromagnetic field emitted by a pulsed current point source above the interface of a nonperfectly conducting Earth. *Radio Sci.*, **31**, 1345–1360.

Kooij, B. J., 1991: The transient electromagnetic field of an electric line source above a plane conducting earth. *IEEE Trans. Electromagn. Compat.*, **EMC-33**, 19–24.

Kuester, E. F., 1984: The transient electromagnetic field of a pulsed line source located above a dispersively reflecting surface. *IEEE Trans. Antennas Propag.*, **AP-32**, 1154–1162.

Langenberg, K. J., 1974: The transient response of a dielectric layer. *Appl. Phys.*, **3**, 179–188.

Seliem, A. A. S. A., and S. T. Bishay, 1996: Transient electromagnetic field of a vertical electric dipole above an atmospheric surface duct. *Radio Sci.*, **31**, 833–840.

Shirai, H., 1995: Transient scattering responses from a plane interface between dielectric half-spaces. *Electron. Comm. Japan, Part 2*, **78**, No. 9, 52–62.

Vlaar, N. J., 1963/64: The transient electromagnetic field from an antenna near the plane boundary of two dielectric halfspaces. *Appl. Sci. Res., Ser. B*, **10**, 353–384.

General

Abramovici, F., 1978: A generalization of the Cagniard method. *J. Comput. Phys.*, **29**, 328–343.

Bleistein, N., and J. K. Cohen, 1992: The Cagniard method in complex time revisited. *Geophys. Prospect.*, **40**, 619–649.

Chapman, C. H., 1974: Generalized ray theory for an inhomogeneous medium. *Geophys. J. R. Astr. Soc.*, **36**, 673–704.

Chapman, C. H., 1976: Exact and approximate generalized ray theory in vertically inhomogeneous media. *Geophys. J. R. Astr. Soc.*, **46**, 201–233.

Craster, R. V., 1996: Wavefront expansions for pulse scattering by a surface inhomogeneity. *Quart. J. Mech. Appl. Math.*, **49**, 657–674.

Drijkoningen, G. G., and C. H. Chapman, 1988: Tunneling rays using the Cagniard–De Hoop method. *Bull. Seism. Soc. Am.*, **78**, 898–907.

Drijkoningen, G. G., 1991: Tunneling and the generalized ray method in piecewise homogeneous media. *Geophys. Prospect.*, **39**, 757–781.

Heyman, E., and L. B. Felsen, 1985: Non-dispersive closed form approximations for transient propagation and scattering of ray fields. *Wave Motion*, **7**, 335–358.

Kraut, E. A., 1965/66: The effect of anisotropy on the integral representation of a cylindrical pulse. *Appl. Sci. Res., Ser. B*, **12**, 308–314.

Geophysical Sciences

Drijkoningen, G. G., and J. T. Fokkema, 1987: The exact seismic response of an ocean and a n-layer configuration. *Geophys. Prospect.*, **35**, 33–61.

Helmberger, D. V., 1968: The crust-mantle transition in the Bering sea. *Bull. Seism. Soc. Am.*, **58**, 179–214.

Richards, P. G., 1971: A theory for pressure radiation from ocean-bottom earthquakes. *Bull. Seism. Soc. Am.*, **61**, 707–721.

Heat Conduction

Shendeleva, M. L., V. V. Zalipaev, N. N. Ljepojevic, 2001: Temperature distribution near the interface in transient thermal processes. *Analytical Sci.*, **17**, S482–S485.

Solid Mechanics

Abramovici, F., L. H. T. Le, and E. R. Kanasewich, 1989: The evanescent wave in Cagniard's problem for a line source generating SH waves. *Bull. Seism. Soc. Am.*, **79**, 1941–1955.

Achenbach, J. D., 1968: Longitudinal force on an embedded filament. *Appl. Sci. Res.*, **19**, 412–425.

Aggarwal, H. R., and C. M. Ablow, 1967: Solution to a class of three-dimensional pulse propagation problems in an elastic half-space. *Int. J. Eng. Sci.*, **5**, 663–679.

Ang, D. D., 1960: Transient motion of a line load on the surface of an elastic half-space. *Quart. Appl. Math.*, **18**, 251–256.

Barclay, D. W., 1990: Isochromatic curves for a dynamically loaded elastic plate. *Acta Mech.*, **84**, 127–137.

Bennett, B. E., and G. Herrmann, 1976: The dynamic response of an elastic half space with an overlying acoustic fluid. *J. Appl. Mech.*, **43**, 39–42.

Ben-Zion, Y., 1989: The response of two joined quarter spaces to SH line sources located at the material discontinuity interface. *Geophys. J. Int.*, **98**, 213–222.

Ben-Zion, Y., 1990: The response of two half spaces to point dislocations at the material interface. *Geophys. J. Int.*, **101**, 507–528.

Brock, L. M., 1978: The effect of a thin layer surface inhomogeneity on dynamic surface response. *J. Appl. Mech.*, **45**, 95–99.

Brock, L. M., 1978: The non-uniform motion of a thin smooth rigid inclusion through an elastic solid. *Quart. Appl. Math.*, **36**, 269–277.

Brock, L. M., 1979: Two basic wave propagation problems for the non-uniform motion of displacement discontinuities in a half-plane. *Int. J. Eng. Sci.*, **17**, 1211–1223.

Brock, L. M., 1979: Two basic wave propagation problems for the non-uniform motion of displacement discontinuities across a bimaterial interface. *Int. J. Eng. Sci.*, **17**, 1289–1302.

Brock, L. M., 1980: The non-uniform motion of a thin smooth rigid wedge into an elastic half-plane. *Quart. Appl. Math.*, **38**, 209–223.

Brock, L. M., 1980: Exact dynamic surface response for sub and through-surface slip. *J. Appl. Mech.*, **47**, 525–530.

Brock, L. M., 1981: A wave propagation problem for non-uniform screw dislocation motion in a viscoelastic half-plane. *J. Elasticity*, **11**, 187–195.

Brock, L. M., 1982: Dynamic solutions for the non-uniform motion of an edge dislocation. *Int. J. Eng. Sci.*, **20**, 113–118.

Brock, L. M., 1983: The dynamic stress intensity factor due to arbitrary screw dislocation motion. *J. Appl. Mech.*, **50**, 383–389.

Brock, L. M., 1983: The dynamic stress intensity factor for a crack due to arbitrary rectilinear screw dislocation motion. *J. Elasticity*, **13**, 429–439.

Brock, L. M., 1986: The transient field under a point force acting on an infinite strip. *J. Appl. Mech.*, **53**, 321–325.

Brock, L. M., 1989: Transient analyses of dislocation emission in the three modes of fracture. *Int. J. Eng. Sci.*, **27**, 1479–1495.

Burridge, R., 1971: Lamb's problem for an isotropic half-space. *Quart. J. Mech. Appl. Math.*, **24**, 81–98.

Bykovtsev, A. S., and D. B. Kramarovskii, 1987: The propagation of a complex fracture area. The exact three-dimensional solution. *J. Appl. Math. Mech.*, **51**, 89–98.

Chandra, U., 1967: Propagation of an SH-torque pulse in a three layered solid half-space. *Pure Appl. Geophys.*, **67**, 54–64.

Chapman, C. H., and J. A. Orcutt, 1985: The computation of body wave synthetic seismograms in laterally homogeneous media. *Rev. Geophys.*, **23**, 105–163.

Cochard, A., and R. Madariaga, 1994: Dynamic faulting under rate-dependent friction. *Pure Appl. Geophys.*, **142**, 419–445.

De Hoop, A. T., and J. H. M. T. van der Hijden, 1985: Seismic waves generated by an impulsive point source in a solid/fluid configuration with a plane boundary. *Geophysics*, **50**, 1083–1090.

De Hoop, A. T., 1988: Large-offset approximations in the modified Cagniard method for computing synthetic seismograms: A survey. *Geophys. Prospect.*, **36**, 465–477.

Fokkema, J. T., and P. M. van den Berg, 1989: Seismic inversion by a RMS Born approximation in the space-time domain. *Geophys. Prospect.*, **37**, 53–72.

Freund, L. B., 1972: Wave motion in an elastic solid due to a nonuniformly moving line load. *Quart. Appl. Math.*, **30**, 271–281.

Freund, L. B., 1987: The stress intensity factor history due to three-dimensional transient loading on the faces of a crack. *J. Mech. Phys. Solids*, **35**, 61–72.

Gakenheimer, D. C., and J. Miklowitz, 1969: Transient excitation of an elastic half space by a point load traveling on the surface. *J. Appl. Mech.*, **36**, 505–515.

Gakenheimer, D. C., 1971: Response of an elastic half space to expanding surface loads. *J. Appl. Mech.*, **38**, 99–110.

Ghosh, M. L., 1971: The axisymmetric problem of propagation of a normal stress discontinuity in a semi-infinite elastic medium. *Appl. Sci. Res.*, **24**, 149–167.

Ghosh, M., 1980/81: Displacement produced in an elastic half-space by the impulsive torsional motion of a circular ring source. *Pure Appl. Geophys.*, **119**, 102–117.

Ghosh, M., and M. L. Ghosh, 1983: Torsional response of an elastic half space to a nonuniformly expanding ring source. *Z. Angew. Math. Mech.*, **63**, 621–629.

Gilbert, F., and L. Knopoff, 1961: The directivity problem for a buried line source. *Geophysics*, **26**, 626–634.

Harkrider, D. G., and D. V. Helmberger, 1978: A note of nonequivalent quadrapole source cylindrical shear potentials which give equal displacements. *Bull. Seism. Soc. Am.*, **68**, 125–132.

Higuchi, N., and Hirashima, K., 1997: Dynamic response of elastic half-space under unsteady buried line moving dislocation (in Japanesse). *Trans. Japan Soc. Mech. Eng.*, **A63**, 1453–1460.

Hwang, L.-F., J. T. Kuo, and Y.-C. Teng, 1982: Three-dimensional elastic wave scattering and diffraction due to a rigid cylinder embedded in an elastic medium by a point source. *Pure Appl. Geophys.*, **120**, 548–576.

Ing, Y.-S., and C.-C. Ma, 1997: Dynamic analysis of a propagating antiplane interface crack. *J. Eng. Mech. Div. (Am. Soc. Civ. Eng.)*, **123**, 783–791.

Ing, Y.-S., and C.-C. Ma, 1997: Transient analysis of a subsonic propagating interface crack subjected to antiplane dynamic loading in dissimilar isotropic materials. *J. Appl. Mech.*, **64**, 546–556.

Jingu, T., H. Matsumoto, and K. Nezu, 1986: The transient stress in an elastic half-space subjected to a semi-infinite line load varying as unit step function on its surface. *Bull. JSME*, **29**, 35–43.

Jingu, T., H. Matsumoto, and K. Nezu, 1986: The transient stress in an elastic half-space excited by impulsive loading over one quarter of its surface. *Bull. JSME*, **29**, 44–51.

Johnson, L. R., 1974: Green's function for Lamb's problem. *Geophys. J. R. Astr. Soc.*, **37**, 99–131.

Kennedy, T. C., and G. Herrmann, 1972: Moving load on a solid-solid interface: Supersonic region. *Arch. Mech.*, **24**, 1023–1028.

Kennedy, T. C., and G. Herrmann, 1973: Moving load on a fluid-solid interface: Supersonic region. *J. Appl. Mech.*, **40**, 137–142.

Kennedy, T. C., and G. Herrmann, 1974: The response of a fluid-solid interface to a moving disturbance. *J. Appl. Mech.*, **41**, 287–288.

Kooij, B. J., and D. Quak, 1988: Three-dimensional scattering of impulsive acoustic waves by a semi-infinite crack in the plane interface of a half-space and a layer. *J. Math. Phys.*, **29**, 1712–1721.

Kraut, E. A., 1963: Advances in the theory of anisotropic elastic wave propagation. *Rev. Geophys.*, **1**, 401–448.

Kraut, E. A., 1968: Diffraction of elastic waves by a rigid 90° wedge. Part II. *Bull. Seism. Soc. Am.*, **58**, 1097–1115.

Laturelle, F. G., 1990: The stresses produced in an elastic half-space by a normal step loading over a circular area: Analytical and numerical results. *Wave Motion*, **12**, 107–127.

Le, L. H. T., 1993: On Cagniard's problem for a qSH line source in transversely isotropic media. *Bull. Seism. Soc. Am.*, **83**, 529–541.

Li, S. F., and P. A. Mataga, 1996: Dynamic crack propagation in piezoelectric materials — Part I. Electrode solution. *J. Mech. Phys. Solids*, **44**, 1799–1830.

Luco, J. E., and M. Georgevich, 1988: A note on the steady-state response of an elastic medium to a moving dislocation of finite width. *Soil Dynam. Earthq. Eng.*, **7**, 170–175.

Ma, C.-C., and S.-K. Chen, 1993: Exact transient analysis of an anti-plane semi-infinite crack subjected to dynamic body forces. *Wave Motion*, **17**, 161–171.

Ma, C.-C., and S.-K. Chen, 1994: Exact transient full-field analysis of an antiplane subsurface crack subjected to dynamic impact loading. *J. Appl. Mech.*, **61**, 649–655.

Ma, C.-C., and Y.-S. Ing, 1995: Transient analysis of dynamic crack propagation with boundary effect. *J. Appl. Mech.*, **62**, 1029–1038.

Ma, C.-C., and L.-R. Hwang, 1996: Dynamic fracture analysis of an inclined subsurface crack subjected to dynamic moving loadings. *Int. J. Fract.*, **80**, 1–18.

Ma, C.-C., and K.-C. Huang, 1996: Analytical transient analysis of layered composite medium subjected to dynamic inplane impact loadings. *Int. J. Solids Struct.*, **33**, 4223–4238.

Ma, C.-C., and K.-C. Huang, 1996: Exact transient solutions of buried dynamic point forces for elastic bimaterials. *Int. J. Solids Struct.*, **33**, 4511–4529.

Ma, C.-C., and Y.-S. Ing, 1997: Transient analysis of crack in composite layered medium subjected to dynamic loadings. *AIAA J.*, **35**, 706–711.

Ma, C.-C., and Y.-S. Ing, 1997: Dynamic crack propagation in a layered medium under antiplane shear. *J. Appl. Mech.*, **64**, 66–72.

Markenscoff, X., and R. J. Clifton, 1981: The nonuniformly moving edge dislocation. *J. Mech. Phys. Solids*, **29**, 253–262.

Markenscoff, X., and L. Ni, 1984: Nonuniform motion of an edge dislocation in an anisotropic solid. I. *Quart. Appl. Math.*, **41**, 475–494.

Mittal, J. P., 1988: Generation of P-SV waves due to a finite line source moving nonuniformly over the surface of an elastic half space. *Proc. Indian Nat. Sci. Acad., Part A*, **54**, 943–957.

Mitra, M., 1963: An SH-point source in a half-space with a layer. *Bull. Seism. Soc. Am.*, **53**, 1031–1037.

Mitra, M., 1964: Disturbance produced in an elastic half-space by impulsive normal pressure. *Proc. Cambridge Philos. Soc.*, **60**, 683–696.

Norwood, F. R., 1969: Exact transient response of an elastic half space loaded over a rectangular region of its surface. *J. Appl. Mech.*, **36**, 516–522.

Nuismer, R. J., and J. D. Achenbach, 1972: Dynamically induced fracture. *J. Mech. Phys. Solids*, **20**, 203–222.

Olsen, D., 1974: Transient waves in two semi infinite viscoelastic media separated by a plane interface. *Int. J. Eng. Sci.*, **12**, 691–712.

Pal, P. C., 1983: Generation of SH-type waves due to non-uniformly moving stress-discontinuity in layered anisotropic elastic half-space. *Acta Mech.*, **49**, 209–220.

Pal, P. C., 1985: On the disturbance produced by an impulsive torsional motion of a circular ring source in a semi-infinite transversely isotropic medium. *Indian J. Pure Appl. Math.*, **16**, 179–188.

Pal, P. C., and L. Kumar, 1993: The effect of inhomogeneity on the torsional impulsive motion over a circular region in transversely isotropic elastic half space. *Indian J. Pure Appl. Math.*, **24**, 133–144.

Pal, S. C., and M. L. Ghosh, 1987: Waves in a semi-infinite elastic medium due to an expanding elliptic ring source on the free surface. *Indian J. Pure Appl. Math.*, **18**, 648–674.

Payton, R. G., 1967: Transient motion of an elastic half-space due to a moving surface line load. *Int. J. Eng. Sci.*, **5**, 49–79.

Ravera, R. J., and G. C. Sih, 1970: Transient analysis of stress waves around cracks under antiplane strain. *J. Acoust. Soc. Am.*, **47**, 875–881.

Richards, P. G., 1973: The dynamic field of a growing plane elliptical shear crack. *Int. J. Solids Struct.*, **9**, 843–861.

Richter, C., and G. Schmid, 1999: A Green's function time-domain boundary element method for the elastodynamic half-plane. *Int. J. Numer. Meth. Eng.*, **46**, 627–648.

Roy, A., 1975: Pulse generation in an elastic half space by normal pressure. *Int. J. Eng. Sci.*, **13**, 641–651.

Salvado, C., and J. B. Minster, 1980: Slipping interfaces: A possible source of S radiation from explosive sources. *Bull. Seism. Soc. Am.*, **70**, 659–670.

Sato, R., 1972: Seismic waves in the near field. *J. Phys. Earth*, **20**, 357–375.

Scott, R. A., and J. Miklowitz, 1969: Transient non-axisymmetric wave propagation in an infinite isotropic elastic plate. *Int. J. Solids Struct.*, **5**, 65–79.

Scott, R. A., and J. Miklowitz, 1969: Near-field transient waves in an anisotropic elastic plates for two and three dimensional problems. *Int. J. Solids Struct.*, **5**, 1059–1075.

Suh, S. L., W. Goldsmith, J. L. Sackman, and R. L. Taylor, 1974: Impact on a transversely anisotropic half-space. *Int. J. Rock Mech. Min. Sci. & Geomech. Abstr.*, **11**, 413–421.

Tanimoto, T., 1982: Cagniard–De Hoop method for a Haskell type vertical fault. *Geophys. J. R. Astr. Soc.*, **70**, 639–646.

Tsai, C.-H., and C.-C. Ma, 1991: Exact transient solutions of buried dynamic point forces and displacement jumps for an elastic half space. *Int. J. Solids Struct.*, **28**, 955–974.

Tsai, C.-H., and C.-C. Ma, 1992: The interaction of two inclined cracks with dynamic stress wave loadings. *Int. J. Fract.*, **58**, 77–91.

Tsai, C.-H., and C.-C. Ma, 1992: Transient analysis of a semi-infinite crack subjected to dynamic concentrated forces. *J. Appl. Mech.*, **59**, 804–811.

Tsai, Y. M., 1968: Stress waves produced by impact on the surface of a plastic medium. *J. Franklin Inst.*, **285**, 204–221.

Van der Hijden, J. H. M. T., 1984: Quantitative analysis of the pseudo-Rayleigh phenomena. *J. Acoust. Soc. Am.*, **75**, 1041–1047.

Verweij, M. D., and A. T. De Hoop, 1990: Determination of seismic wavefields in arbitrarily continuously layered media using the modified Cagniard method. *Geophys. J. Int.*, **103**, 731–754.

Wang, Y.-S., and D. Wang, 1996: Transient motion of an interface dislocation and self-similar propagation of an interface crack: Anti-plane motion. *Eng. Fract. Mech.*, **55**, 717–725.

Watanabe, K., 1981: Transient response of an infinite elastic solid to a moving point load. *Bull. JSME*, **24**, 1115–1122.

Watanabe, K., 1984: Transient response of an elastic solid to a moving torsional load in a cylindrical bore: An approximate solution. *Int. J. Eng. Sci.*, **22**, 277–284.

Ziegler, F., and Y.-H. Pao, 1990: Die Phasenfunktion kugeliger Wellen in der keilförmigen Oberflächenschicht. *Z. Angew. Math. Mech.*, **70**, T222–T223.

Wave Motion

Bakker, M. C. M., M. D. Verweij, B. J. Kooij, and H. A. Dieterman, 1999: The traveling point load revisited. *Wave Motion*, **29**, 119–135.

Barr, G. E., 1967: On the diffraction of a cylindrical pulse by a half-plane. *Quart. Appl. Math.*, **25**, 193–204.

Berg, P., F. If, P. Nielsen, and O. Skovgaard, 1993: Diffraction by a wedge in an acoustic constant density medium. *Geophys. Prospect.*, **41**, 803–831.

Chen, P., and Y.-H., Pao, 1977: The diffraction of sound pulses by a circular cylinder. *J. Math. Phys.*, **18**, 2397–2406.

Chwalczyk, F., J. Rafa, and E. Włodarczyk, 1978: Propagation of three component non-stationary acoustic waves in layer of ideal liquid. *J. Tech. Phys.*, **19**, 467–480.

De Hoop, A. T., and J. H. M. T. van der Hijden, 1983: Generation of acoustic waves by an impulsive line source in a fluid/solid configuration with a plane boundary. *J. Acoust. Soc. Am.*, **74**, 333–342.

De Hoop, A. T., and J. H. M. T. van der Hijden, 1984: Generation of acoustic waves by an impulsive point source in a fluid/solid configuration with a plane boundary. *J. Acoust. Soc. Am.*, **75**, 1709–1715.

De Hoop, A. T., 1990: Acoustic radiation from an impulsive point source in a continuously layered fluid – An analysis based on the Cagniard method. *J. Acoust. Soc. Am.*, **88**, 2376–2388.

Felsen, L. B., 1965: Transient solutions for a class of diffraction problems. *Quart. Appl. Math.*, **23**, 151–169.

Friedland, A. B., and A. D. Pierce, 1969: Reflection of acoustic pulses from stable and unstable interfaces between moving fluids. *Phys. Fluids*, **12**, 1148–1159.

Gopalsamy, K., and B. D. Aggarwala, 1972: Propagation of disturbances from randomly moving sources. *Z. Angew. Math. Mech.*, **52**, 31–35.

Haak, K. F. I., and B. J. Kooij, 1996: Transient acoustic diffraction in a fluid layer. *Wave Motion*, **23**, 139–164.

Harris, J. G., 1980: Diffraction by a crack of a cylindrical longitudinal pulse. *Z. Angew. Math. Phys.*, **31**, 367–383.

Knopoff, L., F. Gilbert, and W. L. Pilant, 1960: Wave propagation in a medium with a single layer. *J. Geophys. Res.*, **65**, 265–278.

Kooij, B. J., and C. Kooij, 2001: The reflected and transmitted acoustic field excited by a supersonic source in a two media configuration. *J. Acoust. Soc. Am.*, **110**, 669–681.

Kurin, V. V., B. E. Nemtsov, and B. Ya. Éidman, 1985: Precursor and lateral waves during pulse reflection from the separation boundary of two media. *Sov. Phys. Usp.*, **28**, 827–841.

Pao, Y.-H., F. Ziegler, and Y.-S. Wang, 1989: Acoustic waves generated by a point source in a sloping fluid layer. *J. Acoust. Soc. Am.*, **85**, 1414–1426.

Rose, L. R. F., 1984: Point-source representation for lasar-generated ultrasound. *J. Acoust. Soc. Am.*, **75**, 723–732.

Rottbrand, K., 1998: Time-dependent plane wave diffraction by a half-plane: Explicit solution for Rawlins' mixed initial boundary value problem. *Z. Angew. Math. Mech.*, **78**, 321–334.

Ryan, R. L., 1971: Pulse propagation in a transversely isotropic half-space. *J. Sound Vib.*, **14**, 511–524.

Sakai, Y., and I. Kawasaki, 1990: Analytic waveforms for a line source in a transversely isotropic medium. *J. Geophys. Res.*, **95**, 11333–11344.

Shmuely, M., 1974: Response of plates to transient sources. *J. Sound Vib.*, **32**, 491–506.

Teng, Y. C., J. T. Kuo, and C. Gong, 1975: Three-dimensional acoustic wave scattering and diffraction by an open-ended vertical soft cylinder in a half-space. *J. Acoust. Soc. Am.*, **57**, 782–790.

Van der Hijden, J. H. M. T., 1987: Radiation from an impulsive line source in an unbounded homogeneous anisotropic medium. *Geophys. J. R. Astr. Soc.*, **91**, 355–372.

Verweij, M. D., 1995: Modeling space-time domain acoustic wave fields in media with attenuation: The symbolic manipulation approach. *J. Acoust. Soc. Am.*, **97**, 831–843.

Watanabe, K., 1978: Transient response of an acoustic half-space to a rotating point load. *Quart. Appl. Math.*, **36**, 39–48.

Williams, D. P., and R. V. Craster, 2000: Cagniard–De Hoop path perturbations with applications to nongeometric wave arrivals. *J. Eng. Math.*, **37**, 253–272.

Zemell, S. H., 1976: New derivation of the exact solution for the diffraction of a cylindrical or spherical pulse on a wedge. *Int. J. Eng. Sci.*, **14**, 845–851.

Zhao, X., 2003: The stress intensity factor history for an advancing crack in a transversely isotropic solid under 3-D loading. *Int. J. Solids Struct.*, **40**, 89–103.

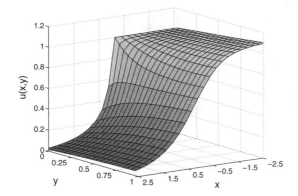

Chapter 7
The Wiener–Hopf Technique

One of the difficulties in solving partial differential equations by Fourier transforms is finding a general transform that applies over the entire spatial domain. The Wiener–Hopf technique is a popular method that avoids this problem by defining Fourier transforms over certain regions and then uses function-theoretic analysis to piece together the complete solution.

Although Wiener and Hopf[1] first devised this method to solve singular integral equations of the form

$$f(x) = \int_0^\infty K(x - y)f(y)\,dy + \varphi(x), \qquad 0 < x < \infty, \qquad (7.0.1)$$

that had arisen in Hopf's 1928 work on the Milne–Schwarzschild equation,[2] it has been in boundary-value problems that it has found its greatest applicability. For example, this technique reduces the problem of diffraction by a semi-infinite plate to the solution of a singular integral equation.[3] The

[1] Wiener, N., and E. Hopf, 1931: Über eine Klasse singulärer Integralgleichungen. *Sitz. Ber. Preuss. Akad. Wiss., Phys.-Math. Kl.*, 696–706.

[2] Shore, S. N., 2002: The evolution of radiative transfer theory from atmospheres to nuclear reactors. *Hist. Math.*, **29**, 463–489.

[3] Magnus, W., 1941: Über die Beugung electromagnetischer Wellen an einer Halbebene. *Z. Phys.*, **117**, 168–179.

Figure 7.0.1: One of the great mathematicians of the twentieth century, Norbert Wiener (1894–1964) graduated from high school at the age of eleven and Tufts at fourteen. Obtaining a doctorate in mathematical logic at 18, he repeatedly traveled to Europe for further education. His work extends over an extremely wide range from stochastic processes to harmonic analysis to cybernetics. (Photo courtesy of the MIT Museum.)

Wiener–Hopf technique[4] then yields the classic result given by Sommerfeld.[5]

Since its original formulation, the Wiener–Hopf technique has undergone simplification by formulating the problem in terms of dual integral equations.[6] The essence of this technique is the process of *factorization* of the Fourier transform of the kernel function into the product of two other Fourier transforms which are analytic and nonzero in certain half planes.

Before we plunge into the use of the Wiener–Hopf technique for solving partial differential equations, let us focus our attention on the mechanics of the method itself. To this end, let us solve the integral equation

$$f(x) = g(x) + \frac{i}{2\kappa_<} \left(\kappa_>^2 - \kappa_<^2 \right) \int_0^\infty e^{i\kappa_< |x-\xi|} f(\xi) \, d\xi, \qquad (7.0.2)$$

where

$$g(x) = \begin{cases} \dfrac{i}{2\kappa_<} e^{-i\kappa_< x}, & x < 0, \\[2ex] \left(x + \dfrac{i}{2\kappa_<} \right) e^{i\kappa_< x}, & x > 0, \end{cases} \qquad (7.0.3)$$

[4] Copson, E. T., 1946: On an integral equation arising in the theory of diffraction. *Quart. J. Math.*, **17**, 19–34.

[5] Sommerfeld, A., 1896: Mathematische Theorie der Diffraction. *Math. Ann.*, **47**, 317–374.

[6] Kaup, S. N., 1950: Wiener–Hopf techniques and mixed boundary value problems. *Comm. Pure Appl. Math.*, **3**, 411–426; Clemmow, P. C., 1951: A method for the exact solution of a class of two-dimensional diffraction problems. *Proc. R. Soc. London, Ser. A*, **205**, 286–308. See Noble, B., 1958: *Methods Based on the Wiener–Hopf Technique for the Solution of Partial Differential Equations*. Pergamon Press, 246 pp.

Figure 7.0.2: Primarily known for his work on topology and ergodic theory, Eberhard Frederich Ferdinand Hopf (1902–1983) received his formal education in Germany. It was during an extended visit to the United States that he worked with Norbert Wiener on what we now know as the Wiener–Hopf technique. Returning to Germany in 1936, he would eventually become an American citizen (1949) and a professor at Indiana University. (Photo courtesy of the MIT Museum.)

and $\mathrm{Im}(\kappa_>)$, $\mathrm{Im}(\kappa_<) \geq \delta > 0$. This integral equation was constructed by Grzesik and Lee[7] to illustrate how various transform methods can be applied to electromagnetic scattering problems.

We intend to solve Equation 7.0.2 via Fourier transforms. An important aspect of the Wiener–Hopf method is the splitting of the Fourier transform into two parts: $F(k) = F_+(k) + F_-(k)$, where

$$F_-(k) = \int_0^\infty f(x)e^{-ikx}\,dx \qquad (7.0.4)$$

and

$$F_+(k) = \int_{-\infty}^0 f(x)e^{-ikx}\,dx. \qquad (7.0.5)$$

---

[7] Grzesik, J. A., and S. C. Lee, 1995: The dielectric half space as a test bed for transform methods. *Radio Sci.*, **30**, 853–862. ©1995 American Geophysical Union. Reproduced/modified by permission of American Geophysical Union.

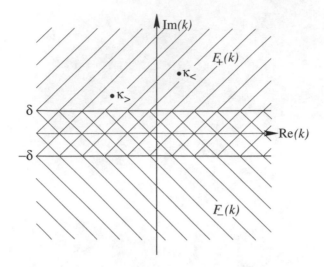

Figure 7.0.3: The location of the half-planes $F_+(k)$ and $F_-(k)$, as well as $\kappa_<$ and $\kappa_>$, in the complex k-plane used in solving Equation 7.0.2 by the Wiener–Hopf technique.

Because the integral in Equation 7.0.4 converges only if $\text{Im}(k) < 0^+$ when $x > 0$, we have added the subscript "$-$" to denote its analyticity in the half-space below $\text{Im}(k) < 0^+$ in the k-plane. Similarly, the integral in Equation 7.0.5 converges only where $\text{Im}(k) > 0^-$ and the plus sign denotes its analyticity in the half-space above $\text{Im}(k) > 0^-$ in the k-plane. We will refine these definitions shortly.

Direct computation of $G(k)$ gives

$$G(k) = \frac{1}{2\kappa_<(k - \kappa_<)} - \frac{1}{2\kappa_<(k + \kappa_<)} - \frac{1}{(k + \kappa_<)^2}. \qquad (7.0.6)$$

Note that this transform is analytic in the strip $|\text{Im}(k)| < \delta$. Next, taking the Fourier transform of Equation 7.0.2, we find that

$$F_+(k) + F_-(k) = G(k) + \frac{\kappa_>^2 - \kappa_<^2}{k^2 - \kappa_<^2} F_-(k), \qquad (7.0.7)$$

or in the more symmetrical form:

$$\frac{k + \kappa_<}{k + \kappa_>} F_+(k) + \frac{k - \kappa_>}{k - \kappa_<} F_-(k) = \frac{k + \kappa_<}{k + \kappa_>} G(k). \qquad (7.0.8)$$

This is permissible as long as $F_+(k)$ is analytic in the half-space $\text{Im}(k) > -\delta$, and $F_-(k)$ is analytic in the half-space $\text{Im}(k) < \delta$. See Figure 7.0.3.

Let us focus on Equation 7.0.8. A quick check shows that the first term on the left side is analytic in the same half-plane as $F_+(k)$. Similarly, the second term on the left side is analytic in the same half-plane as $F_-(k)$. This suggests that it might be advantageous to split the right side into two parts where one term is analytic in the same half plane as $F_+(k)$ while the other part is analytic in the half-plane $F_-(k)$. Our first attempt is

$$\frac{k + \kappa_<}{k + \kappa_>} G(k) = -\frac{1}{2\kappa_<(k + \kappa_>)} + \widetilde{G}(k), \qquad (7.0.9)$$

where

$$\widetilde{G}(k) = \left(\frac{k + \kappa_<}{k + \kappa_>}\right) \left[\frac{1}{2\kappa_<(k - \kappa_<)} - \frac{1}{(k - \kappa_<)^2}\right]. \qquad (7.0.10)$$

The first term on the right side of Equation 7.0.9 is analytic in the same half-plane as $F_+(k)$. However, $\widetilde{G}(k)$ remains unsplit because of the simple pole at $k = -\kappa_>$; otherwise, it would be analytic in the same half-plane as $F_-(k)$. To circumvent this difficulty, we add and subtract out the troublesome singularity:

$$\widetilde{G}(k) = \left\{\widetilde{G}(k) - \frac{\text{Res}\left[\widetilde{G}(k); -\kappa_>\right]}{k + \kappa_>}\right\} + \frac{\text{Res}\left[\widetilde{G}(k); -\kappa_>\right]}{k + \kappa_>} = \widetilde{G}_-(k) + \widetilde{G}_+(k),$$

$$(7.0.11)$$

where

$$\widetilde{G}_+(k) = \left(\frac{\kappa_> - \kappa_<}{k + \kappa_>}\right) \left[\frac{1}{2\kappa_<(\kappa_> + \kappa_<)} + \frac{1}{(\kappa_> + \kappa_<)^2}\right]. \qquad (7.0.12)$$

Substituting Equation 7.0.11 into Equation 7.0.9, we can then rewrite Equation 7.0.8 as

$$\frac{k + \kappa_<}{k + \kappa_>} F_+(k) - \widetilde{G}_+(k) + \frac{1}{2\kappa_<(k + \kappa_>)} = -\left(\frac{k - \kappa_>}{k - \kappa_<}\right) F_-(k) + \widetilde{G}_-(k).$$

$$(7.0.13)$$

Why have we undertaken such an elaborate analysis to obtain Equation 7.0.13? The function on the left side of Equation 7.0.13 is analytic in the half-plane $\text{Im}(k) > -\delta$ while the function on the right side is analytic in the half-plane $\text{Im}(k) < \delta$. By virtue of the principle of analytic continuation, these functions are equal to some function, say $H(k)$, that is analytic over the entire k-plane. We can determine the form of $H(k)$ from the known asymptotic properties of the transform. Here, $H(k)$ must vanish at infinity because (1) $G(k)$ does and (2) both $F_+(k)$ and $F_-(k)$ vanish by the Riemann–Lebesgue theorem. Once we have the asymptotic behavior, we can apply Liouville's theorem:

Liouville's theorem:[8] *If $f(z)$ is analytic for all finite value of z, and as $|z| \to \infty$, $f(z) = O(|z|^m)$, then $f(z)$ is a polynomial of degree $\leq m$.*

---

[8] See Titchmarsh, E. C., 1939: *The Theory of Functions*. 2nd Edition. Oxford University Press, Section 2.52.

Here, for example, because $H(k)$ goes to zero as $|k| \to \infty$, $m = 0$ and $H(k) = 0$. Thus, both sides of Equation 7.0.13 vanish.

The only remaining task is to invert the Fourier transforms and to obtain $f(x)$. Because $F(k) = F_+(k) + F_-(k)$,

$$f(x) = \frac{1}{2\pi} \int_{-\infty}^{\infty} [F_+(k) + F_-(k)]\, e^{ikx}\, dk. \tag{7.0.14}$$

We will evaluate Equation 7.0.14 using the residue theorem, just as we did in Chapter 3 and Chapter 5. Since

$$F_+(k) = \frac{(k + \kappa_>)\widetilde{G}_+(k)}{k + \kappa_<} - \frac{1}{2\kappa_<(k + \kappa_<)} \tag{7.0.15}$$

and

$$F_-(k) = \frac{2\kappa_<(k - \kappa_> - 2\kappa_<)}{(k - \kappa_>)(k - \kappa_<)(\kappa_> + \kappa_<)^2}, \tag{7.0.16}$$

we find that

$$f(x) = \frac{2i\kappa_<}{(\kappa_> + \kappa_<)^2} e^{-i\kappa_< x} \tag{7.0.17}$$

if $x < 0$, while for $x > 0$,

$$f(x) = \frac{2\kappa_<}{i\,(\kappa_>^2 - \kappa_<^2)} \left(\frac{2\kappa_<}{\kappa_> + \kappa_<} e^{i\kappa_> x} - e^{i\kappa_< x} \right). \tag{7.0.18}$$

Having outlined the mechanics behind solving a Wiener–Hopf problem, we are ready to see how this method is used to solve partial differential equations. The crucial step in the procedure is the ability to break all of the Fourier transforms into two functions, one part is analytic on some upper half-plane while the other is analytic on some lower half-plane. For example, we split $G(k)$ into $G_+(k)$ plus $G_-(k)$. Most often, this factor is in the form of a product: $G(k) = G_+(k)G_-(k)$. In the next two sections, we will show various problems which illustrate various types of factorization.

Problems

1. Consider the following equation that appeared in a Wiener–Hopf analysis by Lehner et al.:[9]

$$\sqrt{\omega^2 + \lambda^2}\, U_+(\omega) = \frac{1}{\omega - \kappa} - T_-(\omega), \tag{1}$$

---

[9] Lehner, F. K., V. C. Li, and J. R. Rice, 1981: Stress diffusion along rupturing plate boundaries. *J. Geophys. Res.*, **86**, 6155–6169. ©1981 American Geophysical Union. Reproduced/modified by permission of American Geophysical Union.

where $0 < \lambda$, $\kappa < 0$, $U_+(\omega)$ is analytic in the upper complex ω-plane $0 < \mathrm{Im}(\omega)$ and $T_-(\omega)$ is analytic in the lower half-plane $\mathrm{Im}(\omega) < \tau$, $0 < \tau < \lambda$.

Step 1: Factoring the square root as $\sqrt{\omega + \lambda i}\,\sqrt{\omega - \lambda i}$ so that

$$\sqrt{\omega + \lambda i}\,U_+(\omega) = \frac{1}{(\omega - \kappa)\sqrt{\omega - \lambda i}} - \frac{T_-(\omega)}{\sqrt{\omega - \lambda i}}, \tag{2}$$

show that left side of (2) is analytic in the upper half-plane $0 < \mathrm{Im}(\omega)$, while the second term on the right side is analytic in the lower half-plane $\mathrm{Im}(\omega) < \tau$.

Step 2: Show that the first term on the right side of (2) is neither analytic on the half-plane $0 < \mathrm{Im}(\omega)$ nor on the lower half-plane $\mathrm{Im}(\omega) < \tau$ due to a simple pole that lies in the lower half-plane $\mathrm{Im}(\omega) < 0$.

Step 3: Show that we can split this troublesome term as follows:

$$\frac{1}{(\omega - \kappa)\sqrt{\omega - \lambda i}} = \frac{1}{(\omega - \kappa)\sqrt{\kappa - \lambda i}} + \left(\frac{1}{\sqrt{\omega - \lambda i}} - \frac{1}{\sqrt{\kappa - \lambda i}} \right) \frac{1}{\omega - \kappa}, \tag{3}$$

where the first term on the right side of (3) is analytic in the half-plane $0 < \mathrm{Im}(\omega)$ while the second term is analytic in the half-plane $\mathrm{Im}(\omega) < \tau$.

Step 4: Show that the factorization of (1) is

$$\sqrt{\omega + \lambda i}\,U_+(\omega) - \frac{1}{(\omega - \kappa)\sqrt{\kappa - \lambda i}} = \left(\frac{1}{\sqrt{\omega - \lambda i}} - \frac{1}{\sqrt{\kappa - \lambda i}} \right) \frac{1}{\omega - \kappa} - \frac{T_-(\omega)}{\sqrt{\omega - \lambda i}}.$$

7.1 THE WIENER–HOPF TECHNIQUE WHEN THE FACTORIZATION CONTAINS NO BRANCH POINTS

In the previous section we sketched out the essense of the Wiener–Hopf technique. An important aspect of this technique was the process of factorization. There, we re-expressed several functions as a sum of two parts; one part is analytic in some lower half-plane while the other part is analytic in some upper half-plane. More commonly, the splitting occurs as the *product* of two functions. In this section we illustrate how this factorization arises and how the splitting is accomplished during the solution of a partial differential equation.

• Example 7.1.1

Given that $h, \beta > 0$, let us solve the partial differential equation[10]

$$\frac{\partial^2 u}{\partial x^2} + \frac{\partial^2 u}{\partial y^2} - \beta^2 u = 0, \qquad -\infty < x < \infty, \quad 0 < y < 1, \tag{7.1.1}$$

---

[10] Taken from V. T. Buchwald and F. Viera, Linearized evaporation from a soil of finite depth near a wetted region, *Quart. J. Mech. Appl. Math.*, 1996, **49, No. 1**, 49–64, by permission of Oxford University Press.

with the boundary conditions

$$u_y(x, 1) - \beta u(x, 1) = 0, \qquad -\infty < x < \infty, \qquad (7.1.2)$$

$$u(x, 0) = 1, \qquad x < 0, \qquad (7.1.3)$$

and

$$u_y(x, 0) - (h + \beta)u(x, 0) = 0, \qquad 0 < x. \qquad (7.1.4)$$

We begin by defining

$$U(k, y) = \int_{-\infty}^{\infty} u(x, y)e^{ikx}\, dx \quad \text{and} \quad u(x, y) = \frac{1}{2\pi} \int_{-\infty}^{\infty} U(k, y)e^{-ikx}\, dk.$$
$$(7.1.5)$$

Taking the Fourier transform of Equation 7.1.1, we obtain

$$\frac{d^2 U(k, y)}{dy^2} - m^2 U(k, y) = 0, \qquad 0 < y < 1, \qquad (7.1.6)$$

where $m^2 = k^2 + \beta^2$. The solution to this differential equation is

$$U(k, y) = A(k)\cosh(my) + B(k)\sinh(my). \qquad (7.1.7)$$

Substituting Equation 7.1.7 into Equation 7.1.2 after its Fourier transform has been taken, we find that

$$m\left[A(k)\sinh(m) + B(k)\cosh(m)\right] - \beta\left[A(k)\cosh(m) + B(k)\sinh(m)\right] = 0.$$
$$(7.1.8)$$

The Fourier transform of Equation 7.1.3 is

$$A(k) = \frac{1}{ik} + M_+(k), \qquad (7.1.9)$$

where

$$M_+(k) = \int_0^{\infty} u(x, 0)e^{ikx}\, dx. \qquad (7.1.10)$$

Here, we have assumed that $|u(x, 0)|$ is bounded by $e^{-\epsilon x}$ as $x \to \infty$ with $0 < \epsilon \ll 1$. Consequently, $M_+(k)$ is an analytic function in the half-space $\text{Im}(k) > -\epsilon$. On the other hand, the Fourier transform of Equation 7.1.4 is

$$\int_{-\infty}^{\infty} \left[u_y(x, 0) - (h + \beta)u(x, 0)\right]e^{ikx}\, dx$$

$$= \int_{-\infty}^{0} \left[u_y(x, 0) - (h + \beta)u(x, 0)\right]e^{ikx}\, dx$$

$$+ \int_{0}^{\infty} \left[u_y(x, 0) - (h + \beta)u(x, 0)\right]e^{ikx}\, dx, \qquad (7.1.11)$$

or

$$mB - (h + \beta)A = L_-(k),$$ (7.1.12)

where

$$L_-(k) = \int_{-\infty}^{0} [u_y(x, 0) - (h + \beta)u(x, 0)] e^{ikx} \, dx,$$ (7.1.13)

and $L_-(k)$ is analytic in the half-space $\text{Im}(k) < 0$. We have used Equation 7.1.4 to simplify the right side of Equation 7.1.11. Eliminating $A(k)$ from Equation 7.1.12,

$$mB = L_-(k) + (h + \beta) \left[\frac{1}{ik} + M_+(k) \right].$$ (7.1.14)

Combining Equation 7.1.8, Equation 7.1.9 and Equation 7.1.14, we have that

$$\left[\frac{1}{ik} + M_+(k) \right] \left[hm \cosh(m) - (\beta^2 + h\beta - m^2) \sinh(m) \right]$$
$$+ [m \cosh(m) - \beta \sinh(m)] L_-(k) = 0.$$ (7.1.15)

With Equation 7.1.15, we have reached the point where we must rewrite it so that it is analytic in the half-plane $\text{Im}(k) < 0$ on one side while the other side is analytic in the half-plane $\text{Im}(k) > -\epsilon$. The difficulty arises from the terms $hm \cosh(m) - (\beta^2 + h\beta - m^2) \sinh(m)$ and $[m \cosh(m) - \beta \sinh(m)]$. How can we rewrite them so that we can accomplish our splitting? To do this, we now introduce the *infinite product theorem*:

Infinite Product Theorem:[11] *If $f(z)$ is an entire function of z with simple zeros at $z_1, z_2, \ldots$, then*

$$f(z) = f(0) \exp \left[z f'(0)/f(0) \right] \prod_{n=1}^{\infty} \left(1 - \frac{z}{z_n} \right) e^{z/z_n}.$$ (7.1.16)

Let us apply this theorem to $\cosh(m) - \beta \sinh(m)/m$. We find that

$$\cosh(m) + \frac{\beta}{m} \sinh(m) = e^{-\beta} F(k) F(-k),$$ (7.1.17)

where

$$F(k) = e^{-\gamma ki/\pi} \prod_{n=1}^{\infty} \left(1 - \frac{ki}{\lambda_n} \right) e^{ki/(n\pi)},$$ (7.1.18)

γ is Euler's constant, and $\lambda_n > 0$ is the nth root of $\beta \tan(\lambda) = \lambda$. The reason why Equation 7.1.17 and Equation 7.1.18 is useful lies in the fact that $F(k)$

---

[11] See Titchmarsh, *Ibid.*, Section 3.23.

is analytic and nonzero in the half-plane $\text{Im}(k) > -\epsilon$ while $F(-k)$ is analytic and nonzero in the lower half-plane $\text{Im}(k) < 0$. In a similar vein,

$$\cosh(m) - \frac{\beta^2 + h\beta - m^2}{hm}\sinh(m) = e^{-\beta}G(k)G(-k), \qquad \textbf{(7.1.19)}$$

where

$$G(k) = e^{-\gamma ki/\pi}\prod_{n=1}^{\infty}\left(1 - \frac{ki}{\rho_n}\right)e^{ki/(n\pi)}, \qquad \textbf{(7.1.20)}$$

and ρ_n is the nth root of $\tan(\rho) = h\rho/(\beta^2 + h\beta + \rho^2)$. Here, $G(k)$ is analytic in the half-plane $\text{Im}(k) > -\epsilon$ while $G(-k)$ is analytic in the half-plane $\text{Im}(k) < 0$. Substituting Equation 7.1.17 and Equation 7.1.19 into Equation 7.1.15, we obtain

$$\frac{hG(k)M_+(k)}{F(k)} + \frac{F(-k)L_-(k)}{G(-k)} = -\frac{hG(k)}{ikF(k)} = \frac{h}{ik}\left[1 - \frac{G(k)}{F(k)}\right] - \frac{h}{ik}. \qquad \textbf{(7.1.21)}$$

We observe that the first term on the right side of Equation 7.1.21 is analytic in the upper half-space $\text{Im}(k) > -\epsilon$ while the second term is analytic in the lower half-plane $\text{Im}(k) < 0$. We now rewrite Equation 7.1.21 so that its right side is analytic in the upper half-plane while its left side is analytic in the lower half-plane:

$$\frac{hG(k)M_+(k)}{F(k)} - \frac{h}{ik}\left[1 - \frac{G(k)}{F(k)}\right] = -\frac{F(-k)L_-(k)}{G(-k)} - \frac{h}{ik}. \qquad \textbf{(7.1.22)}$$

At this point we must explore the behavior of both sides of Equation 7.1.22 as $|k| \to \infty$. Applying asymptotic analysis, Buchwald and Viera showed that $G(k)/F(k) \sim k^{1/2}$. Since $M_+(k) \sim k^{-1}$, the first term on the right side of Equation 7.1.21 behaves as $k^{-1/2}$. Because $L_-(k) \sim k^{-1/2}$, the second term behaves as k^{-1}. From Liouville's theorem, both sides of Equation 7.1.22 must equal zero, yielding

$$hM_+(k) = \frac{hF(k)}{ikG(k)}\left[1 - \frac{G(k)}{F(k)}\right] \qquad \textbf{(7.1.23)}$$

and

$$L_-(k) = -\frac{hG(-k)}{ikF(-k)}. \qquad \textbf{(7.1.24)}$$

Therefore, from Equation 7.1.5, we have

$$u(x, y) = \frac{1}{2\pi}\int_{-\infty-\epsilon i}^{\infty-\epsilon i}[A(k)\sinh(mz) + B(k)\cosh(mz)]e^{-ikx}\,dk, \qquad \textbf{(7.1.25)}$$

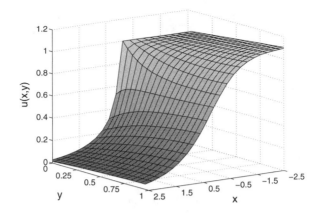

Figure 7.1.1: The solution to Equation 7.1.1 through Equation 7.1.4 obtained via the Wiener–Hopf technique when $h = 2$ and $\beta = 0.1$.

where $A(k)$ is given by a combination of Equation 7.1.9 and Equation 7.1.23 while $B(k)$ follows from Equation 7.1.14, Equation 7.1.23 and Equation 7.1.24. Consequently,

$$ikmA(k) = \frac{F(k)}{G(k)} \quad \text{and} \quad ikmB(k) = (h + \beta)\frac{F(k)}{G(k)} - h\frac{G(-k)}{F(-k)}. \quad \textbf{(7.1.26)}$$

Finally, we apply the residue theorem to evaluate Equation 7.1.25 and find that

$$u(x, y) = e^{\beta y} + he^{-\beta} \sum_{n=1}^{\infty} \frac{\mu_n^2 G(-i\lambda_n) F(i\lambda_n)}{\lambda_n^2 (\lambda_n^2 - \beta) \sin(\mu_n)} \sin(\mu_n y)e^{\lambda_n x}, \qquad x < 0,$$

$$\textbf{(7.1.27)}$$

where $\mu_n^2 = \lambda_n^2 - \beta^2$, and

$$u(x, y) = he^{-\beta} \sum_{n=1}^{\infty} B_n[(h + \beta)\sin(\sigma_n y) + \sigma_n \cos(\sigma_n y)]e^{-\rho_n x}, \qquad 0 < x,$$

$$\textbf{(7.1.28)}$$

where

$$B_n = \frac{\sigma_n(\rho_n^2 + h\beta)F(-i\rho_n)G(i\rho_n)}{\rho_n^2[(\rho_n^2 + h\beta)(\rho_n^2 + h\beta - h) + h(h + 2)\sigma_n^2]\cos(\sigma_n)} \qquad \textbf{(7.1.29)}$$

and $\sigma_n^2 = \rho_n^2 - \beta^2$. Figure 7.1.1 illustrates this solution when $h = 2$ and $\beta = 0.1$.

• Example 7.1.2

The Wiener–Hopf technique is often applied to diffraction problems. To illustrate this in a relatively simple form, consider an infinitely long channel

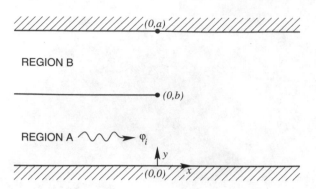

Figure 7.1.2: Schematic of the rotating channel in which Kelvin waves are diffracted.

$-\infty < x < \infty$, $0 < y < a$ rotating on a flat plate[12] filled with an inviscid, homogeneous fluid of uniform depth H. Within the channel, we have another plate of infinitesimal thickness located at $x < 0$ and $y = b$. See Figure 7.1.2. The shallow-water equations govern the motion of the fluid:

$$-i\omega u - fv = -\frac{\partial h}{\partial x}, \tag{7.1.30}$$

$$-i\omega v + fu = -\frac{\partial h}{\partial y}, \tag{7.1.31}$$

and

$$-i\omega h + gH\left(\frac{\partial u}{\partial x} + \frac{\partial v}{\partial y}\right) = 0, \tag{7.1.32}$$

where u and v are the velocities in the x and y directions, respectively, h is the deviation of the free surface from its average height H, g is the gravitational acceleration and f is one half of the angular velocity at which it rotates. All motions within the fluid behave as $e^{-i\omega t}$.

A little algebra shows that we can eliminate u and v and obtain the Helmholtz equation:

$$\frac{\partial^2 h}{\partial x^2} + \frac{\partial^2 h}{\partial y^2} + k^2 h = 0, \tag{7.1.33}$$

where

$$(\omega^2 - f^2)u = -i\omega\frac{\partial h}{\partial x} + f\frac{\partial h}{\partial y}, \tag{7.1.34}$$

$$(\omega^2 - f^2)v = -i\omega\frac{\partial h}{\partial y} - f\frac{\partial h}{\partial x}, \tag{7.1.35}$$

---

[12] Taken from Kapoulitsas, G. M., 1980: Scattering of long waves in a rotating bifurcated channel. *Int. J. Theoret. Phys.*, **19**, 773–788.

and $k^2 = (\omega^2 - f^2)/(gH)$ assuming that $\omega > f$. We solve the problem when an incident wave of the form

$$h = \phi_i = \exp[(i\omega x - fy)/c] \tag{7.1.36}$$
$$= \exp\{k[ix\cosh(\beta) - y\sinh(\beta)]\}, \tag{7.1.37}$$

a so-called "Kelvin wave," propagates toward the origin from $-\infty$ within the lower channel A. See Figure 7.1.2. We have introduced β such that $f = kc\sinh(\beta)$, $\omega = kc\cosh(\beta)$ and $c^2 = gH$.

The first step in the Wiener–Hopf technique is to write the solution as a sum of the incident wave plus some correction ϕ that represents the reflected and transmitted waves. For example, in Region A, h consists of $\phi_i + \phi$ whereas in Region B we have only ϕ. Because ϕ_i satisfies Equation 7.1.33, so must ϕ. Furthermore, ϕ must satisfy certain boundary conditions. Because of the rigid walls, v must vanish along them; this yields

$$\frac{\partial \phi}{\partial y} - i\tanh(\beta)\frac{\partial \phi}{\partial x} = 0 \tag{7.1.38}$$

along $-\infty < x < \infty$, $y = 0, a$ and $x < 0$, $y = b^{\pm}$. Furthermore, because the partition separating Region A from Region B is infinitesimally thin, we must have continuity of v across that boundary; this gives

$$\left[\frac{\partial \phi}{\partial y} - i\tanh(\beta)\frac{\partial \phi}{\partial x}\right]_{y=b^-} = \left[\frac{\partial \phi}{\partial y} - i\tanh(\beta)\frac{\partial \phi}{\partial x}\right]_{y=b^+} \tag{7.1.39}$$

for $-\infty < x < \infty$. Finally, to prevent infinite velocities in the right half of the channel, h must be continuous at $z = b$, or

$$\phi(x, b^-) + \phi_i(x, b) = \phi(x, b^+) \tag{7.1.40}$$

for $x > 0$.

An important assumption in the Wiener–Hopf technique concerns the so-called "edge conditions" at the edge point $(0, b)$; namely, that

$$\phi = O(1) \quad \text{as} \quad x \to 0^{\pm} \quad \text{and} \quad y = b, \tag{7.1.41}$$

and

$$\frac{\partial \phi}{\partial y} = O(x^{-1/2}) \quad \text{as} \quad x \to 0^{\pm} \quad \text{and} \quad y = b. \tag{7.1.42}$$

These conditions are necessary to guarantee the uniqueness of the solution because the edge point is a geometric singularity. Another assumption introduces dissipation by allowing ω to have a small, positive imaginary part. We may either view this as merely reflecting reality or ensuring that we satisfy the Sommerfeld radiation condition that energy must radiate to infinity. As a result of this complex form of ω, k must also be complex with a small, positive imaginary part k_2.

We solve Equation 7.1.33, as well as Equation 7.1.38 through Equation 7.1.40, by Fourier transforms. Let us define the double-sided Fourier transform of $\phi(x, y)$ by

$$\Phi(\alpha, y) = \int_{-\infty}^{\infty} \phi(x, y) e^{i\alpha x} \, dx, \qquad |\tau| < \tau_0, \qquad (7.1.43)$$

as well as the one-sided Fourier transforms

$$\Phi_-(\alpha, y) = \int_{-\infty}^{0} \phi(x, y) e^{i\alpha x} \, dx, \qquad \tau < \tau_0, \qquad (7.1.44)$$

and

$$\Phi_+(\alpha, y) = \int_{0}^{\infty} \phi(x, y) e^{i\alpha x} \, dx, \qquad -\tau_0 < \tau, \qquad (7.1.45)$$

where $\alpha = \sigma + i\tau$ and σ, τ are real. Clearly,

$$\Phi_-(\alpha, y) + \Phi_+(\alpha, y) = \Phi(\alpha, y). \qquad (7.1.46)$$

Note that Equation 7.1.43 through Equation 7.1.45 are analytic in a common strip in the complex α-plane.

Taking the double-sided Fourier transform of Equation 7.1.33,

$$\frac{d^2 \Phi}{dy^2} - \gamma^2 \Phi = 0, \qquad (7.1.47)$$

where $\gamma = \sqrt{\alpha^2 - k^2}$. In general,

$$\Phi(\alpha, y) = \begin{cases} A(\alpha) e^{-\gamma y} + B(\alpha) e^{\gamma y}, & \text{if} \quad 0 \le y \le b, \\ C(\alpha) e^{-\gamma y} + D(\alpha) e^{\gamma y}, & \text{if} \quad b \le y \le a. \end{cases} \qquad (7.1.48)$$

Taking the double-sided Fourier transform of Equation 7.1.38 for the conditions on $y = 0, a$,

$$\Phi'(\alpha, 0) - \alpha \tanh(\beta) \Phi(\alpha, 0) = 0 \qquad (7.1.49)$$

and

$$\Phi'(\alpha, a) - \alpha \tanh(\beta) \Phi(\alpha, a) = 0. \qquad (7.1.50)$$

Similarly, from Equation 7.1.39,

$$\Phi'(\alpha, b^-) - \alpha \tanh(\beta) \Phi(\alpha, b^-) = \Phi'(\alpha, b^+) - \alpha \tanh(\beta) \Phi(\alpha, b^+). \qquad (7.1.51)$$

From the boundary conditions given by Equation 7.1.49 through Equation 7.1.51,

$$B = \lambda A, \qquad (7.1.52)$$

$$C = -A\frac{\sinh(\gamma b)}{\sinh[\gamma(a-b)]}e^{\gamma a}, \tag{7.1.53}$$

and

$$D = -\lambda A\frac{\sinh(\gamma b)}{\sinh[\gamma(a-b)]}e^{-\gamma a}, \tag{7.1.54}$$

where

$$\lambda = \frac{\gamma + \alpha \tanh(\beta)}{\gamma - \alpha \tanh(\beta)}. \tag{7.1.55}$$

By taking the one-sided Fourier transform of Equation 7.1.38 (from $-\infty$ to 0) for the condition $x < 0$, we obtain the two equations:

$$\Phi'_-(\alpha, b^+) - \alpha \tanh(\beta)\Phi_-(\alpha, b^+) = i \tanh(\beta)\phi(0, b), \tag{7.1.56}$$

and

$$\Phi'_-(\alpha, b^-) - \alpha \tanh(\beta)\Phi_-(\alpha, b^-) = i \tanh(\beta)\phi(0, b), \tag{7.1.57}$$

because $\phi(0, b^-) = \phi(0, b^+) = \phi(0, b)$. Using Equation 7.1.56 and Equation 7.1.57 in conjunction with Equation 7.1.51,

$$\begin{aligned}\Phi'_+(\alpha, b^-) &- \alpha \tanh(\beta)\Phi_+(\alpha, b^-) + i \tanh(\beta)\phi(0, b) \\ &= \Phi'_+(\alpha, b^+) - \alpha \tanh(\beta)\Phi_+(\alpha, b^+) + i \tanh(\beta)\phi(0, b) \\ &= P_+(\alpha).\end{aligned} \tag{7.1.58}$$

Finally, the one-sided Fourier transform (from 0 to ∞) of Equation 7.1.40 yields

$$\Phi_+(\alpha, b^-) + \frac{i \exp[-kb \sinh(\beta)]}{\alpha + k \cosh(\beta)} = \Phi_+(\alpha, b^+). \tag{7.1.59}$$

Then, by Equation 7.1.46, Equation 7.1.48, Equation 7.1.56 through Equation 7.1.58 and the definition of B,

$$P_+(\alpha) = 2A[\gamma + \alpha \tanh(\beta)]\sinh(\gamma b). \tag{7.1.60}$$

Let us now introduce the function $Q_-(\alpha)$ defined by

$$Q_-(\alpha) = \tfrac{1}{2}[\Phi_-(\alpha, b^-) - \Phi(\alpha, b^+)]. \tag{7.1.61}$$

From Equation 7.1.48, Equation 7.1.51, Equation 7.1.53, Equation 7.1.55, Equation 7.1.59 and $D = \lambda Ce^{-2\gamma a}$,

$$2Q_-(\alpha) - \frac{i \exp[-kb \sinh(\beta)]}{\alpha + k \cosh(\beta)} = \frac{2A\gamma \sinh(\gamma a)}{[\gamma - \alpha \tanh(\beta)]\sinh[\gamma(a-b)]}. \tag{7.1.62}$$

Eliminating A between Equation 7.1.60 and Equation 7.1.62, we finally obtain a functional equation of the Wiener–Hopf type:

$$2Q_-(\alpha) - \frac{i \exp[-kb \sinh(\beta)]}{\alpha + k \cosh(\beta)} = \frac{P_+(\alpha)}{\gamma^2 - \alpha^2 \tanh(\beta)} \times \frac{\gamma \sinh(\gamma a)}{\sinh(\gamma b)\sinh[\gamma(a-b)]}. \tag{7.1.63}$$

What makes Equation 7.1.63 a functional equation of the Wiener–Hopf type? Note that $Q_-(\alpha)$ is analytic for $\tau < \tau_0$ while $P_+(\alpha)$ is analytic for $\tau > -\tau_0$. In order for Equation 7.1.63 to be true, we must restrict ourselves to the strip $|\tau| < \tau_0$. Thus, the Wiener–Hopf equation contains complex Fourier transforms which are analytic over the common interval of $\tau_- < \tau < \tau_+$, where $Q_-(\alpha)$ is analytic for $\tau < \tau_+$ and $\tau_- < \tau$.

A crucial step in solving the Wiener–Hopf equation Equation 7.1.63 is the process of factorization. In the previous section, we did this by adding and subtracting out a particular singularity. Here, we will rewrite $M(\alpha)$ in terms of the product $M_+(\alpha)M_-(\alpha)$, where $M_+(\alpha)$ and $M_-(\alpha)$ are analytic and free of zeros in an upper and lower half-planes, respectively. These half-planes share a certain strip of the α-plane in common. Applying the infinite product theorem separately to the numerator and denominator of

$$M(\alpha) = \frac{\sinh(\gamma b)\,\sinh[\gamma(a-b)]}{\gamma\sinh(\gamma a)}, \tag{7.1.64}$$

we immediately find that

$$M_+(\alpha) = M_-(-\alpha)$$

$$= \left\{\frac{\sin(kb)\sin[k(a-b)]}{k\sin(ka)}\right\}^{1/2}\exp\left\{\frac{\alpha i}{\pi}\left[b\ln\left(\frac{a}{b}\right)+(a-b)\ln\left(\frac{a}{a-b}\right)\right]\right\}$$

$$\times\prod_{n=1}^{\infty}\left(1+\frac{\alpha}{\alpha_{nb}}\right)e^{ib\alpha/(n\pi)}\prod_{n=1}^{\infty}\left[1+\frac{\alpha}{\alpha_{n(a-b)}}\right]e^{i(a-b)\alpha/(n\pi)}$$

$$\Big/\prod_{n=1}^{\infty}\left(1+\frac{\alpha}{\alpha_{na}}\right)e^{ia\alpha/(n\pi)}, \tag{7.1.65}$$

where $\alpha_{n\ell} = i\sqrt{n^2\pi^2/\ell^2 - k^2}$ and $\ell = a$ or b or $(a-b)$. The square root has a positive real part or a negative imaginary part. Note that in this factorization $M_+(\alpha)$ is analytic and nonzero in the upper half of the α-plane $(-k_2 < \tau)$ while $M_-(\alpha)$ is analytic and nonzero in the lower half of the α-plane $(\tau < k_2)$.

Substituting this factorization into Equation 7.1.63,

$$2[\alpha - k\cosh(\beta)]M_-(\alpha)Q_-(\alpha) - \frac{ie^{-kb\sinh(\beta)}[\alpha - k\cosh(\beta)]M_-(\alpha)}{\alpha + k\cosh(\beta)}$$

$$= \frac{\cosh^2(\beta)}{[\alpha + k\cosh(\beta)]M_+(\alpha)}P_+(\alpha). \tag{7.1.66}$$

Next, we note that

$$\frac{[\alpha - k\cosh(\beta)]M_-(\alpha)}{\alpha + k\cosh(\beta)} = \frac{[\alpha - k\cosh(\beta)]M_-(\alpha) + 2k\cosh(\beta)M_-[-k\cosh(\beta)]}{\alpha + k\cosh(\beta)}$$

$$- \frac{2k\cosh(\beta)M_-[-k\cosh(\beta)]}{\alpha + k\cosh(\beta)}. \tag{7.1.67}$$

Therefore, Equation 7.1.66 becomes

$$2[\alpha - k\cosh(\beta)]M_-(\alpha)Q_-(\alpha) - i\exp[-kb\sinh(\beta)]$$
$$\times \left\{ \frac{[\alpha - k\cosh(\beta)]M_-(\alpha) + 2k\cosh(\beta)M_-[-k\cosh(\beta)]}{\alpha + k\cosh(\beta)} \right\}$$
$$= \frac{\cosh^2(\beta)}{[\alpha + k\cosh(\beta)]M_+(\alpha)}P_+(\alpha)$$
$$- i\exp[-kb\sinh(\beta)]\frac{2k\cosh(\beta)M_-[-k\cosh(\beta)]}{\alpha + k\cosh(\beta)}. \qquad \textbf{(7.1.68)}$$

The primary reason for the factorization and the subsequent algebraic manipulation is the fact that the left side of Equation 7.1.68 is analytic in $-\tau_0 < \tau$ while the right side is analytic in $\tau < \tau_0$. Hence, both sides are analytic on the strip $|\tau| < \tau_0$. Then by analytic continuation it follows that Equation 7.1.68 is defined in the entire α-plane and both sides equal to an entire function $p(\alpha)$. To determine $p(\alpha)$ we examine the asymptotic value of Equation 7.1.68 as $|\alpha| \to \infty$ as well as using the edge conditions, Equation 7.1.41 and Equation 7.1.42. Applying Liouville's theorem, $p(\alpha)$ is a constant. Because in the limit of $|\alpha| \to \infty$, $p(\alpha) \to 0$, then $p(\alpha) = 0$. Therefore, from Equation 7.1.68,

$$P_+(\alpha) = \frac{2ikM_+[k\cosh(\beta)]\exp[-kb\sinh(\beta)]}{\cosh(\beta)}M_+(\alpha). \qquad \textbf{(7.1.69)}$$

Knowing $P_+(\alpha)$, we find from Equation 7.1.52, Equation 7.1.53, Equation 7.1.54 and Equation 7.1.60 that

$$A = \frac{EM_+(\alpha)}{[\gamma + \alpha\tanh(\beta)]\sin(\gamma b)}, \qquad \textbf{(7.1.70)}$$

$$B = \frac{EM_+(\alpha)}{[\gamma - \alpha\tanh(\beta)]\sin(\gamma b)}, \qquad \textbf{(7.1.71)}$$

$$C = -\frac{EM_+(\alpha)e^{\gamma a}}{[\gamma + \alpha\tanh(\beta)]\sin[\gamma(a-b)]}, \qquad \textbf{(7.1.72)}$$

and

$$D = -\frac{EM_+(\alpha)e^{-\gamma a}}{[\gamma - \alpha\tanh(\beta)]\sin[\gamma(a-b)]}, \qquad \textbf{(7.1.73)}$$

where

$$E = \frac{ikM_+[k\cosh(\beta)]\exp[-kb\sinh(\beta)]}{\cosh(\beta)}. \qquad \textbf{(7.1.74)}$$

With these values of A, B, C and D, we have found $\Phi(\alpha, y)$. Therefore, $\phi(x, y)$ follows from the inversion of $\Phi(\alpha, y)$. For example, for $-\infty < x < \infty, 0 \le y \le b$,

$$\phi(x, y) = \frac{E}{2\pi}\int_{-\infty-\epsilon i}^{\infty-\epsilon i} \frac{M_+(\alpha)}{\sinh(\gamma b)}\left[\frac{e^{-\gamma y}}{\gamma + \alpha\tanh(\beta)} + \frac{e^{\gamma y}}{\gamma - \alpha\tanh(\beta)}\right]e^{-i\alpha x}\,d\alpha.$$
$$\textbf{(7.1.75)}$$

For $x < 0$ we evaluate Equation 7.1.75 by closing the integration along the real axis with an infinite semicircle in the upper half of the α-plane by Jordan's lemma and using the residue theorem. The integrand of Equation 7.1.75 has simple poles at $\gamma b = n\pi$, where $n = \pm 1, \pm 2, \ldots$ and the zeros of $\gamma \pm \alpha \tanh(\beta)$. Upon applying the residue theorem,

$$\phi(x, y) = -\frac{k \sinh(\beta) M_+^2[k \cosh(\beta)]}{\sinh[kb \sinh(\beta)]} e^{k[-ix \cosh(\beta) + (y-b) \sinh(\beta)]}$$

$$+ \frac{2\pi iE}{b^2} \sum_{n=1}^{\infty} \frac{(-1)^n n M_+(-\alpha_n)}{\alpha_{nb}[(n\pi/b)^2 + \alpha_{nb}^2 \tanh^2(\beta)]}$$

$$\times \left[\frac{n\pi}{b} \cos\left(\frac{n\pi y}{b}\right) + \alpha_{nb} \tanh(\beta) \sin\left(\frac{n\pi y}{b}\right)\right] e^{-i\alpha_{nb} x}, \quad \textbf{(7.1.76)}$$

where $\alpha_{nb} = i\sqrt{n^2 \pi^2/b^2 - k^2}$. The first term of the right side of Equation 7.1.76 represents the reflected Kelvin wave traveling in the channel ($0 \le y \le b, x < 0$) to the left. The infinite series represents attenuated, stationary modes.

In a similar manner, we apply the residue theorem to obtain the solution in the remaining domains. They are

$$\phi(x, y) = -\frac{\sinh[k(a - b) \sinh(\beta)]}{\sinh[ka \sinh(\beta)]} e^{k[ix \cosh(\beta) - (y+b) \sinh(\beta)]}$$

$$- \frac{2iE}{a} \sum_{n=1}^{\infty} \frac{\sin(n\pi b/a)}{\alpha_{na} M_-(\alpha_{na})[(n\pi/a)^2 + \alpha_{na}^2 \tanh^2(\beta)]}$$

$$\times \left[\frac{n\pi}{a} \cos\left(\frac{n\pi y}{a}\right) - \alpha_{na} \tanh(\beta) \sin\left(\frac{n\pi y}{a}\right)\right] e^{i\alpha_{na} x} \quad \textbf{(7.1.77)}$$

for $0 \le y \le b, 0 < x$, and

$$\phi(x, y) = \frac{k \sinh(\beta) M_+^2[k \cosh(\beta)]}{\sinh[kd \sinh(\beta)]} e^{k[-ix \cosh(\beta) + (y-a-b) \sinh(\beta)]}$$

$$- \frac{2\pi iE}{d^2} \sum_{n=1}^{\infty} \frac{(-1)^n n M_+(\alpha_{nd})}{[(n\pi/d)^2 + \alpha_{nd}^2 \tanh^2(\beta)]}$$

$$\times \left\{\frac{n\pi}{d} \cos\left[\frac{n\pi(y - a)}{d}\right] + \alpha_{nd} \tanh(\beta) \sin\left[\frac{n\pi(y - a)}{d}\right]\right\} e^{-i\alpha_{nd} x}$$

$$\textbf{(7.1.78)}$$

for $b \le y \le a, x < 0$, where $d = a - b$. Finally, for $b \le y \le a, 0 < x$, $\phi(x, y)$ is given by the sum of $\phi_i(x, y)$ and the solution is given by Equation 7.1.77.

Figure 7.1.3 and Figure 7.1.4 illustrate the real and imaginary parts of this solution when $a = 2$, $b = 1$, $k = 1$ and $\beta = 0.5$.

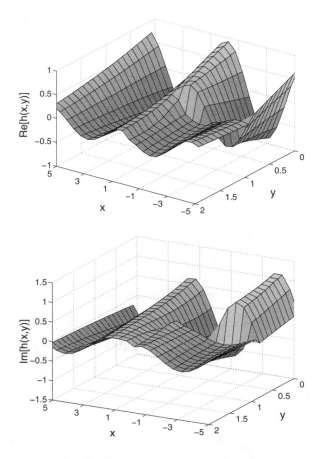

Figure 7.1.3: The real and imaginary parts of the solution to Equation 7.1.33 subject to the boundary conditions given by Equation 7.1.38 through Equation 7.1.40 obtained via the Wiener–Hopf technique when $a = 2$, $b = 1$, $k = 1$ and $\beta = 0.5$.

Problems

1. Use the Wiener–Hopf technique to solve the partial differential equation

$$\frac{\partial^2 u}{\partial x^2} + \frac{\partial^2 u}{\partial y^2} = 0, \qquad -\infty < x < \infty, \quad 0 < y < 1,$$

with the boundary conditions

$$u(x,0) = 0, \qquad -\infty < x < \infty,$$

$$\frac{\partial u(x,1)}{\partial y} = 0, \quad x < 0, \qquad \text{and} \qquad u(x,1) = 1, \quad 0 < x.$$

Step 1: Assuming that $|u(x,1)|$ is bounded by $e^{\epsilon x}$, $0 < \epsilon \ll 1$, as $x \to -\infty$, let us define the following Fourier transforms:

$$U(k,y) = \int_{-\infty}^{\infty} u(x,y)e^{ikx}\,dx, \qquad U_+(k,y) = \int_0^{\infty} u(x,y)e^{ikx}\,dx,$$

and

$$U_-(k,y) = \int_{-\infty}^{0} u(x,y)e^{ikx}\,dx,$$

so that $U(k,y) = U_+(k,y) + U_-(k,y)$. Here, $U_+(k,y)$ is analytic in the half-space $\operatorname{Im}(k) > 0$, while $U_-(k,y)$ is analytic in the half-space $\operatorname{Im}(k) < \epsilon$. Take the Fourier transform of the partial differential equation and the first boundary condition and show that it becomes the boundary-value problem

$$\frac{d^2U}{dy^2} - k^2 U = 0, \qquad 0 < y < 1,$$

with $U(k,0) = 0$.

Step 2: Show that the solution to the boundary-value problem is $U(k,y) = A(k)\sinh(ky)$.

Step 3: From the boundary conditions along $y = 1$, show that

$$\sinh(k)A(k) = L_-(k) + \frac{i}{k} \qquad \text{and} \qquad k\cosh(k)A(k) = M_+(k),$$

where

$$L_-(k) = \int_{-\infty}^{0} u(x,1)e^{ikx}\,dx \quad \text{and} \quad M_+(k) = \int_0^{\infty} u_y(x,1)e^{ikx}\,dx.$$

Step 4: By eliminating $A(k)$ from the equations in Step 3, show that we may factor the resulting equation as

$$L_-(k) + \frac{i}{k} = K(k)M_+(k), \tag{1}$$

where $K(k) = \sinh(k)/[k\cosh(k)]$.

Step 5: Using the infinite product representation[13] for sinh and cosh, show that

$$K(k) = K_+(k)K_-(k), \qquad \text{where} \qquad K_+(k) = K_-(-k) = \prod_{n=1}^{\infty} \frac{1 - ik/(n\pi)}{1 - 2ik/[(2n-1)\pi]}.$$

Step 6: Use the results from Step 5 and show that (1) can be rewritten

$$K_+(k)M_+(k) - \frac{i}{K_-(0)k} = \frac{L_-(k)}{K_-(k)} + \frac{i}{kK_-(k)} - \frac{i}{K_-(0)k}.$$

Note that the left side of the equation is analytic in the upper half-plane $\operatorname{Im}(k) > 0$, while the right side of the equation is analytic in the lower half-plane $\operatorname{Im}(k) < \epsilon$.

---

[13] See, for example, Gradshteyn, I. S., and I. M. Ryzhik, 1965: *Table of Integrals, Series and Products*. Academic Press, Section 1.431, Formulas 2 and 4.

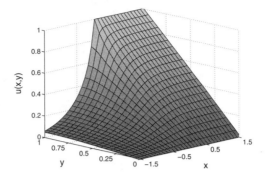

Problem 1

Step 7: Use Liouville's theorem to show that each side of the equation in Step 6 equals zero. Therefore,

$$L_-(k) = \frac{iK_-(k)}{kK_-(0)} - \frac{i}{k}$$

and

$$U(k, y) = \frac{iK_-(k)\sinh(ky)}{k\sinh(k)}.$$

Step 8: Use the inversion integral and show that

$$u(x, y) = \frac{i}{2\pi}\int_{-\infty+\epsilon i}^{\infty+\epsilon i} \frac{K_-(k)\sinh(ky)}{k\sinh(k)}e^{-ikx}\,dk,$$

or

$$u(x, y) = \frac{i}{2\pi}\int_{-\infty+\epsilon i}^{\infty+\epsilon i} \frac{\sinh(ky)}{k^2\,K_+(k)\cosh(k)}e^{-ikx}\,dk.$$

The first integral is best for finding the solution when $x > 0$, while the second integral is best for $x < 0$.

Step 9: Use the residue theorem and show that

$$u(x, y) = y + \sum_{n=1}^{\infty} \frac{(-1)^n}{n\pi} K_-(-n\pi i)\sin(n\pi y)e^{-n\pi x}$$

if $0 < x$, while

$$u(x, y) = \frac{4}{\pi^2}\sum_{n=1}^{\infty} \frac{(-1)^{n+1}}{(2n-1)^2} \frac{\sin[(2n-1)\pi y/2]}{K_+[(2n-1)\pi i/2]} e^{(2n-1)\pi x/2}$$

for $x < 0$. This solution $u(x, y)$ is illustrated in the figure labeled Problem 1.

2. Use the Wiener–Hopf technique to solve the partial differential equation[14]

$$\frac{\partial^2 u}{\partial x^2} + \frac{\partial^2 u}{\partial z^2} - u = 0, \qquad -\infty < x < \infty, \quad 0 < z < 1,$$

---

[14] Taken from Horvay, G., 1961: Temperature distribution in a slab moving from a chamber at one temperature to a chamber at another temperature. *J. Heat Transfer*, **83**, 391–402.

with the boundary conditions $\lim_{|x|\to\infty} u(x,z) \to 0$,

$$\frac{\partial u(x,0)}{\partial z} = 0, \qquad -\infty < x < \infty,$$

$$\frac{\partial u(x,1)}{\partial z} = 0, \quad x < 0, \qquad \text{and} \qquad u(x,1) = e^{-x}, \quad 0 < x.$$

Step 1: Because $u(x,1) = e^{-x}$, we can define the following Fourier transforms:

$$U(k,z) = \int_{-\infty}^{\infty} u(x,z)e^{ikx}\,dx, \qquad U_+(k,z) = \int_0^{\infty} u(x,z)e^{ikx}\,dx,$$

and

$$U_-(k,z) = \int_{-\infty}^0 u(x,z)e^{ikx}\,dx,$$

so that $U(k,z) = U_+(k,z) + U_-(k,z)$. Therefore, $U_+(k,z)$ is analytic in the half-plane $\text{Im}(k) > -1$ while $U_-(k)$ is analytic in the half-plane $\text{Im}(k) < 0$. Show that we may write the partial differential equation and boundary conditions

$$\frac{d^2U}{dz^2} - m^2 U = 0, \qquad 0 < z < 1,$$

with $U'(k,0) = 0$, $U_+(k,1) = 1/(1-ki)$ and $U'_-(k,1) = 0$, where $m^2 = k^2 + 1$.

Step 2: Show that we can write the solution to Step 1 as

$$U(k,z) = A(k)\frac{\cosh(mz)}{\cosh(m)},$$

with $U'_+(k,1) = m\tanh(m)A(k)$ and $A(k) = U_-(k,1) + i/(k+i)$.

Step 3: By eliminating $A(k)$ from the last two equations in Step 2, show that we may factor the resulting equation as

$$K_+(k)U'_+(k,1) = \frac{m^2 A(k)}{K_-(k)} = \frac{(k-i)[(k+i)U_-(k,1) + i]}{K_-(k)} = J,$$

where $m\coth(m) = K_+(k)K_-(k)$. Note that the left side of the equation is analytic in the upper half-plane $\text{Im}(k) > -1$, while the right side of the equation is analytic in the lower half-plane $\text{Im}(k) < 0$.

Step 4: It may be shown that $K_-(k) \sim |k|^{1/2}$ as $|k| \to \infty$. Show that $m^2 A(k)/K_-(k)$ cannot increase faster than $|k|^{1/2}$. Then use Liouville's theorem to show that each side equals a constant value J.

Step 5: Use the results from Step 4 to show that $J = 2/K_-(-i)$.

Step 6: From the infinite product theorem we have $K_+(-k) = K_-(k) = \Omega(ik)$, where

$$\Omega(z) = \prod_{n=1}^{\infty} \frac{\dfrac{z}{(n-1/2)\pi} + \sqrt{1 + \dfrac{1}{(n-1/2)^2\pi^2}}}{\dfrac{z}{n\pi} + \sqrt{1 + \dfrac{1}{n^2\pi^2}}}.$$

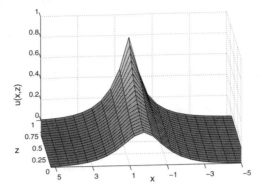

Problem 2

Use this information and show that

$$u(x,z) = \frac{1}{\pi\Omega(1)} \int_{-\infty-i}^{\infty-i} \frac{\Omega(ik)\,\cosh(mz)}{(k^2+1)\cosh(m)} e^{-ikx}\,dk.$$

Step 7: Use the residue theorem and show that

$$u(x,z) = e^{-x} - \frac{2i}{\Omega(1)} \sum_{n=0}^{\infty} \frac{\Omega(i\alpha_n)\,\cosh(\mu_n z)}{\mu_n \alpha_n \sinh(\mu_n)} e^{-i\alpha_n x},$$

where $0 < x$, $\mu_n = (n+1/2)\pi i$ and $\alpha_n = -i\sqrt{1+(2n+1)^2\pi^2/4}$, and

$$u(x,z) = \frac{e^x}{\Omega^2(1)} + \frac{2i}{\Omega(1)} \sum_{n=1}^{\infty} \frac{\cosh(\mu_n z)e^{-i\alpha_n x}}{\alpha_n \Omega(-i\alpha_n)\cosh(\mu_n)},$$

where $x < 0$, $\mu_n = n\pi i$ and $\alpha_n = i\sqrt{1+n^2\pi^2}$. This solution $u(x,z)$ is illustrated in the figure labeled Problem 2.

3. Use the Wiener–Hopf technique to solve the partial differential equation

$$\frac{\partial^2 u}{\partial x^2} + \frac{\partial^2 u}{\partial y^2} - \beta^2 u = 0, \qquad -\infty < x < \infty, \quad 0 < y < 1,$$

with the boundary conditions

$$u(x,1) = 0, \qquad -\infty < x < \infty,$$

$$u(x,0) = 0, \quad x < 0, \qquad \text{and} \qquad \frac{\partial u(x,0)}{\partial y} - hu(x,0) = 1, \quad 0 < x,$$

where $h > 0$.

Step 1: Assuming that $|u(x,0)|$ is bounded by $e^{\epsilon x}$, $0 < \epsilon \ll 1$, as $x \to -\infty$, let us introduce the Fourier transforms

$$U(k,y) = \int_{-\infty}^{\infty} u(x,y)e^{ikx}\,dx, \qquad U_{+}(k,y) = \int_{0}^{\infty} u(x,y)e^{ikx}\,dx,$$

and

$$U_{-}(k,y) = \int_{-\infty}^{0} u(x,y)e^{ikx}\,dx,$$

so that $U(k,y) = U_{+}(k,y) + U_{-}(k,y)$. Here $U_{+}(k,y)$ is analytic in the half-plane $\operatorname{Im}(k) > 0$ while $U_{-}(k,y)$ is analytic in the half-plane $\operatorname{Im}(k) < \epsilon$. Show that we may write the partial differential equation and boundary conditions as the boundary-value problem

$$\frac{d^2 U}{dy^2} - m^2 U = 0, \qquad 0 < y < 1,$$

with $U'(k,1) = 0$, $U'_{+}(k,0) - hU_{+}(k,0) = i/k$ and $U'_{-}(k,0) = 0$, where $m^2 = k^2 + \beta^2$.

Step 2: Show that the solution to Step 1 is $U(k,y) = A(k)\sinh[m(1-y)]$,

$$U'_{+}(k,0) = \sinh(m)A(k) \quad \text{and} \quad -\left[m\cosh(m) + h\sinh(m)\right]A(k) = U'_{-}(k,0) + \frac{i}{k}, \qquad (1)$$

where

$$U'_{-}(k,0) = \int_{-\infty}^{0} u_y(x,0)e^{ikx}\,dx.$$

Step 3: It may be shown[15] that $m\coth(m) + h$ can be factorized as follows:

$$m\coth(m) + h = K(0)P(k)P(-k), \qquad P(k) = \prod_{n=1}^{\infty} \left(1 - \frac{ik}{\rho_n}\right)\left(1 - \frac{ik}{\sqrt{n^2\pi^2 + \beta^2}}\right)^{-1},$$

where $K(0) = h + \beta\coth(\beta)$, $\rho_n = \sqrt{\beta^2 + \lambda_n^2}$ and λ_n is the nth root of $\lambda + h\tan(\lambda) = 0$. Note that $P(k)$ is analytic in the half-plane $\operatorname{Im}(k) > 0$ while $P(-k)$ is analytic in the half-plane $\operatorname{Im}(k) < \epsilon$. By eliminating $A(k)$ from (1) in Step 2 and using this factorization, show that we have the Wiener–Hopf equation

$$K(0)P(k)U'_{+}(k,0) + \frac{i}{k} = \frac{i}{k}\left[1 - \frac{1}{P(-k)}\right] - \frac{U'_{-}(k,0)}{P(-k)}. \qquad (2)$$

Note that the left side of (2) is analytic in the upper half-plane $\operatorname{Im}(k) > 0$, while the right side is analytic in the lower half-plane $\operatorname{Im}(k) < \epsilon$.

Step 4: It may be shown that $P(k) \sim |k|^{1/2}$. Show that $U_{+}(k,0) \sim k^{-1}$ and $U'_{-}(k,0) \sim \ln(k)$ as $|k| \to \infty$. Then use Liouville's theorem to show that each side of (2) equals zero.

---

[15] See Appendix A in Buchwald, V. T., and F. Viera, 1998: Linearized evaporation from a soil of finite depth above a water table. *Austral. Math. Soc., Ser. B*, **39**, 557–576.

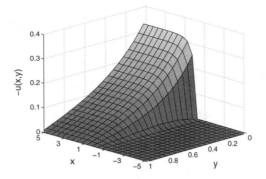

Problem 3

Step 5: Use the results from Step 4 to show that

$$K(0)U_+(k,0) = \frac{1}{ki\,P(k)}.$$

Therefore,

$$u(x,y) = \frac{1}{2\pi i K(0)} \int_{-\infty+\epsilon i}^{\infty+\epsilon i} \frac{\sinh[m(1-y)]}{k\sinh(m)P(k)}e^{-ikx}\,dk$$

$$= \frac{1}{2\pi i} \int_{-\infty+\epsilon i}^{\infty+\epsilon i} \frac{P(-k)\sinh[m(1-y)]}{k[m\cosh(m)+h\sinh(m)]}e^{-ikx}\,dk.$$

The first integral is best for computations when $x < 0$, while the second integral is best for $x > 0$.

Step 6: Use the results from Step 5 to show that

$$u(x,y) = -\frac{\pi}{K(0)}\sum_{n=1}^{\infty} \frac{n\,\sin(n\pi y)}{\xi_n^2 P(i\xi_n)}e^{\xi_n x}, \qquad x < 0,$$

where $\xi_n = \sqrt{n^2\pi^2+\beta^2}$, and

$$u(x,y) = \frac{e^{-\beta(1-y)}-e^{\beta(1-y)}}{(\beta+h)e^\beta+(\beta-h)e^{-\beta}} + \sum_{n=1}^{\infty} \frac{\lambda_n^2 P(i\rho_n)\,\sin[\lambda_n(1-y)]}{\rho_n^2[h(1+h)+\lambda_n^2]\sin(\lambda_n)}e^{-\lambda_n x}, \qquad 0 < x.$$

This solution $u(x,y)$ is illustrated in the figure labeled Problem 3 when $h = 1$ and $\beta = 2$.

4. Use the Wiener–Hopf technique to solve the partial differential equation

$$\frac{\partial^2 u}{\partial x^2} + \frac{\partial^2 u}{\partial y^2} - u = 0, \qquad -\infty < x < \infty, \quad 0 < y < 1,$$

with the boundary conditions

$$\frac{\partial u(x,1)}{\partial y} = 0, \qquad -\infty < x < \infty,$$

$$u(x,0) = 1, \quad x < 0, \quad \text{and} \quad \frac{\partial u(x,0)}{\partial y} = 0, \quad 0 < x.$$

Step 1: Assuming that $|u(x,0)|$ is bounded by $e^{-\epsilon x}$, $0 < \epsilon \ll 1$, as $x \to \infty$, let us define the following Fourier transforms:

$$U(k,y) = \int_{-\infty}^{\infty} u(x,y)e^{ikx}\,dx, \qquad U_+(k,y) = \int_{0}^{\infty} u(x,y)e^{ikx}\,dx,$$

and

$$U_-(k,y) = \int_{-\infty}^{0} u(x,y)e^{ikx}\,dx,$$

so that $U(k,y) = U_+(k,y) + U_-(k,y)$. Here, $U_+(k,y)$ is analytic in the half-space $\mathrm{Im}(k) > -\epsilon$, while $U_-(k,y)$ is analytic in the half-space $\mathrm{Im}(k) < 0$. Then show that the partial differential equation becomes

$$\frac{d^2 U}{dy^2} - m^2 U = 0, \qquad 0 < y < 1,$$

with $U'(k,1) = 0$, where $m^2 = k^2 + 1$.

Step 2: Show that the solution to Step 1 is $U(k,y) = A(k)\cosh[m(1-y)]$.

Step 3: From the boundary conditions along $y = 0$, show that

$$\cosh(m)A(k) = M_+(k) - \frac{i}{k} \qquad \text{and} \qquad -m\sinh(m)A(k) = L_-(k),$$

where

$$M_+(k) = \int_{0}^{\infty} u(x,0)e^{ikx}\,dx \quad \text{and} \quad L_-(k) = \int_{-\infty}^{0} u_y(x,0)e^{ikx}\,dx.$$

Step 4: By eliminating $A(k)$ from the equations in Step 3, show that we may factor the resulting equation as

$$-m^2\left[M_+(k) - \frac{i}{k}\right] = m\coth(m)L_-(k).$$

Step 5: Using the results that $m\coth(m) = K_+(k)K_-(k)$, where $K_+(k)$ and $K_-(k)$ are defined in Step 6 of Problem 2, show that

$$\frac{i}{K_+(k)} + \frac{K_+(k) - K_+(0)}{k\,K_+(0)K_+(k)} - \frac{(k+i)M_+(k)}{K_+(k)} = \frac{K_-(k)L_-(k)}{k-i} + \frac{1}{k\,K_+(0)}.$$

Note that the left side of the equation is analytic in the upper half-plane $\mathrm{Im}(k) > -\epsilon$, while the right side of the equation is analytic in the lower half-plane $\mathrm{Im}(k) < 0$.

Step 6: Use Liouville's theorem to show that each side of the equation in Step 5 equals zero. Therefore,

$$M_+(k) = \frac{i}{k} + \frac{K_+(k)}{k(k+i)K_+(0)},$$

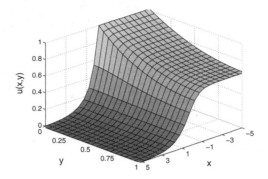

Problem 4

and

$$U(k,y) = \frac{K_+(k)\,\cosh[m(1-y)]}{k(k+i)\,K_+(0)\,\cosh(m)}.$$

Step 7: Use the inversion integral and show that

$$u(x,y) = \frac{1}{2\pi}\int_{-\infty-\epsilon i}^{\infty-\epsilon i}\frac{K_+(k)\,\cosh[m(1-y)]}{k(k+i)\,K_+(0)\,\cosh(m)}e^{-ikx}\,dk,$$

or

$$u(x,y) = \frac{K_-(0)}{2\pi\,\coth(1)}\int_{-\infty-\epsilon i}^{\infty-\epsilon i}\frac{m\,\cosh[m(1-y)]}{k(k+i)\,\sinh(m)}e^{-ikx}\,dk.$$

The first integral is best for finding the solution when $x < 0$, while the second integral is best for $0 < x$.

Step 8: Use the residue theorem and show that

$$u(x,y) = \tanh(1)\frac{K_-(0)}{K_-(-i)}e^{-x} + \tanh(1)\sum_{n=1}^{\infty}\frac{(1+\kappa_n)\,K_-(0)}{\kappa_n^2\,K_-(-i\kappa_n)}\cos(n\pi y)e^{-\kappa_n x}$$

if $0 < x$, where $\kappa_n = \sqrt{1+n^2\pi^2}$, while

$$u(x,y) = \frac{\cosh(1-y)}{\cosh(1)} + \sum_{n=1}^{\infty}\frac{(1-k_n)K_+(ik_n)}{m_n k_n^2 K_+(0)}\sin(m_n y)e^{k_n x}$$

for $x < 0$, where $k_n = \sqrt{1+(2n-1)^2\pi^2/4}$ and $m_n = (2n-1)\pi/2$. This solution $u(x,y)$ is illustrated in the figure labeled Problem 4.

7.2 THE WIENER–HOPF TECHNIQUE WHEN THE FACTORIZATION CONTAINS BRANCH POINTS

In the previous section, the product factors $K_+(k)$ and $K_-(k)$ were always meromorphic, resulting in a solution that consisted of a sum of residues. This occurred because $K(k)$ contained terms such as $m\sinh(m)$ and $\cosh(m)$, whose power series expansion consists only of powers of m^2, and there were no branch points. The form of $K(k)$ was due, in turn, to the presence of a finite domain in one of the spatial domains.

In this section we consider infinite or semi-infinite domains where $K(k)$ will become multivalued. As one might expect, the sum of residues becomes a branch cut integral just as it did in the case of Fourier transforms. There we found that single-valued Fourier transform yielded inverses that were a sum of residues whereas the inverses of multivalued Fourier transforms contained branch cut integrals.

• Example 7.2.1

An insightful example arises from a heat conduction problem[16] in the upper half-plane $y > 0$:

$$\frac{\partial u}{\partial t} = \frac{\partial^2 u}{\partial x^2} + \frac{\partial^2 u}{\partial y^2}, \qquad -\infty < x < \infty, \quad 0 < t, y, \qquad (7.2.1)$$

with the boundary conditions

$$u(x, 0, t) = e^{-\epsilon x}, \quad 0 < \epsilon, x, \qquad \frac{\partial u(x, 0, t)}{\partial y} = 0, \quad x < 0, \qquad (7.2.2)$$

and

$$\lim_{y \to \infty} u(x, y, t) \to 0, \qquad (7.2.3)$$

while the initial condition is $u(x, y, 0) = 0$. Eventually we will consider the limit $\epsilon \to 0$.

What makes this problem particularly interesting is the boundary condition that we specify along $y = 0$; it changes from a Dirichlet condition when $x < 0$ to a Neumann boundary condition when $x > 0$. The Wiener–Hopf technique is commonly used to solve these types of boundary-value problems where the nature of the boundary condition changes along a given boundary – the so-called *mixed boundary-value problem*.

We begin by introducing the Laplace transform in time

$$U(x, y, s) = \int_0^\infty u(x, y, t) e^{-st} \, dt, \qquad (7.2.4)$$

and the Fourier transform in the x-direction

$$\overline{U}_+(k, y, s) = \int_0^\infty U(x, y, s) e^{ikx} \, dx, \qquad (7.2.5)$$

---

[16] Simplified version of a problem solved by Huang, S. C., 1985: Unsteady-state heat conduction in semi-infinite regions with mixed-type boundary conditions. *J. Heat Transfer*, **107**, 489–491.

$$\overline{U}_-(k,y,s) = \int_{-\infty}^{0} U(x,y,s)e^{ikx}\,dx, \tag{7.2.6}$$

and

$$\overline{U}(k,y,s) = \overline{U}_+(k,y,s) + \overline{U}_-(k,y,s) = \int_{-\infty}^{\infty} U(x,y,s)e^{ikx}\,dx. \tag{7.2.7}$$

Here, we have assumed that $|u(x,y,t)|$ is bounded by $e^{-\epsilon x}$ as $x \to \infty$ while $|u(x,y,t)|$ is $O(1)$ as $x \to -\infty$. For this reason, the subscripts "+" and "−" denote that $\overline{U}_+$ is analytic in the upper half-plane $\text{Im}(k) > -\epsilon$ while $\overline{U}_-$ is analytic in the lower half-plane $\text{Im}(k) < 0$.

Taking the joint transform of Equation 7.2.1, we find that

$$\frac{d^2\overline{U}(k,y,s)}{dy^2} - (k^2 + s)\overline{U}(k,y,s) = 0, \qquad 0 < y < \infty, \tag{7.2.8}$$

with the transformed boundary conditions

$$\overline{U}_+(k,0,s) = \frac{1}{s(\epsilon - ki)}, \qquad \frac{d\overline{U}_-(k,0,s)}{dy} = 0, \tag{7.2.9}$$

and $\lim_{y\to\infty} \overline{U}(k,y,s) \to 0$. The general solution to Equation 7.2.8 is

$$\overline{U}(k,y,s) = A(k,s)e^{-y\sqrt{k^2+s}}. \tag{7.2.10}$$

Consequently,

$$A(k,s) = \frac{1}{s(\epsilon - ki)} + \overline{U}_-(k,0,s) \tag{7.2.11}$$

and

$$-\sqrt{k^2 + s}\, A(k,s) = \frac{d\overline{U}_+(k,0,s)}{dy}. \tag{7.2.12}$$

Note that we have a multivalued function $\sqrt{k^2 + s}$ with branch points $k = \pm\sqrt{s}\,i$. Eliminating $A(k,s)$ between Equation 7.2.11 and Equation 7.2.12, we obtain the Wiener–Hopf equation:

$$\frac{d\overline{U}_+(k,0,s)}{dy} = -\sqrt{k^2 + s}\left[\overline{U}_-(k,0,s) + \frac{1}{s(\epsilon - ki)}\right]. \tag{7.2.13}$$

Our next goal is to rewrite Equation 7.2.13 so that the left side is analytic in the upper half-plane $\text{Im}(k) > -\epsilon$ while the right side is analytic in the lower half-plane $\text{Im}(k) < 0$. We begin by factoring $\sqrt{k^2 + s} = \sqrt{k - i\sqrt{s}}\sqrt{k + i\sqrt{s}}$, where the branch cuts lie along the imaginary axis in the k-plane from $(-\infty i, -\sqrt{s}i]$ and $[\sqrt{s}i, \infty i)$. Equation 7.2.13 can then be rewritten

$$\frac{1}{\sqrt{k + i\sqrt{s}}}\frac{d\overline{U}_+(k,0,s)}{dy} = -\sqrt{k - i\sqrt{s}}\left[\overline{U}_-(k,0,s) + \frac{1}{s(\epsilon - ki)}\right]. \tag{7.2.14}$$

The left side of Equation 7.2.14 is what we want; the same is true of the first term on the right side. However, the second term on the right side falls short. At this point we note that

$$\frac{\sqrt{k - i\sqrt{s}}}{\epsilon - ki} = \frac{\sqrt{k - i\sqrt{s}} - \sqrt{-i\epsilon - i\sqrt{s}}}{\epsilon - ki} + \frac{\sqrt{-i\epsilon - i\sqrt{s}}}{\epsilon - ki}. \qquad (7.2.15)$$

Substituting Equation 7.2.15 into Equation 7.2.14 and rearranging terms, we obtain

$$\frac{1}{\sqrt{k + i\sqrt{s}}} \frac{d\overline{U}_+(k, 0, s)}{dy} + \frac{\sqrt{-i\epsilon - i\sqrt{s}}}{s(\epsilon - ki)}$$

$$= -\sqrt{k - i\sqrt{s}}\,\overline{U}_-(k, 0, s) - \frac{\sqrt{k - i\sqrt{s}} - \sqrt{-i\epsilon - i\sqrt{s}}}{s(\epsilon - ki)}. \qquad (7.2.16)$$

In this form the right side of Equation 7.2.16 is analytic in the lower half-plane $\mathrm{Im}(k) < 0$ while the left side is analytic in the upper half-plane $\mathrm{Im}(k) > -\epsilon$. Since they share a common strip of analyticity $-\epsilon < \mathrm{Im}(k) < 0$, they are analytic continuations of each other and equal some entire function. Using Liouville's theorem and taking the limit as $|k| \to \infty$, we see that both sides of Equation 7.2.16 equal zero. Therefore,

$$\overline{U}_-(k, 0, s) = -\frac{\sqrt{k - i\sqrt{s}} - \sqrt{-i\epsilon - i\sqrt{s}}}{s(\epsilon - ki)\sqrt{k - i\sqrt{s}}}, \qquad (7.2.17)$$

$$A(k, s) = \frac{\sqrt{-i\epsilon - i\sqrt{s}}}{s(\epsilon - ki)\sqrt{k - i\sqrt{s}}}, \qquad (7.2.18)$$

and

$$U(x, y, s) = \frac{\sqrt{-i\epsilon - i\sqrt{s}}}{2\pi s} \int_{-\infty}^{\infty} \frac{\exp\left(-ikx - y\sqrt{k^2 + s}\right)}{(\epsilon - ki)\sqrt{k - i\sqrt{s}}}\, dk. \qquad (7.2.19)$$

Upon taking the limit $\epsilon \to 0$ and introducing $x = r\cos(\theta)$, $y = r\sin(\theta)$ and $k = \sqrt{s}\,\eta$, Equation 7.2.19 becomes

$$U(r, \theta, s) = \frac{\sqrt{i}}{2\pi s} \int_{-\infty}^{\infty} \frac{\exp\left\{-r\sqrt{s}\left[i\eta\cos(\theta) + \sin(\theta)\sqrt{\eta^2 + 1}\right]\right\}}{\eta\sqrt{\eta - i}}\, d\eta. \qquad (7.2.20)$$

To invert Equation 7.2.20, we employ the technique shown in Example 5.1.4. Introducing

$$\cosh(\tau) = i\eta\cos(\theta) + \sin(\theta)\sqrt{\eta^2 + 1}, \qquad (7.2.21)$$

we find that $\eta = \sin(\theta)\sinh(\tau) - i\cos(\theta)\cosh(\tau) = -i\cos(\theta - i\tau)$. We now deform the original contour along the real axis to the one defined by η. Particular care must be exercised in the case of $x > 0$ as we deform the contour

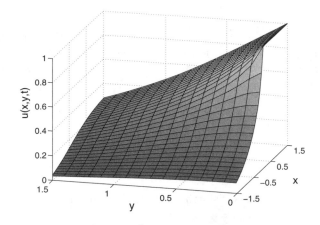

Figure 7.2.1: The solution to Equation 7.2.1 through Equation 7.2.3 obtained via the Wiener–Hopf technique.

into the lower half of the k-plane since $0 < \theta < \pi/2$. During this deformation, we pass over the singularity at $k = 0^-$ and must consequently add its contribution to the inverse. Thus, the Laplace transform of the solution now reads

$$U(r, \theta, s) = \frac{e^{-y\sqrt{s}}}{s} H(x) - \frac{1}{\sqrt{2}\,\pi s} \int_{-\infty}^{\infty} \frac{\sin[(\theta - i\tau)/2]}{\cos(\theta - i\tau)} e^{-r\sqrt{s}\,\cosh(\tau)} \, d\tau.$$

(**7.2.22**)

Taking the inverse Laplace transform of Equation 7.2.22, we obtain

$$u(r, \theta, t) = \operatorname{erfc}\left(\frac{y}{\sqrt{4t}}\right) H(x)$$
$$- \frac{\sqrt{2}}{\pi} \int_0^{\infty} \operatorname{Re}\left\{ \frac{\sin[(\theta - i\tau)/2]}{\cos(\theta - i\tau)} \right\} \operatorname{erfc}\left[\frac{r\,\cosh(\tau)}{\sqrt{4t}}\right] d\tau. \quad (\textbf{7.2.23})$$

Figure 7.2.1 illustrates this solution when $t = 1$.

• Example 7.2.2

As we saw in the previous section, the Wiener–Hopf technique is useful in solving mixed boundary-value problems involving Laplace's and Helmholtz's equations. For example, let us solve

$$\frac{\partial^2 u}{\partial x^2} + \frac{\partial^2 u}{\partial y^2} - \alpha^2 u = 0, \qquad -\infty < x < \infty, \quad 0 < y. \quad (\textbf{7.2.24})$$

At infinity, we have that $\lim_{y \to \infty} u(x, y) \to 0$ while along $y = 0$,

$$u(x, 0) = 1, \quad x < 0, \qquad u(x, 0) = 1 + \lambda u_y(x, 0), \quad 0 < x, \quad (\textbf{7.2.25})$$

where $0 < \alpha, \lambda$.

Before we solve the general problem, let us find the solution at large $|x|$. In these regions, the solution becomes essentially independent of x and Equation 7.2.24 becomes an ordinary differential equation in y. The solution as $x \to -\infty$ is

$$u(x, y) = e^{-\alpha y}, \tag{7.2.26}$$

while for $x > 0$ the solution approaches

$$u(x, y) = \frac{e^{-\alpha y}}{1 + \alpha \lambda}. \tag{7.2.27}$$

Note how these solutions satisfy the differential equation and boundary conditions as $y \to \infty$ and at $y = 0$.

Why did we find these limiting cases? From Equation 7.2.26 and Equation 7.2.27 we see that $u(x, 0)$ tends to constant, nonzero values as $x \to -\infty$ and $x \to \infty$. Because $u(x, 0)$ does not tend to zero at one (or both) of the end points, the Wiener–Hopf technique would not work as the problem is currently posed. To circumvent this difficulty we introduce the intermediate variable $v(x, y)$ via

$$u(x, y) = \frac{e^{-\alpha y}}{1 + \alpha \lambda} + v(x, y). \tag{7.2.28}$$

Substituting Equation 7.2.28 into Equation 7.2.24, we obtain

$$\frac{\partial^2 v}{\partial x^2} + \frac{\partial^2 v}{\partial y^2} - \alpha^2 v = 0, \qquad -\infty < x < \infty, \quad 0 < y. \tag{7.2.29}$$

The boundary condition at infinity now reads $\lim_{y \to \infty} v(x, y) \to 0$ while

$$v(x, 0) = \frac{\alpha \lambda}{1 + \alpha \lambda}, \quad x < 0, \qquad v(x, 0) = \lambda v_y(x, 0), \quad 0 < x, \tag{7.2.30}$$

Therefore, by introducing $v(x, y)$, we have subtracted out the value of $u(x, 0)$ as $x \to \infty$ and we now assume that $|v(x, y)|$ is bounded by $e^{-\epsilon x}$, $0 < \epsilon \ll 1$, as $x \to \infty$ while $|v(x, y)|$ is $O(1)$ as $x \to -\infty$.

Applying the Wiener–Hopf technique to Equation 7.2.29 and Equation 7.2.30, we introduce the Fourier transform in the x-direction:

$$V_+(k, y) = \int_0^\infty v(x, y) e^{ikx} \, dx, \tag{7.2.31}$$

$$V_-(k, y) = \int_{-\infty}^0 v(x, y) e^{ikx} \, dx, \tag{7.2.32}$$

and

$$V(k, y) = V_+(k, y) + V_-(k, y) = \int_{-\infty}^\infty v(x, y) e^{ikx} \, dx. \tag{7.2.33}$$

The subscripts "+" and "−" denote the fact that V_+ is analytic in the upper half-space $\text{Im}(k) > -\epsilon$ while V_- is analytic in the lower half-space $\text{Im}(k) < 0$.

Taking the Fourier transform of Equation 7.2.29, we find that

$$\frac{d^2 V(k,y)}{dy^2} - m^2 V(k,y) = 0, \qquad m^2 = k^2 + \alpha^2, \qquad 0 < y < \infty, \quad \textbf{(7.2.34)}$$

along with the transformed boundary condition $\lim_{y\to\infty} V(k,y) \to 0$. The general solution to Equation 7.2.34 is $V(k,y) = A(k)e^{-my}$. Note that m is multivalued with branch points $k = \pm\alpha i$.

Turning to the boundary conditions given by Equation 7.2.30, we obtain

$$A(k) = \frac{\alpha\lambda}{ik(1+\alpha\lambda)} + M_+(k) \quad \text{and} \quad A(k) + m\lambda A(k) = L_-(k), \quad \textbf{(7.2.35)}$$

where

$$M_+(k) = \int_0^\infty v(x,0)e^{ikx}\, dx \text{ and } L_-(k) = \int_{-\infty}^0 [v(x,0) - \lambda v_y(x,0)]\, e^{ikx}\, dx.$$
$$\textbf{(7.2.36)}$$

Eliminating $A(k)$ in Equation 7.2.35, we obtain the Wiener–Hopf equation:

$$\frac{\alpha\lambda}{ik(1+\alpha\lambda)} + M_+(k) = \frac{L_-(k)}{1+m\lambda}. \qquad \textbf{(7.2.37)}$$

Our next goal is to rewrite Equation 7.2.37 so that the left side is analytic in the upper half-plane $\text{Im}(k) > -\epsilon$ while the right side is analytic in the lower half-plane $\text{Im}(k) < 0$. We begin by factoring $P(k) = 1 + \lambda m = P_+(k)P_-(k)$, where $P_+(k)$ and $P_-(k)$ are analytic in the upper and lower half-planes, respectively. We will determine them shortly. Equation 7.2.37 can then be rewritten

$$P_+(k)M_+(k) + \frac{\alpha\lambda P_+(k)}{ik(1+\alpha\lambda)} = \frac{L_-(k)}{P_-(k)}. \qquad \textbf{(7.2.38)}$$

The right side of Equation 7.2.38 is what we want; it is analytic in the half-plane $\text{Im}(k) < 0$. The first term on the left side is analytic in the half-plane $\text{Im}(k) > -\epsilon$. The second term, unfortunately, is not analytic in either half-planes. However, we note that

$$\frac{P_+(k)}{ik} = \frac{P_+(k)}{ik} - \frac{P_+(0)}{ik} + \frac{P_+(0)}{ik}. \qquad \textbf{(7.2.39)}$$

Substituting Equation 7.2.39 into Equation 7.2.38 and rearranging terms, we obtain

$$P_+(k)M_+(k) + \frac{\alpha\lambda}{1+\alpha\lambda}\left[\frac{P_+(k)}{ik} - \frac{P_+(0)}{ik}\right] = \frac{L_-(k)}{P_-(k)} - \frac{\alpha\lambda P_+(0)}{ik(1+\alpha\lambda)}. \quad \textbf{(7.2.40)}$$

In this form the left side of Equation 7.2.40 is analytic in the upper half-plane $\text{Im}(k) > -\epsilon$ while the right side is analytic in the lower half-plane $\text{Im}(k) < 0$. Since both sides share a common strip of analyticity $-\epsilon < \text{Im}(k) < 0$, they are analytic continuations of each other and equal some entire function. Using Liouville's theorem and taking the limit as $|k| \to \infty$, we see that both sides of Equation 7.2.40 equal zero. Therefore,

$$M_+(k) = \frac{\alpha\lambda P_+(0)}{ik(1+\alpha\lambda)P_+(k)} - \frac{\alpha\lambda}{ik(1+\alpha\lambda)}, \qquad (7.2.41)$$

$$A(k) = \frac{\alpha\lambda P_+(0)}{ik(1+\alpha\lambda)P_+(k)}, \qquad (7.2.42)$$

and

$$u(x,y) = \frac{e^{-\alpha y}}{1+\alpha\lambda} + \frac{\alpha\lambda P_+(0)}{2\pi i(1+\alpha\lambda)} \int_{-\infty-\epsilon i}^{\infty-\epsilon i} \frac{e^{-my-ikx}}{k\,P_+(k)}\,dk \qquad (7.2.43)$$

$$= \frac{e^{-\alpha y}}{1+\alpha\lambda} + \frac{\alpha\lambda}{2\pi i} \int_{-\infty-\epsilon i}^{\infty-\epsilon i} \frac{P_-(k)}{P_-(0)} \frac{e^{-my-ikx}}{k(1+\lambda m)}\,dk. \qquad (7.2.44)$$

Equation 7.2.43 is best for computing $u(x,y)$ when $x < 0$. Using the contour shown in Figure 5.1.1, we deform the original contour to the contour $AGFEDCB$ shown there. During the deformation, we cross the simple pole at $k = 0$ and must add its contribution. Therefore,

$$\begin{aligned}
u(x,y) = {} & \frac{e^{-\alpha y}}{1+\alpha\lambda} + \frac{\alpha\lambda e^{-\alpha y}}{1+\alpha\lambda} + \frac{\alpha\lambda P_+(0)}{2\pi i(1+\alpha\lambda)} \int_{\infty}^{\alpha} \frac{e^{iy\sqrt{\eta^2-\alpha^2}+\eta x}}{P_+(i\eta)}\,\frac{d\eta}{\eta} \\
& + \frac{\alpha\lambda P_+(0)}{2\pi i(1+\alpha\lambda)} \int_{\alpha}^{\infty} \frac{e^{-iy\sqrt{\eta^2-\alpha^2}+\eta x}}{P_+(i\eta)}\,\frac{d\eta}{\eta} \qquad\qquad (7.2.45) \\[4pt]
= {} & e^{-\alpha y} - \frac{\alpha\lambda P_+(0)}{\pi(1+\alpha\lambda)} \int_{\alpha}^{\infty} e^{x\eta}\sin\!\left(y\sqrt{\eta^2-\alpha^2}\right) \frac{d\eta}{\eta\,P_+(i\eta)} \qquad (7.2.46) \\[4pt]
= {} & e^{-\alpha y} - \frac{\alpha\lambda P_+(0)}{\pi(1+\alpha\lambda)} \int_{1}^{\infty} e^{\alpha x\xi}\sin\!\left(\alpha y\sqrt{\xi^2-1}\right) \frac{d\xi}{\xi\,P_+(i\alpha\xi)}. \qquad (7.2.47)
\end{aligned}$$

The contribution from the arcs GA and BC vanish as the radius of the semicircle in the upper half-plane becomes infinite.

Turning to the case $x > 0$, we deform the original contour to the contour $A'G'F'E'D'C'B'$ in Figure 5.1.1. During the deformation we do not cross any singularities. Therefore,

$$\begin{aligned}
u(x,y) = {} & \frac{e^{-\alpha y}}{1+\alpha\lambda} + \frac{\alpha\lambda}{2\pi i} \int_{\infty}^{\alpha} \frac{P_-(-i\eta)}{P_-(0)} \frac{e^{-iy\sqrt{\eta^2-\alpha^2}-\eta x}}{1+i\lambda\sqrt{\eta^2-\alpha^2}}\,\frac{d\eta}{\eta} \\
& + \frac{\alpha\lambda}{2\pi i} \int_{\alpha}^{\infty} \frac{P_-(-i\eta)}{P_-(0)} \frac{e^{iy\sqrt{\eta^2-\alpha^2}-\eta x}}{1-i\lambda\sqrt{\eta^2-\alpha^2}}\,\frac{d\eta}{\eta} \qquad\qquad (7.2.48)
\end{aligned}$$

$$u(x,y) = \frac{e^{-\alpha y}}{1+\alpha\lambda} + \frac{\alpha\lambda}{\pi}\int_\alpha^\infty \frac{P_-(-i\eta)}{P_-(0)} e^{-x\eta}\, \frac{d\eta}{\eta} \tag{7.2.49}$$

$$\times\; \frac{\sin\left(y\sqrt{\eta^2-\alpha^2}\right) + \lambda\sqrt{\eta^2-\alpha^2}\cos\left(y\sqrt{\eta^2-\alpha^2}\right)}{1+\lambda^2(\eta^2-\alpha^2)}$$

$$= \frac{e^{-\alpha y}}{1+\alpha\lambda} + \frac{\alpha\lambda}{\pi}\int_1^\infty \frac{P_-(-i\alpha\xi)}{P_-(0)} e^{-\alpha x\xi}\, \frac{d\xi}{\xi} \tag{7.2.50}$$

$$\times\; \frac{\sin\left(\alpha y\sqrt{\xi^2-1}\right) + \alpha\lambda\sqrt{\xi^2-1}\cos\left(\alpha y\sqrt{\xi^2-1}\right)}{1+\alpha^2\lambda^2(\xi^2-1)}.$$

The final task is the factorization. We begin by introducing the functions

$$\varphi(z) = \varphi_+(z) + \varphi_-(z) = \frac{P'(z)}{P(z)} = \frac{z\lambda}{m(1+\lambda m)}, \quad m^2 = z^2 + \alpha^2, \tag{7.2.51}$$

which is analytic in the region $-\alpha < \mathrm{Im}(z) < \alpha$ and $\varphi_\pm(z) = P'_\pm(z)/P_\pm(z)$. By definition, $\varphi_\pm(z)$ is analytic and nonzero in the same half-planes as $P_\pm(z)$. Furthermore,

$$P_\pm(z) = P_\pm(0)\exp\left[\int_0^z \varphi_\pm(\zeta)\, d\zeta\right]. \tag{7.2.52}$$

From Cauchy's integral theorem,

$$\varphi_+(\zeta) = \frac{1}{2\pi i}\int_{-\infty+\epsilon i}^{\infty+\epsilon i} \frac{\varphi(z)}{z-\zeta}\, dz \quad \text{and} \quad \varphi_-(\zeta) = -\frac{1}{2\pi i}\int_{-\infty+\delta i}^{\infty+\delta i} \frac{\varphi(z)}{z-\zeta}\, dz, \tag{7.2.53}$$

where $-\alpha < \epsilon < \delta < \alpha$. We will now evaluate the line integrals in Equation 7.2.53 by converting them into the closed contours shown in Figure 5.1.1. In particular, for φ_+, we employ the closed contour $A'B'C'D'E'F'G'A'$ with the branch cut running from $-\alpha i$ to $-\infty i$, while for the evaluation of φ_-, we use $ABCDEFGA$ with the branch cut running from αi to ∞i. For example,

$$\varphi_+(\zeta) = \frac{\lambda}{2\pi i}\int_{-\infty+\epsilon i}^{\infty+\epsilon i} \frac{z}{m(1+\lambda m)(z-\zeta)}\, dz \tag{7.2.54}$$

$$= \frac{\lambda}{2\pi i}\int_\infty^\alpha \frac{(-i\eta)(-i\, d\eta)}{\left(i\sqrt{\eta^2-\alpha^2}\right)\left(1+i\lambda\sqrt{\eta^2-\alpha^2}\right)(-i\eta-\zeta)}$$

$$+ \frac{\lambda}{2\pi i}\int_\alpha^\infty \frac{(-i\eta)(-i\, d\eta)}{\left(-i\sqrt{\eta^2-\alpha^2}\right)\left(1-i\lambda\sqrt{\eta^2-\alpha^2}\right)(-i\eta-\zeta)} \tag{7.2.55}$$

$$= \frac{\lambda}{\pi}\int_\alpha^\infty \frac{\eta\, d\eta}{(\zeta+i\eta)\sqrt{\eta^2-\alpha^2}\left[1+\lambda^2(\eta^2-\alpha^2)\right]} \tag{7.2.56}$$

$$= \frac{\rho}{\pi}\int_0^\infty \frac{ds}{(s^2+\rho^2)\left(\zeta+i\sqrt{s^2+\alpha^2}\right)}. \tag{7.2.57}$$

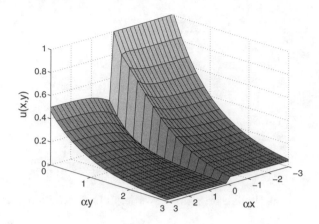

Figure 7.2.2: The solution to Equation 7.2.24 and Equation 7.2.25 obtained via the Wiener–Hopf technique.

where $s^2 = \eta^2 - \alpha^2$ and $\rho = 1/\lambda$. A similar analysis of $\varphi_-(\zeta)$ shows that $P_-(-\zeta)/P_-(0) = P_+(\zeta)/P_+(0)$. Figure 7.2.2 illustrates $u(x,y)$ when $\alpha\lambda = 1$. The numerical calculations begin with a computation of $\varphi_+(\zeta)$ using Simpson's rule. Then, $P_+(\zeta)/P_+(0)$ is found using the trapezoidal rule. Finally, Equation 7.2.47 and Equation 7.2.50 are evaluated using Simpson's rule.

Problems

1. Use the Wiener–Hopf technique to solve

$$\frac{\partial^2 u}{\partial x^2} + \frac{\partial^2 u}{\partial y^2} - u = -2\rho(x)\delta(y), \qquad -\infty < x, y < \infty,$$

subject to the boundary conditions

$$\lim_{|x|\to\infty} u(x,y) \to 0, \qquad -\infty < y < \infty,$$

$$\lim_{|y|\to\infty} u(x,y) \to 0, \qquad -\infty < x < \infty,$$

and

$$u(x,0) = e^{-x}, \qquad 0 < x < \infty.$$

The function $\rho(x)$ is only nonzero for $0 < x < \infty$.

Step 1: Assuming that $|u(x,y)|$ is bounded by $e^{\epsilon x}$, where $0 < \epsilon \ll 1$, as $x \to -\infty$, let us introduce

$$U(k,y) = \int_{-\infty}^{\infty} u(x,y)e^{-ikx}\,dx \qquad \text{and} \qquad \overline{U}(k,\ell) = \int_{-\infty}^{\infty} U(k,y)e^{-i\ell y}\,dy.$$

Use the differential equation and first two boundary conditions to show that

$$U(k,y) = R(k)\frac{e^{-|y|\sqrt{k^2+1}}}{\sqrt{k^2+1}},$$

where $R(k)$ is the Fourier transform of $\rho(x)$.

Step 2: Taking the Fourier transform of the last boundary condition, show that

$$\frac{R(k)}{\sqrt{k^2+1}} = \frac{1}{1+ik} + F_+(k), \tag{1}$$

where

$$F_+(k) = \int_{-\infty}^0 u(x,0)e^{-ikx}\,dx.$$

Note that $R(k)$ is analytic in the lower half-space where $\text{Im}(k) < \epsilon$. Why?

Step 3: Show that (1) can be rewritten

$$\frac{R(k)}{\sqrt{1+ik}} - \frac{\sqrt{2}}{1+ik} = \frac{\sqrt{1-ik}-\sqrt{2}}{1+ik} + \sqrt{1-ik}\,F_+(k).$$

Note that the left side of this equation is analytic in the lower half-plane $\text{Im}(k) < \epsilon$, while the right side is analytic in the upper half-plane $\text{Im}(k) > 0$.

Step 4: Use Liouville's theorem and deduce that $R(k) = \sqrt{2}/\sqrt{1+ik}$.

Step 5: Show that

$$U(k,y) = \frac{\sqrt{2}}{\sqrt{1+ik}}\frac{e^{-|y|\sqrt{k^2+1}}}{\sqrt{k^2+1}}.$$

Step 6: Using integral tables, show that

$$\mathcal{F}^{-1}\left(\frac{e^{-|y|\sqrt{k^2+1}}}{\sqrt{k^2+1}}\right) = \frac{1}{2\pi}\int_{-\infty}^\infty \frac{e^{ikx-|y|\sqrt{k^2+1}}}{\sqrt{k^2+1}}\,dk = \frac{K_0(r)}{\pi},$$

where $r^2 = x^2 + y^2$.

Step 7: Using contour integration, show that

$$\mathcal{F}^{-1}\left(\frac{\sqrt{2}}{\sqrt{1+ik}}\right) = e^{-x}\sqrt{\frac{2}{\pi x}}\,H(x).$$

Step 8: Using the results from Step 6 and Step 7 and applying the convolution theorem, show that

$$u(x,y) = e^{-x}\sqrt{\frac{2}{\pi x}}\,H(x) * \frac{K_0(r)}{\pi}.$$

Step 9: Complete the problem and show that

$$u(x,y) = \sqrt{\frac{2}{\pi^3}}\int_0^\infty \frac{e^{-\chi}}{\sqrt{\chi}}K_0\left[\sqrt{(x-\chi)^2+y^2}\right]d\chi$$

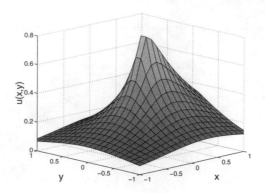

Problem 1

or

$$u(x,y) = \sqrt{\frac{8}{\pi^3}} \int_0^\infty e^{-\eta^2} K_0\left[\sqrt{(x-\eta^2)^2 + y^2}\right] d\eta.$$

This solution is illustrated in Problem 1.

2. Use the Wiener–Hopf technique to solve

$$\frac{\partial^2 u}{\partial x^2} + \frac{\partial^2 u}{\partial y^2} - u = 0, \qquad -\infty < x < \infty, \quad 0 < y,$$

subject to the boundary conditions

$$\lim_{y \to \infty} u(x,y) \to 0, \qquad -\infty < x < \infty,$$

$$u_y(x,0) = 0, \quad x < 0, \qquad \text{and} \qquad u(x,0) = e^{-x}, \quad 0 < x.$$

Step 1: Introducing

$$U(k,y) = \int_{-\infty}^\infty u(x,y)e^{ikx}\, dx,$$

use the differential equation and first two boundary conditions to show that $U(k,y) = A(k)e^{-|y|\sqrt{k^2+1}}$.

Step 2: Taking the Fourier transform of the boundary condition along $y = 0$, show that

$$A(k) = U_-(k) + \frac{1}{1-ki} \qquad \text{and} \qquad U'_+(k) = -\sqrt{k^2+1}\, A(k), \tag{1}$$

where

$$U_-(k) = \int_{-\infty}^0 u(x,0)e^{ikx}\, dx \qquad \text{and} \qquad U'_+(k) = \int_0^\infty u_y(x,0)e^{ikx}\, dx.$$

Here we have assumed that $|u(x,0)|$ is bounded by $e^{-\epsilon x}$, $0 < \epsilon \ll 1$, as $x \to \infty$ so that U_+ is analytic in the upper half-plane $\text{Im}(k) > -\epsilon$, while U_- is analytic in the lower half-plane $\text{Im}(k) < 0$.

Step 3: Show that (1) can be rewritten

$$-\frac{U_+(k)}{\sqrt{1-ki}} + \frac{\sqrt{2}}{1-ki} = \sqrt{1-ki}\, U_-(k) + \frac{\sqrt{1+ki} - \sqrt{2}}{1-ki}.$$

Note that the right side of this equation is analytic in the lower half-plane $\text{Im}(k) < 0$ while the left side is analytic in the upper half-plane $\text{Im}(k) > -\epsilon$.

Step 4: Use Liouville's theorem and deduce that

$$U_-(k) = \frac{\sqrt{2}}{(1-ki)\sqrt{1+ki}} - \frac{1}{1-ki}.$$

Step 5: Show that

$$U(k,y) = \frac{\sqrt{2}}{\sqrt{1-ki}}\,\frac{e^{-|y|\sqrt{k^2+1}}}{\sqrt{k^2+1}}.$$

Step 6: Finish the problem by retracing Step 6 through Step 9 of the previous problem and show that you recover the same solution.

3. Use the Wiener–Hopf technique to solve the partial differential equation

$$\frac{\partial^2 u}{\partial x^2} + \frac{\partial^2 u}{\partial y^2} - u = 0, \qquad -\infty < x < \infty, \quad 0 < y,$$

with the boundary conditions $\lim_{y\to\infty} u(x,y) \to 0$,

$$u(x,0) = 1, \quad x < 0, \qquad \text{and} \qquad \frac{\partial u(x,0)}{\partial y} = 0, \quad 0 < x.$$

Step 1: Assuming that $|u(x,0)|$ is bounded by $e^{-\epsilon x}$ as $x \to \infty$, where $0 < \epsilon \ll 1$, let us define the following Fourier transforms:

$$U(k,y) = \int_{-\infty}^{\infty} u(x,y)e^{ikx}\,dx, \qquad U_+(k,y) = \int_{0}^{\infty} u(x,y)e^{ikx}\,dx,$$

and

$$U_-(k,y) = \int_{-\infty}^{0} u(x,y)e^{ikx}\,dx,$$

so that $U(k,y) = U_+(k,y) + U_-(k,y)$. Here, $U_+(k,y)$ is analytic in the half-space $\text{Im}(k) > -\epsilon$, while $U_-(k,y)$ is analytic in the half-space $\text{Im}(k) < 0$. Then show that the partial differential equation becomes

$$\frac{d^2U}{dy^2} - m^2 U = 0, \qquad 0 < y,$$

with $\lim_{y\to\infty} U(k,y) \to 0$, where $m^2 = k^2 + 1$.

Step 2: Show that the solution to Step 1 is $U(k,y) = A(k)e^{-my}$.

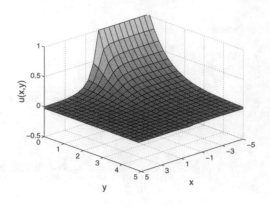

Problem 3

Step 3: From the boundary conditions along $x = 0$, show that

$$A(k) = M_+(k) - \frac{i}{k} \quad \text{and} \quad -mA(k) = L_-(k),$$

where

$$M_+(k) = \int_0^\infty u(x,0)e^{ikx}\,dx \quad \text{and} \quad L_-(k) = \int_{-\infty}^0 u_y(x,0)e^{ikx}\,dx.$$

Note that $M_+(k)$ is analytic in the half-space $\text{Im}(k) > -\epsilon$, while $L_-(k)$ is analytic in the half-space $\text{Im}(k) < 0$.

Step 4: By eliminating $A(k)$ from the equations in Step 3, show that we may factor the resulting equation as

$$-\sqrt{k+i}\,M_+(k) + i\frac{\sqrt{k+i} - \sqrt{i}}{k} = \frac{L_-(k)}{\sqrt{k-i}} - \frac{i^{3/2}}{k}.$$

Note that the left side of the equation is analytic in the upper half-plane $\text{Im}(k) > -\epsilon$, while the right side of the equation is analytic in the lower half-plane $\text{Im}(k) < 0$.

Step 5: Use Liouville's theorem to show that each side of the equation in Step 4 equals zero. Therefore,

$$M_+(k) = \frac{i}{k} - \frac{i^{3/2}}{k\sqrt{k+i}} \quad \text{and} \quad U(k,y) = -\frac{i^{3/2}}{k\sqrt{k+i}}e^{-y\sqrt{k^2+1}}.$$

Step 6: Use the inversion integral and show that

$$u(x,y) = -\frac{i^{3/2}}{2\pi}\int_{-\infty-\epsilon i}^{\infty-\epsilon i}\frac{\exp\left(-ikx - y\sqrt{k^2+1}\right)}{k\sqrt{k+i}}\,dk.$$

Step 7: Using the techniques from Section 5.1 and Figure 5.1.1, show that

$$u(x,y) = \frac{1}{\pi}\int_1^\infty \cos\left(y\sqrt{\eta^2-1}\right)e^{-x\eta}\frac{d\eta}{\eta\sqrt{\eta-1}}$$

if $x > 0$, and

$$u(x,y) = e^{-y} - \frac{1}{\pi}\int_1^\infty \sin\left(y\sqrt{\eta^2-1}\right)e^{x\eta}\frac{d\eta}{\eta\sqrt{\eta+1}}$$

if $x < 0$. This solution $u(x,y)$ is illustrated in the figure labeled Problem 3.

Papers Using the Wiener–Hopf Technique

Aquifers, Reservoirs and Porous Media

Merkin, J. H., and D. J. Needham, 1987: The natural convection flow above a heated wall in a saturated porous medium. *Quart. J. Mech. Appl. Math.*, **40**, 559–574.

Rice, J. R., and D. A. Simons, 1976: The stabilization of spreading shear faults by coupled deformation-diffusion effects in fluid-infiltrated porous materials. *J. Geophys. Res.*, **81**, 5322–5334.

Diffusion

Boersma, J., J. J. E. Indenkleef, and H. K. Kuiken, 1984: A diffusion problem in semiconductor technology. *J. Eng. Math.*, **18**, 315–333.

Brykina, I. G., 1988: An analytical solution of the problem of convective diffusion in the neighborhood of a discontinuity of the catalytic properties of a surface. *J. Appl. Math. Mech.*, **52**, 191–197.

Dua, S. S., and C. L. Tien, 1976: Two-dimensional analysis of conduction-controlled rewetting with precursory cooling. *J. Heat Transfer*, **98**, 407–413.

Wijngaarden, L. van, 1966: Asymptotic solution of a diffusion problem with mixed boundary conditions. *Koninkl. Ned. Akad. Weterschap.*, **B69**, 263–276.

Electromagnetism

Albertsen, N. Chr., and P. Skov-Madsen, 1983: A compact septum polarizer. *IEEE Trans. Microwave Theory Tech.*, **MTT-31**, 654–660.

Amatuni, A. Ts., 1965: Solution of the problem of transitional radiation in a plasma-like medium. *Sov. Phys. Tech. Phys.*, **9**, 1049–1056.

Anderson, I., 1979: Plane wave diffraction by a thin dielectric half-plane. *IEEE Trans. Antennas Propag.*, **AP-27**, 584–589.

Angulo, C. M., and W. S. C. Chang, 1959: The launching of surface waves by a parallel plate waveguide. *IEEE Trans. Antennas Propag.*, **AP-7**, 359–368.

Aoki, K., T. Miyazaki, K. Uchida, and Y. Shimada, 1982: On junction of two semi-infinite dielectric guides. *Radio Sci.*, **17**, 11–19.

Arora, R. K., 1965: Bifurcation of a parallel-plate waveguide by a unidirectionally conducting screen. *Proc. R. Soc. Edinburgh, Ser. A*, **67**, 50–68.

Arora, R. K., and S. Vijayaraghavan, 1970: Scattering of a shielded surface wave by a wall-impedance discontinuity. *IEEE Trans. Microwave Theory Tech.*, **MTT-18**, 734–736.

Bailey, R. C., 1977: Electromagnetic induction over the edge of a perfectly conducting ocean: The *H*-polarization case. *Geophys. J. Astron. Soc.*, **48**, 385–392.

Baksht, F. G., 1970: Electron energy distribution in the electrode sheath in a weakly ionized plasma. *Sov. Phys. Tech. Phys.*, **14**, 1196–1204.

Bates, C. P., and R. Mittra, 1968: Waveguide excitation of dielectric and plasma slabs. *Radio Sci.*, **3**, 251–266.

Bates, C. P., and R. Mittra, 1969: A factorization procedure for Wiener–Hopf kernels. *IEEE Trans. Antennas Propag.*, **AP-17**, 102–103.

Bates, C. P., 1970: Internal reflections at the open end of a semi-infinite waveguide. *IEEE Trans. Antennas Propag.*, **AP-18**, 230–235.

Beilis, A., J. W. Dash, and A. Farrow, 1987: Analytic "α-pole" approach to an electromagnetic scattering problem. *IEEE Trans. Electromagn. Compat.*, **EMC-29**, 175–185.

Breithaupt, R. W., 1963: Diffraction of a cylindrical surface wave by a discontinuity in surface reactance. *Proc. IEEE*, **51**, 1455–1463.

Breithaupt, R. W., 1966: Diffraction of a general surface wave mode by a surface reactance discontinuity. *IEEE Trans. Antennas Propag.*, **AP-14**, 290–297.

Broome, N. L., 1979: Improvements to nonnumerical methods for calculating the transient behavior of linear and aperture antennas. *IEEE Trans. Antennas Propag.*, **AP-27**, 51–62.

Buchwald, V. T., and H. E. Doran, 1965: Eigenfunctions of plane elastostatics. II. A mixed boundary value problem of the strip. *Proc. R. Soc. London, Ser. A*, **284**, 69–82.

Büyükaksoy, A., 1985: Diffraction coefficients related to cylindrically curved soft-hard surfaces. *Ann. Télécommun.*, **40**, 402–410.

Büyükaksoy, A., G. Uzgören, and A. H. Serbest, 1989: Diffraction of an obliquely incident plane wave by the discontinuity of a two part thin dielectric plane. *Int. J. Eng. Sci.*, **27**, 701–710.

Carlson, J. F., and A. E. Heins, 1946: The reflection of an electromagnetic plane wave by an infinite set of plates. I. *Quart. Appl. Math.*, **4**, 313–329.

Chakrabarti, A., 1977: Diffraction by a uni-directionally conducting strip. *Indian J. Pure Appl. Math.*, **8**, 702–717.

Chakrabarti, A., and S. Dowerah, 1984: Traveling waves in a parallel plate waveguide with periodic wall perturbations. *Can. J. Phys.*, **62**, 271–284.

Chakrabarti, A., 1986: Diffraction by a dielectric half-plane. *IEEE Trans. Antennas Propag.*, **AP-34**, 830–833.

Chang, D. C., 1969: VLF wave propagation along a mixed path in the curved earth-ionosphere waveguide. *Radio Sci.*, **4**, 335–345.

Chang, D. C., 1973: Radiation of a buried dipole in the presence of a semi-infinite metallic tube. *Radio Sci.*, **8**, 147–154.

Chew, W. C., and J. A. Kong, 1981: Asymptotic formula for the capacitance of two oppositely charged discs. *Math. Proc. Cambridge Philos. Soc.*, **89**, 373–384.

Chew, W. C., and J. A. Kong, 1982: Microstrip capacitance for a circular disk through matched asymptotic expansions. *SIAM J. Appl. Math.*, **42**, 302–317.

Chuang, C. A., C. S. Liang, and S.-W. Lee, 1975: High frequency scattering from an open-ended semi-infinite cylinder. *IEEE Trans. Antennas Propag.*, **AP-23**, 770–776.

Clausert, H., 1968: A parallel-plate waveguide, asymmetrically bifurcated by a unidirectionally conducting half screen. *Proc. Cambridge Philos. Soc.*, **64**, 559–563.

Clemmow, P. C., 1953: Radio propagation over a flat earth across a boundary separating two different media. *Philos. Trans. R. Soc. London, Ser. A*, **246**, 1–55.

Comstock, C., 1966: Transmission and reflection of electromagnetic waves normally incident on a warm plasma. *Phys. Fluids*, **9**, 1514–1521.

Daniele, V. G., I. Montrosset, and R. S. Zich, 1979: Wiener–Hopf solution for the junction between a smooth and a corrugated cylindrical waveguide. *Radio Sci.*, **14**, 943–955.

Dawson, T. W., and J. T. Weaver, 1979: H-polarization induction in two thin half-sheets. *Geophys. J. R. Astr. Soc.*, **56**, 419–438.

Dawson, T. W., J. W. Weaver, and U. Raval, 1982: B-polarization induction in two generalized thin sheets at the surface of a conducting half-space. *Geophys. J. R. Astr. Soc.*, **69**, 209–234.

Dawson, T. W., 1983: E-polarization induction in two thin half-sheets. *Geophys. J. R. Astr. Soc.*, **73**, 83–107.

Dawson, T. W., 1996: B-polarization induction in two thin half-sheets coupled to the mantle by a conducting crust – I. Solution by the Wiener–Hopf technique, with limiting forms. *Geophys. J. Int.*, **124**, 556–576.

Delogne, P., 1969: Problèmes de diffraction sur guide unifilaire. *Ann. Télécommunic.*, **24**, 405–426.

Dobrott, D. R., 1966: Propagation of VLF waves past a coastline. *Radio Sci.*, **1**, 1411–1424.

Dorfman, L. G., and V. V. Filatov, 1966: Scattering coefficients caused by abrupt change in electrical properties of the narrow wall of a rectangular waveguide. *Radio Eng. Electron. Phys.*, **11**, 170–177.

Dowerah, S., and A. Chakrabarti, 1985: Diffraction by two staggered half-planes with periodic wall perturbations. *Indian J. Pure Appl. Math.*, **16**, 411–447.

Du Cloux, R., 1984: Pulsed electromagnetic radiation from a line source in the presence of a semi-infinite screen in the plane interface of two different media. *Wave Motion*, **6**, 459–476.

Eaves, R. E., and D. M. Bolle, 1970: Modes of shielded slot lines. *Arch. Elektr. Übertrag.*, **24**, 389–394.

Elmoazzen, Y. E., and L. Shafai, 1973: Mutual coupling between parallel-plate waveguides. *IEEE Trans. Microwave Theory Tech.*, **MTT-21**, 825–833.

Elmoazzen, Y., and L. Shafai, 1974: Mutual coupling between two circular waveguides. *IEEE Trans. Antennas Propag.*, **AP-22**, 751–760.

Fernandes, C. A., and A. M. Barbosa, 1990: Radiation from a sheath helix excited by a circular waveguide: a Wiener–Hopf analysis. *IEE Proc., Part H*, **137**, 269–275.

Fong, T. T., 1972: Radiation from an open-ended waveguide with extended dielectric loading. *Radio Sci.*, **7**, 965–972.

Göksu, Ç., and F. Güneş, 1991: High frequency surface currents induced on a soft-hard cylindrical strip. *IEE Proc., Part A*, **138**, 113–118.

Grzesik, J. A., and S. C. Lee, 1995: The dielectric half space as a test bed for transform methods. *Radio Sci.*, **30**, 853–862.

Hallén, E., 1956: Exact treatment of antenna current wave reflection at the end of a tube-shaped cylindrical antenna. *IEEE Antennas Propag.*, **AP-3**, 479–491.

Hayat, T., S. Asghar, and B. Asghar, 1999: Field of a point source within perfectly conducting parallel-plates in a homogeneous biisotropic medium. *Int. J. Infrared Millimeter Waves*, **20**, 1169–1194.

Hazeltinc, R. D., M. N. Rosenbluth, and A. M. Sessler, 1971: Diffraction radiation by a line charge moving past a comb: A model of radiation losses in an electron ring accelerator. *J. Math. Phys.*, **12**, 502–514.

Heins, A. E., and J. F. Carlson, 1947: The reflection of an electromagnetic plane wave by an infinite set of plates. II. *Quart. Appl. Math.*, **5**, 82–88.

Heins, A. E., 1950: The reflection of an electromagnetic plane wave by an infinite set of plates. III. *Quart. Appl. Math.*, **8**, 281–291.

Heitman, W. G., and P. M. van den Berg, 1975: Diffraction of electromagnetic waves by a semi-infinite screen in a layered medium. *Can. J. Phys.*, **53**, 1305–1317.

Hemp, J., 1987: The virtual current in electromagnetic flowmeters with end effects. *Quart. J. Mech. Appl. Math.*, **40**, 507–525.

Hurd, R. A., 1960: Diffraction by a unidirectionally conducting half-plane. *Can. J. Phys.*, **38**, 168–175.

Hurd, R. A., 1965: The admittance of a linear antenna in a uniaxial medium. *Can. J. Phys.*, **43**, 2276–2309.

Hurd, R. A., 1966: Admittance of a long linear antenna. *Can. J. Phys.*, **44**, 1723–1744.

Hurd, R. A., 1972: Mutual admittance of two collinear antennas. *J. Appl. Phys.*, **43**, 3701–3707.

Igarashi, A., 1964: Simultaneous Wiener–Hopf equations and their application to diffraction problems in electromagnetic theory. I. *J. Phys. Soc. Japan*, **19**, 1213–1221.

Igarashi, A., 1968: Simultaneous Wiener–Hopf equations and their application to diffraction problems in electromagnetic theory. II. *J. Phys. Soc. Japan*, **25**, 260–271.

Igarashi, A., 1969: Simultaneous Wiener–Hopf equations and their application to diffraction problems in electromagnetic theory. IV. *J. Phys. Soc. Japan*, **26**, 549–560.

Igushkin, L. P., and É. I. Urazakov, 1967: Motion of symmetric current-carrying plasmoids along the axis of a semiinfinite cylindrical waveguide. *Sov. Phys. Tech. Phys.*, **12**, 27–32.

Ittipiboon, A., and M. Hamid, 1981: Application of the Wiener–Hopf technique to dielectric slab waveguide discontinuities. *IEE Proc., Part H*, **128**, 188–196.

Johansen, E. L., 1962: Scattering coefficients for wall impedance changes in waveguides. *IEEE Trans. Microwave Theory Tech.*, **10**, 26–29.

Johansen, E. L., 1965: The radiation properties of a parallel-plane waveguide in a transversely magnetized, homogeneous plasma. *IEEE Trans. Microwave Theory Tech.*, **MTT -13**, 77–83.

Johansen, E. L., 1967: Surface wave scattering by a step. *IEEE Trans. Antennas Propag.*, **AP-15**, 442–448.

Johansen, E. L., 1968: Surface wave radiation from a thick, semi-infinite plane with a reactive surface. *IEEE Trans. Antennas Propag.*, **AP-16**, 391–398.

Jull, E. V., 1967: Aperture fields in an anisotropic medium. *Radio Sci.*, **2**, 837–852.

Jull, E. V., 1968: Diffraction by a wide unidirectionally conducting strip. *Can. J. Phys.*, **46**, 2107–2117.

Kane, J., 1960: The efficiency of launching surface waves on a reactive half plane by an arbitrary antenna. *IEEE Trans. Antennas Propag.*, **AP-8**, 500–507.

Kashyap, S. C., and M. A. K. Hamid, 1971: Diffraction characteristics of a slit in a thick conducting screen. *IEEE Trans. Antennas Propag.*, **AP-19**, 499–507.

Kay, A. F., 1959: Scattering of a surface wave by a discontinuity in reactance. *IEEE Trans. Antennas Propag.*, **AP-7**, 22–31.

Kobayashi, K., 1984: On the factorization of certain kernels arising in functional equations of the Wiener–Hopf type. *J. Phys. Soc. Japan*, **53**, 2885–2898.

Kobayashi, K., and T. Inoue, 1988: Diffraction of a plane wave by an inclined parallel plate grating. *IEEE Trans. Antennas Propag.*, **AP-36**, 1424–1434.

Koshikawa, S., and K. Kobayashi, 1997: Diffraction by a terminated semi-infinite parallel-plate waveguide with three-layer material loading. *IEEE Trans. Antennas Propag.*, **AP-45**, 949–959.

Kostelnicek, R. J., and R. Mittra, 1971: Radiation from a parallel-plate waveguide into a dielectric or plasma layer. *Radio Sci.*, **6**, 981–990.

Kurushin, E. P., and E. I. Nefedov, 1969: Diffraction of H-waves on a wide slit in a plane multiwave waveguide. *Radio Eng. Electron. Phys.*, **14**, 176–182.

Lam, J., 1967: Radiation of a point charge moving uniformly over an infinite array of conducting half-planes. *J. Math. Phys.*, **8**, 1053–1060.

Latham, R. W., and K. S. H. Lee, 1968: Magnetic field leakage into a semi-infinite pipe. *Can. J. Phys.*, **46**, 1455–1462.

Lee, S. W., 1967: Radiation from an infinite array of parallel-plate waveguides with thick walls. *IEEE Trans. Microwave Theory Tech.*, **MTT-15**, 364–371.

Lee, S. W., 1969: Cylindrical antenna in uniaxial resonant plasmas. *Radio Sci.*, **4**, 179–189.

Lee, S. W., and R. Mittra, 1969: Admittance of a solid cylindrical antenna. *Can. J. Phys.*, **47**, 1959–1970.

Lee, S.-W., V. Jamnejad, and R. Mittra, 1973: Near field of scattering by a hollow semi-infinite cylinder and its application to sensor booms. *IEEE Trans. Antennas Propag.*, **AP-21**, 182–188.

Lehman, G. W., 1970: Diffraction of electromagnetic waves by planar dielectric structures. I. Transverse electric excitation. *J. Math. Phys.*, **11**, 1522–1535.

Levinson, Y. B., and E. V. Sukhorukov, 1991: Bending of electron edge states in a magnetic field. *J. Phys., Condens. Matter*, **3**, 7291–7306.

Lovis, D., 1971: The relationship between the electric and magnetic fields with an arbitrary field distribution on an open-ended parallel-plate waveguide and its radiation characteristic. *Arch. Electron. Übertrag.*, **25**, 502–508.

Marin, L., 1969: On the coupling between two adjacent rectangular waveguides through a slot in the common wall. *Acta Polytech. Scandinavica, Electr. Eng. Ser. No. 20*, 39 pp.

Marin, L., 1971: Analysis of a slotted circular waveguide. *Acta Polytech. Scandinavica, Electr. Eng. Ser. No. 28*, 23 pp.

Mason, R. J., 1968: Electric-field penetration into a plasma with a fractionally accommodating boundary. *J. Math. Phys.*, **9**, 868–874.

Millar, R. F., 1964: Plane wave spectra in grating theory. III. Scattering by a semiinfinite grating of identical cylinders. *Can. J. Phys.*, **42**, 1149–1184.

Mitsioulis, G., 1990: A Wiener–Hopf theory for a semi-infinite dielectric slab. *Can. J. Phys.*, **68**, 1348–1351.

Nasalski, W., 1983: Leaky and surface wave diffraction by an asymmetric impedance half plane. *Can. J. Phys.*, **61**, 906–918.

Papadopoulos, V. M., 1956: Scattering by a semi-infinite resistive strip of dominant-mode propagation in an infinite rectangular wave-guide. *Proc. Cambridge Philos. Soc.*, **52**, 553–563.

Papadopoulos, V. M., 1957: The scattering effect of a junction between two circular wave-guides. *Quart. J. Mech. Appl. Math.*, **10**, 191–209.

Pathak, P. H., and R. G. Kouyoumjian, 1979: Surface wave diffraction by a truncated dielectric slab recessed in a perfectly conducting surface. *Radio Sci.*, **14**, 405–417.

Pearson, J. D., 1953: The diffraction of electro-magnetic waves by a semi-infinite circular wave guide. *Proc. Cambridge Philos. Soc.*, **49**, 659–667.

Poddar, M., 1982: Very low-frequency electromagnetic response of a perfectly conducting half-plane in a layered half-space. *Geophysics*, **47**, 1059–1067.

Rao, T. C. K., and M. A. K. Hamid, 1980: On the coaxial excitation of the modified Goubau line. *Radio Sci.*, **15**, 17–24.

Ricoy, M. A., and J. L. Volakis, 1991: Diffraction by a multilayer slab recessed in a ground plane via generalized impedance boundary conditions. *Radio Sci.*, **26**, 313–327.

Rojas, R. G., H. C. Ly and P. H. Pathak, 1991: Electromagnetic plane wave diffraction by a planar junction of two thin dielectric/ferrite half planes. *Radio Sci.*, **26**, 641–660.

Rulf, B., 1967: Diffraction by a halfplane in a uniaxial medium. *SIAM J. Appl. Math.*, **15**, 120–127.

Senior, T. B. A., 1960: Diffraction by an imperfectly conducting half-plane at oblique incidence. *Appl. Sci. Res., Ser. B*, **8**, 35–61.

Serbest, A. H., G. Uzgören, and A. Büyükaksoy, 1991: Diffraction of plane waves by a resistive strip residing between two impedance half-planes. *Ann. Télécommunic.*, **46**, 359–366.

Seshadri, S. R., and T. T. Wu, 1960: High-frequency diffraction of plane waves by an infinite slit for grazing incidence. *IEEE Trans. Antennas Propag.*, **AP-8**, 37–42.

Seshadri, S. R., 1961: Diffraction of a plane wave by an infinite slit in a unidirectionally conducting screen. *IEEE Trans. Antennas Propag.*, **AP-9**, 199–207.

Seshadri, S. R., and T. T. Wu, 1963: Diffraction by a circular aperture in a unidirectionally conducting screen. *IEEE Trans. Antennas Propag.*, **AP-11**, 56–67.

Shen, H.-M., and T. T. Wu, 1989: The universal current distribution near the end of a tubular antenna. *J. Math. Phys.*, **30**, 2721–2729.

Takahashi, K., 1967: The electromagnetic field of a coaxial antenna. *Elect. Eng. Japan*, **87**, No. 12, 35–44.

Uchida, K., and K. Aoki, 1984: Scattering of surface waves on transverse discontinuities in symmetrical three-layer dielectric waveguides. *IEEE Trans. Microwave Theory Tech.*, **MTT-32**, 11–19.

Uzgören, G., A. Büyükaksoy, and A. H. Serbest, 1989: Diffraction coefficient related to a discontinuity formed by impedance and resistive half-planes. *IEE Proc., Part H*, **136**, 19–23.

Uzgören, G., A. Büyükaksoy, and A. H. Serbert, 1990: Plane wave diffraction by the discontinuity formed by resistive and impedance half-planes: Oblique incident case. *Ann. Télécommunic.*, **45**, 410–418.

Vanblaricum, G. F., and R. Mittra, 1969: A modified residue-calculus technique for solving a class of boundary value problems. Part I: Waveguide discontinuities. *IEEE Trans. Microwave Theory Tech.*, **MTT-17**, 302–309.

Van der Pauw, L. J., 1973: Diffraction of a Bleustein-Gulyaev wave by a conductive semi-infinite surface layer. *J. Acoust. Soc. Am.*, **53**, 1107–1115.

Venkataratnam, K., and T. P. Rao, 1982: Analysis of end effects in a linear induction motor by the Wiener–Hopf technique. *IEE Proc., Part B*, **129**, 364–372.

Vijayaraghavan, S., and R. K. Arora, 1971: Scattering of a shielded surface wave in a coaxial waveguide by a wall impedance discontinuity. *IEEE Trans. Microwave Theory Tech.*, **MTT-19**, 736–739.

Volakis, J. L., 1987: Scattering by a thick impedance half plane. *Radio Sci.*, **22**, 13–25.

Volakis, J. L., and M. A. Ricoy, 1987: Diffraction by a thick perfectly conducting half-plane. *IEEE Trans. Antennas Propag.*, **AP-35**, 62–72.

Volakis, J. L., and M. A. Ricoy, 1989: H-polarization diffraction by a thick metal-dielectric join. *IEEE Trans. Antennas Propag.*, **AP-37**, 1453–1462.

Volakis, J. L., and J. D. Collins, 1990: Electromagnetic scattering from a resistive half plane on a dielectric interface. *Wave Motion*, **12**, 81–96.

Wait, J. R., 1970: On launching an azimuthal surface wave on a cylindrical impedance boundary. *Acta Phys. Austr.*, **32**, 122–130.

Wait, J. R., 1970: Propagation of electromagnetic waves over a smooth multisection curved earth – an exact theory. *J. Math. Phys.*, **11**, 2851–2860.

Wait, J. R., 1970: Factorization method applied to electromagnetic wave propagation in a curved waveguide with nonuniform walls. *Radio Sci.*, **5**, 1059–1068.

Weidelt, P., 1971: The electromagnetic induction in two thin half-sheets. *Z. Geophys.*, **37**, 649–665.

Weidelt, P., 1983: The harmonic and transient electromagnetic response of a thin dipping dike. *Geophysics*, **48**, 934–952.

Wenger, N. C., 1965: The launching of surface waves on an axial-cylindrical reactive surface. *IEEE Trans. Antennas Propag.*, **AP-13**, 126–134.

Wenger, N. C., 1967: Resonant frequency of open-ended cylindrical cavity. *IEEE Trans. Microwave Theory Tech.*, **MTT-15**, 334–340.

Williams, W. E., 1957: Step discontinuities in waveguides. *IEEE Antennas Propag.*, **AP-5**, 191–198.

Williams, W. E., and A. Chakrabarti, 1982: Reflection at a discontinuity in a transmission line. *IMA J. Appl. Math.*, **28**, 185–195.

Wu, C. P., 1967: Diffraction of a plane electromagnetic wave by an infinite set of parallel metallic plates in an anisotropic plasma. *Can. J. Phys.*, **45**, 1911–1923.

Wu, T. T., 1961: Theory of the dipole antenna and the two-wire transmission line. *J. Math. Phys.*, **2**, 550–574.

Yoshidomi, K., and K. Aoki, 1988: Scattering of an E-polarized plane wave by two parallel rectangular impedance cylinders. *Radio Sci.*, **23**, 471–480.

Fluid Dynamics

Abrahams, I. D., and G. R. Wickham, 1991: The scattering of water waves by two semi-infinite opposed vertical walls. *Wave Motion*, **14**, 145–168.

Adamczyk, J. J., and M. E. Goldstein, 1978: Unsteady flow in a supersonic cascade with subsonic leading-edge locus. *AIAA J.*, **16**, 1248–1254.

Aizin, L. B., 1992: Sound generation by a Tollmien-Schlichting wave at the end of a plate in a flow. *J. Appl. Mech. Tech. Phys.*, **33**, 355–362.

Alblas, J. B., 1957: On the diffraction of sound waves in a viscous medium. *Appl. Sci. Res., Ser. A*, **6**, 237–262.

Barcilon, V., and D. R. MacAyeal, 1993: Steady flow of a viscous ice stream across a no-slip/free-slip transition at the bed. *J. Glaciol.*, **39**, 167–185.

Bodstein, G. C. R., A. R. George, and C.-Y. Hui, 1996: The three-dimensional interaction of a streamwise vortex with a large-chord lifting surface: Theory and experiment. *J. Fluid Mech.*, **322**, 51–79.

Budden, P. J., and J. Norbury, 1984: Stability of a subcritical flow under a sluice gate. *Quart. J. Mech. Appl. Math.*, **37**, 293–310.

Burke, J. E., 1964: Scattering of surface waves on an infinitely deep fluid. *J. Math. Phys.*, **5**, 805–819.

Cargill, A. M., 1982: Low frequency sound radiation and generation due to the interaction of unsteady flow with a jet pipe. *J. Fluid Mech.*, **121**, 59–105.

Case, K. M., 1960: Edge effects and the stability of plane Couette flow. *Phys. Fluids*, **3**, 432–435.

Clarke, J. F., 1967: The laminar diffusion flame in Oseen flow: The stoichiometric Burke-Schumann flame and frozen flow. *Proc. R. Soc. London, Ser. A*, **296**, 519–545.

Crighton, D. G., and F. G. Leppington, 1974: Radiative properties of the semi-infinite vortex sheet: The initial-value problem. *J. Fluid Mech.*, **64**, 393–414.

Durbin, P. A., 1989: Stokes flow near a moving contact line with yield-stress boundary condition. *Quart. J. Mech. Appl. Math.*, **42**, 99–113.

Englert, G. W., 1984: Interaction of upstream flow distortions with high-Mach-number cascades. *J. Eng. Gas Turbines Power*, **106**, 260–270.

Evans, D. V., 1972: The application of a new source potential to the problem of the transmission of water waves over a shelf of arbitrary profile. *Proc. Cambridge Philos. Soc.*, **71**, 391–410.

Foster, M. R., 1980: A flat plate in a rotating, stratified flow. *J. Eng. Math.*, **14**, 117–132.

Gautesen, A. K., 1971: Symmetric Oseen flow past a semi-infinite plate and two force singularities. *Z. Angew. Math. Phys.*, **22**, 144–155.

Gautesen, A. K., 1971: Oseen flow past a semi-infinite flat plate and a force singularity. *Z. Angew. Math. Phys.*, **22**, 247–257.

Goldstein, M. E., W. Braun, and J. J. Adamczyk, 1977: Unsteady flow in a supersonic cascade with strong in-passage shocks. *J. Fluid Mech.*, **83**, 569–604.

Goldstein, M. E., 1978: Characteristics of the unsteady motion on transversely sheared mean flow. *J. Fluid Mech.*, **84**, 305–329.

Goldstein, M. E., 1979: Scattering and distortion of the unsteady motion on transversely sheared mean flows. *J. Fluid Mech.*, **91**, 601–632.

Goldstein, M. E., 1981: The coupling between flow instabilities and incident disturbances at a leading edge. *J. Fluid Mech.*, **104**, 217–246.

Graebel, W. P., 1965: Slow viscous shear flow past a plate in a channel. *Phys. Fluids*, **8**, 1929–1935.

Greene, T. R., and A. E. Heins, 1953: Water waves over a channel of infinite depth. *Quart. Appl. Math.*, **11**, 201–214.

Hamblin, P. F., 1980: An analysis of advective diffusion in branching channels. *J. Fluid Mech.*, **99**, 101–110.

Heins, A. E., and H. Feshbach, 1947: The coupling of two acoustical ducts. *J. Math. Phys. (Cambridge, MA)*, **26**, 143–155.

Holford, R. L., 1964: Short surface waves in the presence of a finite dock. *Proc. Cambridge Philos. Soc.*, **60**, 957–983, 985–1011.

Jeong, J.-T., and M.-U. Kim, 1983: Slow viscous flow around an inclined fence on a plane. *J. Phys. Soc. Japan*, **52**, 2356–2363.

Jeong, J.-T., and M.-U. Kim, 1985: Slow viscous flow due to sliding of a semi-infinite plate over a plane. *J. Phys. Soc. Japan*, **54**, 1789–1799.

Jeong, J.-T., and M.-U. Kim, 1988: Two-dimensional slow viscous flow from a converging nozzle. *J. Phys. Soc. Japan*, **57**, 856–865.

Jeong, J.-T., and M.-U. Kim, 1990: Two-dimensional slow viscous flow due to the rotation of a body with cusped edges. *J. Phys. Soc. Japan*, **59**, 3194–3202.

Jeong, J.-T., and M.-U. Kim, 1993: Slow viscous flow around a vertical plate above a plane wall. *J. Phys. Soc. Japan*, **62**, 4243–4254.

Jeong, J.-T., and C.-G. Park, 1997: Slow viscous flow past a lattice of equal flat plates. *J. Phys. Soc. Japan*, **66**, 1660–1667.

Jeong, J.-T., 1998: Two-dimensional Stokes flow through a slit in a vertical plate on a plane wall. *J. Phys. Soc. Japan*, **67**, 4074–4079.

Jeong, J.-T., 2000: Two-dimensional stagnation flow around a vertical plate above a plane wall. *Phys. Fluids*, **12**, 511–517.

Jeong, J.-T., 2001: Slow viscous flow in a partitioned channel. *Phys. Fluids*, **13**, 1577–1582.

Jeong, J.-T., 2001: Slip boundary condition on an idealized porous wall. *Phys. Fluids*, **13**, 1884–1890.

Kempton, A. J., 1976: Heat diffusion as a source of aerodynamic sound. *J. Fluid Mech.*, **78**, 1–31.

Kim, D. H., and L. L. Koss, 1990: Sound radiation from a circular duct with axial temperature gradients. *J. Sound Vib.*, **141**, 1–16.

Kim, M.-U., and M. K. Chung, 1984: Two-dimensional slow viscous flow past a plate midway between an infinite channel. *J. Phys. Soc. Japan*, **53**, 156–166.

Kim, M.-U., J.-T. Jeong, and T.-H. Lee, 1986: Slow viscous flow due to the rotation of two finite hinged plates. *J. Phys. Soc. Japan*, **55**, 115–127.

Koch, W., and G. S. S. Ludford, 1970: Diffusion in shear flow past a semi-infinite flat plate. *Acta Mech.*, **10**, 229–250.

Koch, W., G. S. S. Ludford, and A. R. Seebass, 1971: Diffusion in shear flow past a semi-infinite flat plate. Part II: Viscous effects. *Acta Mech.*, **12**, 99–120.

Koch, W., 1975: Heat transfer from two hot plates in laterally bounded uniform flow as an example of an n-part Wiener–Hopf-problem. *Z. Angew. Math. Phys.*, **26**, 187–198.

Koch, W., and W. Möhring, 1983: Eigensolutions for liners in uniform mean flow ducts. *AIAA J.*, **21**, 200–213.

Kuiken, H. K., 1986: On the influence of longitudinal diffusion in time-dependent convective-diffusive systems. *J. Fluid Mech.*, **165**, 147–162.

Laurmann, J. A., 1961: Linearized slip flow past a semi-infinite flat plate. *J. Fluid Mech.*, **11**, 82–96.

Leib, S. J., and M. E. Goldstein, 1986: The generation of capillary instabilities on a liquid jet. *J. Fluid Mech.*, **168**, 479–500.

Lennox, S. C., and D. C. Pack, 1963: The flow in compound jets. *J. Fluid Mech.*, **15**, 513–526.

Ludford, G. S. S., and V. O. S. Olunloyo, 1972: Further results concerning the forces on a flat plate in a Couette flow. *Z. Angew. Math. Phys.*, **23**, 729–744.

Mahmood-Ul-Hassan and A. D. Rawlins, 1998: Two problems of waveguides carrying mean fluid flow. *J. Sound Vib.*, **216**, 713–738.

Martin, P. A., and R. A. Dalrymple, 1994: On the propagation of water waves along a porous-walled channel. *Proc. R. Soc. London, Ser. A*, **444**, 411–428.

Martinez, R., and S. E. Widnall, 1980: Unified aerodynamic-acoustic theory for a thin rectangular wing encountering a gust. *AIAA J.*, **18**, 636–645.

McCue, S. W., and D. M. Stump, 2000: Linear stern waves in finite depth channels. *Quart. J. Mech. Appl. Math.*, **53**, 629–643.

Meister, E., 1965: Zur Theorie der ebenen, instationären Unterschallströmung um ein schwingendes Profil im Kanal. *Z. Angew. Math. Phys.*, **16**, 770-780.

Meister, E., 1969: Theorie instationärer Unterschallströmungen durch ein schwingendes Gitter im Windkanal. *Z. Angew. Math. Mech.*, **49**, 481–494.

Miles, J. W., 1951: The oscillating rectangular airfoil at supersonic speeds. *Quart. Appl. Math.*, **9**, 47–65.

Morioka, S., 1980: Steady two-dimensional jet-flow of a dispersive compressible fluid. *J. Phys. Soc. Japan*, **48**, 1009–1017.

Munt, R. M., 1977: The interaction of sound with a subsonic jet issuing from a semi-infinite cylindrical pipe. *J. Fluid Mech.*, **83**, 609–640.

Munt, R. M., 1990: Acoustic transmission properties of a jet pipe with subsonic jet flow: I. The cold jet reflection coefficient. *J. Sound Vib.*, **142**, 413–436.

Olmstead, W. E., 1966: An exact solution for Oseen flow past a half plane and a horizontal force singularity. *J. Math. Phys. (Cambridge, MA)*, **45**, 156–161.

Orszag, S. A., and S. C. Crow, 1970: Instability of a vortex sheet leaving a semi-infinite plate. *Stud. Appl. Math.*, **49**, 167–181.

Osborne, T. J., and D. M. Stump, 2001: Capillary waves on a Eulerian jet emerging from a channel. *Phys. Fluids*, **13**, 616–623.

Ranta, M. A., 1964: An application of integral transformations and the Wiener–Hopf techniques to the supersonic flow past an oscillating, nearly circular, slender body and past an elastic, two-dimensional, thin wing. *Acta Polytech. Scandinavica, Mech. Eng. Ser. No. 18*, 69 pp.

Ranta, M. A., 1969: On the application of perturbation techniques to the study of missile flight with special references to internal mass flow, thrust malalignment, and the effect of gravity. *Acta Polytech. Scandinavica, Mech. Eng. Ser. No. 43*, 41 pp.

Richardson, S., 1970: A 'stick-slip' problem related to the motion of a free jet at low Reynolds numbers. *Proc. Cambridge Philos. Soc.*, **67**, 477–489.

Rienstra, S. W., 1984: Acoustic radiation from a semi-infinite annular duct in a uniform subsonic mean flow. *J. Sound Vib.*, **94**, 267–288.

Schmidt, G. H., 1981: Linearized stern flow of a two-dimensional shallow-draft ship. *J. Ship Res.*, **25**, 236–242.

Schorr, B., 1967: Das Anfangs-Rantwertproblem eines dünnen Profils in kompressibler Unterschallströmung. *Z. Angew. Math. Phys.*, **18**, 149–164.

Seebass, R., K. Tamada, and T. Miyagi, 1966: Oseen flow past a finite flat plate. *Phys. Fluids*, **9**, 1697–1703.

Sharma, U., K. Viswanathan, K. Singh, and M. L. Gogna, 1991: Diffraction of obliquely incident surface waves by a soft vertical plane barrier of finite depth. *Indian J. Pure Appl. Math.*, **22**, 337–353.

Shaw, D. C., 1985: The asymptotic behavior of a curved line source plume within an enclosure. *IMA J. Appl. Math.*, **35**, 71–89.

Sparenberg, J. A., 1974: On the linear theory of an actuator disk in a viscous fluid. *J. Ship Res.*, **18**, 16–21.

Surampudi, S. P., and J. J. Adamczyk, 1986: Unsteady transonic flow over cascade blades. *AIAA J.*, **24**, 293–302.

Takahashi, K., and S. Kaji, 1991: Analytical study on plate edge noise (Trailing edge noise caused by vorticity waves). *JSME Int. J., Ser. 2*, **34**, 431–438.

Trogden, S. A., and D. D. Joseph, 1980: The stick-slip problem for a round jet. I. Large surface tension. *Rheol. Acta*, **19**, 404–420.

Vitiuk, V. F., 1970: Diffraction of surface waves at a dock of finite width. *J. Appl. Math. Mech.*, **34**, 27–35.

Weitz, M., and J. B. Keller, 1950: Reflection of water waves from floating ice in water of finite depth. *Commun. Pure Appl. Math.*, **3**, 305–318.

Williams, M. M. R., 1969: The temperature distribution in a radiating fluid flowing over a flat plate. *Quart. J. Mech. Appl. Math.*, **22**, 487–500.

General

Abrahams, I. D., and G. R. Wickham, 1990: General Wiener–Hopf factorization of matrix kernels with exponential phase factors. *SIAM J. Appl. Math.*, **50**, 819–838.

Atkinson, C., 1975: On the stress intensity factors associated with cracks interacting with an interface between two elastic media. *Int. J. Eng. Sci.*, **13**, 489–504.

Atkinson, C., and C. R. Champion, 1984: Some boundary-value problems for the equation $\nabla \bullet (|\nabla \phi|^N \nabla \phi) = 0$. *Quart. J. Mech. Appl. Math.*, **37**, 401–419.

Bierman, G. J., 1971: A particular class of singular integral equations. *SIAM J. Appl. Math.*, **20**, 99–109.

Bolotovskiĭ, B. M., and G. V. Voskresenskiĭ, 1966: Diffraction radiation. *Sov. Phys. Uspekhi*, **9**, 73–96.

Boyd, W. G. C., 1974: High-frequency scattering in a certain stratified medium. The two-part problem. *Proc. R. Soc. Edinburgh, Ser. A*, **72**, 149–178.

Boyd, W. G. C., 1977: High-frequency scattering in a certain stratified medium: The three-part problem. *Proc. R. Soc. London, Ser. A*, **356**, 315–343.

Chakrabarti, A., 1979: Diffraction by a strip under mixed boundary conditions. *J. Indian Inst. Sci.*, **B61**, 163–176.

Chakrabarti, A., and S. Dowerah, 1984: Diffraction by a periodically corrugated strip. *J. Tech. Phys.*, **25**, 113–126.

Clemmow, P. C., 1951: A method for the exact solution of a class of two-dimensional diffraction problems. *Proc. R. Soc. London, Ser. A*, **205**, 286–308.

Davidson, R. F., 1988: Waves below first cutoff in a duct. *J. Austr. Math. Soc.*, **B29**, 448–460.

Davies, B., 1978: *Integral Transforms and Their Applications.* Springer-Verlag, Chapters 18–19.

DeSanto, J. A., 1971: Scattering from a periodic corrugated structure: Thin comb with soft boundaries. *J. Math. Phys.*, **12**, 1913–1923.

Evans, D. V., 1972: The application of a new source potential to the problem of the transmission of water waves over a shelf of arbitrary profile. *Proc. Cambridge Philos. Soc.*, **71**, 391–410.

Friedrich, N., 1980: Zur Faktorisierung einiger Kerne, die bei Ausstrahlungsproblemen der Elektrodynamik und Aeroelastizität auftreten. *J. Reine Angew. Math.*, **316**, 15–30.

Georgiadis, H. G., 1992: A correspondence principle connecting IBVPs of wave propagation and heat conduction. *ZAMP*, **43**, 742–755.

Haj-Hariri, H., 1994: Generalized modal expansion: A Wiener–Hopf problem. *J. Acoust. Soc. Am.*, **95**, 3039–3048.

Heins, A. E., 1956: The scope and limitations of the method of Wiener and Hopf. *Commun. Pure Appl. Math.*, **9**, 447–466.

Kobayashi, K., 1984: On the factorization of certain kernels arising in functional equations of the Wiener–Hopf type. *J. Phys. Soc. Japan*, **53**, 2885–2898.

Levine, H., 1982: On a mixed boundary value problem of diffusion type. *Appl. Sci. Res.*, **39**, 261–276.

Marshak, R. E., 1947: The Milne problem for a large plane slab with constant source and anisotropic scattering. *Phys. Rev., Ser. 2*, **72**, 47–50.

Mitsioulis, G., 1991: Renormalization of the energies stored around a Wiener–Hopf structure: I. *Can. J. Phys.*, **69**, 875–890.

Mittra, R., and S. W. Lee, 1970: On the solution of a generalized Wiener–Hopf equation. *J. Math. Phys.*, **11**, 775–783.

Noble, B., 1958: *Methods Based on the Wiener–Hopf Technique for the Solution of Partial Differential Equations.* Pergamon Press, 246 pp.

Pridmore-Brown, D. C., 1968: A Wiener–Hopf solution of a radiation problem in conical geometry. *J. Phys. Math. (Cambridge, MA)*, **47**, 79–94.

Ramakrishnan, R., 1982: A note on the calculation of Wiener–Hopf split functions. *J. Sound Vib.*, **81**, 592–595.

Rawlins, A. D., 1980: Simultaneous Wiener–Hopf equations. *Can. J. Phys.*, **58**, 420–428.

Shirai, H., and L. B. Felsen, 1986: Spectral method for multiple edge diffraction by a flat strip. *Wave Motion*, **8**, 499–524.

Vajnshtejn, L. A., 1948: Rigorous solution of the problem of an open-ended parallel-plate waveguide (in Russian). *Ivz. Akad. Nauk. USSR, Ser. Fiz.*, **12**, 144–165.

Vajnshtejn, L. A., 1948: On the theory of diffraction by two parallel half-planes (in Russian). *Ivz. Akad. Nauk. USSR, Ser. Fiz.*, **12**, 166–180.

Vajnshtejn, L. A., 1948: Theory of symmetric waves in a cylindrical waveguide with an open end (in Russian). *Zh. Tekh. Fiz.*, **18**, 1543–1564.

Williams, W. E., 1954: Diffraction by two parallel planes of finite length. *Proc. Cambridge Philos. Soc.*, **50**, 309–318.

Williams, W. E., 1956: Diffraction by a cylinder of finite length. *Proc. Cambridge Philos. Soc.*, **52**, 322–335.

Geophysical Sciences

Asghar, S., and G. H. Zahid, 1985: Diffraction of Kelvin waves by a finite barrier. *ZAMP*, **36**, 712–722.

Asghar, S., and F. D. Zaman, 1986: Diffraction of Love waves by a finite rigid barrier. *Bull. Seism. Soc. Am.*, **76**, 241–257.

Buchwald, V. T., 1968: The diffraction of Kelvin waves at a corner. *J. Fluid Mech.*, **31**, 193–205.

Buchwald, V. T., and F. Viera, 1996: Linearised evaporation from a soil of finite depth near a wetted region. *Quart. J. Mech. Appl. Math.*, **49**, 49–64.

Buchwald, V. T., and F. Viera, 1998: Linearised evaporation from a soil of finite depth above a water table. *J. Aust. Math. Soc., Ser. B*, **39**, 557–576.

Chao, S.-Y., L. J. Pietrafesa, and G. S. Janowitz, 1979: The scattering of continental shelf waves by an isolated topographic irregularity. *J. Phys. Oceanogr.*, **9**, 687–695.

Clovis, L. F., and H. G. Pinsent, 1983: The incidence of internal waves onto a thin submerged barrier. *Geophys. Astrophys. Fluid Dyn.*, **25**, 191–212.

Crease, J., 1956: Long waves on a rotating earth in the presence of a semi-infinite barrier. *J. Fluid Mech.*, **1**, 86–96.

Davis, A. M. J., 1987: Continental shelf wave scattering by a semi-infinite coastline. *Geophys. Astrophys. Fluid Dyn.*, **39**, 25–55.

Davis, A. M. J., 1990: Continental shelf wave scattering by a semi-infinite coastline; partial removal of the "rigid lid." *Geophys. Astrophys. Fluid Dyn.*, **50**, 175–194.

Deshwal, P. S., and K. K. Mann, 1988: Rayleigh wave scattering at coastal regions. *J. Acoust. Soc. Am.*, **84**, 286–291.

Deshwal, P. S., and N. Mohan, 1988: P-wave scattering at a coastal region in a shallow ocean. *Indian J. Pure Appl. Math.*, **19**, 1020–1030.

Deshwal, P. S., and S. Rathi, 1996: Rayleigh wave scattering at the edge of pack-ice in shallow oceans. *Indian J. Pure Appl. Math.*, **27**, 1043–1056.

Deshwal, P. S., and S. Mudgal, 1998: Scattering of Love waves due to the presence of a rigid barrier of finite depth in the crustal layer of the earth. *Proc. Indian Acad. Sci., Math. Sci.*, **108**, 81–94.

Diprima, R. C., 1957: On the diffusion of tides into permeable rock of finite depth. *Quart. Appl. Math.*, **15**, 329–339.

Ghosh, M. L., 1973: On reflection and diffraction of pressure waves from floating ice. *Pure Appl. Geophys.*, **111**, 2163–2176.

Gol'dshtein, R. V., and A. V. Marchenko, 1989: The diffraction of plane gravitational waves by the edge of an ice cover. *J. Appl. Math. Mech.*, **53**, 731–736.

Haines, C. R., 1981: A Wiener–Hopf approach to Kelvin wave generation by a semi-infinite barrier and a depth discontinuity. *Quart. J. Mech. Appl. Math.*, **34**, 139–151.

Hsieh, W. W., and V. T. Buchwald, 1984: The scattering of a continental shelf wave by a semi-infinite barrier located along the outer edge of a step shelf. *Geophys. Astrophys. Fluid Dyn.*, **28**, 257–276.

Hutter, K., and V. O. S. Olunloyo, 1980: On the distribution of stress and velocity in an ice strip, which is partly sliding over and partly adhering to its bed, by using a Newtonian viscous approximation. *Proc. R. Soc. London, Ser. A*, **373**, 385–403.

Kapoulitsas, G. M., 1977: Diffraction of long waves by a semi-infinite vertical barrier on a rotating earth. *Int. J. Theoret. Phys.*, **16**, 763–773.

Kapoulitsas, G. M., 1979: Diffraction of Kelvin waves from a rotating channel with an infinite and a semi-infinite barrier. *J. Phys. A*, **12**, 733–742.

Kapoulitsas, G. M., 1980: Scattering of long waves in a rotating, bifurcated channel. *Int. J. Theoret. Phys.*, **19**, 773–788.

Kapoulitsas, G. M., 1984: Diffraction of long waves by a step. *Acta Mech.*, **52**, 77–91.

Kapoulitsas, G. M., 1984: Propagation of long waves into a set of parallel vertical barriers on a rotating earth. *Wave Motion*, **6**, 1–14.

Lehner, F. K., V. C. Li, and J. R. Rice, 1981: Stress diffusion along rupturing plate boundaries. *J. Geophys. Res.*, **86**, 6155–6169.

Mann, K. K., and P. S. Deshwal, 1986: Rayleigh wave scattering by a plane barrier in a shallow ocean. *Indian J. Pure Appl. Math.*, **17**, 1056–1066.

Murphy, D. G., and A. J. Willmott, 1991: Rossby wave scattering by a meridional line barrier in an infinitely long zonal channel. *J. Phys. Oceanogr.*, **21**, 621–634.

Nicoll, M. A., and J. T. Weaver, 1977: H-polarization induction over an ocean edge coupled to the mantle by a conducting crust. *Geophys. J. R. Astr. Soc.*, **49**, 427–441.

Packham, B. A., 1969: Reflexion of Kelvin waves at the open end of a rotating semi-infinite channel. *J. Fluid Mech.*, **39**, 321–328.

Pinsent, H. G., 1971: The effect of a depth discontinuity on Kelvin wave diffraction. *J. Fluid Mech.*, **45**, 747–758.

Plis, A. I., and V. I. Plis, 1980: Diffraction of Kelvin waves at the open end of a plane-parallel channel. *J. Appl. Math. Mech.*, **44**, 45–50.

Plis, V. I., 1981: Propagation of Kelvin waves from a channel into a semibounded tank. *J. Appl. Math. Mech.*, **45**, 785–790.

Plis, V. I., 1986: Diffraction of Kelvin waves in a rotating semibounded basin containing a semi-infinite wall. *J. Appl. Math. Mech.*, **50**, 286–290.

Raval, U., J. T. Weaver, and T. W. Dawson, 1981: The ocean-coast effect re-examined. *Geophys. J. R. Astr. Soc.*, **67**, 115–123.

Singh, J., and M. Deshwal, 2003: Rayleigh waves scattering due to mountain of finite depth at the coastal region. *Indian J. Pure Appl. Math.*, **34**, 785–797.

Zyryanov, V. N., 1974: On wind-driven currents in a strait. *Oceanol.*, **14**, 386–391.

Heat Conduction

Avilov, V. V., and I. Decker, 1994: Heat transfer to laminar flow across plates and cylinders of various cross sections. *JETP Lett.*, **59**, 890–895.

Evans, D. V., 1984: A note on the cooling of a cylinder entering a fluid. *IMA J. Appl. Math.*, **33**, 49–54.

Georgiadis, H. G., J. R. Barber, and F. Ben Ammar, 1991: An asymptotic solution for short-time transient heat conduction between two dissimilar bodies in contact. *Quart. J. Mech. Appl. Math.*, **44**, 303–322.

Horvay, G., and M. DaCosta, 1964: Temperature distribution in a cylindrical rod moving from a chamber at one temperature to a chamber at another temperature. *J. Heat Transfer*, **86**, 265–266.

Huang, S. C., 1985: Unsteady-state heat conduction in semi-infinite regions with mixed-type boundary conditions. *J. Heat Transfer*, **107**, 489–491.

Koch, W., 1970: On the heat transfer from a finite plate in channel flow for vanishing Prandtl number. *Z. Angew. Math. Phys.*, **21**, 910–918.

Koch, W., 1974: Radiation – convection interaction for a laterally bounded flow past a hot plate. *Int. J. Heat Mass Transfer*, **17**, 915–931.

Ma, S. W., and F. M. Gerner, 1993: Forced convection heat transfer from microstructures. *J. Heat Transfer*, **115**, 872–880.

Olek, S., 1989: Rewetting of a solid cylinder with precursory cooling. *Appl. Sci. Res.*, **46**, 347–364.

Springer, S. G., and T. J. Pedley, 1973: The solution of heat-transfer problems by the Wiener–Hopf technique. *Proc. R. Soc. London, Ser. A*, **333**, 347–362.

Tien, C. L., and L. S. Yao, 1975: Analysis of conduction-controlled rewetting of a vertical surface. *J. Heat Transfer*, **97**, 161–165.

Magnetohydrodynamics

Fouad, A. A., 1966: End losses in a magnetohydrodynamic channel: DC channel with fluid having large magnetic Reynolds number. *IEEE Trans. Electron Devices*, **ED-13**, 554–561.

Greenspan, H. P., and L. A. Peletier, 1962: Some exact solutions of magneto-hydrodynamic viscous flow problems. *J. Phys. Math. (Cambridge, MA)*, **41**, 116–131.

Hector, D. L., 1967: On the linearized M. H. D. flow past a semi-infinite flat plate in the presence of a transverse magnetic field. *J. Phys. Math. (Cambridge, MA)*, **46**, 408–424.

Kapila, A. K., and G. S. S. Ludford, 1977: MHD with inertia: Flow over blunt obstacles in channels. *Int. J. Eng. Sci.*, **15**, 465–480.

Kudriavtsev, B. A., V. Z. Parton, and B. D. Rubinskii, 1980: Magnetothermoelastic field in a body with a semi-infinite cut. *J. Appl. Math. Mech.*, **44**, 646–650.

Radlow, J., and W. B. Ericson, 1962: Transverse magnetohydrodynamic flow past a semi-infinite plate. *Phys. Fluids*, **5**, 1428–1434.

Waechter, R. T., 1968: Steady electrically driven flows. *Proc. Cambridge Philos. Soc.*, **64**, 871–894.

Wolfersdorf, L. von, 1964: Eine magnetohydrodynamische Kanalströmung bei inhomogenem elecktrischem Verhalten der Kanalwände. *Rev. Roum. Sci. Tech., Ser. Mec. Appl.*, **9**, 963–976.

Other

Aoki, K., and J. Osteryoung, 1981: Diffusion-controlled current at the stationary finite disk electrode. *J. Electroanal. Chem.*, **122**, 19–35.

Aoki, K., K. Tokuda, and H. Matsuda, 1987: Hydrodynamic voltammetry at channel electrodes. Part IX. Edge effects at rectangular channel flow microelectrodes. *J. Electroanal. Chem.*, **217**, 33–47.

Aoki, K., K. Tokuda, and H. Matsuda, 1987: Theory of chrono-amperometric curves at microband electrodes. *J. Electroanal. Chem.*, **225**, 19–32.

Aoki, K., K. Tokuda, and H. Matsuda, 1987: Theory of stationary current-potential curves at microdisk electrodes for quasi-reversible and totally irreversible electrode reactions. *J. Electroanal. Chem.*, **235**, 87–96.

Atkinson, C., 1981: The growth kinetics of individual ledges during solid-solid phase transformation. *Proc. R. Soc. London, Ser. A*, **378**, 351–368.

Atkinson, C., 1982: Diffusion controlled ledge growth in a medium of finite extent. *J. Appl. Phys.*, **53**, 5689–5696.

Avdeyev, Ye. V., and G. V. Voskresenskiy, 1967: The radiation accompanying the uniform motion of a charged filament in the vicinity of a comb structure. General solution. *Radio Eng. Electron. Phys.*, **12**, 432–440.

Baksht, F. G., V. G. Ivanov, and B. Ya. Moizhes, 1972: Calculation of the electron emission probability from dielectrics in the case of secondary electron emission and the external photoeffect. *Sov. Phys. Solid State*, **13**, 2436–2443.

Bera, R., and B. Patra, 1979: On the shear of an elastic wedge as a mixed boundary value problem. *Acta Mech.*, **33**, 307–316.

Blatter, G., 1985: Scattering of atomic beams off a single surface step: An exact solution. *Ann. Physics*, **162**, 100–131.

Bolotovskiǐ, B. M., and G. V. Voskresenskii, 1968: Emission from charged particles in periodic structures. *Sov. Phys. Uspekhi*, **11**, 143–162.

Eid, Y., S. Morsy, A. Abboud, S. El-Konsol, A. Hussein, and I. Hamouda, 1976: Investigation of the space-dependent energy and angular neutron spectrum in two adjacent moderating media. *Atomkernenergie*, **28**, 259–264.

Elliott, J. P., 1955: Milne's problem with a point source. *Proc. R. Soc. London, Ser. A*, **228**, 424–433.

Fainstein-Pedraza, D., and G. F. Bolling, 1975: Superdendritic growth. II. A mathematical analysis. *J. Crystal Growth*, **28**, 319–333.

Kuiken, H. K., 1984: Etching: A two-dimensional mathematical approach. *Proc. R. Soc. London, Ser. A*, **392**, 199–225.

Kuiken, H. K., 1985: Edge effects in crystal growth under intermediate diffusive-kinetic control. *IMA J. Appl. Math.*, **35**, 117–129.

Misiakos, K., C. H. Wang, A. Neugroschel, and F. A. Lindholm, 1990: Simultaneous extraction of minority-carrier transport parameters in crystalline semiconductors by lateral photocurrent. *J. Appl. Phys.*, **67**, 321–333.

Morsy, S., and I. Kamha, 1976: Flux distortion in three-dimensional reactors due to plane absorbers. *Atomkernenergie*, **28**, 265–270.

Popova, V. A., and G. A. Shneerson, 1978: Field and current density near the edge of a semiinfinite thin solenoid. *Sov. Phys. Tech. Phys.*, **22**, 1169–1173.

Sander, L. M., 1985: Exact solution for the peripheral photoresponse of a p-n junction. *J. Appl. Phys.*, **57**, 2057–2059.

Shaw, D. C., 1985: The asymptotic behavior of a curved line source plume within an enclosure. *IMA J. Appl. Math.*, **35**, 71–89.

Sparenberg, J. A., 1958: On a shrink-fit problem. *Appl. Sci. Res., Ser. A*, **7**, 109–120.

Stakhanov, I. P., 1968: Nonequilibrium ionization in a low-voltage discharge. II. *Sov. Phys. Tech. Phys.*, **12**, 935–942.

Thomas, R. M., 1988: Methods for calculating the conduction-controlled rewetting of a cladded rod. *Nucl. Eng. Des.*, **110**, 1–16.

Vaisleib, Yu. V., 1971: Perfectly conducting finite cone in the field of a point charge and related electrostatic problems. *Sov. Phys. Tech. Phys.*, **15**, 1395–1404.

Von Roos, O., 1978: Analysis of the interaction of an electron beam with a solar cell. II. *Solid-State Electron.*, **21**, 1101–1108.

Vorotnytsev, M. A., 1981: Distribution of the potential in the electric double layer at the contact between two different semiinfinite planar electrodes. *Sov. Electrochem.*, **17**, 472–479.

Wichman, I. S., 1983: Flame spread in an opposed flow with a linear velocity gradient. *Combust. Flame*, **50**, 287–304.

Williams, M. M. R., 1965: Neutron flux perturbations due to infinite plane absorbers. II: Exponential flux. *Brit. J. Appl. Phys.*, **16**, 1841–1852.

Williams, M. M. R., 1966: An exact solution of the two group Milne problem by the method of Wiener and Hopf. *J. Math. Phys. (Cambridge, MA)*, **45**, 64–76.

Williams, M. M. R., 1976: The energy spectrum of sputtered atoms. *Philos. Mag., Ser. 8*, **34**, 669–683.

Williams, M. M. R., 1979: The spatial dependence of the energy spectrum of slowing down particles – I. Applications to reactor physics and atomic sputtering. *Ann. Nucl. Energy*, **6**, 145–173.

Solid Mechanics

Abrahams, I. D., 1986: Scattering of sound by a semi-infinite elastic plate with a soft backing; a matrix Wiener-Hopf problem. *IMA J. Appl. Math.*, **37**, 227–245.

Achenbach, J. D., and A. K. Gautesen, 1977: Elastodynamic stress-intensity factors for a semi-infinite crack under 3-D loading. *J. Appl. Mech.*, **44**, 243–249.

Achenbach, J. D., and A. K. Gautesen, 1977: Geometrical theory of diffraction for three-D elastodynamics. *J. Acoust. Soc. Am.*, **61**, 413–421.

Alblas, J. B., 1957: On the diffraction of sound waves in a viscous medium. *Appl. Sci. Res., Ser. A*, **6**, 237–262.

Alblas, J. B., 1958: On the generation of water waves by a vibrating strip. *Appl. Sci. Res., Ser. A*, **7**, 224–236.

Angel, Y. C., 1990: Singular integral equation for antiplane-wave scattering by a semi-infinite crack. *J. Elasticity*, **23**, 53–67.

Atkinson, C., 1975: On the stress intensity factors associated with cracks interacting with an interface between two elastic media. *Int. J. Eng. Sci.*, **13**, 489–504.

Atkinson, C., and R. D. List, 1978: Steady state crack propagation into media with spatially varying elastic properties. *Int. J. Eng. Sci.*, **16**, 717–730.

Atkinson, C., 1979: A note on some dynamic crack problems in linear viscoelasticity. *Arch. Mech.*, **31**, 829–849.

Austin, D. M., and R. D. Gregory, 1988: On the bending and flexure of a plate restrained by smooth-rigid clamps. *Quart. J. Mech. Appl. Math.*, **41**, 549–562.

Baker, B. R., 1966: Ductile yielding and brittle fracture at the ends of parallel cracks in a stretched orthotropic sheet. *Int. J. Fract. Mech.*, **2**, 576–596.

Banks-Sills, L., and Y. Benveniste, 1983: Steady interface crack propagation between two viscoelastic standard solids. *Int. J. Fract.*, **21**, 243–260.

Bera, R., and B. Patra, 1979: On the shear of an elastic wedge as a mixed boundary value problem. *Acta Mech.*, **33**, 307–316.

Block, L. M., 1982: Shear and normal impact loadings on one face of a narrow slit. *Int. J. Solids Struct.*, **18**, 467–477.

Block, L. M., 1992: Transient thermal effects in edge dislocation generation near a crack edge. *Int. J. Solids Struct.*, **29**, 2217–2234.

Block, L. M., and J. P. Thomas, 1992: Thermal effects in rudimentary crack edge inelastic zone growth under stress wave loading. *Acta Mech.*, **93**, 223 239.

Bose, S. K., 1968: On the sudden subsidence of half the surface of an elastic half-space. *Bull. Calcutta Math. Soc.*, **60**, 117–128.

Cannell, P. A., 1975: Edge scattering of aerodynamic sound by a lightly loaded elastic half-plane. *Proc. R. Soc. London, Ser. A*, **347**, 213–238.

Champion, C. R., 1988: The stress intensity factor history for an advancing crack under three-dimensional loading. *Int. J. Solids Struct.*, **24**, 285–300.

Chang, S.-J., 1971: Diffraction of plane dilatational waves by a finite crack. *Quart. J. Mech. Appl. Math.*, **24**, 423–443.

Chang, S.-J., and S. M. Ohr, 1984: Inclined pileup of screw dislocations at the crack tip without a dislocation-free zone. *J. Appl. Phys.*, **55**, 3505–3513.

Chattopadhyay, A., and U. Bandyopadhyay, 1988: Propagation of a crack due to shear waves in a medium of monoclinic type. *Acta Mech.*, **71**, 145–156.

Cherepanov, G. P., and V. D. Kuliev, 1975: On crack twinning. *Int. J. Fract.*, **11**, 29–38.

Cherepanov, G. P., 1976: Equilibrium of a slope with a tectonic crack. *J. Appl. Math. Mech.*, **40**, 119–133.

Cherepanov, G. P., 1976: Plastic rupture lines at the tip of a crack. *J. Appl. Math. Mech.*, **40**, 666–674.

Choi, S. R., and Y. Y. Earmme, 1991: Green's function of semi-infinite kinked crack under anti-plane shear. *Int. J. Fract.*, **51**, R3–R11.

Craster, R. V., and C. Atkinson, 1992: Interfacial fracture in elastic diffusive media. *Int. J. Solids Struct.*, **29**, 1463–1498.

Crighton, D. G., and D. Innes, 1984: The modes, resonances and forced response of elastic structures under heavy fluid loading. *Philos. Trans. R. Soc. London, Ser. A*, **312**, 295–341.

Fonseca, J. G., J. D. Eshelby, and C. Atkinson, 1971: The fracture mechanics of flint-knapping and allied processes. *Int. J. Fract.*, **7**, 421–433.

Fredricks, R. W., and L. Knopoff, 1960: The reflection of Rayleigh waves by a high imped-ance obstacle on a half-space. *Geophysics*, **25**, 1195–1202.

Freund, L. B., and J. D. Achenbach, 1967: Diffraction of a plane pulse by a semi-infinite barrier at a fluid-solid interface. *J. Appl. Mech.*, **34**, 571–578.

Freund, L. B., and J. D. Achenbach, 1968: Diffraction of a plane pulse by a closed crack at the interface of elastic solids. *Z. Angew. Math. Mech.*, **48**, 173–185.

Freund, L. B., and J. D. Achenbach, 1968: Waves in a semi-infinite plate in smooth contact with a harmonically disturbed half-space. *Int. J. Solids Struct.*, **4**, 605–621.

Georgiadis, H. G., 1987: Moving punch on a highly orthotropic elastic layer. *Acta Mech.*, **68**, 193–202.

Ghosh, M. L., 1962: Reflection of Love wave from a rigid obstacle. *Z. Geophys.*, **28**, 223–230.

Ghosh, M. L., 1974: On the propagation of Love's waves in an elastic layer in the presence of a vertical crack. *Proc. Vib. Prob.*, **15**, 147–165.

Ghosh, S. C., 1968: Diffraction of SH-waves originating from a moving point source by a rigid quarter space. *Pure Appl. Geophys.*, **70**, 22–27.

Gregory, R. D., 1966: The attenuation of a Rayleigh wave in a half-space by a surface impedance. *Proc. Cambridge Philos. Soc.*, **62**, 811–827.

Gregory, R. D., 1977: A circular disc containing a radial edge crack opened by a constant internal pressure. *Math. Proc. Cambridge Philos. Soc.*, **81**, 497–521.

Gregory, R. D., 1979: The edge-cracked circular disc under symmetric pin-loading. *Math. Proc. Cambridge Philos. Soc.*, **85**, 523–538.

Gregory, R. D., 1989: The spinning circular disc with a radial edge crack; an exact solution. *Int. J. Fract.*, **41**, 39–50.

Gupta, S. D., 1969: Diffraction of Love waves in layered media by half-plane. *J. Technol.*, **14**, 21–35.

Hanzawa, H., M. Kishida, and M. Asano, 1981: Dynamic interference between a crack and a plane boundary: Dynamic stress intensity factor induced by a plane harmonic SH-wave. *Bull. JSME*, **24**, 895–901.

Hebenstreit, H., 1976: Oberflächenwellen bei verschiedenen Anregungsformen. *Acta Phys. Austr.*, **44**, 259–277.

Herrmann, J. M., and L. Schovanec, 1990: Quasi-static mode III fracture in a nonhomoge-neous viscoelastic body. *Acta Mech.*, **85**, 235–249.

Ing, Y.-S., and C.-C. Ma, 1997: Transient analysis of a subsonic propagating interface crack subjected to antiplane dynamic loading in dissimilar isotropic materials. *J. Appl. Mech.*, **64**, 546–556.

Kazi, M. H., 1975: Diffraction of Love waves by perfectly rigid and perfectly weak half-plane. *Bull. Seism. Soc. Am.*, **65**, 1461–1479.

Keogh, P. S., 1985: High-frequency scattering by a Griffith crack. I: A crack Green's function. *Quart. J. Mech. Appl. Math.*, **38**, 185–204.

Keogh, P. S., 1985: High-frequency scattering by a Griffith crack. II: Incident plane and cylindrical waves. *Quart. J. Mech. Appl. Math.*, **38**, 205–232.

Keogh, P. S., 1986: High-frequency scattering of a normally incident plane compressional wave by a penny-shaped crack. *Quart J. Mech. Appl. Math.*, **39**, 535–566.

Kovalenko, G. P., 1974: Pressure pulse on the boundary of an elastic homogeneous half-plane. *J. Appl. Mech. Tech. Phys.*, **15**, 832–836.

Kraut, E. A., 1968: Diffraction of elastic waves by a rigid 90° wedge. Part I. *Bull. Seism. Soc. Am.*, **58**, 1083–1096.

Kuhn, G., and M. Matczyński, 1974: Beitrag zum gemischten Randwertproblem am Streifen. *Z. Angew. Math. Mech.*, **54**, T88–T91.

Kuhn, G., and M. Matczyński, 1975: Elastic strip with a crack under periodic loading. *Arch. Mech.*, **27**, 459–472.

Kuhn, G., and M. Matczyński, 1975: Analytische Ermittlung des Spannungsintensitäts-faktors eines ebenen Rißproblems unter periodischer Belastung. *Z. Angew. Math. Mech.*, **55**, T99–T102.

Kuo, M. K., and S. H. Cheng, 1991: Elastodynamic responses due to anti-plane point impact loadings on the faces of an interface crack along dissimilar anisotropic materials. *Int. J. Solids Struct.*, **28**, 751–768.

Lawrie, J. B., 1986: Vibrations of a heavily loaded, semi-infinite cylindrical elastic shell. I. *Proc. R. Soc. London, Ser. A*, **408**, 103–128.

Li, S.-F., and P. A. Mataga, 1996: Dynamic crack propagation in piezoelectric materials – Part I. Electrode solution. *J. Mech. Phys. Solids*, **44**, 1799–1830.

Linton, C. M., and D. V. Evans, 1991: Trapped modes above a submerged horizontal plate. *Quart. J. Mech. Appl. Math.*, **44**, 487–506.

Liu, C. T., and X. P. Li, 1994: A half plane crack under three-dimensional combined mode impact loading. *Acta Mech. Sinica*, **10**, 40–48.

Low, R. D., and H. J. Weiss, 1962: On a mixed boundary value problem for an infinite elastic cone. *Z. Angew. Math. Phys.*, **13**, 232–242.

Lyamshev, L. M., 1967: Scattering of sound by a semiinfinite elastic tube in a moving medium. *Sov. Phys. Acoust.*, **13**, 71–77.

Ma, C. C., and P. Burgers, 1986: Mode-III crack kinking with delay time: An analytical approximation. *Int. J. Solids Struct.*, **22**, 883–899.

Ma, C.-C., 1990: A dynamic crack model with viscous resistance to crack opening: Antiplane shear mode. *Eng. Fract. Mech.*, **37**, 127–144.

Mandal, S. C., and M. L. Ghosh, 1990: Moving punch on a viscoelastic semi-infinite medium. *Indian J. Pure Appl. Math.*, **21**, 847–864.

Matczyński, M., 1965: Axi-symmetric problem for a partly clamped elastic rod. *Arch. Mech. Stosow.*, **17**, 43–64.

Matczyński, M., 1973: Quasi-static problem of a crack in an elastic strip subject to antiplane state of strain. *Arch. Mech.*, **25**, 851–860.

Matczyński, M., 1974: Quasistatic problem of a non-homogeneous elastic layer containing a crack. *Acta Mech.*, **19**, 153–168.

Maue, A.-W., 1953: Die Beugung elastischer Wellen an der Halbebene. *Z. Angew. Math. Mech.*, **33**, 1–10.

Meijers, P., 1968: The contact problem of a rigid cylinder on an elastic layer. *Appl. Sci. Res.*, **18**, 353–383.

Mohan, N., and P. S. Deshwal, 1989: Scattering of a compressional wave at the corner of a quarter space. *Indian J. Pure Appl. Math.*, **20**, 386–394.

Mukhopadhyay, D., 1977: Stresses in semi-infinite strip under symmetric load on the lateral sides. *Indian J. Pure Appl. Math.*, **8**, 663–671.

Norris, A. N., and J. D. Achenbach, 1984: Elastic wave diffraction by a semi-infinite crack in a transversely isotropic material. *Quart. J. Mech. Appl. Math.*, **37**, 565–580.

Nuismer, R. J., and J. D. Achenbach, 1972: Dynamically induced fracture. *J. Mech. Phys. Solids*, **20**, 203–222.

Pal, A., and L. W. Pearson, 1986: A new approach to the electromagnetic diffraction problem of a perfectly conducting half-plane screen. *IEEE Trans. Antennas Propag.*, **AP-34**, 1281–1287.

Pal, S. C., and M. L. Ghosh, 1990: High frequency scattering of antiplane shear waves by an interface crack. *Indian J. Pure Appl. Math.*, **21**, 1107–1124.

Ping, L. X., and L. C. Tu, 1995: Elastodynamic stress-intensity factors for a semi-infinite crack under 3-D combined mode loading. *Int. J. Fract.*, **69**, 319–339.

Popelar, C. H., and C. Atkinson, 1980: Dynamic crack propagation in a viscoelastic strip. *J. Mech. Phys. Solids*, **28**, 79–93.

Przeździecki, S., and R. A. Hurd, 1977: Diffraction by a half-plane perpendicular to the distinguished axis of a general gyrotropic medium. *Can. J. Phys.*, **55**, 305–324.

Przeździecki, S., and R. A. Hurd, 1981: Diffraction by a half plane perpendicular to the distinguished axis of a gyrotropic medium (oblique incidence). *Can. J. Phys.*, **59**, 403–424.

Ramirez, J.-C., 1987: The three-dimensional stress intensity factor due to the motion of a load on the faces of a crack. *Quart. Appl. Math.*, **45**, 361–375.

Rokhlin, S., 1980: Diffraction of Lamb waves by a finite crack in an elastic layer. *J. Acoust. Soc. Am.*, **67**, 1157–1165.

Rose, L. R. F., 1976: An approximate (Wiener–Hopf) kernel for dynamic crack problems in linear elasticity and viscoelasticity. *Proc. R. Soc. London, Ser. A*, **349**, 497–521.

Rosenbaum, S., 1967: Edge diffraction in an arbitrary anisotropic medium. I. *Can. J. Phys.*, **45**, 3479–3502.

Roy, A., and N. Visalakshi, 1982: Diffraction of shear waves by an edge crack in an elastic wedge. *Int. J. Eng. Sci.*, **20**, 553–563.

Ryvkin, M., and L. Banks-Sills, 1993: Steady-state mode III propagation of an interface crack in an inhomogeneous viscoelastic strip. *Int. J. Solids Struct.*, **30**, 483–498.

Shindo, Y., 1983: Dynamic singular stresses for a Griffith crack in a soft ferromagnetic elastic solid subjected to a uniform magnetic field. *J. Appl. Mech.*, **50**, 50–56.

Shindo, Y., 1984: Diffraction of waves and singular stresses in a soft ferromagnetic elastic solid with two coplanar Griffith cracks. *J. Acoust. Soc. Am.*, **75**, 50-57.

Sills, L. B., and Y. Benveniste, 1981: Steady state propagation of a mode III interface crack between dissimilar viscoelastic media. *Int. J. Eng. Sci.*, **19**, 1255–1268.

Simons, D. A., 1975: Scattering of a Love wave by the edge of a thin surface layer. *J. Appl. Mech.*, **42**, 842–846.

Simons, D. A., 1976: Scattering of a Rayleigh wave by the edge of a thin surface layer of negligible inertia. *J. Acoust. Soc. Am.*, **59**, 12–18.

Sinha, T. K., 1980: Propagation of Love waves in elastic layer of variable thickness resting on a rigid half space. *J. Tech. Phys.*, **21**, 275–289.

Skelton, E. A., 1991: Acoustic scattering by a rigid barrier between two fluid-loaded parallel elastic plates. *Proc. R. Soc. London, Ser. A*, **435**, 217–232.

Stakhanov, I. P., 1968: Nonequilibrium ionization in a low-voltage discharge. II. *Sov. Phys. Tech. Phys.*, **12**, 935–942.

Thau, S. A., and Y.-H. Lu, 1970: Diffraction of transient horizontal shear waves by a finite crack and a finite rigid ribbon. *Int. J. Eng. Sci.*, **8**, 857–874.

Wesenberg, D. L., and L. M. Murphy, 1974: Steady-state response of an infinite plate to an exponential edge load along a fusing crack – A model for the dynamic edge fusion of two semi-infinite plates. *Int. J. Mech. Sci.*, **16**, 91–103.

Wickham, G. R., 1980: Short-wave radiation from a rigid strip in smooth contact with a semi-infinite elastic solid. *Quart. J. Mech. Appl. Math.*, **33**, 409–433.

Willis, J. R., 1967: Crack propagation in viscoelastic media. *J. Mech. Phys. Solids*, **15**, 229–240.

Yamada, T., and R. Sato, 1976: SH wave propagation in a medium having step-shaped discontinuity. *J. Phys. Earth*, **24**, 105–130.

Yoneyama, T., and S. Nishida, 1976: Reflection and transmission of Rayleigh waves by the edge of a deposited thin film. *J. Acoust. Soc. Am.*, **55**, 738–743.

Yoneyama, T., and S. Nishida, 1974: Diffraction of elastic surface wave at the edge of deposited thin films. *Electron. Communic. Japan*, **A57, No. 6**, 35–44.

Yoneyama, T., and S. Nishida, 1974: Transmission and reflection of Rayleigh waves by a high impedance obstacle of finite length. *J. Acoust. Soc. Am.*, **60**, 90–94.

Zhao, X.-H., 2003: The stress intensity factor history for an advancing crack in a transversely isotropic solid under 3-D loading. *Int. J. Solids Struct.*, **40**, 89–103.

Zaman, F. D., S. Asghar and M. Ahmad, 1987: Diffraction of SH-waves in a layered plate. *J. Tech. Phys.*, **28**, 143–152.

Zaman, F. D., 2001: Diffraction of SH-waves across a mixed boundary in a plate. *Mech. Res. Commun.*, **28**, 171–178.

Zhong, Z., W. Yang and S.-W. Yu, 1990: Further result on symmetric bending of cracked Reissner plate. *Theoret. Appl. Fract. Mech.*, **12**, 231–240.

Thermoelasticity

Brock, L. M., and M. J. Rodgers, 1997: Steady-state response of a thermoelastic half-space to the rapid motion of surface thermal/mechanical loads. *J. Elasticity*, **47**, 225–240.

Das, S. C., and S. C. Prasad, 1972: Thermal stresses in an elastic cone due to a mixed thermal boundary condition. *Indian J. Pure Appl. Math.*, **3**, 248–262.

Wave Equation

Abrahams, I. D., 1982: Scattering of sound by large finite geometries. *IMA J. Appl. Math.*, **29**, 79–97.

Abrahams, I. D., 1987: Scattering of sound by two parallel semi-infinite screens. *Wave Motion*, **9**, 289–300.

Abrahams, I. D., and G. R. Wickham, 1988: On the scattering of sound by two semi-infinite parallel staggered plates. I. Explicit matrix Wiener-Hopf factorization. *Proc. R. Soc. London, Ser. A*, **420**, 131–156.

Abrahams, I. D., and G. R. Wickham, 1990: Acoustic scattering by two parallel slightly staggered rigid plates. *Wave Motion*, **12**, 281–297.

Aivazyan, Yu. M., and D. M. Sedrakyan, 1965: Radiation from the open end of a planar semiinfinite waveguide. *Sov. Phys. Tech. Phys.*, **10**, 358–361.

Alblas, J. B., 1957: On the diffraction of sound waves in a viscous medium. *Appl. Sci. Res., Ser. A*, **6**, 237–262.

Ando, Y., 1969/70: On the sound radiation from semi-infinite circular pipe of certain wall thickness. *Acustica*, **22**, 219–225.

Ando, Y., and T. Koizumi, 1976: Sound radiation from a semi-infinite circular pipe having an arbitrary profile of orifice. *J. Acoust. Soc. Am.*, **59**, 1033–1039.

Asghar, S., 1988: Acoustic diffraction by an absorbing finite strip in a moving liquid. *J. Acoust. Soc. Am.*, **83**, 812–816.

Asghar, S., B. Ahmad, and M. Ayub, 1991: Point-source diffraction by an absorbing half-plane. *IMA J. Appl. Math.*, **46**, 217–224.

Bailin, L. L., 1951: An analysis of the effect of the discontinuity in a bifurcated circular guide upon plane longitudinal waves. *J. Res. NBS*, **47**, 315–335.

Baldwin, G. L., and A. E. Heins, 1954: On the diffraction of a plane wave by an infinite plane grating. *Math. Scand.*, **2**, 103–118.

Candel, S. M., 1973: Diffraction of a plane wave by a half plane in a subsonic and supersonic medium. *J. Acoust. Soc. Am.*, **54**, 1008–1016.

Candel, S. M., and C. Crance, 1981: Direct Fourier synthesis of waves: Application to acoustic source radiation. *AIAA J.*, **19**, 290–295.

Cannell, P. A., 1975: Edge scattering of aerodynamic sound by a lightly loaded elastic half-plane. *Proc. R. Soc. London, Ser. A*, **347**, 213–238.

Chester, W., 1950: The propagation of sound waves in an open-ended channel. *Philos. Mag., Ser. 7*, **41**, 11–33.

Dahl, P. H., and G. V. Frisk, 1991: Diffraction from the juncture of pressure release and locally reacting half-planes. *J. Acoust. Soc. Am.*, **90**, 1093–1100.

Das Gupta, S. P., 1970: Diffraction by a corrugated half-plane. *Proc. Vib. Prob.*, **11**, 413–424.

Davies, H. G., 1974: Natural motion of a fluid-loaded semi-infinite membrane. *J. Acoust. Soc. Am.*, **55**, 213–219.

Davies, H. G., 1975: Edge-mode radiation from fluid-loaded semi-infinite membranes. *J. Sound Vib.*, **40**, 179–189.

DeBruijn, A., 1973: A mathematical analysis concerning the edge effect of sound absorbing materials. *Acustica*, **28**, 33–44.

DeSanto, J. A., 1973: Scattering from a periodic corrugated surface: Semi-infinite alternately filled plates. *J. Acoust. Soc. Am.*, **53**, 719–734.

Deshwal, P. S., 1971: Diffraction of a compressional wave by a rigid barrier in a liquid half-space. *Pure Appl. Geophys.*, **85**, 107–124.

Deshwal, P. S., 1971: Diffraction of a compressional wave by a rigid barrier in a liquid layer - Part I. *Pure Appl. Geophys.*, **88**, 12–28.

Deshwal, P. S., 1971: Diffraction of a compressional wave by a rigid barrier in a liquid layer - Part II. *Pure Appl. Geophys.*, **91**, 14–33.

Deshwal, P. S., 1981: Diffraction of compressional waves by a strip in a liquid halfspace. *J. Math. Phys. Sci.*, **15**, 263–281.

Deshwal, P. S., 1987: Rayleigh wave scattering due to a surface impedance in a liquid layer. *J. Math. Phys. Sci.*, **21**, 399–421.

Dowerah, S., and A. Chakrabarti, 1985: Diffraction by two staggered half-planes with periodic wall perturbations. *Indian J. Pure Appl. Math.*, **16**, 411–447.

Dowerah, S., and A. Chakrabarti, 1987: Double knife-edge diffraction with mixed boundary conditions on one of the scatterers. *J. Sound Vib.*, **116**, 49–70.

Faulkner, T. R., 1966: The diffraction of an obliquely incident surface wave by a vertical barrier of finite depth. *Proc. Cambridge Philos. Soc.*, **62**, 829–838.

Fredricks, R. W., and L. Knopoff, 1960: The reflection of Rayleigh waves by a high impedance obstacle on a half-space. *Geophysics*, **25**, 1195–1202.

Gazarian, Iu. L., 1957: Waveduct propagation of sound for one particular class of laminarly-inhomogeneous media. *Sov. Phys. Acoust.*, **3**, 135–149.

Haak, K. F. I., and B. J. Kooij, 1996: Transient acoustic diffraction in a fluid layer. *Wave Motion*, **23**, 139–164.

Heins, A. E., 1948: The radiation and transmission properties of a pair of semi-infinite parallel plates. *Quart. Appl. Math.*, **6**, 157–166, 215–220.

Heins, A. E., 1950: Water waves over a channel of finite depth with a submerged plane barrier. *Can. J. Math.*, **2**, 210–222.

Heins, A. E., 1951: Some remarks on the coupling of two ducts. *J. Math. Phys. (Cambridge, MA)*, **30**, 164–169.

Heins, A. E., 1957: The Green's function for periodic structures in diffraction theory with an application to parallel plate media. I. *J. Math. Mech.*, **6**, 401–426.

Idemen, M., 1975: On the scalar scattering by a strip in a dissipative medium. *J. Eng. Math.*, **9**, 93–102.

Johnson, G. W., and K. Ogimoto, 1980: Sound radiation from a finite length unflanged circular duct with uniform axial flow. I. Theoretical analysis. *J. Acoust. Soc. Am.*, **68**, 1858–1870.

Jones, D. S., 1952: Diffraction by a wave-guide of finite length. *Proc. Cambridge Philos. Soc.*, **48**, 118–134.

Jones, D. S., 1952: A simplifying technique in the solution of a class of diffraction problems. *Quart. J. Math.*, Ser. 2, **3**, 189–196.

Jones, D. S., 1953: Diffraction by a thick semi-infinite plate. *Proc. R. Soc. London, Ser. A*, **217**, 153–175.

Jones, D. S., 1955: The scattering of a scalar wave by a semi-infinite rod of circular cross section. *Philos. Trans. R. Soc. London, Ser. A*, **247**, 499–528.

Jones, D. S., 1973: Double knife-edge diffraction and ray theory. *Quart. J. Mech. Appl. Math.*, **26**, 1–18.

Kashyap, S. C., and M. A. K. Hamid, 1971: Diffraction characteristics of a slit in a thick conducting screen. *IEEE Trans. Antennas Propag.*, **AP-19**, 499–507.

Kim, D. H., and L. L. Koss, 1990: Sound radiation from a circular duct with axial temperature gradients. *J. Sound Vib.*, **141**, 1–16.

Koch, W., 1971: On the transmission of sound waves through a blade row. *J. Sound Vib.*, **18**, 111–128.

Koch, W., 1977: Attenuation of sound in multi-element acoustically lined rectangular ducts in the absence of mean flow. *J. Sound Vib.*, **52**, 459–496.

Koch, W., 1977: Radiation of sound from a two-dimensional acoustically lined duct. *J. Sound Vib.*, **55**, 255–274.

Konovalyuk, I. P., 1969: Diffraction of a plane sound wave by a plate reinforced with stiffness members. *Sov. Phys. Acoust.*, **14**, 465–469.

Korovkin, A. N., and D. D. Plakhov, 1973: Sound diffraction by a baffle coupled with an elastic plate. *Sov. Phys. Acoust.*, **19**, 459–462.

Kouzov, D. P., 1969: Diffraction of a cylindrical hydroacoustic wave at the joint of two semi-infinite plates. *J. Appl. Math. Mech.*, **33**, 225–234.

Lamb, G. L., 1959: Diffraction of a plane sound wave by a semi-infinite thin elastic plate. *J. Acoust. Soc. Am.*, **31**, 929–935.

Lawrie, J. B., 1988: Axisymmetric radiation from a finite gap in an infinite, rigid, circular duct. *IMA J. Appl. Math.*, **40**, 113–128.

Lawrie, J. B., 1988: Scattering of sound by a heavily loaded finite cylindrical elastic shell. *Quart. J. Mech. Appl. Math.*, **41**, 445–467.

Lee, S. W., 1970: Ray theory of diffraction by open-ended waveguides. I. Field in waveguides. *J. Math. Phys.*, **11**, 2830–2850.

Lee, S. W., 1972: Ray theory of diffraction by open-ended waveguides. II. Applications. *J. Math. Phys.*, **13**, 656–664.

Lee, S. W., and J. Boersma, 1975: Ray-optical analysis of fields on shadow boundaries of two parallel plates. *J. Math. Phys.*, **16**, 1746–1764.

Leppington, F. G., 1983: Travelling waves in a dielectric slab with an abrupt change in thickness. *Proc. R. Soc. London, Ser. A*, **386**, 443–460.

Levine, H., and J. Schwinger, 1948: On the radiation of sound from an unflanged circular pipe. *Phys. Rev., Series 2*, **73**, 383–406.

Levine, H., 1954: On the theory of sound reflection in an open-ended cylindrical tube. *J. Acoust. Soc. Am.*, **26**, 200–211.

Luk'yanov, V. D., 1997: The radiation of acoustic waves by a piston radiator in a rigid screen partially covering a waveguide cross-section. *J. Appl. Math. Mech.*, **61**, 609–618.

Mani, R., 1973: Refraction of acoustic duct waveguide modes by exhaust jets. *Quart. Appl. Math.*, **30**, 501–520.

Matsui, E., 1971: Free-field correction for laboratory standard microphones mounted on a semiinfinite rod. *J. Acoust. Soc. Am.*, **49**, 1475–1483.

Meister, E., 1962: Zum Dirichlet-Problem der Helmholtzschen Schwingungsgleichung für ein gestaffeltes Streckengitter. *Arch. Rat. Mech. Anal.*, **10**, 67–100.

Meister, E., 1962: Zum Neumann-Problem der Helmholtzschen Schwingungsgleichung für ein gestaffeltes Streckengitter. *Arch. Rat. Mech. Anal.*, **10**, 127–148.

Millar, R. F., 1964: Plane wave spectra in grating theory. III. Scattering by a semiinfinite grating of identical cylinders. *Can. J. Phys.*, **42**, 1149–1184.

Millar, R. F., 1966: Plane wave spectra in grating theory. V. Scattering by a semi-infinite grating of isotropic scatterers. *Can. J. Phys.*, **44**, 2839–2874.

Pierucci, M., 1978: Acoustic radiation due to a fluid loading discontinuity on an infinite membrane. *J. Acoust. Soc. Am.*, **64**, 223–231.

Rawlins, A. D., 1975: Acoustic diffraction by an absorbing semi-infinite half plane in a moving fluid. *Proc. R. Soc. Edinburgh, Ser. A*, **72**, 337–357.

Rawlins, A. D., 1975: The solution of a mixed boundary value problem in the theory of diffraction by a semi-infinite plane. *Proc. R. Soc. London, Ser. A*, **346**, 469–484.

Rawlins, A. D., 1976: Diffraction of sound by a rigid screen with a soft or perfectly adsorbing edge. *J. Sound Vib.*, **45**, 53–67.

Rawlins, A. D., 1976: Diffraction of sound by a rigid screen with absorbent edge. *J. Sound Vib.*, **47**, 523–541.

Rawlins, A. D., 1977: The engine-over-the-wing noise problem. *J. Sound Vib.*, **50**, 553–569.

Rawlins, A. D., 1977: Diffraction by an acoustically penetrable or an electromagnetically dielectric half plane. *Int. J. Eng. Sci.*, **15**, 569–578.

Rawlins, A. D., 1978: Radiation of sound from an unflanged rigid cylindrical duct with an acoustically absorbing internal surface. *Proc. R. Soc. London, Ser. A*, **361**, 65–91.

Rawlins, A. D., 1984: The solution of a mixed boundary value problem in the theory of diffraction. *J. Eng. Math.*, **18**, 37–62.

Rawlins, A. D., E. Meister, and F.-O. Speck, 1991: Diffraction by an acoustically transmissive or an electromagnetic dielectric half-plane. *Math. Meth. App. Sci.*, **14**, 387–402.

Rawlins, A. D., 1995: A bifurcated circular waveguide problem. *IMA J. Appl. Math.*, **54**, 59–81.

Reuter, G. E. H., and E. H. Sondheimer, 1948: The theory of the anomalous skin effect in metals. *Proc. R. Soc. London, Ser. A*, **195**, 336–364.

Rienstra, S. W., 1983: A small Strouhal number analysis for acoustic wave-jet flow-pipe interactions. *J. Sound Vib.*, **86**, 539–556.

Rulf, B., 1967: Diffraction by a halfplane in a uniaxial medium. *SIAM J. Appl. Math.*, **15**, 120–127.

Savkar, S. D., 1975: Radiation of cylindrical duct acoustic modes with flow mismatch. *J. Sound Vib.*, **42**, 363–386.

Senior, T. B. A., 1952: Diffraction by a semi-infinite metallic sheet. *Proc. R. Soc. London, Ser. A*, **213**, 436–458.

Vaisleib, Yu. V., 1971: Sound scattering by a finite cone. *Sov. Phys. Acoust.*, **17**, 26–33.

Vajnshtejn, L. A., 1949: The theory of sound waves in open tubes (in Russian). *Zh. Tekh. Fiz.*, **19**, 911–930.

Zaitseva, I. A., and A. A. Zolotarev, 1990: Propagation and attenuation of waves generated in a layer by a plane source. *J. Appl. Mech. Tech. Phys.*, **31**, 696–701.

Worked Solutions
to Some of the Problems

Section 1.3

Section 1.3, Problem 1. A quick check shows that the general solution is

$$y(x) = A \sinh\left[(1-x)\sqrt{a^2+s}\right] + B \cosh\left[(1-x)\sqrt{a^2+s}\right].$$

We have written the solution in this form because $y(1) = B = 0$. Therefore,

$$y(-1) = A \sinh\left(2\sqrt{a^2+s}\right) = \frac{1}{s} \quad \text{and} \quad y(x) = \frac{\sinh\left[(1-x)\sqrt{a^2+s}\right]}{s \sinh\left(2\sqrt{a^2+s}\right)}.$$

Section 1.6

Section 1.6, Problem 1. Because

$$\oint_{|z|=1} a^z \, dz = \oint_{|z|=1} \exp[z \log(a)] \, dz,$$

$$\oint_{|z|=1} a^z \, dz = \int_0^{2\pi} \exp\left[\log(a)e^{\theta i}\right] i e^{\theta i} \, d\theta = \left. \frac{\exp\left[\log(a)e^{\theta i}\right]}{\log(a)} \right|_0^{2\pi}$$

$$= \frac{\exp\left[\log(a)e^{2\pi i}\right] - \exp\left[\log(a)e^0\right]}{\log(a)} = 0.$$

627

Section 1.6, Problem 2.

$$\oint_{|z|=1} z^\alpha \, dz = i \int_0^{2\pi} e^{\alpha\theta i} e^{\theta i} \, d\theta = i \int_0^{2\pi} e^{(\alpha+1)\theta i} \, d\theta.$$

If $\alpha \neq -1$,

$$\oint_{|z|=1} z^\alpha \, dz = \frac{e^{(\alpha+1)\theta i}}{1+\alpha} \bigg|_0^{2\pi} = \frac{e^{2\alpha\pi i} - 1}{1+\alpha}.$$

One the other hand, if $\alpha = -1$,

$$\oint_{|z|=1} z^\alpha \, dz = i \int_0^{2\pi} d\theta = 2\pi i.$$

Section 1.6, Problem 3. Using the given contour,

$$\oint_C \frac{dz}{z \log(z)} = \int_{C_\infty} \frac{dz}{z \log(z)} + \int_{-\infty+0+i}^{0+i} \frac{dz}{z \log(z)} + \int_{C_\epsilon} \frac{dz}{z \log(z)} + \int_{0-i}^{-\infty+0-i} \frac{dz}{z \log(z)}.$$

Now,

$$\int_{C_\epsilon} \frac{dz}{z \log(z)} = \lim_{\epsilon \to 0} \int_\pi^{-\pi} \frac{\epsilon i e^{\theta i} \, d\theta}{\epsilon e^{\theta i}[\ln(\epsilon) + \theta i]} = \lim_{\epsilon \to 0} \log[\ln(\epsilon) + \theta i] \bigg|_\pi^{-\pi}$$

$$= \lim_{\epsilon \to 0} \left\{ \log[\ln(\epsilon) - \pi i] - \log[\ln(\epsilon) + \pi i] \right\} = -2\pi i,$$

because we crossed the branch cut. On the other hand,

$$\int_{C_\infty} \frac{dz}{z \log(z)} = \lim_{r \to \infty} \int_{-\pi}^\pi \frac{r i e^{\theta i} \, d\theta}{r e^{\theta i}[\ln(r) + \theta i]} = \lim_{r \to \infty} \log[\ln(r) + \theta i] \bigg|_{-\pi}^\pi$$

$$= \lim_{r \to \infty} \left\{ \log[\ln(r) + \pi i] - \log[\ln(r) - \pi i] \right\} = 0,$$

because we did not cross the branch cut. Along the branch cut,

$$\int_{-\infty+0+i}^{0+i} \frac{dz}{z \log(z)} = \int_\infty^0 \frac{(-dx)}{(-x)[\ln(x) + \pi i]}$$

and

$$\int_{0-i}^{-\infty+0-i} \frac{dz}{z \log(z)} = \int_0^\infty \frac{(-dx)}{(-x)[\ln(x) - \pi i]}.$$

Therefore,

$$\int_0^\infty \frac{dx}{x[\ln(x) - \pi i]} - \int_0^\infty \frac{dx}{x[\ln(x) + \pi i]} = 2\pi i, \quad \text{or} \quad \int_0^\infty \frac{dx}{x[\ln(x)^2 + \pi^2]} = 1.$$

Section 1.6, Problem 4. From the residue theorem,

$$\oint_C \frac{dz}{(z+1)\sqrt{z}} = \int_{C_\infty} \frac{dz}{(z+1)\sqrt{z}} + \int_{0+i}^{\infty+0^+i} \frac{dz}{(z+1)\sqrt{z}}$$

$$+ \int_{C_\epsilon} \frac{dz}{(z+1)\sqrt{z}} + \int_{\infty+0-i}^{0^-i} \frac{dz}{(z+1)\sqrt{z}} = 2\pi.$$

Now,

$$\int_{C_\epsilon} \frac{dz}{(z+1)\sqrt{z}} = \lim_{\epsilon \to 0} \int_{2\pi}^{0} \frac{\epsilon i e^{\theta i}\, d\theta}{\sqrt{\epsilon e^{\theta i/2}}(1+\epsilon e^{\theta i})} \to 0,$$

and

$$\int_{C_\infty} \frac{dz}{(z+1)\sqrt{z}} = \lim_{r \to \infty} \int_0^{2\pi} \frac{r i e^{\theta i}\, d\theta}{\sqrt{r}e^{\theta i/2}(1+r e^{\theta i})} \to 0.$$

Along the branch cut, $z = xe^{0i}$ so that

$$\int_{0+i}^{\infty+0^+i} \frac{dz}{(z+1)\sqrt{z}} = \int_0^\infty \frac{dx}{\sqrt{x}(1+x)},$$

while along the other side $z = xc^{2\pi i}$,

$$\int_{\infty+0-i}^{0^-i} \frac{dz}{(z+1)\sqrt{z}} = \int_\infty^0 \frac{dx}{\sqrt{x}(1+x)e^{\pi i}}.$$

Therefore,

$$\int_0^\infty \frac{dx}{\sqrt{x}\,(1+x)} = \pi.$$

Section 1.6, Problem 5. From the residue theorem,

$$\oint_C \frac{dz}{z\sqrt{z^2-1}} = \int_{C_\infty} \frac{dz}{z\sqrt{z^2-1}} + \int_{Br_1} \frac{dz}{z\sqrt{z^2-1}} + \int_{Br_2} \frac{dz}{z\sqrt{z^2-1}}$$

$$+ \int_{1+0+i}^{\infty+0^+i} \frac{dz}{z\sqrt{z^2-1}} + \int_{\infty+0-i}^{1+0^-i} \frac{dz}{z\sqrt{z^2-1}}$$

$$+ \int_{-1+0-i}^{-\infty+0^-i} \frac{dz}{z\sqrt{z^2-1}} + \int_{-\infty+0+i}^{-1+0^+i} \frac{dz}{z\sqrt{z^2-1}} = 2\pi.$$

Now,

$$\int_{C_\infty} \frac{dz}{z\sqrt{z^2-1}} = \lim_{r \to \infty} \int_0^{2\pi} \frac{r i e^{\theta i}\, d\theta}{r e^{\theta i}\sqrt{r^2 e^{2\theta i}+1}} \to 0.$$

Along the branch cut, $z - 1 = (x - 1)e^{0i}$ or $(x - 1)e^{2\pi i}$, and $z + 1 = (x + 1)e^{0i}$ with $1 < x < \infty$ so that

$$\int_{1+0+i}^{\infty+0^+i} \frac{dz}{z\sqrt{z^2-1}} = \int_1^\infty \frac{dx}{x\sqrt{x^2-1}}$$

and

$$\int_{\infty+0^-i}^{1+0^-i} \frac{dz}{z\sqrt{z^2-1}} = -\int_{\infty}^{1} \frac{dx}{x\sqrt{x^2-1}},$$

while along the other side, $z - 1 = (x+1)e^{\pi i}$ and $z + 1 = (x-1)e^{\pm\pi i}$ so that

$$\int_{-1+0^-i}^{-\infty+0^-i} \frac{dz}{z\sqrt{z^2-1}} = \int_{1}^{\infty} \frac{(-dx)}{(-x)\sqrt{x^2-1}}$$

and

$$\int_{-1+0^+i}^{-\infty+0^+i} \frac{dz}{z\sqrt{z^2-1}} = -\int_{\infty}^{1} \frac{(-dx)}{(-x)\sqrt{x^2-1}}.$$

Therefore,

$$\int_{1}^{\infty} \frac{dx}{x\sqrt{x^2-1}} = \frac{\pi}{2}.$$

Section 1.6, Problem 6. (a) From the residue theorem,

$$\oint_C \frac{dz}{(z^2+1)\sqrt{z^2-1}} = 2\pi i \left\{ \mathrm{Res}\left[\frac{1}{(z^2+1)\sqrt{z^2-1}}; i\right] + \mathrm{Res}\left[\frac{1}{(z^2+1)\sqrt{z^2-1}}; -i\right] \right\},$$

where

$$\mathrm{Res}\left[\frac{1}{(z^2+1)\sqrt{z^2-1}}; i\right] = -\frac{1}{2\sqrt{2}} \quad \text{and} \quad \mathrm{Res}\left[\frac{1}{(z^2+1)\sqrt{z^2-1}}; -i\right] = -\frac{1}{2\sqrt{2}}.$$

Now, along the top of the branch cut,

$$\int_{\substack{top \\ branch\ cut}} \frac{dz}{(z^2+1)\sqrt{z^2-1}} = \int_{-1}^{1} \frac{dx}{i(x^2+1)\sqrt{1-x^2}},$$

because $z - 1 = (1-x)e^{\pi i}$ and $z + 1 = (1+x)e^{0i}$. Along the bottom, however,

$$\int_{\substack{bottom \\ branch\ cut}} \frac{dz}{(z^2+1)\sqrt{z^2-1}} = \int_{1}^{-1} \frac{dx}{(-i)(x^2+1)\sqrt{1-x^2}},$$

because $z - 1 = (1-x)e^{\pi i}$ and $z + 1 = (1+x)e^{2\pi i}$. Therefore, because the sum of the residues equals the branch cut integrals,

$$\int_{-1}^{1} \frac{dx}{(x^2+1)\sqrt{1-x^2}} = \frac{\pi}{\sqrt{2}}.$$

For part (b),

$$\oint_C \frac{dz}{(z^2+1)^2\sqrt{z^2-1}} = 2\pi i \left\{ \mathrm{Res}\left[\frac{1}{(z^2+1)^2\sqrt{z^2-1}}; i\right] + \mathrm{Res}\left[\frac{1}{(z^2+1)^2\sqrt{z^2-1}}; -i\right] \right\}.$$

This time, however, $z = \pm i$ are second-order poles so that

$$\text{Res}\left[\frac{1}{(z^2+1)^2\sqrt{z^2-1}};i\right] = \lim_{z\to i}\frac{d}{dz}\left[\frac{1}{(z+i)^2\sqrt{z^2-1}}\right] = -\frac{3}{8\sqrt{2}},$$

and

$$\text{Res}\left[\frac{1}{(z^2+1)^2\sqrt{z^2-1}};-i\right] = \lim_{z\to-i}\frac{d}{dz}\left[\frac{1}{(z-i)^2\sqrt{z^2-1}}\right] = -\frac{3}{8\sqrt{2}}.$$

Now, along the top of the branch cut,

$$\int_{\substack{top\\branch\ cut}}\frac{dz}{(z^2+1)^2\sqrt{z^2-1}} = \int_{-1}^{1}\frac{dx}{i(x^2+1)^2\sqrt{1-x^2}},$$

because $z - 1 = (1-x)e^{\pi i}$ and $z + 1 = (1+x)e^{0i}$. Along the bottom, however,

$$\int_{\substack{bottom\\branch\ cut}}\frac{dz}{(z^2+1)^2\sqrt{z^2-1}} = \int_{1}^{-1}\frac{dx}{(-i)(x^2+1)^2\sqrt{1-x^2}},$$

because $z - 1 = (1-x)e^{\pi i}$ and $z + 1 = (1+x)e^{2\pi i}$. Therefore, because the sum of the residues equals the branch cut integrals,

$$\int_{-1}^{1}\frac{dx}{(x^2+1)^2\sqrt{1-x^2}} = \frac{3\pi}{4\sqrt{2}}.$$

Section 1.6, Problem 7. From the residue theorem,

$$\oint_C \frac{1}{z^2}\sqrt{\frac{z-1}{z+1}}\,dz = 2\pi i\,\text{Res}\left(\frac{1}{z^2}\sqrt{\frac{z-1}{z+1}};0\right),$$

where

$$\text{Res}\left(\frac{1}{z^2}\sqrt{\frac{z-1}{z+1}};0\right) = \lim_{z\to 0}\frac{d}{dz}\left(\sqrt{\frac{z-1}{z+1}}\right) = \frac{1}{i}.$$

Now,

$$\int_{1+0^+i}^{\infty+0^+i}\frac{1}{z^2}\sqrt{\frac{z-1}{z+1}}\,dz = \int_{1}^{\infty}\frac{dx}{x^2}\sqrt{\frac{x-1}{x+1}},$$

because $z - 1 = (x-1)e^{0i}$ and $z + 1 = (x+1)e^{0i}$.

$$\int_{\infty+0^-i}^{1+0^-i}\frac{1}{z^2}\sqrt{\frac{z-1}{z+1}}\,dz = e^{\pi i}\int_{\infty}^{1}\frac{dx}{x^2}\sqrt{\frac{x-1}{x+1}},$$

because $z - 1 = (x-1)e^{2\pi i}$ and $z + 1 = (x+1)e^{0i}$.

$$\int_{-\infty+0^+i}^{-1+0^+i}\frac{1}{z^2}\sqrt{\frac{z-1}{z+1}}\,dz = \int_{\infty}^{1}\frac{(-dx)}{x^2}\sqrt{\frac{x+1}{x-1}},$$

because $z - 1 = (x + 1)e^{\pi i}$ and $z + 1 = (x - 1)e^{\pi i}$.

$$\int_{-1+0^- i}^{-\infty+0^- i} \frac{1}{z^2} \sqrt{\frac{z-1}{z+1}}\, dz = e^{\pi i} \int_1^\infty \frac{(-dx)}{x^2} \sqrt{\frac{x+1}{x-1}},$$

because $z - 1 = (x + 1)e^{\pi i}$ and $z + 1 = (x - 1)e^{-\pi i}$. Adding the branch cut integrals together,

$$\int_1^\infty \frac{dx}{x^2} \sqrt{\frac{x-1}{x+1}} = \pi - \int_1^\infty \frac{dx}{x^2} \sqrt{\frac{x+1}{x-1}}.$$

Section 1.6, Problem 8. If

$$f(z) = \frac{z^{\alpha-1}}{1-z} = \frac{\exp[(\alpha-1)\log(z)]}{1-z},$$

then,

$$\oint_C f(z)\, dz = \int_{C_\infty} f(z)\, dz + \int_{C_1} f(z)\, dz + \int_{C_{\delta_b}} f(z)\, dz + \int_{C_2} f(z)\, dz$$

$$+ \int_{C_{\epsilon_0}} f(z)\, dz + \int_{C_3} f(z)\, dz + \int_{C_{\delta_t}} f(z)\, dz + \int_{C_4} f(z)\, dz = 0.$$

Along C_1 and C_2, $z = x\, e^{2\pi i}$ and $dz = dx\, e^{2\pi i}$. Therefore,

$$\int_{C_1} f(z)\, dz = \int_\infty^{1+\delta} \frac{e^{(\alpha-1)[\ln(x)+2\pi i]}}{1-x}\, dx, \quad \text{and} \quad \int_{C_2} f(z)\, dz = \int_{1-\delta}^\epsilon \frac{e^{(\alpha-1)[\ln(x)+2\pi i]}}{1-x}\, dx.$$

On the other hand, along C_3 and C_4, $z = x\, e^{0i}$ and $dz = dx\, e^{0i}$. Therefore,

$$\int_{C_3} f(z)\, dz = \int_\epsilon^{1-\delta} \frac{e^{(\alpha-1)\ln(x)}}{1-x}\, dx, \quad \text{and} \quad \int_{C_4} f(z)\, dz = \int_{1+\delta}^\infty \frac{e^{(\alpha-1)\ln(x)}}{1-x}\, dx.$$

Around the singularity at $z = 0$,

$$\int_{C_{\epsilon_0}} f(z)\, dz = \lim_{\epsilon\to 0} \int_{2\pi}^0 \frac{\exp\{(\alpha-1)[\ln(\epsilon)+\theta i]\}}{1-\epsilon e^{\theta i}}\, i\epsilon e^{\theta i}\, d\theta$$

$$= \lim_{\epsilon\to 0} \int_{2\pi}^0 \frac{\epsilon^{\alpha-1}e^{(\alpha-1)\theta i}}{1-\epsilon e^{\theta i}}\, i\epsilon e^{\theta i}\, d\theta = \lim_{\epsilon\to 0} \int_{2\pi}^0 \frac{i\epsilon^\alpha e^{\alpha\theta i}}{1-\epsilon e^{\theta i}}\, d\theta = 0.$$

Around the singularity at $z = 1 + \delta e^{\theta i}$,

$$\int_{C_{\delta_b}} f(z)\, dz = \lim_{\delta\to 0} \int_{2\pi}^\pi \frac{\exp\left[(\alpha-1)\delta e^{\theta i} + 2\pi i(\alpha-1)\right]}{-\delta e^{\theta i}}\, i\delta e^{\theta i}\, d\theta = \pi i e^{2\alpha\pi i},$$

since $\log(z) = \log\left(1 + \delta e^{\theta i}\right) \approx \delta e^{\theta i}$. Similarly,

$$\int_{C_{\delta_t}} f(z)\, dz = \lim_{\delta\to 0} \int_\pi^0 \frac{\exp\left[(\alpha-1)\delta e^{\theta i}\right]}{-\delta e^{\theta i}}\, i\delta e^{\theta i}\, d\theta = \pi i.$$

Substitution leads to

$$-e^{2\alpha\pi i} \int_0^\infty \frac{x^{\alpha-1}}{1-x}\,dx + \int_0^\infty \frac{x^{\alpha-1}}{1-x}\,dx = -\pi i\left(1 + e^{2\alpha\pi i}\right).$$

Simplification yields

$$\int_0^\infty \frac{x^{\alpha-1}}{1-x}\,dx = \pi i\,\frac{e^{\alpha\pi i} + e^{-\alpha\pi i}}{e^{\alpha\pi i} - e^{-\alpha\pi i}} = \pi\cot(\alpha\pi).$$

Section 1.6, Problem 9. From the residue theorem,

$$\oint_C \frac{dz}{(z^3 - z^2)^{1/3}} = 2\pi i.$$

Now, along the top of the branch cut,

$$\int_{\substack{top \\ branch\ cut}} \frac{dz}{(z^3 - z^2)^{1/3}} = e^{-\pi i/3} \int_1^0 \frac{dx}{(x^2 - x^3)^{1/3}},$$

because $z = xe^{0i}$ and $z - 1 = (1 - x)e^{\pi i}$. Along the bottom, however,

$$\int_{\substack{bottom \\ branch\ cut}} \frac{dz}{(z^3 - z^2)^{1/3}} = e^{-5\pi i/3} \int_0^1 \frac{dx}{(x^2 - x^3)^{1/3}},$$

because $z = xe^{2\pi i}$ and $z - 1 = (1-x)e^{\pi i}$. Therefore, because the sum of the residues equals the branch cut integrals,

$$\int_0^1 \frac{dx}{(x^2 - x^3)^{1/3}} = \frac{2\pi}{\sqrt 3}.$$

Section 1.6, Problem 10. The integration comprises four parts: an integration along the top of the branch cut, an integration along the bottom of the branch cut and integrations around the branch points at $z = 0$ and $z = -1$. The integrations around the branch points vanish. Along the top of the branch cut, $z = xe^{\pi i}$ and $z + 1 = (1 - x)e^{0i}$ so that

$$\int_{\substack{top \\ branch\ cut}} \left(\frac{z+1}{z}\right)^\alpha z^{n-1}\,dz = -\int_{0+}^1 \left(\frac{1-x}{x}\right)^\alpha e^{-\pi\alpha i} x^{n-1} e^{(n-1)\pi i}\,dx,$$

while along the bottom of the branch cut, $z = xe^{-\pi i}$ and $z + 1 = (1 - x)e^{0i}$ so that

$$\int_{\substack{bottom \\ branch\ cut}} \left(\frac{z+1}{z}\right)^\alpha z^{n-1}\,dz = -\int_1^{0+} \left(\frac{1-x}{x}\right)^\alpha e^{\pi\alpha i} x^{n-1} e^{-(n-1)\pi i}\,dx.$$

Combining the two integrals, dividing by $2\pi i$ and simplifying, the desired result is obtained.

Section 1.6, Problem 11. Because

$$\oint_\Gamma \left(\frac{z-c}{z+c}\right)^{1/2} \frac{dz}{(z-\lambda)(\xi - z)} = 0,$$

we have

$$2\pi i \operatorname{Res}\left[\left(\frac{z-c}{z+c}\right)^{1/2}\frac{1}{(z-\lambda)(\xi-z)};\lambda\right]+2\pi i\operatorname{Res}\left[\left(\frac{z-c}{z+c}\right)^{1/2}\frac{1}{(z-\lambda)(\xi-z)};\xi\right]$$

$$+\int_{C_1}\left(\frac{z-c}{z+c}\right)^{1/2}\frac{dz}{(z-\lambda)(\xi-z)}+\int_{C_2}\left(\frac{z-c}{z+c}\right)^{1/2}\frac{dz}{(z-\lambda)(\xi-z)}$$

$$+\int_{C_3}\left(\frac{z-c}{z+c}\right)^{1/2}\frac{dz}{(z-\lambda)(\xi-z)}+\int_{C_4}\left(\frac{z-c}{z+c}\right)^{1/2}\frac{dz}{(z-\lambda)(\xi-z)}=0.$$

For C_1,

$$\int_{C_1}\left(\frac{z-c}{z+c}\right)^{1/2}\frac{dz}{(z-\lambda)(\xi-z)}$$

$$=\lim_{\epsilon\to 0}\int_0^{2\pi}\frac{\left(-2c+\epsilon e^{\theta i}\right)^{1/2}}{\epsilon^{1/2}e^{i\theta/2}}\frac{i\epsilon e^{\theta i}}{\left(-c-\lambda+\epsilon e^{\theta i}\right)\left(c+\xi-\epsilon e^{\theta i}\right)}\,d\theta=0.$$

For C_2, $z-c=(c-x)e^{-\pi i}$ and $z+c=(c+x)e^{0i}$ so that

$$\int_{C_2}\left(\frac{z-c}{z+c}\right)^{1/2}\frac{dz}{(z-\lambda)(\xi-z)}=-i\int_{-c}^{c}\left(\frac{c-x}{c+x}\right)^{1/2}\frac{dx}{(x-\lambda)(\xi-x)}.$$

For C_3,

$$\int_{C_3}\left(\frac{z-c}{z+c}\right)^{1/2}\frac{dz}{(z-\lambda)(\xi-z)}$$

$$=\lim_{\epsilon\to 0}\int_{-\pi}^{\pi}\frac{\epsilon^{1/2}e^{\theta i/2}}{\left(2c+\epsilon e^{i\theta}\right)^{1/2}}\frac{i\epsilon e^{\theta i}}{\left(c+\epsilon e^{\theta i}-\lambda\right)\left(\xi-c-\epsilon e^{\theta i}\right)}\,d\theta=0.$$

For C_4, $z-c=(c-x)e^{\pi i}$ and $z+c=(c+x)e^{0i}$ so that

$$\int_{C_4}\left(\frac{z-c}{z+c}\right)^{1/2}\frac{dz}{(z-\lambda)(\xi-z)}=i\int_{c}^{-c}\left(\frac{c-x}{c+x}\right)^{1/2}\frac{dx}{(x-\lambda)(\xi-x)}.$$

Direct substitution yields

$$-2i\int_{-c}^{c}\left(\frac{c-x}{c+x}\right)^{1/2}\frac{dx}{(x-\lambda)(\xi-x)}=-2\pi i\left[\left(\frac{\lambda-c}{\lambda+c}\right)^{1/2}\frac{1}{\xi-\lambda}+\left(\frac{\xi-c}{\xi+c}\right)^{1/2}\frac{-1}{\xi-\lambda}\right].$$

Simplification leads to the final result.

Section 1.6, Problem 12. Let

$$f(z)=\left(\frac{z-c}{z+c}\right)^{1/2}\frac{1}{(z-\sigma)(\xi-z)}.$$

Because $zf(z)$ tends to zero as $|z|\to\infty$, $\oint_\Gamma f(z)\,dz=0$.

We evaluate the integral around the contour $C = C_1 \cup \cdots \cup C_8$ by noting that an integration around the contour Γ is equivalent to integrations along C and around the pole at $z = \xi$. If the phase of $z - c$ is $\pm\pi$ while the phase for $z + c$ is always 0, then the residue theorem yields

$$\oint_C f(z)\,dz = \oint_\Gamma f(z)\,dz - 2\pi i\,\mathrm{Res}[f(z);\xi] = 2\pi i\left(\frac{\xi-c}{\xi+c}\right)^{1/2}\frac{1}{\xi-\sigma}.$$

Now,

$$\oint_C f(z)\,dz = \int_{C_1} f(z)\,dz + \int_{C_2} f(z)\,dz + \int_{C_3} f(z)\,dz + \int_{C_4} f(z)\,dz$$
$$+ \int_{C_5} f(z)\,dz + \int_{C_6} f(z)\,dz + \int_{C_7} f(z)\,dz + \int_{C_8} f(z)\,dz.$$

At C_1, $z = -c + \epsilon e^{\theta i}$ and

$$\int_{C_1} f(z)\,dz = \lim_{\epsilon \to 0}\int_0^{2\pi} \frac{(-2c+\epsilon e^{\theta i})^{1/2}}{\epsilon^{1/2}e^{\theta i/2}}\,\frac{i\epsilon e^{\theta i}}{\left(-c-\sigma+\epsilon e^{\theta i}\right)\left(\xi+c-\epsilon e^{\theta i}\right)}\,d\theta = 0,$$

while for C_5, $z - c = \epsilon e^{\theta i}$ and $z + c = 2c + \epsilon e^{\theta i}$, resulting in

$$\int_{C_5} f(z)\,dz = \lim_{\epsilon \to 0}\int_{-\pi}^{\pi} \left(\frac{\epsilon e^{\theta i}}{2c+\epsilon e^{\theta i}}\right)^{1/2}\frac{i c e^{\theta i}}{\left(c-\sigma+\epsilon e^{\theta i}\right)\left(\xi-c-\epsilon e^{\theta i}\right)}\,d\theta = 0.$$

At $z = \sigma$, $z - \sigma = \epsilon e^{\theta i}$, $z + c = (c+\sigma)e^{0i} + \epsilon e^{\theta i}$, $z - c = (c-\sigma)e^{\pi i}$ $\quad c e^{\theta i}$ on C_7 and $z - c = (c+\sigma)e^{-\pi i} + \epsilon e^{\theta i}$ on C_3, so that

$$\int_{C_3} f(z)\,dz = -i\lim_{\epsilon \to 0}\left(\frac{c+\sigma}{c-\sigma}\right)^{1/2}\frac{1}{\xi-\sigma}\int_{-\pi}^0 \frac{i\epsilon e^{\theta i}}{\epsilon e^{\theta i}}\,d\theta = \frac{\pi}{\xi-\sigma}\left(\frac{c+\sigma}{c-\sigma}\right)^{1/2}$$

and

$$\int_{C_7} f(z)\,dz = i\lim_{\epsilon \to 0}\left(\frac{c+\sigma}{c-\sigma}\right)^{1/2}\frac{1}{\xi-\sigma}\int_0^{\pi} \frac{i\epsilon e^{\theta i}}{\epsilon e^{\theta i}}\,d\theta = -\frac{\pi}{\xi-\sigma}\left(\frac{c+\sigma}{c-\sigma}\right)^{1/2}.$$

We evaluate the remaining contour integrals by noting that $z + c = (c+x)e^{0i}$ and $z - c = (c-x)e^{-\pi i}$ along C_2 and C_4, and $z - c = (c-x)e^{\pi i}$ along C_6 and C_8, where $-c < x < c$. Upon substituting these definitions into the integrals and simplifying,

$$PV \int_{C_2 \cup C_4} f(z)\,dz = -i\int_{-c}^c \left(\frac{c-x}{c+x}\right)^{1/2}\frac{dx}{(x-\sigma)(\xi-x)}$$

and

$$PV \int_{C_6 \cup C_8} f(z)\,dz = i\int_c^{-c} \left(\frac{c-x}{c+x}\right)^{1/2}\frac{dx}{(x-\sigma)(\xi-x)}.$$

Substituting these results into the second equation, our final answer is

$$PV \int_{-c}^c \left(\frac{c-x}{c+x}\right)^{1/2}\frac{dx}{(x-\sigma)(\xi-x)} = \frac{\pi}{\sigma-\xi}\left(\frac{\xi-c}{\xi+c}\right)^{1/2}.$$

Section 1.6, Problem 13. We have two poles within the contour which are located at $z = \alpha \pm i\sqrt{1-\alpha^2}$. Therefore, the residues are

$$\text{Res}\left(\frac{z^{p-1}}{z^2 - 2\alpha z + 1}; \alpha + i\sqrt{1-\alpha^2}\right) = \lim_{z \to \alpha + i\sqrt{1-\alpha^2}} \left[\frac{z^{p-1}\left(z - \alpha - i\sqrt{1-\alpha^2}\right)}{z^2 - 2\alpha z + 1}\right]$$

$$= \frac{\left(\alpha + i\sqrt{1-\alpha^2}\right)^{p-1}}{2i\sqrt{1-\alpha^2}}$$

and

$$\text{Res}\left(\frac{z^{p-1}}{z^2 - 2\alpha z + 1}; \alpha - i\sqrt{1-\alpha^2}\right) = -\frac{\left(\alpha - i\sqrt{1-\alpha^2}\right)^{p-1}}{2i\sqrt{1-\alpha^2}}.$$

The contribution from the integration around the branch point and at infinity vanish. Along the top of the branch cut, $z = xe^{0i}$ and

$$\int_{\substack{top \\ branch\ cut}} \frac{z^{p-1}}{z^2 - 2\alpha z + 1}\,dz = \int_0^\infty \frac{x^{p-1}}{x^2 - 2\alpha x + 1}\,dx,$$

while along the bottom of the branch cut, we have $z = xe^{2\pi i}$ and

$$\int_{\substack{bottom \\ branch\ cut}} \frac{z^{p-1}}{z^2 - 2\alpha z + 1}\,dz = e^{2\pi i(p-1)} \int_\infty^0 \frac{x^{p-1}}{x^2 - 2\alpha x + 1}\,dx.$$

From the residue theorem,

$$\left[1 - e^{2\pi i(p-1)}\right] \int_0^\infty \frac{x^{p-1}}{x^2 - 2\alpha x + 1}\,dx = \frac{\pi}{\sqrt{1-\alpha^2}}\left[\left(\alpha + i\sqrt{1-\alpha^2}\right)^{p-1}\right.$$
$$\left. - \left(\alpha - i\sqrt{1-\alpha^2}\right)^{p-1}\right],$$

or

$$\int_0^\infty \frac{x^{p-1}}{x^2 - 2\alpha x + 1}\,dx = -\frac{\pi}{\sqrt{1-\alpha^2}} \frac{\exp[i(p-1)\arccos(\alpha)] - \exp[-i(p-1)\arccos(\alpha)]}{\{\exp[(p-1)\pi i] - \exp[-(p-1)\pi i]\}\exp[(p-1)\pi i]}$$

$$= \frac{\pi}{\sqrt{1-\alpha^2}}\left\{\cos[(p-1)\arccos(\alpha)]\right.$$

$$\left. - \cot[(p-1)\pi]\sin[(p-1)\arccos(\alpha)]\right\}.$$

Section 1.6, Problem 14. Let $\omega < 0$ and $z = e^{\theta i}$ with $dz = iz\,d\theta$. Then,

$$\int_0^{2\pi} \frac{e^{-i\omega\theta}}{\cos(\theta) - a}\,d\theta = \int_{-\pi}^\pi \frac{e^{-i\omega\theta}}{\cos(\theta) - a}\,d\theta = \oint_{C_2} \frac{\left(e^{i\theta}\right)^{-\omega}}{\frac{1}{2}\left(e^{\theta i} + e^{-\theta i}\right) - a}\,d\theta$$

$$= \frac{2}{i}\oint_{C_2} \frac{z^{|\omega|}}{z^2 - 2az + 1}\,dz.$$

Similarly, if $\omega > 0$, $z = e^{-\theta i}$ and $dz = -iz\,d\theta$. The analysis follows as before once we realize that the line integration runs in the negative sense.

The singularities within the contour occur at

$$z = \begin{cases} a \pm i\sqrt{1-a^2}, & 0 < a < 1, \\ a - \sqrt{a^2-1}, & 1 < a < \infty. \end{cases}$$

If $1 < a < \infty$, then

$$\mathrm{Res}\left(\frac{z^{|\omega|}}{z^2 - 2az + 1}; a - \sqrt{a^2-1}\right) = -\frac{\left(a - \sqrt{a^2-1}\right)^{|\omega|}}{2\sqrt{a^2-1}}.$$

On the other hand, if $0 < a < 1$, then

$$\mathrm{Res}\left(\frac{z^{|\omega|}}{z^2 - 2az + 1}; a \pm i\sqrt{1-a^2}\right) = \pm\frac{\left(a \pm i\sqrt{1-a^2}\right)^{|\omega|}}{2i\sqrt{1-a^2}} = \pm\frac{e^{\pm|\omega|\theta i}}{2i\sqrt{1-a^2}},$$

where $\theta = \arccos(a)$. Therefore,

$$\mathrm{Res}\left(\frac{z^{|\omega|}}{z^2 - 2az + 1}; a + i\sqrt{1-a^2}\right) + \mathrm{Res}\left(\frac{z^{|\omega|}}{z^2 - 2az + 1}; a - i\sqrt{1-a^2}\right)$$

$$= \frac{e^{|\omega|\theta i} - e^{|\omega|\theta i}}{2i\sqrt{1-a^2}} = \frac{\sin\left(|\omega|\theta\right)}{\sqrt{1-a^2}} = \frac{\sin\left[|\omega|\arccos(a)\right]}{\sqrt{1-a^2}}.$$

Turning to the branch cut integrals,

$$\int_{C_1} \frac{z^{|\omega|}}{z^2 - 2az + 1}\,dz = -\int_{0+}^{1} \frac{\left(xe^{-\pi i}\right)^{|\omega|}}{x^2 + 2ax + 1}\,dx = -e^{\pi|\omega|i}\int_{0+}^{1} \frac{x^{|\omega|}}{x^2 + 2ax + 1}\,dx,$$

$$\int_{C_3} \frac{z^{|\omega|}}{z^2 - 2az + 1}\,dz = -\int_{1}^{0+} \frac{\left(xe^{\pi i}\right)^{|\omega|}}{x^2 + 2ax + 1}\,dx = e^{\pi|\omega|i}\int_{0+}^{1} \frac{x^{|\omega|}}{x^2 + 2ax + 1}\,dx,$$

and

$$\int_{C_4} \frac{z^{|\omega|}}{z^2 - 2az + 1}\,dz = \lim_{\epsilon \to 0}\int_{\pi}^{-\pi} \frac{\epsilon^{|\omega|}e^{|\omega|\theta i}}{\epsilon^2 e^{2\theta i} - 2a\epsilon e^{\theta i} + 1} i\epsilon e^{\theta i}\,d\theta = 0.$$

Noting that the contour integral along C_2 equals to $2\pi i$ times the sum of the residues minus the branch cut integrals C_1 to C_3, we substitute the contour integral into the first equation and simplify.

Section 1.7

Section 1.7, Problem 1. The general solution is

$$y(r) = AI_0\left(r\sqrt{s}\right) + BK_0\left(r\sqrt{s}\right) + \frac{1}{s}.$$

Because $K_0(z) \to \infty$ as $z \to 0$, $B = 0$. Using the condition at $r = a$, $A\sqrt{s}\,I_0'\left(a\sqrt{s}\right) = -1/s$. However, because $I_0'(z) = I_1(z)$,

$$y(r) = \frac{1}{s} - \frac{I_0\left(r\sqrt{s}\right)}{s^{3/2} I_1\left(a\sqrt{s}\right)}.$$

Section 1.7, Problem 2.

$$\mathcal{H}[\delta(r)] = \int_0^\infty \delta(r) J_0(kr) \, r \, dr = \frac{1}{2\pi} \int_{-\infty}^\infty \delta(x) \left[\int_{-\infty}^\infty \delta(y) J_0\left(k\sqrt{x^2 + y^2} \right) dy \right] dx$$

$$= \frac{1}{2\pi} \int_{-\infty}^\infty \delta(x) J_0(kx) \, dx = \frac{1}{2\pi}.$$

Section 2.1

Section 2.1, Problem 1. From Bromwich's integral,

$$f(t) = \frac{1}{2\pi i} \int_{c-\infty i}^{c+\infty i} \frac{e^{tz}}{z^3(z+1)^2} \, dz = \frac{1}{2\pi i} \oint_C \frac{e^{tz}}{z^3(z+1)^2} \, dz.$$

Computing the residues,

$$\text{Res}\left[\frac{e^{tz}}{z^3(z+1)^2}; 0 \right] = \lim_{z \to 0} \frac{1}{2} \frac{d^2}{dz^2} \left[z^3 \frac{e^{tz}}{z^3(z+1)^2} \right]$$

$$= \lim_{z \to 0} \frac{1}{2} \left[\frac{t^2 e^{tz}}{(z+1)^2} - \frac{4te^{tz}}{(z+1)^3} + \frac{6e^{tz}}{(z+1)^4} \right]$$

$$= (t^2 - 4t + 6)/2,$$

and

$$\text{Res}\left[\frac{e^{tz}}{z^3(z+1)^2}; -1 \right] = \lim_{z \to -1} \frac{d}{dz} \left[(z+1)^2 \frac{e^{tz}}{z^3(z+1)^2} \right]$$

$$= \lim_{z \to -1} \frac{te^{tz}}{z^3} - \frac{3e^{tz}}{z^4} = -te^{-t} - 3e^{-t}.$$

Therefore, the inverse equals $f(t) = t^2/2 - 2t + 3 - (t+3)e^{-t}$.

Section 2.1, Problem 2. From Bromwich's integral,

$$f(t) = \frac{1}{2\pi i} \int_{c-\infty i}^{c+\infty i} \frac{(z+1)e^{tz}}{(z+2)^2(z+3)} \, dz = \frac{1}{2\pi i} \oint_C \frac{(z+1)e^{tz}}{(z+2)^2(z+3)} \, dz.$$

Computing the residues,

$$\text{Res}\left[\frac{(z+1)e^{tz}}{(z+2)^2(z+3)}; -2 \right] = \lim_{z \to -2} \frac{d}{dz} \left[(z+2)^2 \frac{(z+1)e^{tz}}{(z+2)^2(z+3)} \right]$$

$$= \lim_{z \to -2} \left[\frac{t(z+1)e^{tz}}{z+3} + \frac{e^{tz}}{z+3} - \frac{(z+1)e^{tz}}{(z+3)^2} \right]$$

$$= 2e^{-2t} - te^{-2t},$$

and

$$\text{Res}\left[\frac{(z+1)e^{tz}}{(z+2)^2(z+3)}; -3 \right] = \lim_{z \to -3} \left[(z+3) \frac{(z+1)e^{tz}}{(z+2)^2(z+3)} \right] = -2e^{-3t}.$$

Therefore, the inverse equals $f(t) = (2 - t)e^{-2t} - 2e^{-3t}$.

Section 2.1, Problem 3. From Bromwich's integral,

$$f(t) = \frac{1}{2\pi i} \int_{c-\infty i}^{c+\infty i} \frac{(z+2)e^{tz}}{z(z-a)(z^2+4)} \, dz = \frac{1}{2\pi i} \oint_C \frac{(z+2)e^{tz}}{z(z-a)(z^2+4)} \, dz.$$

Computing the residues,

$$\text{Res}\left[\frac{(z+2)e^{tz}}{z(z-a)(z^2+4)}; 0\right] = \lim_{z \to 0} \frac{z(z+2)e^{tz}}{z(z-a)(z^2+4)} = \frac{2}{-4a} = -\frac{1}{2a},$$

$$\text{Res}\left[\frac{(z+2)e^{tz}}{z(z-a)(z^2+4)}; a\right] = \lim_{z \to a} \frac{(z-a)(z+2)e^{tz}}{z(z-a)(z^2+4)} = \frac{(a+2)e^{at}}{a(a^2+4)},$$

$$\text{Res}\left[\frac{(z+2)e^{tz}}{z(z-a)(z^2+4)}; -2i\right] = \lim_{z \to -2i} \frac{(z+2i)(z+2)e^{tz}}{z(z-a)(z^2+4)} = \frac{(2a-4-2ai-4i)e^{-2it}}{8(a^2+4)},$$

and

$$\text{Res}\left[\frac{(z+2)e^{tz}}{z(z-a)(z^2+4)}; 2i\right] = \lim_{z \to 2i} \frac{(z-2i)(z+2)e^{tz}}{z(z-a)(z^2+4)} = \frac{(-2a+4-2ia-4i)e^{2it}}{-8(a^2+4)}.$$

Then the inverse is

$$f(t) = \frac{a+2}{a(a^2+4)}e^{at} - \frac{1}{2a} + \frac{(a-2)\cos(2t) - (a+2)\sin(2t)}{2(a^2+4)}.$$

Section 2.1, Problem 4. From Bromwich's integral,

$$f(t) = \frac{1}{2\pi i} \int_{c-\infty i}^{c+\infty i} \frac{e^{tz}}{z\sinh(az)} \, dz = \frac{1}{2\pi i} \oint_C \frac{e^{tz}}{z\sinh(az)} \, dz.$$

Some of the poles are located where $\sinh(az) = i\sin(-iaz) = 0$ or $z_n = \pm n\pi i/a$, where $n = 1, 2, 3, \ldots$. On the other hand, at $z = 0$,

$$F(z) = \frac{1}{az^2(1 + a^2z^2/3! + a^4z^4/5! + \cdots)} = \frac{1}{az^2}\left(1 - \frac{a^2z^2}{3!} + \cdots\right)$$

and we have a second-order pole at $z = 0$. Therefore,

$$\text{Res}\left[\frac{e^{tz}}{z\sinh(az)}; 0\right] = \lim_{z \to 0} \frac{d}{dz}\left[\frac{ze^{tz}}{\sinh(az)}\right] = \lim_{z \to 0} \frac{d}{dz}\left[\frac{ze^{tz}}{az(1 + a^2z^2/3! + a^4z^4/5! + \cdots)}\right]$$

$$= \lim_{z \to 0} \frac{t}{a}\left(\frac{e^{tz}}{1 + a^2z^2/3! + a^4z^4/5! + \cdots}\right)$$

$$- \lim_{z \to 0} \frac{e^{tz}(2a^2z/3! + 4a^4z^3/5! + \cdots)}{a(1 + a^2z^2/3! + a^4z^4/5! + \cdots)^2} = \frac{t}{a}.$$

At the remaining poles,

$$\text{Res}\left[\frac{e^{tz}}{z\sinh(az)};\frac{n\pi i}{a}\right] = \lim_{z\to n\pi i/a}\frac{(z-n\pi i/a)\,e^{tz}}{z\sinh(az)} = \frac{\exp(n\pi it/a)}{n\pi i\cosh(n\pi i)} = (-1)^n\frac{\exp(n\pi it/a)}{n\pi i}.$$

Therefore, the inverse equals

$$f(t) = \frac{t}{a} + \frac{1}{\pi}\sum_{\substack{n=-\infty\\n\neq 0}}^{\infty}(-1)^n\frac{\exp(n\pi it/a)}{ni} = \frac{t}{a} + \frac{2}{\pi}\sum_{n=1}^{\infty}\frac{(-1)^n}{n}\sin\left(\frac{n\pi t}{a}\right).$$

Section 2.1, Problem 5. From Bromwich's integral,

$$f(t) = \frac{1}{2\pi i}\int_{c-\infty i}^{c+\infty i}\frac{\sinh(az/2)e^{tz}}{z\cosh(az/2)}\,dz = \frac{1}{2\pi i}\oint_C\frac{\sinh(az/2)e^{tz}}{z\cosh(az/2)}\,dz.$$

Some of the poles are located where $\cosh(az/2) = \cos(-iaz/2) = 0$ or $z_n = \pm(2n-1)\pi i/a$, where $n = 1,2,3,\ldots$. On the other hand, at $z = 0$,

$$F(z) = \frac{(az/2)[1 + a^2z^2/(3!4) + a^4z^4/(5!16) + \cdots]}{z[1 + a^2z^2/(2!4) + a^4z^4/(4!16) + \cdots]}$$

$$= \frac{a}{2}\left(1 + \frac{a^2z^2}{3!4} + \cdots\right)\left(1 - \frac{a^2z^2}{2!4} + \cdots\right)$$

and $z = 0$ is a removable singularity. Therefore, at the remaining poles,

$$\text{Res}\left[\frac{\sinh(az/2)e^{tz}}{z\cosh(az/2)};z_n\right] = \lim_{z\to z_n}\frac{(z-z_n)\sinh(az/2)e^{tz}}{z\cosh(az/2)} = \pm\frac{2\exp[\pm(2n-1)\pi it/a]}{(2n-1)\pi i}.$$

Therefore, the inverse is

$$f(t) = \frac{2}{\pi}\sum_{n=1}^{\infty}\frac{\exp[(2n-1)\pi it/a] - \exp[-(2n-1)\pi it/a]}{(2n-1)i} = \frac{4}{\pi}\sum_{n=1}^{\infty}\frac{1}{2n-1}\sin\left[\frac{(2n-1)\pi t}{a}\right].$$

Section 2.1, Problem 6. From Bromwich's integral,

$$f(t) = \frac{1}{2\pi i}\int_{c-\infty i}^{c+\infty i}\frac{e^{tz}}{z^2\cosh\left(\sqrt{az}\right)}\,dz = \frac{1}{2\pi i}\oint_C\frac{e^{tz}}{z^2\cosh\left(\sqrt{az}\right)}\,dz.$$

Some of the poles are located where $\cosh\left(\sqrt{az}\right) = \cos\left(-i\sqrt{az}\right) = 0$ or $z_n = -(2n+1)^2\pi^2/4a$, where $n = 0,1,2,3,\ldots$ and $\sqrt{az_n} = (2n+1)\pi i/2$. On the other hand, at $z = 0$, we have a second-order pole. Therefore, because

$$\frac{e^{tz}}{z^2\cosh\left(\sqrt{az}\right)} = \frac{1}{z^2} + \frac{t - a/2}{z} + \cdots, \qquad \text{Res}\left[\frac{e^{tz}}{z^2\cosh\left(\sqrt{az}\right)};0\right] = t - \frac{a}{2}.$$

At the remaining singularities,

$$\text{Res}\left[\frac{e^{tz}}{z^2 \cosh(\sqrt{az})}; z_n\right] = \lim_{z \to z_n} \frac{e^{tz}}{z^2} \lim_{z \to z_n} \frac{z - z_n}{\cosh(\sqrt{az})} = \frac{16a(-1)^n}{(2n+1)^3 \pi^3}.$$

Therefore, the inverse equals

$$f(t) = t - \frac{a}{2} + \frac{16a}{\pi^3} \sum_{n=1}^{\infty} \frac{(-1)^n}{(2n+1)^3} \exp\left[-\frac{(2n+1)^2 \pi^2 t}{4a}\right].$$

§2.1, Problem 7. From Bromwich's integral,

$$f(t) = \frac{1}{2\pi i} \int_{c-\infty i}^{c+\infty i} \frac{\sinh(a\sqrt{z}) e^{tz}}{z \sinh(\sqrt{z})} \, dz = \frac{1}{2\pi i} \oint_C \frac{\sinh(a\sqrt{z}) e^{tz}}{z \sinh(\sqrt{z})} \, dz.$$

Some of the poles are located where $\sinh(\sqrt{z}) = i \sin(-i\sqrt{z}) = 0$ or $z_n = -n^2\pi^2$ with $\sqrt{z_n} = n\pi i$, where $n = 1, 2, 3, \ldots$. On the other hand, at $z = 0$,

$$F(z) = \frac{az^{1/2} + a^3 z^{3/2}/3! + a^5 z^{5/3}/5! + \cdots}{z(z^{1/2} + z^{3/2}/3! + z^{5/2}/5! + \cdots)} = \frac{a}{z}\left(1 + \frac{a^2 z}{3!} + \cdots\right)\left(1 - \frac{z}{3!} + \cdots\right)$$

and $z = 0$ is a simple pole. The easiest way to compute its residue is by constructing its Laurent expansion:

$$\frac{\sinh(a\sqrt{z}) e^{tz}}{z \sinh(\sqrt{z})} = \frac{(az^{1/2} + a^3 z^{3/2}/3! + a^5 z^{5/2}/5! + \cdots)(1 + tz + \cdots)}{z(z^{1/2} + z^{3/2}/3! + z^{5/2}/5! + \cdots)}$$

$$= \frac{a}{z}\left(1 + \frac{a^2 z}{3!} + \cdots\right)(1 + tz + \cdots)\left(1 - \frac{z}{3!} + \cdots\right)$$

$$= \frac{a}{z}\left(1 + \frac{a^2 z}{3!} + tz - \frac{z}{3!} + \cdots\right)$$

and the residue equals a. The remaining poles yield the residues

$$\text{Res}\left[\frac{\sinh(a\sqrt{z}) e^{tz}}{z \sinh(\sqrt{z})}; z_n\right] = \lim_{z \to z_n} \frac{(z - z_n) \sinh(a\sqrt{z}) e^{tz}}{z \sinh(\sqrt{z})} = \lim_{z \to z_n} \frac{2\sqrt{z} \sinh(a\sqrt{z}) e^{tz}}{z \cosh(\sqrt{z})}$$

$$= \frac{2 \sinh(n\pi a i)(n\pi i) \exp(-n^2\pi^2 t)}{-n^2\pi^2 \cosh(n\pi i)} = 2(-1)^n \sin(n\pi a) \frac{e^{-n^2\pi^2 t}}{n\pi}.$$

Summing the residues gives

$$f(t) = a + \frac{2}{\pi} \sum_{n=1}^{\infty} \frac{(-1)^n}{n} \sin(n\pi a) e^{-n^2\pi^2 t}.$$

Section 2.1, Problem 8. From Bromwich's integral,

$$f(t) = \frac{1}{2\pi i} \int_{c-\infty i}^{c+\infty i} \frac{\sinh(z)e^{tz}}{z^2 \cosh(z)}\, dz = \frac{1}{2\pi i} \oint_C \frac{\sinh(z)e^{tz}}{z^2 \cosh(z)}\, dz.$$

Some of the poles are located where $\cosh(z) = \cos(-iz) = 0$ or $z_n = \pm(2n-1)\pi i/2$, where $n = 1, 2, 3, \ldots$. On the other hand, at $z = 0$,

$$F(z) = \frac{\sinh(z)}{z^2 \cosh(z)} = \frac{z(1 + z^2/3! + z^4/5! + \cdots)}{z^2(1 + z^2/2! + z^4/4! + \cdots)} = \frac{1}{z}\left(1 + \frac{z^2}{3!} + \cdots\right)\left(1 - \frac{z^2}{2!} + \cdots\right)$$

and $z = 0$ is a simple pole. Therefore, its residue is

$$\mathrm{Res}\left[\frac{\sinh(z)e^{tz}}{z^2 \cosh(z)}; 0\right] = \lim_{z \to 0} \frac{z\,\sinh(z)e^{tz}}{z^2 \cosh(z)} = 1.$$

The remaining poles yield the residues

$$\mathrm{Res}\left[\frac{\sinh(z)e^{tz}}{z^2 \cosh(z)}; z_n\right] = \lim_{z \to z_n} \frac{(z-z_n)\sinh(z)e^{tz}}{z^2 \cosh(z)} = \lim_{z \to z_n} \frac{e^{tz}}{z^2} = -\frac{4\exp[\pm(2n-1)\pi it/2]}{(2n-1)^2\pi^2}.$$

Summing the residues gives

$$f(t) = 1 - \frac{4}{\pi^2}\sum_{n=1}^{\infty} \frac{e^{(2n-1)\pi it/2} + e^{-(2n-1)\pi it/2}}{(2n-1)^2} = 1 - \frac{8}{\pi^2}\sum_{n=1}^{\infty} \frac{\cos[(2n-1)\pi t/2]}{(2n-1)^2}.$$

Section 2.1, Problem 9. From Bromwich's integral,

$$f(t) = \frac{1}{2\pi i} \int_{c-\infty i}^{c+\infty i} \frac{e^{tz}}{Rz + M^2/\alpha}\left[1 - \frac{\cosh\left(a\sqrt{Rz + M^2/\alpha}\right)}{\cosh\left(\sqrt{Rz + M^2/\alpha}\right)}\right] dz$$

$$= \frac{1}{2\pi i} \oint_C \frac{e^{tz}}{Rz + M^2/\alpha}\left[1 - \frac{\cosh\left(a\sqrt{Rz + M^2/\alpha}\right)}{\cosh\left(\sqrt{Rz + M^2/\alpha}\right)}\right] dz.$$

Some of the poles are located where

$$\cosh\left(\sqrt{Rz + M^2/\alpha}\right) = \cos\left(-i\sqrt{Rz + M^2\alpha}\right) = 0,$$

or $z_n = -M^2/(\alpha R) - (2n-1)^2\pi^2/(4R)$ and $\sqrt{Rz + M^2\alpha} = (2n-1)\pi i/2$, where $n = 1, 2, 3, \ldots$. On the other hand, at $z = -M^2/(\alpha R)$,

$$\mathrm{Res}\left\{\frac{e^{tz}}{Rz + M^2/\alpha}\left[1 - \frac{\cosh\left(a\sqrt{Rz + M^2\alpha}\right)}{\cosh\left(\sqrt{Rz + M^2/\alpha}\right)}\right]; -\frac{M^2}{\alpha R}\right\}$$

$$= \lim_{z \to -M^2/(\alpha R)} \frac{[z + M^2/(\alpha R)]e^{tz}}{Rz + M^2/\alpha}\left[1 - \frac{\cosh\left(a\sqrt{Rz + M^2\alpha}\right)}{\cosh\left(\sqrt{Rz + M^2/\alpha}\right)}\right] = 0,$$

while the remaining poles give the residues

$$
\text{Res}\left\{\frac{e^{tz}}{Rz + M^2/\alpha}\left[1 - \frac{\cosh\left(a\sqrt{Rz + M^2\alpha}\right)}{\cosh\left(\sqrt{Rz + M^2/\alpha}\right)}\right]; z_n\right\}
$$

$$
= -\lim_{z \to z_n}\frac{(z - z_n)\,e^{tz}}{Rz + M^2/\alpha}\frac{\cosh\left(a\sqrt{Rz + M^2/\alpha}\right)}{\cosh\left(\sqrt{Rz + M^2/\alpha}\right)}
$$

$$
= \frac{\exp\{-[M^2/\alpha + (2n - 1)^2\pi^2/4](t/R)\}}{R[M^2/(\alpha R) + (2n - 1)^2\pi^2/(4R) - M^2/(\alpha R)]}
$$

$$
\times\frac{2\cosh[a(2n - 1)\pi i/2][(2n - 1)\pi i/2]}{R\sinh[(2n - 1)\pi i/2]}
$$

$$
= -\frac{4(-1)^n}{R\pi(2n - 1)}\exp\left\{-\left[\frac{M^2}{\alpha} + \frac{(2n - 1)^2\pi^2}{4}\right]\frac{t}{R}\right\}\cos\left[\frac{(2n - 1)\pi a}{2}\right].
$$

Adding the residues yields

$$
f(t) = -\frac{4}{\pi R}\sum_{n=1}^{\infty}\frac{(-1)^n}{(2n - 1)}\cos\left[\frac{(2n - 1)\pi a}{2}\right]\exp\left\{-\left[\frac{(2n - 1)^2\pi^2}{4} + \frac{M^2}{\alpha}\right]\frac{t}{R}\right\}.
$$

Section 2.1, Problem 10. From Bromwich's integral,

$$
f(t) = \frac{1}{2\pi i}\int_{c-\infty i}^{c+\infty i}\frac{e^{tz}}{2za - \sqrt{z}\tanh\left(\sqrt{z}\right)}\,dz = \frac{1}{2\pi i}\oint_C\frac{e^{tz}}{2za - \sqrt{z}\tanh\left(\sqrt{z}\right)}\,dz.
$$

Some of the poles are located where $2a - \tanh(\sqrt{z})/\sqrt{z}$ or $\tan(\alpha_n) = 2a\alpha_n$, $z_n = -\alpha_n^2$, and $\sqrt{z_n} = \alpha_n i$ with $n = 1, 2, 3, \ldots$. We also have a simple pole at $z = 0$. At that point, its residue is

$$
\text{Res}\left[\frac{e^{tz}}{2za - \sqrt{z}\tanh(\sqrt{z})}; 0\right] = \lim_{z \to 0}\frac{z\,e^{tz}}{2za - \sqrt{z}\tanh(\sqrt{z})}
$$

$$
= \lim_{z \to 0}\frac{e^{tz}}{2a - \tanh(\sqrt{z})/\sqrt{z}} = \frac{1}{2a - 1},
$$

while the remaining poles have the residues

$$
\text{Res}\left[\frac{e^{tz}}{2za - \sqrt{z}\tanh(\sqrt{z})}; -\alpha_n^2\right] = \lim_{z \to -\alpha_n^2}\frac{(z + \alpha_n^2)e^{tz}}{2za - \sqrt{z}\tanh(\sqrt{z})}
$$

$$
= \frac{e^{-\alpha_n^2 t}}{2a - i\tan(\alpha_n)/(2i\alpha_n) - i/[2i\sec^2(\alpha_n)]}
$$

$$
= \frac{2e^{-\alpha_n^2 t}}{2a - 1 + 4a^2\alpha_n^2}.
$$

Adding the residues gives

$$f(t) = \frac{1}{2a-1} + 2\sum_{n=1}^{\infty} \frac{2e^{-\alpha_n^2 t}}{2a-1+4a^2\alpha_n^2}.$$

Section 2.1, Problem 11. Using the Taylor expansion for hyperbolic tangent, we see that $F(s)$ is single-valued. Clearly, there is a singularity at $z=0$. Expanding about that point,

$$\frac{\tanh\left(\alpha\sqrt{z}\right)e^{tz}}{z\left[m\sqrt{z}+\tanh\left(\alpha\sqrt{z}\right)\right]} = \frac{(\alpha-\alpha^3 z/3+\cdots)(1+tz+\cdots)}{z(m+\alpha-\alpha^3 z/3+\cdots)}$$

and we have a simple pole there. The residue equals $\alpha/(m+\alpha)$.

The remaining poles occur where $m\sqrt{z}+\tanh(\alpha\sqrt{z})=0$. If $\alpha\sqrt{z}=i\lambda$, then these simple poles occur where $\tan(\lambda)=-b\lambda$ and $z_n=-\lambda_n^2/\alpha^2$. The corresponding residues are

$$\operatorname{Res}\left[\frac{\tanh\left(\alpha\sqrt{z}\right)e^{tz}}{z\left[m\sqrt{z}+\tanh\left(\alpha\sqrt{z}\right)\right]};z_n\right] = \lim_{z\to z_n}\frac{(z-z_n)\tanh\left(\alpha\sqrt{z}\right)e^{tz}}{z[m\sqrt{z}+\tanh\left(\alpha\sqrt{z}\right)]}$$

$$= \frac{1}{b\lambda_n}\left[\frac{\sin(2\lambda_n)\exp(-\lambda_n^2 t/b^2)}{1/b+\cos^2(\lambda_n)}\right].$$

Summing the residues gives the inverse.

Section 2.1, Problem 12. From Bromwich's integral

$$f(t) = \frac{1}{2\pi i}\oint_C \frac{I_0\left(r\sqrt{z/\kappa}\right)e^{tz}}{I_0\left(a\sqrt{z/\kappa}\right)}\,dz,$$

where C includes all of the singularities. Let $\sqrt{z/\kappa}=-\alpha i$. Then, $I_0(-\alpha a i)=J_0(\alpha a)=0$. Let us denote the nth zero by $\alpha_n a$. Then, $z_n=-\kappa\alpha_n^2$ and the poles lie along the negative real axis. Using the residue theorem,

$$\operatorname{Res}\left[\frac{I_0\left(r\sqrt{z/\kappa}\right)e^{tz}}{I_0\left(a\sqrt{z/\kappa}\right)};-\kappa\alpha_n^2\right] = \lim_{z\to-\kappa\alpha_n^2}\frac{(z+\kappa\alpha_n^2)I_0\left(r\sqrt{z/\kappa}\right)e^{tz}}{I_0\left(a\sqrt{z/\kappa}\right)}$$

$$= \frac{I_0(-r\alpha_n i)\exp(-\alpha_n^2\kappa t)}{I_1(-a\alpha_n i)[a/(-2\kappa\alpha_n i)]} = \frac{2\kappa}{a}\frac{\alpha_n J_0(r\alpha_n)\exp(-\alpha_n^2\kappa t)}{J_1(\alpha_n a)}.$$

Summing the residues,

$$f(t) = \frac{2\kappa}{a}\sum_{n=1}^{\infty}\frac{\alpha_n J_0(r\alpha_n)\exp(-\alpha_n^2\kappa t)}{J_1(\alpha_n a)},$$

where $J_0(\alpha_n a)=0$ for $n=1,2,3,\dots$.

Section 2.1, Problem 13. $F(s)$ has simple poles located at $s = 0$, $s = k_1^2/\kappa$ and $\sqrt{s/\kappa} = -\alpha i$, where $I_0(-\alpha a i) = J_0(\alpha a) = 0$. Therefore, the last group of poles is at $s_n = -\kappa \alpha_n^2$, where α_n is the nth root of $J_0(\alpha a) = 0$. From Bromwich's integral,

$$f(t) = \frac{1}{2\pi i} \oint_C \frac{I_0\left(r\sqrt{z/\kappa}\right) e^{tz}}{z(k_1^2/\kappa - z) I_0\left(a\sqrt{z/\kappa}\right)} \, dz,$$

where C includes all of the singularities. Now, the residues are

$$\text{Res}\left[\frac{I_0\left(r\sqrt{z/\kappa}\right) e^{tz}}{z(k_1^2/\kappa - z) I_0\left(a\sqrt{z/\kappa}\right)}; 0\right] = \lim_{z \to 0} \frac{I_0\left(r\sqrt{z/\kappa}\right) e^{tz}}{(k_1^2/\kappa - z) I_0\left(a\sqrt{z/\kappa}\right)} = \frac{\kappa}{k_1^2},$$

$$\text{Res}\left[\frac{I_0\left(r\sqrt{z/\kappa}\right) e^{tz}}{z(k_1^2/\kappa - z) I_0\left(a\sqrt{z/\kappa}\right)}; k_1^2/\kappa\right] = \lim_{z \to k_1^2/\kappa} -\frac{I_0\left(r\sqrt{z/\kappa}\right) e^{tz}}{z I_0\left(a\sqrt{z/\kappa}\right)}$$

$$= -\frac{\kappa}{k_1^2} \frac{I_0(k_1 r/\kappa)}{I_0(k_1 a/\kappa)} \exp(k_1^2 t/\kappa),$$

and

$$\text{Res}\left[\frac{I_0\left(r\sqrt{z/\kappa}\right) e^{tz}}{z(k_1^2/\kappa - z) I_0\left(a\sqrt{z/\kappa}\right)}; -\kappa \alpha_n^2\right] = \lim_{z \to -\kappa \alpha_n^2} \frac{(z + \kappa \alpha_n^2) I_0\left(r\sqrt{z/\kappa}\right) e^{tz}}{z(k_1^2/\kappa - z) I_0\left(a\sqrt{z/\kappa}\right)}$$

$$= \frac{I_0(-r\alpha_n i) \exp(-\alpha_n^2 \kappa t)}{I_1(-a\alpha_n i)(-a\alpha_n i/2)(k_1^2/\kappa + \kappa \alpha_n^2)}$$

$$= \frac{2\kappa}{a} \frac{J_0(r\alpha_n) \exp(-\alpha_n^2 \kappa t)}{\alpha_n (k_1^2 + \kappa^2 \alpha_n^2) J_1(\alpha_n a)}.$$

Summing the residues,

$$f(t) = \frac{\kappa}{k_1^2} - \frac{\kappa}{k_1^2} \frac{I_0(k_1 r/\kappa)}{I_0(k_1 a/\kappa)} \exp(k_1^2 t/\kappa) + \frac{2\kappa}{a} \sum_{n=1}^{\infty} \frac{J_0(r\alpha_n) \exp(-\alpha_n^2 \kappa t)}{\alpha_n (k_1^2 + \kappa^2 \alpha_n^2) J_1(\alpha_n a)},$$

where $J_0(\alpha_n a) = 0$ for $n = 1, 2, 3, \dots$.

Section 2.1, Problem 14. Because

$$I_0(z) = 1 + \frac{z^2}{4} + \frac{z^4}{64} + \cdots \qquad \text{and} \qquad I_1(z) = \frac{z}{2} + \frac{z^3}{16} + \frac{z^5}{384} + \cdots,$$

then

$$F(s) = \frac{I_0\left(r\sqrt{s}\right)}{s^{3/2} I_1\left(a\sqrt{s}\right)}$$

is single-valued and has a second-order pole at $s = 0$. Furthermore, if $a\sqrt{s} = -i\alpha$, then $I_1(-\alpha i) = iJ_1(\alpha) = 0$. The residue for $s = 0$ is

$$\text{Res}\left[\frac{I_0(r\sqrt{z})\,e^{tz}}{z^{3/2}I_1(a\sqrt{z})};0\right] = 1 - \lim_{z\to 0}\frac{d}{dz}\left[\frac{(1+r^2z/4+r^4z^2/64+\cdots)(1+tz+t^2z^2/2+\cdots)}{(a/2)(1+a^2z/8+a^4z^2/192+\cdots)}\right]$$

$$= 1 - \frac{2}{a}\left(t + \frac{r^2}{4} - \frac{a^2}{8}\right).$$

On the other hand, the residue at $z_n = -\alpha_n^2/a^2$ is

$$\text{Res}\left[\frac{I_0(r\sqrt{z}\,e^{tz})}{z^{3/2}I_1(a\sqrt{z})};-\alpha_n^2/a^2\right] = \lim_{z\to -\alpha_n^2/a^2}\frac{(z+\alpha_n^2/a^2)I_0(r\sqrt{z})\,e^{tz}}{z^{3/2}I_1(a\sqrt{z})}$$

$$= \frac{I_0(-i\alpha_n r/a)\exp(-\alpha_n^2 t/a^2)}{I_1'(-i\alpha_n)(a/2)(-\alpha_n^2/a^2)}$$

$$= 2a\frac{J_0(\alpha_n r/a)\exp(-\alpha_n^2 t/a^2)}{\alpha_n^2 J_0(\alpha_n)},$$

because $2I_1'(z) = I_0(z) + I_2(z)$, $I_0(-i\alpha_n) = J_0(\alpha_n)$ and $I_2(-i\alpha_n) = -J_2(\alpha_n)$. Finally, $J_2(\alpha_n) = J_0(\alpha_n)$ because $J_2(\alpha_n) + J_0(\alpha_n) = 2J_1(\alpha_n)/\alpha_n = 0$. Therefore, summing the residues gives

$$f(t) = 1 - a\left[\frac{2t}{a^2} + \frac{r^2}{2a^2} - \frac{1}{4} - 2\sum_{n=1}^{\infty}\frac{J_0(\alpha_n r/a)}{\alpha_n^2 J_0(\alpha_n)}e^{-\alpha_n^2 t/a^2}\right].$$

Section 2.1, Problem 15. From Bromwich's integral,

$$f(t) = \frac{1}{2\pi i}\int_{c-\infty i}^{c+\infty i}\frac{I_n(a\sqrt{z})}{\sqrt{z}\,I_{n+1}(b\sqrt{z})}e^{zt}\,dz.$$

Closing the line integral with a semicircle of infinite radius in the left side of the z-plane, there are simple poles at $z = 0$ and $z_m = -\alpha_m^2/b^2$ with $\sqrt{z_m} = -i\alpha_m/b$, where $J_{n+1}(\alpha_m) = 0$. The residue at $z = 0$ is

$$\text{Res}\left[\frac{I_n(a\sqrt{z})\,e^{tz}}{\sqrt{z}\,I_{n+1}(b\sqrt{z})};0\right] = \lim_{z\to 0}\frac{\sqrt{z}\,I_n(a\sqrt{z})}{I_{n+1}(b\sqrt{z})}e^{zt} = 2(n+1)\frac{a^n}{b^{n+1}},$$

because $I_n(z) \approx (z/2)^n/n!$ as $z \to 0$. On the other hand,

$$\text{Res}\left[\frac{I_n(a\sqrt{z})\,e^{tz}}{\sqrt{z}\,I_{n+1}(b\sqrt{z})};z_m\right] = \lim_{z\to z_m}\frac{(z-z_m)I_n(a\sqrt{z})}{\sqrt{z}\,I_{n+1}(b\sqrt{z})}e^{zt}$$

$$= \frac{4e^{-n\pi i/2}J_n(a\alpha_m/b)e^{-\alpha_m^2 t/b^2}}{b[e^{-n\pi i/2}J_n(\alpha_m) + e^{(n+2)\pi i/2}J_{n+2}(\alpha_m)]}.$$

Summing the residues, the inverse is

$$f(t) = 2(n+1)\frac{a^n}{b^{n+1}} + \frac{4}{b}\sum_{m=1}^{\infty}\frac{J_n(a\alpha_m/b)}{J_n(\alpha_m) - J_{n+2}(\alpha_m)}e^{-\alpha_m^2 t/b^2}.$$

Section 2.1, Problem 16. Deforming Bromwich's integral along the imaginary axis,

$$f(t) = \frac{1}{2\pi i}\left[\int_{-\infty i}^{-0i} e^{-r\sqrt{z}\tanh(\sqrt{z})}\frac{dz}{z} + \int_{C_\epsilon} e^{-r\sqrt{z}\tanh(\sqrt{z})}\frac{dz}{z} + \int_{+0i}^{\infty i} e^{-r\sqrt{z}\tanh(\sqrt{z})}\frac{dz}{z}\right],$$

where C_ϵ is a very small semicircle passing to the right of the singularity at $z = 0$. From a straightforward integration with $z = \epsilon e^{\theta i}$ with $\epsilon \to 0$ and $-\pi/2 < \theta < \pi/2$,

$$\int_{C_\epsilon} e^{-r\sqrt{z}\tanh(\sqrt{z})}\frac{dz}{z} = \pi i.$$

Along the imaginary axis, we have $s = \eta^2 e^{\pm \pi i/2}/2$ and

$$\tanh(\sqrt{s}) = \frac{\sinh(\eta) \pm i\sin(\eta)}{\cosh(\eta) + \cos(\eta)}.$$

Then,

$$\int_{-\infty i}^{0^- i} e^{-r\sqrt{z}\tanh(\sqrt{z}}\frac{dz}{z} = \int_{\infty}^{0^+}\frac{\eta\, d\eta}{\eta^2/2}\exp\left[-\frac{r\eta}{2}(1-i)\frac{\sinh(\eta) - i\sin(\eta)}{\cosh(\eta) + \cos(\eta)} - i\frac{\eta^2 t}{2}\right]$$

and

$$\int_{0+i}^{\infty i} e^{-r\sqrt{z}\tanh(\sqrt{z}}\frac{dz}{z} = \int_{0+}^{\infty}\frac{\eta\, d\eta}{\eta^2/2}\exp\left[-\frac{r\eta}{2}(1+i)\frac{\sinh(\eta) + i\sin(\eta)}{\cosh(\eta) + \cos(\eta)} + i\frac{\eta^2 t}{2}\right]$$

so that

$$\int_{-\infty i}^{0^- i} e^{-r\sqrt{z}\tanh(\sqrt{z})}\frac{dz}{z} + \int_{0+i}^{\infty i} e^{-r\sqrt{z}\tanh(\sqrt{z})}\frac{dz}{z}$$

$$= 4i\int_{0+}^{\infty}\frac{d\eta}{\eta}\exp\left[-\frac{r\eta}{2}\frac{\sinh(\eta) - \sin(\eta)}{\cosh(\eta) + \cos(\eta)}\right]\sin\left[\frac{\eta^2 t}{2} - \frac{r\eta}{2}\frac{\sinh(\eta) + \sin(\eta)}{\cosh(\eta) + \cos(\eta)}\right].$$

Adding the above result to πi gives the Bromwich integral after dividing the sum by $2\pi i$.

Section 3.1

Section 3.1, Problem 1. The poles are located at $z = 0$ and $z = a^2 i$. Therefore, the inversion integral is

$$f(t) = \frac{1}{2\pi}\int_{-\infty-\epsilon i}^{\infty-\epsilon i}\frac{e^{i\omega t}}{\omega(\omega - a^2 i)}\,d\omega.$$

Because there are no singularities in the lower half-plane, $f(t) = 0$ for $t < 0$. In the limit of $\epsilon \to 0$,

$$\text{Res}\left[\frac{e^{itz}}{z(z - a^2 i)}; 0\right] = \frac{1}{-a^2 i} \quad \text{and} \quad \text{Res}\left[\frac{e^{itz}}{z(z - a^2 i)}; a^2 i\right] = \frac{e^{-a^2 t}}{a^2 i}.$$

Thus, the residue theorem gives $f(t) = \left(e^{-a^2 t} - 1\right) H(t)/a^2$.

Section 3.1, Problem 2. The inversion formula is

$$f(t) = \frac{1}{2\pi} \int_{-\infty}^{\infty} \frac{e^{i\omega t}}{\omega^2 + a^2} \, d\omega.$$

The poles are located at $\omega = \pm ai$. For $t < 0$, we take the pole in the lower half-plane and

$$f(t) = \frac{1}{2\pi}\left[-2\pi i \, \text{Res}\left(\frac{e^{itz}}{z^2 + a^2}; -ai\right)\right] = -i \lim_{z \to -ia} \frac{(z + ai)e^{itz}}{(z + ai)(z - ai)} = \frac{e^{at}}{2a}.$$

The negative sign arises from taking the contour in the negative sense. For $t > 0$, we use the pole $\omega = ai$ and

$$f(t) = \frac{1}{2\pi}\left[2\pi i \, \text{Res}\left(\frac{e^{itz}}{z^2 + a^2}; ai\right)\right] = i \lim_{z \to ia} \frac{(z - ai)e^{itz}}{(z + ai)(z - ai)} = \frac{e^{-at}}{2a}.$$

Therefore, the complete answer, using the absolute value sign, is $f(t) = e^{-a|t|}/(2a)$.

Section 3.1, Problem 3. The inversion formula is

$$f(t) = \frac{1}{2\pi} \int_{-\infty}^{\infty} \frac{\omega e^{i\omega t}}{\omega^2 + a^2} \, d\omega.$$

The poles are located at $\omega = \pm ai$. For $t < 0$, we take the pole in the lower half-plane and

$$f(t) = \frac{1}{2\pi}\left[-2\pi i \, \text{Res}\left(\frac{ze^{itz}}{z^2 + a^2}; -ai\right)\right] = -i \lim_{z \to -ai} \frac{(z + ai)ze^{itz}}{(z + ai)(z - ai)} = \frac{-ie^{at}}{2}.$$

The negative sign arises from taking the contour in the negative sense. For $t > 0$, we use the pole $\omega = ai$ and

$$f(t) = \frac{1}{2\pi}\left[2\pi i \, \text{Res}\left(\frac{ze^{itz}}{z^2 + a^2}; ai\right)\right] = i \lim_{z \to ai} \frac{(z - ai)ze^{itz}}{(z + ai)(z - ai)} = \frac{ie^{-at}}{2}.$$

Therefore, the complete answer, using the absolute value sign, is $f(t) = i \,\text{sgn}(t) \, e^{-a|t|}/2$.

Section 3.1, Problem 4. The inversion formula is

$$f(t) = \frac{1}{2\pi} \int_{-\infty}^{\infty} \frac{\omega e^{i\omega t}}{(\omega^2 + a^2)^2} \, d\omega.$$

The poles are located at $\omega = \pm ai$ and are second order. For $t < 0$, we take the pole in the lower half-plane and

$$f(t) = \frac{1}{2\pi}\left\{-2\pi i \operatorname{Res}\left[\frac{ze^{itz}}{(z^2+a^2)^2};-ai\right]\right\} = -i \lim_{z\to -ai}\frac{d}{dz}\left[\frac{(z+ai)^2 ze^{itz}}{(z+ai)^2(z-ai)^2}\right]$$

$$= -i \lim_{z\to -ai}\left[\frac{e^{itz}}{(z-ai)^2}+\frac{itze^{itz}}{(z-ai)^2}-\frac{2ze^{itz}}{(z-ai)^3}\right]$$

$$= -i\left(-\frac{e^{at}}{4a^2}-\frac{te^{at}}{4a}+\frac{e^{at}}{4a^2}\right) = \frac{ite^{at}}{4a}.$$

The negative sign arises from taking the contour in the negative sense. For $t > 0$, we use the pole $\omega = ai$ and

$$f(t) = \frac{1}{2\pi}\left\{2\pi i \operatorname{Res}\left[\frac{ze^{itz}}{(z^2+a^2)^2};ai\right]\right\} = i \lim_{z\to ai}\frac{d}{dz}\left[\frac{(z-ai)^2 ze^{itz}}{(z+ai)^2(z-ai)^2}\right]$$

$$= i \lim_{z\to ai}\left[\frac{e^{itz}}{(z+ai)^2}+\frac{itze^{itz}}{(z+ai)^2}-\frac{2ze^{itz}}{(z+ai)^3}\right]$$

$$= i\left(-\frac{e^{-at}}{4a^2}+\frac{te^{-at}}{4a}+\frac{e^{-at}}{4a^2}\right) = \frac{ite^{-at}}{4a}.$$

Therefore, the inverse, using the absolute value sign, is $f(t) = ite^{-a|t|}/(4a)$.

Section 3.1, Problem 5. The inversion formula is

$$f(t) = \frac{1}{2\pi}\int_{-\infty}^{\infty}\frac{\omega^2 e^{i\omega t}}{(\omega^2+a^2)^2}\,d\omega.$$

The poles are located at $\omega = \pm ai$ and are second order. For $t < 0$, we take the pole in the lower half-plane and

$$f(t) = \frac{1}{2\pi}\left\{-2\pi i \operatorname{Res}\left[\frac{z^2 e^{itz}}{(z^2+a^2)^2};-ai\right]\right\} = -i \lim_{z\to -ai}\frac{d}{dz}\left[\frac{(z+ai)^2 z^2 e^{itz}}{(z+ai)^2(z-ai)^2}\right]$$

$$= -i \lim_{z\to -ia}\left[\frac{2ze^{itz}}{(z-ai)^2}+\frac{itz^2 e^{itz}}{(z-ai)^2}-\frac{2z^2 e^{itz}}{(z-ai)^3}\right] = \frac{e^{at}}{2a}+\frac{te^{at}}{4}-\frac{e^{at}}{4a} = \frac{ate^{at}}{4a}+\frac{e^{at}}{4a}.$$

The negative sign arises from taking the contour in the negative sense. For $t > 0$, we use the pole $\omega = ai$ and

$$f(t) = \frac{1}{2\pi}\left\{2\pi i \operatorname{Res}\left[\frac{z^2 e^{itz}}{(z^2+a^2)^2};ai\right]\right\} = i \lim_{z\to ai}\frac{d}{dz}\left[\frac{(z-ai)^2 z^2 e^{itz}}{(z+ai)^2(z-ai)^2}\right]$$

$$= i \lim_{z\to ai}\left[\frac{2ze^{itz}}{(z+ai)^2}+\frac{itz^2 e^{itz}}{(z+ai)^2}-\frac{2z^2 e^{itz}}{(z+ai)^3}\right] = \frac{e^{-at}}{2a}-\frac{te^{-at}}{4}-\frac{e^{-at}}{4a}$$

$$= \frac{e^{-at}}{4a}-\frac{ate^{-at}}{4a}.$$

Therefore, the inverse, using the absolute value sign, is $f(t) = (1-a|t|)e^{-a|t|}/(4a)$.

Section 3.1, Problem 6. The inversion formula is

$$f(t) = \frac{1}{2\pi} \int_{-\infty}^{\infty} \frac{e^{i\omega t}}{\omega^2 - 3i\omega - 3} \, d\omega.$$

The simple poles are located at $\omega = \pm\sqrt{3}/2 + 3i/2$. Because there are no singularities in the lower half-plane, $f(t) = 0$ for $t < 0$. For $t > 0$,

$$f(t) = \frac{1}{2\pi} \left[2\pi i \operatorname{Res}\left(\frac{e^{itz}}{z^2 - 3iz - 3}; \frac{\sqrt{3}}{2} + \frac{3i}{2} \right) + 2\pi i \operatorname{Res}\left(\frac{e^{itz}}{z^2 - 3iz - 3}; -\frac{\sqrt{3}}{2} + \frac{3i}{2} \right) \right]$$

$$= i \left[\lim_{z \to \sqrt{3}/2 + 3i/2} \frac{\left(z - \sqrt{3}/2 - 3i/2\right) e^{itz}}{z^2 - 3iz - 3} + \lim_{z \to -\sqrt{3}/2 + 3i/2} \frac{\left(z + \sqrt{3}/2 - 3i/2\right) e^{itz}}{z^2 - 3iz - 3} \right]$$

$$= \frac{i e^{-3t/2} \left[\cos(\sqrt{3}t/2) + i\sin(\sqrt{3}t/2) \right]}{2\left(\sqrt{3}/2 + 3i/2\right) - 3i} + \frac{i e^{-3t/2} \left[\cos(\sqrt{3}t/2) - i\sin(\sqrt{3}t/2) \right]}{2\left(-\sqrt{3}/2 + 3i/2\right) - 3i}$$

$$= -\frac{2}{\sqrt{3}} e^{-3t/2} \sin\left(\frac{\sqrt{3}t}{2} \right).$$

Therefore, the inverse is $f(t) = -2e^{-3t/2} \sin(\sqrt{3}\,t/2)\, H(t)/\sqrt{3}$.

Section 3.1, Problem 7. The inversion formula is

$$f(t) = \frac{1}{2\pi} \int_{-\infty}^{\infty} \frac{e^{i\omega t}}{(\omega - ai)^{2n+2}} \, d\omega.$$

The pole is located at $\omega = ai$ and it is a $(2n + 2)$th-order pole. Because there are no singularities in the lower half-plane, $f(t) = 0$ for $t < 0$. For $t > 0$,

$$f(t) = \frac{1}{2\pi} \left\{ 2\pi i \operatorname{Res}\left[\frac{e^{itz}}{(z - ai)^{2n+2}}; ai \right] \right\} = \lim_{z \to ai} \left\{ \frac{1}{(2n+1)!} \frac{d^{2n+1}}{dz^{2n+1}} \left[\frac{(z - ai)^{2n+2} e^{itz}}{(z - ai)^{2n+2}} \right] \right\}$$

$$= \frac{i}{(2n+1)!} (i)^{2n+2} t^{2n+1} e^{-at} = \frac{(-1)^{n+1}}{(2n+1)!} t^{2n+1} e^{-at}.$$

Therefore, the inverse is $f(t) = (-1)^{n+1} t^{2n+1} e^{-at} H(t)/(2n+1)!$.

Section 3.1, Problem 8. The inversion formula is

$$f(t) = \frac{1}{2\pi} \int_{-\infty}^{\infty} \frac{2i\sin(\omega h/2) e^{i\omega t}}{\omega^2 + a^2} \, d\omega = \frac{1}{2\pi} \int_{-\infty}^{\infty} \frac{e^{i(t+h/2)\omega} - e^{i(t-h/2)\omega}}{\omega^2 + a^2} \, d\omega$$

$$= \frac{1}{2\pi} \int_{-\infty}^{\infty} \frac{e^{i(t+h/2)\omega}}{\omega^2 + a^2} \, d\omega - \frac{1}{2\pi} \int_{-\infty}^{\infty} \frac{e^{i(t-h/2)\omega}}{\omega^2 + a^2} \, d\omega.$$

The simple poles are located at $\omega = \pm ai$. The corresponding residues are

$$\operatorname{Res}\left[\frac{e^{i(t+h/2)z}}{z^2 + a^2}; ai \right] = \frac{e^{-a(t+h/2)}}{2ia}, \qquad \operatorname{Res}\left[\frac{e^{i(t-h/2)z}}{z^2 + a^2}; ai \right] = \frac{e^{-a(t-h/2)}}{2ia},$$

$$\operatorname{Res}\left[\frac{e^{i(t+h/2)z}}{z^2 + a^2}; -ai \right] = \frac{e^{a(t+h/2)}}{-2ia}, \quad \text{and} \quad \operatorname{Res}\left[\frac{e^{i(t-h/2)z}}{z^2 + a^2}; -ai \right] = \frac{e^{a(t-h/2)}}{-2ia}.$$

If $t > h/2$,

$$f(t) = \frac{1}{2\pi} \left\{ 2\pi i \operatorname{Res}\left[\frac{e^{i(t+h/2)z}}{z^2+a^2}; ai \right] - 2\pi i \operatorname{Res}\left[\frac{e^{i(t-h/2)z}}{z^2+a^2}; ai \right] \right\}$$

$$= \frac{1}{2a} \left[e^{-a(t+h/2)} - e^{-a(t-h/2)} \right] = -\frac{e^{-at}}{a} \sinh\left(\frac{ah}{2} \right).$$

If $-h/2 < t < h/2$,

$$f(t) = \frac{1}{2\pi} \left\{ 2\pi i \operatorname{Res}\left[\frac{e^{i(t+h/2)z}}{z^2+a^2}; ai \right] - 2\pi i \operatorname{Res}\left[\frac{e^{i(t-h/2)z}}{z^2+a^2}; -ai \right] \right\}$$

$$= \frac{1}{2a} \left[e^{-a(t+h/2)} - e^{-a(t-h/2)} \right] = -\frac{e^{-ah/2}}{a} \sinh(at).$$

If $t < -h/2$,

$$f(t) = \frac{1}{2\pi} \left\{ 2\pi i \operatorname{Res}\left[\frac{e^{i(t+h/2)z}}{z^2+a^2}; -ai \right] - 2\pi i \operatorname{Res}\left[\frac{e^{i(t-h/2)z}}{z^2+a^2}; -ai \right] \right\}$$

$$= \frac{1}{2a} \left[e^{a(t+h/2)} - e^{a(t-h/2)} \right] = \frac{e^{at}}{a} \sinh\left(\frac{ah}{2} \right).$$

Therefore,

$$f(t) = \begin{cases} -e^{-at} \sinh(ah/2)/a, & \text{if} & h/2 < t, \\ -e^{-ah/2} \sinh(at)/a, & \text{if} & -h/2 < t < h/2, \\ e^{at} \sinh(ah/2)/a, & \text{if} & t < -h/2. \end{cases}$$

Section 3.1, Problem 9. The inversion formula is

$$f(t) = \frac{1}{2\pi} \int_{-\infty}^{\infty} \frac{\omega^2 e^{i\omega t}}{(\omega^2 - 1)^2 + 4a^2\omega^2} \, d\omega.$$

The simple poles are located at $\omega = \pm\sqrt{1 - a^2} \pm ai$. Let us assume that $a < 1$. Then,

$$\operatorname{Res}\left[\frac{z^2 e^{itz}}{(z^2-1)^2 + 4a^2 z^2}; \sqrt{1-a^2} + ai \right] = \lim_{z \to \sqrt{1-a^2}+ai} \left[\frac{\left(z - \sqrt{1-a^2} - ai\right) z^2 e^{itz}}{(z^2-1)^2 + 4a^2 z^2} \right]$$

$$= \lim_{z \to \sqrt{1-a^2}+ai} \frac{z e^{itz}}{4(z^2-1) + 8a^2}$$

$$= \frac{e^{-at+it\sqrt{1-a^2}}}{8ia\sqrt{1-a^2}} \left(\sqrt{1-a^2} + ai \right),$$

and

$$\operatorname{Res}\left[\frac{z^2 e^{itz}}{(z^2-1)^2 + 4a^2 z^2}; -\sqrt{1-a^2} + ai \right] = \lim_{z \to -\sqrt{1-a^2}+ai} \left[\frac{\left(z + \sqrt{1-a^2} - ai\right) z^2 e^{itz}}{(z^2-1)^2 + 4a^2 z^2} \right]$$

$$= \lim_{z \to -\sqrt{1-a^2}+ai} \frac{z e^{itz}}{4(z^2-1) + 8a^2}$$

$$= \frac{e^{-at-it\sqrt{1-a^2}}}{-8ia\sqrt{1-a^2}} \left(-\sqrt{1-a^2} + ai \right).$$

Therefore, for $t > 0$, the residue theorem gives

$$f(t) = \frac{e^{-at}}{8a\sqrt{1-a^2}} \left[\sqrt{1-a^2}\,\cos\!\left(t\sqrt{1-a^2}\right) - a\sin\!\left(t\sqrt{1-a^2}\right) \right.$$

$$+ ia\cos\!\left(t\sqrt{1-a^2}\right) + i\sqrt{1-a^2}\,\sin\!\left(t\sqrt{1-a^2}\right)$$

$$+ \sqrt{1-a^2}\,\cos\!\left(t\sqrt{1-a^2}\right) - a\sin\!\left(t\sqrt{1-a^2}\right)$$

$$\left. - ia\cos\!\left(t\sqrt{1-a^2}\right) - i\sqrt{1-a^2}\,\sin\!\left(t\sqrt{1-a^2}\right) \right]$$

$$= \frac{e^{-at}}{4a}\cos\!\left(t\sqrt{1-a^2}\right) - \frac{e^{-at}}{4\sqrt{1-a^2}}\sin\!\left(t\sqrt{1-a^2}\right).$$

Similarly, for the lower half-plane,

$$\mathrm{Res}\!\left[\frac{z^2 e^{itz}}{(z^2-1)^2 + 4a^2 z^2}; \sqrt{1-a^2} - ai \right] = \lim_{z \to \sqrt{1-a^2}-ai} \left[\frac{\left(z - \sqrt{1-a^2} + ai\right) z^2 e^{itz}}{(z^2-1)^2 + 4a^2 z^2} \right]$$

$$= \lim_{z \to \sqrt{1-a^2}-ai} \frac{z e^{itz}}{4(z^2-1) + 8a^2}$$

$$= \frac{e^{it\sqrt{1-a^2}} e^{at}}{-8ia\sqrt{1-a^2}} \left(\sqrt{1-a^2} - ai \right),$$

and

$$\mathrm{Res}\!\left[\frac{z^2 e^{itz}}{(z^2-1)^2 + 4a^2 z^2}; -\sqrt{1-a^2} - ai \right] = \lim_{z \to -\sqrt{1-a^2}-ai} \left[\frac{\left(z + \sqrt{1-a^2} + ai\right) z^2 e^{itz}}{(z^2-1)^2 + 4a^2 z^2} \right]$$

$$= \lim_{z \to -\sqrt{1-a^2}-ai} \frac{z e^{itz}}{4(z^2-1) + 8a^2}$$

$$= \frac{e^{at-it\sqrt{1-a^2}}}{8ia\sqrt{1-a^2}} \left(-\sqrt{1-a^2} - ai \right).$$

Therefore, for $t < 0$, the residue theorem gives

$$f(t) = -\frac{e^{at}}{8a\sqrt{1-a^2}} \left[-\sqrt{1-a^2}\,\cos\!\left(t\sqrt{1-a^2}\right) - a\sin\!\left(t\sqrt{1-a^2}\right) \right.$$

$$- ia\cos\!\left(t\sqrt{1-a^2}\right) + i\sqrt{1-a^2}\,\sin\!\left(t\sqrt{1-a^2}\right)$$

$$- \sqrt{1-a^2}\,\cos\!\left(t\sqrt{1-a^2}\right) - a\sin\!\left(t\sqrt{1-a^2}\right)$$

$$\left. + ia\cos\!\left(t\sqrt{1-a^2}\right) - i\sqrt{1-a^2}\,\sin\!\left(t\sqrt{1-a^2}\right) \right]$$

$$= \frac{e^{at}}{4a}\cos\!\left(t\sqrt{1-a^2}\right) + \frac{e^{at}}{4\sqrt{1-a^2}}\sin\!\left(t\sqrt{1-a^2}\right).$$

Now, let us consider $a > 1$. The easiest method is to use our earlier results with $\sqrt{1 - a^2} = i\sqrt{a^2 - 1}$ or $-i\sqrt{a^2 - 1}$. Therefore, the inverse is

$$
f(t) = \begin{cases} \dfrac{e^{-a|t|}\cosh\left(\sqrt{a^2 - 1}\,|t|\right)}{4a} - \dfrac{e^{-a|t|}\sinh\left(\sqrt{a^2 - 1}\,|t|\right)}{4\sqrt{a^2 - 1}}, & \text{if} \quad 1 < a, \\[4mm] \dfrac{e^{-a|t|}\cos\left(\sqrt{1 - a^2}\,|t|\right)}{4a} - \dfrac{e^{-a|t|}\sin\left(\sqrt{1 - a^2}\,|t|\right)}{4\sqrt{a^2 - 1}}, & \text{if} \quad 0 < a < 1. \end{cases}
$$

Section 3.1, Problem 10. The inversion formula is

$$
f(t) = \frac{1}{2\pi}\int_{-\infty}^{\infty} \frac{e^{i\omega t}}{I_0(\omega)}\, d\omega.
$$

Because $I_0(\omega) = J_0(\omega i)$ if $-\pi < \arg(\omega) \leq \pi/2$, then the poles are located at $\omega = \pm i\alpha_n$, where $J_0(\alpha_n) = 0$ and $n = 1, 2, 3, \ldots$. For $t < 0$,

$$
f(t) = \frac{1}{2\pi}(-2\pi i)\left\{\sum_{n=1}^{\infty} \text{Res}\left[\frac{e^{itz}}{I_0(z)}; -\alpha_n i\right]\right\}
$$

$$
= -i\sum_{n=1}^{\infty} \lim_{z \to -\alpha_n i}\left[\frac{(z + \alpha_n i)e^{itz}}{I_0(z)}\right] = -i\sum_{n=1}^{\infty} \frac{e^{\alpha_n t}}{I_0'(-i\alpha_n)}.
$$

The negative sign arises from taking the contour in the negative sense. Now, $I_0'(-i\alpha_n) = I_1(-i\alpha_n) = -iJ_1(\alpha_n)$. Therefore,

$$
f(t) = \sum_{n=1}^{\infty} \frac{e^{\alpha_n t}}{J_1(\alpha_n)}.
$$

For $t > 0$,

$$
f(t) = \frac{1}{2\pi}(2\pi i)\left\{\sum_{n=1}^{\infty} \text{Res}\left[\frac{e^{itz}}{I_0(z)}; \alpha_n i\right]\right\} = i\sum_{n=1}^{\infty} \lim_{z \to \alpha_n i}\left[\frac{(z - \alpha_n i)e^{itz}}{I_0(z)}\right] = i\sum_{n=1}^{\infty} \frac{e^{-\alpha_n t}}{I_0'(i\alpha_n)}.
$$

Now, $I_0'(i\alpha_n) = -iJ_1(-\alpha_n)$. Consequently,

$$
f(t) = -\sum_{n=1}^{\infty} \frac{e^{-\alpha_n t}}{J_1(-\alpha_n)} = \sum_{n=1}^{\infty} \frac{e^{-\alpha_n t}}{J_1(\alpha_n)}.
$$

Therefore, the inverse, using absolute value signs, is

$$
f(t) = \sum_{n=1}^{\infty} \frac{\exp(-z_n|t|)}{J_1(z_n)}, \quad \text{where} \quad J_0(z_n) = 0, \quad n = 1, 2, 3, \ldots.
$$

Section 3.1, Problem 11. The poles are located at $\sinh(\omega h) = -i\sin(i\omega h) = 0$ or $\omega = \pm n\pi i/h$, with $n = 0, 1, 2, \ldots$. Therefore, the inversion integral is

$$f(t) = \frac{1}{2\pi} \int_{-\infty-\epsilon i}^{\infty-\epsilon i} \frac{\cosh[\omega(x-h)]\cosh(\omega a)e^{i\omega t}}{i\sinh(\omega h)} \, d\omega.$$

In the limit of $\epsilon \to 0$,

$$\mathrm{Res}\left\{ \frac{\cosh[(x-h)z]\cosh(az)e^{itz}}{i\sinh(hz)} ; 0 \right\} = \lim_{z\to 0} \left\{ z\frac{\cosh[(x-h)z]\cosh(az)e^{itz}}{i\sinh(hz)} \right\} = \frac{1}{ih}$$

and

$$\mathrm{Res}\left\{ \frac{\cosh[(x-h)z]\cosh(az)e^{itz}}{i\sinh(hz)} ; \frac{n\pi i}{h} \right\} = \lim_{z\to n\pi i/h} \frac{(z-n\pi i/h)\cosh[(x-h)z]\cosh(az)e^{itz}}{i\sinh(hz)}$$

$$= \frac{\cosh[n\pi(x-h)i/h]\cosh(n\pi ai/h)\exp(-n\pi t/h)}{ih\cosh(n\pi i)}$$

$$= \frac{\cos[n\pi(x-h)/h]\cos(n\pi a/h)\exp(-n\pi t/h)}{ih(-1)^n}.$$

Therefore, for $t > 0$,

$$f(t) = \frac{1}{h} + \frac{1}{h}\sum_{n=1}^{\infty} \cos\left(\frac{n\pi x}{h}\right)\cos\left(\frac{n\pi a}{h}\right)e^{-n\pi t/h}.$$

On the other hand, for $t < 0$,

$$f(t) = -\frac{1}{h}\sum_{n=1}^{\infty} \cos\left(\frac{n\pi x}{h}\right)\cos\left(\frac{n\pi a}{h}\right)e^{n\pi t/h}.$$

Thus, the most general form of the inverse is

$$f(t) = \frac{H(t)}{h} + \frac{\mathrm{sgn}(t)}{h}\sum_{n=1}^{\infty} \cos\left(\frac{n\pi x}{h}\right)\cos\left(\frac{n\pi a}{h}\right)\exp\left(-\frac{n\pi |t|}{h}\right).$$

Section 3.1, Problem 12. The inversion integral is

$$f(t) = \frac{1}{2\pi} \int_{-\infty-\epsilon i}^{\infty-\epsilon i} \frac{\sin(\omega x/a)}{\omega\cos(\omega h/a)} e^{i\omega t} \, d\omega.$$

The integrand has poles at $\cos(h\omega/a) = 0$ or $\omega = \pm(2n-1)\pi a/(2h)$, where $n = 1, 2, 3, \ldots$. The pole $\omega = 0$ is removable. Because there are no singularities in the lower half of the ω-plane, $f(t) = 0$ for $t < 0$. On the other hand, in the limit of $\epsilon \to 0$,

$$\mathrm{Res}\left[\frac{\sin(xz/a)e^{itz}}{z\cos(hz/a)} ; \frac{(2n-1)\pi a}{2h} \right] = \lim_{z\to(2n-1)\pi a/h} \left\{ \left[z - \frac{(2n-1)\pi a}{2h} \right] \frac{\sin(xz/a)e^{itz}}{z\cos(hz/a)} \right\}$$

$$= \frac{\sin[(2n-1)\pi x/(2h)]}{(2n-1)\pi/2} \times \frac{\exp[(2n-1)\pi ati/(2h)]}{-\sin[(2n-1)\pi/2]}$$

$$= (-1)^{n+1}\sin\left[\frac{(2n-1)\pi x}{2h} \right] \frac{\exp[(2n-1)\pi ati/(2h)]}{(2n-1)\pi/2}.$$

A similar result is found for $\omega = -(2n+1)\pi a/(2h)$:

$$\text{Res}\left[\frac{\sin(xz/a)e^{itz}}{z\cos(hz/a)}; -\frac{(2n-1)\pi a}{2h}\right] = (-1)^{n+1}\sin\left[\frac{(2n-1)\pi x}{2h}\right]\frac{\exp[-(2n-1)\pi ati/(2h)]}{(2n-1)\pi/2}.$$

Therefore, for $t > 0$,

$$f(t) = \frac{4}{\pi}\sum_{n=1}^{\infty}\frac{(-1)^{n+1}}{2n-1}\sin\left[\frac{(2n-1)\pi x}{2h}\right]\sin\left[\frac{(2n-1)\pi at}{2h}\right].$$

Thus, the most general form of the inverse is

$$f(t) = \frac{4}{\pi}\sum_{n=1}^{\infty}\frac{(-1)^{n+1}}{2n-1}\cos\left[\frac{(2n-1)\pi x}{2h}\right]\sin\left[\frac{(2n-1)\pi at}{2h}\right]H(t).$$

Section 3.1, Problem 13. The inversion integral is

$$f(t) = \frac{1}{2\pi}\int_{-\infty-\epsilon i}^{\infty-\epsilon i}\frac{m\pi\sinh(\omega a)e^{i\omega t}}{2\omega(\omega^2-m^2\pi^2)\cosh(\omega a)}[1+(-1)^m\cos(\omega)-i(-1)^m\sin(\omega)]\,d\omega.$$

The integrand has poles located at $\cosh(a\omega) = 0$ or $\omega = \pm(2n-1)\pi i/(2a)$, where $n = 1, 2, 3, \ldots$. The pole $\omega = 0$ is removable. There are also simple poles at $\omega = \pm m\pi$. In the limit of $\epsilon \to 0$,

$$\text{Res}\left\{\frac{m\pi\sinh(az)e^{itz}}{2z(z^2-m^2\pi^2)\cosh(az)}[1+(-1)^m\cos(z)-i(-1)^m\sin(z)]; m\pi\right\}$$

$$= \frac{m\pi\sinh(m\pi a)e^{im\pi t}}{2m\pi(2m\pi)\cosh(m\pi a)}[1+(-1)^m\cos(m\pi)-i(-1)^m\sin(m\pi)]$$

$$= \frac{\tanh(m\pi a)}{2m\pi}e^{im\pi t},$$

$$\text{Res}\left\{\frac{m\pi\sinh(az)e^{itz}}{2z(z^2-m^2\pi^2)\cosh(az)}[1+(-1)^m\cos(z)-i(-1)^m\sin(z)]; -m\pi\right\}$$

$$= \frac{m\pi\sinh(-m\pi a)e^{-im\pi t}}{(-2m\pi)(-2m\pi)\cosh(-m\pi a)}[1+(-1)^m\cos(-m\pi)-i(-1)^m\sin(-m\pi)]$$

$$= -\frac{\tanh(m\pi a)}{2m\pi}e^{-im\pi t},$$

$$\text{Res}\left\{\frac{m\pi\sinh(az)[1+(-1)^m\cos(z)-i(-1)^m\sin(z)]\,e^{itz}}{2z(z^2-m^2\pi^2)\cosh(az)}; \frac{(2n-1)\pi i}{a}\right\}$$

$$= \frac{m\pi\sinh[(2n-1)\pi i/2]\exp[-(2n-1)\pi t/(2a)]}{2a\sinh[(2n-1)\pi i/2][(2n-1)\pi i/(2a)][-m^2\pi^2-(2n-1)^2\pi^2/(4a^2)]}$$

$$\times\left\{1+(-1)^m\cos\left[\frac{(2n-1)\pi i}{2a}\right]-i(-1)^m\sin\left[\frac{(2n-1)\pi i}{2a}\right]\right\}$$

$$= -\frac{m\exp[-(2n-1)\pi t/(2a)]}{[m^2\pi^2+(2n-1)^2\pi^2/(4a^2)](2n-1)i}$$

$$\times\left\{1+(-1)^m\cosh\left[\frac{(2n-1)\pi}{2a}\right]+(-1)^m\sinh\left[\frac{(2n-1)\pi}{2a}\right]\right\},$$

and

$$\text{Res}\left\{\frac{m\pi\sinh(az)\left[1+(-1)^m\cos(z)-i(-1)^m\sin(z)\right]e^{itz}}{2z(z^2-m^2\pi^2)\cosh(az)}; -\frac{(2n-1)\pi i}{a}\right\}$$

$$=\frac{m\pi\sinh[-(2n-1)\pi i/2]\exp[(2n-1)\pi t/(2a)]}{2a\sinh[-(2n-1)\pi i/2][-(2n-1)\pi i/(2a)][-m^2\pi^2-(2n-1)^2\pi^2/(4a^2)]}$$

$$\times\left\{1+(-1)^m\cos\left[\frac{-(2n-1)\pi i}{2a}\right]-i(-1)^m\sin\left[\frac{-(2n-1)\pi i}{2a}\right]\right\}$$

$$\text{Res}\left\{\frac{m\pi\sinh(az)\left[1+(-1)^m\cos(z)-i(-1)^m\sin(z)\right]e^{itz}}{2z(z^2-m^2\pi^2)\cosh(az)}; -\frac{(2n-1)\pi i}{a}\right\}$$

$$=\frac{m\exp[(2n-1)\pi t/(2a)]}{[m^2\pi^2+(2n-1)^2\pi^2/(4a^2)](2n-1)i}$$

$$\times\left\{1+(-1)^m\cosh\left[\frac{(2n-1)\pi}{2a}\right]-(-1)^m\sinh\left[\frac{(2n-1)\pi}{2a}\right]\right\}.$$

Thus, the most general form of the inverse is

$$f(t)=-\frac{\tanh(m\pi)}{m\pi}H(t)$$

$$-\sum_{n=1}^{\infty}\frac{m\exp\left[-(2n-1)\pi|t|/(2a)\right]\left\{1-(-1)^m\exp[(2n-1)\pi\,\text{sgn}(t)/(2a)]\right\}}{(2n-1)\left[m^2\pi^2+(2n-1)^2\pi^2/(4a^2)\right]}.$$

Section 3.1, Problem 14. Here, the inversion integral is

$$f(t)=\frac{1}{2\pi i}\int_{-\infty-\epsilon i}^{\infty-\epsilon i}\left[\frac{e^{i(t+x)\omega}}{\omega\cosh(\omega)}+\frac{e^{i(t-x)\omega}}{2\omega\cosh(\omega)}-\frac{e^{it\omega}}{\omega}\right]d\omega.$$

The poles are located at $\omega=0$, $\omega=z_n=(2n-1)\pi i/2a$ and $\omega=-z_n$, where $n=1,2,3,\dots$. They are all simple poles.

For $0\le x<t$, we close all of the contours in the upper half plane. For $0<t<x$, however, we must evaluate the second term with a contour in the lower half plane. Now, for the third term,

$$\text{Res}\left(\frac{e^{izt}}{z};0\right)=-1.$$

For the first term,

$$\text{Res}\left[\frac{e^{i(t+x)z}}{z\cosh(z)};0\right]=1$$

and

$$\text{Res}\left[\frac{e^{i(t+x)z}}{z\cosh(z)};z_n\right]=\lim_{z\to z_n}\frac{(z-z_n)e^{i(t+x)z}}{z\cosh(z)}=\frac{2a\exp[-(2n-1)\pi(x+t)/(2a)]}{(2n-1)\pi ai\sinh[(2n-1)\pi i/2]}$$

$$=-\frac{2(-1)^n}{(2n-1)\pi}\exp[-(2n-1)\pi(x+t)/(2a)].$$

For the second term,

$$\text{Res}\left[\frac{e^{i(t-x)z}}{2z\cosh(z)};0\right]=-1,$$

$$\text{Res}\left[\frac{e^{i(t+x)z}}{z\cosh(z)};z_n\right] = -\lim_{z\to z_n}\frac{(z-z_n)e^{i(t-x)z}}{z\cosh(z)} = \frac{2a\exp[-(2n-1)\pi(t-x)/(2a)]}{(2n-1)\pi a i \sinh[(2n-1)\pi i/2]}$$

$$= \frac{2(-1)^n}{(2n-1)\pi}\exp[-(2n-1)\pi(t-x)/(2a)],$$

and

$$\text{Res}\left[\frac{e^{i(t+x)z}}{z\cosh(z)};-z_n\right] = -\lim_{z\to z_n}\frac{(z-z_n)e^{i(t-x)z}}{z\cosh(z)} = -\frac{2a\exp[(2n-1)\pi(t-x)/(2a)]}{(2n-1)\pi a i \sinh[(2n-1)\pi i/2]}$$

$$= -\frac{2(-1)^n}{(2n-1)\pi}\exp[(2n-1)\pi(t-x)/(2a)].$$

Thus, for $0 \le x < t$, $f(t) = -1 + g(-x,t) - g(x,t)$, while for $0 < t < x$, $f(t) = -g(x,t) + g(x,-t)$, where

$$g(x,t) = \frac{2}{\pi}\sum_{n=1}^{\infty}\frac{(-1)^n}{2n-1}\exp\left[-\frac{(2n-1)\pi t}{2a}\right]\exp\left[-\frac{(2n-1)\pi x}{2a}\right].$$

Section 3.1, Problem 15. From the inversion formula,

$$f(t) = -\frac{Pa}{2\pi r\rho c}\int_{-\infty}^{\infty}\frac{\exp[-\omega^2 b^2(r-a)/(2c^3)+i\omega t]}{(\omega+B-Ai)(\omega-B-Ai)}\,d\omega.$$

The poles are located at $\omega = \pm B + Ai$. Therefore, for $t < 0$, $f(t) = 0$. For $t > 0$,

$$f(t) = -\frac{Pai}{r\rho c}\left(\text{Res}\left\{\frac{\exp[-z^2b^2(r-a)/(2c^3)+itz]}{(z+B-Ai)(z-B-Ai)};B+Ai\right\}\right.$$

$$\left. + \text{Res}\left\{\frac{\exp[-z^2b^2(r-a)/(2c^3)+itz]}{(z+B-Ai)(z-B-Ai)};-B+Ai\right\}\right).$$

Now,

$$\text{Res}\left\{\frac{\exp[-z^2b^2(r-a)/(2c^3)+itz]}{(z+B-Ai)(z-B-Ai)};B+Ai\right\}$$

$$= \frac{\exp[-b^2(B^2-A^2+2iAB)(r-a)/(2c^3)+iBt-At]}{2B}$$

and

$$\text{Res}\left\{\frac{\exp[-z^2b^2(r-a)/(2c^3)+itz]}{(z+B-Ai)(z-B-Ai)};-B+Ai\right\}$$

$$= \frac{\exp[-b^2(B^2-A^2-2iAB)(r-a)/(2c^3)-iBt-At]}{-2B}.$$

Adding the residues and simplifying, the inverse is

$$f(t) = \frac{Pa}{r\rho cB}\exp\left[-\frac{b^2(B^2-A^2)(r-a)}{2c^3}-At\right]\sin\left[Bt-\frac{ABb^2}{c^3}(r-a)\right].$$

Section 3.1, Problem 16. Because P_n is a real positive number, the poles at $z_n = i\bar{z}_n = i(P_n^2 + \lambda^2)$ lie along the positive imaginary axis. Therefore, $f(t)$ and $g(t)$ are zero for $t < 0$. For $t > 0$, the residue for computing $f(t)$ is

$$\text{Res}\left\{\frac{[P\cot(P) + \lambda]e^{itz}}{2\lambda P \cot(P) - P^2 + \lambda^2}; z_n\right\} = \frac{[P_n \cot(P_n) + \lambda]e^{iz_n t}}{\{2\lambda \cot(P_n) - 2\lambda P_n[1 + \cot^2(P_n)] - 2P_n\}[-i/(2P_n)]}$$

$$= \frac{[(2P_n^2 - \lambda^2)/2\lambda + \lambda]e^{-\bar{z}_n t}}{[(P_n^2 - \lambda^2)/P_n - \bar{z}_n^2/(2\lambda P_n) - 2P_n][-i/(2P_n)]}$$

$$= \frac{2(\bar{z}_n - \lambda^2)\exp(-\bar{z}_n t)}{i(\bar{z}_n + 2\lambda)},$$

because $\cot(P_n) = (P_n^2 - \lambda^2)/(2\lambda_n P_n)$, $\cot^2(P_n) = (P_n^2 - \lambda^2)^2/(4\lambda^2 P_n^2)$ and $1 + \cot^2(P_n) = \bar{z}_n^2/(4\lambda^2 P_n^2)$. Therefore,

$$f(t) = 2\sum_{n=1}^{\infty} \frac{\bar{z}_n - \lambda^2}{\bar{z}_n + 2\lambda}e^{-\bar{z}_n t}.$$

On the other hand, the residue used in computing $g(t)$ is

$$\text{Res}\left[\frac{e^{itz}}{P\cot(P) + \lambda}; z_n\right] = \frac{e^{iz_n t}}{\{\cot(P_n) - P_n[1 + \cot^2(P_n)]\}[-i/(2P_n)]}$$

$$= \frac{2(\bar{z}_n - \lambda^2)\exp(-\bar{z}_n t)}{i(\bar{z}_n + \lambda)},$$

because $\cot(P_n) = -\lambda/P_n$, $\cot^2(P_n) = \lambda^2/P_n^2$ and $1 + \cot^2(P_n) = \bar{z}_n/P_n^2$. Therefore,

$$g(t) = 2\sum_{n=1}^{\infty} \frac{\bar{z}_n - \lambda^2}{\bar{z}_n + \lambda}e^{-\bar{z}_n t}.$$

Section 3.1, Problem 17. The poles of the transform are located at $\zeta = 0$, a second-order pole, and $\zeta_n = n\pi i$, where $n = 1, 2, 3, \ldots$. To express these poles in terms of z, we solve for z and find that $z_n^{\pm} = \pm\sqrt{k^2 + \zeta_n^2} + i\delta = \pm\sqrt{k^2 - n^2\pi^2} + i\delta$. For sufficiently large n, say $N + 1$, z_n becomes purely imaginary. Consequently, there are $N + 1$ poles that lie along and above the real axis; the vast majority lie along the imaginary axis.

To find the inverse for $t > 0$, we convert the line integral into a closed contour by adding an infinite semicircle in the upper half-plane. Therefore,

$$f(t) = \frac{1}{2\pi}\int_{-\infty}^{\infty} \frac{\cosh(\zeta)}{\zeta\sinh(\zeta)}e^{i\omega t}\,d\omega = \frac{1}{2\pi}\oint_C \frac{\cosh(\zeta)}{\zeta\sinh(\zeta)}e^{izt}\,dz.$$

Upon applying the residue theorem,

$$f(t) = \lim_{z\to k+i\delta}\frac{d}{dz}\left[\frac{(z - k - \delta i)^2\cosh(\zeta)e^{itz}}{2\zeta\sinh(\zeta)}\right] + \lim_{z\to -k+i\delta}\frac{d}{dz}\left[\frac{(z + k - \delta i)^2\cosh(\zeta)e^{itz}}{2\zeta\sinh(\zeta)}\right]$$

$$+ i\sum_{n=0}^{N} \frac{\sinh(\zeta)e^{-\delta t}}{\frac{d}{d\zeta}[\zeta\cosh(\zeta)]\frac{d\zeta}{dz}\big|_{z=z_n^+}}\exp\left(it\sqrt{k^2 - n^2\pi^2}\right)$$

$$+ i\sum_{n=0}^{N} \frac{\sinh(\zeta)e^{-\delta t}}{\frac{d}{d\zeta}[\zeta\cosh(\zeta)]\frac{d\zeta}{dz}\big|_{z=z_n^-}}\exp\left(-it\sqrt{k^2 - n^2\pi^2}\right)$$

$$+ i\sum_{n=N+1}^{\infty} \frac{\sinh(\zeta)e^{-\delta t}}{\frac{d}{d\zeta}[\zeta\cosh(\zeta)]\frac{d\zeta}{dz}\big|_{z=z_n^+}}\exp\left(-t\sqrt{n^2\pi^2 - k^2}\right),$$

or

$$f(t) = -\frac{e^{-\delta t}\sin(kt)}{2} - 2\sum_{n=0}^{N}\frac{e^{-\delta t}\sin\left(t\sqrt{k^2 - n^2\pi^2}\right)}{\sqrt{k^2 - n^2\pi^2}} + \sum_{n=N+1}^{\infty}\frac{\exp\left(-\delta t - t\sqrt{n^2\pi^2 - k^2}\right)}{\sqrt{n^2\pi^2 - k^2}}.$$

Section 3.1, Problem 18. The transform has singularities where $\sinh\left(d\sqrt{zi}\right) = i\sin\left(-id\sqrt{zi}\right)$ $= 0$, or $\sqrt{z_n i} = n\pi i/d$ and $z_n = n^2\pi^2 i/d^2$ with $n = 1, 2, 3, \ldots$. The singularity $z = 0$ is a removable singularity. From the Taylor expansion of hyperbolic sine, we find that $F(z)$ is single-valued. Therefore,

$$f(t) = \frac{1}{2\pi}\oint_C \frac{\sqrt{zi}}{\sinh(d\sqrt{zi}\,)}e^{itz}\,dz,$$

where C is a semicircle of infinite radius in the upper half-plane if $t > 0$ and a semicircle of infinite radius in the lower half-plane if $t < 0$. Because there are no singularities in the lower half-plane, $f(t) = 0$. In the upper half-plane,

$$\text{Res}\left[\frac{\sqrt{zi}\,e^{itz}}{\sinh\left(d\sqrt{zi}\right)};z_n\right] = \lim_{z \to z_n}\frac{(z - z_n)\sqrt{zi}}{\sinh\left(d\sqrt{zi}\right)}e^{itz} = -\frac{2n^2\pi^2}{d^3 i}e^{-n^2\pi^2 t/d^2}.$$

Multiplying by $2\pi i$ and summing the residues, we obtain the inverse for $t > 0$.

Section 3.1, Problem 19. From the definition of the Fourier transform,

$$f(t) = \frac{1}{2\pi}\int_{-\infty}^{\infty}\frac{\cosh\left(y\sqrt{\omega^2 + 1}\right)e^{it\omega}}{\sqrt{\omega^2 + 1}\,\sinh(p\sqrt{\omega^2 + 1}/2)}\,d\omega = \frac{1}{2\pi}\oint_C \frac{\cosh\left(y\sqrt{z^2 + 1}\right)e^{itz}}{\sqrt{z^2 + 1}\,\sinh(p\sqrt{z^2 + 1}/2)}\,dz,$$

where we have closed the line integral with an infinite semicircle in the upper half-plane if $t > 0$. Simple poles are located at $z = i$ and $p\sqrt{z^2 + 1} = 2n\pi i$, $n = 0, 1, 2, \ldots$, or $z_n = i\sqrt{1 + 4n^2\pi^2/p^2}$. Now, the residue at $z = i$ is

$$\text{Res}\left[\frac{\cosh\left(y\sqrt{z^2 + 1}\right)e^{izt}}{\sqrt{z^2 + 1}\,\sinh\left(p\sqrt{z^2 + 1}/2\right)};i\right] = \lim_{z \to i}\frac{(z - i)\cosh\left(y\sqrt{z^2 + 1}\right)e^{izt}}{(z^2 + 1)p[1 + p^2(z^2 + 1)/24 + \cdots]/2} = \frac{e^{-t}}{ip},$$

and

$$\text{Res}\left[\frac{\cosh\left(y\sqrt{z^2 + 1}\right)e^{izt}}{\sqrt{z^2 + 1}\,\sinh\left(p\sqrt{z^2 + 1}/2\right)};z_n\right] = \lim_{z \to z_n}\frac{(z - z_n)\cosh\left(y\sqrt{z^2 + 1}\right)e^{izt}}{\sqrt{z^2 + 1}\,\sinh\left(p\sqrt{z^2 + 1}/2\right)}$$

$$= \lim_{z \to z_n}\frac{\cosh\left(y\sqrt{z^2 + 1}\right)e^{izt}}{(pz/2)\cosh\left(p\sqrt{z^2 + 1}/2\right)}$$

$$= \frac{2\cos(2n\pi y/p)\exp\left(-t\sqrt{1 + 4n^2\pi^2/p^2}\right)}{ip(-1)^n\sqrt{1 + 4n^2\pi^2/p^2}}.$$

Therefore, the inverse equals i times the sum of the residues, or

$$f(t) = \frac{e^{-t}}{p} + \frac{2}{p}\sum_{n=1}^{\infty}(-1)^n\cos\left(\frac{2n\pi y}{p}\right)\frac{\exp\left(-t\sqrt{1 + 4n^2\pi^2/p^2}\right)}{\sqrt{1 + 4n^2\pi^2/p^2}}.$$

For $t < 0$, we close the contour in the lower half-plane and evaluate the poles at $z = -i$ and $z_n = -i\sqrt{1 + 4n^2\pi^2/p^2}$. We find that

$$\text{Res}\left[\frac{\cosh\left(y\sqrt{z^2+1}\right)e^{izt}}{\sqrt{z^2+1}\,\sinh\left(p\sqrt{z^2+1}/2\right)}; -i\right] = \lim_{z \to -i}\frac{(z+i)\cosh\left(y\sqrt{z^2+1}\right)e^{izt}}{(z^2+1)p[1+p^2(z^2+1)/24+\cdots]/2} = -\frac{e^t}{ip},$$

and

$$\text{Res}\left[\frac{\cosh\left(y\sqrt{z^2+1}\right)e^{izt}}{\sqrt{z^2+1}\,\sinh\left(p\sqrt{z^2+1}/2\right)}; z_n\right] = \lim_{z \to z_n}\frac{(z-z_n)\cosh\left(y\sqrt{z^2+1}\right)e^{izt}}{\sqrt{z^2+1}\,\sinh\left(p\sqrt{z^2+1}/2\right)}$$

$$= \lim_{z \to z_n}\frac{\cosh\left(y\sqrt{z^2+1}\right)e^{izt}}{h(pz/2)\cosh\left(p\sqrt{z^2+1}/2\right)}$$

$$= -\frac{2\cos(2n\pi y/p)\exp\left(t\sqrt{1+4n^2\pi^2/p^2}\right)}{ip(-1)^n\sqrt{1+4n^2\pi^2/p^2}}.$$

Summing the residues and multiplying by $-i$,

$$f(t) = \frac{e^t}{p} + \frac{2}{p}\sum_{n=1}^{\infty}(-1)^n\cos\left(\frac{2n\pi y}{p}\right)\frac{\exp\left(t\sqrt{1+4n^2\pi^2/p^2}\right)}{\sqrt{1+4n^2\pi^2/p^2}}.$$

Section 3.1, Problem 20. From the definition of the Fourier transform,

$$f(t) = \frac{1}{2\pi}\int_{-\infty}^{\infty}\frac{e^{it\omega}}{\cos(\omega L/\{\beta[1+i\gamma\,\text{sgn}(\omega)]\})}\,d\omega = \frac{1}{2\pi}\oint_C\frac{e^{itz}}{\cos(zL/\{\beta[1+i\gamma\,\text{sgn}(z)]\})}\,dz.$$

The integral has singularities at $z_n = \pm(2n-1)\beta\pi/(2L) + (2n-1)i\beta\gamma\pi/(2L)$, where $n = 1, 2, 3, \ldots$. Because the poles are only in the upper half-plane, $f(t) = 0$ for $t < 0$. For $t > 0$,

$$f(t) = i\left\{\sum_{n=1}^{\infty}\text{Res}\left[\frac{e^{itz}}{\cos(zL/\{\beta[1+i\gamma\,\text{sgn}(z)]\})}; z_{n+}\right]\right.$$

$$\left. + \text{Res}\left[\frac{e^{itz}}{\cos(zL/\{\beta[1+i\gamma\,\text{sgn}(z)]\})}; z_{n-}\right]\right\},$$

where

$$\text{Res}\left[\frac{e^{itz}}{\cos(zL/\{\beta[1+i\gamma\,\text{sgn}(z)]\})}; z_n\right] = \lim_{z \to z_n}\frac{(z-z_n)e^{itz}}{\cos(zL/\{\beta[1+i\gamma\,\text{sgn}(z)]\})}$$

$$= \pm\frac{e^{itz_n}}{(-1)^n(L/\{\beta[1+i\gamma\,\text{sgn}(z_n)]\})}$$

$$= \pm\frac{(-1)^n\beta}{L}e^{\pm(2n-1)\beta\pi it/(2L)}$$

$$\times[1+i\gamma\,\text{sgn}(z_n)]\,e^{-(2n-1)\beta\gamma\pi t/(2L)}.$$

Therefore,

$$f(t) = \frac{i\beta}{L} \sum_{n=1}^{\infty} (-1)^n e^{-(2n-1)\beta\gamma\pi t/(2L)}$$

$$\times \left[e^{(2n-1)\beta\pi it/(2L)} + i\gamma e^{\pm(2n-1)\beta\pi it/(2L)} \right.$$

$$\left. - e^{-(2n-1)\beta\pi it/(2L)} + i\gamma e^{\pm(2n-1)\beta\pi it/(2L)} \right]$$

$$= -\frac{2\beta}{L} \sum_{n=1}^{\infty} (-1)^n e^{-(2n-1)\beta\gamma\pi t/(2L)} \left\{ \gamma \cos\left[\frac{(2n-1)\beta\pi t}{2L} \right] + \sin\left[\frac{(2n-1)\beta\pi t}{2L} \right] \right\}.$$

Section 3.1, Problem 21.

$$\int_{-\infty}^{\infty} \frac{\cosh(hx) - 1}{x\sinh(hx)} \cos(ax)\, dx = \text{Re}\left(\int_{-\infty}^{\infty} \frac{\cosh(hx) - 1}{x\sinh(hx)} e^{aix}\, dx \right)$$

$$= \text{Re}\left(\oint_C \frac{\cosh(hz) - 1}{z\sinh(hz)} e^{aiz}\, dz \right),$$

if $a > 0$ and C is a semicircle of infinite radius in the upper half-plane. Within the contour, there is a removal singularity at $z = 0$ and simple poles at $hz_n = n\pi i$ with $n = 1, 2, 3, \ldots$. Therefore,

$$\oint_C \frac{\cosh(hz) - 1}{z\sinh(hz)} e^{aiz}\, dz = 2\pi i \sum_{n=1}^{\infty} \text{Res}\left[\frac{\cosh(hz) - 1}{z\sinh(hz)} e^{aiz}; \frac{n\pi i}{h} \right]$$

with

$$\text{Res}\left[\frac{\cosh(hz) - 1}{z\sinh(hz)} e^{aiz}; \frac{n\pi i}{h} \right] = \lim_{z\to n\pi i/h} \frac{\cosh(hz) - 1}{z} \times \lim_{z\to n\pi i/h} \frac{e^{aiz}}{h\cosh(hz)}$$

$$= \frac{1 - (-1)^n}{n\pi i} e^{-n\pi a/h}.$$

Thus,

$$\int_{-\infty}^{\infty} \frac{\cosh(hx) - 1}{x\sinh(hx)} \cos(ax)\, dx = 4 \sum_{m=1}^{\infty} \frac{\exp[-(2m-1)\pi a/h]}{2m - 1} = \ln\left[\coth\left(\frac{\pi a}{h} \right) \right].$$

Because we obtain the same integral if we replace a by $-a$, we will get the same answer except that we must place an absolute value sign around a in the final answer.

Section 3.1, Problem 22. (a) Poles are located where $(U\omega - i\epsilon)^2 \cosh(\omega h) - g\omega\sinh(\omega h) = 0$. Let $\omega = \pm\omega_0 + \epsilon\omega_1 + \epsilon^2\omega_2 + \cdots$. Taking the positive sign

$$[U\omega_0 + (U\omega_1 - 1)\epsilon + \omega_2\epsilon^2 + \cdots]\cosh[h(\omega_0 + \epsilon\omega_1 + \epsilon^2\omega_2 + \cdots)]$$

$$- g(\omega_0 + \epsilon\omega_1 + \epsilon^2\omega_2 + \cdots)\sinh[h(\omega_0 + \epsilon\omega_1 + \epsilon^2\omega_2 + \cdots)] = 0.$$

Equating powers of ϵ, the $O(1)$ terms give $\tanh(\omega_0 h) = U^2\omega_0/g$ and $\omega_0 = 0$. We obtain the same results when we take $-\omega_0$. Taking the $O(\epsilon)$ terms,

$$[2U^2\omega_0\cosh(\omega_0 h) + U_0^2\omega_0^2 h\sinh(\omega_0 h)$$

$$- g\sinh(\omega_0 h) - g\omega_0 h\cosh(\omega_0 h)]\omega_1 = 2iU\omega_0\cosh(\omega_0 h).$$

Using $\tanh(\omega_0 h) = U^2\omega_0/g$,

$$\omega_1 = -\frac{2iU\omega_0}{g[\omega_0 h\,\mathrm{sech}^2(\omega_0 h) - \tanh(\omega_0 h)]}.$$

We obtain the same result for $-\omega_0$. Now, the $O(\epsilon)$ terms do not give us any information when $\omega_0 = 0$ and we must go the $O(\epsilon^2)$ terms. The $O(\epsilon^2)$ terms are

$$[2\omega_0\omega_2 U + (U\omega_1 - i)^2 + U^2\omega_0^2\omega_1^2/2 - gh\omega_0\omega_2 - gh\omega_1^2]\cosh(\omega_0 h)$$

$$+ [\omega_0^2\omega_2 U^2 + 2\omega_0\omega_1 U(U\omega_1 - i) - gh\omega_0\omega_1^2/2 - gh\omega_2]\sinh(\omega_0 h) = 0.$$

For $\omega_0 = 0$, $\omega_1 = i/\left(U \pm \sqrt{gh}\right)$.

(b)

$$f(t) = \frac{1}{2\pi}\int_{-\infty}^{\infty}\frac{\cosh(\omega h)e^{it\omega}}{(U\omega - i\epsilon)^2\cosh(\omega h) - g\omega\sinh(\omega)}\,d\omega.$$

Now,

$$\mathrm{Res}\left[\frac{\cosh(hz)e^{itz}}{(Uz - i\epsilon)^2\cosh(hz) - gz\sinh(z)};\omega_n\right] = \lim_{z\to\omega_n}\frac{(z - \omega_n)\cosh(hz)e^{itz}}{(Uz - i\epsilon)^2\cosh(hz) - gz\sinh(z)}$$

$$= \frac{\cosh(\omega_n h)e^{it\omega_n}}{D(\omega_n)},$$

where

$$D(\omega) = 2U(U\omega - i\epsilon)\cosh(\omega h) + (U\omega - i\epsilon)^2 h\sinh(\omega h) - g\sinh(\omega h) - gh\omega\cosh(\omega h).$$

For $t < 0$, $\omega_I = i\epsilon/\left(U - \sqrt{gh}\right)$ is the only pole. Therefore,

$$\mathrm{Res}\left[\frac{\cosh(hz)e^{itz}}{(Uz - i\epsilon)^2\cosh(hz) - gz\sinh(z)};\omega_I\right]$$

$$= \frac{\exp\left[-\epsilon t/\left(U - \sqrt{gh}\right)\right]}{2Ui\epsilon\left[U/\left(U - \sqrt{gh}\right) - 1\right] - 2i\epsilon gh/\left(U - \sqrt{gh}\right)}$$

$$= \frac{1}{2i\epsilon\sqrt{gh}} - \frac{t}{2i\sqrt{gh}\left(U - \sqrt{gh}\right)} + O(\epsilon).$$

Multiplying by i and noting that the contour is taken in the negative sense,

$$f(t) = -\frac{1}{2\epsilon\sqrt{gh}} - \frac{t}{2i\sqrt{gh}\left(\sqrt{gh} - U\right)} + O(\epsilon).$$

For $t > 0$,

$$\text{Res}\left[\frac{\cosh(hz)e^{itz}}{(Uz - i\epsilon)^2 \cosh(hz) - gz \sinh(z)}; \omega_{II}\right] = -\frac{1}{2i\epsilon\sqrt{gh}} + \frac{t}{2i\sqrt{gh}\left(U + \sqrt{gh}\right)} + O(\epsilon),$$

$$\text{Res}\left[\frac{\cosh(hz)e^{itz}}{(Uz - i\epsilon)^2 \cosh(hz) - gz \sinh(z)}; \omega_{III}\right] = \frac{\cosh(\omega_0 h)e^{it\omega_0}}{D(\omega_0)},$$

and

$$\text{Res}\left[\frac{\cosh(hz)e^{itz}}{(Uz - i\epsilon)^2 \cosh(hz) - gz \sinh(z)}; \omega_{IV}\right] = -\frac{\cosh(\omega_0 h)e^{it\omega_0}}{D(\omega_0)},$$

where

$$D(\omega_0) = 2U^2\omega_0 \cosh(\omega_0 h) + U^2\omega_0^2 h \sinh(\omega_0 h) - g \sinh(\omega_0 h) - gh\omega_0 \cosh(\omega_0 h).$$

Therefore,

$$i\left\{\text{Res}\left[\frac{\cosh(hz)e^{itz}}{(Uz - i\epsilon)^2 \cosh(hz) - gz \sinh(z)}; \omega_{III}\right]\right.$$

$$\left. + \text{Res}\left[\frac{\cosh(hz)e^{itz}}{(Uz - i\epsilon)^2 \cosh(hz) - gz \sinh(z)}; \omega_{IV}\right]\right\}$$

$$= -\frac{2\cosh(\omega_0 h)\sin(\omega_0 t)}{D(\omega_0)} + O(\epsilon)$$

$$= \frac{2\cosh^2(\omega_0 h)\sin(\omega_0 t)}{g\tanh(\omega_0 h)[gh/U^2 - \cosh^2(\omega_0 h)]} + O(\epsilon).$$

In summary,

$$f(t) = \begin{cases} -\dfrac{1}{2\epsilon\sqrt{gh}} - \dfrac{t}{2\sqrt{gh}\left(\sqrt{gh} - U\right)} + O(\epsilon), & \text{if } t < 0, \\[4mm] -\dfrac{1}{2\epsilon\sqrt{gh}} + \dfrac{t}{2\sqrt{gh}\left(\sqrt{gh} + U\right)} \\[2mm] \quad + \dfrac{2\cosh^2(\omega_0 h)\sin(\omega_0 t)}{g\tanh(\omega_0 h)[gh/U^2 - \cosh^2(\omega_0 h)]} + O(\epsilon), & \text{if } t > 0. \end{cases}$$

Section 3.1, Problem 23. (a) Poles are at $(U\omega - i\epsilon)^2\{\omega^2 + [nN/(U\kappa)]^2\} - N^2\omega^2 = 0$. Let $\omega = \pm\omega_0 + \epsilon\omega_1 + \epsilon^2\omega_2 + \cdots$. Taking the positive sign,

$$[U\omega_0 + (U\omega_1 - i)\epsilon + U\omega_2\epsilon^2 + \cdots]^2\{(\omega_0 + \epsilon\omega_1 + \epsilon^2\omega_2 + \cdots)^2$$

$$+ [nN/(U\kappa)]^2\} - N^2(\omega_0 + \epsilon\omega_1 + \epsilon^2\omega_2 + \cdots)^2 = 0.$$

Equating powers of ϵ, the $O(1)$ terms give $\omega_0 = N\sqrt{1 - (n/\kappa)^2}/U$ and $\omega_0 = 0$. We get the same results when we take $-\omega_0$. Taking the $O(\epsilon)$ terms,

$$2U\omega_0(U\omega_1 - i)\{\omega_0^2 + [nN/(U\kappa)]^2\} + 2\omega_1\omega_0^3U^2 - 2\omega_0\omega_1 N^2 = 0.$$

Simplifying, $\omega_1 = i/\{U[1 - (n/\kappa)^2]\}$. We get the same result for $-\omega_0$. Now, the $O(\epsilon)$ terms do not give us any information when $\omega_0 = 0$ and we must go to the $O(\epsilon^2)$ terms. For $\omega_0 = 0$, the $O(\epsilon^2)$ terms are $(U\omega_1 - i)^2[nN/(U\kappa)]^2 - N^2\omega_1^2 = 0$, or $\omega_1 = [ni/(U\kappa)]/(n/\kappa \pm 1)$.

(b)

$$f(t) = -\frac{1}{2\pi} \int_{-\infty}^{\infty} \frac{iU\omega + \epsilon}{(U\omega - i\epsilon)^2\{\omega^2 + [nN/(U\kappa)]^2\} - N^2\omega^2} \, d\omega.$$

Now, for $n/\kappa < 1$, we have three poles in the upper half plane:

$$\omega_I = \omega_0 + \frac{\epsilon i}{U(1 - n^2/\kappa^2)} + O(\epsilon^2), \qquad \omega_{II} = -\omega_0 + \frac{\epsilon i}{U(1 - n^2/\kappa^2)} + O(\epsilon^2),$$

$$\omega_{III} = \frac{n\epsilon i/(U\kappa)}{n/\kappa + 1} + O(\epsilon^2),$$

and one pole in the lower half-plane:

$$\omega_{IV} = \frac{n\epsilon i/(U\kappa)}{n/\kappa - 1} + O(\epsilon^2),$$

where $\omega_0 = N\sqrt{1 - (n/\kappa)^2}/U$. These poles are all simple and the corresponding residues are

$$\text{Res}\left\{\frac{iUz + \epsilon}{(Uz - i\epsilon)^2\{z^2 + [nN/(U\kappa)]^2\} - N^2z^2}; \omega_I\right\} = -\frac{iU\exp\left[itN\sqrt{1 - (n/\kappa)^2}/U\right]}{2N^2[1 - (n/\kappa)^2]},$$

$$\text{Res}\left\{\frac{iUz + \epsilon}{(Uz - i\epsilon)^2\{z^2 + [nN/(U\kappa)]^2\} - N^2z^2}; \omega_{II}\right\} = -\frac{iU\exp\left[-itN\sqrt{1 - (n/\kappa)^2}/U\right]}{2N^2[1 - (n/\kappa)^2]},$$

$$\text{Res}\left\{\frac{iUz + \epsilon}{(Uz - i\epsilon)^2\{z^2 + [nN/(U\kappa)]^2\} - N^2z^2}; \omega_{III}\right\} = \frac{U}{2N^2i(n/\kappa)(1 + n/\kappa)},$$

and

$$\text{Res}\left\{\frac{iUz + \epsilon}{(Uz - i\epsilon)^2\{z^2 + [nN/(U\kappa)]^2\} - N^2z^2}; \omega_{IV}\right\} = \frac{U}{2N^2i(n/\kappa)(n/\kappa - 1)}.$$

Using the residue theorem and recalling that for $t < 0$, we have a contour in the negative sense,

$$f(t) = \frac{U}{2N^2}\left\{\frac{1}{(n/\kappa)(1 + n/\kappa)} + \frac{2\cos\left[Nt\sqrt{1 - (n/\kappa)^2}/U\right]}{1 - (n/\kappa)^2}\right\}$$

for $t > 0$, and

$$f(t) = \frac{U}{2N^2(n/\kappa)(1 - n/\kappa)}$$

for $t < 0$. On the other hand, for $n/\kappa > 1$, we have three poles in the upper half plane:

$$\omega_I = i\omega_0 - \frac{\epsilon i}{U(n^2/\kappa^2 - 1)} + O(\epsilon^2), \qquad \omega_{II} = \frac{n\epsilon i/(U\kappa)}{n/\kappa + 1} + O(\epsilon^2),$$

$$\omega_{III} = \frac{n\epsilon i/(U\kappa)}{n/\kappa - 1} + O(\epsilon^2),$$

and one pole in the lower half plane:

$$\omega_{IV} = -i\omega_0 - \frac{\epsilon i}{U(n^2/\kappa^2 - 1)} + O(\epsilon^2),$$

where $\omega_0 = N\sqrt{(n/\kappa)^2 - 1}/U$. These poles are all simple and the corresponding residues are

$$\text{Res}\left\{ \frac{iUz + \epsilon}{(Uz - i\epsilon)^2\{z^2 + [nN/(U\kappa)]^2\} - N^2 z^2} ; \omega_I \right\} = -\frac{iU \exp\left[-tN\sqrt{(n/\kappa)^2 - 1}/U\right]}{2N^2[1 - (n/\kappa)^2]},$$

$$\text{Res}\left\{ \frac{iUz + \epsilon}{(Uz - i\epsilon)^2\{z^2 + [nN/(U\kappa)]^2\} - N^2 z^2} ; \omega_{II} \right\} = \frac{U}{2N^2 i(n/\kappa)(1 + n/\kappa)},$$

$$\text{Res}\left\{ \frac{iUz + \epsilon}{(Uz - i\epsilon)^2\{z^2 + [nN/(U\kappa)]^2\} - N^2 z^2} ; \omega_{III} \right\} = \frac{U}{2N^2 i(n/\kappa)(n/\kappa - 1)},$$

and

$$\text{Res}\left\{ \frac{iUz + \epsilon}{(Uz - i\epsilon)^2\{z^2 + [nN/(U\kappa)]^2\} - N^2 z^2} ; \omega_{IV} \right\} = -\frac{iU \exp\left[tN\sqrt{(n/\kappa)^2 - 1}/U\right]}{2N^2[1 - (n/\kappa)^2]}.$$

Using the residue theorem and recalling that we have a contour in the negative sense if $t < 0$,

$$f(t) = \frac{U}{2N^2}\left\{ \frac{2 - \exp\left[-Nt\sqrt{(n/\kappa)^2 - 1}/U\right]}{(n/\kappa)^2 - 1} \right\}$$

for $t < 0$, and

$$f(t) = \frac{U}{2N^2} \frac{\exp\left[Nt\sqrt{(n/\kappa)^2 - 1}/U\right]}{(n/\kappa)^2 - 1}$$

for $t > 0$.

Section 4.1

Section 4.1, Problem 1. From Bromwich's integral,

$$f(t) = \frac{1}{2\pi i} \oint_C \frac{e^{zt}}{\sqrt{z}\left(\sqrt{z} - b\right)} \, dz,$$

where our closed contour is the same as Figure 4.1.3 with $0 < b^2 < c$. Because $|\arg(z)| < \pi$ and $\left|\arg\left(\sqrt{z}\right)\right| < \pi/2$, there is a simple pole at $z = b^2$. Therefore,

$$f(t) = \text{Res}\left[\frac{e^{zt}}{\sqrt{z}\left(\sqrt{z} - b\right)}; b^2\right] + \frac{1}{2\pi i}\int_{DC}\frac{e^{zt}}{\sqrt{z}\left(\sqrt{z} - b\right)} \, dz + \frac{1}{2\pi i}\int_{GF}\frac{e^{zt}}{\sqrt{z}\left(\sqrt{z} - b\right)} \, dz.$$

The value of residue is

$$\text{Res}\left[\frac{e^{zt}}{\sqrt{z}\left(\sqrt{z} - b\right)}; b^2\right] = \lim_{z \to b^2}\frac{\sqrt{z} + b}{\sqrt{z}} e^{tz} = 2e^{b^2 t}.$$

Along the top of the branch cut, $z = \eta^2 e^{\pi i}$ and

$$\int_{DC} \frac{e^{zt}}{\sqrt{z}\left(\sqrt{z}-b\right)}\, dz = \frac{2}{i}\int_0^\infty \frac{(b-i\eta)e^{-t\eta^2}}{b^2+\eta^2}\, d\eta,$$

while along the bottom of the branch cut, $z = \eta^2 e^{-\pi i}$ and

$$\int_{GF} \frac{e^{zt}}{\sqrt{z}\left(\sqrt{z}-b\right)}\, dz = \frac{2}{i}\int_0^\infty \frac{(b+i\eta)e^{-t\eta^2}}{b^2+\eta^2}\, d\eta.$$

Thus, the total branch cut contribution is

$$\int_{DC\cup GF} \frac{e^{zt}}{\sqrt{z}\left(\sqrt{z}-b\right)}\, dz = \frac{4b}{i}\int_0^\infty \frac{e^{-t\eta^2}}{b^2+\eta^2}\, d\eta.$$

The inverse equals the sum of the residue plus the branch cut integral after we divide it by $2\pi i$.

Section 4.1, Problem 2. From Bromwich's integral,

$$f(t) = \frac{1}{2\pi i}\oint_C \frac{e^{zt}}{1+a^p z^p}\, dz,$$

where our closed contour is the same as Figure 4.1.3. Along the top of the branch cut, $z = \eta e^{\pi i}$ and

$$\int_{DC} \frac{e^{zt}}{1+a^p z^p}\, dz = -\int_0^\infty \frac{e^{-t\eta}}{1+a^p\eta^p e^{p\pi i}}\, d\eta,$$

while along the bottom of the branch cut, $z = \eta e^{-\pi i}$ and

$$\int_{GF} \frac{e^{zt}}{1+a^p z^p}\, dz = -\int_\infty^0 \frac{e^{-t\eta}}{1+a^p\eta^p e^{-p\pi i}}\, d\eta.$$

Thus, the total branch cut contribution is

$$\int_{DC\cup GF} \frac{e^{zt}}{1+a^p z^p}\, dz = 2ia^p\sin(p\pi)\int_0^\infty \frac{\eta^p e^{-t\eta}}{1+a^{2p}\eta^{2p}+2a^p\eta^p\cos(p\pi)}\, d\eta.$$

The inverse equals the branch cut integral after we divide it by $2\pi i$. Note that integrand has simple poles located at $s_{1,2} = [\cos(\pi/p) \pm i\sin(\pi/p)]/a$. However, they are located on different Riemann surfaces.

Section 4.1, Problem 3. From Bromwich's integral,

$$f(t) = \frac{1}{2\pi i}\int_{c-\infty i}^{c+\infty i} \frac{a}{z(z^{1/n}+a)}e^{zt}\, dz.$$

If we take the branch cut along the negative real axis, we close Bromwich's integral as shown in Figure 4.1.3. The integral has poles at $z = a^n e^{2n(2m-1)\pi i}$, where m is an integer. Because we restrict our phase between $-\pi$ to π, these poles lie on another Riemann surface and we may exclude them from further consideration. Integration around the branch point $z = 0$ yields

$$\int_{FED} \frac{a}{z(z^{1/n} + a)} e^{zt}\, dz = -2\pi i.$$

Along the top of the branch cut, $z = a^n \eta^n e^{\pi i}$ so that

$$\int_{DC} \frac{a}{z(z^{1/n} + a)} e^{zt} dz = n \int_{\infty}^{0} \frac{\exp(-a^n \eta^n t)}{1 + \eta \cos(\pi/n) + i\eta \sin(\pi/n)} \frac{d\eta}{\eta},$$

while, along the bottom of the branch cut, $z = a^n \eta^n e^{-\pi i}$ so that

$$\int_{GF} \frac{a}{z(z^{1/n} + a)} e^{zt}\, dz = n \int_{0}^{\infty} \frac{\exp(-a^n \eta^n t)}{1 + \eta \cos(\pi/n) - i\eta \sin(\pi/n)} \frac{d\eta}{\eta}.$$

Combining the two branch cuts integrals,

$$\int_{GF \cup FED \cup DC} \frac{a}{z(z^{1/n} + a)} e^{zt}\, dz = 2in \sin\left(\frac{\pi}{n}\right) \int_{0}^{\infty} \frac{\exp(-a^n \eta^n t)}{1 + \eta^2 + 2\eta \cos(\pi/n)} \frac{d\eta}{\eta}.$$

The inverse equals the branch cut integral after dividing the sum by $2\pi i$.

Section 4.1, Problem 4. From Bromwich's integral,

$$f(t) = \frac{1}{2\pi i} \int_{c-\infty i}^{c+\infty i} \frac{e^{tz}}{a + z + b\sqrt{z}} e^{zt}\, dz.$$

If we take the branch cut along the negative real axis, we close Bromwich's integral as shown in Figure 4.1.3. Because $|\arg(z)| < \pi$ and $|\arg(\sqrt{s})| < \pi/2$, there are no poles if $a, b > 0$ that lie on the proper Riemann surface. Integration around the branch point $z = 0$ yields

$$\int_{FED} \frac{e^{tz}}{a + z + b\sqrt{z}} e^{zt}\, dz = 0.$$

Along the top of the branch cut $z = \eta e^{\pi i}$,

$$\int_{DC} \frac{e^{tz}}{a + z + b\sqrt{z}} e^{zt}\, dz = -\int_{0}^{\infty} \frac{(a - \eta - ib\sqrt{\eta})e^{-t\eta}}{(a - \eta)^2 + b^2 \eta}\, d\eta,$$

whereas along the bottom of the branch cut, $z = \eta e^{-\pi i}$ and

$$\int_{GF} \frac{e^{tz}}{a + z + b\sqrt{z}} e^{zt}\, dz = -\int_{\infty}^{0} \frac{(a - \eta + ib\sqrt{\eta})e^{-t\eta}}{(a - \eta)^2 + b^2 \eta}\, d\eta.$$

Combining the two branch cuts integrals,

$$\int_{GF \cup FED \cup DC} \frac{e^{tz}}{a + z + b\sqrt{z}} e^{zt}\, dz = 2bi \int_{0}^{\infty} \frac{\sqrt{\eta}\, e^{-t\eta}}{(a - \eta)^2 + b^2 \eta}\, d\eta.$$

The inverse equals the branch cut and branch point integrals after dividing the sum by $2\pi i$.

Section 4.1, Problem 5. Closing Bromwich's integral as shown in Figure 4.1.3,

$$f(t) = \frac{1}{2\pi i} \oint_C \frac{\sqrt{z}\, e^{tz}}{z\sqrt{z} + a^3}\, dz.$$

Inside the contour, there are simple poles at $z_{1,2} = a^2 e^{\pm 2\pi i/3}$ with $\sqrt{z_{1,2}} = a e^{\pm \pi i/3}$. Therefore,

$$\mathrm{Res}\left(\frac{\sqrt{z}\, e^{tz}}{z\sqrt{z} + a^3}; s_1\right) = \lim_{z \to s_1} \frac{(z - s_1)\sqrt{z}\, e^{tz}}{z\sqrt{z} + a^3} = 2e^{-a^2 t/2} \exp\!\left(ia^2 t\sqrt{3}/2\right) /3$$

and

$$\mathrm{Res}\left(\frac{\sqrt{z}\, e^{tz}}{z\sqrt{z} + a^3}; s_2\right) = \lim_{z \to s_2} \frac{(z - s_2)\sqrt{z}\, e^{tz}}{z\sqrt{z} + a^3} = 2e^{-a^2 t/2} \exp\!\left(-ia^2 t\sqrt{3}/2\right) /3.$$

Therefore,

$$\sum_{j=1}^{2} \mathrm{Res}\left(\frac{\sqrt{z}\, e^{tz}}{z\sqrt{z} + a^3}; s_j\right) = 4e^{-a^2 t/2} \cos\!\left(a^2 t\sqrt{3}/2\right) /3.$$

Along the top of the branch cut, $z = x^2 e^{\pi i}$ and $\sqrt{z} = x e^{\pi i/2}$ so that

$$\int_{CD} \frac{\sqrt{z}\, e^{tz}}{z\sqrt{z} + a^3}\, dz = -\int_{\infty}^{0} \frac{xi\, e^{-x^2 t}}{a^3 - ix^3} (2x\, dx),$$

while along the bottom of the branch cut, $z = x^2 e^{-\pi i}$ and $\sqrt{z} = x e^{-\pi i/2}$ so that

$$\int_{GF} \frac{\sqrt{z}\, e^{tz}}{z\sqrt{z} + a^3}\, dz = \int_{0}^{\infty} \frac{xi\, e^{-x^2 t}}{a^3 + ix^3} (2x\, dx).$$

Consequently, the contribution from the branch cut is

$$\int_{CD \cup GF} \frac{\sqrt{z}\, e^{tz}}{z\sqrt{z} + a^3}\, dz = 4ia^3 \int_{0}^{\infty} \frac{x^2}{x^6 + a^6} e^{-tx^2}\, dx.$$

The inverse equals the sum of the residues minus the branch cut integral after we divide the integral by $2\pi i$.

Section 4.1, Problem 6. From Bromwich's integral,

$$f(t) = \frac{1}{2\pi i} \oint_C \frac{z\sqrt{1 + z^2/k^2}\, e^{tz}}{(z^2 + 2)^2 - 4\sqrt{(1 + z^2/k^2)(z^2 + 1)}}\, dz.$$

We close Bromwich's integral in a manner similar to Figure 4.1.18. From the expansion about $z = 0$, the residue at $z = 0$ equals $k^2/[2(k^2 - 1)]$. At $z = i/\gamma$,

$$\mathrm{Res}\left[\frac{z\sqrt{1 + z^2/k^2}\, e^{tz}}{(z^2 + 2)^2 - 4\sqrt{(1 + z^2/k^2)(z^2 + 1)}}; i/\gamma\right] = \lim_{z \to i/\gamma} \frac{(z - i/\gamma)z\sqrt{1 + z^2/k^2}}{g(z)} e^{tz}$$

$$= \frac{(i/\gamma)\sqrt{1 - 1/(k\gamma)^2}}{g'(i/\gamma)} e^{it/\gamma}.$$

Similarly, at $z = -i/\gamma$,

$$\text{Res}\left[\frac{z\sqrt{1+z^2/k^2}\,e^{tz}}{(z^2+2)^2 - 4\sqrt{(1+z^2/k^2)(z^2+1)}}; -i/\gamma\right] = \lim_{z \to -i/\gamma} \frac{(z+i/\gamma)z\sqrt{1+z^2/k^2}}{g(z)} e^{tz}$$

$$= \frac{(-i/\gamma)\sqrt{1-1/(k\gamma)^2}}{g'(-i/\gamma)} e^{-it/\gamma}.$$

Because $g'(i/\gamma) = -g'(-i/\gamma)$, the sum of the residues equals

$$\frac{k^2}{2(k^2-1)} + \frac{2\sqrt{1-1/(k\gamma)^2}}{[-i\gamma g'(-i/\gamma)]} \cos(t/\gamma),$$

which is purely real.

The integrations around the branch points equal zero. On both sides of the branch cut, from $z = i$ to ki, we have

$$z - i = (\eta - 1)e^{\pi i/2} \quad \text{or} \quad (\eta-1)e^{-3\pi i/2}, \quad z+i = (\eta+1)e^{\pi i/2},$$

$$z - ki = (k-\eta)e^{-\pi i/2}, \qquad z + ki = (k+\eta)e^{\pi i/2},$$

so that

$$\int_{C_1} \frac{z\sqrt{1+z^2/k^2}\,e^{tz}}{(z^2+2)^2 - 4\sqrt{(1+z^2/k^2)(z^2+1)}}\,dz$$

$$= \int_k^1 \frac{i\eta\sqrt{k^2-\eta^2}}{k(2-\eta^2)^2 - 4i\sqrt{(k^2-\eta^2)(\eta^2-1)}} e^{it\eta}\,i\,d\eta$$

$$+ \int_1^k \frac{i\eta\sqrt{k^2-\eta^2}}{k(2-\eta^2)^2 + 4i\sqrt{(k^2-\eta^2)(\eta^2-1)}} e^{it\eta}\,i\,d\eta.$$

On both sides of the branch, from $z = -i$ to $-ki$, we have

$$z - i = (\eta-1)e^{-\pi i/2}, \quad z+i = (\eta-1)e^{-\pi i/2} \quad \text{or} \quad (\eta-1)e^{3\pi i/2},$$

$$z - ki = (k+\eta)e^{-\pi i/2}, \qquad z+ki = (k-\eta)e^{\pi i/2},$$

and

$$\int_{C_2} \frac{z\sqrt{1+z^2/k^2}\,e^{tz}}{(z^2+2)^2 - 4\sqrt{(1+z^2/k^2)(z^2+1)}}\,dz$$

$$= \int_k^1 \frac{-i\eta\sqrt{k^2-\eta^2}}{k(2-\eta^2)^2 - 4i\sqrt{(k^2-\eta^2)(\eta^2-1)}} e^{-it\eta}\,(-i\,d\eta)$$

$$+ \int_1^k \frac{-i\eta\sqrt{k^2-\eta^2}}{k(2-\eta^2)^2 + 4i\sqrt{(k^2-\eta^2)(\eta^2-1)}} e^{-it\eta}\,(-i\,d\eta).$$

Combining the integrals and simplifying,

$$\int_{C_1 \cup C_2} \frac{z\sqrt{1+z^2/k^2}\, e^{tz}}{(z^2+2)^2 - 4\sqrt{(1+z^2/k^2)(z^2+1)}}\, dz$$

$$= 16i \int_1^k \frac{\eta(1-\eta^2/k^2)\sqrt{\eta^2-1}}{(2-\eta^2)^4 + 16(1-\eta^2/k^2)(\eta^2-1)} \cos(\eta t)\, d\eta.$$

On the other hand, both sides of the branch from $z = ki$ to ∞i can be parameterized as follows:

$$z - i = (\eta-1)e^{\pi i/2} \quad \text{or} \quad (\eta-1)e^{-3\pi i/2}, \quad z + i = (\eta+1)e^{\pi i/2},$$

$$z - ki = (\eta-k)e^{\pi i/2} \quad \text{or} \quad (\eta-k)e^{-3\pi i/2}, \quad z + ki = (\eta+k)e^{\pi i/2},$$

and

$$\int_{C_3} \frac{z\sqrt{1+z^2/k^2}\, e^{tz}}{(z^2+2)^2 - 4\sqrt{(1+z^2/k^2)(z^2+1)}}\, dz$$

$$= \int_\infty^k \frac{i\eta\left(i\sqrt{\eta^2-k^2}\right)}{k(2-\eta^2)^2 - 4\left(i\sqrt{\eta^2-k^2}\right)\left(i\sqrt{\eta^2-1}\right)} e^{it\eta}\, i\, d\eta$$

$$+ \int_k^\infty \frac{i\eta\left(-i\sqrt{k^2-\eta^2}\right)}{k(2-\eta^2)^2 - 4\left(-i\sqrt{\eta^2-k^2}\right)\left(-i\sqrt{\eta^2-1}\right)} e^{it\eta}\, i\, d\eta.$$

Finally, from $z = -ki$ to $-\infty i$,

$$z - i = (\eta-1)e^{-\pi i/2}, \quad z + i = (\eta-1)e^{-\pi i/2} \quad \text{or} \quad (\eta-1)e^{3\pi i/2},$$

$$z - ki = (k+\eta)e^{-\pi i/2}, \quad z + ki = (k-\eta)e^{-\pi i/2} \quad \text{or} \quad (k-\eta)e^{3\pi i/2},$$

and

$$\int_{C_4} \frac{z\sqrt{1+z^2/k^2}\, e^{tz}}{(z^2+2)^2 - 4\sqrt{(1+z^2/k^2)(z^2+1)}}\, dz$$

$$= \int_\infty^k \frac{(-i\eta)\left(i\sqrt{\eta^2-k^2}\right)}{k(2-\eta^2)^2 - 4\left(i\sqrt{\eta^2-k^2}\right)\left(i\sqrt{\eta^2-1}\right)} e^{-it\eta}\, (-i\, d\eta)$$

$$+ \int_k^\infty \frac{-i\eta\left(-i\sqrt{k^2-\eta^2}\right)}{k(2-\eta^2)^2 - 4\left(-i\sqrt{\eta^2-k^2}\right)\left(-i\sqrt{\eta^2-1}\right)} e^{-it\eta}(-i\, d\eta).$$

Combining the integrals and simplifying,

$$\int_{C_3 \cup C_4} \frac{z\sqrt{1+z^2/k^2}\, e^{tz}}{(z^2+2)^2 - 4\sqrt{(1+z^2/k^2)(z^2+1)}}\, dz$$

$$= 4i \int_k^\infty \frac{\eta\sqrt{\eta^2/k^2-1}}{(2-\eta^2)^2 + 4\sqrt{(\eta^2/k^2-1)(\eta^2-1)}} \cos(\eta t)\, d\eta.$$

The inverse equals the sum of the residues minus the branch cut integrals after we divide it by $2\pi i$.

Section 4.1, Problem 7. The inverse consists of two parts. There is a simple pole at $z = 0$ and branch cut integrals along the negative real axis. The value of the residue at $z = 0$ is

$$\text{Res}\left[\sqrt{\frac{v}{c}\left(z + \frac{\beta}{2}\right) + \sqrt{z(z + \beta)}} + \frac{v^2\beta^2}{4c^2}\frac{e^{tz}}{z}; 0\right] = \sqrt{\frac{v\beta}{c}}.$$

Let us rewrite the transform as

$$F(s) = \frac{1}{s}\sqrt{\frac{v}{c}\left(s + \frac{\beta}{2}\right) + \sqrt{(s - \alpha_1)(s - \alpha_2)}}.$$

In the interval from $z = \alpha_2$ to $z = \alpha_1$,

$$z - \alpha_1 = \eta e^{\pm\pi i}, \qquad z - \alpha_2 = \left(\beta\sqrt{1 - v^2/c^2} - \eta\right)e^{0i},$$

and

$$\sqrt{(s - \alpha_1)(s - \alpha_2)} = \sqrt{\eta\left(\beta\sqrt{1 - v^2/c^2} - \eta\right)}e^{\pm\pi i/2}.$$

Therefore,

$$\sqrt{\frac{v}{c}\left(s + \frac{\beta}{2}\right) + \sqrt{(s - \alpha_1)(s - \alpha_2)}}$$

$$= \sqrt{\frac{v}{c}\left(\frac{\beta}{2}\sqrt{1 - \frac{v^2}{c^2}} - \eta\right) \pm i\sqrt{\eta\left(\beta\sqrt{1 - v^2/c^2} - \eta\right)}}$$

$$= \sqrt{\frac{r + x}{2} \pm i\sqrt{\frac{r - x}{2}}},$$

where $0 < \eta < \beta\sqrt{1 - v^2/c^2}$. Consequently, the branch cut integral's contribution for this segment is

$$I_1 = 2i \int_0^{\beta\sqrt{1 - v^2/c^2}} \frac{e^{(\alpha_1 - \eta)t}}{\alpha_1 - \eta}\sqrt{\frac{r - x}{2}}\, d\eta.$$

In the interval from $z = -\infty$ to $z = \alpha_2$, $z - \alpha_1 = \eta e^{\pm\pi i}$,

$$z - \alpha_2 = \left(\eta - \beta\sqrt{1 - v^2/c^2}\right)e^{\pm\pi i},$$

and

$$\sqrt{(s - \alpha_1)(s - \alpha_2)} = \sqrt{\eta\left(\eta - \beta\sqrt{1 - v^2/c^2}\right)}e^{\pm\pi i}.$$

Therefore,

$$
\sqrt{\frac{v}{c}\left(s+\frac{\beta}{2}\right)+\sqrt{(s-\alpha_1)(s-\alpha_2)}}
$$

$$
=\sqrt{\frac{v}{c}\left(\eta-\frac{\beta}{2}\sqrt{1-\frac{v^2}{c^2}}\right)+\sqrt{\eta\left(\eta-\beta\sqrt{1-v^2/c^2}\right)}}\,e^{\pm\pi i/2},
$$

where $\beta\sqrt{1-v^2/c^2}<\eta<\infty$. Thus,

$$
I_2=2i\int_{\beta\sqrt{1-v^2/c^2}}^{\infty}\frac{e^{(\alpha_1-\eta)t}}{\alpha_1-\eta}\,d\eta\sqrt{\frac{v}{c}\left(\eta-\frac{\beta}{2}\sqrt{1-\frac{v^2}{c^2}}\right)+\sqrt{\eta\left(\eta-\beta\sqrt{1-v^2/c^2}\right)}}.
$$

The inverse equals the residue minus the sum of the branch cut integrals after we divide them by $2\pi i$.

Section 4.1, Problem 8. We may close Bromwich's integral with a semicircle in the left half of the z-plane with a branch cut from $z=-c$ to $z=0$. Because the arcs at infinity vanish,

$$
f(t)=\operatorname{Res}\left\{\frac{e^{tz}}{z[a+b\log(1+c/z)]};z_0\right\}-\frac{1}{2\pi i}\int_{\substack{branch\\cut}}\frac{e^{tz}}{z[a+b\log(1+c/z)]}\,dz,
$$

since a simple pole is located at $z_0=-c/(1-e^{-a/b})$. The residue there is

$$
\operatorname{Res}\left\{\frac{e^{tz}}{z[a+b\log(1+c/z)]};z_0\right\}=\lim_{z\to z_0}\frac{(z-z_0)e^{zt}}{z[a+b\log(1+c/z)]}
$$

$$
=\frac{\exp[-ct/(1-e^{-a/b})]}{b(e^{a/b}-1)}.
$$

There is no contribution from the integrations around the branch points. Along the top of the branch cut, $z=c\eta e^{\pi i}$ and $z+c=c(1-\eta)e^{0i}$ so that

$$
\int_{\substack{top\\branch\ cut}}\frac{e^{tz}}{z[a+b\log(1+c/z)]}\,dz=\int_1^0\frac{e^{-ct\eta}}{(-c\eta)[a+b\ln(1/\eta-1)-b\pi i]}(-c\,d\eta),
$$

while along the bottom of the branch cut, $z=c\eta e^{\pi i}$ and $z+c=c(1-\eta)e^{2\pi i}$ so that

$$
\int_{\substack{bottom\\branch\ cut}}\frac{e^{tz}}{z[a+b\log(1+c/z)]}\,dz=\int_0^1\frac{e^{-ct\eta}}{(-c\eta)[a+b\ln(1/\eta-1)+b\pi i]}(-c\,d\eta).
$$

Therefore, the branch cut equals

$$
\int_{\substack{branch\\cut}}\frac{e^{tz}}{z[a+b\log(1+c/z)]}\,dz=-2b\pi i\int_0^1\frac{e^{-ct\eta}}{\eta\{[a+b\ln(1/\eta-1)]^2+b^2\pi^2\}}\,d\eta.
$$

Substitution of the residue and branch cut integral into our first equation yields the inverse.

Section 4.1, Problem 9. If we use the contour shown in Figure 4.1.19, the inverse equals to the sum of the residues corresponding to the singularities inside the contour plus the branch cut integral. Here we must consider three separate cases: (a) $0 < a < 1$, (b) $a = 1$ and (c) $1 < a < \infty$. In all of these cases, we restrict ourselves to the principal value of the logarithm so that $z^2 + a^2 = 1$ at the singularity.

In case (a), the singularities are located at $z_\pm = \pm\sqrt{1 - a^2}$ and the residue equals

$$\operatorname{Res}\left[\frac{e^{tz}}{\log(z^2 + a^2)}; z_\pm\right] = \lim_{z \to z_\pm} \frac{(z - z_\pm)e^{tz}}{\log(z^2 + a^2)} = \frac{e^{\pm t\sqrt{1-a^2}}}{\pm 2\sqrt{1-a^2}}.$$

The sum of the residues for z_- and z_+ equals $\sinh\left(t\sqrt{1-a^2}\right)/\sqrt{1-a^2}$.

In case (b), we note that $\ln(1 + z^2) \approx z^2$ as $z \to 0$. Therefore, the residue equals t in this case.

Finally, in case (c), the singularities are located at $z_\pm = \pm i\sqrt{a^2 - 1}$ and the residue equals

$$\operatorname{Res}\left[\frac{e^{tz}}{\log(z^2 + a^2)}; z_\pm\right] = \lim_{z \to z_\pm} \frac{(z - z_\pm)e^{tz}}{\log(z^2 + a^2)} = \frac{e^{\pm it\sqrt{a^2-1}}}{\pm 2i\sqrt{a^2 - 1}}.$$

The sum of the residues for z_- and z_+ equals $\sin\left(t\sqrt{a^2 - 1}\right)/\sqrt{a^2 - 1}$.

Turning to the branch cut integrals, our choice of branch cuts yields

$$z + ai = (\eta - a)e^{\theta i}, \qquad -\pi/2 < \theta < 3\pi/2,$$

and

$$z - ai = (\eta + a)e^{\theta i}, \qquad -3\pi/2 < \theta < \pi/2.$$

From these definitions, it follows that

$$\int_{C_1} \frac{e^{tz}}{\log(z^2 + a^2)} \, dz = \int_\infty^a \frac{e^{-it\eta}}{\ln(\eta^2 - a^2) - \pi i}(-i \, d\eta),$$

$$\int_{C_2} \frac{e^{tz}}{\log(z^2 + a^2)} \, dz = \int_a^\infty \frac{e^{-it\eta}}{\ln(\eta^2 - a^2) + \pi i}(-i \, d\eta),$$

$$\int_{C_3} \frac{e^{tz}}{\log(z^2 + a^2)} \, dz = \int_\infty^a \frac{e^{it\eta}}{\ln(\eta^2 - a^2) - \pi i}(i \, d\eta),$$

and

$$\int_{C_4} \frac{e^{tz}}{\log(z^2 + a^2)} \, dz = \int_a^\infty \frac{e^{it\eta}}{\ln(\eta^2 - a^2) + \pi i}(i \, d\eta).$$

Therefore,

$$\int_{C_1 \cup C_2} \frac{e^{tz}}{\log(z^2 + a^2)} \, dz = -2\pi \int_a^\infty \frac{e^{-it\eta}}{\ln^2(\eta^2 - a^2) + \pi^2} \, d\eta,$$

and

$$\int_{C_3 \cup C_4} \frac{e^{tz}}{\log(z^2 + a^2)} \, dz = 2\pi \int_a^\infty \frac{e^{it\eta}}{\ln^2(\eta^2 - a^2) + \pi^2} \, d\eta.$$

Finally,

$$\int_{C_1 \cup C_2 \cup C_3 \cup C_4} \frac{e^{tz}}{\log(z^2 + a^2)} \, dz = 4\pi i \int_a^\infty \frac{\sin(t\eta)}{\ln^2(\eta^2 - a^2) + \pi^2} \, d\eta.$$

The inverse equals the sum of the residues plus the branch cut integral after dividing it by $2\pi i$.

Section 4.1, Problem 10. If we use the contour shown in Figure 4.1.19, the inverse equals to the sum of the residues corresponding to the singularities inside the contour plus the branch cut integrals. If we restrict ourselves to the principal value of the logarithm, there are only two singularities where $z^2 + a^2 = 1$ or $z_\pm = \pm\sqrt{1 - a^2}$. Expanding $\log(z^2 + a^2)$ as a Taylor expansion about $z = z_\pm$,

$$\log(z^2 + a^2) = 2z_\pm(z - z_\pm) - (1 - 2a^2)(z - z_\pm)^2 + \cdots$$

and

$$\log^2(z^2 + a^2) = 4(1 - a^2)(z - z_\pm)^2 - 4(1 - 2a^2)z_\pm(z - z_\pm)^3 + \cdots,$$

since $z_\pm^2 = 1 - a^2$. Therefore, because

$$\frac{e^{tz}}{\log^2(z^2 + a^2)} = \frac{e^{tz_\pm}}{4(1 - a^2)(z - z_\pm)^2} + \frac{t e^{tz_\pm}}{4(1 - a^2)(z - z_\pm)} + \frac{(1 - 2a^2)e^{tz_\pm}}{4(1 - a^2)z_\pm(z - z_\pm)} + \cdots,$$

$$\mathrm{Res}\left[\frac{e^{tz}}{\log^2(z^2 + a^2)}; z_\pm\right] = \frac{t e^{tz_\pm}}{4(1 - a^2)} + \frac{(1 - 2a^2)e^{tz_\pm}}{4(1 - a^2)z_\pm}$$

and the sum of the residues for z_- and z_+ equals

$$\frac{t \cosh\left(t\sqrt{1 - a^2}\right)}{2(1 - a^2)} + \frac{(1 - 2a^2)\sinh\left(t\sqrt{1 - a^2}\right)}{2(1 - a^2)^{3/2}}.$$

Turning to the branch cut integrals, our choice of branch cuts yields

$$z + ai = (\eta - a)e^{\theta i}, \qquad -\pi/2 < \theta < 3\pi/2,$$

and

$$z - ai = (\eta + a)e^{\theta i}, \qquad -3\pi/2 < \theta < \pi/2.$$

From these definitions, it follows

$$\int_{C_1} \frac{e^{tz}}{\log^2(z^2 + a^2)}\, dz = \int_\infty^a \frac{e^{-it\eta}}{[\ln(\eta^2 - a^2) - \pi i]^2}(-i\, d\eta),$$

$$\int_{C_2} \frac{e^{tz}}{\log^2(z^2 + a^2)}\, dz = \int_a^\infty \frac{e^{-it\eta}}{[\ln(\eta^2 - a^2) + \pi i]^2}(-i\, d\eta),$$

$$\int_{C_3} \frac{e^{tz}}{\log^2(z^2 + a^2)}\, dz = \int_\infty^a \frac{e^{it\eta}}{[\ln(\eta^2 - a^2) - \pi i]^2}(i\, d\eta),$$

and

$$\int_{C_4} \frac{e^{tz}}{\log^2(z^2 + a^2)}\, dz = \int_a^\infty \frac{e^{it\eta}}{[\ln(\eta^2 - a^2) + \pi i]^2}(i\, d\eta).$$

Therefore,

$$\int_{C_1 \cup C_2} \frac{e^{tz}}{\log^2(z^2 + a^2)}\, dz = -4\pi \int_a^\infty \frac{\ln(\eta^2 - a^2)e^{-it\eta}}{\left[\ln^2(\eta^2 - a^2) + \pi^2\right]^2}\, d\eta$$

and

$$\int_{C_3 \cup C_4} \frac{e^{tz}}{\log^2(z^2 + a^2)} \, dz = 4\pi \int_a^\infty \frac{\ln(\eta^2 - a^2)e^{it\eta}}{\left[\ln^2(\eta^2 - a^2) + \pi^2\right]^2} \, d\eta.$$

Finally,

$$\int_{C_1 \cup C_2 \cup C_3 \cup C_4} \frac{e^{tz}}{\log^2(z^2 + a^2)} \, dz = 8\pi i \int_a^\infty \frac{\ln(\eta^2 - a^2)\sin(t\eta)}{\left[\ln^2(\eta^2 - a^2) + \pi^2\right]^2} \, d\eta.$$

The inverse equals the sum of the residue plus the branch cut integral after dividing it by $2\pi i$.

Section 4.1, Problem 11. We use a closed contour similar to Figure 4.1.12 except that the branch cut runs from $z = -ai$ to $z = ai$. There is only one singularity located at $a - \arctan(a/z) = 0$, or $z_0 = a/\tan(a)$. We choose the principal branch of the inverse tangent; therefore, $0 < z_0 < \pi/2$. The residue there is

$$\operatorname{Res}\left[\frac{e^{tz}}{a - \arctan(a/z)}; z_0\right] = \lim_{z \to z_0} \frac{(z - z_0)e^{tz}}{a - \arctan(a/z)} = \frac{a \exp[at/\tan(a)]}{\sin^2(a)}.$$

For the branch cut, we have

$$z - ai = a(1 - \eta)e^{\theta i}, \qquad -3\pi/2 < \theta < \pi/2,$$

and

$$z + ai = a(1 + \eta)e^{\theta i}, \qquad -3\pi/2 < \theta < \pi/2.$$

Along the right side of the branch cut,

$$\int_{RHS} \frac{e^{tz}}{a - \arctan(a/z)} \, dz = \int_{-1}^1 \frac{e^{iat\eta}}{a - \log[(1 + \eta)e^{\pi i}/(1 - \eta)]/(2i)} \, d\eta$$

$$= i \int_{-1}^1 \frac{e^{iat\eta}}{(a - \pi/2) + i\ln[(1 + \eta)/(1 - \eta)]/2} \, d\eta,$$

while, on the other side,

$$\int_{LHS} \frac{e^{tz}}{a - \arctan(a/z)} \, dz = \int_1^{-1} \frac{e^{iat\eta}}{a - \log[(1 + \eta)e^{-\pi i}/(1 - \eta)]/(2i)} \, d\eta$$

$$= -i \int_{-1}^1 \frac{e^{iat\eta}}{(a + \pi/2) + i\ln[(1 + \eta)/(1 - \eta)]/2} \, d\eta.$$

The inverse equals the sum of the residue plus the branch cut integral after dividing it by $2\pi i$.

Section 4.1, Problem 12. We close Bromwich's integral with an infinite semicircle in the left side of the z-plane, except for a cut running along the negative real axis. See Figure 4.1.3. By Jordan's lemma, the arcs at infinity vanish. Along the cut from $-\infty$ to $-a$, both $\sqrt{z}$ and $\sqrt{z + a}$ have the same arguments of π or $-\pi$ along the top and bottom of the branch cut, respectively. The integration along the top of the branch cut will cancel the integration along the bottom of the branch cut. Consequently, the only contribution arises from that portion of the branch cut integral from the segment from $-a$ to 0 along the top and bottom of the branch cut as well as integrations around the branch points $z = 0$ and $z = -a$. The contribution from the branch point $z = 0$ is $2\pi i$ while the contribution from the branch point $z = -a$ is zero. Along the top of the branch cut, $z = \eta e^{\pi i}$ and $z + a = (a - \eta)e^{0i}$. Therefore,

$$\int_{\substack{top \\ branch}} \exp\left(tz - x\sqrt{\frac{z}{z+a}}\,\right) \frac{dz}{z} = \int_a^0 \frac{d\eta}{\eta} \exp\left(-t\eta - xi\sqrt{\frac{\eta}{a-\eta}}\,\right).$$

Along the bottom of the branch cut, $z = \eta e^{-\pi i}$ and $z + a = (a - \eta)e^{0i}$. Therefore,

$$\int_{\substack{bottom \\ branch}} \exp\left(tz - x\sqrt{\frac{z}{z+a}}\,\right) \frac{dz}{z} = \int_0^a \frac{d\eta}{\eta} \exp\left(-t\eta + xi\sqrt{\frac{\eta}{a-\eta}}\,\right).$$

Combining the integrals yields

$$f(t) = 1 - \frac{1}{\pi} \int_0^a e^{-t\eta} \sin\left(x\sqrt{\frac{\eta}{a-\eta}}\,\right) \frac{d\eta}{\eta}.$$

Finally, we substitute $\eta = a\sigma^2/(1 + \sigma^2)$ to obtain the inverse.

Section 4.1, Problem 13. Figure 4.1.3 shows the closed Bromwich integral. The integration around the branch point yields $-2\pi i$. Along the top of the branch cut, $z = \eta e^{\pi i}$ so that

$$\int_{CD} \exp\left(tz - \lambda\sqrt{\frac{z^\alpha}{z^\alpha + a}}\,\right) \frac{dz}{z} = \int_\infty^0 \exp\left(-t\eta - \lambda\sqrt{\frac{\eta^\alpha e^{\pi\alpha i}}{\eta^\alpha e^{\pi\alpha i} + a}}\,\right) \frac{d\eta}{\eta}.$$

Setting $\eta^\alpha e^{\pi\alpha i} + a = \rho e^{\phi i}$,

$$\int_{CD} \exp\left(tz - \lambda\sqrt{\frac{z^\alpha}{z^\alpha + a}}\,\right) \frac{dz}{z} = -\int_0^\infty \exp\left[-t\eta - \lambda\sqrt{\frac{\eta^\alpha}{\rho}}\, e^{(\pi\alpha - \phi)i/2}\right] \frac{d\eta}{\eta}$$

$$= -\int_0^\infty e^{-t\eta} \exp\left[-\lambda\sqrt{\frac{\eta^\alpha}{\rho}} \cos\left(\frac{\pi\alpha - \phi}{2}\right)\right]$$

$$\times \exp\left[-i\lambda\sqrt{\frac{\eta^\alpha}{\rho}} \sin\left(\frac{\pi\alpha - \phi}{2}\right)\right] \frac{d\eta}{\eta}.$$

Along the bottom of the branch cut, $z = \eta e^{-\pi i}$ and

$$\int_{GF} \exp\left(tz - \lambda\sqrt{\frac{z^\alpha}{z^\alpha + a}}\,\right) \frac{dz}{z} = \int_0^\infty \exp\left[-t\eta - \lambda\sqrt{\frac{\eta^\alpha}{\rho}}\, e^{-(\pi\alpha - \phi)i/2}\right] \frac{d\eta}{\eta}$$

$$= \int_0^\infty e^{-t\eta} \exp\left[-\lambda\sqrt{\frac{\eta^\alpha}{\rho}} \cos\left(\frac{\pi\alpha - \phi}{2}\right)\right]$$

$$\times \exp\left[i\lambda\sqrt{\frac{\eta^\alpha}{\rho}} \sin\left(\frac{\pi\alpha - \phi}{2}\right)\right] \frac{d\eta}{\eta}.$$

Consequently, the branch cut integral is

$$\int_{branch\ cut} \exp\!\left(tz - \lambda\sqrt{\frac{z^\alpha}{z^\alpha + a}}\,\right) \frac{dz}{z} = -2\pi i$$

$$+ 2i \int_0^\infty e^{-t\eta} \exp\!\left[-\lambda\sqrt{\frac{\eta^\alpha}{\rho}}\,\cos\!\left(\frac{\pi\alpha - \phi}{2}\right)\right] \sin\!\left[\lambda\sqrt{\frac{\eta^\alpha}{\rho}}\,\sin\!\left(\frac{\pi\alpha - \phi}{2}\right)\right] \frac{d\eta}{\eta}.$$

The inverse equals the negative of the branch cut integral after we divide it by $2\pi i$.

Section 4.1, Problem 14. Deforming Bromwich's contour to one along the imaginary z-axis, except for a small semicircle to the right of the singularity at $z = 0$, we have

$$f(t) = \frac{1}{2\pi i} \int_{-\infty i}^{\infty i} \exp\!\left[tz - r\sqrt{\frac{z(1+z)}{(az+1)}}\,\right] \frac{dz}{z}$$

$$= \frac{1}{2\pi i}\left\{ \int_{-\infty i}^{-0+i} \exp\!\left[tz - r\sqrt{\frac{z(1+z)}{(az+1)}}\,\right] \frac{dz}{z} + \int_{C_\epsilon} \exp\!\left[tz - r\sqrt{\frac{z(1+z)}{(az+1)}}\,\right] \frac{dz}{z} \right.$$

$$\left. + \int_{0+i}^{\infty i} \exp\!\left[tz - r\sqrt{\frac{z(1+z)}{(az+1)}}\,\right] \frac{dz}{z} \right\}$$

Now,

$$\int_{C_\epsilon} \exp\!\left[tz - r\sqrt{\frac{z(1+z)}{(az+1)}}\,\right] \frac{dz}{z} = \lim_{\epsilon\to 0} \int_{-\pi/2}^{\pi/2} \frac{\exp[-r\sqrt{\epsilon e^{\theta i}(1+\epsilon e^{\theta i})/(1+a\epsilon e^{\theta i})}\,]}{\epsilon e^{\theta i}} i\epsilon e^{\theta i}\, d\theta$$

$$= \pi i.$$

Along the contour from $(-\infty i, -0+i]$,

$$\int_{-\infty i}^{-0+i} \exp\!\left[tz - r\sqrt{\frac{z(1+z)}{(az+1)}}\,\right] \frac{dz}{z} = \int_\infty^{0+} \frac{du}{u} \exp\{-iut - r\sqrt{u/2}\,M(1-i)$$

$$\times [\cos(\theta) - i\sin(\theta)]\}$$

$$= -\int_{0+}^\infty \frac{du}{u} \exp\{-iut - r\sqrt{u/2}\,M$$

$$\times [\cos(\theta) - i\cos(\theta) - i\sin(\theta) - \sin(\theta)]\}$$

$$= -\int_{0+}^\infty \frac{du}{u} \exp\{-r\sqrt{u/2}\,M[\cos(\theta) - \sin(\theta)]\}$$

$$\times \exp\{-iut + ir\sqrt{u/2}\,M[\cos(\theta) + \sin(\theta)]\},$$

because $z = -ui$ with $0+ < u < \infty$, $z+1 = \sqrt{1+u^2}\,e^{-2\theta_1 i}$ where $2\theta_1 = \arctan(u)$, $1+az = \sqrt{1+a^2u^2}\,e^{-2\theta_2 i}$ where $2\theta_2 = \arctan(au)$, $\theta = \theta_1 - \theta_2$ and $M = \sqrt{(1+u^2)/(1+a^2u^2)}$. Similarly, along $[0+i, \infty i)$,

$$\int_{0+i}^{\infty i} \exp\!\left[tz - r\sqrt{\frac{z(1+z)}{(az+1)}}\,\right] \frac{dz}{z} = \int_{0+}^\infty \frac{du}{u} \exp\{-r\sqrt{u/2}\,M[\cos(\theta) - \sin(\theta)]\}$$

$$\times \exp\{iut - ir\sqrt{u/2}\,M[\cos(\theta) + \sin(\theta)]\}.$$

Therefore,

$$\frac{1}{2\pi i}\left\{\int_{-\infty i}^{-0+i}\exp\left[tz-r\sqrt{\frac{z(1+z)}{(az+1)}}\right]\frac{dz}{z}+\int_{0+i}^{\infty i}\exp\left[tz-r\sqrt{\frac{z(1+z)}{(az+1)}}\right]\frac{dz}{z}\right\}$$

$$=\frac{1}{\pi}\int_{0+}^{\infty}\frac{du}{u}\exp\{-r\sqrt{u/2}\,M[\cos(\theta)-\sin(\theta)]\}\sin\{ut-r\sqrt{u/2}\,M[\cos(\theta)+\sin(\theta)]\}.$$

Section 4.1, Problem 15. The closed contour for the inversion is similar to Figure 4.1.3, where the point $z = k$ lies on the positive real axis between the origin and the Bromwich contour. The arcs at infinity vanish by Jordan's lemma as does the integration around the branch point at $z = 0$. Therefore,

$$f(t) = \frac{\exp\left(kt - a\sqrt{k}\right)}{\sqrt{k}} - \frac{1}{2\pi i}\int_{\substack{branch\\cut}}\frac{\exp\left(tz - a\sqrt{z}\right)}{(z-k)\sqrt{z}}\,dz,$$

where the first term on the right side is due to the residue at $z = k$. Along the branch cut, $z = \eta^2 e^{\pm\pi i}$. Therefore,

$$\int_{\substack{branch\\cut}}\frac{\exp\left(tz - a\sqrt{z}\right)}{(z-k)\sqrt{z}}\,dz = -\int_{\infty}^{0}\frac{\exp(-a\eta i - \eta^2 t)}{(i\eta)(-k-\eta^2)}\,2\eta\,d\eta - \int_{0}^{\infty}\frac{\exp(a\eta i - \eta^2 t)}{(-i\eta)(-k-\eta^2)}\,2\eta\,d\eta$$

$$= 4i\int_{0}^{\infty}\frac{\cos(a\eta)}{k+\eta^2}e^{-\eta^2 t}\,d\eta.$$

Substitution into the first equation gives the inverse.

Section 4.1, Problem 16. From Bromwich's integral,

$$f(t) = \frac{a}{2\pi i}\int_{c-\infty i}^{c+\infty i}\frac{\exp\left(tz - x\sqrt{z}\right)}{z\left[a + (2-a)\sqrt{\pi z/2}\right]}\,dz.$$

Closing the line integral as shown in Figure 4.1.3, we only have an integration along the branch cut. The integration around the branch point $z = 0$ yields

$$\int_{FED}\frac{\exp(tz - x\sqrt{z})}{z\left[a + (2-a)\sqrt{\pi z/2}\right]}\,dz = -2\pi i.$$

Along the branch cut, we have $z = \eta^2 e^{\pm\pi i}$. Therefore, along the top of the branch cut,

$$\int_{DC}\frac{\exp\left(tz - x\sqrt{z}\right)}{z[a + (2-a)\sqrt{\pi z/2}]}\,dz = a\int_{\infty}^{0}\frac{\exp(-\eta^2 t - x\eta i)}{a + (2-a)\eta i\sqrt{\pi/2}}\frac{2\eta\,d\eta}{\eta^2}$$

$$= -2a\int_{0}^{\infty}e^{-\eta^2 t}\frac{(a - c\eta i)e^{-x\eta i}}{a^2 + \eta^2 c^2}\frac{d\eta}{\eta},$$

while along the bottom of the branch cut,

$$\int_{GF} \frac{\exp(tz - x\sqrt{z})}{z[a + (2-a)\sqrt{\pi z/2}]}\, dz = 2a \int_0^\infty \frac{\exp(-\eta^2 t + x\eta i)}{a - c\eta i}\, \frac{d\eta}{\eta}$$

$$= 2a \int_0^\infty e^{-\eta^2 t} \frac{(a + c\eta i)e^{-x\eta i}}{a^2 + \eta^2 c^2}\, \frac{d\eta}{\eta}.$$

The inverse equals the negative of the sum of the branch cut and branch point integrals after dividing the sum by $2\pi i$ and replacing the complex exponentials with sines and cosines.

Section 4.1, Problem 17. From Bromwich's integral,

$$f(t) = \frac{1}{2\pi i} \int_{c-\infty i}^{c+\infty i} \frac{\exp\left[tz - x\sqrt{z(z+b)}\right]}{z + a\sqrt{z(z+b)}}\, dz.$$

If $t < x$, we close Bromwich's integral in the right half of the s-plane and find that $f(t) = 0$. If $t > x$, we close the line integral as shown in Figure 4.1.3 and have an integration along the branch cut. The integration around the branch point $z = 0$ yields

$$\int_{FED} \frac{\exp\left[tz - x\sqrt{z(z+b)}\right]}{z + a\sqrt{z(z+b)}}\, dz = 0.$$

Along the branch cut, the integration consists of two parts along the intervals $(-\infty, -b)$ and $[-b, 0]$. Over the first interval, the contribution to the inversion integral along the top of the branch cut cancels the contribution from the integration along the bottom of the branch cut. Over the second interval, $z = b\eta^2 e^{\pm \pi i}$ and $z + b = b(1 - \eta^2)e^{0i}$. Therefore,

$$f(t) = \frac{1}{\pi i} \int_1^0 \frac{\exp\left(-bt\eta^2 + bxi\eta\sqrt{1-\eta^2}\right)}{b\eta + abi\sqrt{1-\eta^2}}\, d\eta - \frac{1}{\pi i} \int_0^1 \frac{\exp\left(-bt\eta^2 - bxi\eta\sqrt{1-\eta^2}\right)}{-b\eta + abi\sqrt{1-\eta^2}}\, d\eta$$

$$= -\frac{1}{\pi i} \int_0^1 \frac{\left(\eta - ai\sqrt{1-\eta^2}\right)\exp\left(-bt\eta^2 + bxi\eta\sqrt{1-\eta^2}\right)}{b\eta^2 + a^2 b(1-\eta^2)}\, d\eta$$

$$- \frac{1}{\pi i} \int_0^1 \frac{\left(-\eta - ai\sqrt{1-\eta^2}\right)\exp\left(-bt\eta^2 - bxi\eta\sqrt{1-\eta^2}\right)}{b\eta^2 + a^2 b(1-\eta^2)}\, d\eta$$

$$= \frac{2a}{\pi b} \int_0^1 \frac{\sqrt{1-\eta^2}\, \cos\left(bx\eta\sqrt{1-\eta^2}\right)}{a^2 + (1-a^2)\eta^2} e^{-bt\eta^2}\, d\eta$$

$$- \frac{2}{\pi b} \int_0^1 \frac{\eta \sin\left(bx\eta\sqrt{1-\eta^2}\right)}{a^2 + (1-a^2)\eta^2} e^{-bt\eta^2}\, d\eta.$$

Section 4.1, Problem 18. Using the contour shown in Figure 4.1.3, the inversion consists of two parts. From the integration around the branch point $z = 0$, we have $-2\pi i$. The second part arises from the integration along the branch cut. Along the top of the branch cut, $z = \eta^2 e^{\pi i}$ and $\sqrt{z} = i\eta$, while along the bottom of the branch cut, $z = \eta^2 e^{-\pi i}$ and $\sqrt{z} = -i\eta$. Therefore, the branch cut integration is

$$
\int_{GF \cup DC} e^{-(r-a)\sqrt{z}} \frac{1 + r\sqrt{z}}{z\left(1 + a\sqrt{z}\right)} \, dz = -\int_{\infty}^{0} \left(\frac{1}{-\eta^2}\right) \frac{1 + r\eta i}{1 + a\eta i} e^{-(r-a)\eta i} e^{-\eta^2 t} \, 2\eta \, d\eta
$$

$$
- \int_{0}^{\infty} \left(\frac{1}{-\eta^2}\right) \frac{1 - r\eta i}{1 - a\eta i} e^{(r-a)\eta i} e^{-\eta^2 t} \, 2\eta \, d\eta
$$

$$
= 4i \int_{0}^{\infty} \left\{ \frac{1 + ar\eta^2}{1 + a^2\eta^2} \sin[(r-a)\eta] \right.
$$

$$
\left. - \frac{r\eta - a\eta}{1 + a^2\eta^2} \cos[(r-a)\eta] \right\} e^{-\eta^2 t} \frac{d\eta}{\eta}.
$$

Thus, the inverse equals the negative of the integration around the branch point and the branch cut integral after we divide them by $2\pi i$.

Section 4.1, Problem 19. The inversion contour is identical to Figure 4.1.3. The contribution from the integration around the branch point at $z = 0$ is $2\pi i$. Along CD, $z = \eta e^{\pi i}$ and $\sqrt{z} = i\sqrt{\eta}$ so that

$$
\int_{CD} \exp\left(tz - a\sqrt{z} \frac{1 - ce^{-b\sqrt{z}}}{1 + ce^{-b\sqrt{z}}}\right) \frac{dz}{z} = \int_{0}^{\infty} \exp\left(-\eta t - ai\sqrt{\eta} \frac{1 - ce^{-bi\sqrt{\eta}}}{1 + ce^{-bi\sqrt{\eta}}}\right) \frac{d\eta}{\eta}.
$$

Along FG, $z = \eta e^{-\pi i}$ and $\sqrt{z} = -i\sqrt{\eta}$ so that

$$
\int_{FG} \exp\left(tz - a\sqrt{z} \frac{1 - ce^{-b\sqrt{z}}}{1 + ce^{-b\sqrt{z}}}\right) \frac{dz}{z} = \int_{\infty}^{0} \exp\left(-\eta t + ai\sqrt{\eta} \frac{1 - ce^{bi\sqrt{\eta}}}{1 + ce^{bi\sqrt{\eta}}}\right) \frac{d\eta}{\eta}.
$$

Using Euler's formula,

$$
\frac{1 - c\exp(-bi\sqrt{\eta})}{1 + c\exp(-bi\sqrt{\eta})} = \frac{1 - c^2 + 2ic\sin(b\sqrt{\eta})}{1 + c^2 + 2c\cos(b\sqrt{\eta})} = v(\eta) + iu(\eta)
$$

and

$$
\frac{1 - c\exp(bi\sqrt{\eta})}{1 + c\exp(bi\sqrt{\eta})} = \frac{1 - c^2 - 2ic\sin(b\sqrt{\eta})}{1 + c^2 + 2c\cos(b\sqrt{\eta})} = v(\eta) - iu(\eta).
$$

Therefore,

$$
\int_{CD} \exp\left(tz - a\sqrt{z} \frac{1 - ce^{-b\sqrt{z}}}{1 + ce^{-b\sqrt{z}}}\right) \frac{dz}{z} = \int_{0}^{\infty} \exp\{-\eta t - a\sqrt{\eta} \, [iv(\eta) - u(\eta)]\} \frac{d\eta}{\eta}
$$

$$
= \int_{0}^{\infty} e^{-\eta t + a\sqrt{\eta} \, u(\eta)} e^{-ai\sqrt{\eta} \, v(\eta)} \frac{d\eta}{\eta}
$$

and

$$
\int_{FG} \exp\left(tz - a\sqrt{z} \frac{1 - ce^{-b\sqrt{z}}}{1 + ce^{-b\sqrt{z}}}\right) \frac{dz}{z} = -\int_{0}^{\infty} e^{-\eta t + a\sqrt{\eta} \, u(\eta)} e^{ai\sqrt{\eta} \, v(\eta)} \frac{d\eta}{\eta}.
$$

Finally,

$$f(t) = 1 + \frac{1}{2\pi i} \int_{FG} \exp\left(tz - a\sqrt{z}\,\frac{1 - ce^{-b\sqrt{z}}}{1 + ce^{-b\sqrt{z}}}\right) \frac{dz}{z}$$

$$+ \frac{1}{2\pi i} \int_{CD} \exp\left(tz - a\sqrt{z}\,\frac{1 - ce^{-b\sqrt{z}}}{1 + ce^{-b\sqrt{z}}}\right) \frac{dz}{z}$$

$$= 1 - \frac{1}{\pi} \int_0^\infty e^{-\eta t + a\sqrt{\eta}\,u(\eta)} \sin[a\sqrt{\eta}\,v(\eta)] \frac{d\eta}{\eta}.$$

Section 4.1, Problem 20. Taking the branch cut to run between $z = -i$ and $z = i$, we have a closed contour similar to the one shown in Figure 4.1.12. We have simple poles at $z = \pm a$. For $t < |x|$, we close Bromwich's contour in the right half-plane and find that $f(t) = 0$. If $t > |x|$, we close Bromwich contour in the left half-plane. There we have contributions from the residues

$$\mathrm{Res}\left[\frac{\exp\left(tz - |x|\sqrt{z^2 + 1}\right)}{z^2 - a^2}; a\right] = \lim_{z \to a} \frac{(z - a)\exp\left(tz - |x|\sqrt{z^2 + 1}\right)}{(z - a)(z + a)} = \frac{e^{at - |x|\sqrt{1 + a^2}}}{2a}$$

and

$$\mathrm{Res}\left[\frac{\exp\left(tz - |x|\sqrt{z^2 + 1}\right)}{z^2 - a^2}; -a\right] = \lim_{z \to -a} \frac{(z + a)\exp\left(tz - |x|\sqrt{z^2 + 1}\right)}{(z - a)(z + a)}$$

$$= -\frac{e^{-at + |x|\sqrt{1 + a^2}}}{2a}.$$

Therefore, the sum of the residues equals $\sinh\left(at - |x|\sqrt{1 + a^2}\right)/a$.

For the branch cut integration, we define the cuts as follows:

$$z + i = (1 + \eta)e^{\theta i}, \quad -1 < \eta < 1, \quad -3\pi/2 < \theta < \pi/2,$$

and

$$z - i = (1 - \eta)e^{\theta i}, \quad -1 < \eta < 1, \quad -3\pi/2 < \theta < \pi/2.$$

Therefore, along the right side of the branch cut,

$$\int_{RHS} \frac{\exp\left(tz - |x|\sqrt{z^2 + 1}\right)}{z^2 - a^2}\, dz = \int_{-1}^1 \frac{e^{it\eta - |x|\sqrt{1 - \eta^2}}}{-\eta^2 - a^2}\, i\, d\eta.$$

Along the left side of the branch cut,

$$\int_{LHS} \frac{\exp\left(tz - |x|\sqrt{z^2 + 1}\right)}{z^2 - a^2}\, dz = \int_1^{-1} \frac{e^{it\eta + |x|\sqrt{1 - \eta^2}}}{-\eta^2 - a^2}\, i\, d\eta.$$

Therefore,

$$\int_{RHS \cup LHS} \frac{\exp\left(tz - |x|\sqrt{z^2 + 1}\right)}{z^2 - a^2}\, dz = 2i \int_{-1}^1 \frac{\sinh\left(|x|\sqrt{1 - \eta^2}\right)e^{it\eta}}{\eta^2 + a^2}\, d\eta$$

$$= 4i \int_0^1 \frac{\sinh\left(|x|\sqrt{1 - \eta^2}\right)\cos(t\eta)}{-\eta^2 - a^2}\, d\eta.$$

The inverse equals the sum of the residues plus the branch cut integral after dividing it by $2\pi i$.

Section 4.1, Problem 21. Taking the branch cut to run between $z = -i$ and $z = i$, the closed contour is similar to the one shown in Figure 4.1.12. We have simple poles at $z = \pm a$. For $t < |x|$, we close Bromwich's contour in the right half-plane and find that $f(t) = 0$. If $t > |x|$, we close Bromwich contour in the left half-plane. There we have contributions from the residues

$$\mathrm{Res}\left[\frac{z\exp\left(tz - |x|\sqrt{z^2+1}\right)}{(z^2-a^2)\sqrt{z^2+1}}; a\right] = \lim_{z \to a} \frac{z(z-a)\exp\left(tz - |x|\sqrt{z^2+1}\right)}{(z-a)(z+a)\sqrt{z^2+1}} = \frac{e^{at-|x|\sqrt{1+a^2}}}{2\sqrt{1+a^2}}$$

and

$$\mathrm{Res}\left[\frac{z\exp\left(tz - |x|\sqrt{z^2+1}\right)}{(z^2-a^2)\sqrt{z^2+1}}; -a\right] = \lim_{z \to -a} \frac{z(z+a)\exp\left(tz - |x|\sqrt{z^2+1}\right)}{(z-a)(z+a)\sqrt{z^2+1}}$$

$$= \frac{e^{-at+|x|\sqrt{1+a^2}}}{-2\sqrt{1+a^2}}.$$

Therefore, the sum of the residues equals $\sinh\left(at - |x|\sqrt{1+a^2}\right)/\sqrt{1+a^2}$.

For the branch cut integration, we define the cuts as follows:

$$z + i = (1+\eta)e^{\theta i}, \qquad -1 < \eta < 1, \qquad -3\pi/2 < \theta < \pi/2,$$

and

$$z - i = (1-\eta)e^{\theta i}, \qquad -1 < \eta < 1, \qquad -3\pi/2 < \theta < \pi/2.$$

Therefore, along the right side of the branch cut,

$$\int_{RHS} \frac{z\exp\left(tz - |x|\sqrt{z^2+1}\right)}{(z^2-a^2)\sqrt{z^2+1}}\,dz = \int_{-1}^{1} \frac{\eta\, e^{it\eta - |x|\sqrt{1-\eta^2}}}{(\eta^2+a^2)\sqrt{1-\eta^2}}\,d\eta.$$

Along the left side of the branch cut,

$$\int_{LHS} \frac{z\exp\left(tz - |x|\sqrt{z^2+1}\right)}{(z^2-a^2)\sqrt{z^2+1}}\,dz = \int_{1}^{-1} \frac{\eta\, e^{it\eta + |x|\sqrt{1-\eta^2}}}{(-\eta^2-a^2)\sqrt{1-\eta^2}}\,d\eta.$$

Therefore,

$$\int_{RHS\cup LHS} \frac{z\exp\left(tz - |x|\sqrt{z^2+1}\right)}{(z^2-a^2)\sqrt{1+z^2}}\,dz = 2\int_{-1}^{1} \frac{\cosh\left(x\sqrt{1-\eta^2}\right)e^{it\eta}}{(\eta^2+a^2)\sqrt{1-\eta^2}}\,\eta\,d\eta$$

$$= 4i\int_{0}^{1} \frac{\cosh\left(x\sqrt{1-\eta^2}\right)\sin(t\eta)}{(\eta^2+a^2)\sqrt{1-\eta^2}}\,\eta\,d\eta$$

$$= 4i\int_{0}^{1} \frac{\cosh(x\chi)\sin\left(t\sqrt{1-\chi^2}\right)}{1+a^2-\chi^2}\,d\chi.$$

The inverse equals the sum of the residues plus the branch cut integral after dividing it by $2\pi i$.

Section 4.1, Problem 22. From Bromwich's integral

$$f(t) = \frac{a}{2\pi i} \int_{c-\infty i}^{c+\infty i} \frac{\sqrt{z}\exp\left[tz - x\sqrt{z(z+1)}\right]}{z\left[a\sqrt{z+1} + (2-a)\sqrt{\pi z/2}\right]}\, dz.$$

By Jordan's lemma, we close the contour in the right half-plane if $t < x$; in the left half-plane, if $t > x$. If we take the branch cuts associated with $\sqrt{z}$ and $\sqrt{z+1}$ along the negative real axis so that the phase lies between $-\pi$ and π, the integration along the top of the branch cut cancels the integration along the bottom of the branch cut in the interval $(-\infty, -1)$. Consequently, the branch cut for $\sqrt{z(z+1)}$ results in a dumbbell-shaped contour along the negative real axis from -1 to 0.

Integrations around the branch point $z = 0$ and $z = -1$ equal zero. Along the branch cut, $z = \eta^2 e^{\pm\pi i}$ and $z + 1 = (1 - \eta^2)e^{0i}$ with $0 < \eta < 1$. Along the top of the branch cut,

$$\int_{\substack{top\ of \\ branch\ cut}} \frac{\sqrt{z}\exp\left[tz - x\sqrt{z(z+1)}\right]}{z\left[a\sqrt{z+1} + (2-a)\sqrt{\pi z/2}\right]}\, dz$$

$$= \int_1^0 \left(\frac{2a}{\eta}\right) \frac{\eta i \exp\left(-\eta^2 t - ix\eta\sqrt{1-\eta^2}\right)}{a\sqrt{1-\eta^2} + (2-a)i\eta\sqrt{\pi/2}}\, d\eta$$

$$= -2ai \int_0^1 e^{-\eta^2 t} \frac{\left[a\sqrt{1-\eta^2} - (2-a)i\eta\sqrt{\pi/2}\right]}{a^2 + b^2\eta^2} \exp\left(-ix\eta\sqrt{1-\eta^2}\right)\, d\eta,$$

while along the bottom of the branch cut,

$$\int_{\substack{bottom\ of \\ branch\ cut}} \frac{\sqrt{z}\exp\left[tz - x\sqrt{z(z+1)}\right]}{z\left[a\sqrt{z+1} + (2-a)\sqrt{\pi z/2}\right]}\, dz$$

$$= \int_0^1 \left(\frac{2a}{\eta}\right) \frac{-\eta i \exp\left(-\eta^2 t + ix\eta\sqrt{1-\eta^2}\right)}{a\sqrt{1-\eta^2} - (2-a)i\eta\sqrt{\pi/2}}\, d\eta$$

$$= -2ai \int_0^1 e^{-\eta^2 t} \frac{\left[a\sqrt{1-\eta^2} + (2-a)i\eta\sqrt{\pi/2}\right]}{a^2 + b^2\eta^2} \exp\left(ix\eta\sqrt{1-\eta^2}\right)\, d\eta.$$

Combining the results, we find

$$\int_{\substack{branch \\ cut}} \frac{\sqrt{z}\exp\left[tz - x\sqrt{z(z+1)}\right]}{z\left[a\sqrt{z+1} + (2-a)\sqrt{\pi z/2}\right]}\, dz$$

$$= -4a^2 i \int_0^1 e^{-t\eta^2} \frac{\sqrt{1-\eta^2}}{a^2 + b^2\eta^2} \cos\left(x\eta\sqrt{1-\eta^2}\right)\, d\eta$$

$$+ 4a(2-a)\sqrt{\frac{\pi}{2}} \int_0^1 e^{-t\eta^2} \frac{\eta}{a^2 + b^2\eta^2} \sin\left(x\eta\sqrt{1-\eta^2}\right)\, d\eta.$$

The inverse equals the negative of the branch cut integral after it has been divided by $2\pi i$.

Section 4.1, Problem 23. With the branch cut taken along the negative real axis, Bromwich's integral reduces to residues from the poles $s = \pm\omega i$ plus a contour integral along two semicircles of infinite radius in the second and third quadrants and a line integration along the branch cut. From Jordan's lemma, the contributions from the arcs vanish. Furthermore, the integration around the branch point vanishes. Computing the residues first,

$$\mathrm{Res}\left(\frac{\omega\, e^{-x\sqrt{z/\kappa}}}{z^2+\omega^2}; \omega i\right) = \lim_{z\to\omega i}\frac{\omega\,(z-\omega i)e^{-x\sqrt{z/\kappa}}}{z^2+\omega^2} = \frac{1}{2i}\exp\left(i\omega t - x e^{\pi i/4}\sqrt{\omega/\kappa}\right)$$

$$= \frac{1}{2i}\exp\left(-x\sqrt{\omega/2\kappa}\right)\exp\left(i\omega t - ix\sqrt{\omega/2\kappa}\right)$$

and

$$\mathrm{Res}\left(\frac{\omega\, e^{-x\sqrt{z/\kappa}}}{z^2+\omega^2}; -\omega i\right) = \lim_{z\to-\omega i}\frac{\omega\,(z+\omega i)e^{-x\sqrt{z/\kappa}}}{z^2+\omega^2}$$

$$= -\frac{1}{2i}\exp\left(-i\omega t - x e^{-\pi i/4}\sqrt{\omega/\kappa}\right)$$

$$= -\frac{1}{2i}\exp\left(-x\sqrt{\omega/2\kappa}\right)\exp\left(-i\omega t + ix\sqrt{\omega/2\kappa}\right).$$

The sum of the residues equals

$$\exp\left(-x\sqrt{\omega/2\kappa}\right)\sin\left(\omega t - x\sqrt{\omega/2\kappa}\right).$$

Along the top of the branch cut, $z = \chi e^{\pi i}$, where $0 < \chi < \infty$. Therefore,

$$\int_{\substack{top\ of \\ branch\ cut}}\frac{\omega\, e^{-x\sqrt{z/\kappa}}}{z^2+\omega^2}\,dz = -\int_{\infty}^{0}\frac{\omega}{\omega^2+\chi^2}e^{-t\chi - xi\sqrt{\chi/\kappa}}\,d\chi.$$

Along the bottom of the branch cut, $z = \chi e^{-\pi i}$, where $0 < \chi < \infty$. Consequently,

$$\int_{\substack{bottom\ of \\ branch\ cut}}\frac{\omega\, e^{-x\sqrt{z/\kappa}}}{z^2+\omega^2}\,dz = -\int_{0}^{\infty}\frac{\omega}{\omega^2+\chi^2}e^{-t\chi + xi\sqrt{\chi/\kappa}}\,d\chi.$$

Therefore,

$$\frac{1}{2\pi i}\int_{\substack{branch \\ cut}}\frac{\omega\, e^{-x\sqrt{z/\kappa}}}{z^2+\omega^2}\,dz = -\frac{1}{\pi}\int_{0}^{\infty}\frac{\omega}{\omega^2+\chi^2}e^{-t\chi}\sin\left(x\sqrt{\chi/\kappa}\right)\,d\chi.$$

Introducing $\chi = \kappa\eta^2$, this integral becomes

$$\frac{1}{2\pi i}\int_{\substack{branch \\ cut}}\frac{\omega\, e^{-x\sqrt{z/\kappa}}}{z^2+\omega^2}\,dz = -\frac{2\kappa}{\pi}\int_{0}^{\infty}\frac{\omega\eta}{\omega^2+\kappa^2\eta^4}e^{-\kappa t\eta^2}\sin(x\eta)\,d\eta.$$

The inverse equals the sum of the residues minus the branch cut integral.

Section 4.1, Problem 24. With the branch cut taken along the negative real axis (see Figure 4.1.3), Bromwich's integral reduces to residues from the poles $z = \pm i$ plus the branch cut integral. Computing the residues first,

$$\operatorname{Res}\left[\frac{1}{z^2+1}\exp\left(tz - \frac{zx}{\sqrt{z+1}}\right);i\right] = \frac{1}{2i}\exp\left(it - ix/\sqrt{1+i}\right)$$

$$= \frac{1}{2i}\exp\left\{it - ix[\cos(\pi/8) - i\sin(\pi/8)]/\sqrt[4]{2}\right\}$$

$$= \frac{1}{2i}\exp\left[ti - xi\cos(\pi/8)/\sqrt[4]{2}\right]\exp\left[-x\sin(\pi/8)/\sqrt[4]{2}\right]$$

and

$$\operatorname{Res}\left[\frac{1}{z^2+1}\exp\left(tz - \frac{zx}{\sqrt{z+1}}\right);-i\right] = -\frac{1}{2i}\exp\left(-it + ix/\sqrt{1-i}\right)$$

$$= -\frac{1}{2i}\exp\left\{-it + ix[\cos(\pi/8) + i\sin(\pi/8)]/\sqrt[4]{2}\right\}$$

$$= -\frac{1}{2i}\exp\left[-ti + xi\cos(\pi/8)/\sqrt[4]{2}\right]e^{-x\sin(\pi/8)/\sqrt[4]{2}}.$$

The sum of the residues equals $\exp\left[-x\sin(\pi/8)/\sqrt{2}\right]\sin\left[t - x\cos(\pi/8)/\sqrt{2}\right]$.

Along the top of the branch cut, $z + 1 = \eta e^{\pi i}$, where $0 < \eta < \infty$. Therefore,

$$\int_{\substack{top\ of\\branch\ cut}} \frac{\exp\left(tz - zx/\sqrt{s+1}\right)}{z^2+1}\,dz = \int_\infty^0 \frac{-d\eta}{1+(-1-\eta)^2}\exp\left[-\frac{(-1-\eta)x}{i\sqrt{\eta}} + t(-1-\eta)\right].$$

Along the bottom of the branch cut, $z + 1 = \eta e^{-\pi i}$, where $0 < \eta < \infty$. Therefore,

$$\int_{\substack{bottom\ of\\branch\ cut}} \frac{\exp\left(tz - zx/\sqrt{s+1}\right)}{z^2+1}\,dz = \int_0^\infty \frac{-d\eta}{1+(-1-\eta)^2}\exp\left[-\frac{(-1-\eta)x}{-i\sqrt{\eta}} + t(-1-\eta)\right].$$

Consequently, the contribution to Bromwich's integral from the branch cut integral is

$$\int_{\substack{branch\\cut}} \frac{\exp\left(tz - zx/\sqrt{s+1}\right)}{z^2+1}\,dz = -2i\int_0^\infty \frac{\exp[-t(1+\eta)]}{\eta^2 + 2\eta + 2}\sin\left[\frac{(1+\eta)x}{\sqrt{\eta}}\right]d\eta.$$

The inverse equals the sum of the residues plus the branch cut integral after it has been divided by $2\pi i$.

Section 4.1, Problem 25. We have simple poles at $z = \pm 2\pi i$. Therefore, the residues are

$$\operatorname{Res}\left\{\frac{2\pi\exp\left[tz - x\sqrt{z(1+az)/(1+bz)}\right]}{s^2+4\pi^2};2\pi i\right\} = \frac{1}{2i}\exp\left[2\pi ti - x\sqrt{\frac{2\pi i(1+2\pi ai)}{1+2\pi bi}}\right]$$

$$= \frac{1}{2i}\exp\left[2\pi ti - x\sqrt{d_3(d_2 + id_1)}\right]$$

$$= \frac{1}{2i}\exp\left[2\pi ti - x\Delta e^{\psi i/2}\right],$$

and

$$\text{Res}\left\{\frac{2\pi \exp\left(tz - x\sqrt{z(1+az)/(1+bz)}\right)}{s^2 + 4\pi^2}; -2\pi i\right\} = -\frac{1}{2i}\exp\left[-2\pi ti - x\Delta e^{-\psi i/2}\right].$$

Thus, the sum of the residues is $\exp[-x\Delta\cos(\psi/2)]\sin[2\pi t - x\Delta\sin(\psi/2)]$.

We can break the integration along the branch cuts into three parts; each part has a contribution from above and below the negative real axis. Above the negative real axis from $-1/b$ to 0,

$$z = \eta e^{\pi i}, \quad z + 1/b = (1/b - \eta)e^{0i}, \quad z + 1/a = (1/a - \eta)e^{0i},$$

for $0 < \eta < 1/b$. Below the negative real axis, the only change is $z = \eta e^{-\pi i}$. Above the negative real axis from $-1/a$ to $-1/b$,

$$z = \eta e^{\pi i}, \quad z + 1/b = (\eta - 1/b)e^{\pi i}, \quad z + 1/a = (1/a - \eta)e^{0i},$$

for $1/a < \eta < 1/b$. Below the negative real axis, we have two changes:

$$z = \eta e^{-\pi i}, \quad \text{and} \quad z + 1/b = (\eta - 1/b)e^{-\pi i}.$$

Finally, from $-\infty$ to $-1/a$ above the negative real axis,

$$z = \eta e^{\pi i}, \quad z + 1/b = (\eta - 1/b)e^{\pi i}, \quad z + 1/a = (\eta - 1/a)e^{\pi i},$$

for $1/b < \eta < \infty$, while below the negative real axis,

$$z = \eta e^{-\pi i}, \quad z + 1/b = (\eta - 1/b)e^{-\pi i}, \quad z + 1/a = (\eta - 1/a)e^{-\pi i}.$$

Therefore, direct substitution yields the branch cut integral

$$2\pi \int_{\substack{branch \\ cut}} \frac{\exp\left[tz - x\sqrt{z(1+az)/(1+bz)}\right]}{z^2 + 4\pi^2}\, dz$$

$$= -\int_{\infty}^{1/a} \frac{2\pi\, e^{-\eta t - xi\sqrt{\eta(a\eta-1)/(b\eta-1)}}}{4\pi^2 + \eta^2}\, d\eta - \int_{1/a}^{\infty} \frac{2\pi\, e^{-\eta t + xi\sqrt{\eta(a\eta-1)/(b\eta-1)}}}{4\pi^2 + \eta^2}\, d\eta$$

$$-\int_{1/a}^{1/b} \frac{2\pi\, e^{-\eta t - x\sqrt{\eta(a\eta-1)/(b\eta-1)}}}{4\pi^2 + \eta^2}\, d\eta - \int_{1/b}^{1/a} \frac{2\pi\, e^{-\eta t - x\sqrt{\eta(a\eta-1)/(b\eta-1)}}}{4\pi^2 + \eta^2}\, d\eta$$

$$-\int_{1/b}^{0} \frac{2\pi\, e^{-\eta t - xi\sqrt{\eta(a\eta-1)/(b\eta-1)}}}{4\pi^2 + \eta^2}\, d\eta - \int_{0}^{1/b} \frac{2\pi\, e^{-\eta t + xi\sqrt{\eta(a\eta-1)/(b\eta-1)}}}{4\pi^2 + \eta^2}\, d\eta$$

$$= -4\pi i \int_{0}^{1/b} F(x, t, \eta)\, d\eta - 4\pi i \int_{1/a}^{\infty} F(x, t, \eta)\, d\eta,$$

where

$$F(x, t, \eta) = \frac{e^{-\eta t}}{4\pi^2 + \eta^2}\sin\left[x\sqrt{\frac{\eta(1 - a\eta)}{1 - b\eta}}\right].$$

The inverse equals

$$f(t) = e^{-x\Delta\cos(\psi/2)}\sin[2\pi t - x\Delta\sin(\psi/2)] - \frac{1}{i}\int_{\substack{branch\\cut}} \frac{e^{tz-x\sqrt{z(1+az)/(1+bz)}}}{z^2 + 4\pi^2}\,dz.$$

Section 4.1, Problem 26. Closing Bromwich's integral as shown in Figure 4.1.3,

$$f(t) = \frac{1}{2\pi i}\oint_C \left[(1+i)e^{tz-(1+i)a\sqrt[4]{z}} + (1-i)e^{tz-(1-i)a\sqrt[4]{z}}\right]\frac{dz}{z}.$$

There are no singularities inside the contour. The integration around the branch point yields

$$\int_{\substack{branch\\point}} \left[(1+i)e^{tz-(1+i)a\sqrt[4]{z}} + (1-i)e^{tz-(1-i)a\sqrt[4]{z}}\right]\frac{dz}{z} = 4\pi i.$$

Along the top of the branch cut, $z = \eta^4 e^{\pi i}$, while along the bottom of the branch cut, $z = \eta^4 e^{-\pi i}$. Therefore,

$$\int_{\substack{top\\branch\ cut}} \left[(1+i)e^{tz-(1+i)a\sqrt[4]{z}} + (1-i)e^{tz-(1-i)a\sqrt[4]{z}}\right]\frac{dz}{z}$$

$$= 4\int_0^\infty (1+i)e^{-t\eta^4}\exp\left[-(1+i)^2\frac{a\eta}{\sqrt{2}}\right]\frac{d\eta}{\eta}$$

$$+ 4\int_0^\infty (1-i)e^{-t\eta^4}\exp\left[-(1+i)(1-i)\frac{a\eta}{\sqrt{2}}\right]\frac{d\eta}{\eta}$$

and

$$\int_{\substack{bottom\\branch\ cut}} \left[(1+i)e^{tz-(1+i)a\sqrt[4]{z}} + (1-i)e^{tz-(1-i)a\sqrt[4]{z}}\right]\frac{dz}{z}$$

$$= 4\int_\infty^0 (1+i)e^{-t\eta^4}\exp\left[-(1+i)(1-i)\frac{a\eta}{\sqrt{2}}\right]\frac{d\eta}{\eta}$$

$$+ 4\int_0^\infty (1-i)e^{-t\eta^4}\exp\left[-(1-i)^2\frac{a\eta}{\sqrt{2}}\right]\frac{d\eta}{\eta}.$$

Adding these three integrals together, simplifying, dividing by $2\pi i$, and using Euler's identity, we obtain the inverse.

Section 4.1, Problem 27. Closing Bromwich's integral as shown in Figure 4.1.3,

$$f(t) = \frac{1}{2\pi i}\oint_C \frac{i}{z\sqrt[4]{z}}\left[e^{tz-(1+i)a\sqrt[4]{z}} - e^{tz-(1-i)a\sqrt[4]{z}}\right]dz.$$

There are no singularities inside the contour. The integration around the branch point yields

$$\int_{\substack{branch\\point}} \frac{i}{z\sqrt[4]{z}}\left[e^{tz-(1+i)a\sqrt[4]{z}} - e^{tz-(1-i)a\sqrt[4]{z}}\right]dz = 4\pi ai.$$

Along the top of the branch cut, $z = \eta^4 e^{\pi i}$, while along the bottom of the branch cut, $z = \eta^4 e^{-\pi i}$. Therefore,

$$\int_{\substack{top \\ branch\ cut}} \frac{i}{z\sqrt[4]{z}} \left[e^{tz - (1+i)a\sqrt[4]{z}} - e^{tz - (1-i)a\sqrt[4]{z}} \right] dz$$

$$= i \int_0^\infty \frac{\exp\left[-(1+i)^2 a\eta/\sqrt{2} \right] - \exp\left[-(1-i)(1+i)a\eta/\sqrt{2} \right]}{(1+i)\eta^5/\sqrt{2}} e^{-t\eta^4} 4\eta^3 \, d\eta$$

$$= 2\sqrt{2}\, i (1-i) \int_0^\infty \left[\exp\left(-\sqrt{2}\, ia\eta \right) - \exp\left(-\sqrt{2}\, a\eta \right) \right] e^{-t\eta^4} \frac{d\eta}{\eta^2}$$

and

$$\int_{\substack{bottom \\ branch\ cut}} \frac{i}{z\sqrt[4]{z}} \left[e^{tz - (1+i)a\sqrt[4]{z}} - e^{tz - (1-i)a\sqrt[4]{z}} \right] dz$$

$$= i \int_\infty^0 \frac{\exp\left[-(1+i)(1-i)a\eta/\sqrt{2} \right] - \exp\left[-(1-i)^2 a\eta/\sqrt{2} \right]}{(1-i)\eta^5/\sqrt{2}} e^{-t\eta^4} 4\eta^3 \, d\eta$$

$$= -2\sqrt{2}\, i (1+i) \int_0^\infty \left[\exp\left(-\sqrt{2}\, a\eta \right) - \exp\left(\sqrt{2}\, ia\eta \right) \right] e^{-t\eta^4} \frac{d\eta}{\eta^2}.$$

Adding these three integrals together, simplifying, dividing by $2\pi i$, and using Euler's identity, we obtain the inverse.

Section 4.1, Problem 28. For $t < a - b$, we close the contour on the right side of the plane by Jordan's lemma. Because there are no singularities, $f(t) = 0$. For $t > a - b$, we close Bromwich's integral as shown in Figure 4.1.3,

$$f(t) = \frac{1}{2\pi i} \oint_C \frac{K_1(az)e^{tz}}{z\, K_1(bz)} \, dz.$$

There are no singularities inside the contour. The integration around the branch point yields

$$\int_{\substack{branch \\ point}} \frac{K_1(az)e^{tz}}{z\, K_1(bz)} \, dz = -\frac{2\pi ib}{a},$$

because $K_n(z) \approx (n-1)!(2/z)^n/2$ as $z \to 0$. Along the top of the branch cut, $z = \eta e^{\pi i}$ and $K_1(\eta e^{\pi i}) = K_1(\eta) - \pi i I_1(\eta)$, while along the bottom of the branch cut, $z = \eta e^{-\pi i}$ and $K_1(\eta e^{-\pi i}) = K_1(\eta) + \pi i I_1(\eta)$. Therefore,

$$\int_{\substack{bottom \\ branch\ cut}} \frac{K_1(az)e^{tz}}{z\, K_1(bz)} \, dz = \int_\infty^0 e^{-\eta t} \left[\frac{K_1(a\eta) + \pi i I_1(a\eta)}{K_1(b\eta) + \pi i I_1(b\eta)} \right] \frac{d\eta}{\eta},$$

$$\int_{\substack{top \\ branch\ cut}} \frac{K_1(az)e^{tz}}{z\, K_1(bz)} \, dz = \int_0^\infty e^{-\eta t} \left[\frac{K_1(a\eta) - \pi i I_1(a\eta)}{K_1(b\eta) - \pi i I_1(b\eta)} \right] \frac{d\eta}{\eta},$$

and

$$\int_{\substack{branch \\ cut}} \frac{K_1(az)e^{tz}}{z\, K_1(bz)} \, dz = -2\pi i \int_0^\infty e^{-\eta t} \left[\frac{I_1(a\eta)K_1(b\eta) - K_1(a\eta)I_1(b\eta)}{K_1^2(b\eta) + \pi^2 I_1^2(b\eta)} \right] \frac{d\eta}{\eta}.$$

The inverse equals the negative of the branch cut and branch point integrals after we divide them by $2\pi i$.

Section 4.1, Problem 29. For $t < r/c$, we close the contour on the right side of the plane by Jordan's lemma. Because there are no singularities, $f(t) = 0$. For $t > r/c$, we close it on the left side. With the branch cut along the negative real axis, Figure 4.1.3 shows the contour for the inversion. If $sa/c = \eta e^{\pm \pi i}$, it follows that

$$f(t) = -\frac{1}{2\pi i} \int_\infty^0 \frac{\exp(-\eta\tau)}{(-\eta)[K_0(\eta) + i\pi I_0(\eta)]}\, d\eta - \frac{1}{2\pi i} \int_0^\infty \frac{\exp(-\eta\tau)}{(-\eta)[K_0(\eta) - i\pi I_0(\eta)]}\, d\eta$$

$$= \int_0^\infty \frac{I_0(\eta)}{K_0^2(\eta) + \pi^2 I_0^2(\eta)} e^{-\eta\tau}\, \frac{d\eta}{\eta}.$$

Section 4.1, Problem 30. Here

$$f(t) = \frac{1}{2\pi i} \int_{c-\infty i}^{c+\infty i} \frac{K_0\left(a\sqrt{z}\right) e^{tz}}{\sqrt{z} - b}\, dz.$$

Converting this integral into the contour shown in 4.1.3, we have

$$f(t) = \operatorname{Res}\left[\frac{K_0\left(a\sqrt{z}\right) e^{tz}}{\sqrt{z} - b}; b^2\right] + \frac{1}{2\pi i} \int_{GF} \frac{K_0(a\sqrt{z}) e^{tz}}{\sqrt{z} - b}\, dz + \frac{1}{2\pi i} \int_{DC} \frac{K_0(a\sqrt{z}) c^{tz}}{\sqrt{z} - b}\, dz.$$

The residue at $z = b^2$ is

$$\operatorname{Res}\left[\frac{K_0(a\sqrt{z}) e^{tz}}{\sqrt{z} - b}; b^2\right] = \lim_{z \to b^2} (\sqrt{z} + b) K_0\left(a\sqrt{z}\right) e^{tz} = 2b K_0(ab) e^{b^2 t}.$$

On the other hand, along the bottom of the branch cut GF,

$$\frac{1}{2\pi i} \int_{GF} \frac{K_0(a\sqrt{z}) e^{tz}}{\sqrt{z} - b}\, dz = \frac{1}{2} \int_\infty^0 \frac{J_0(a\eta) + i Y_0(a\eta)}{b + i\eta} \eta e^{-t\eta^2}\, d\eta,$$

while along the top of the branch cut DC,

$$\frac{1}{2\pi i} \int_{DC} \frac{K_0(a\sqrt{z}) e^{tz}}{\sqrt{z} - b}\, dz = \frac{1}{2} \int_0^\infty \frac{-J_0(a\eta) + i Y_0(a\eta)}{b - i\eta} \eta e^{-t\eta^2}\, d\eta.$$

Combining the two integrals together,

$$\frac{1}{2\pi i} \int_{GF \cup DC} \frac{K_0(a\sqrt{z}) e^{tz}}{\sqrt{z} - b}\, dz = -\int_0^\infty \frac{b J_0(a\eta) + \eta Y_0(a\eta)}{b^2 + \eta^2} \eta e^{-t\eta^2}\, d\eta.$$

Combining these results, the inverse is

$$f(t) = 2b K_0(ab) e^{b^2 t} - \int_0^\infty \frac{b J_0(a\eta) + \eta Y_0(a\eta)}{b^2 + \eta^2} \eta e^{-t\eta^2}\, d\eta.$$

Section 4.1, Problem 31. From Figure 4.1.3,

$$\int_{GF} I_\nu(az)K_\nu(bz)e^{tz}\,dz = \int_0^\infty \left[I_\nu(a\eta)K_\nu(b\eta) + \pi i e^{-\nu\pi i}I_\nu(a\eta)I_\nu(b\eta)\right]e^{-t\eta}\,d\eta$$

and

$$\int_{CD} I_\nu(az)K_\nu(bz)e^{tz}\,dz = -\int_0^\infty \left[I_\nu(a\eta)K_\nu(b\eta) - \pi i e^{\nu\pi i}I_\nu(a\eta)I_\nu(b\eta)\right]e^{-t\eta}\,d\eta.$$

Therefore,

$$\int_{GF\cup CD} I_\nu(az)K_\nu(bz)e^{tz}\,dz = 2\pi i\,\cos(\nu\pi)\int_0^\infty I_\nu(a\eta)I_\nu(b\eta)e^{-t\eta}\,d\eta.$$

Therefore, the inverse equals the branch cut integral divided by $2\pi i$.

Section 4.1, Problem 32. Here

$$f(t) = \frac{1}{2\pi i}\int_{c-\infty i}^{c+\infty i} I_{-\nu}\!\left(a\sqrt{z}\right)K_\nu\!\left(b\sqrt{z}\right)e^{tz}\,dz.$$

Rewriting $f(t)$ using the contours shown in Figure 4.1.3, we have

$$f(t) = \frac{1}{2\pi i}\int_{FED} I_{-\nu}\!\left(a\sqrt{z}\right)K_\nu\!\left(b\sqrt{z}\right)e^{tz}\,dz + \frac{1}{2\pi i}\int_{GF} I_{-\nu}\!\left(a\sqrt{z}\right)K_\nu\!\left(b\sqrt{z}\right)e^{tz}\,dz$$

$$+ \frac{1}{2\pi i}\int_{DC} I_{-\nu}\!\left(a\sqrt{z}\right)K_\nu\!\left(b\sqrt{z}\right)e^{tz}\,dz.$$

Now,

$$\int_{FED} I_{-\nu}\!\left(a\sqrt{z}\right)K_\nu\!\left(b\sqrt{z}\right)e^{tz}\,dz = 0.$$

On the other hand, along the bottom of the branch cut GF,

$$\int_{GF} I_{-\nu}\!\left(a\sqrt{z}\right)K_\nu\!\left(b\sqrt{z}\right)e^{tz}\,dz = -\int_\infty^0 I_{-\nu}(-ia\sqrt{\eta})\,K_\nu(-ib\sqrt{\eta})\,e^{-t\eta}\,d\eta,$$

while along the top of the branch cut DC,

$$\int_{DC} I_{-\nu}\!\left(a\sqrt{z}\right)K_\nu\!\left(b\sqrt{z}\right)e^{tz}\,dz = -\int_0^\infty I_{-\nu}(ia\sqrt{\eta})\,K_\nu(ib\sqrt{\eta})\,e^{-t\eta}\,d\eta.$$

Combining the integrals together,

$$\int_{GF\cup FED\cup DC} I_{-\nu}\!\left(a\sqrt{z}\right)K_\nu\!\left(b\sqrt{z}\right)e^{tz}\,dz$$

$$= \int_0^\infty \left[I_{-\nu}(-ia\sqrt{\eta})\,K_\nu(-ib\sqrt{\eta}) - I_{-\nu}(ia\sqrt{\eta})\,K_\nu(ib\sqrt{\eta})\right]e^{-t\eta}\,d\eta$$

$$= \frac{\pi}{2\sin(\nu\pi)}\int_0^\infty I_{-\nu}(ia\sqrt{\eta})\,I_{-\nu}(ib\sqrt{\eta})\left(e^{2\nu\pi i} - 1\right)e^{-t\eta}\,d\eta$$

$$= \pi i\int_0^\infty J_{-\nu}(a\sqrt{\eta})\,J_{-\nu}(b\sqrt{\eta})\,e^{-t\eta}\,d\eta.$$

Upon using the result stated in the text and dividing by $2\pi i$, we obtain the inverse.

Section 4.1, Problem 33. There are no singularities, just the branch cut. Therefore, the inverse equals the branch cut integrals using the contours shown on Figure 4.1.3:

$$f(t) = \frac{1}{2\pi i} \int_0^\infty \frac{K_0(i\eta r)\,e^{-\eta^2 t}}{(i\eta)K_1(i\eta a)}\, 2\eta\, d\eta + \frac{1}{2\pi i} \int_\infty^0 \frac{K_0(-i\eta r)\,e^{-\eta^2 t}}{(-i\eta)K_1(-i\eta a)}\, 2\eta\, d\eta$$

$$= \frac{1}{\pi i} \int_0^\infty \frac{J_0(\eta r) - iY_0(\eta r)}{J_1(\eta a) - iY_1(\eta a)} e^{-\eta^2 t}\, d\eta - \frac{1}{\pi i} \int_0^\infty \frac{J_0(\eta r) + iY_0(\eta r)}{J_1(\eta a) + iY_1(\eta a)} e^{-\eta^2 t}\, d\eta,$$

because $K_0(iz) = -\pi i[J_0(z) - iY_0(z)]/2$, $K_0(-iz) = \pi i[J_0(z) + iY_0(z)]/2$, $K_1(iz) = -\pi[J_1(z) - iY_1(z)]/2$, $K_1(-iz) = -\pi[J_1(z) + iY_1(z)]/2$ and $z = -\eta^2$. The first integral arises from the top of the branch cut, while the second integral arises from the bottom. After a little bit of complex algebra, we obtain the inverse.

Section 4.1, Problem 34. The contour is shown in Figure 4.1.3. Because

$$K_0(z) \approx \log(2/z) \quad \text{and} \quad K_n(z) \approx \frac{(n-1)!}{2} \left(\frac{z}{2}\right)^{-n} \quad \text{as } z \to 0,$$

where $n = 1, 2, 3, \ldots$,

$$\int_{\substack{branch \\ point}} \frac{K_n(\sqrt{z})\,e^{tz}}{\sqrt{z}\,K_{n-1}(\sqrt{z})}\, dz = 4\pi i \begin{cases} 0, & n = 1, \\ n-1, & n \geq 2. \end{cases}$$

Along GF,

$$\int_{GF} \frac{K_n(\sqrt{z})\,e^{tz}}{\sqrt{z}\,K_{n-1}(\sqrt{z})}\, dz = -\int_\infty^0 \frac{K_n(-i\eta)e^{-t\eta^2}}{(-i\eta)K_{n-1}(-i\eta)}\, 2\eta\, d\eta$$

$$= -2\int_0^\infty \frac{J_n(\eta) + iY_n(\eta)}{J_{n-1}(\eta) + iY_{n-1}(\eta)} e^{-t\eta^2}\, d\eta,$$

because $K_n(-i\eta) = \pi i e^{n\pi i/2}[J_n(\eta) + iY_n(\eta)]/2$. Along DC,

$$\int_{DC} \frac{K_n(\sqrt{z})\,e^{tz}}{\sqrt{z}\,K_{n-1}(\sqrt{z})}\, dz = -\int_\infty^0 \frac{K_n(i\eta)e^{-t\eta^2}}{(i\eta)K_{n-1}(i\eta)}\, 2\eta\, d\eta$$

$$= 2\int_0^\infty \frac{J_n(\eta) - iY_n(\eta)}{J_{n-1}(\eta) - iY_{n-1}(\eta)} e^{-t\eta^2}\, d\eta,$$

because $K_n(-i\eta) = -\pi i e^{-n\pi i/2}[J_n(\eta) - iY_n(\eta)]/2$. Combining the GF and DC integrals,

$$\int_{GF \cup DC} \frac{K_n(\sqrt{z})\,e^{tz}}{\sqrt{z}\,K_{n-1}(\sqrt{z})}\, dz = \frac{8i}{\pi} \int_0^\infty \frac{e^{-t\eta^2}}{J_{n-1}^2(\eta) + Y_{n-1}^2(\eta)} \frac{d\eta}{\eta}.$$

The inverse equals the sum of the branch point and cut integrals after they are divided by $2\pi i$.

Section 4.1, Problem 35. Following Figure 4.1.3, Bromwich's integral is

$$f(t) = \frac{K_0(r)}{K_0(1)} + \frac{1}{2\pi i} \int_{\substack{branch \\ cut}} \frac{K_0\big(r\sqrt{z+1}\big)\,e^{tz}}{z\,K_0\big(\sqrt{z+1}\big)}\,dz,$$

where the first term on the right side is from the residue at $z = 0$. Now, the branch cut integral is

$$\int_{\substack{branch \\ cut}} \frac{K_0\big(r\sqrt{z+1}\big)\,e^{tz}}{z\,K_0\big(\sqrt{z+1}\big)}\,dz$$

$$= -\int_0^\infty \frac{K_0(r\eta i)e^{-(1+\eta^2)t}}{(-1-\eta^2)K_0(\eta i)}\,2\eta\,d\eta - \int_\infty^0 \frac{K_0(-r\eta i)e^{-(1+\eta^2)t}}{(-1-\eta^2)K_0(-\eta i)}\,2\eta\,d\eta$$

$$= 2\int_0^\infty \frac{J_0(r\eta) + iY_0(r\eta)}{(1+\eta^2)[J_0(\eta) + iY_0(\eta)]}e^{-(1+\eta^2)t}\,\eta\,d\eta$$

$$- 2\int_0^\infty \frac{J_0(r\eta) - iY_0(r\eta)}{(1+\eta^2)[J_0(\eta) - iY_0(\eta)]}e^{-(1+\eta^2)t}\,\eta\,d\eta$$

$$= 4i\int_0^\infty \frac{Y_0(r\eta)J_0(\eta) - J_0(r\eta)Y_0(\eta)}{J_0^2(\eta) + Y_0^2(\eta)}e^{-(1+\eta^2)t}\,\frac{\eta}{1+\eta^2}\,d\eta,$$

because $z+1 = \eta^2 e^{\pm\pi i}$, $K_0(iz) = -\pi i[J_0(z)-iY_0(z)]/2$ and $K_0(-iz) = \pi i[J_0(z)+iY_0(z)]/2$. After substituting the branch cut integral into the first equation, we obtain the inverse.

§4.1, Problem 36. The contour used to invert the transform is similar to Figure 4.1.1. The arcs at infinity vanish by Jordan's lemma. Therefore,

$$f(t) = \text{Res}\left[\frac{K_0\big(r\sqrt{z/\nu + 1/b^2}\big)\,e^{tz}}{z\,K_0\big(a\sqrt{z/\nu + 1/b^2}\big)};0\right] - \frac{1}{2\pi i}\int_{BC \cup IH} \frac{K_0\big(r\sqrt{z/\nu + 1/b^2}\big)\,e^{tz}}{z\,K_0\big(a\sqrt{z/\nu + 1/b^2}\big)}\,dz.$$

Now, the residue at $z = 0$ is

$$\text{Res}\left[\frac{K_0\big(r\sqrt{z/\nu + 1/b^2}\big)\,e^{tz}}{z\,K_0\big(a\sqrt{z/\nu + 1/b^2}\big)};0\right] = \lim_{z\to 0}\frac{K_0(r\sqrt{z/\nu + 1/b^2})}{K_0(a\sqrt{z/\nu + 1/b^2})} = \frac{K_0(r/b)}{K_0(a/b)}.$$

The branch cut consists of two parts. Along the top, $z + \nu/b^2 = \nu\chi^2 e^{\pi i}$, $0 < \chi < \infty$. Therefore,

$$\int_{BC} \frac{K_0\big(r\sqrt{z/\nu + 1/b^2}\big)\,e^{tz}}{z\,K_0\big(a\sqrt{z/\nu + 1/b^2}\big)}\,dz = 2\int_\infty^0 \frac{K_0(ir\chi)\exp(-\nu t/b^2 - \nu t\chi^2)}{(\chi^2 + 1/b^2)K_0(ia\chi)}\,\chi\,d\chi$$

$$= -2e^{-\nu t/b^2}\int_0^\infty \frac{\exp(-\nu t\chi^2)[J_0(r\chi) - iY_0(r\chi)]}{(\chi^2 + 1/b^2)[J_0(a\chi) - iY_0(a\chi)]}\,\chi\,d\chi,$$

because $K_0(ir\chi) = -\pi i[J_0(r\chi)-iY_0(r\chi)]/2$. Along the bottom of the branch cut, $z+\nu/b^2 = \nu\chi^2 e^{-\pi i}$, $0 < \chi < \infty$. Therefore,

$$\int_{IH} \frac{K_0\big(r\sqrt{z/\nu + 1/b^2}\big)\,e^{tz}}{z\,K_0\big(a\sqrt{z/\nu + 1/b^2}\big)}\,dz = 2e^{-\nu t/b^2}\int_0^\infty \frac{\exp(-\nu t\chi^2)[J_0(r\chi) + iY_0(r\chi)]}{(\chi^2 + 1/b^2)[J_0(a\chi) + iY_0(a\chi)]}\,\chi\,d\chi.$$

Therefore,

$$\int_{BC \cup IH} \frac{K_0\!\left(r\sqrt{z/\nu + 1/b^2}\right) e^{tz}}{z\, K_0\!\left(a\sqrt{z/\nu + 1/b^2}\right)}\, dz$$

$$= 4ie^{-\nu t/b^2} \int_0^\infty \frac{\exp(-\nu t\chi^2)}{(\chi^2 + 1/b^2)}\, \frac{Y_0(r\chi)J_0(a\chi) - J_0(r\chi)Y_0(a\chi)}{J_0^2(a\chi) + Y_0^2(a\chi)} \chi\, d\chi.$$

Substituting $\chi = a\eta$ into the branch cut integral and then introducing this integral into the first equation, we obtain the inverse.

Section 4.1, Problem 37. Because there are no poles, the only contribution arise from the branch cut integral. Along the top of the negative real axis, there are three regions. In Region 1, $-\infty < \eta e^{\pi i} < -b$, $\sqrt{z} = i\sqrt{\eta}$, $\sqrt{z+a} = i\sqrt{\eta - a}$ and $\sqrt{z+b} = i\sqrt{\eta - b}$ so that

$$K_0\!\left[r\sqrt{\frac{z(z+b)}{c(z+a)}}\right] = -\frac{\pi i}{2}\left\{ J_0\!\left[r\sqrt{\frac{\eta(\eta-b)}{c(\eta-a)}}\right] - iY_0\!\left[r\sqrt{\frac{\eta(\eta-b)}{c(\eta-a)}}\right] \right\}.$$

In Region 2, $-b < \eta e^{\pi i} < -a$, $\sqrt{z} = i\sqrt{\eta}$, $\sqrt{z+a} = i\sqrt{\eta - a}$ and $\sqrt{z+b} = \sqrt{b - \eta}$ so that

$$K_0\!\left[r\sqrt{\frac{z(z+b)}{c(z+a)}}\right] = K_0\!\left[r\sqrt{\frac{\eta(b-\eta)}{c(\eta-a)}}\right].$$

In Region 3, $-a < \eta e^{\pi i} < 0$, $\sqrt{z} = i\sqrt{\eta}$, $\sqrt{z+a} = \sqrt{a - \eta}$ and $\sqrt{z+b} = \sqrt{b - \eta}$ so that

$$K_0\!\left[r\sqrt{\frac{z(z+b)}{c(z+a)}}\right] = -\frac{\pi i}{2}\left\{ J_0\!\left[r\sqrt{\frac{\eta(b-\eta)}{c(a-\eta)}}\right] - iY_0\!\left[r\sqrt{\frac{\eta(b-\eta)}{c(a\quad \eta)}}\right] \right\}.$$

Similarly, below the branch cut in Region 1, $-\infty < \eta e^{-\pi i} < -b$, $\sqrt{z} = -i\sqrt{\eta}$, $\sqrt{z+a} = -i\sqrt{\eta - a}$ and $\sqrt{z+b} = -i\sqrt{\eta - b}$ so that

$$K_0\!\left[r\sqrt{\frac{z(z+b)}{c(z+a)}}\right] = \frac{\pi i}{2}\left\{ J_0\!\left[r\sqrt{\frac{\eta(\eta-b)}{c(\eta-a)}}\right] + iY_0\!\left[r\sqrt{\frac{\eta(\eta-b)}{c(\eta-a)}}\right] \right\}.$$

In Region 2, $-b < \eta e^{-\pi i} < -a$, $\sqrt{z} = -i\sqrt{\eta}$, $\sqrt{z+a} = -i\sqrt{\eta - a}$ and $\sqrt{z+b} = \sqrt{b - \eta}$ so that

$$K_0\!\left[r\sqrt{\frac{z(z+b)}{c(z+a)}}\right] = K_0\!\left[r\sqrt{\frac{\eta(b-\eta)}{c(\eta-a)}}\right].$$

In Region 3, $-a < \eta e^{-\pi i} < 0$, $\sqrt{z} = -i\sqrt{\eta}$, $\sqrt{z+a} = \sqrt{a - \eta}$ and $\sqrt{z+b} = \sqrt{b - \eta}$ so that

$$K_0\!\left[r\sqrt{\frac{z(z+b)}{c(z+a)}}\right] = \frac{\pi i}{2}\left\{ J_0\!\left[r\sqrt{\frac{\eta(b-\eta)}{c(a-\eta)}}\right] + iY_0\!\left[r\sqrt{\frac{\eta(b-\eta)}{c(a-\eta)}}\right] \right\}.$$

Now,

$$f(t) = \frac{1}{2\pi i}\left\{ \int_{-\infty-0i}^{-b-0i} K_0\!\left[r\sqrt{\frac{z(z+b)}{c(z+a)}}\right] e^{tz}\, dz + \int_{-b-0i}^{-a-0i} K_0\!\left[r\sqrt{\frac{z(z+b)}{c(z+a)}}\right] e^{tz}\, dz \right.$$

$$+ \int_{-a-0i}^{0-0i} K_0\!\left[r\sqrt{\frac{z(z+b)}{c(z+a)}}\right] e^{tz}\, dz + \int_{0+0i}^{-a+0i} K_0\!\left[r\sqrt{\frac{z(z+b)}{c(z+a)}}\right] e^{tz}\, dz$$

$$\left. + \int_{-a+0i}^{-b+0i} K_0\!\left[r\sqrt{\frac{z(z+b)}{c(z+a)}}\right] e^{tz}\, dz + \int_{-b+0i}^{-\infty+0i} K_0\!\left[r\sqrt{\frac{z(z+b)}{c(z+a)}}\right] e^{tz}\, dz \right\}.$$

Upon substituting, we find that the second and fifth integrals cancel, leaving the inverse.

Section 4.1, Problem 38. The closed Bromwich integral is similar to Figure 4.1.3. Within the contour, we have singularities at $z = \pm \omega i$. Their residues are

$$\text{Res}\left[\frac{\omega\, K_1\!\left(r\sqrt{z}\right) e^{tz}}{(z^2 + \omega^2) K_1\!\left(\sqrt{z}\right)}; \omega i \right] = \frac{1}{2i} \frac{K_1\!\left(r\sqrt{\omega i}\right)}{K_1\!\left(\sqrt{\omega i}\right)} e^{i\omega t}$$

and

$$\text{Res}\left[\frac{\omega\, K_1\!\left(r\sqrt{z}\right) e^{tz}}{(z^2 + \omega^2) K_1\!\left(\sqrt{z}\right)}; -\omega i \right] = -\frac{1}{2i} \frac{K_1\!\left(r\sqrt{-\omega i}\right)}{K_1\!\left(\sqrt{-\omega i}\right)} e^{-i\omega t}.$$

Because

$$K_1\!\left(z e^{\pm \pi i/4}\right) e^{\mp \pi i/2} = \ker_1(z) \pm i \kei_1(z),$$

and defining

$$\ker_1(z) + i\,\kei_1(z) = N_1(z) e^{-i\phi_1(z)},$$

the sum of the residues yields

$$\frac{N_1\!\left(r\sqrt{\omega}\right)}{N_1\!\left(\sqrt{\omega}\right)} \sin\!\left[\omega t + \phi_1\!\left(r\sqrt{\omega}\right) - \phi_1\!\left(\sqrt{\omega}\right)\right].$$

Along the top of the branch cut, $z = \eta^2 e^{\pi i}$ and

$$\int_{DC} \frac{\omega\, K_1\!\left(r\sqrt{z}\right) e^{tz}}{(z^2 + \omega^2) K_1\!\left(\sqrt{z}\right)}\, dz = -2 \int_0^\infty \frac{\omega}{\omega^2 + \eta^4} \frac{K_1(r\eta i)}{K_1(\eta i)} e^{-\eta^2 t}\, \eta\, d\eta,$$

while along the bottom of the branch cut, $z = \eta^2 e^{-\pi i}$ and

$$\int_{GF} \frac{\omega\, K_1\!\left(r\sqrt{z}\right) e^{tz}}{(z^2 + \omega^2) K_1\!\left(\sqrt{z}\right)}\, dz = -2 \int_\infty^0 \frac{\omega}{\omega^2 + \eta^4} \frac{K_1(-r\eta i)}{K_1(-\eta i)} e^{-\eta^2 t}\, \eta\, d\eta.$$

Using the relationship

$$K_1\!\left(z e^{\pm i/2}\right) = \pm \pi i e^{\mp i/2} [-J_1(z) \pm i Y_1(z)]/2,$$

the branch cut integral reduces to

$$-4i \int_0^\infty \frac{\omega \eta}{\omega^2 + \eta^4} \frac{J_1(r\eta) Y_1(\eta) - Y_1(r\eta) J_1(\eta)}{J_1^2(\eta) + Y_1^2(\eta)} e^{-\eta^2 t}\, d\eta.$$

The inverse $f(t)$ equals the sum of the residues plus the branch cut integral after we divide it by $2\pi i$.

Section 4.1, Problem 39. We close Bromwich's contour in a manner *similar* to Figure 4.1.3. If we take the branch cuts along the negative real axis of the z-plane, then for $-\infty < z < -b^2$, the phase of $z + a^2$ and $z + b^2$ equals π above the branch cut and equals $-\pi$ below the branch cut. Therefore, the argument of the modified Bessel function is the same on either side of the cut and the value of the integration along the top of the branch cut equals the negative of the integration just below the branch cut. Thus, there is no contribution from the integration between $z = -\infty$ and $z = -b^2$.

However, between $z = -b^2$ and $z = -a^2$, $z = -\eta$, $z + a^2 = (\eta - a^2)e^{\pm\pi i}$ and $z + b^2 = (b^2 - \eta)e^{0i}$. Therefore,

$$f(t) = \frac{1}{2\pi i} \int_{b^2}^{a^2} \frac{e^{-\eta t}}{b^2 - \eta} K_0\left(rci\sqrt{\frac{\eta - a^2}{b^2 - \eta}}\right) d\eta - \frac{1}{2\pi i} \int_{a^2}^{b^2} \frac{e^{-\eta t}}{b^2 - \eta} K_0\left(-rci\sqrt{\frac{\eta - a^2}{b^2 - \eta}}\right) d\eta.$$

Because $K_0(\pm xi) = \mp\pi i[J_0(x) \mp iY_0(x)]/2$ when $x > 0$,

$$f(t) = \frac{1}{2} \int_{a^2}^{b^2} \frac{1}{b^2 - \eta} J_0\left(rc\sqrt{\frac{\eta - a^2}{b^2 - \eta}}\right) e^{-t\eta} d\eta.$$

Section 4.1, Problem 40. Let us first find the inverse of

$$G(s) = \frac{K_1\left(a\sqrt{s}\right)}{bK_1\left(\sqrt{s}\right) + \sqrt{s}K_0\left(\sqrt{s}\right)}.$$

This transform has no poles, so

$$\int_{DC \cup GF} \frac{K_1\left(a\sqrt{s}\right)e^{tz}}{bK_1\left(\sqrt{s}\right) + \sqrt{s}K_0\left(\sqrt{s}\right)} dz$$

$$= -\int_0^\infty \frac{K_1(\eta ai)\, e^{-\eta^2 t}}{bK_1(\eta i) + \eta i K_0(\eta i)} (2\eta\, d\eta) - \int_\infty^0 \frac{K_1(-\eta ai)\, e^{-\eta^2 t}}{bK_1(-\eta i) - \eta i K_0(-\eta i)} (2\eta\, d\eta)$$

$$= 2\int_0^\infty \frac{\eta\,[J_1(\eta a) + iY_1(\eta a)]e^{-\eta^2 t}}{b[J_1(\eta) + iY_1(\eta)] - \eta[J_0(\eta) + iY_0(\eta)]} d\eta$$

$$- 2\int_0^\infty \frac{\eta\,[J_1(\eta a) - iY_1(\eta a)]e^{-\eta^2 t}}{b[J_1(\eta) - iY_1(\eta)] - \eta[J_0(\eta) - iY_0(\eta)]} d\eta$$

$$= 4i\int_0^\infty \eta e^{-\eta^2 t}\, d\eta\, \frac{Y_1(\eta a)[bJ_1(\eta) - \eta J_0(\eta)] - J_1(\eta a)[bY_1(\eta) - \eta Y_0(\eta)]}{[bJ_1(\eta) - \eta J_0(\eta)]^2 + [bY_1(\eta) - \eta Y_0(\eta)]^2},$$

because $z = \eta^2 e^{\pm\pi i}$, $K_0(zi) = -\pi i[J_0(z) - iY_0(z)]/2$, $K_0(-zi) = \pi i[J_0(z) + iY_0(z)]/2$, $K_1(zi) = -\pi[J_1(z) - iY_1(z)]/2$ and $K_1(-zi) = -\pi[J_1(z) + iY_1(z)]/2$. Therefore,

$$g(t) = \frac{2}{\pi} \int_0^\infty e^{-\eta^2 t} \frac{J_1(\eta a)[bY_1(\eta) - \eta Y_0(\eta)] - Y_1(\eta a)[bJ_1(\eta) - \eta J_0(\eta)]}{[bJ_1(\eta) - \eta J_0(\eta)]^2 + [bY_1(\eta) - \eta Y_0(\eta)]^2} \eta\, d\eta.$$

Finally,

$$f(t) = \int_0^\infty g(\tau)\, d\tau = \frac{2}{\pi} \int_0^\infty \left(1 - e^{-\eta^2 t}\right) \frac{J_1(\eta a)[bY_1(\eta) - \eta Y_0(\eta)] - Y_1(\eta a)[bJ_1(\eta) - \eta J_0(\eta)]}{[bJ_1(\eta) - \eta J_0(\eta)]^2 + [bY_1(\eta) - \eta Y_0(\eta)]^2} \frac{d\eta}{\eta}.$$

Section 4.1, Problem 41. First, we note that

$$\left.\frac{\partial K_1\left(r\sqrt{z}\right)}{\partial r}\right|_{r=a} = -\frac{\sqrt{z}}{2}\left[K_0\left(a\sqrt{z}\right) + K_2\left(a\sqrt{z}\right)\right].$$

The contour integral is identical to Figure 4.1.3. The integration around the branch point is

$$\int_{FED} \frac{K_1\!\left(r\sqrt{z}\right) e^{tz}}{(Iz + 2\pi\mu a^2)K_1\!\left(a\sqrt{z}\right) - 2\pi\mu a^3 \partial K_1\!\left(r\sqrt{z}\right)/\partial r|_{r=a}} \frac{dz}{z} = -\frac{i}{2a\mu r},$$

where we have used the asymptotic forms of $K_0(\cdot)$, $K_1(\cdot)$ and $K_2(\cdot)$ as $a\sqrt{z} \to 0$.

Turning to the branch cut integral, $z = \eta^2 e^{\pi i}$, $K_0\!\left(a\sqrt{z}\right) = -\pi i[J_0(a\eta) - iY_0(a\eta)]/2$, $K_1\!\left(a\sqrt{z}\right) = -\pi[J_1(a\eta) - iY_1(a\eta)]/2$ and $K_2\!\left(a\sqrt{z}\right) = \pi i[J_2(a\eta) - iY_2(a\eta)]/2$ along the top of the branch cut so that

$$\int_{CD} \frac{K_1\!\left(r\sqrt{z}\right) e^{tz}}{(Iz + 2\pi\mu a^2)K_1\!\left(a\sqrt{z}\right) - 2\pi\mu a^3 \partial K_1\!\left(r\sqrt{z}\right)/\partial r|_{r=a}} \frac{dz}{z}$$

$$= -\frac{2}{I}\int_0^\infty \frac{[J_1(r\eta) - iY_1(r\eta)]\exp(-\eta^2 t)}{[\eta J_1(a\eta) - \chi J_2(a\eta)] - i[\eta Y_1(a\eta) - \chi Y_2(a\eta)]} \frac{d\eta}{\eta^2}.$$

Along the bottom of the branch cut, $z = \eta^2 e^{-\pi i}$, $K_0\!\left(a\sqrt{z}\right) = \pi i[J_0(a\eta) + iY_0(a\eta)]/2$, $K_1\!\left(a\sqrt{z}\right) = -\pi[J_1(a\eta) + iY_1(a\eta)]/2$ and $K_2\!\left(a\sqrt{z}\right) = -\pi i[J_2(a\eta) + iY_2(a\eta)]/2$, so that

$$\int_{GF} \frac{K_1\!\left(r\sqrt{z}\right) e^{tz}}{(Iz + 2\pi\mu a^2)K_1\!\left(a\sqrt{z}\right) - 2\pi\mu a^3 \partial K_1\!\left(r\sqrt{z}\right)/\partial r|_{r=a}} \frac{dz}{z}$$

$$= \frac{2}{I}\int_0^\infty \frac{[J_1(r\eta) + iY_1(r\eta)]\exp(-\eta^2 t)}{[\eta J_1(a\eta) - \chi J_2(a\eta)] + i[\eta Y_1(a\eta) - \chi Y_2(a\eta)]} \frac{d\eta}{\eta^2}.$$

We have used the relationships that $J_0(a\eta) = 2J_1(a\eta)/(a\eta) - J_2(a\eta)$ and $Y_0(a\eta) = 2Y_1(a\eta)/(a\eta) - Y_2(a\eta)$ to eliminate $J_0(\cdot)$ and $Y_0(\cdot)$ from the integrals. The inverse equals the branch cut integral after we divide it by $2\pi i$.

Section 5.1

Section 5.1, Problem 1. The cut ω-plane is similar to Figure 5.1.1 with $\epsilon = 0$ except that there is no branch cut in the lower half plane. Because the singularity is in the upper half-plane, $f(t) = 0$ if $t < 0$. Along the branch cut, $z - i = \chi e^{\pi i/2}$ along CD and $z - i = \chi e^{-3\pi i/2}$ along GF. Therefore, for $t > 0$,

$$f(t) = \frac{1}{2\pi}\int_{-\infty}^\infty \frac{e^{it\omega}}{\sqrt{1 + i\omega}}\, d\omega = -\frac{1}{2\pi}\int_{CD} \frac{e^{itz}}{\sqrt{1 + iz}}\, dz - \frac{1}{2\pi}\int_{FG} \frac{e^{itz}}{\sqrt{1 + iz}}\, dz$$

$$= -\frac{1}{2\pi}\int_\infty^0 \frac{e^{-t-t\chi}}{i\sqrt{\chi}}(i\,d\chi) - \frac{1}{2\pi}\int_0^\infty \frac{e^{-t-t\chi}}{-i\sqrt{\chi}}(i\,d\chi) = \frac{e^{-t}}{\pi}\int_0^\infty \frac{e^{-t\chi}}{\sqrt{\chi}}\, d\chi$$

$$= \frac{2e^{-t}}{\pi}\int_0^\infty e^{-t\eta^2}\, d\eta = \frac{e^{-t}}{\sqrt{\pi t}},$$

where $\chi = \eta^2$.

Section 5.1, Problem 2. For $t > t_0$, we close the contour in the upper half-plane, while for $t < t_0$, we close the contour in the lower half-plane. Because the singularities are only in the upper half-plane, $f(t) = 0$ if $t < t_0$. Along the branch cut on the positive real axis, $z - a = (\eta - a)e^{\theta i}$, where $\theta = 0$ or 2π, and $z + a = (\eta + a)e^{\pi i}$. Along the branch cut on the negative real axis, $z - a = (\eta + a)e^{\pi i}$ and $z + a = (\eta - a)e^{\theta i}$, where $\theta = -\pi$ or π. The nonvanishing portions of the line integrals arise from the branch cut edge below the positive real axis, the branch cut edge above the positive real axis, the branch cut edge above the negative real axis and the branch cut edge below the negative real axis. In that order,

$$
f(t) = \frac{-i}{2\pi} \int_\infty^a \frac{e^{it\eta} \exp\left(ito\sqrt{\eta^2 - a^2}\right)}{-\sqrt{\eta^2 - a^2}}\, d\eta + \frac{-i}{2\pi} \int_a^\infty \frac{e^{it\eta} \exp\left(-ito\sqrt{\eta^2 - a^2}\right)}{\sqrt{\eta^2 - a^2}}\, d\eta
$$

$$
+ \frac{-i}{2\pi} \int_\infty^a \frac{e^{-it\eta} \exp\left(ito\sqrt{\eta^2 - a^2}\right)}{-\sqrt{\eta^2 - a^2}}\, (-d\eta)
$$

$$
+ \frac{-i}{2\pi} \int_a^\infty \frac{e^{-it\eta} \exp\left(-ito\sqrt{\eta^2 - a^2}\right)}{\sqrt{\eta^2 - a^2}}\, (-d\eta)
$$

$$
= \frac{2}{\pi} \int_a^\infty \frac{\cos(t\eta)\cos\left(to\sqrt{\eta^2 - a^2}\right)}{\sqrt{\eta^2 - a^2}}\, d\eta = J_0\left(a\sqrt{t^2 - t_0^2}\right).
$$

Section 5.1, Problem 3. We can rewrite

$$
F(\omega) = \frac{\exp\left[-\omega i\sqrt{r^2 + (z + a/\omega i)^2}\right]}{\sqrt{r^2 + (z + a/\omega i)^2}}
$$

as

$$
F(\omega) = \omega i \frac{\exp\left[-\sqrt{(r\omega i)^2 + (z\omega i + a)^2}\right]}{\sqrt{(r\omega i)^2 + (z\omega i + a)^2}} = \omega i G(\omega).
$$

Therefore,

$$
f(t) - \frac{dg(t)}{dt}.
$$

Now, let

$$
G(\omega) = \frac{\exp\left[-ir_i\sqrt{(\omega - \omega_1)(\omega - \omega_2)}\right]}{ir_i\sqrt{(\omega - \omega_1)(\omega - \omega_2)}},
$$

where $\omega_1 = \beta + \alpha i$, $\omega_2 = -\beta + \alpha i$, $\alpha = az/r_i^2$, $\beta = ar/r_i^2$ and $r_i^2 = r^2 + z^2$. In order to satisfy the conditions on the square root, $z - \omega_{1,2} = re^{\theta i}$ with $2\pi < \theta < 4\pi$. The integration is similar to Figure 5.1.3; the only nonzero contribution arises from the integration along the branch cuts. Therefore,

$$
g(t) = \frac{1}{2\pi} \int_{-\infty}^\infty G(\omega)e^{i\omega t}\, d\omega = \frac{e^{-\alpha t}}{2\pi} \int_{-\infty - i\epsilon}^{\infty - i\epsilon} \frac{e^{-ir_i\sqrt{(\omega + \beta)(\omega - \beta)}}e^{it\omega}}{ir_i\sqrt{(\omega + \beta)(\omega - \beta)}}\, d\omega
$$

$$
= \frac{e^{-\alpha t}}{2\pi r_i}\left(\int_0^\beta \frac{e^{r_i\sqrt{\beta^2 - \eta^2}}e^{i\eta t}}{\sqrt{\beta^2 - \eta^2}}\, d\eta - \int_\beta^0 \frac{e^{-r_i\sqrt{\beta^2 - \eta^2}}e^{i\eta t}}{\sqrt{\beta^2 - \eta^2}}\, d\eta \right.
$$

$$
\left. + \int_0^\beta \frac{e^{-r_i\sqrt{\beta^2 - \eta^2}}e^{-i\eta t}}{\sqrt{\beta^2 - \eta^2}}\, d\eta - \int_\beta^0 \frac{e^{r_i\sqrt{\beta^2 - \eta^2}}e^{-i\eta t}}{\sqrt{\beta^2 - \eta^2}}\, d\eta \right)
$$

$$g(t) = \frac{e^{-\alpha t}}{\pi r_i} \left[\int_0^\beta \frac{e^{r_i \sqrt{\beta^2 - \eta^2}} \cos(\eta t)}{\sqrt{\beta^2 - \eta^2}} \, d\eta + \int_0^\beta \frac{e^{-r_i \sqrt{\beta^2 - \eta^2}} \cos(\eta t)}{\sqrt{\beta^2 - \eta^2}} \, d\eta \right]$$

$$= \frac{2e^{-\alpha t}}{\pi r_i} \int_0^\beta \frac{\cosh\left(r_i \sqrt{\beta^2 - \eta^2}\right) \cos(\eta t)}{\sqrt{\beta^2 - \eta^2}} \, d\eta = \frac{2e^{-\alpha t}}{\pi r_i} J_0\left(\beta \sqrt{t^2 - r_i^2}\right) H(t - r_i).$$

Section 5.1, Problem 4. For $t > 0$, we close *both* integrals with a semicircle of infinite radius in the upper half of the z-plane. Within the first integral, we have a second-order pole at $z = 0$; its residue is

$$\text{Res}\left(\frac{e^x \sqrt{z^2 + k^2}}{k - \sqrt{z^2 + k^2}}; 0 \right) = \lim_{z \to 0} \frac{d}{dz} \left[\frac{z^2 \exp\left(x\sqrt{z^2 + k^2} + itz\right)}{k - \sqrt{z^2 + k^2}} \right] = -2ikte^{kx}.$$

From our choice of branch cuts, $z - ki = re^{\theta i}$, where $-3\pi/2 < \theta < \pi/2$, and $z + ki = re^{\theta i}$, where $-\pi/2 < \theta < 3\pi/2$. Therefore, the integration along the branch cut in the upper half-plane is

$$\int_{\substack{branch \\ cut}} \frac{e^x \sqrt{z^2 + k^2}}{k - \sqrt{z^2 + k^2}} \, dz = \int_\infty^k \frac{\exp\left(ix\sqrt{\eta^2 - k^2} - t\eta\right)}{k - i\sqrt{\eta^2 - k^2}} \, i \, d\eta$$

$$+ \int_k^\infty \frac{\exp\left(-ix\sqrt{\eta^2 - k^2} - t\eta\right)}{k + i\sqrt{\eta^2 - k^2}} \, i \, d\eta$$

$$= 2 \int_k^\infty \frac{e^{-t\eta}}{\eta^2} \left[k \sin\left(x\sqrt{\eta^2 - k^2}\right) \right.$$

$$\left. + \sqrt{\eta^2 - k^2} \, \cos\left(x\sqrt{\eta^2 - k^2}\right) \right] d\eta.$$

Thus,

$$\frac{1}{2\pi} \int_{-\infty - \epsilon i}^{\infty - \epsilon i} \frac{e^x \sqrt{z^2 + k^2}}{k - \sqrt{z^2 + k^2}} \, dz = 2kte^{kz} - \frac{1}{2\pi} \int_{\substack{branch \\ cut}} \frac{e^x \sqrt{z^2 + k^2}}{k - \sqrt{z^2 + k^2}} \, dz.$$

On the other hand,

$$\frac{1}{2\pi} \int_{-\infty + \epsilon i}^{\infty + \epsilon i} \frac{e^x \sqrt{z^2 + k^2}}{k - \sqrt{z^2 + k^2}} \, dz = -\frac{1}{2\pi} \int_{\substack{branch \\ cut}} \frac{e^x \sqrt{z^2 + k^2}}{k - \sqrt{z^2 + k^2}} \, dz.$$

We now substitute these results into the formula.

For $t < 0$, we close both integrals in the lower half-plane and repeat the previous analysis. Recall that the contour is now in the negative sense so we must take the negative of the residue.

Section 5.1, Problem 5.

$$f(t) = -\frac{1}{2\pi} \int_{-\infty}^\infty \frac{|\omega| \exp(-|\omega|s + i\omega t)}{|\omega| - a_1} \, d\omega,$$

where we pass over the singularities at $\omega = \pm a_1$. The residues for these singularities are

$$\mathrm{Res}\left(\frac{|z|\,e^{-s|z|+itz}}{|z|-a_1};a_1\right) = a_1 e^{-a_1 s + i a_1 t} \quad \text{and} \quad \mathrm{Res}\left(\frac{|z|\,e^{-s|z|+itz}}{|z|-a_1};-a_1\right) = -a_1 e^{-a_1 s - i a_1 t}.$$

Therefore, the sum of the residues is $2i a_1 e^{-a_1 s} \sin(a_1 t)$.

For $t < 0$, we close the contour in the lower half-plane. This includes two simple poles. Recalling that the closed contour in the lower half-plane is in the negative sense,

$$f(t) = i\left[2i a_1 e^{-a_1 s}\sin(a_1 t)\right] - \frac{1}{2\pi}\int_{\substack{branch\\cut}} \frac{|z|\,e^{-s|z|+itz}}{|z|-a_1}\,dz.$$

Now,

$$\int_{\substack{branch\\cut}} \frac{|z|\,e^{-s|z|+itz}}{|z|-a_1}\,dz = -\int_\infty^0 \frac{(-i\eta)e^{i\eta s + \eta t}}{-i\eta - a_1}(-i\,d\eta) - \int_0^\infty \frac{(i\eta)e^{-i\eta s + \eta t}}{i\eta - a_1}(-i\,d\eta)$$

$$= 2\int_0^\infty \frac{\eta e^{t\eta}}{\eta^2 + a_1^2}\left[a_1\cos(s\eta) + \eta\sin(s\eta)\right]d\eta.$$

For $t > 0$, we close the contour in the upper half-plane. There are no poles in this case and

$$f(t) = -\frac{1}{2\pi}\int_{\substack{branch\\cut}} \frac{|z|\,e^{-s|z|+itz}}{|z|-a_1}\,dz,$$

where

$$\int_{\substack{branch\\cut}} \frac{|z|\,e^{-s|z|+itz}}{|z|-a_1}\,dz = -\int_\infty^0 \frac{(i\eta)e^{-i\eta s - \eta t}}{i\eta - a_1}(i\,d\eta) - \int_0^\infty \frac{(-i\eta)e^{i\eta s - \eta t}}{-i\eta - a_1}(i\,d\eta)$$

$$= 2\int_0^\infty \frac{\eta e^{-t\eta}}{\eta^2 + a_1^2}\left[a_1\cos(s\eta) + \eta\sin(s\eta)\right]d\eta.$$

Section 5.1, Problem 6.

$$f(t) = \frac{1}{2\pi}\int_{-\infty}^\infty \frac{e^{i\omega t}}{\omega^2 + a^2}\log\left(\frac{b+i\omega}{b-i\omega}\right)d\omega.$$

For $t > 0$, we close the line integral in the upper half of the ω-plane. Within the contour, there is a simple pole at $\omega = ai$. Its residue is

$$\mathrm{Res}\left[\frac{e^{itz}}{z^2 + a^2}\log\left(\frac{b+iz}{b-iz}\right);ai\right] = \frac{e^{-at}}{2ai}\ln\left(\frac{b-a}{b+a}\right).$$

There is also a branch point at $\omega = bi$. If we define $\omega - bi = (\eta - b)e^{\theta i}$ with $\theta = -3\pi/2$ or $\pi/2$ and $\omega + bi = (\eta + b)e^{\pi i/2}$, then the branch cut integrals along the sides of the branch cut are

$$\int_{\substack{RHS\ of\\branch\ cut}} \frac{e^{itz}}{z^2 + a^2}\log\left(\frac{b+iz}{b-iz}\right)dz = \int_\infty^b \frac{e^{-t\eta}}{a^2 - \eta^2}\left[\ln\left(\frac{\eta - b}{\eta + b}\right) - \pi i\right]i\,d\eta$$

and

$$\int_{\substack{LHS\ of \\ branch\ cut}} \frac{e^{itz}}{z^2 + a^2} \log\left(\frac{b + iz}{b - iz}\right) dz = \int_b^\infty \frac{e^{-t\eta}}{a^2 - \eta^2} \left[\ln\left(\frac{\eta - b}{\eta + b}\right) - 3\pi i\right] i\, d\eta.$$

Therefore,

$$\int_{\substack{branch \\ cut}} \frac{e^{itz}}{z^2 + a^2} \log\left(\frac{b + iz}{b - iz}\right) dz = 2\pi \int_b^\infty \frac{e^{-t\eta}}{a^2 - \eta^2}\, d\eta.$$

The inverse equals i times the residue minus the branch cut integral after it has been divided by 2π.

For $t < 0$, we close the contour in the lower half of the ω-plane. Here the residue equals

$$\text{Res}\left[\frac{e^{itz}}{z^2 + a^2} \log\left(\frac{b + iz}{b - iz}\right); -ai\right] = \frac{e^{at}}{2ai} \ln\left(\frac{b - a}{b + a}\right).$$

There is also a branch point at $\omega = -bi$. If we define $\omega - bi = (\eta + b)e^{-\pi i/2}$ and $\omega + bi = (\eta - b)e^{\theta i}$ with $\theta = -3\pi/2$ or $\pi/2$, then the branch cut integrals along the sides of the branch cut are

$$\int_{\substack{RHS\ of \\ branch\ cut}} \frac{e^{itz}}{z^2 + a^2} \log\left(\frac{b + iz}{b - iz}\right) dz = \int_\infty^b \frac{e^{t\eta}}{a^2 - \eta^2} \left[\ln\left(\frac{\eta + b}{\eta - b}\right) - \pi i\right] (-i\, d\eta)$$

and

$$\int_{\substack{LHS\ of \\ branch\ cut}} \frac{e^{itz}}{z^2 + a^2} \log\left(\frac{b + iz}{b - iz}\right) dz = \int_b^\infty \frac{e^{t\eta}}{a^2 - \eta^2} \left[\ln\left(\frac{\eta + b}{\eta - b}\right) - 3\pi i\right] (-i\, d\eta).$$

Therefore,

$$\int_{\substack{branch \\ cut}} \frac{e^{itz}}{z^2 + a^2} \log\left(\frac{b + iz}{b - iz}\right) dz = -2\pi \int_b^\infty \frac{e^{t\eta}}{a^2 - \eta^2}\, d\eta.$$

The inverse equals $-i$ times the residue, because we are traveling along the contour in the negative sense, minus the branch cut integral after it has been divided by 2π.

Section 5.1, Problem 7. The inverse is

$$f(t) = \frac{1}{2\pi} \int_{-\infty - \epsilon i}^{\infty - \epsilon i} \left\{ \frac{V(z)i}{z[1 - cV(z)]} - \frac{V(z)zi}{(z^2 + a^2)[1 - cV(z)]} \right\} e^{izt}\, dz.$$

For $t > 0$, we close the contour with an infinite semicircle in the upper half of the z-plane, except for a branch cut along the imaginary axis from i to ∞i. Within the closed contour, there are three poles: $z = 0$, $z = \alpha i$ and $z = \omega_0 i$ if $cV(\omega_0 i) = 1$. The corresponding residues are

$$\text{Res}\left\{ \frac{iV(z)e^{izt}}{z[1 - cV(z)]} - \frac{ziV(z)e^{izt}}{(z^2 + a^2)[1 - cV(z)]}; 0 \right\} = \frac{i}{1 - c},$$

$$\text{Res}\left\{ \frac{iV(z)e^{izt}}{z[1 - cV(z)]} - \frac{ziV(z)e^{izt}}{(z^2 + a^2)[1 - cV(z)]}; \alpha i \right\} = -\frac{iV(\alpha i)}{2[1 - cV(\alpha i)]} e^{-\alpha t},$$

and

$$\text{Res}\left\{\frac{iV(z)e^{izt}}{z[1-cV(z)]}-\frac{ziV(z)e^{izt}}{(z^2+a^2)[1-cV(z)]};\omega_0 i\right\}=\frac{ia^2(1-\omega_0^2)e^{-\omega_0 t}}{c(a^2-\omega_0^2)(\omega_0^2+c-1)}.$$

Along the branch cut, we have $z-i=(\chi-1)e^{\pi i/2}$ or $z-i=(\chi-1)e^{-3\pi i/2}$, and $z+i=(\chi+1)e^{\pi i/2}$. Then,

$$\int_{\substack{branch\\cut}}\left\{\frac{V(z)i}{z[1-cV(z)]}-\frac{V(z)zi}{(z^2+a^2)[1-cV(z)]}\right\}e^{izt}\,dz$$

$$=\frac{1}{2\pi}\int_\infty^1\left\{\frac{V_+(\chi)}{\chi[1-cV_+(\chi)]}-\frac{V_+(\chi)(-\chi)}{(a^2-\chi^2)[1-cV_+(\chi)]}\right\}e^{-\chi t}i\,d\chi$$

$$+\frac{1}{2\pi}\int_1^\infty\left\{\frac{V_-(\chi)}{\chi[1-cV_-(\chi)]}-\frac{V_-(\chi)(-\chi)}{(a^2-\chi^2)[1-cV_-(\chi)]}\right\}e^{-\chi t}i\,d\chi,$$

where

$$V_+(\chi)=-\frac{1}{2\chi}\ln\left(\frac{\chi-1}{\chi+1}\right)-\frac{\pi i}{2\chi}\quad\text{and}\quad V_-(\chi)=-\frac{1}{2\chi}\ln\left(\frac{\chi-1}{\chi+1}\right)+\frac{\pi i}{2\chi}.$$

A little algebra gives

$$\int_{\substack{branch\\cut}}\left\{\frac{V(z)i}{z[1-cV(z)]}-\frac{V(z)zi}{(z^2+a^2)[1-cV(z)]}\right\}e^{izt}\,dz$$

$$=\frac{\pi a^2}{2\pi}\int_1^\infty\left\{\left[1+\frac{c}{2\chi}\ln\left(\frac{\chi-1}{\chi+1}\right)\right]^2+\frac{\pi^2 c^2}{4\chi^2}\right\}^{-1}\frac{e^{-\chi t}}{\chi^2(\chi^2-a^2)}\,d\chi.$$

The inverse equals the residue times i minus the branch cut integral after we divide it by 2π. Note that we have made the substitution $\chi=1/\eta$ in the branch cut integral.

Section 5.1, Problem 8. The inverse is

$$f(t)=\frac{1}{2\pi}\int_{-\infty}^\infty\frac{e^{it\omega}}{2\alpha\sqrt{k^2+i\omega}+i\omega}\,d\omega.$$

The integrand has simple poles at $2\alpha\sqrt{k^2+i\omega}+i\omega=0$, or

$$\omega_{1,2}=-2i\alpha^2\pm 2i\alpha\sqrt{\alpha^2+k^2}.$$

For $t>0$, we close the contour with an infinite semicircle in the upper half of the ω-plane, except for a branch cut along the imaginary axis from $k^2 i$ to ∞i. Within the closed contour, there is a single pole, ω_1. The residue there is

$$\text{Res}\left(\frac{e^{itz}}{2\alpha\sqrt{k^2+iz}+iz};\omega_1\right)=\lim_{z\to\omega_1}\frac{(z-\omega_1)e^{itz}}{2\alpha\sqrt{k^2+iz}+iz}$$

$$=\frac{\left(-\alpha+\sqrt{\alpha^2+k^2}\right)\exp\left(2\alpha^2 t-2\alpha t\sqrt{\alpha^2+k^2}\right)}{i\sqrt{\alpha^2+k^2}},$$

since

$$\sqrt{k^2 + \omega_1 i} = -\frac{\omega_1 i}{2\alpha} = -\alpha + \sqrt{\alpha^2 + k^2}.$$

For the branch cut, $z - k^2 i = \eta^2 e^{\pi i/2}$ on the right side of the branch cut and $\eta^2 e^{-3\pi i/2}$ on the left. Therefore,

$$\int_{\substack{\text{right side of} \\ \text{branch cut}}} \frac{e^{itz}}{2\alpha\sqrt{k^2 + iz} + iz}\, dz = \int_\infty^0 \frac{e^{-(k^2 + \eta^2)t}}{-2\alpha\eta i + k^2 + \eta^2}\,(2\eta\,d\eta)$$

and

$$\int_{\substack{\text{left side of} \\ \text{branch cut}}} \frac{e^{itz}}{2\alpha\sqrt{k^2 + iz} + iz}\, dz = \int_0^\infty \frac{e^{-(k^2 + \eta^2)t}}{-2\alpha\eta i - k^2 - \eta^2}\,(2\eta\,d\eta),$$

so that the branch cut integral is

$$\int_{\substack{\text{branch} \\ \text{cut}}} \frac{e^{itz}}{2\alpha\sqrt{k^2 + iz} + iz}\, dz = -8\alpha \int_0^\infty \frac{\eta^2 e^{-(k^2 + \eta^2)t}}{(k^2 + \eta^2)^2 + 4\alpha^2\eta^2}\, d\eta.$$

The inverse for $t > 0$ equals i times the residue minus the branch cut integral divided by 2π.

For $t < 0$, we close the contour with a semicircle of infinite radius in the lower half of the ω-plane. There is no branch cut; we only have the pole at $z = \omega_2$. In this case,

$$\text{Res}\left(\frac{e^{itz}}{2\alpha\sqrt{k^2 + iz} + iz}; \omega_2\right) = \lim_{z \to \omega_2} \frac{(z - \omega_2)e^{itz}}{2\alpha\sqrt{k^2 + iz} + iz}$$

$$= \frac{\left(-\alpha - \sqrt{\alpha^2 + k^2}\right)\exp\left(2\alpha^2 t + 2\alpha t\sqrt{\alpha^2 + k^2}\right)}{-i\sqrt{\alpha^2 + k^2}}.$$

Here, the inverse equals $-i$ times the residue; the minus sign is due to the negative sense in which the countour is taken.

Section 5.1, Problem 9. From the definition of the bilateral Laplace transform, the inverse is

$$f(t) = \frac{1}{2\pi} \int_{-\infty - \epsilon i}^{\infty - \epsilon i} \frac{\log(i\omega\tau)}{(\lambda + \sqrt{\lambda^2 + \omega i})^2}\, e^{i\omega t}\, d\omega,$$

where $\epsilon > 0$. For $t > 0$, we close the contour with a semicircle in the upper half of the ω-plane, except for a cut along the imaginary axis. For $t < 0$, we close the contour with a semicircle in the lower half of the ω-plane. Within the closed contours there are no singularities. Therefore, $f(t) = 0$ if $t < 0$. For $t > 0$, we have four line integrals that comprise the branch cut integration: (1) from $0^+ + \infty i$ to $0^+ + \lambda^2 i$, (2) from $0^+ + \lambda^2 i$ to $0^+ + 0^+ i$, (3) from $0^- + 0^+ i$ to $0^- + \lambda^2 i$ and (4) from $0^- + \lambda^2 i$ to $0^- + \infty i$. The integrations around the branch points equal zero.

For the first contour, $z = \lambda^2 \eta i$, $z - \lambda^2 i = \lambda^2(\eta - 1)e^{\pi i/2}$ and $\log(iz\tau) = \ln(\lambda^2\tau\eta) + \pi i$, so that

$$\int_{C_1} \frac{\log(iz\tau)\, e^{itz}}{(\lambda + \sqrt{\lambda^2 + zi})^2}\, dz = \int_\infty^1 \frac{[\ln(\lambda^2\tau\eta) + \pi i]}{\left(\lambda + \lambda i\sqrt{\eta - 1}\right)^2}\, e^{-\lambda^2 t\eta}\, i\lambda^2\, d\eta,$$

while along the fourth contour $z = \lambda^2 \eta i$, $z - \lambda^2 i = \lambda^2(\eta - 1)e^{-3\pi i/2}$ and $\log(iz\tau) = \ln(\lambda^2\tau\eta) - \pi i$, so that

$$\int_{C_4} \frac{\log(iz\tau)\, e^{itz}}{(\lambda + \sqrt{\lambda^2 + zi})^2}\, dz = \int_1^\infty \frac{[\ln(\lambda^2\tau\eta) - \pi i]}{\left(\lambda - \lambda i\sqrt{\eta - 1}\right)^2}\, e^{-\lambda^2 t\eta}\, i\lambda^2\, d\eta.$$

Consequently, the sum of the integrals equals

$$\int_{C_1 \cup C_4} \frac{\log(iz\tau)\, e^{itz}}{(\lambda + \sqrt{\lambda^2 + zi})^2}\, dz = -2 \int_1^\infty \left[2\sqrt{\eta - 1}\, \ln(\lambda^2 \tau \eta) - \pi(2 - \eta) \right] e^{-\lambda^2 t\eta}\, \frac{d\eta}{\eta^2}.$$

Along the second contour, $z = \lambda^2 \eta i$, $z - \lambda^2 i = \lambda^2(1-\eta)e^{-\pi i/2}$ and $\log(iz\tau) = \ln(\lambda^2 \tau \eta) + \pi i$ so that

$$\int_{C_2} \frac{\log(iz\tau)\, e^{itz}}{(\lambda + \sqrt{\lambda^2 + zi})^2}\, dz = 2 \int_1^0 \frac{[\ln(\lambda^2 \tau \eta) + \pi i]}{\lambda^2 \left(1 + \sqrt{1-\eta}\right)^2} e^{-\lambda^2 t\eta}\, i\lambda^2\, d\eta,$$

while along the third contour, $z = \lambda^2 \eta i$, $z - \lambda^2 i = \lambda^2(1-\eta)e^{-\pi i/2}$ and $\log(iz\tau) = \ln(\lambda^2 \tau \eta) - \pi i$ so that

$$\int_{C_3} \frac{\log(iz\tau)\, e^{itz}}{(\lambda + \sqrt{\lambda^2 + zi})^2}\, dz = \int_0^1 \frac{[\ln(\lambda^2 \tau \eta) - \pi i]}{\lambda^2 \left(1 + \sqrt{1-\eta}\right)^2} e^{-\lambda^2 t\eta}\, i\lambda^2\, d\eta.$$

Consequently, the sum of the integrals equals

$$\int_{C_2 \cup C_3} \frac{\log(iz\tau)\, e^{itz}}{\left(\lambda + \sqrt{\lambda^2 + zi}\right)^2}\, dz = 2\pi \int_0^1 \frac{e^{-\lambda^2 t\eta}}{\left(1 + \sqrt{1-\eta}\right)^2}\, d\eta.$$

The inverse equals the negative of the sum of all of the integrals after they have been divided by 2π.

Section 5.1, Problem 11. From the inversion integral,

$$f(x,y) = \frac{1}{2\pi} \int_{-\infty}^\infty \frac{\exp\!\left(ikx - y\sqrt{k^2 - ikV/c}\,\right)}{\sqrt{k^2 - ikV/c}}\, dk$$

$$= \frac{1}{2\pi} \int_{-\infty}^\infty \frac{\exp\!\left[ikx - y\sqrt{(k - iV/2c)^2 + V^2/4c^2}\,\right]}{\sqrt{(k - iV/2c)^2 + V^2/4c^2}}\, dk$$

$$= \frac{e^{-Vx/2c}}{2\pi} \int_{-\infty - Vi/2c}^{\infty - Vi/2c} \frac{\exp\!\left[i\alpha x - y\sqrt{\alpha^2 + V^2/4c^2}\,\right]}{\sqrt{\alpha^2 + V^2/4c^2}}\, d\alpha,$$

if $\alpha = k - Vi/2c$. Let $i\alpha x - y\sqrt{\alpha^2 + V^2/4c^2} = -s$ so that

$$\alpha_\pm = \frac{ixs}{r^2} \pm \frac{z}{r^2}\sqrt{s^2 - (rV/2c)^2},$$

where α_- gives the contour in the second (third) quadrant and α_+ is the contour in the first (fourth) quadrant if $x > 0$ ($x < 0$). We can deform our contour because there are no singularities between the original contour and the new contour. Straightforward substitution and algebra leads to

$$f(x,y) = \frac{e^{-Vx/2c}}{\pi} \int_{rV/2c}^\infty \frac{e^{-s}}{\sqrt{s^2 - (rV/2c)^2}}\, ds.$$

Now, let $s = rV\eta/2c$. Then,

$$f(x,y) = \frac{e^{-Vx/2c}}{\pi} \int_1^\infty \frac{e^{-rV\eta/2c}}{\sqrt{\eta^2 - 1}}\, d\eta.$$

Index